Reviewing Earth Science

The Physical Setting

With Sample Examinations

Third Edition

Includes the 2011 Edition of the Earth Science Reference Tables

Thomas McGuire
Earth Science Educator

Amsco School Publications, Inc.,
a division of Perfection Learning®

The publisher wishes to thank the following people who acted as reviewers for this project:

Joseph Andrews
Earth Science Teacher
Paul V. Moore School
Central Square, NY

Michelle Ebert
Earth Science Teacher
Greece Arcadia High School
Greece, NY

Lee Posniack
Earth Science Teacher
Haldane High School
Cold Spring, NY

Greg Wagner
Department Chair/Earth Science Teacher
Island Trees Middle School
Long Island, NY

Cover Design: Meghan J. Shupe
Text Design: Howard Petlack/A Good Thing, Inc.
Composition: Northeastern Graphic, Inc.
Art: Hadel Studio, Perfect Pix Studios, and Eric Heiber, artist
Front Cover Photo: Horseshoe Bend, Glen Canyon, Arizona, Thomas McGuire
Back Cover Photos: (l. to r.) Delicate Arch, Arches National Park, Utah; Star Trails, New York; Hanging Glacier, Denali National Park, Alaska; Chert Pebbles, Rodeo Beach, California—all by Thomas McGuire

Readings in *Science Technology and Society* written by **Jonathan Kolleeny**

When ordering this book, please specify: *Either* **1344201** *or*
REVIEWING EARTH SCIENCE: The Physical Setting, Third Edition

Please visit our Web sites at: ***www.amscopub.com*** and ***www.perfectionlearning.com***

ISBN: 978-1-56765-942-9

Printed in the United States of America

25 26 27 28 EBM 21 20 19 18 17

To the Student

You are embarking on an adventure. Adventures are known for the challenges and accomplishments they bring. However, hardships are also part of the process. The challenge will be to understand Earth's systems. As you read, you will notice **bold-faced** words within the text. These words are defined in the text where you see them. In addition, you will find them in the Glossary. Words printed in *italics* are words that have been defined in an earlier chapter, or other science-related words you should know. Rather than just memorizing numerous facts, you will be expected to use information sources, such as the 2010 edition of the *Earth Science Reference Tables*, to help you gain an understanding of our planet. These tables are printed near the back of this book. They should be used in class, during labs, and for reference during tests.

Reviewing Earth Science: The Physical Setting, Third Edition, will help you place factual information in context. This skill will allow you to answer challenging questions. Each section of text is followed by a set of questions divided into Part A and Part B. In addition, there are Chapter Review questions, which are divided into the Part A, Part B, and Part C types that you will find in the Regents Exam. Each chapter also has a Problem Solving Activity. This wealth of questions provides practice in answering multiple-choice items and questions that require longer answers. In many cases, these questions test your ability to apply your knowledge and skills in new ways.

The best learning takes place as a cooperative venture. You will work with your teacher and this book to learn the principles of Earth science. In my 30 years of teaching, I have seen that the majority of students in late middle school or high school can succeed in this course. Remember that success will require effort on your part. Adventures are seldom easy. But, the rewards can make the effort worthwhile. I wish you success in this adventure.

To the Teacher

Every course of study needs guiding principles. The Physical Setting: Earth Science Core Curriculum is based on the Learning Standards for Mathematics, Science, and Technology. Most teachers find the ordering of content in the State Education Department "Physical Setting/Earth Science Core Curriculum" document unsuitable for an instructional sequence. There are, in fact, many good "storylines" (content sequences) for Regents Earth science. The introduction to Earth science, with some relatively simple measurements, along with density activities and calculations, is a good way to start. Models and map reading are important to all content areas.

Some teachers like to move into meteorology next so they can highlight weather events as they occur during the year. Some prefer to have the astronomy content in the winter when nights are long and skies are more likely to be clear for star viewing. This book can be used for any of those sequences.

However, there is a very different reason the author has chosen the sequence in this book; not for the science content, but for the student. Many students will mature in their thinking skills as they progress through the course. In general, geology is the most concrete content area. Weather is less so. A significant amount of the astronomy content is more abstract. The sequence in this book is an attempt to match the growth of your students during the academic year.

You are free to use any sequence of content you wish. You should find that if you use a sequence other than that chosen by the author, this book should still work nicely. In fact, some teachers try a different sequence every year. The choice is yours.

Reviewing Earth Science: The Physical Setting, Third Edition, is abundantly illustrated with clearly labeled drawings and diagrams that illuminate and reinforce the subject matter. Important science terms are **boldfaced** and defined in the text. Other terms that may be unfamiliar to students are *italicized* for emphasis. The large format makes this book easy for students to read. Within each chapter, there are several sets of multiple-choice questions that are divided into Part A and Part B. The Chapter Review breaks out questions into Part A, Part B, and Part C, as does the Regents Exam. The questions test students' knowledge and reasoning, while provoking thought. These questions can be used throughout the year for review, exams, and homework assignments.

No book by itself can impart the skills your students will need in this course. The most important part of instruction will be the learning activities that you provide in your classes. Your classroom learning activities should be physically and intellectually engaging, and stress an inquiry/problem-solving style of learning. Your students may have had little experience in courses that develop these higher-level skills. Patience, planning, and keen professional judgment will be needed to guide students in this important mode of learning.

Contents

Introduction
Welcome to Earth Science

Vocabulary at a Glance

astronomy	geology	observation
coordinate system	inference	percent deviation
density	instrument	scientific notation
Earth science	mass	sense
engineering	measurement	volume
exponential notation	meteorology	

WHAT IS EARTH SCIENCE?

Earth science is the study of our planet, its changing systems, and its place in the universe. Like all scientists, Earth scientists use standard methods to make observations, as well as to record, analyze, test, and apply these observations. Many Earth scientists often reach similar conclusions based on of the results of different investigations. They share their knowledge, and it becomes a part of Earth science. Earth science is a method of investigating our physical surroundings that yields a commonly accepted body of knowledge. One of the most important characteristics of science is that it is dynamic, which means it changes. As scientists improve their methods and as they apply new ideas and technology, they make new discoveries.

Throughout their investigations, scientists encounter new questions that lead them to a constantly expanding understanding of our physical surroundings. Inquiry is the process by which people investigate questions, and problem solving is the application of those investigations to address our needs. Problem solving and inquiry are the driving forces of Earth science. Scientists, like other people, have personal beliefs. However, science should independent of these beliefs or any other bias.

There are a several branches of learning that fall within the Earth sciences. **Geology** is the study of rock materials and how the solid Earth changes. **Meteorology** is the study of the changing conditions of Earth's atmosphere, which we know as weather and climate. **Astronomy** is study of Earth's motions and the nature of other objects in space. Each of these branches of learning includes specialties. For example, some geologists focus on the recovery of commercial Earth resources and are therefore known as economic geologists. Petroleum geologists are even more specialized because they concentrate on the geological and environmental aspects of locating and drilling for oil. In this course, you will learn about many career and academic opportunities in Earth science. In addition, you will learn about the efforts to conserve resources and protect the environment.

Some astronauts are Earth scientists. Figure I-1 on page 2 shows astronaut Karen Nyberg making remote observations of our home planet. As you study Earth science, you will see many ways to apply what you learn to protect the natural environment and improve your life.

Engineering is the application of science to solving problems. For example, a civil engineer might be assigned to find ways to improve water flow over your school grounds. The engineer

Figure I-1. Anyone who uses the methods of science to investigate Earth systems is an Earth scientist.

might try to find ways to keep rainwater away from buildings and playing fields and to avoid problems associated with poor drainage.

Observations and Measurements

Observations and measurement are basic to all sciences. **Observations** are information received directly from any of our five **senses**: sight, touch, smell, hearing, and taste. In some cases, we use **instruments** to extend our senses. In other cases, instruments, such as balance scales and thermometers, are used to make our observations more accurate. Instruments can be as simple as a meterstick, or as complicated as the Hubble Space Telescope. The oceanographers in Figure I-2 are repairing instruments on an ocean buoy that help them make remote observations even when no human is present.

Figure I-2. When scientists need precise observations or observations that are difficult to make, they use instruments such as those on this ocean buoy.

Measurements are observations made using an instrument. They are expressed with numbers and units of measure, such as 25 meters or 4 light-years. (Meters and light-years are units of length.) These units give meaning to observations. Without the unit, how would you know what the measurement stands for? For example, if your friend said that a certain distance was five, you might ask five meters, kilometers, or inches?

You may be more familiar with the units of the U.S. Customary Measures. These units include the foot, the pound, and the degree Fahrenheit. Scientists use the *International System of Units, SI*, based on the *metric system*. In the SI/metric system, prefixes relate larger and smaller units by factors of 10 (0.1, 100, etc.) The SI/metric system is used in most countries of the world. Converting from one unit to another within the metric system is often as easy as changing the position of the decimal point. For example, 3.5 meters = 350 centimeters = 0.0035 kilometer. Table I-1 shows some of the units used in metric measures and similar U.S. Customary Measures. Table I-2 includes the most common metric prefixes.

Table I-1. Basic Units of Measurement Used in Earth Science

Physical Quantity	Metric Basic Unit	Metric Symbol	U.S. Customary Measure
Length	Meter	m	Inch, foot, mile
Mass	Gram	g	Ounce, pound, ton
Time	Second	s	(Same as metric)
Temperature	Kelvin Degree Celsius	K °C	Degree Fahrenheit (°F) Degree Fahrenheit (°F)

Let us apply the information in this table to the meter. A distance of 1 ($\times 10^0$) meter is 1 meter (1 m). A distance of 1 thousandth ($\times 10^{-3}$) of a meter is 1 millimeter (1 mm). A distance of 1 hundredth ($\times 10^{-2}$) of a meter is 1 centimeter

Table I-2. Common Metric Prefixes

Prefix	Meaning	Exponential Notation
milli-	thousandth	$\times 10^{-3}$
centi-	hundredth	$\times 10^{-2}$
(Basic unit, for example meter)	one	$\times 10^{0}$
kilo-	thousand	$\times 10^{3}$

(1 cm). A distance of 1 thousand ($\times 10^3$) meters is 1 kilometer (1 km).

Exponential Notation

Scientists often work with very large or very small numbers. These numbers are easier to read, write, and use in calculations if they are expressed in the mathematical shorthand known as **exponential**, or **scientific**, **notation**. A number written in scientific notation has the form

$$M \times 10^n$$

where M is a number from 1 to nearly 10 (but not 10), and 10^n is a power of 10.

The number 2500 written in scientific notation is 2.5×10^3. The first part, 2.5, is a number between 1 and 10; and the second part, 10^3, is a power of 10. So 2.5×10^3 really means 2.5×1000. When expressed in exponential notation, Earth's volume is approximately 1.1×10^{21} cubic meters.

Sample Problems

1. Write 27,508 in scientific notation.

Solution

Step 1: Determine M by moving the decimal point in the original number to the left or right so that only one nonzero digit is to the left of the decimal.

2.7508

Step 2: Determine n, the exponent of 10, by counting the number of places the decimal point has been moved. If moved to the left, n is positive. If moved to the right, n is negative.

4 3 2 1
2.7508
← 4 places to the left
$27{,}508 = 2.7508 \times 10^4$

2. Write 0.00875 in scientific notation.

Step 1: 0.008.75

Step 2:
1 2 3
0.008.75
3 places to the right →
$0.00875 = 8.75 \times 10^{-3}$

A negative power of 10 does not mean a negative number: $8.75 \times 10^{-3} = 0.00875$. A negative power of 10 mean is a number less than 1.

Observations and Inferences

Scientists use observations to draw conclusions, or **inferences**, which may include generalities, predictions, or extensions of their theories to situations where observations are difficult or impossible to make. Inferences that are based on a wide range of observations can be nearly as reliable as the best observations.

For example, if you found a small, rounded, shiny yellow rock that felt very dense and was easily scratched, you might infer from these observations that you have found a gold nugget. If you also observe that you are in a location where gold has been known to occur, this would strengthen your inference.

If observations do not support a scientific theory, it is the theory that must change. For example, in Chapter 3 you will learn how a variety of observations disagreed with the theory that the positions of the continents are permanent. Over a period of half a century, evidence accumulated that led to the modern paradigm (scientific model) of plate tectonics. Plate tectonics theory has helped scientists to understand observations that were difficult to explain under the former paradigms. It also has guided new scientific inquiry to expand our understanding of planet Earth. In fact, many of the most important discoveries of science came about when traditional beliefs failed to account for new observations.

QUESTIONS

Part A

1. Which of the following would most likely require the skills of an Earth scientist? (1) finding a cure for liver cancer (2) calculating the weight-bearing capacity of a bridge (3) predicting the local effects of global climate change (4) creating a new kind of plastic

2. A student observed and recorded the time of sunset at his home in New York State over a period of one year. Which of the following statements is most likely to be an inference? (1) the sunset time will continue as an annual cycle (2) the sunset time increased from January through May (3) it is difficult to establish the sunset time on cloudy days (4) after sunset, the sky became darker

3. Which of the following conclusions probably required the use of one or more scientific instruments? (1) Hot coffee spills cause painful burns. (2) The sun appears red as it sets. (3) Hurricanes have dangerous winds. (4) Water freezes at 0°Celsius.

4. Two scientists conduct experiments to find the composition of Earth's oceans. Good scientific investigation by both persons would probably result in the two scientists (1) having exactly the same data (2) working in the same country (3) reaching similar conclusions (4) being friends who communicate regularly

5. The number 5.1×10^{-3} is equal to (1) –5100 (2) 0.0051 (3) 5.13 (4) 5100

6. Which is a scientific instrument widely used by astronomers? (1) meterstick (2) thermometer (3) telescope (4) textbook

7. Which of the following measures would be most likely to be reported in exponential (scientific) notation? (1) the diameter of an atom in centimeters (2) the height of a tall tree in meters (3) the mass of a coin in grams (4) the volume of a bathtub in cubic meters

8. The mass of Earth is approximately 100 times the mass of our moon. Earth's mass is approximately 6×10^{24} kg. What is the approximate mass of the moon? (1) 600 kg (2) 6×10^{22} kg (3) 600×10^{24} kg (4) 6×10^{100} kg

9. Why do scientists usually prefer to use observations rather than inferences?
(1) Inferences are usually incorrect.
(2) Inferences are not scientific.
(3) Observations are never in error.
(4) Observations are less subject to interpretation.

10. Which of the following is the largest distance? (1) 2×10^{5} meters (2) 5×10^{-2} meter (3) 2×10^{-5} meter (4) 5×10^{2} meters

11. A prediction is always a/an (1) observation (2) inference (3) measurement (4) fact

Determining Percent Deviation

All measurements can be made with more care or by using better instruments. However, no measurement is perfect. Therefore, we have no way to determine any value exactly and totally free of error.

Percent deviation, or percent error, is a convenient way of comparing a measurement to the commonly accepted value for that measurement. The advantage of percent deviation is that it compares the size of the error to the value being measured. To calculate percent deviation, divide the difference between the measured and accepted value by the accepted value, and then multiply by 100. (*Note:* You always subtract the smaller of the two values from the larger value.)

$$\text{deviation (\%)} = \frac{\text{difference from accepted value}}{\text{accepted value}} \times 100$$

Sample Problem

In measuring a table, a student found its length to be 1.9 meters (m). If the accepted value is 2.0 meters, what is the percent deviation of the student's measurement?

Solution

$$\text{deviation (\%)} = \frac{\text{accepted value} - \text{measured value}}{\text{accepted value}} \times 100$$

$$= \frac{2.0 \text{ m} - 1.9 \text{ m}}{2.0 \text{ m}} \times 100$$

$$= \frac{0.1 \text{ m}}{2.0 \text{ m}} \times 100 = 5.0\%$$

QUESTIONS

Part A

12. A student used string to determine the length of a building that was known to be 10 meters. Unfortunately, the string stretched so the student's measurement was 10.5 meters. What was the student's percent deviation? (1) 5% (2) 10% (3) 10.5% (4) 59%

13. If a student estimates the mass of a rock as 200 grams when the actual mass is 100 grams, what is the student's percent error? (1) 100 grams (2) 200 grams (3) 100% (4) 200%

14. A scientist determined the half-life of carbon-14 to be 5000 years. Using the value given on the first page of the *Earth Science Reference Tables*, what was the scientist's approximate percent deviation? (1) 0.122% (2) 7.0% (3) 12.2% (4) 12.3%

15. A student claimed that he made a measurement that had an error of just 1 millimeter. However, this would be a large percent deviation if he were measuring the (1) height of a chair (2) the thickness of a coin (3) the length of a school building (4) the distance from Earth to the moon

TYPES OF GRAPHS

A graph is a way to organize and present data visually. Instead of reading paragraphs of information or studying columns of figures, you can see the data in a graph and make comparisons between variables almost at a glance. Unlike a data table, a graph helps you to visualize changes in data, to understand relationships between variables within the data, and to picture trends or patterns in data.

Line Graph

A line graph, such as the one in Figure I-3, shows how a measured quantity changes with respect to time, distance, or some other variable.

Line graphs are constructed by plotting data on a **coordinate system** that is set up on vertical and horizontal axes. The horizontal (x) axis is usually used for the independent variable. For example, time is generally shown on the horizontal (x) axis. The vertical (y) axis is

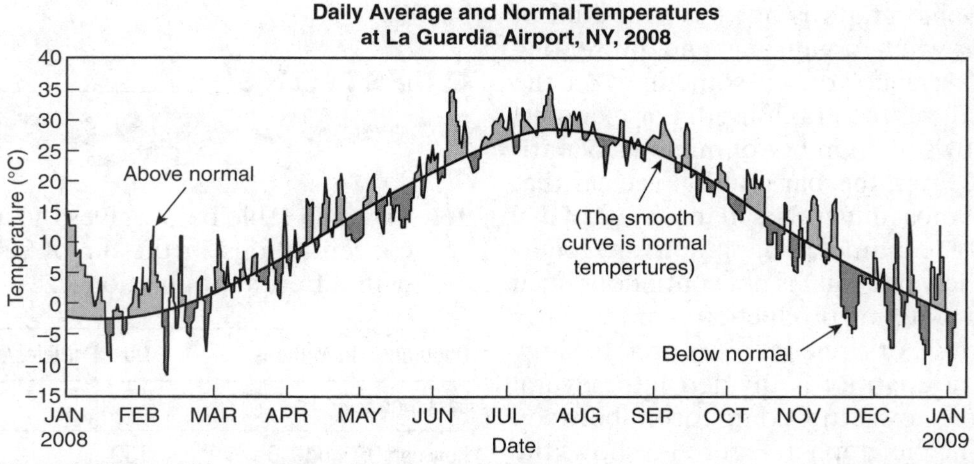

Figure I-3. The irregular line on this graph shows measured daily temperatures at La Guardia Airport during 2008. The smooth curve is the normal daily temperatures averaged over many years.

used for the dependent variable. It usually indicates the amount of the measured quantity being studied, such as temperature, height, or population. Normally, you know in advance the regular change expected in the independent variable. The values of the dependent variable are what you are trying to find or to know more about. You draw the graph to see how the dependent variable changes with respect to the independent variable.

When a line graph trends upward to the right, it represents a continuous increase. When the line graph trends downward to the right, it indicates a continuous decrease. A horizontal line represents no change. The steeper the line segment in a graph rises to the right, the greater the slope of the segment and the greater the increase in the dependent variable. Likewise, the steeper the line segment falls to the right, the greater the decrease in the dependent variable.

Line graphs allow you to read values even when they are not in the set of data points. That is, you can read values anywhere along the graph line. You can also extend a line graph to values outside the range of data if the trend of the change is clear.

Not all line graphs are straight lines. Some line graphs are curved or have an irregular shape. For example, Figure I-3 shows the normal and actual daily temperatures at La Guardia Airport in New York City.

Bar Graphs and Pie Graphs

Sometimes a line graph is not the best kind of graph to use when organizing and presenting data. In Earth science, you will sometimes see the bar graph and the pie graph used. For example, Figure I-4 shows the number of minerals of various densities. From this bar graph it is clear that the density of most minerals is in the range of 2–9 grams per cubic centimeter. Bar graphs are generally used to show data that is not continuous, such as the United States oil production year by year.

The pie graph, or circle graph, is used to show how an entire quantity is divided into several parts. It also shows the comparison between these parts. The pie graph in Figure I-5 shows the sources of electric energy used in New York in a typical year.

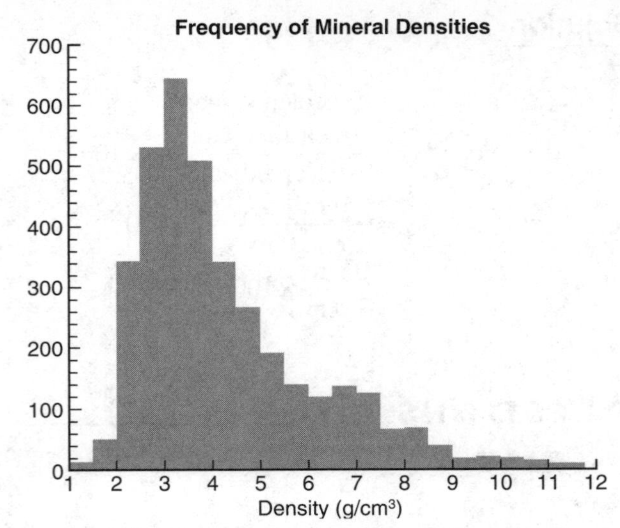

Figure I-4. The density of most minerals falls within the range 2 to 9 g/cm^3.

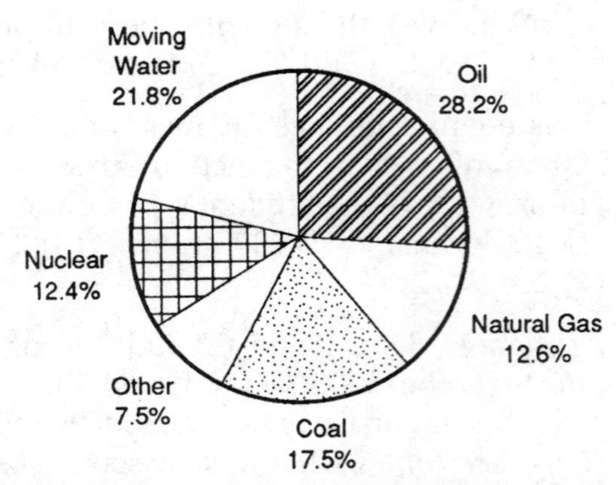

Figure I-5. This pie chart shows that more than half of New York's electric energy comes from fossil fuels. Renewable energy sources (hydroelectric, solar, wind, etc.) account for only about one-third of our electrical power.

QUESTIONS

Part A

16. The data table below gives the average dust concentrations in the air of selected cities with different population sizes.

Population in Millions	Dust Particles/Meter^3
Less than 0.7	110
Between 0.7 and 1.0	150
Greater than 1.0	190

Based on this data table, which graph best represents the general relationship between population and concentration of dust particles?

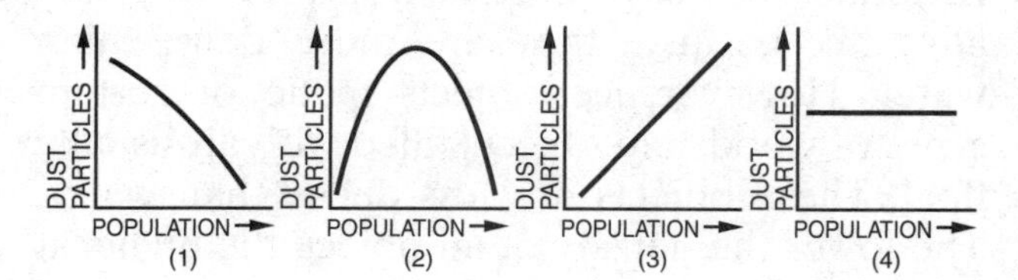

17. Why do scientists often publish their data in graphs? (1) data tables often contain errors (2) graphs can make data easier to understand (3) most scientific journals will not accept articles without graphs (4) graphs make it clear that scientific methods were used to obtain the data

Base your answers to questions 18 through 20 on the following graphs, which show the approximate elemental composition by mass of Earth's crust.

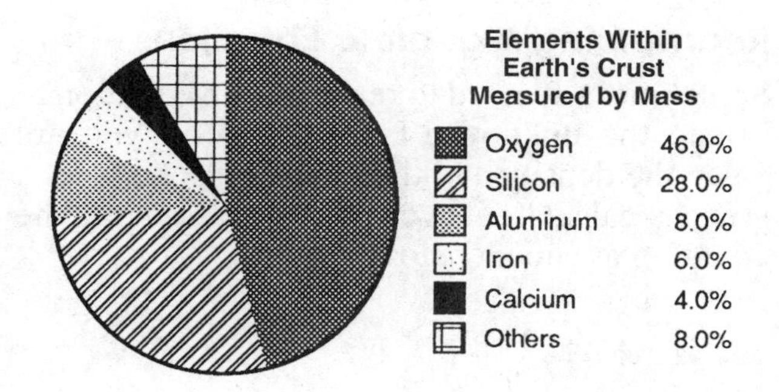

18. According to the graph, what one element makes up almost half the rocks of Earth's crust? (1) iron (2) aluminum (3) silicon (4) oxygen

19. What is the fourth most common element in Earth's crust? (1) iron (2) aluminum (3) silicon (4) oxygen

20. Limestone, which contains atoms of calcium, carbon, and oxygen, is a common rock on Earth's surface. According to the graph above, what is the percentage of carbon in Earth's crust? (1) less than 8% (2) approximately 10% (3) 33.3% (4) more than 50%

Base your answers to questions 21 through 23 on the graph below, which shows normal and actual precipitation for Albany, New York.

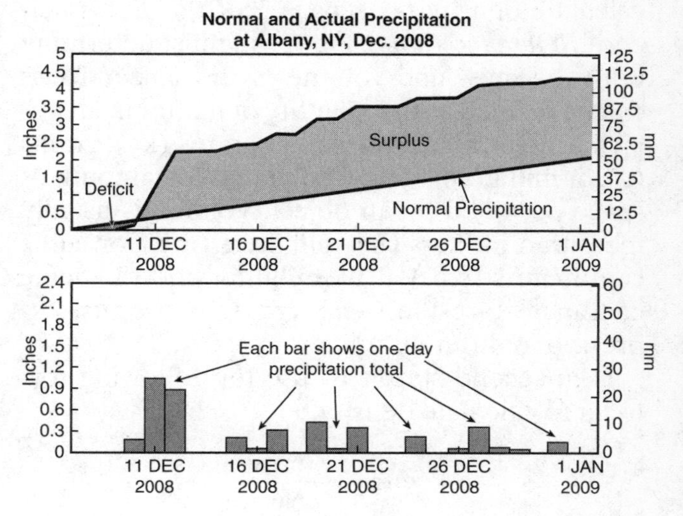

21. Approximately how much rain fell in Albany during the 3-day rain event of December 10–12? (1) 0.2 inch (2) 1.0 inch (3) 1.2 inches (4) 2.0 inches

22. A deficit is a shortage. Why was there a deficit for the first few days of December? (1) December was a dry month in Albany. (2) The first precipitation did not occur until December 10. (3) Normal precipitation does not begin until half way through December. (4) All the precipitation fell as snow.

23. Which scenario is best supported by the graph above? (1) December 2008 was a relatively dry month in Albany. (2) After several dry days, Albany had more rain than normal in December 2008. (3) In spite of a big rain event early in December 2008, this was a relatively dry month. (4) December 2008 was the wettest month ever in Albany.

DENSITY

Density is the property of matter that relates mass to volume for any particular material. In general, no matter how much or how little of that substance you have, the density of the substance will always be the same. For example, the density of 1000 grams of gold is 19.3 g/cm^3 and the density of 1.0 gram of gold is also 19.3 g/cm^3.

Determining Density

You must complete two steps to determine the density of an object. The equation you will use to calculate density is on page 1 of the *Earth Science Reference Tables.* First, you must determine both the mass and volume of the object. **Mass** can be defined as the quantity of matter in an object. Mass is measured in kilograms (kg), grams (g), or milligrams (mg). **Volume** is the amount of space occupied by an object. Volume is usually measured in liters (L), milliliters (mL), or cubic centimeters (cm^3). (The milliliter and the cubic centimeter are the same size; they are just expressed in different units.)

The second step is to use the following formula to calculate density:

$$\text{density} = \frac{\text{mass}}{\text{volume}}$$

Sample Problem

What is the density of an object that has a mass of 20 grams and a volume of 10 milliliters?

Solution

$$\text{density} = \frac{\text{mass}}{\text{volume}}$$
$$= \frac{20 \text{ g}}{10 \text{ mL}}$$
$$= 2 \text{ g/mL}$$ That is the same as 2 g/cm^3.

Dividing 20 grams (mass) by 10 milliliters (volume) yields a density of 2 grams per milliliter. Note that dividing these quantities results in the unit grams per milliliter (g/mL). You can also express the result as grams per cubic centimeter (g/cm^3):

Water is a special substance. (Because there is so much water present, Earth is sometimes called the water planet.) Since water is matter, it has mass, volume, and density. One gram of liquid water at 3.98°C occupies a volume of 1 milliliter. Therefore the density of liquid water is 1.0 g/mL. Since 1 milliliter is equal to 1 cubic centimeter (cc), the density of water also can be expressed as 1.0 g/cm^3. Solid water, or ice, is less dense than liquid water. The properties of water are listed on page 1 of the *Earth Science Reference Tables*.

Water gives us a simple way to test the density of a substance. Density determines an object's ability to sink or float in water, as illustrated in Figure I-6. When placed in water, materials such as a piece of iron or a rock usually sink because they are more dense than water. However, ice, objects made of certain types of wood, and objects filled with air usually float. These objects are less dense than water. The lower the density of an object, the higher it floats in water.

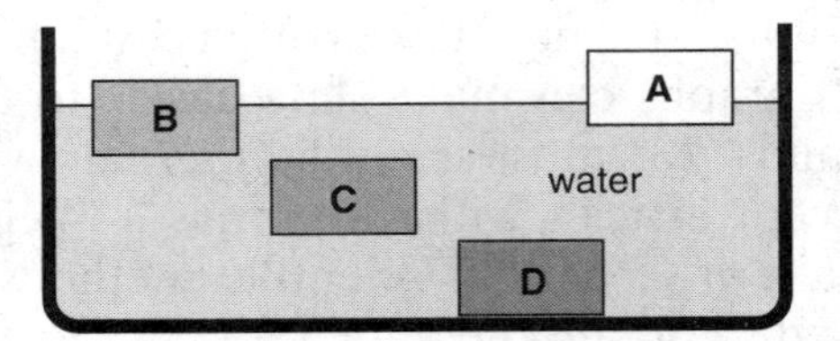

Figure I-6. Objects that are less dense than water such as A and B float in water. The lower the density, the greater the portion the object that floats above the level of the water. C is suspended because its density is equal to that of water. D sinks as do all objects that are more dense than water.

Rearranging Algebraic Equations

Sometimes you need to rearrange an equation to isolate the unknown. For example, if you are given the density and the mass of a solid object, you can calculate its volume by rearranging the equation as shown below:

$$\text{density} = \frac{\text{mass}}{\text{volume}}$$
$$d = \frac{m}{v}$$

Multiplying both sides of the equation by v (volume) gives you

$$d \times v = \frac{m \times \cancel{v}}{\cancel{v}}$$
$$d \times v = m$$

Divide both sides of the equation by d (density)

$$\frac{\cancel{d} \times v}{\cancel{d}} = \frac{m}{d}$$
$$v = \frac{m}{d}$$

Any equation can be rearranged in this way, which will allow you substitute the appropriate values to calculate the unknown.

QUESTIONS

PART A

24. The ill-fated steamship *Titanic* is shown below as it was towed from port in Southampton, England in 1912.

Why can such a large, steel ship float on the ocean? (1) The density of the ship is greater than the density of water. (2) Freshwater is more dense than seawater. (3) The steel ship is mostly filled with air. (4) Dense objects usually float on water.

25. The density of a 100-g sample of lead is approximately 11 grams per cubic centimeter. If the sample is cut into two 50-g pieces, what is the density of each half? (1) 11 g/cm^3 (2) 5.5 g/cm^3 (3) 1.1 g/cm^3 (4) 0.5 g/cm^3

26. The mass and dimensions of a rectangular solid are shown below.

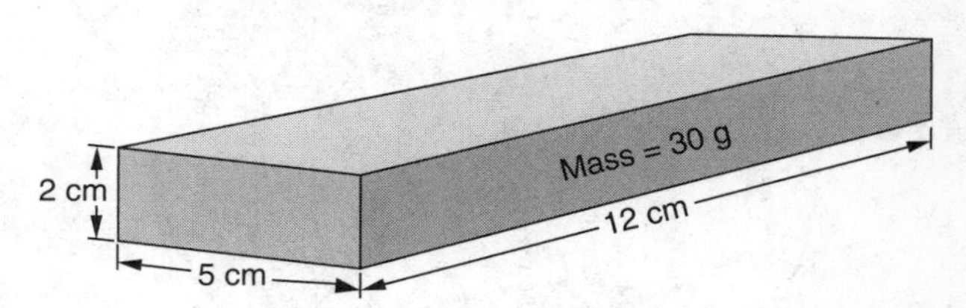

What is the density of this object? (1) 0.25 g/cm^3 (2) 0.5 g/cm^3 (3) 25 g/cm^3 (4) 50 g/cm^3

27. Granite rock has a density of 2.7 g/cm^3. What is the approximate volume of a 100-g sample of granite? (1) 2.7 cm^3 (2) 37 cm^3 (3) 270 cm^3 (4) 370 cm^3

28. How do you know that ice is less dense than water? (1) Ice is a solid. (2) Ice is transparent. (3) Ice is usually light in color. (4) Ice floats in water.

29. A student examined 100-g samples of iron, lead, gold, and ice. According to the figure below, which sequence shows these four samples in proper order of volume from smallest to largest?

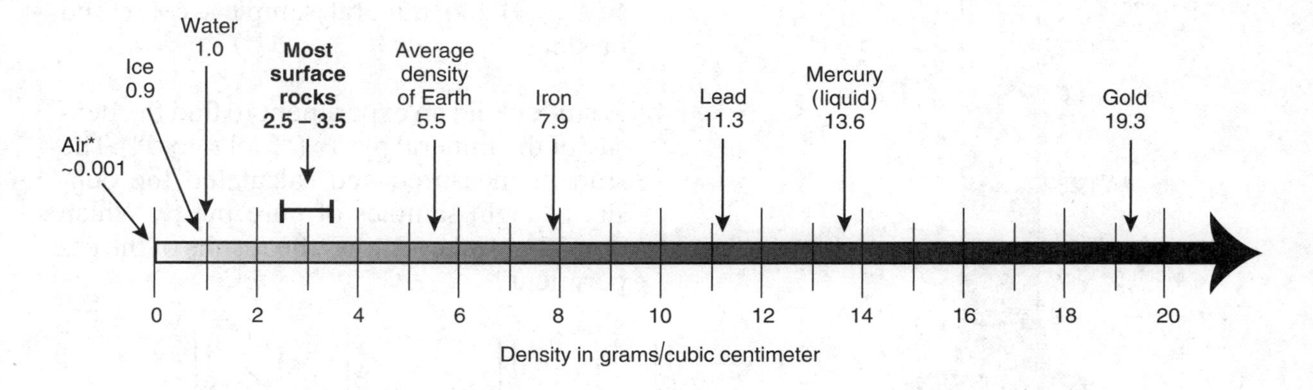

(1) lead–gold–iron–ice (2) gold–lead– iron–ice (3) iron–ice–lead–gold (4) ice– iron–lead–gold

30. The diagram below shows layers of Earth as revealed by earthquake waves.

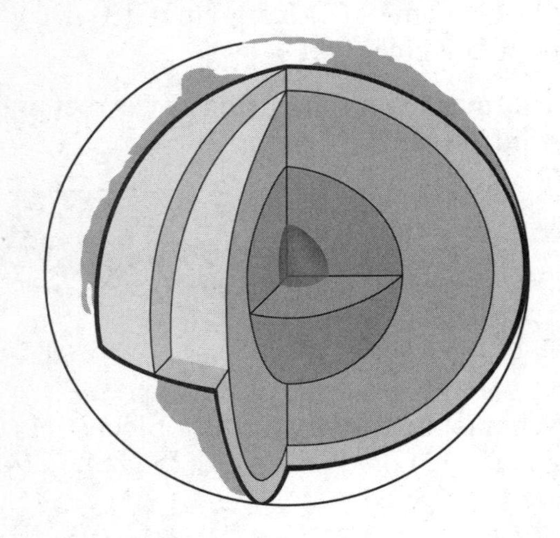

What could have caused our planet to separate into these layers? (1) Gravity pulled the hardest substances toward the Earth's center. (2) The hardest substances floated toward Earth' surface. (3) Gravity pulled the most dense substances toward Earth's center. (4) The most dense substances floated toward Earth's surface.

31. The mineral magnetite has a density of 5.17 g/cm^3 and does not dissolve in any of the liquids in the diagram below. A 1-cm^3 sample of magnetite is dropped into this cylinder.

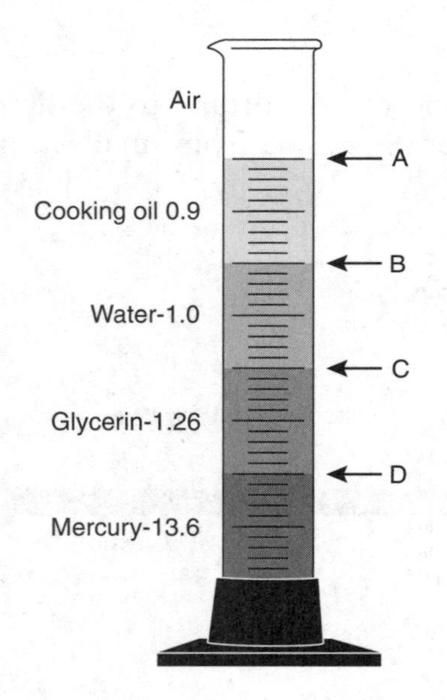

Where in the cylinder is the magnetite sample most likely to come to rest? (1) A (2) B (3) C (4) D

Chapter Review

PART A

1. A scientist observing the light given off by a star discovers that this star emits a kind of starlight never before observed by other scientists. Which of the following reactions would *not* be good science? (1) He contacts other astronomers to help him observe and understand this unusual star. (2) He extends his idea of what a star is to include this newly discovered object. (3) He changes his observations to conform to what he has learned from textbooks. (4) He repeats his observations to be sure that he has not made some kind of error.

2. An observer listed the following properties of a mineral sample.

A. The object is a dark gray rectangular solid.
B. The object rusts when exposed to rain.
C. The object most likely was purchased from a science catalog.
D. The object is more dense than water.

Which of these statements is an inference? (1) A (2) B (3) C (4) D

3. What values would be needed to calculate the mass of a solid mineral sample? (1) the mineral sample's hardness and volume (2) the mineral sample's density and volume (3) the mineral sample's hardness and density (4) the mineral sample's color and hardness

4. A student did an experiment to find the density of the mineral pyrite ("fool's gold"). The student measured and calculated the density of eight samples of pure pyrite. Which graph below best shows the results of the experiment?

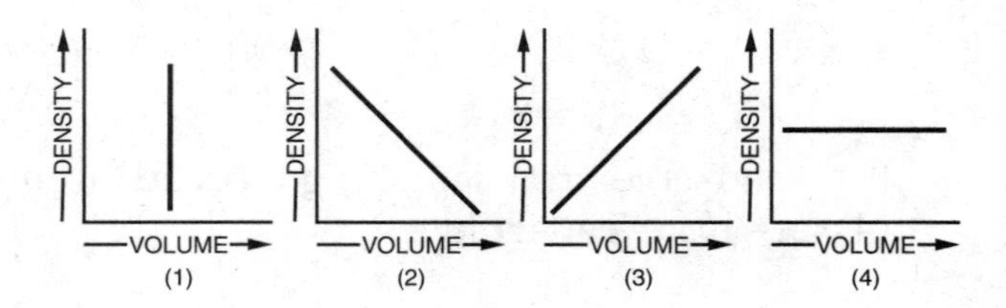

5. Which of the following is most likely to be an inference? (1) Earth's circumference is approximately 40,000 km (2) Earth's center is composed mainly of iron (3) Earth is shaped like a sphere (4) Earth is the third planet from the sun

6. Quartz is a mineral with a density of 2.63 g/cm^3. If a sample of quartz was dropped into the graduated cylinder shown below, where would the quartz sample come to rest in the cylinder?

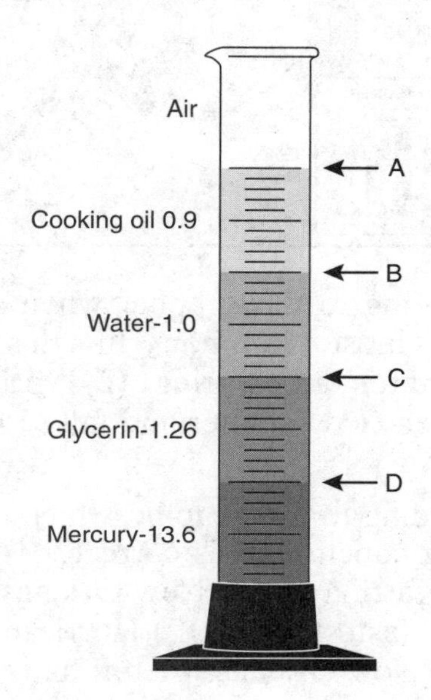

(1) A (2) B (3) C (4) D

7. A student estimated the mass of a mineral sample to be 50 g. Measurement showed it was actually 40 grams. What was the student's percent deviation? (1) 20% (2) 25% (3) 50% (4) 100%

8. Which is the greatest distance? (1) 3×10^{-8} km (2) 8×10^{-3} km (3) 8×10^{3} km (4) 3×10^{8} km

9. Why do objects float higher in seawater than they do in fresh water? (1) Seawater is less dense than freshwater. (2) Seawater is more dense than freshwater (3) Seawater is more pure than freshwater (4) All freshwater is polluted

10. When aluminum foil is formed into a loose ball, it can float on water. But when the ball of foil is pounded flat with a hammer, the foil sinks. Why? (1) Aluminum is less dense than water. (2) The act of pounding with a hammer increases the mass of the ball of foil. (3) The act of pounding with a hammer increases the volume of the ball of foil. (4) Air caught in the ball of foil makes the ball less dense than water.

Base your answers to questions 11 and 12 on the graph below, which was constructed after a student measured the mass and volume of five samples of the mineral pyrite.

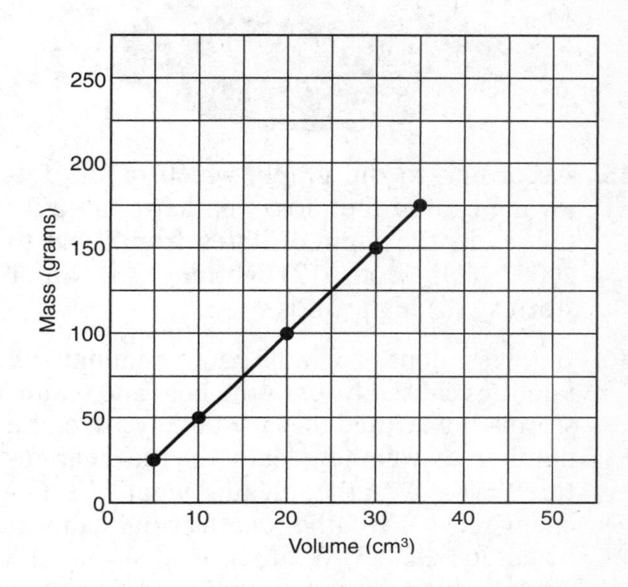

11. What is the approximate volume of 75 grams of pyrite? (1) 15 cm^3 (2) 50 cm^3 (3) 72 cm^3 (4) 150 cm^3

12. What is the approximate density of pyrite? (1) 0.2 g/cm^3 (2) 0.5 g/cm^3 (3) 2 g/cm^3 (4) 5 g/cm^3

13. Lillian looked up into the sky and estimated the clouds to be 1 kilometer above the ground. But measurements by weather scientists showed them to actually be 10 kilometers above the ground. What was Lillian's percent error? (1) 9% (2) 10% (3) 90% (4) 100%

14. Which of the following professionals would be *least* likely to need training in the Earth sciences for his or her occupation? (1) a water rights lawyer (2) a plastic surgeon (3) a mining engineer (4) an airline pilot

Base your answers to questions 15 and 16 on the graph below

15. According to the graph, which of the following caused the most weather-related fatalities in the United States from 1988 to 2005? (1) floods (2) lightning (3) winter storms (4) high winds

16. If government programs could eliminate all fatalities caused by extreme heat and winter storms, how would this most likely affect the number of weather deaths in the near future? (1) Weather deaths would be unchanged. (2) Weather deaths would almost be cut in half. (3) Weather deaths would be cut by a little more than half. (4) Weather deaths would be cut by about 80%.

PART B

Base your answers to questions 17 through 22 on the graphic below, which provides information about Hurricane Katrina, which devastated the New Orleans area on August 29, 2005. The associated graphs show when Atlantic hurricanes generally occur, and the central air pressure within Hurricane Katrina.

17. What was the general direction of movement of Hurricane Katrina from August 25 to August 28? (1) north (2) south (3) east (4) west

18. The map shows that Hurricane Katrina came ashore in the continental United States how many times? (1) 1 (2) 2 (3) 3 (4) 4

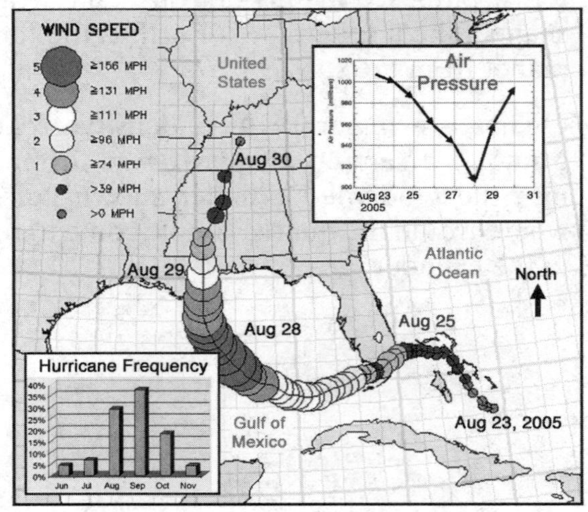

19. According to this graphic, when are hurricanes most common in this region? (1) winter and spring (2) spring and summer (3) summer and fall (4) fall and winter

20. According to this graphic, which of the following conclusions is correct? (1) Hurricane Katrina struck New Orleans before it had its fastest winds. (2) Hurricane Katrina struck New Orleans when it had the fastest winds. (3) Hurricane Katrina struck New Orleans after it had its fastest winds. (4) Hurricane Katrina struck New Orleans in September.

21. What was the relationship between the air pressure in Hurricane Katrina and the speed of the hurricane winds? (1) As the air pressure went down, the wind speed increased. (2) As the air pressure went down, the wind speed decreased. (3) The air pressure was not related to wind speed. (4) The air pressure remained constant from August 23 to August 30.

22. How did the strength of Hurricane Katrina change from August 23 to August 30? (1) The hurricane increased in wind speed for the whole period. (2) The hurricane decreased in wind speed for the whole period. (3) The hurricane became weaker over water and stronger over land. (4) The hurricane became stronger over water and weaker over land.

Base your answers to questions 23 through 25 on the passage below, which discusses Earth's weight and mass.

Earth's Weight and Mass

Earth orbits the sun at a relatively constant distance. This indicates that there is a balance between inertia (the tendency of Earth to move in a straight line) and the gravity of the sun (which pulls Earth into a curved path). This balance means Earth is weightless in space. However, Earth clearly has a very large mass. Scientists have calculated that the mass of Earth is 6×10^{24} kg, a large number.

23. What is the mass of Earth? (1) Because it is weightless, Earth has no mass. (2) 0.000 000 000 000 000 000 000 000 6 kg (3) 600,000,000,000,000,000,000,000 kg (4) 6,000,000,000,000,000,000,000,000 kg

24. From the paragraph above, we can conclude that (1) Earth has no mass. (2) Earth is the only planet in the solar system. (3) Scientists have been able to determine Earth's mass. (4) All objects that have mass also have weight.

25. How do we know that Earth has no weight? (1) The moon orbits Earth at a constant distance. (2) Earth orbits the sun at a constant distance. (3) Earth does not move with respect to the sun and moon. (4) Earth has been placed on a very sensitive mass scale.

26. A metric ton is 1000 kg. According to the passage above, what is Earth's mass in metric tons? (1) 6×10^{21} (2) 6×10^{24} (3) 1000×10^{24} (4) 6×10^{27}

PART C

Base your answers to questions 27 through 29 on the paragraph below, which describes the dwarf planet Eris.

Eris is a dwarf planet that orbits the sun at about twice Pluto's distance. Eris has been measured to have about 27 percent more mass than Pluto. The mass was calculated by timing the orbit of Eris' moon Dysnomia. Images taken with the ground-based Keck telescope were combined with images taken by the Hubble Space Telescope. These images show that Dysnomia has a nearly circular orbit lasting about 16 days. Infrared images also show that Eris is larger in diameter than Pluto. No space missions are currently planned to Eris, although the robotic *New Horizons* spacecraft bound for Pluto has recently passed Jupiter.

27. If Eris and Pluto have the same average density, how do the sizes (volumes) of Eris and Pluto compare?

28. What type of Earth scientist most likely discovered Eris?

29. If the mass of Pluto is 1.3×10^{22} kilograms, what is the approximate mass of Eris?

30. The following data represent four samples of the same material. Show this data as a line graph. Graph the mass on the vertical (up-and-down) axis and the volume on the horizontal axis. Be sure to follow the graphing rules you learned in class.

Four Samples of the Mineral Magnetite

Mass (g)	Volume (cm^3)
15	3
40	8
50	10
25	5

Follow the steps below and use the grid to plot and graph the data in the table.

(a) Mark each data point as an X on the grid.

(b) Connect the four points with a line.

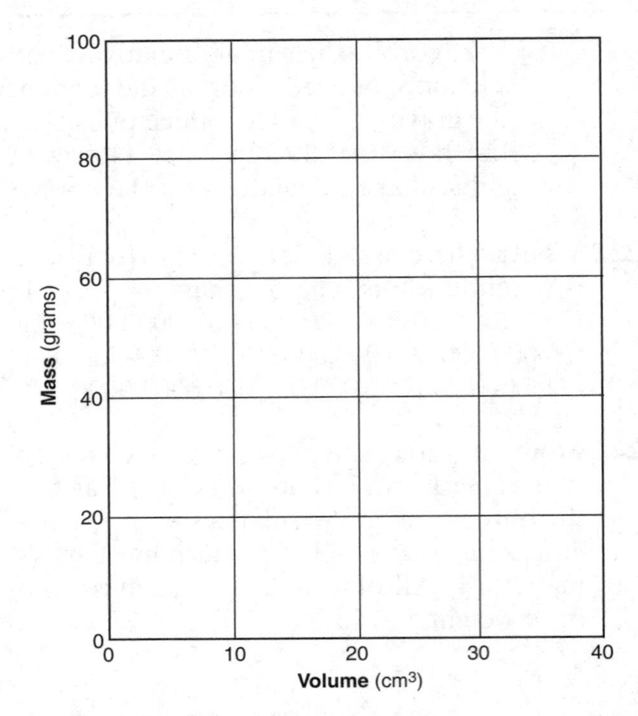

31. How many of the objects shown below are less dense than water?

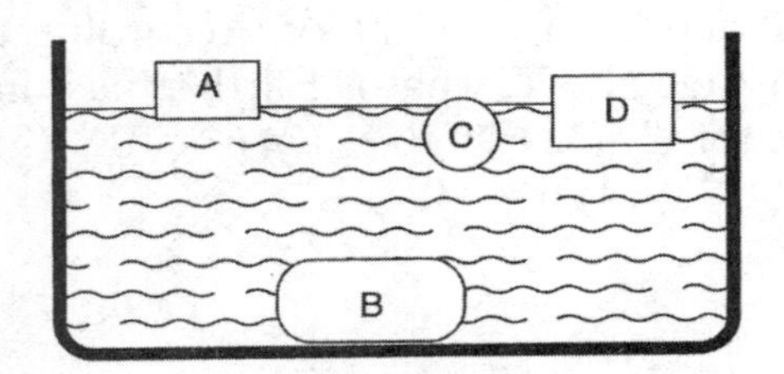

32. If you know the mass of Earth, what other single value would you need to know to calculate Earth's average density?

33. The data table below shows a student's measurements of five different samples of aluminum.

Five Samples of Aluminum

Sample Number	Mass (g)	Volume (cm^3)
1	21.6	8
2	32.4	12
3	54.0	20
4	94.5	35

34. What is the mass of a solid sample of aluminum that has a volume of 30 cm^3?

35. What is the density of aluminum?

36. Based on the map in the *Earth Science Reference Tables*, student estimated the distance from Syracuse, NY to Watertown, NY to be 150 kilometers. What is the percent deviation of the student's estimate? (Please solve this problem in three steps: (1) Write the appropriate algebraic equation, (2) substitute the appropriate values, (3) solve the equation to find the percent deviation.)

37. Page 15 of the *Earth Science Reference Tables* lists the density of each of the eight planets. Use this information to construct a bar graph on the grid below.

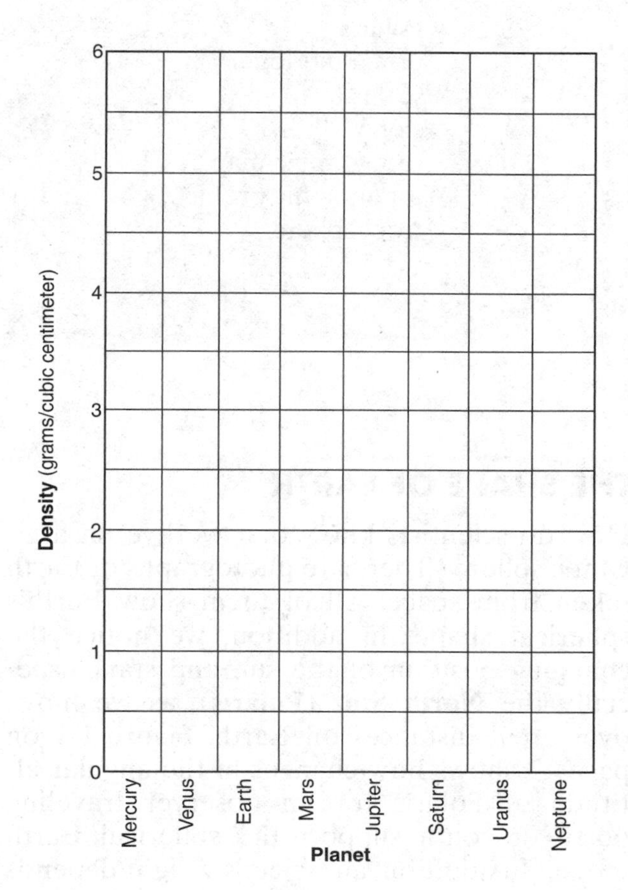

38. The diagram below represents four solid objects of the same size and shape (*A*, *B*, *C*, and *D*) resting in a bowl of water.

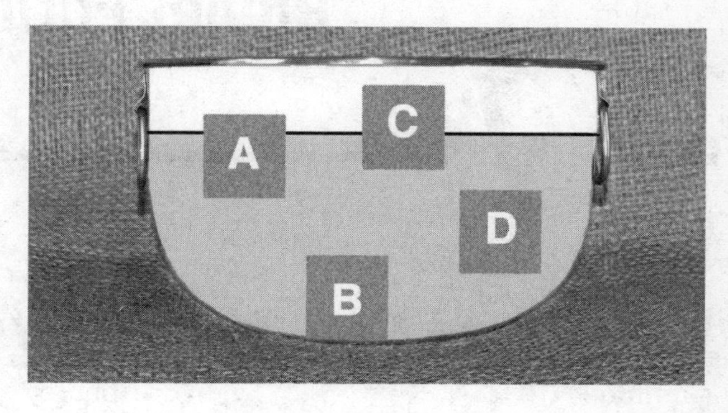

(a) List these four objects in order of their density, from the least dense to the most dense.

(b). What is the density of object *D*?

CHAPTER

Planet Earth

Vocabulary at a Glance

altitude	hydrosphere	oblate spheroid
atmosphere	isobar	parallel
circumference	isoline	Polaris
contour interval	isotherm	prime meridian
contour line	latitude	scale
contour map	lithosphere	slope
diameter	longitude	topographic map
equator	meridian	topographic profile
field	model	troposphere
gradient	navigation	
Greenwich Mean Time	North Star	

MODELS

A **model** is a representation of an object or a natural event. The most accurate model of Earth is a globe. Globes make Earth appear to be shaped like a ball or sphere. For most purposes, these models are excellent representations of Earth's shape. Earth looks flat to us who live on its surface because it is very large, and we usually see only a very small part of its surface. We live on an almost perfect sphere nearly 13,000 kilometers (8000 miles) in diameter. Even the tallest mountains do not seem very large when compared with the size of our planet. From space, Earth appears round and smooth from every direction. However, due to its daily rotation on its axis (spin), Earth bulges about 40 kilometers at the equator. Our planet is actually an **oblate spheroid**, slightly flattened at the poles and bulging a little at the equator. However, you must be careful not to exaggerate Earth's "out-of-roundness."

THE SHAPE OF EARTH

How do scientists know that we live on a gigantic sphere? There are photographs of Earth taken from space. All of them show Earth's spherical shape. In addition, we notice the changing position of the sun and stars, especially the **North Star** (**Polaris**), as we move over great distances on Earth. Figure 1-1 on page 17 shows how changes in the angular altitude of Polaris for an observer traveling north or south support the spherical Earth model. In addition, an object's weight depends on its distance from Earth's center. The farther it is from Earth's center the less it weighs. On a sphere, all points on the surface are equidistant from the center. Only on a spherical planet would we expect an object to have the same weight everywhere on the planet's surface. On Earth an object's weight changes only slightly from pole to equator, indicating a nearly spherical shape.

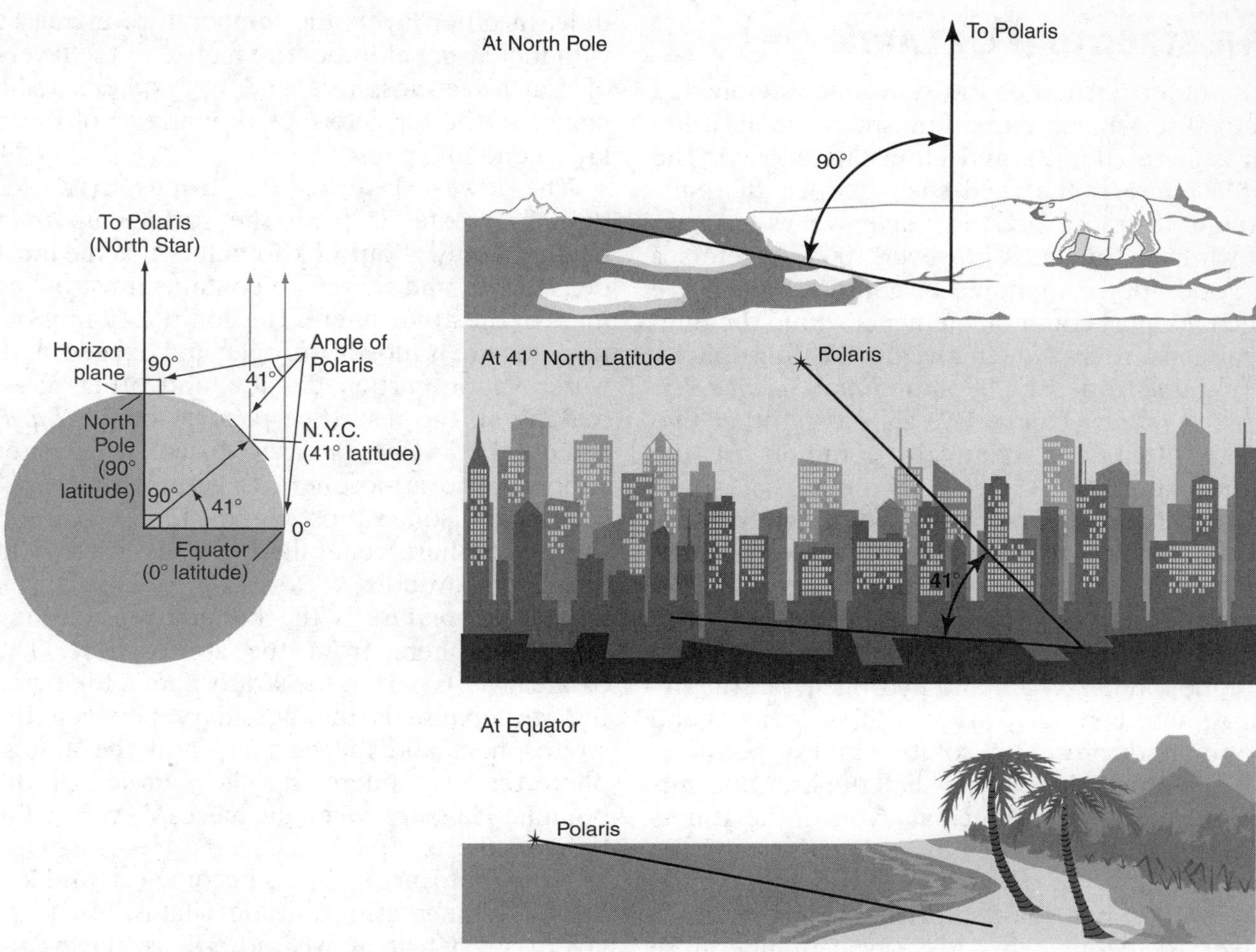

Figure 1-1. Earth's shape is apparent to any observer who travels north from the equator to the North Pole. This diagram shows how the position of Polaris (the North Star) changes along that journey. This change would happen only on a spherical planet or nearly spherical planet.

Table 1-1. Clues to Earth's Shape

How Do We Know that Earth Is a Nearly Perfect Sphere?
1. Ships seem to "sink" below the horizon.
2. The force of gravity is nearly the same everywhere on Earth's surface.
3. When seen from Earth at the same time but from different places on Earth, the sun and stars are in different positions.
4. Earth's shadow on the moon in a lunar eclipse is round.
5. Photographs taken from space show Earth is a globe.

Diameter and Circumference

Earth's **circumference** (distance around a sphere) and **diameter** (distance through the center of an object) have been measured with great accuracy. If our planet were a perfect sphere, the polar and equatorial values would be equal. As seen in the table below, the small difference between these values shows that Earth is very slightly out of round.

	Diameter (km)	Circumference (km)
Polar	12,714	40,008
Equatorial	12,756	40,076

THE STRUCTURE OF EARTH

The outer portion of Earth is generally divided into three major parts: lithosphere (solid), hydrosphere (liquid), and atmosphere (gas). The **lithosphere** is the rigid shell of rock that surrounds the more fluid inner layers and solid core. It varies in thickness, however, 100 kilometers is a good approximation. Oxygen and silicon are the two most common elements within the minerals and rocks that make up the lithosphere. This table from page 11 of the *Earth Science Reference Tables* (Figure 1-2) shows six other elements that are relatively common in the lithosphere.

The **hydrosphere** consists of Earth's liquid water—the oceans, lakes, rivers, and water in the ground. The average depth of the oceans is between 3 and 5 kilometers. The oceans cover about 70 percent of Earth's surface. Water is a compound of oxygen and hydrogen, but the hydrosphere contains various other elements and compounds present in solution and suspension.

The **atmosphere** is the shell of gases that surrounds Earth. The principal layers of the atmosphere, listed in order of increasing altitude (height), are the troposphere, the stratosphere, the mesosphere, and the thermosphere. Each layer is defined by a vertical temperature change, as shown in the *Earth Science Reference Tables*, page 14, and Figure 1-3. In some layers, the temperature decreases with increasing altitude. In other layers the temperature increases with increasing altitude. The names of the layers of Earth's atmosphere end in "-sphere." The names of the top interface (boundary) of those layers end in "-pause."

The lowest layer is the **troposphere**. Although it extends from the surface up to a height of only about 12 kilometers, it is the most dense layer and therefore contains most of the mass of the atmosphere. The composition of the troposphere is mostly nitrogen and oxygen, with water vapor, carbon dioxide, and other gases making up the rest. (See page 11 of the *Earth Science Reference Tables*.) Most of the water vapor in the atmosphere is within the troposphere, and so are most clouds. Temperature in the troposphere generally becomes lower with increasing altitude.

The *tropopause* is the boundary separating the troposphere from the stratosphere. The ozone layer is part of the stratosphere. Likewise, the *stratopause* is the boundary between the stratosphere and the next layer of the atmosphere, the mesosphere. And the *mesopause* is the boundary layer between the mesosphere and the thermosphere.

Note that Earth's layers become less and less dense with increasing altitude. That is, less-dense layers always sit above more-dense layers. In fact, the thermosphere is so thin that it gradually changes into the vacuum of outer space with no clear upper boundary.

ELEMENT (symbol)	CRUST		HYDROSPHERE	TROPOSPHERE
	Percent by mass	Percent by volume	Percent by volume	Percent by volume
Oxygen (O)	46.10	94.04	33.0	21.0
Silicon (Si)	28.20	0.88		
Aluminum (Al)	8.23	0.48		
Iron (Fe)	5.63	0.49		
Calcium (Ca)	4.15	1.18		
Sodium (Na)	2.36	1.11		
Magnesium (Mg)	2.33	0.33		
Potassium (K)	2.09	1.42		
Nitrogen (N)				78.0
Hydrogen (H)			66.0	
Other	0.91	0.07	1.0	1.0

Figure 1-2. Average chemical composition of Earth's crust, hydrosphere, and troposphere.

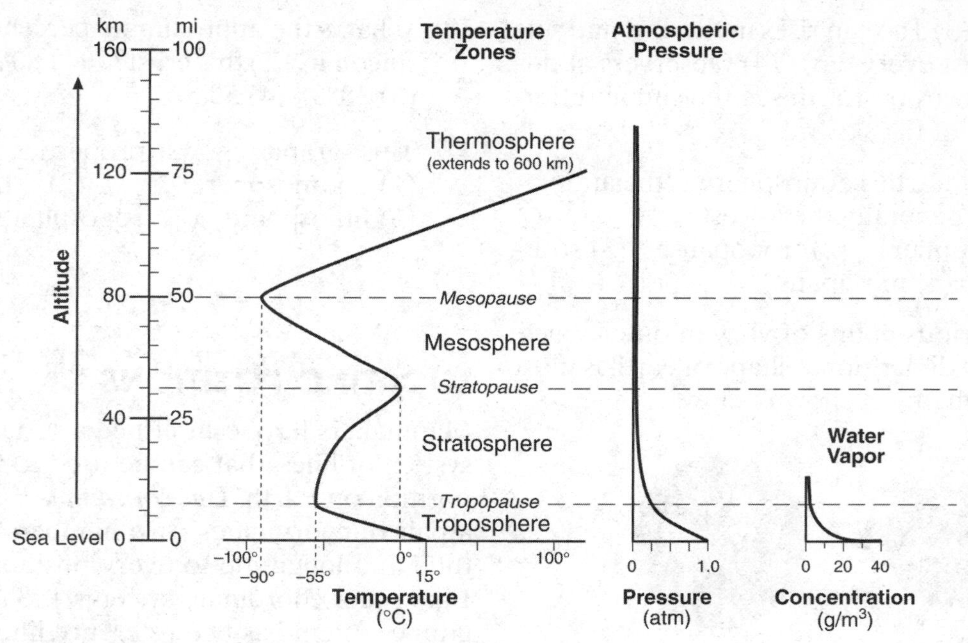

Figure 1-3. Selected properties of Earth's atmosphere.

QUESTIONS

Note that to answer many of the questions that follow, you will need to use information from the Earth Science Reference Tables, which are printed near the back of this book, following page 336. You should have the Reference Tables handy whenever you answer questions in this book or when you take quizzes and tests, including the Regents examination.

Part A

1. In what layer is most of the mass of Earth's atmosphere? (1) troposphere (2) stratosphere (3) mesosphere (4) thermosphere

2. Which best indicates the exact shape of Earth? (1) a perfect sphere (2) very slightly flattened at the equator (3) very slightly flattened at the poles (4) bulging outward at the poles

3. If you could travel upward from the middle of the stratosphere to the middle of the mesosphere, how would the air temperature change? (1) The temperature would decrease constantly. (2) The temperature would increase constantly. (3) The temperature would decrease and then increase. (4) The temperature would increase and then decrease.

4. Which pie graph below best shows the most abundant gases in Earth's atmosphere?

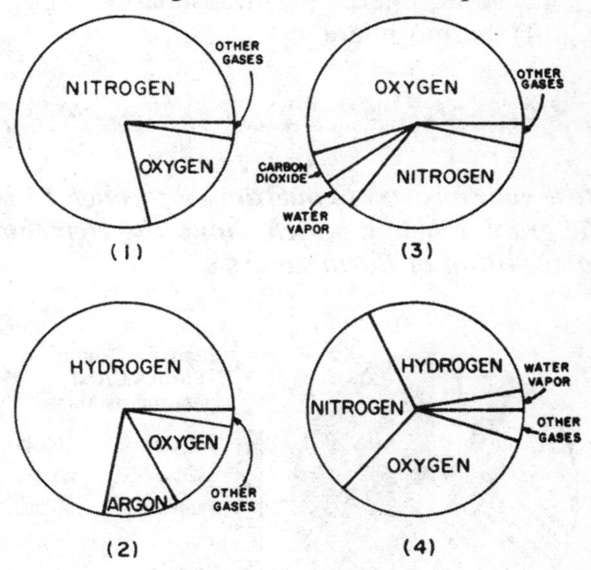

5. Which of the following observations is the best evidence that Earth is round? (1) Mountains are found over many areas of Earth. (2) The Great Plains of the American Midwest are the flattest area on the continents. (3) The sun rises in the east and sets in the west every day. (4) Observers at different places on Earth see the sun in different parts of the sky.

6. Where in Earth's atmosphere is the air temperature generally the lowest?
(1) troposphere (2) tropopause (3) stratosphere (4) mesopause

7. From various points of view in outer space, the three-dimensional shape of earth is most like which object shown below.

8. Nearly all the water vapor in Earth's atmosphere is located in the (1) troposphere (2) stratosphere (3) mesosphere (4) thermosphere

Part B

Base your answers to questions 9 through 11 on the graphic below, which shows the elemental composition of Earth's crust.

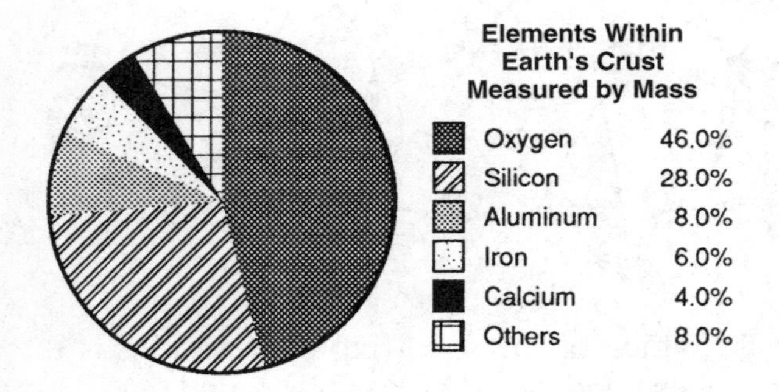

9. Which of the following elements is more abundant in Earth's crust? (1) silicon (2) calcium (3) aluminum (4) iron

10. What is the approximate percent by mass of silicon in Earth's crust? (1) 80% (2) 50% (3) 30% (4) 5%

11. This graph shows properties of Earth's (1) atmosphere (2) hydrosphere (3) lithosphere (4) troposphere

LOCATING POSITIONS ON EARTH

Mapmakers have established a surface grid, or a system of lines, that can be used to identify any position on Earth. This *coordinate system* assigns a pair of coordinates (two numbers) called latitude and longitude to every position on Earth. Like other coordinate systems, the latitude-longitude system has two reference lines: the equator and the prime meridian.

The reference line for latitude is based on the rotation of Earth. The spin axis of Earth determines the positions of its north and south poles. The **equator** is an imaginary line that circles Earth halfway between the North Pole and the South Pole. Angular distance in degrees north or south of the equator is **latitude**. Imaginary lines drawn around Earth parallel to the equator represent lines of equal latitude and are often called **parallels** since they never meet. The latitude of the equator is zero degrees (0°). The highest degrees of latitude are 90°N at the North Pole and 90°S at the South Pole, as you can see in Figure 1-4.

Unfortunately, there is no natural location for the line from which to begin measuring longitude. The **prime meridian** is an imaginary line (semicircle) that runs through Greenwich, England, from the North Pole to the South Pole. The prime meridian was located there because the observatory used by the British Navy to set their clocks to the apparent motion of the sun was there. Angular distance in degrees east or west of the prime meridian is called **longitude**. See Figure 1-5. Imaginary semicircles (**meridians**) drawn around Earth from the North Pole to the South Pole represent lines of equal longitude. The longitude of the prime meridian is zero degrees (0°). If you move east or west away from the prime meridian, the farthest you can go is 180°. The 180th meridian is half the distance around Earth. The half of the

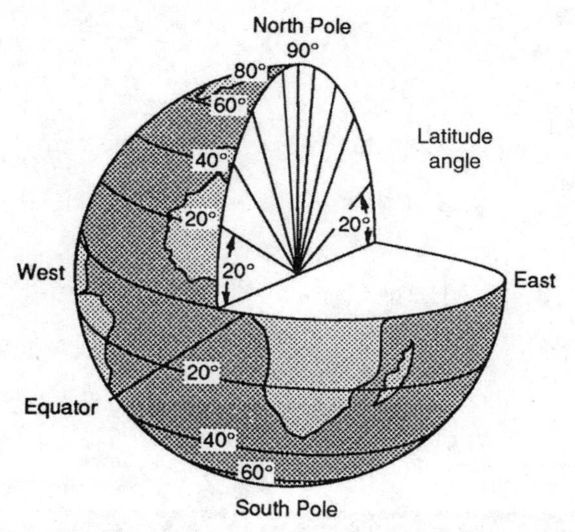

Figure 1-4. Latitude is the angular distance north or south of the equator.

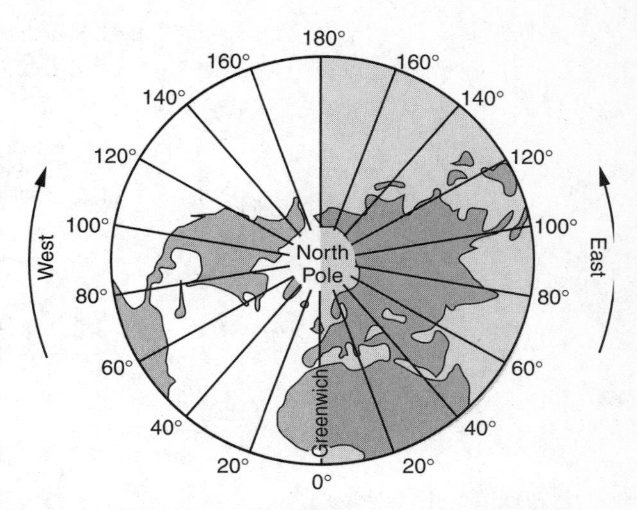

Figure 1-6. Earth is divided into the Eastern and Western hemispheres by the prime meridian and the 180° meridian. In this diagram, you are looking down from a position high above the North Pole.

world that is west of the prime meridian has west longitude. The half of the world that is east of the prime meridian has east longitude as illustrated in Figure 1-6. The Eastern and Western Hemispheres meet at 180° longitude. The International Date Line generally follows the 180° line of longitude.

Many students think of latitude as lines that circle Earth, parallel to the equator and longitude as lines that circle north and south. In this course, however, you will need a deeper understanding. Remember that latitude is the angle north or south of the equator, and longitude is the angle east or west of the prime meridian. Terrestrial coordinates are actually angles, not lines.

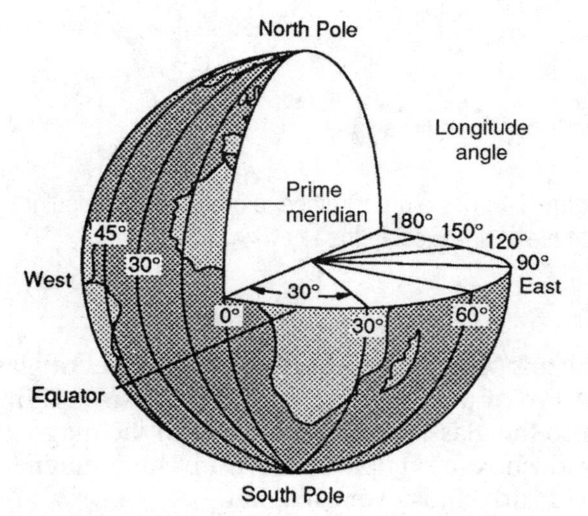

Figure 1-5. Longitude is the angular distance east or west of the prime meridian.

Terrestrial Navigation

Navigation is the science of identifying your position on Earth. The identity of any location can be expressed as the measure of the angle north or south of the equator (latitude) and the measure of the angle east or west of the prime meridian (longitude). In other words, each place on Earth has its own unique coordinates of latitude and longitude. The latitude of any location north of the equator is basically equal to the angle of Polaris (altitude of the North Star) above the horizon. For example, the latitude of New York City's is 41°N, so the angle of Polaris above the horizon at New York City is 41°.

To find Polaris in the night sky, look for the Big Dipper. If you have a clear northern horizon, the Big Dipper will always be visible in the northern part of the sky. However, at times it may be tipped on its side or upside down. Figure 1-7 on page 22 illustrates how to locate Polaris in the Northern sky. The two stars at the end of the bowl of the dipper are known as the pointer stars. A line connecting the pointer stars always points to Polaris, no matter where the Big Dipper is in the sky. There are many stars brighter than Polaris.

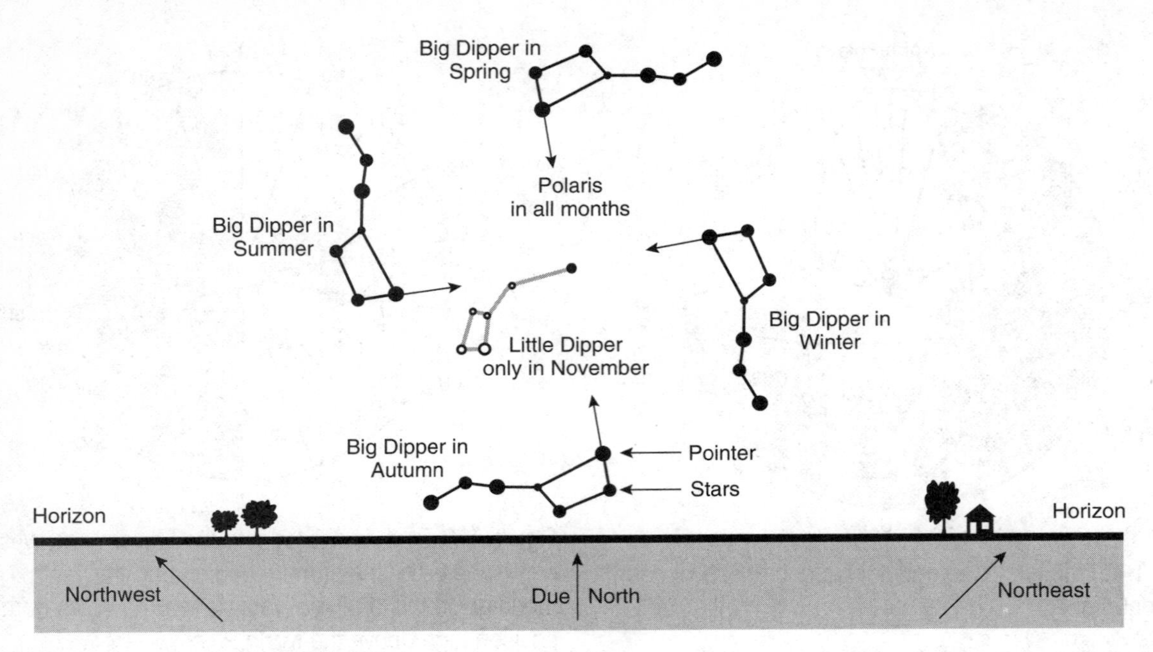

Figure 1-7. The easiest way to find Polaris is to locate the Big Dipper and follow the pointer stars at the end of the bowl as they point toward Polaris. This diagram shows the way the Big Dipper looks in the evening sky at the middle of each season. The Little Dipper also rotates around Polaris, but it is shown only in its autumn position.

However, it is one of the brighter stars in the Little Dipper, which seems to rotate around Polaris with the Big Dipper. Polaris is located at the end of the handle of the Little Dipper. You can use Figure 1-7 to help find the North Star in the night sky. You will probably need to look an hour or more after sunset from a dark location with a good view of the northern sky.

People living in the Northern Hemisphere are fortunate to have a bright star located almost exactly above Earth's North Pole. Residents of the Southern Hemisphere, however, are not as fortunate. They must use a number of stars to locate the South celestial Pole.

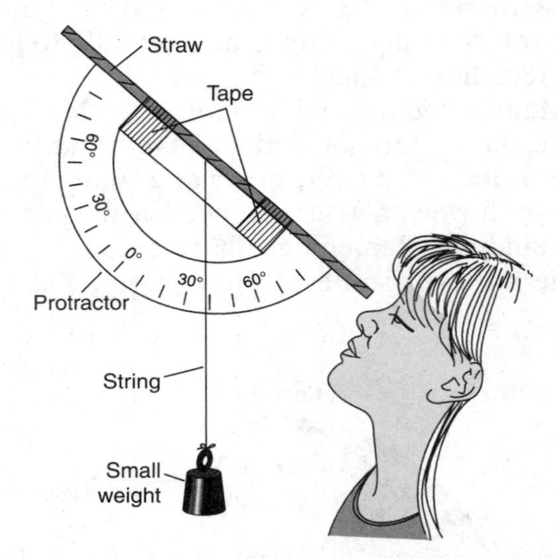

Figure 1-8. Using an astrolabe to measure the angular altitude of an object in the sky.

Finding the Altitude of a Star With an Astrolabe The **altitude** of an object in the sky is its angular height above the observer's horizon. The altitude of Polaris can be measured using an astrolabe. You can construct your own astrolabe by suspending a weight from a protractor as you see in Figure 1-8.

When you sight any star along the edge of the astrolabe, the weight hangs straight down toward Earth's center, or center of mass. Therefore, the string's position on the protractor indicates the angle of the star above the horizon.

Earth's Magnetic Field If you lay a sheet of glass on top of a bar magnet and sprinkle iron filings onto the glass, the filings align with the magnetic field, showing its pattern. An invisible magnetic field surrounds every magnet.

Earth itself has a magnetic field that lines up within about 12° of Earth's spin axis. In that way,

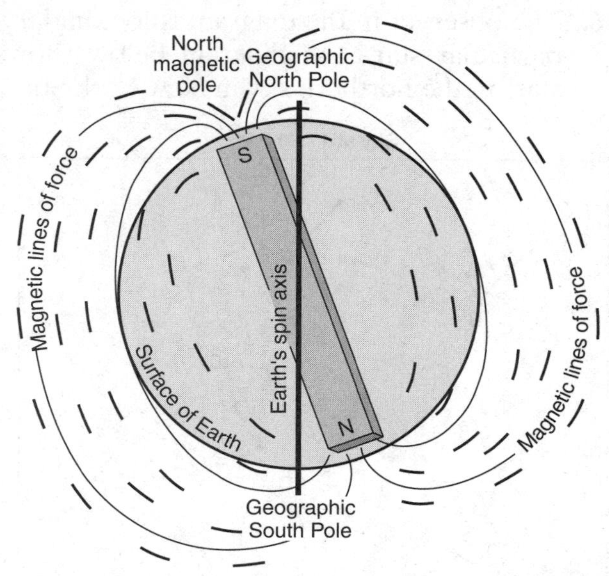

Figure 1-9. Our planet has a magnetic field. It is as if a giant bar magnet were at Earth's core. The magnet's north and south poles are reversed because the north-pointing end of a compass needle points to geographic north. (Opposite poles attract.)

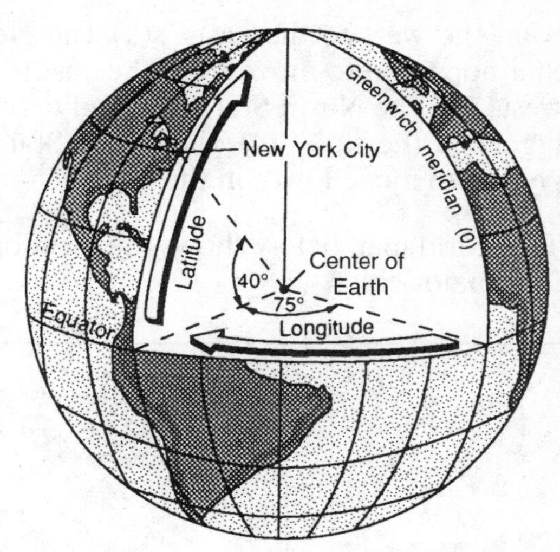

Figure 1-10. The parallels of latitude and the meridians of longitude allow us to locate positions. The position of New York City (approximate latitude 40°N, longitude 74°W) is illustrated here.

it is like Earth has an imaginary gigantic bar magnet at its core, as shown in Figure 1-9. The needle of a magnetic compass points toward the magnetic pole that is located relatively close to Earth's geographic North Pole. (You may recall that the geographic pole is where Earth's spin axis intercepts Earth's surface.) A magnetic compass is a convenient way to find directions if clouds cover the sun and stars.

Time Our system of time is based on observations of the sun. Noon can be defined as the time the sun reaches its highest point in the sky. Note that for all locations in the continental United States the noon sun is never directly overhead. For us in the Northern Hemisphere, it is always in the southern half of the sky. The day is divided into 24 divisions called hours. Each hour is further divided into 60 minutes, with each minute divided into 60 seconds.

You can calculate your approximate longitude if you know the present time along the prime meridian at Greenwich (**Greenwich Mean Time**) and your local time. (Greenwich is a suburb of London, England, where observations were made to establish the prime meridian.) To find longitude, you must first find the time difference, in hours, between local time and Greenwich time. This time difference multiplied by 15° per hour equals your longitude. (The rate at which the sun appears to move from east to west is 15° per hour.) If local time is earlier than Greenwich time, your position is west of the prime meridian or west longitude. If local time is later than Greenwich time, your position is east of the prime meridian or east longitude.

For example, there is a time difference of 5 hours between the time in Greenwich, which is located at 0° longitude, and the time in New York City. Multiplying 15° by 5 (hours), you get a difference of 75° of longitude. The fact that New York time is 5 hours earlier than Greenwich time indicates that New York is west of Greenwich. Therefore, the longitude of New York is approximately 75° west. See Figure 1-10.

QUESTIONS

Part A

12. On a clear night, Ruby traveled from New York City 120 miles north to Albany, New York. She made observations of the North Star every 30 miles. How did the position of the North Star appear to change during her trip? (1) The North Star appeared to move from the west to the east. (2) The North Star appeared to move from the east to the west. (3)The North Star appeared to move higher in the sky. (4) The North Star appeared to move lower in the sky.

13. The world map below shows the location of four major cities.

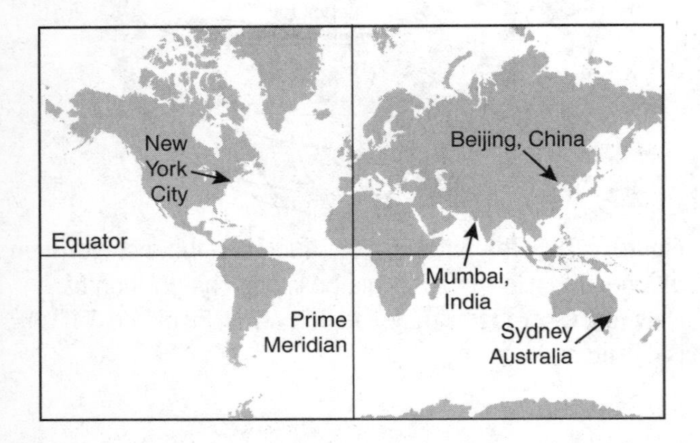

Polaris is never visible for observers in

(1) New York City (3) Beijing, China
(2) Mumbai, India (4) Sydney, Australia

Base your answers to questions 14 through 16 on the picture below, which shows a person using an astrolabe.

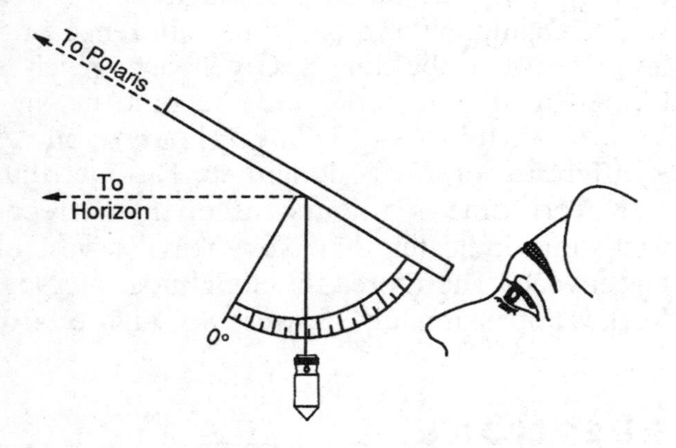

14. What is the position of the person? (1) 34° west longitude 2) 60° west longitude (3) 34° north latitude (4) 60° north latitude

15. The person in this diagram is making an observation to find his angular distance along Earth's surface from (1) the equator (2) the Prime Meridian (3) Polaris (4) the North Star

16. The observer in the diagram is looking at a particular star. The diagram below shows stars in the northern sky in New York State.

Sky at 11:00 P.M.

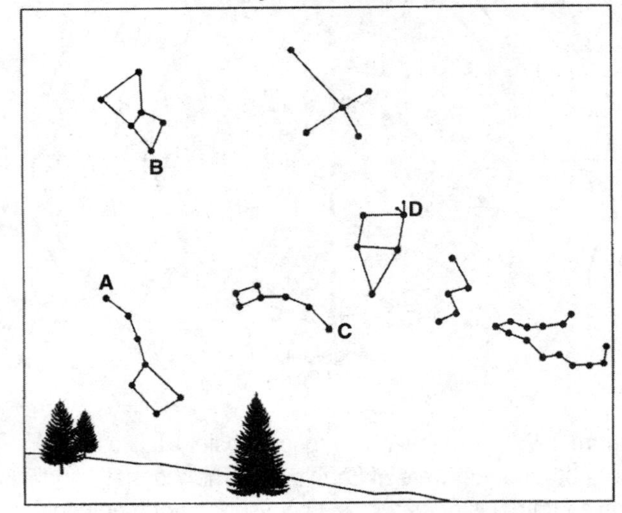

At which star on the figure above is this observer looking? (1) *A* (2) *B* (3) *C* (4) *D*

17. A compass used to find directions on Earth's surface does not point northward in some locations. Where would this problem most likely occur? (1) over a geologic deposit of magnetic iron (2) where strong winds always blow from the same direction (3) a region that has not been explored and mapped (4) at the South Pole

18. What is the longitude of the Yellowstone Hot Spot? (Hint: Use the *Earth Science Reference Tables*.) (1) 120° east (2) 120° west (3) 45° north (4) 45° south

19. The diagram below shows the Big Dipper in the evening in spring. What is the correct position of Polaris?

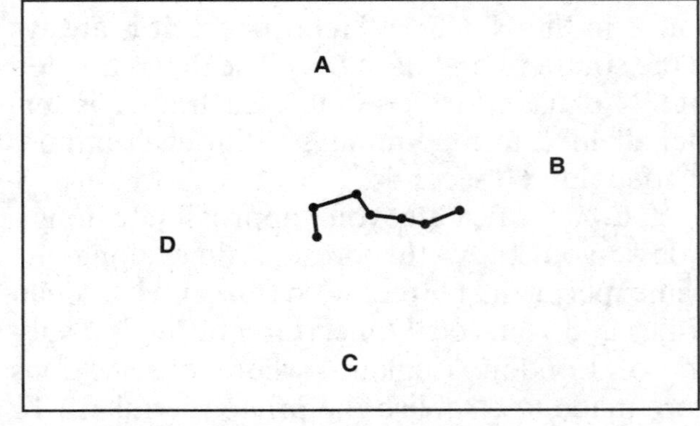

(1) *A* (2) *B* (3) *C* (4) *D*

20. On a clear night, observers in Albany, New York, and Buffalo, New York, attempt to measure the altitude of Polaris. How do their observations compare?

(1) They see Polaris at the same angular altitude. (2) The observer in Albany sees Polaris about 10° higher in the sky (3) The observer in Buffalo sees Polaris about 10° higher in the sky. (4) Polaris is never visible in Buffalo.

21. Which New York location below would see the sun rise earlier in the morning than the other three locations? (1) Watertown (2) Albany (3) Elmira (4) Buffalo

Part B

22. The antipode of any location on Earth is the point on the opposite side of the planet. So the antipode of New York City is in the Indian Ocean off Perth, Western Australia. What are the terrestrial coordinates of this Indian Ocean location? (1) 41° north, 74° west (2) 41° north, 74° east (3) 41° south, 74° west (4) 41° south, 74° east

23. London, England, is on the prime meridian. Massena, New York, is located at approximately 45° north latitude, 75° west longitude. If it is noon in London, what time is it in Massena? (1) 7 A.M (2) 9 A.M (3) 3 P.M (4) 5 P.M.

24. At which location would Polaris be visible on the horizon? (1) North Pole (2) New York State (3) equator (4) South Pole

25. Longitude is best determined experimentally by (1) observations of the North Star (2) comparing local times (3) recording times of sunrise (4) using a magnetic compass

26. New York City is at 41° north latitude, 74° west longitude. Osomo is a city in Chile (South America) that is located at approximately 41° south latitude, 74° west longitude. At New York City and Osomo (1) Polaris is at the same angle above the horizon. (2) The calendars are six months apart. (3) The time is the same. (4) The moon is never visible at the same time.

27. St. Louis, Missouri, is 90° west. Calcutta, India is 90° east. When the sun is just setting in St. Louis, in Calcutta (1) it is midnight (2) the sun is just rising (3) it is noon (4) the sun is also setting.

28. The diagram below indicates the altitude of Polaris as measured at three locations on Earth's Northern Hemisphere.

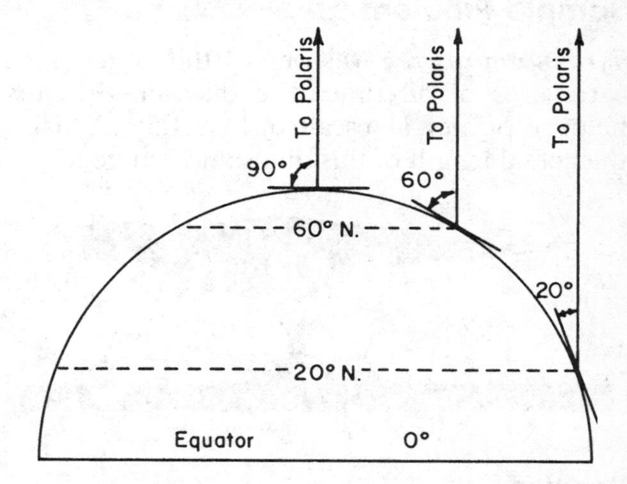

These observations could lead to the inference that Earth (1) rotates 15° per hour (2) has an elliptical orbit (3) has a curved surface (4) revolves around the sun

USING SCALE

Physical models are usually made to **scale**. That is, the dimensions of the real object are related to the model by a specific ratio. A scale is usually based on linear measurement. For example, a one-to-eight (1:8) model of a truck would mean that the real vehicle is 8 times longer than the scale model. So the model would probably be about 1 meter long, because the truck is about 8 meters long.

The United States Geological Survey (USGS) publishes maps at many scales, but the 1:24 000 maps are popular with hikers and others who enjoy the outdoors. So places about 240 m (24,000 cm) apart would be 1 cm apart on the map. If no scale is specified for a particular model, sometimes it can be determined by making measurements of length on the model and the length of the object.

Some models are smaller than the real object, for example a road map. Others are larger, such as diagrams that represent single atoms. In many cases, you will see drawings that are not to scale. If a true scale drawing showing the distance from Earth to the sun were printed on this page, Earth would probably be a dot too small to see without a magnifier. So sometimes a diagram that is not to scale is useful.

Sample Problem

Tyrannosaurus rex was one of the largest land carnivores of all time. The diagram of T-rex below is printed at a scale of 1 cm:1.5 m. What is the actual length of this dinosaur in meters?

Solution

$$\frac{\text{scale length (1 cm)}}{\text{actual length (1.5 m)}} = \frac{\text{length of drawing (6.8 cm)}}{\text{length of dinosaur } (x)}$$

$$\frac{1 \text{ cm}}{1.5 \text{ m}} = \frac{6.8 \text{ cm}}{x}$$

$$x = 10.2 \text{ m}$$

(That is probably 7 to 8 times as long as you are tall!)

Part A: Field values (elevations in meters)

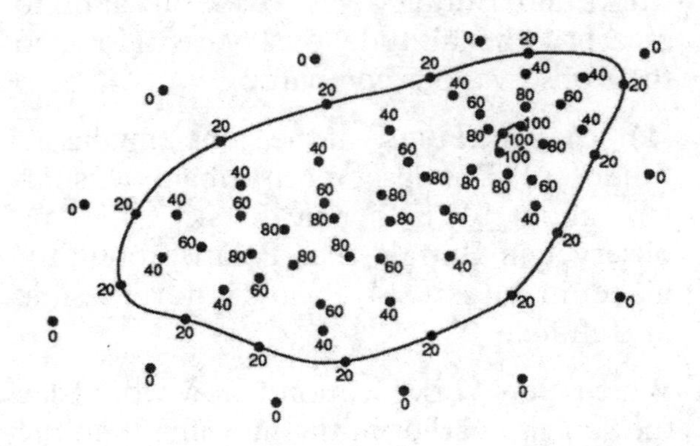

Part B: Isoline map

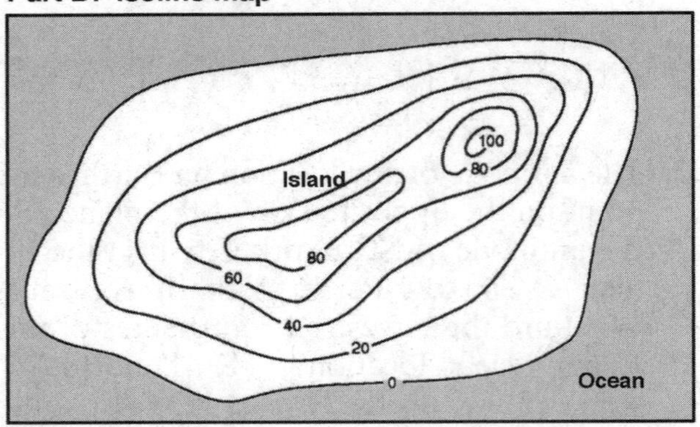

Figure 1-11. The numbers on Part A of this figure are land elevations in meters above sea level on an island. The 0 locations are at sea level and therefore describe the shoreline of the island. Two lines have been drawn: the 20-meter level and the 100-meter level. These are isolines of elevation, or contour lines. Part B shows the completed isolines without the data points. The ocean area is shaded.

FIELDS

A **field** is a region in which a similar quantity can be measured at every point or location, for example temperature in a room or the depth of snow. **Isolines** connect points of equal value on a field map as in Figure 1-11. Types of isolines include isotherms, isobars, and contour lines. **Isotherms** connect points of equal temperature. **Isobars** connect points of equal air pressure. **Contour lines** connect points of equal elevation.

Topographic Maps

A **topographic**, or **contour**, **map** shows the shape of Earth's surface. On a topographic map, contour lines connect places with the same elevation, as you can see in Figure 1-12. Each contour line is separated from the next by a uniform difference in elevation known as the **contour interval**. In other words, the contour interval is the difference in height between two adjacent contour lines. Where the contour lines are close, the slope of the ground is steep. Where the contour lines are far apart, the slope of the ground is gentle. *Note:* Do not confuse a contour interval, which shows a difference in height, with horizontal distance along the ground. The horizontal distance between two positions is determined by using the map scale.

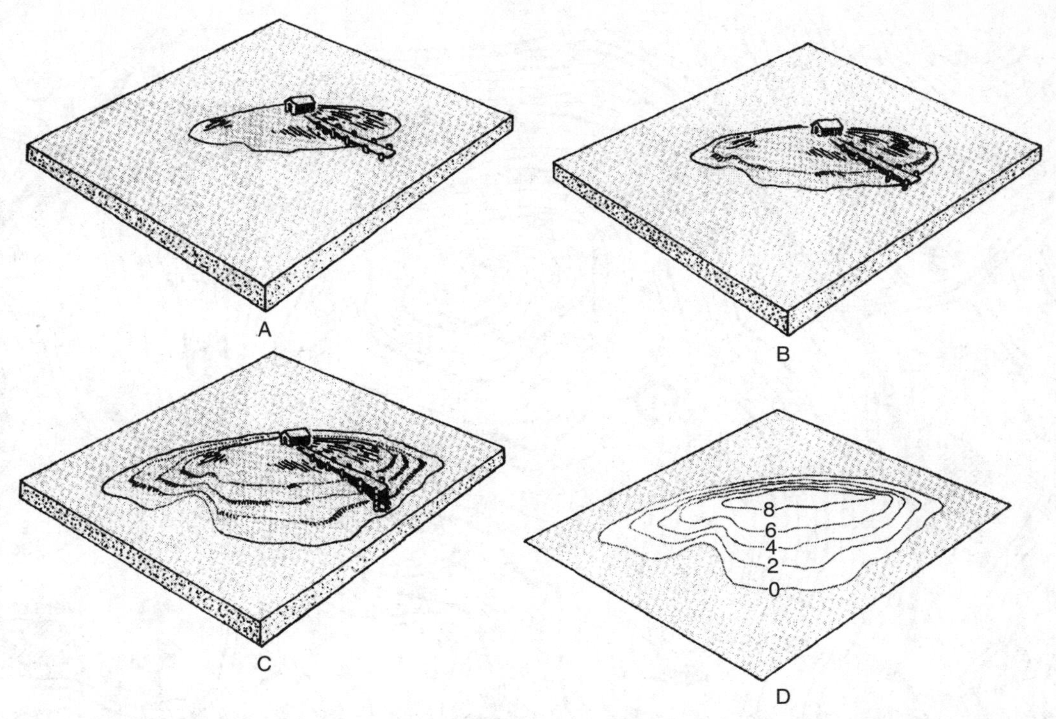

Figure 1-12. To better understand contour maps, think about a receding flood. As the water level falls, successive shorelines form. Each new shoreline can be thought of as a topographic contour line.

From the shape of a set of contours you can recognize landscape features, such as hills, valleys, depressions, *escarpments* (cliffs), etc., as illustrated in Figure 1-13 on page 28. A topographic map may also use other symbols or colors to represent bodies of water, vegetation, buildings, and roads. The maps usually have a key or legend describing the symbols.

Note the following numbered features of the topographic contour map in Figure 1-13:

1. The contour lines connect places with the same elevation. If you can visualize the shapes represented on this map, you have acquired an important skill in map reading. The darker lines are the index contours. Some of the index contours are labeled with their elevation. The line labeled 1 is 450 feet above sea level. What is the contour interval on this map? (*Ans.* 10 ft.)
2. This is a stream valley. Streams are located in the bottom of their valleys. Notice that where contour lines cross streams, the contour lines make a V shape. Also note that the Vs always point upstream. In this case, the Vs are upside down; this shows that the stream flows to the west.
3. The slope (gradient) is steepest where contour lines are closest.
4. The highest elevations on this map are found where the contour lines make small closed curves. On this map it is at least 570 feet, but not as high as 580 feet. (Otherwise, there would be a 580-feet contour line.)
5. The gradient is the least (flattest) where the contour lines are farthest apart.
6. A closed depression is shown by a contour line with small marks (hachures) pointing toward the lowest part inside the depression. On this map, the first depression contour must have an elevation of 260 feet, the same elevation as the next contour line to the north. The second depression contour is marked 250 feet. Therefore the elevation at the center of this closed depression must be a little lower than 250 feet above sea level.

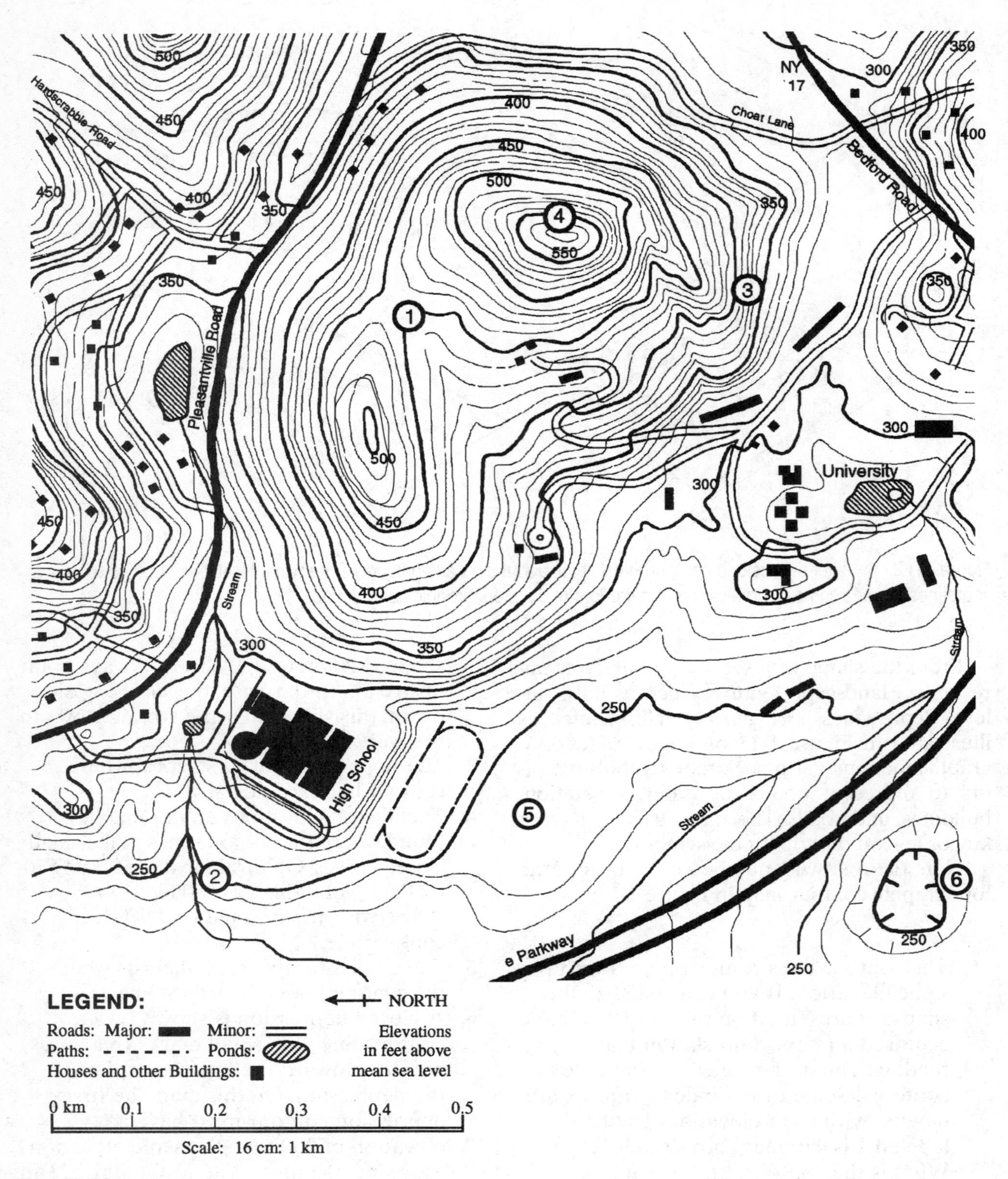
500
450
400
350
300
250
550
NY
17
Choat Lane
Bedford Road
Hardscrabble Road
Pleasantville Road
Stream
University
High School
e Parkway
1
2
3
4
5
6
LEGEND:
NORTH
Roads: Major:
Minor:
Paths:
Ponds:
Houses and other Buildings:
Elevations in feet above mean sea level
0 km
0.1
0.2
0.3
0.4
0.5
Scale: 16 cm: 1 km

Figure 1-13.

QUESTIONS

Part A

29. The mineral sample in the picture below is 9 centimeters long. The actual sample is how many times the size of this image?

(1) 1 (2) 2 (3) 3 (4) 4

30. Unlike most road maps, topographic maps can be used to find (1) the best route to a town 20 kilometers away (2) the temperature at any place on the map (3) the elevation above sea level (4) the distance between two locations

Part B

Base your answers to questions 31 through 38 on the York Towne Regional Topographic map below.

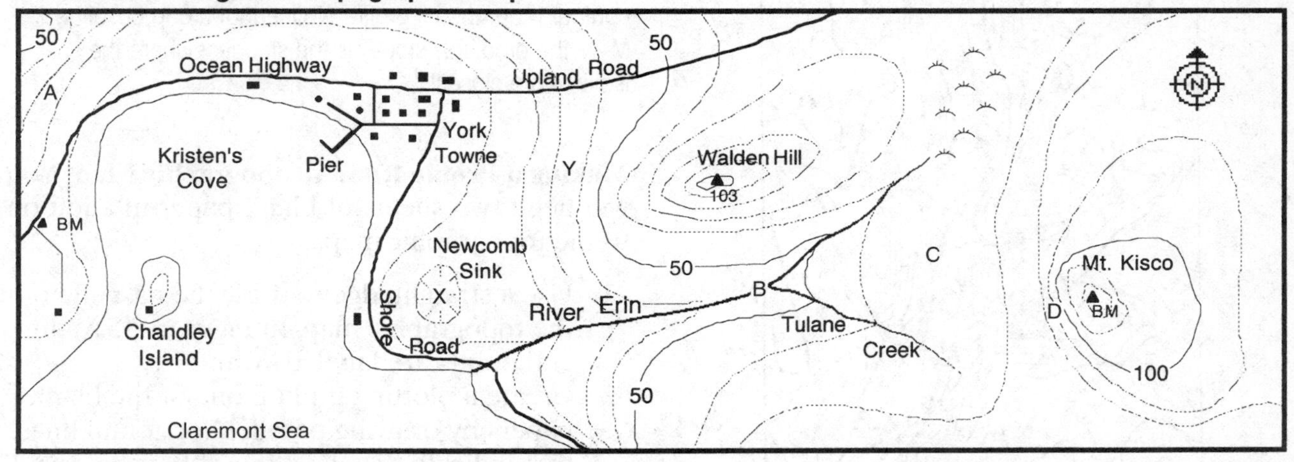

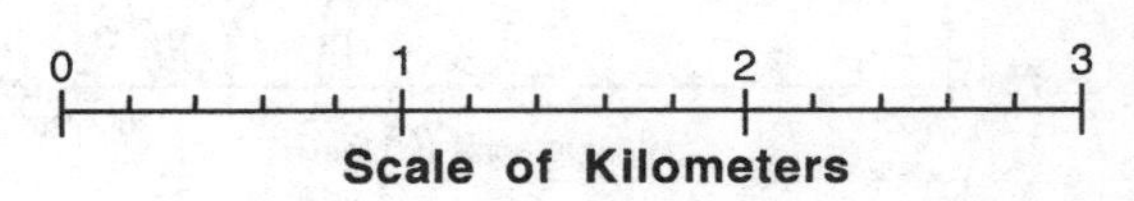

31. Based on the scale given on the map, the distance from Walden Hill to Mt. Kisco is (1) 0.7 km (2) 1 km (3) 1.3 km (4) 2.35 km

32. In what direction does the River Erin flow? (1) north (2) south (3) southwest (4) northwest

33. What is the vertical contour interval on this map? (1) 5 m (2) 10 m (3) 50 m (4) 100 m

34. What is a possible elevation for the bench mark (BM) on the top of Mt. Kisco? (1) 100 ft (2) 125 ft (3) 130 ft (4) 140 ft

35. If you were to walk from Walden Hill to Mt. Kisco, how would your elevation change? (1) Your elevation would increase only. (2) Your elevation would decrease only. (3) Your elevation would decrease, and then increase. (4) Your elevation would increase, and then decrease.

36. What is the elevation of the house on Chandley Island? (1) 0 m (2) 10 m (3) 20 m (4) 30 m

37. Where is the slope (gradient) of the land greatest? (1) *A* (2) *B* (3) *C* (4) *D*

38. What is the lowest point on this map? (1) shore of Kristen's Cove (2) Newcomb Sink (3) point *A* (4) point *C*

Base your answers to questions 39 and 40 on the map below. Points A, B, C, X, and Y are locations on the topographic map.

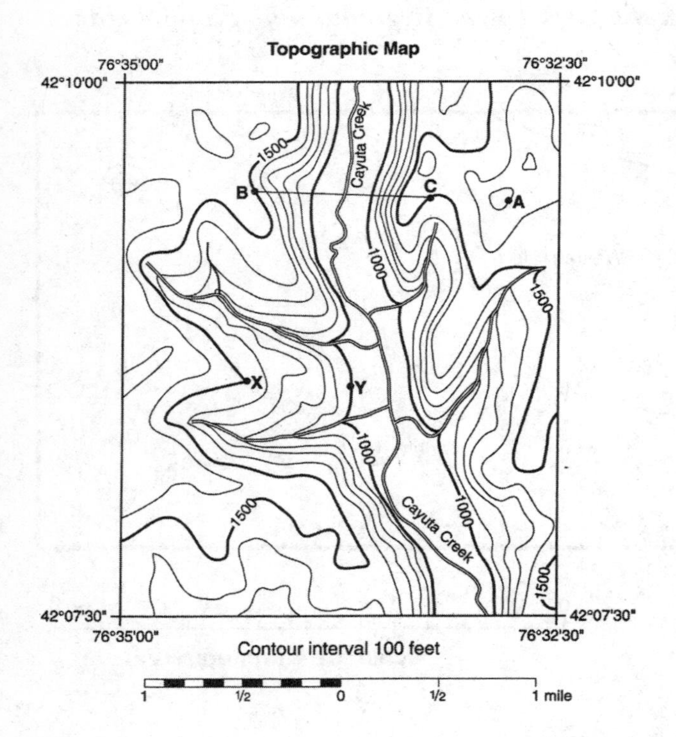

39. What is the elevation of point *A*? (1) 1700 ft (2) 1650 ft (3) 1600 ft (4) 1550 ft

40. What is the approximate gradient between points *X* and *Y*? (1) 100 ft/mi (2) 250 ft/mi (3) 500 ft/mi (4) 1000 ft/mi

Topographic Profiles

Topographic maps, which include information about length, width, and height of surface features, are used to draw topographic profiles. A **topographic profile** is a cross-sectional view that shows the elevation of the land along a particular baseline. Wherever the baseline crosses a contour line, the exact height above sea level can be plotted on a vertical scale as shown at the bottom of Figure 1-14. A topographic profile is a vertical slice of the topography. It shows the ups and downs along a specific route.

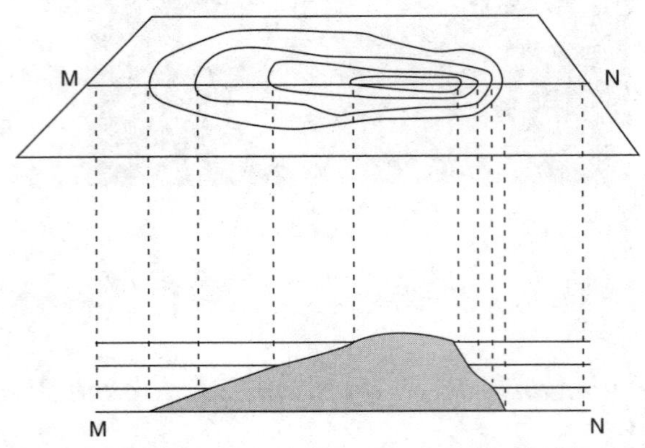

Figure 1-14. This figure shows the relationship between a contour map and its profile. Notice that along baseline M–N, the gradient (slope) is the steepest where the isolines are closest.

Making a Profile from a Topographic Map You will need two sheets of blank paper in addition to the topographic map.

1. Use a straightedge to draw the baseline on the topographic map. In Figure 1-15A, the end points are labeled W and E.
2. Create a plotting grid on one of the blank papers by drawing parallel horizontal lines that indicate the elevation of the contour lines along your profile. These lines need to be as long as the profile line, W–E. To see this, look at the bottom of Figure 1-15A.
3. Label the elevation of each horizontal line at the appropriate contour interval: one line for each elevation crossed by the profile line. (The profile may cross the same elevation several times.)
4. Place an edge of the second sheet of blank paper on the topographic map along the profile line, W–E. Mark W and E at the

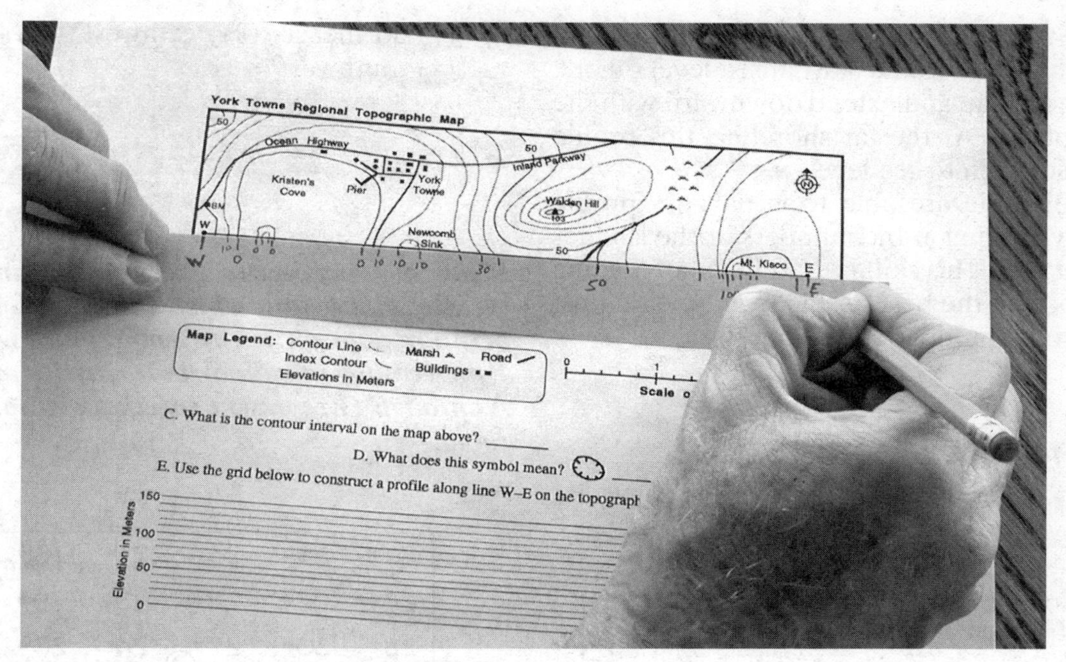

Figure 1-15A. Marking the contour lines on the blank sheet.

appropriate places along this blank edge. Then mark the same edge where the profile crosses each contour line. Label these marks with the elevation of each contour line. See Figure 1-15A.

5. Now move the sheet with the marked edge to the bottom of the parallel lines. Use the marks and elevation labels to project where the profile crosses each horizontal line. See Figure 1-15B.
6. Connect the points on the profile to show a vertical cross section along line W–E. See Figure 1-15B.

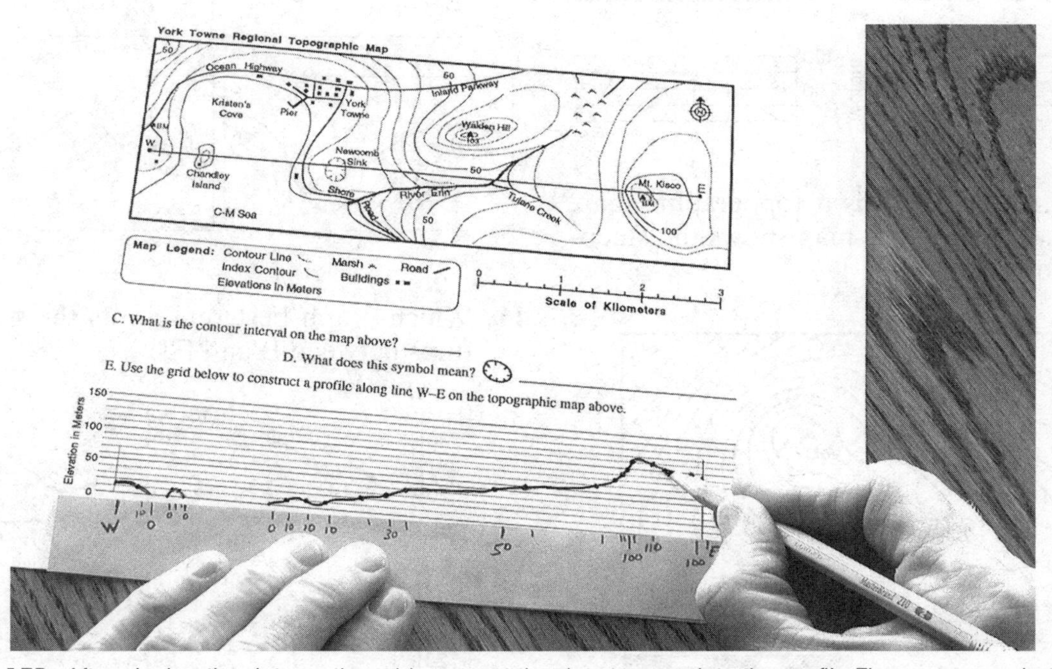

Figure 1-15B. After placing the dots on the grid, connect the dots to complete the profile. There are gaps in this profile along the water surface between the shore and Chandley Island in Kristen's Cove

Imagine a profile of the bottom of an ocean. An ocean profile would start at sea level on one side of the ocean and extend downward with the ocean bottom. At the far shoreline, this profile would rise again to sea level.

Some people are able to visualize a profile simply by looking at the numbers or the isolines on a field map. This skill is useful in map reading or in choosing the best profile from several possibilities in a multiple-choice question.

QUESTIONS

Part A

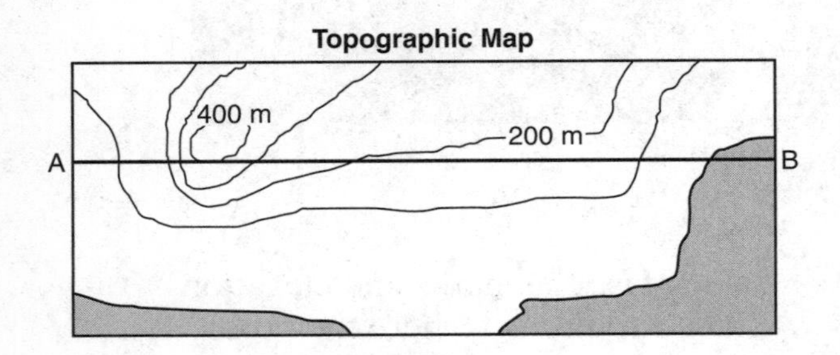

41. Which profile from *A* to *B* below best represents the isoline map above?

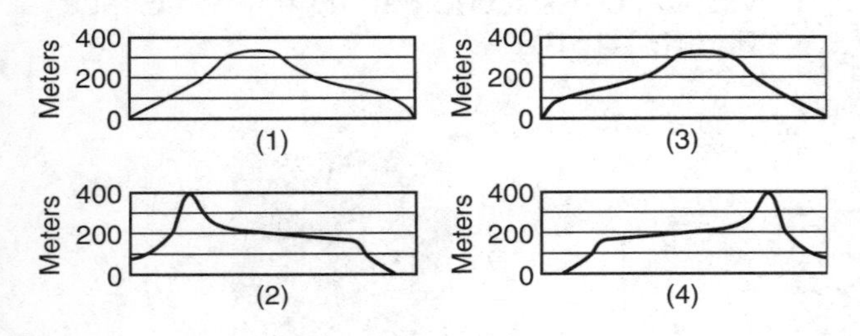

42. The diagram below is a topographic map. Which section of the map shows the steepest gradient?

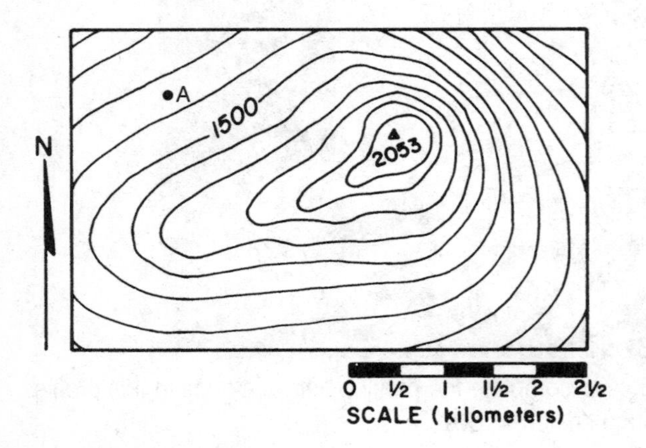

(1) northeast (2) northwest (3) southeast (4) southwest

Part B

Base your answers to questions 43 and 44 on the diagram below, which shows a plot of temperature data (°C). The temperature was taken 1 meter above the floor in a classroom. Letters A through G represent specific locations in the room.

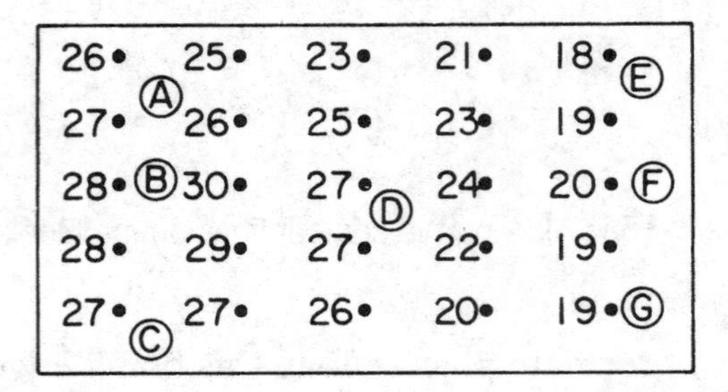

43. Which of the four diagrams below best represents the 26°C isotherm?

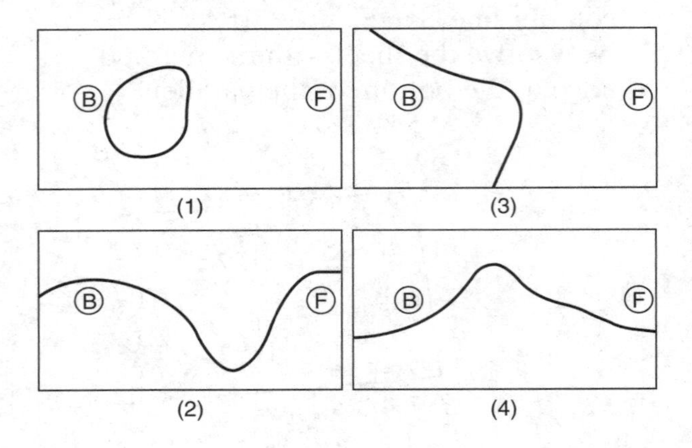

44. Which graph best represents the temperatures between B and F?

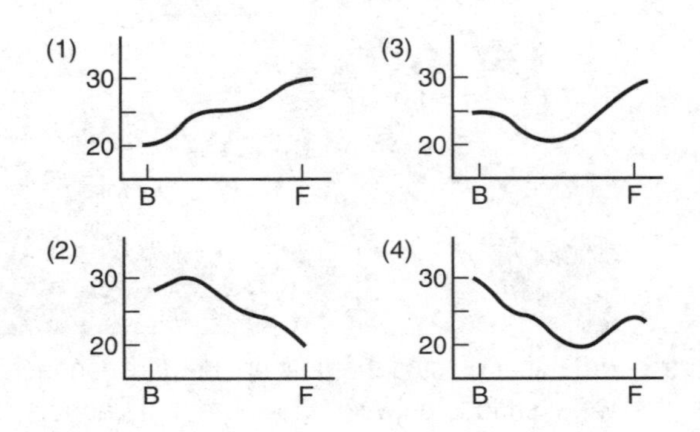

Determining a Gradient, or Slope You can determine the average rate of change of a value within a field by finding the gradient between two points. **Gradient**, or **slope**, is the rate of change in a field. It is calculated by dividing the difference in field values at two points by the distance between those two points. For example, the average slope, or gradient, between any two points (A and B) on a mountain can easily be determined from a contour map. If you know how many meters the mountain drops or rises between A and B in a given distance, you can calculate the average gradient by using the following equation (also found on page 1 of the *Earth Science Reference Tables*):

$$\text{gradient} = \frac{\text{change in field value}}{\text{distance}}$$

$$= \frac{\text{difference in elevation between } A \text{ and } B \text{ (m)}}{\text{distance between } A \text{ and } B \text{ (km)}}$$

Please notice that the units of gradient will be the unit of the field value divided by unit of distance. Since the field value in the example above is elevation in meters, the unit of gradient is meters/kilometer (m/km). If the field value were the temperatures in a classroom, the unit might be °Celsius/meter (°C/m).

Sample Problem

Calculate the average slope of a mountain trail from the 980-meter contour line to the 480-meter contour. The distance between these two elevations measures 4 kilometers.

Solution

$$\text{gradient} = \frac{\text{change in field value}}{\text{distance}}$$

$$= \frac{\text{difference in elevation (m)}}{\text{distance between points (km)}}$$

$$= \frac{980.\ \text{m} - 480.\ \text{m}}{4.00\ \text{km}}$$

$$= \frac{500.\ \text{m}}{4.00\ \text{km}} = 125\ \text{m/km}$$

QUESTIONS

Part A

45. The top of a hill is 1000. meters above a point at the bottom of the hill that is 3000. meters away in horizontal distance. What is the average gradient between these two points? (1) 0.3333 m/m (2) 30.00 m/m (3) 333.0 m/m (4) 3000. m/m

46. The source of a stream is at an elevation of 1861 meters. The stream ends in a lake at an elevation of 361 meters. If the lake is 100. kilometers away from the source, what is the average gradient of the stream?
(1) 1.50 m/km (3) 10.0 m/km
(2) 0.50 m/km (4) 15.0 m/km

PART B

47. The distance from point X at the road intersection to point Y near the top of the hill on the map below is 0.2 mile. What is the average gradient from X to Y?

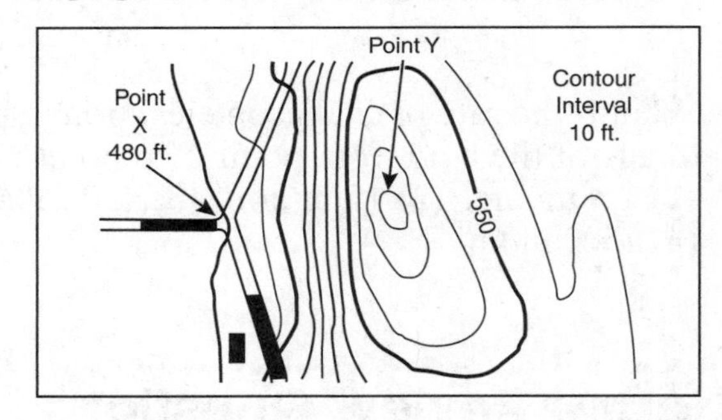

(1) 100 ft/mi (2) 200 ft/mi (3) 400 ft/mi
(4) 500 ft/mi

Base your answers to questions 48 through 52 on the topographic map below, which represents a coastal landscape. The contour lines show the elevations in meters.

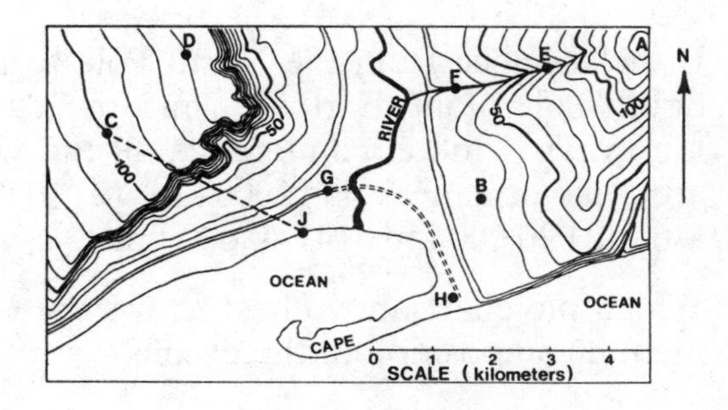

48. What is the approximate distance from point *G* to point *B*? (1) 2.1 km (2) 2.6 km (3) 3.2 km (4) 4.0 km

49. Points *F* and *E* are along a small stream. In what general direction does that stream flow? (1) north (2) south (3) east (4) west

50. What is the elevation of the highest contour line shown on the map? (1) 100 m (2) 140 m (3) 155 m (4) 180 m

51. Which diagram best represents the profile along a straight line between points *C* and *J*?

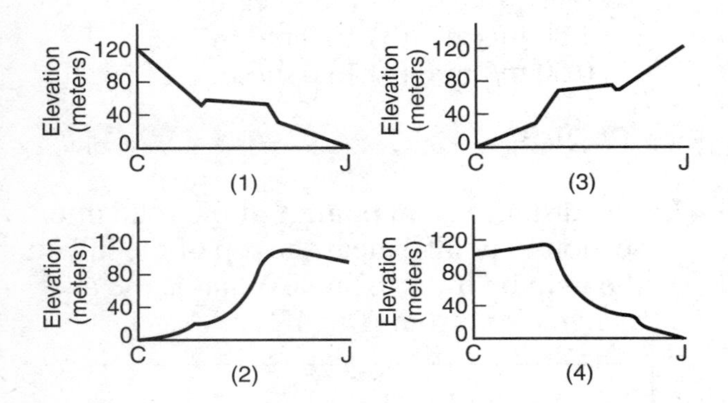

52. What is the rate of change in elevation (gradient) of the land from point *C* to point *D*? (1) 1.5 m/km (2) 15 m/km (3) 150 m/km (4) 1500 m/km

Chapter Review Questions

PART A

1. The International Space Station revolves around Earth once every 91.34 minutes. Through how many degrees of longitude does it travel in this time? (1) 90° (2) 180° (3) 270° (4) 360°

2. If the distance from the North Pole to the South Pole along Earth's surface is 20,000 km, what is the circumference all the way around our planet? (1) 5000 km (2) 10.000 km (3) 20,000 km (4) 40,000 km

3. Which pie chart above best shows the elemental composition of the oceans?

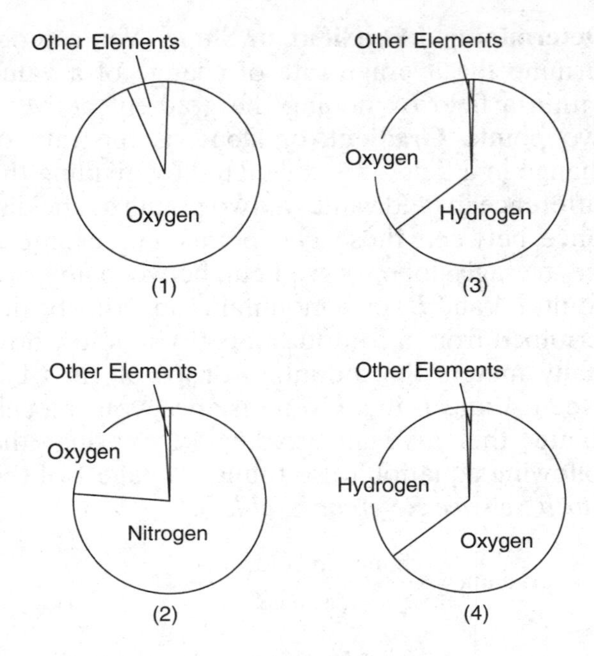

4. Which city is the farthest from Earth's center? (That is, at which city is Earth's spherical radius greatest?) (1) Anchorage, Alaska (2) New York, New York (3) Atlanta, Georgia (4) Miami, Florida

5. How high above Earth's surface are you likely to find the coldest part of Earth's atmosphere? (1) at sea level (2) 12 kilometers (3) 50 kilometers (4) 82 kilometers

Base your answers to questions 6 through 9 on the diagram below, which gives temperature field data collected in a room that contains an operating heater.

Temperature Field in a Room (°C)

N

25 26 27 28 28 29 **B** 29

A

26 27 28 29 30 30 30

27 28 30 30 31 31 31

C

25 30 33 36 34 32 **D** 32

Scale: Distance in meters

0 5 10 15 20

6. Along which wall is the heater located? (1) north (2) south (3) east (4) west

7. Near which letter is the temperature gradient greatest? (1) *A* (2) *B* (3) *C* (4) *D*

8. Which diagram below best shows the position of the 34°C isotherm in the temperature field?

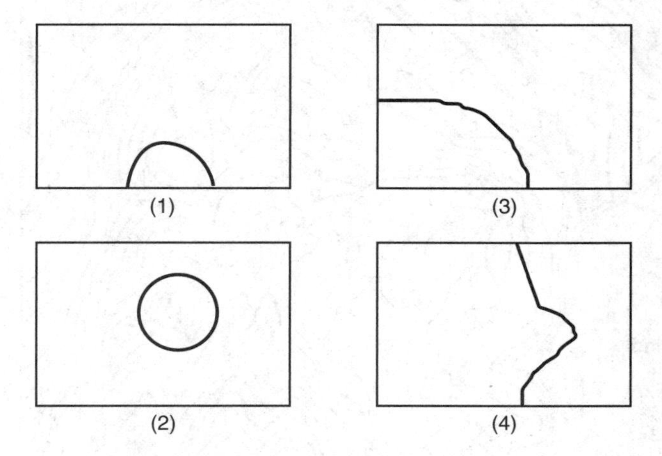

9. What is the gradient between points *B* and *D*? (1) 0.1 °C/m (2) 0.2 °C/m (3) 3 °C/m (4) 15 °C/m

10. At what altitude in the night sky will observers in Niagara Falls, New York, see Polaris? (1) 43° (2) 45° (3) 79° (4) 90°

11. Which of the following units could be used to show the land gradient on a topographic map? (1) °C/kilometer (2) miles/hour (3) degrees of angle (4) meters/kilometer

12. What are the most common units of measure for expressing longitude? (1) km east or west (2) km north or south (3) degrees east or west (4) degrees north or south

13. Astronauts far enough away in space can see our entire planet. If they carefully measure the shape of planet Earth, from what position would an astronaut be able to measure the most flattening?

(1) above the North Pole (2) above the South Pole (3) above the equator (4) above New York City

14. Students in New York City and Buffalo, New York, observe Polaris on the same clear night. How will their observations differ? (1) Polaris will appear higher in the sky in New York City (2) Polaris will appear higher in the sky in Buffalo (3) Polaris is never visible in New York City (4) Polaris is never visible in Buffalo

PART B

Base your answers to questions 15 through 17 on the diagram below, which shows the positions of stars in the northern part of an observer's night sky.

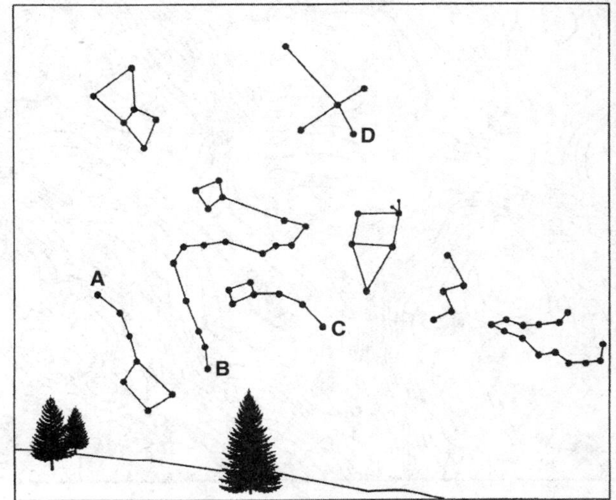

15. Which letter represents Polaris (the North Star)? (1) *A* (2) *B* (3) *C* (4) *D*

16. What is the common name of the group of stars that includes star *A*? (1) the Little Dipper (2) the Big Dipper (3) the Hunter (4) the Scorpion

17. Star *C* is 30° above the horizon at the time shown in this diagram. Which is the most likely Earth location where these observations were made? (1) 30°N (2) 30°S (3) 90°N (4) 90°S

Base your answers to questions 18 through 27 on the topographic map below, which shows a

location in Harriman State Park, about 30 miles northwest of New York City.

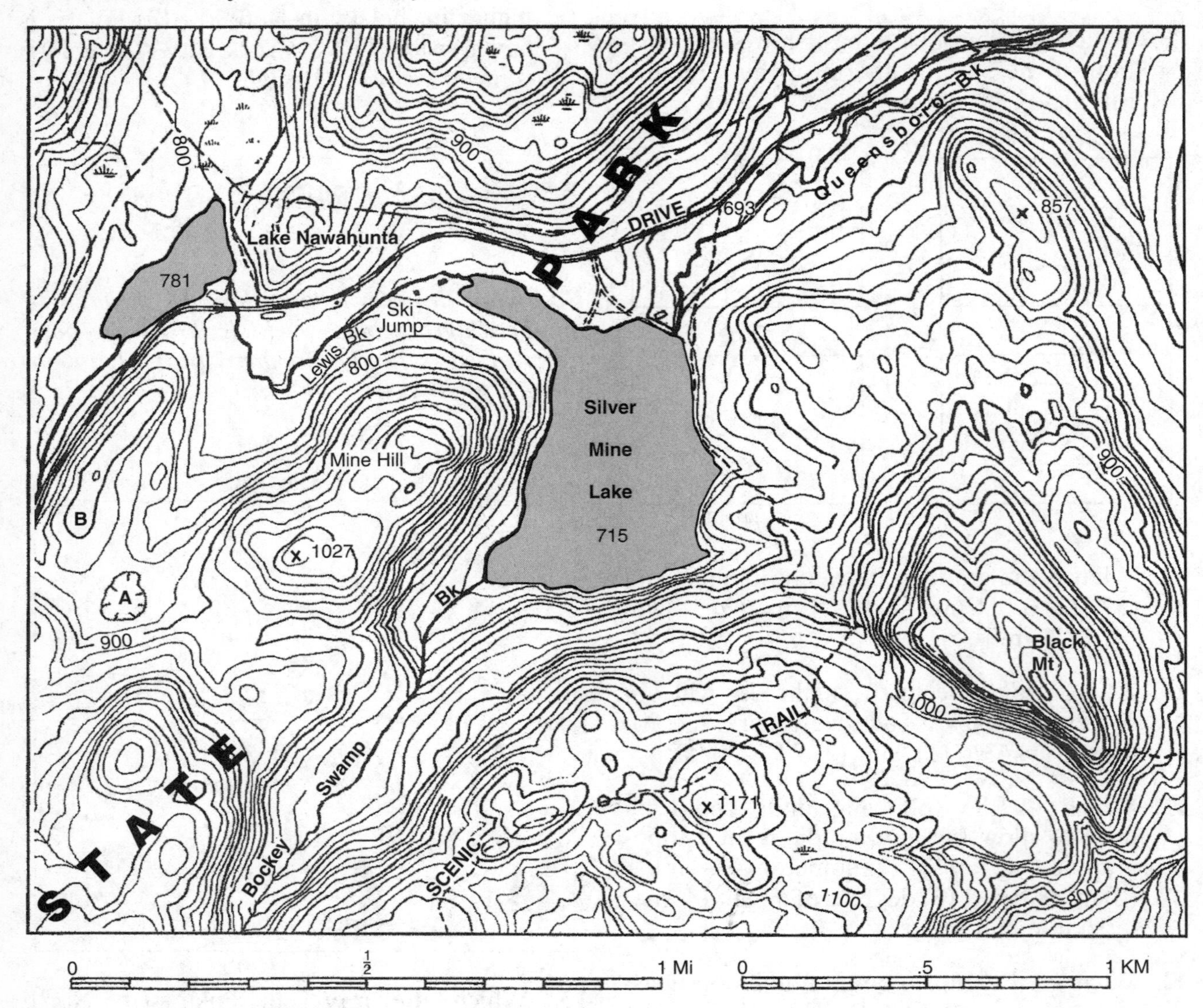

18. What is the length of Lake Nawahunta? (1) 0.2 km (2) 0.3 km (3) 0.4 km (4) 0.5 km

19. Which side of Black Mountain is the steepest? (1) northeast (2) northwest (3) southeast (4) southwest

20. In what general direction does Lewis Brook flow? (1) northward (2) southward (3) eastward (4) westward

21. What is the contour interval on this map? (1) 10 feet (2) 20 feet (3) 50 feet (4) 100 feet

22. What is the approximate elevation of top of Mine Hill? (1) 1000 feet (2) 1010 feet (3) 1020 feet (4) 1030 feet

23. In the southwestern part of the map are letters *A* and *B*. If you were to walk from *A* straight to *B*, how would your elevation change? (1) You would walk uphill, continuously. (2) You would walk downhill, continuously. (3) You would walk uphill, then downhill. (4) You would walk downhill, then uphill.

24. What is the approximate elevation at location *A*? (1) 850 feet above sea level (2) 870 feet above sea level (3) 890 feet above sea level (4) 910 feet above sea level

25. What feature occupies the largest area of relatively level land that is not completely covered by water? (1) the trail near the southern edge of the map (2) Mine Hill (3) Silver Mine Lake (4) Lewis Brook

26. What value is common to every location along a single contour line?

27. Near where will you find the lowest elevation on this map ?

PART C

28. The diagram below represents Celsius temperatures in a schoolroom. Draw isotherms at an interval of 2°C on this field map.

25	26	27	27	26
24	25	27	28	27
25	26	28	29	28
25	26	26	25	26

Base your answers to questions 29 and 30 on the topographic map of Bear Creek, Alaska, and a profile grid below.

29. Draw a profile from *A* to *B* on the grid below the map.

Bear Cheek, AK

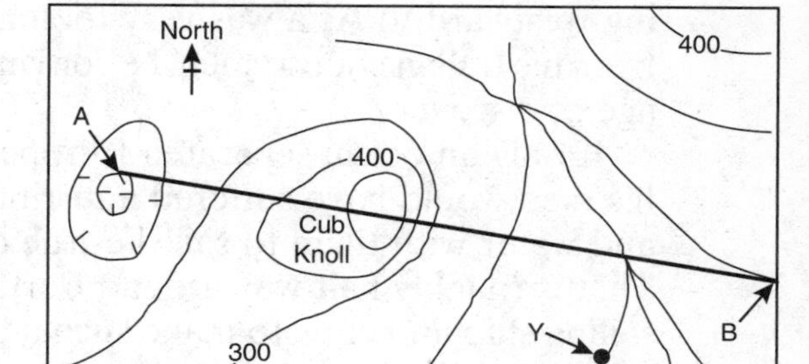

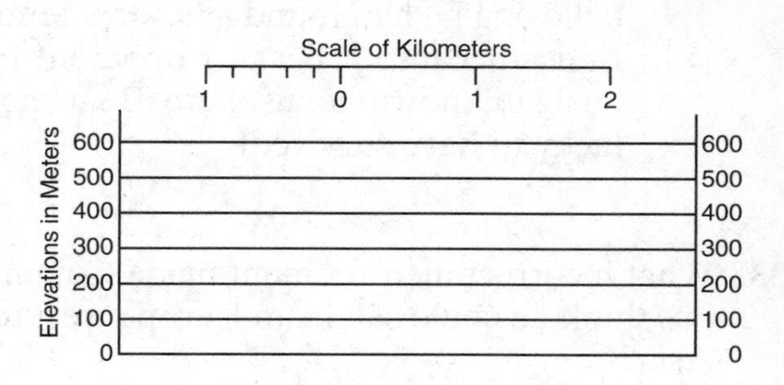

30. Use the Bear Creek map to calculate the average gradient between points *X* and *Y*.

31. To find your longitude without using a map or printed materials, what two time measurements do you need?

32. State two different kinds of observations you could make on Earth's surface to show that Earth is a sphere.

Base your answers to questions 33 through 37 on the passage below and on your knowledge of Earth science.

Just What Did Columbus Prove?

Some people say Christopher Columbus proved that Earth is round. However, when they say this they are probably confusing Columbus with Ferdinand Magellan, the first captain whose ship sailed all the way around Earth. This was 15 years after the death of Columbus. (Unfortunately, hostile natives killed Magellan before his ship returned to Europe.)

Columbus accepted the (incorrect) estimate of Posidonius for Earth's circumference. The Posidonius estimate was only about half Earth's true size. Columbus had an idea how far east Asia was, so he figured that a relatively short trip westward across the Atlantic Ocean would take him to Asia. If he had accepted the correct calculations of Eratosthenes, he would have known that sailing westward to Asia was very foolhardy. Unable to obtain a ship and funding from his Italian countrymen, he convinced the Queen of Spain to finance his voyage of discovery.

If only an ocean separated Europe from Asia to the west, Columbus and all his men would have suffered a terrible fate. They could not bring enough food and water with them to sail the true distance to Asia, about 20,000 kilometers. That is roughly half way around Earth. In the late 1400s it was impossible for a sailing ship and crew to make such a long journey.

We now know that Columbus landed in the West Indies near modern Cuba and Puerto Rico. They are called the "Indies" because Columbus thought he had sailed to Asia (India). In fact, he made four voyages to the Americas and he died believing he had found a new route to Asia. So we now call this area the "West" Indies because it became apparent later that Columbus had sailed to a previously unknown hemisphere. Being so wrong, Columbus and his men were very lucky to have survived!

33. What incorrect measurement made Columbus think he could sail from Europe west to Asia?

34. According to the *Earth Science Reference Tables*, what is Earth's diameter?

35. From your knowledge of celestial navigation, why was Columbus unable to tell how far west (in longitude) he had actually sailed?

36. Posidonius estimated Earth's circumference to be roughly 25,000 kilometers. But today we know Earth's circumference is 40,000 km. What was the percent error of the estimate of Posidonius? Please show your calculation starting with the formula for percent deviation.

37. It was relatively easy for navigators to find their latitude at the time of Columbus. They needed only one measurement. What is that measurement?

CHAPTER

Minerals, Rocks, and Resources

Vocabulary at a Glance

banding	foliation	natural resource
bioclastic	fossil	nonrenewable resource
bond	fossil fuel	organic rock
cement	fracture	origin
clastic	fragmental	plutonic
cleavage	hardness	regional metamorphism
cleavage planes	igneous rock	renewable resource
conglomerate	inorganic	rock
contact metamorphism	intrusive	rock cycle
crystal	lava	sedimentary rock
crystalline sedimentary rock	luster	silicates
crystallize	mafic	specific gravity
double refraction	magma	streak
evaporite	metamorphic rock	texture
extrusive	mineral	vesicular
felsic	Mohs scale	

MINERAL RESOURCES

Earth's mineral resources include minerals, rocks, and fossil fuels (coal, oil, and natural gas). Minerals are an important part of our lives. Iron, copper, aluminum, gold, and other metals come from minerals or combinations of minerals mined from Earth. Minerals are also essential ingredients in computer chips, concrete, and even foods. Rocks, which are made up of minerals, make up the planet on which we live. Rocks weather, or wear away, forming soil and sediments. Although fossil fuels are found in naturally occurring underground deposits, they are not minerals. This is because they are organic, that is, formed from the remains of living things.

What Are Minerals?

Minerals are naturally occurring, inorganic, homogeneous (uniform) solid substances. **Inorganic** means that the substance was not formed by or from living things such as plants or animals. As a result, inorganic substances such as minerals do not contain the complex carbon compounds made by living things.

Some minerals are chemical elements while others are compounds or mixtures. Graphite and diamond are different forms of elemental carbon (C). Quartz (SiO_2, also called silica), and calcite ($CaCO_3$) are chemical compounds composed of elements in set ratios. A third group of minerals are mixtures, for which exact chemical formulas cannot be written. For example, minerals in the feldspar family contain potassium, calcium, and/or sodium in varying proportions along with silicon, oxygen, and aluminum.

Although scientists have identified and classified about 3000 minerals, fewer than 12 are found commonly. Most of the rocks found near Earth's surface are made up of clay, feldspar, quartz, and calcite. Pyroxene, amphibole, and olivine are common deep underground, but less common at or near the surface. The elements oxygen and silicon make up more than half of the mass of these minerals (**silicates**). With just another dozen elements, you can account for about 98 percent of Earth's mass. Figure 2-1 shows the chemical composition of Earth's crust by mass and by volume.

You can usually identify minerals by their physical and chemical properties. Always use a fresh, unweathered sample or an unweathered surface when you try to identify a mineral. Table 2-1 can help you identify about two dozen of the most common minerals. This table is from page 16 of the *Earth Science Reference Tables*. You should become very familiar with its use. The next few pages will help you understand the properties on this table that are most useful in mineral identification.

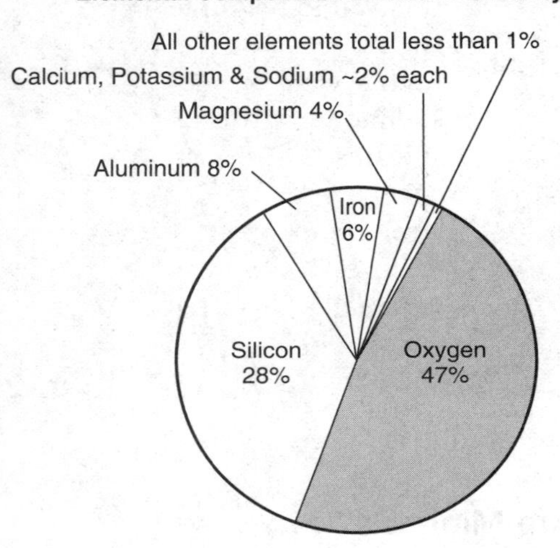

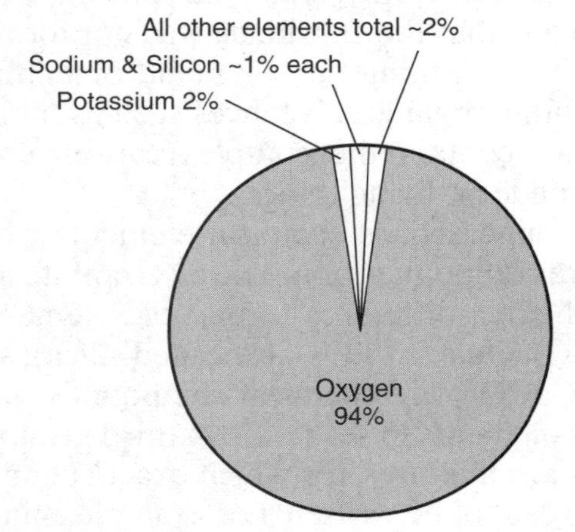

Figure 2-1. Oxygen and silicon make up most of Earth's crust by mass and by volume. The difference in these graphs is due to the relatively large size of oxygen ions.

Color Many minerals have a characteristic color. Pyrite ("fool's gold") is usually a brassy yellow, like gold. Almandine garnet is valued for its deep red color. Many colorless or white minerals are often colored by impurities. Quartz and calcite are usually white or colorless. However, they also may be pink, green, or even black, depending on the impurities they contain. Dark-colored minerals, such as the amphibole and pyroxene families, are easier to identify by color because impurities are unlikely to cause color variations.

Luster The way light is reflected from the freshly cut surface of a mineral produces the mineral's **luster**. Minerals with a metallic luster have a hard, shiny look, like polished metal. This is because light cannot penetrate the surface, so almost all the light is reflected. Minerals with nonmetallic luster can also be shiny, but nonmetallic luster differs from a metallic luster because some of the light is transmitted into or through the mineral while some is reflected. Nonmetallic lusters include glassy, waxy, pearly, and earthy, which is dull.

Streak The **streak** is the powdered form of the mineral. Some metallic minerals, for example hematite, leave behind a powder that is not the same color as the mineral sample. The test for streak is performed by rubbing a fresh corner of the mineral across a white, unglazed porcelain streak plate. Figure 2-2 on page 42 illustrates the test for streak.

Crystal Structure A **crystal** is a regularly shaped solid formed by an ordered pattern of atoms. Minerals form characteristically shaped crystals. Figure 2-3 on page 42 shows characteristic crystal forms of five minerals. Calcite and quartz can look very similar. Both are very common and are often white or colorless with a glassy luster.

Table 2-1. Common Rock-forming Minerals

LUSTER	HARDNESS	CLEAVAGE	FRACTURE	COMMON COLORS	DISTINGUISHING CHARACTERISTICS	USE(S)	COMPOSITION*	MINERAL NAME
Metallic luster	1–2	✔		silver to gray	black streak, greasy feel	pencil lead, lubricants	C	**Graphite**
Metallic luster	2.5	✔		metallic silver	gray-black streak, cubic cleavage, density = 7.6 g/cm^3	ore of lead, batteries	PbS	**Galena**
Metallic luster	5.5–6.5		✔	black to silver	black streak, magnetic	ore of iron, steel	Fe_3O_4	**Magnetite**
Metallic luster	6.5		✔	brassy yellow	green-black streak, (fool's gold)	ore of sulfur	FeS_2	**Pyrite**
Either	5.5 – 6.5 or 1		✔	metallic silver or earthy red	red-brown streak	ore of iron, jewelry	Fe_2O_3	**Hematite**
Nonmetallic luster	1	✔		white to green	greasy feel	ceramics, paper	$Mg_3Si_4O_{10}(OH)_2$	**Talc**
Nonmetallic luster	2		✔	yellow to amber	white-yellow streak	sulfuric acid	S	**Sulfur**
Nonmetallic luster	2	✔		white to pink or gray	easily scratched by fingernail	plaster of paris, drywall	$CaSO_4 \cdot 2H_2O$	**Selenite gypsum**
Nonmetallic luster	2–2.5	✔		colorless to yellow	flexible in thin sheets	paint, roofing	$KAl_3Si_3O_{10}(OH)_2$	**Muscovite mica**
Nonmetallic luster	2.5	✔		colorless to white	cubic cleavage, salty taste	food additive, melts ice	$NaCl$	**Halite**
Nonmetallic luster	2.5–3	✔		black to dark brown	flexible in thin sheets	construction materials	$K(Mg,Fe)_3AlSi_3O_{10}(OH)_2$	**Biotite mica**
Nonmetallic luster	3	✔		colorless or variable	bubbles with acid, rhombohedral cleavage	cement, lime	$CaCO_3$	**Calcite**
Nonmetallic luster	3.5	✔		colorless or variable	bubbles with acid when powdered	building stones	$CaMg(CO_3)_2$	**Dolomite**
Nonmetallic luster	4	✔		colorless or variable	cleaves in 4 directions	hydrofluoric acid	CaF_2	**Fluorite**
Nonmetallic luster	5–6	✔		black to dark green	cleaves in 2 directions at 90°	mineral collections, jewelry	$(Ca,Na)(Mg,Fe,Al)(Si,Al)_2O_6$	**Pyroxene** (commonly augite)
Nonmetallic luster	5.5	✔		black to dark green	cleaves at 56° and 124°	mineral collections, jewelry	$CaNa(Mg,Fe)_4(Al,Fe,Ti)_3Si_6O_{22}(O,OH)_2$	**Amphibole** (commonly hornblende)
Nonmetallic luster	6	✔		white to pink	cleaves in 2 directions at 90°	ceramics, glass	$KAlSi_3O_8$	**Potassium feldspar** (commonly orthoclase)
Nonmetallic luster	6	✔		white to gray	cleaves in 2 directions, striations visible	ceramics, glass	$(Na,Ca)AlSi_3O_8$	**Plagioclase feldspar**
Nonmetallic luster	6.5		✔	green to gray or brown	commonly light green and granular	furnace bricks, jewelry	$(Fe,Mg)_2SiO_4$	**Olivine**
Nonmetallic luster	7		✔	colorless or variable	glassy luster, may form hexagonal crystals	glass, jewelry, electronics	SiO_2	**Quartz**
Nonmetallic luster	6.5–7.5		✔	dark red to green	often seen as red glassy grains in NYS metamorphic rocks	jewelry (NYS gem), abrasives	$Fe_3Al_2Si_3O_{12}$	**Garnet**

*Chemical symbols: Al = aluminum, C = carbon, Ca = calcium, Cl = chlorine, F = fluorine, Fe = iron, H = hydrogen, K = potassium, Mg = magnesium, Na = sodium, O = oxygen, Pb = lead, S = sulfur, Si = silicon, Ti = titanium

✔ = dominant form of breakage

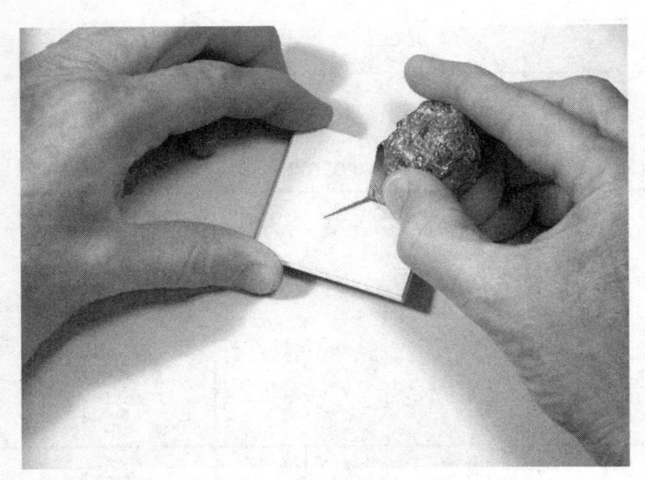

Figure 2-2. The streak test is used to find the color of the powder of a mineral with a metallic luster. A corner of the sample is rubbed across an unglazed porcelain streak plate.

However, their crystal shapes are very different. Quartz crystals are hexagonal (six-sided) in cross section, whereas calcite usually forms rhombohedral crystals. A rhombohedral solid resembles a rectangular solid that has been pushed over to one corner so that all the faces have parallel edges but none meet at right angles, such as the calcite crustal shown in Figure 2-3. Halite, pyrite, and galena form cubic crystals. The crystals of mica grow in "books" of thin, flexible sheets. Many minerals form distinctive and beautiful crystals.

Hardness Minerals also differ in **hardness**, the resistance to being scratched. The stronger the **bonds** between the atoms and molecules in the mineral, the harder it is. You test for hardness by scratching the unknown mineral with the edge or point of other minerals of known hardness. Quartz can scratch most other minerals. Because calcite is softer than quartz, calcite cannot scratch quartz, but quartz can scratch calcite. The geologist Friedrich Mohs created a scale of hardness that is based on relatively common minerals. On the **Mohs scale**, talc is the softest mineral, with a hardness of 1. The hardest natural substance is diamond. Its hardness is 10. The hardness of a fingernail is 2.5, calcite is 3, window glass is 5.5, and quartz has a hardness of 7. See Figure 2-4 for the hardness of other common materials.

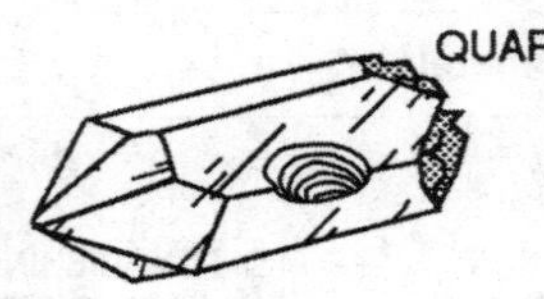

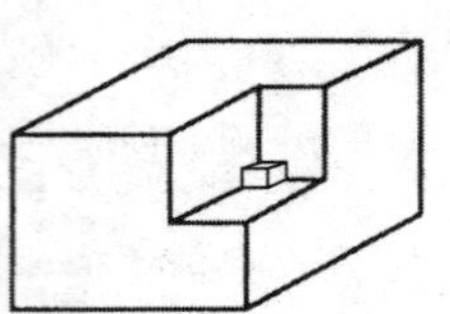

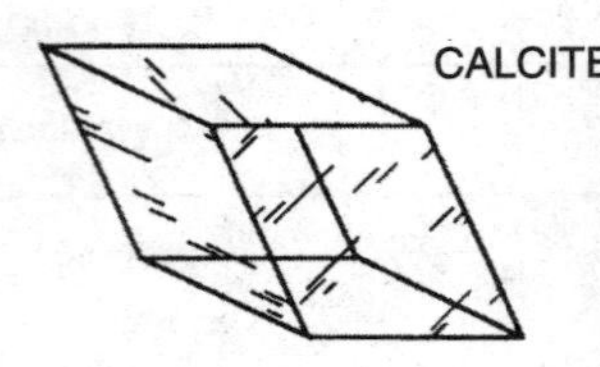

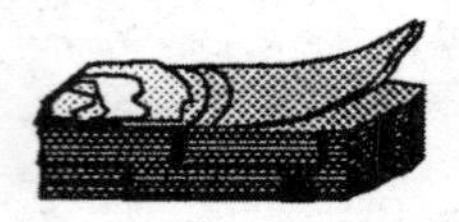

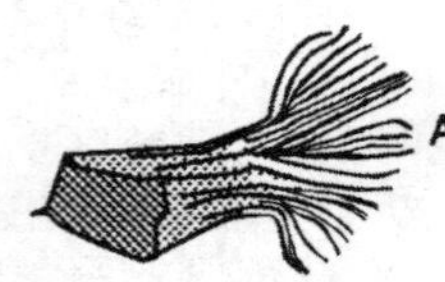

Figure 2-3. Crystal form, fracture, and cleavage are keys to mineral identification.

Cleavage and Fracture If a mineral breaks at specific angles, it shows cleavage. **Cleavage** comes from a Latin word meaning to cut, or slice. Many minerals break along flat surfaces called **cleavage planes.** These cleavage planes are sometimes, but not always, parallel to the sides of crystals. For example, halite (rock salt) breaks into small cubes and rectangular solids, the same shape as its crystals. However, quartz, which forms hexagonal crystals with flat faces, does not break along parallel planes. Minerals that break along even surfaces that do not follow the atomic arrangement (crystal faces) are said to show **fracture**. Quartz breaks along curved surfaces, a property called conchoidal fracture. Figure 2-5 shows cleavage directions in the mica family of minerals and in halite and galena.

Density Minerals also vary in density and specific gravity. The density of many minerals falls

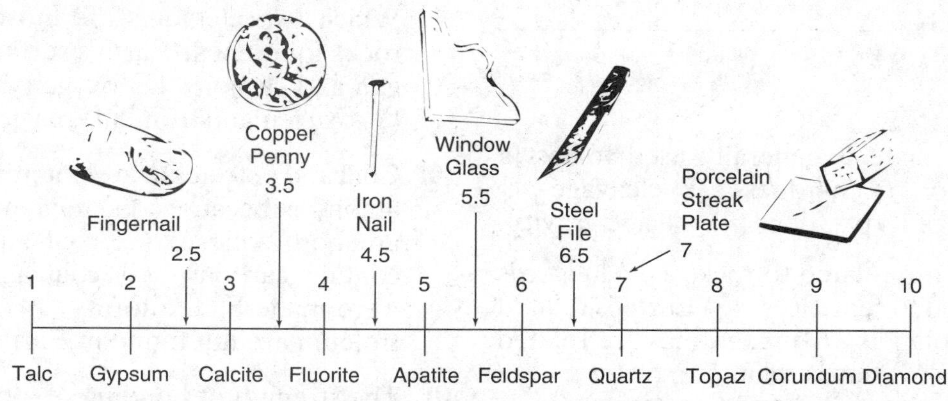

Figure 2-4. Mohs scale of hardness is a relative index based upon the hardness of selected common minerals. This is not an absolute scale. The changes in hardness from step to step are not always uniform. For example, diamond is much harder than corundum, but it is only one step above corundum on the Mohs scale.

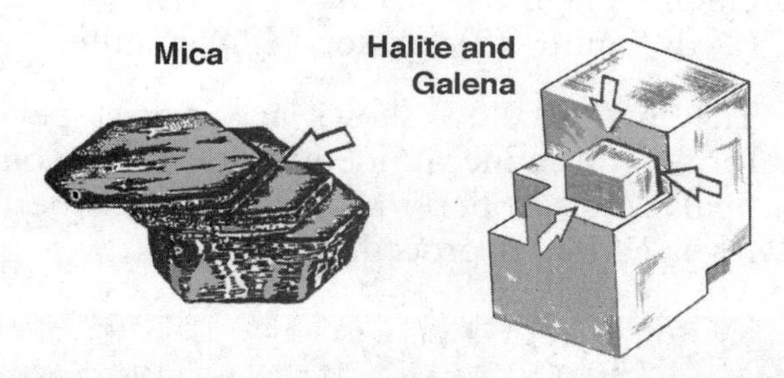

Figure 2-5. Mica minerals (left) show cleavage in one direction only (sheets). Both galena and halite have cubic cleavage.

within the range 2.5 to 3.5 g/cm³. Gold, the densest substance we see commonly, has a density of about 19 g/cm³, and the density of lead is about 12 g/cm³. We can determine density by dividing the mass of a sample by its volume or by flotation. A substance will sink in a liquid that is less dense than the substance and float in a liquid that is more dense than the substance. **Specific gravity** is a ratio of the density of a substance to the density of water. Water's density is 1.0 g/cm³. Therefore, a mineral with a density of 4.0 g/cm³ has a specific gravity of 4.0 (no units).

Unusual Properties Figure 2-6 illustrates some of the special properties of certain minerals that make identification nearly certain. For example, when a printed line or word is looked at through a clear calcite crystal (A), the line or word will appear doubled (**double refraction**). Calcite also bubbles when acid is dropped on it. Uranium (B) is radioactive, which can be detected by a Geiger counter. Some samples of magnetite (C) are so magnetic they will pick up paper clips or small nails.

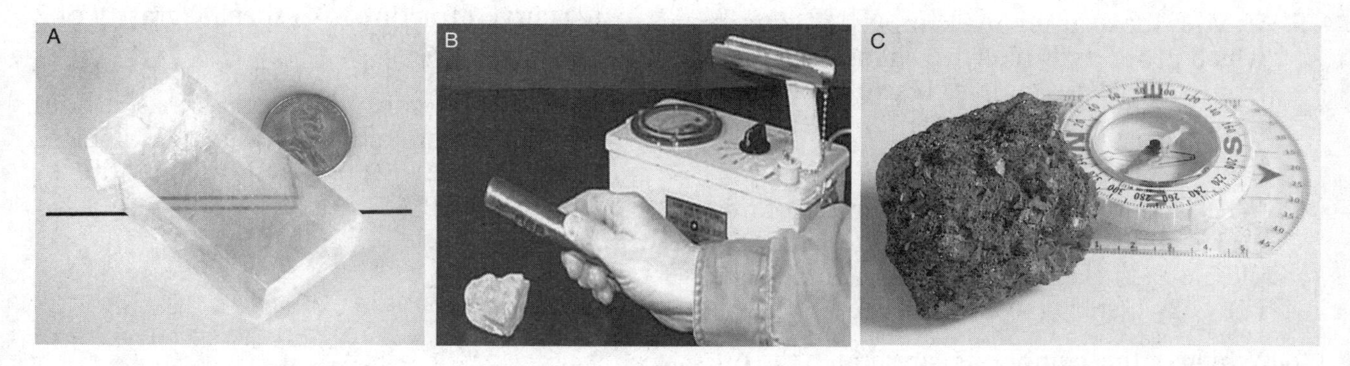

Figure 2-6. Some minerals have unusual properties that make them easy to identify. Calcite crystals (A) show a single line as two lines. Uranium (B) is radioactive. Magnetite (C) attracts iron or deflects a compass needle.

QUESTIONS

Part A

1. A streak plate is generally used to test a mineral for (1) hardness (2) cleavage (3) color of its powder (4) specific gravity

2. Minerals are related to rocks as (1) words are related to sentences (2) Earth is related to other planets (3) magazines are related to books (4) day is related to night

3. What is the best way to determine if a mineral sample is calcite or quartz? (1) Observe the color of the mineral. (2) Place the mineral near a magnetic compass. (3) Place a drop of acid on the mineral. (4) Measure the mass of the mineral.

4. What mineral property is illustrated in the image below?

(1) streak (2) double refraction (3) hardness (4) cleavage

5. If you look at many samples of quartz, which property is likely to change the most, from sample to sample, because of minor impurities? (1) luster (2) color (3) streak (4) density

6. When light falling on a fresh surface of mineral is almost completely reflected, the sample has (1) glassy luster (2) metallic luster (3) waxy luster (4) earthy luster

7. Which of these minerals can scratch the others? (1) galena (2) pyroxene (3) hematite (4) garnet

8. Which two elements are most common in rocks found near Earth's surface? (1) oxygen and carbon (2) oxygen and hydrogen (3) oxygen and iron (4) oxygen and silicon

9. Coal and petroleum are not minerals, but diamond is because (1) coal and petroleum are both solids (2) coal and petroleum contain carbon (3) coal and petroleum were made by life forms (4) coal and petroleum are not found in Earth's crust

10. The strength of bonding of atoms and molecules determine which property of a mineral? (1) luster (2) streak (3) hardness (4) crystal shape

11. Which mineral is a source of the metallic element magnesium (Mg)? (1) pyrite (2) dolomite (3) calcite (4) magnetite

12. The image below shows a geologist performing a mineral identification test on opal. What property of this mineral is best revealed in this procedure?

(1) metallic luster (2) curved fracture (3) hardness of 7 (4) specific gravity of 2

13. Which of the following is a dark-colored mineral that can scratch glass, has a nonmetallic luster, and cleaves at right angles? (1) magnetite (2) pyroxene (3) biotite (4) amphibole

Part B

14. Based on the information in the table, how is olivine unlike the other minerals?

Elemental Composition of Selected Minerals

	O	Si	Al	Fe	Ca	Na	C
Quartz	✓	✓					
Feldspar	✓	✓	✓		✓	✓	
Olivine	✓	✓		✓	✓		
Diamond							✓

(1) olivine is a silicate (2) olivine contains carbon (3) olivine is the hardest (4) olivine contains iron

Base your answers to questions 15 and 16 on the diagram below, which shows three minerals each of which is undergoing a different physical test, A, B, or C.

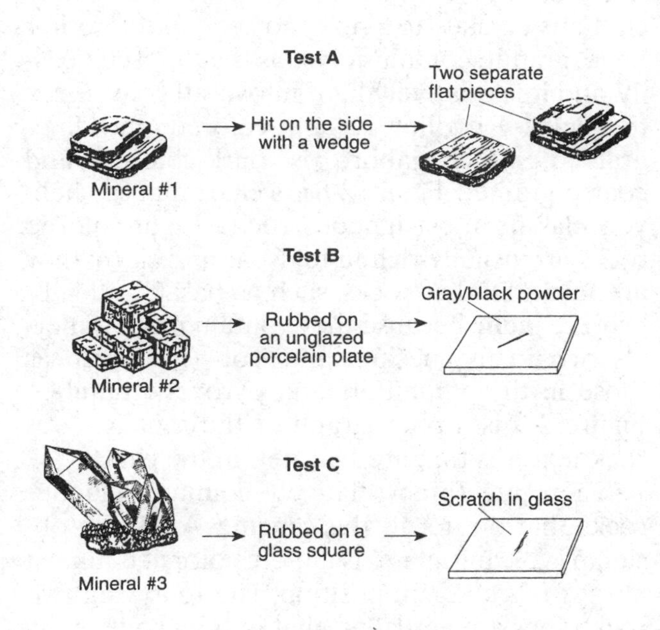

15. Which sequence correctly matches each test, *A*, *B*, and *C*, with the mineral property tested?

(1) *A*–cleavage; *B*–streak; *C*–hardness
(2) *A*–cleavage; *B*–hardness; *C*–streak
(3) *A*–streak; *B*–cleavage; *C*–hardness
(4) *A*–streak; *B*–hardness; *C*–cleavage

16. The results of these three physical tests are most useful in determining the (1) rate of weathering of the minerals (2) identity of the minerals (3) environment where the minerals formed (4) geologic period when the minerals formed

17. Diamond is the hardest substance on Mohs scale. This means that (1) Nothing is more dense than a diamond. (2) It is not possible to break a diamond. (3) Diamonds have a glassy luster. (4) Diamonds can scratch any other mineral.

ROCKS

Nearly all **rocks** are made up of one or more minerals. Limestone is made of calcite, and rock salt is made of the mineral halite. However, most rocks, such as granite, basalt, and sandstone, contain a few minerals. The most common minerals in granite—feldspar, quartz, amphibole, and biotite mica—are shown in Figure 2-7 on page 46. Coal and organic limestone, which form from the remains of plants and animals, respectively, are rocks that are not composed of primarily minerals.

Geologists classify rocks according to their **origin** (how they were formed). **Igneous** rocks form when molten rock, or **magma** (or **lava**, when it reaches the surface), cools and solidifies. **Sedimentary** rocks usually form by the compaction and cementation of layers of sediment. The third kind of rock is metamorphic. **Metamorphic** rocks form when igneous, sedimentary, or other metamorphic rocks are changed by heat and/or pressure.

Texture, the look and feel of surface patterns on a mineral, is the result of the size and arrangement of particles in a rock. We can often infer the way a rock formed (its origin) by examining its texture.

Igneous Rocks

Igneous rocks are always the result of the solidification of magma or lava. Furthermore, only igneous rocks form directly from a hot liquid. When magma cools from 1000°C and 600°C, it **crystallizes**, or becomes a solid. At the temperature of solidification, molecules arrange themselves into an ordered pattern, forming crystals.

The size of the crystals provides a clue to how quickly a rock solidified. Slow cooling allows the molecules enough time to form large crystals.

Primary Minerals in Granite

Figure 2-7. Granite, like most rocks, is made of a variety of minerals. These five minerals are especially common in granite.

Granite and gabbro may have crystals 1 to 10 millimeter across. In general, igneous rocks that cool slowly deep within Earth, where temperatures are just below the melting temperature, are composed of large crystals. These are **plutonic (intrusive)** rocks that are usually composed of large, easily seen mineral crystals. Pegmatite is an igneous rock with very large crystals (10 millimeters or larger) that does not follow this rule. (See Figure 2-8.) Scientists think the large crystals in pegmatite are due to the large amount of water in the magma, and not just because of slow cooling.

Rocks that form at or near Earth's surface usually cool quickly, and therefore have small crystals or no visible crystals. Rapid cooling, such as occurs at the surface of a volcano, produces rock, such as basalt or rhyolite. These rocks have crystals too small to be seen without a magnifier. In fact, volcanic glass (obsidian) cools so quickly that crystals do not have time to form. Fine-grained igneous rocks that cool quickly at or near Earth's surface are called volcanic or **extrusive** rocks. The magma may have been extruded as lava at the surface before it crystallized.

Another characteristic of most igneous rocks is their general lack of layering. However, successive lava flows can form a layered igneous rock structure.

Classification The most common igneous rocks are classified using just two characteristics: crystal size and color. For example, basalt is relatively dark in color (black to gray) and usually contains crystals too small to see. On the other hand, granite contains crystals that are seen easily, and it is relatively light in overall color (gray to pink). Rhyolite is light-colored and fine-grained, while gabbro is dark-colored and coarse-grained. Figure 2-8 is a chart that can help you classify most igneous rocks. Light-colored rocks are usually rich in <u>fel</u>dspar and <u>sil</u>i<u>c</u>a; they are **felsic**. Darker rocks, such as basalt and gabbro, are **mafic** because they contain more minerals rich in <u>ma</u>gnesium and iron (<u>F</u>e), such as those in the amphibole and pyroxene families. Figure 2-9 is a photograph of the four igneous rock textures that are featured in the chart.

There are three relatively common igneous rocks that do not fit this scheme. Although obsidian (volcanic glass) is dark in color, it is almost always felsic in composition. The following two rock types are **vesicular**; that is, they have many holes. As the magma rose to Earth's surface, these holes were formed by expanding gas. Scoria looks like cinders, full of large air pockets. Pumice contains smaller holes that make it look foamy. Sometimes pumice contains so many air pockets it can float in water.

You can read the typical mineral compositions of various igneous rocks by looking directly below them on the Scheme for Igneous Rock Identification on page 6 of the *Earth Science Reference Tables*. For example, granite and rhyolite are about 25 percent potassium feldspar, 40 per-

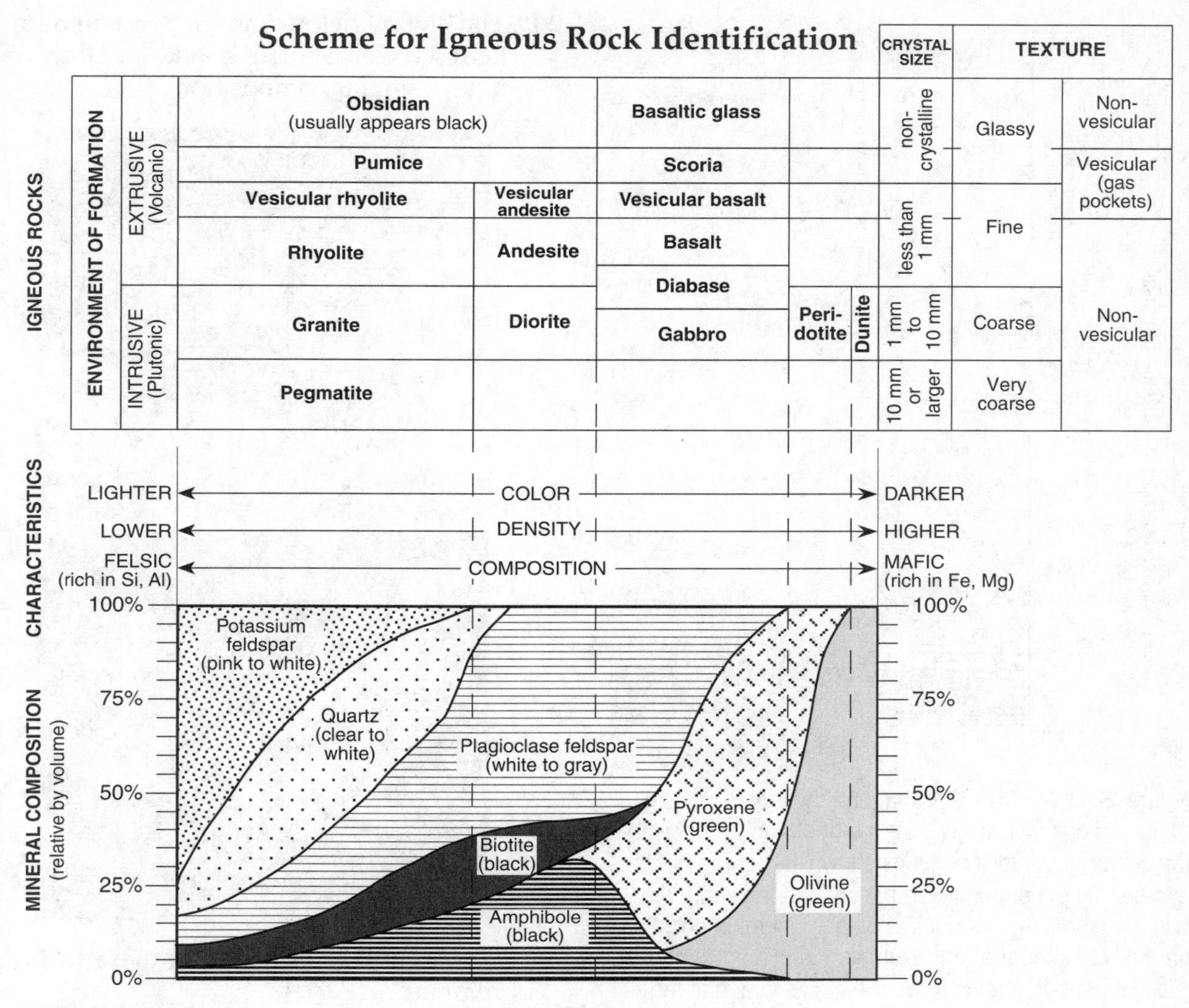

Figure 2-8. This table is also found on page 6 of the *Earth Science Reference Tables.*

Figure 2-9. Note that the four igneous rocks in this image are positioned as they are in the Igneous Rock Identification Chart in Figure 2-8. You can easily see the different textures of these rocks.

cent quartz, 15 percent plagioclase, and 10 percent each of biotite and amphibole. That is a total of 100 percent. Each vertical slice represents a different mineral composition. Figure 2-10 on page 48 illustrates how this is done.

A wide variety of other igneous rocks is found in the Scheme for Igneous Rock Identification in the *Earth Science Reference Tables* (Figure 2-8). From a rock's position on the chart, it is easy to identify its properties and approximate mineral composition.

It is important to remember that igneous rock is the only kind of rock that forms from hot magma or lava, and that all rocks that crystallize from melted rock are igneous.

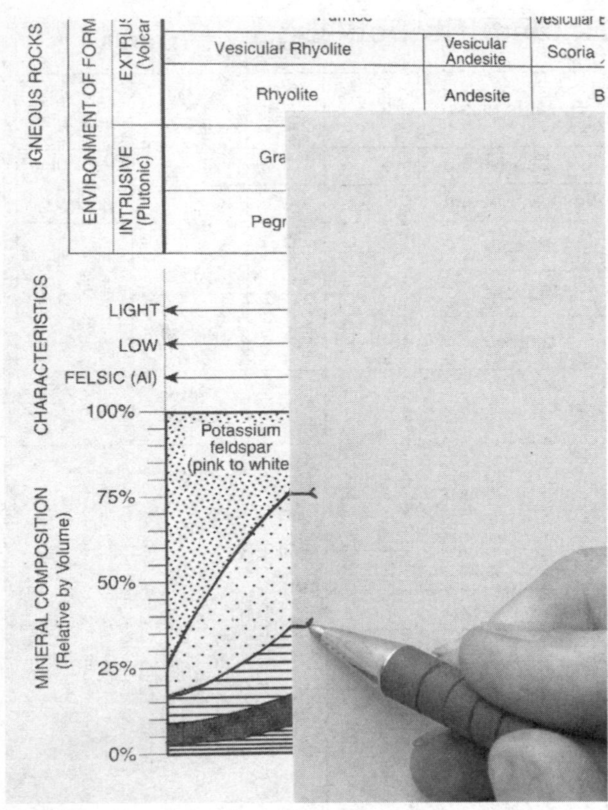

Figure 2-10. To find the percentage mineral composition of an igneous rock, align the edge of a blank paper with the center of the rock name as shown. Be sure to hold the edge of the paper parallel to the edge of the chart. You can then mark the boundaries of the mineral in question on the blank paper. In this example the mineral is quartz. (See Figure 2-8) Finally, move the marked edge to the scale on the left. This example shows that rhyolite, granite, and pegmatite are typically about 40% quartz.

QUESTIONS

Part A

18. Which rock feature is the best indication that a rock has an igneous origin? (1) rounded grains of sediment (2) rounded holes (3) composition of more than one mineral (4) fossils

19. How do igneous rocks differ from other kinds of rock? (1) Igneous rocks formed from a liquid. (2) Igneous rocks contain oxygen and silicon. (3) Igneous rocks contain the mineral feldspar. (4) Igneous rocks are usually found in layers.

20. The figure below shows two common igneous rocks labeled *X* and *Y*. Both rocks have a similar composition.

How does the formation of these two rocks compare?
(1) *X* formed from a more mafic magma.
(2) *Y* formed from a more felsic magma.
(3) *X* cooled quickly, and *Y* cooled slowly.
(4) *X* cooled slowly, and *Y* cooled quickly.

21. Which mineral is generally found in granite, but is not common in basalt? (1) potassium feldspar (2) plagioclase feldspar (3) amphibole (4) olivine

22. The famous green sand beaches on the island of Hawaii are the result of the physical weathering of basaltic rocks. Which two minerals are most likely found in this sand? (1) potassium feldspar and quartz (2) quartz and biotite (3) pyroxene and olivine (4) biotite and potassium feldspar

23. What is the approximate range of the percentage of potassium feldspar in granite? (1) 0% to 25% (2) 0% to 75% (3) 0% to 100% (4) 25% to 100%

24. Which of these igneous rocks has a composition similar to gabbro? (1) black obsidian (2) basaltic glass (3) rhyolite (4) dunite

25. In what property does pumice differ most from pegmatite? (1) mineral composition (2) overall color (3) size of the mineral crystals (4) pegmatite is made of rounded grains

26. According to the *Earth Science Reference Tables*, which of these is a fine-grained, dark-colored igneous rock? (1) rhyolite (2) granite (3) diorite (4) basalt

27. We can learn the most about where and how an igneous rock formed by studying its (1) mineral composition (2) texture (3) color (4) density

Part B

Base your answers to questions 28 and 29 on the diagram below, which shows a very thin slice of a coarse-grained igneous rock.

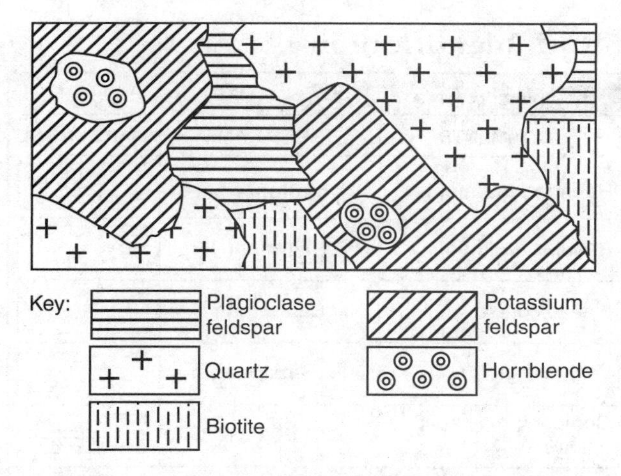

28. What kind of rock is this? (1) granite (2) rhyolite (3) gabbro (4) basalt

29. What is the most probable origin of the rock in the diagram above? (1) rapid cooling at earth's surface (2) rapid cooling deep underground (3) slow cooling at earth's surface (4) slow cooling deep underground

30. Different samples of basalt and gabbro can have different mineral compositions. Which mineral is most likely to be found in all of them? (1) plagioclase feldspar (2) biotite (3) pyroxene (4) olivine

31. Which of the graphs below best shows the sizes of the crystals in basalt, granite, and rhyolite?

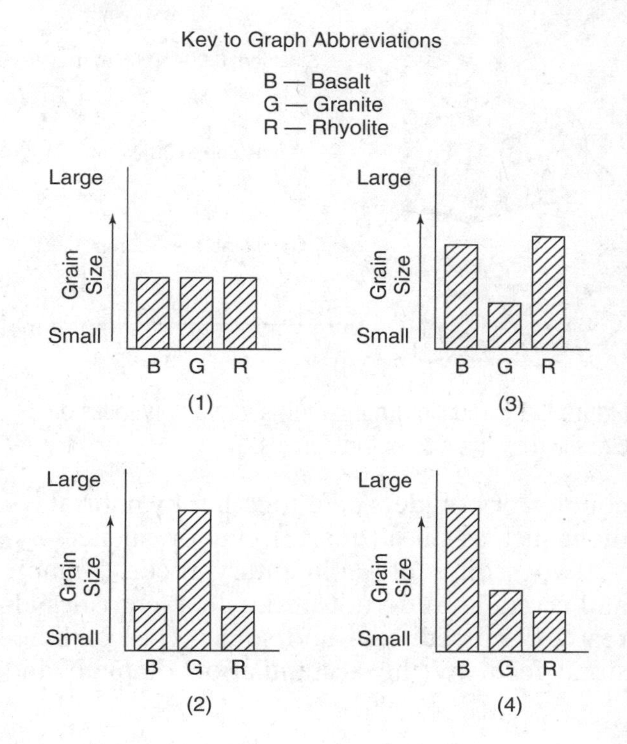

Sedimentary Rocks

Most sedimentary rocks are composed mainly of the weathered remains of other rocks that are called sediment. Sedimentary rocks usually form by the compression and cementation of sediment particles, or grains. The individual grains, which are usually rounded by abrasion, are deposited in layers. These characteristics (rounded grains in layers) will help you recognize most sedimentary rocks. Sedimentary rocks are the most common rocks at the surface. However, they generally form a relatively thin layer over metamorphic and igneous rocks, which are more abundant within Earth.

Characteristics The most common group of sedimentary rocks is the **fragmental (clastic)** rocks, which are made up of different sized particles as you can see Figure 2-11 on page 50. Shale is made of tiny clay particles that cannot be seen without magnification. Siltstone is composed of particles between the size of sand and clay. Sandstone has particles of sand large enough to be seen. These particles give sandstone a gritty feel. **Conglomerate** is composed of large pebbles,

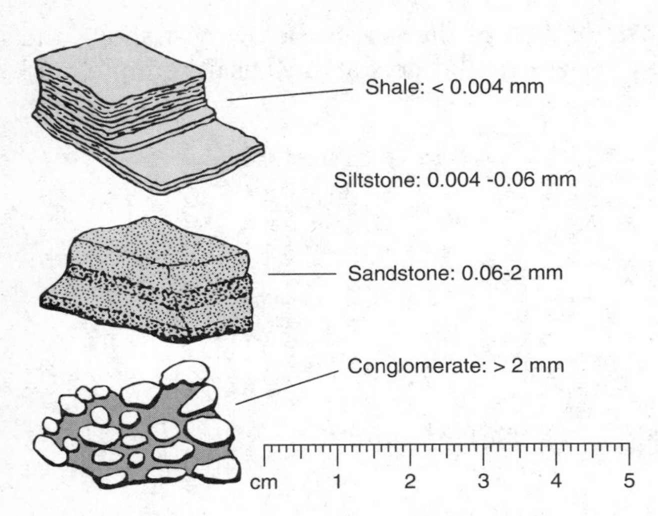

Figure 2-11. Clastic (fragmental) sedimentary rocks are classified by the size of their grains.

cobbles, or boulders held together by natural **cement** such as silica (quartz), clay, or calcite.

Two groups of sedimentary rocks (organic and crystalline) do not fit the characteristics already discussed: coal and some types of limestone form by the accumulation of plant and animal remains. For this reason, they are called **organic**, or **bioclastic**, sedimentary rocks. **Fossils**, the remains of prehistoric life, are almost always found in sedimentary rocks. Organic sedimentary rocks are especially rich in fossils.

Crystalline sedimentary rocks are deposited from seawater by chemical precipitation (settling) of materials. This occurs when seawater evaporates and chemical reactions in the water form insoluble compounds that settle. Although rock salt is the most common example of the crystalline **evaporite** rocks, there are others, such as gypsum. Figure 2-12 shows how sedimentary rocks are classified.

QUESTIONS

Part A

32. A sedimentary rock composed of rounded particles about 0.1 cm in diameter is best classified as (1) sandstone (2) rock salt (3) shale (4) conglomerate

Scheme for Sedimentary Rock Identification

INORGANIC LAND-DERIVED SEDIMENTARY ROCKS					
TEXTURE	**GRAIN SIZE**	**COMPOSITION**	**COMMENTS**	**ROCK NAME**	**MAP SYMBOL**
Clastic (fragmental)	Pebbles, cobbles, and/or boulders embedded in sand, silt, and/or clay	Mostly quartz, feldspar, and clay minerals; may contain fragments of other rocks and minerals	Rounded fragments	**Conglomerate**	
			Angular fragments	**Breccia**	
	Sand (0.006 to 0.2 cm)		Fine to coarse	**Sandstone**	
	Silt (0.0004 to 0.006 cm)		Very fine grain	**Siltstone**	
	Clay (less than 0.0004 cm)		Compact; may split easily	**Shale**	
CHEMICALLY AND/OR ORGANICALLY FORMED SEDIMENTARY ROCKS					
TEXTURE	**GRAIN SIZE**	**COMPOSITION**	**COMMENTS**	**ROCK NAME**	**MAP SYMBOL**
Crystalline	Fine to coarse crystals	Halite	Crystals from chemical precipitates and evaporites	**Rock salt**	
		Gypsum		**Rock gypsum**	
		Dolomite		**Dolostone**	
Crystalline or bioclastic	Microscopic to very coarse	Calcite	Precipitates of biologic origin or cemented shell fragments	**Limestone**	
Bioclastic		Carbon	Compacted plant remains	**Bituminous coal**	

Figure 2-12. This table is also found on page 7 of the *Earth Science Reference Tables.*

33. A student found two samples of fine-grained, dark-colored sedimentary rocks. One was coal and the other was shale. Which classification best matches these two rock samples? (1) Coal and shale are both organic in origin. (2) Coal and shale are both clastic (fragmental) rocks. (3) Coal is organic and shale is clastic (fragmental). (4) Coal is clastic (fragmental) and shale is organic.

Base your answers to questions 34 and 35 on the figure below, which shows two sedimentary rocks A and B.

Rock A

Rock B

34. How are these two rocks most clearly different? (1) mineral composition (2) grain size (3) depth within Earth where they formed (4) shape of the particles of which they are composed

35. Rock *A* is most likely (1) sandstone (2) rock salt (3) shale (4) conglomerate

36. Which sedimentary rock forms through the accumulation of the remains of living organisms? (1) limestone (2) rock salt (3) sandstone (4) conglomerate

37. Which of the four objects shown below is most likely a sedimentary rock?

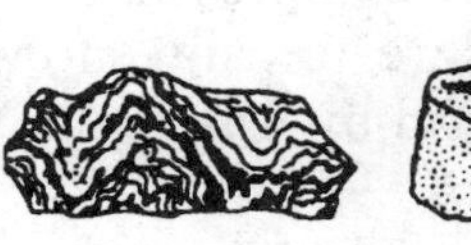

Sample A

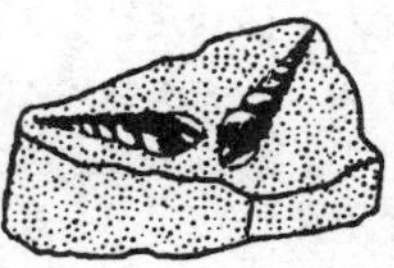

Sample C

Sample B

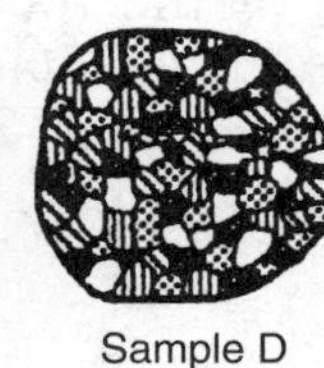

Sample D

(1) sample *A* (2) sample *B* (3) sample *C* (4) sample *D*

38. Which sedimentary rock is formed when seawater evaporates? (1) rock gypsum (2) shale (3) conglomerate (4) breccia

39. Which process would *not* form a sedimentary rock? (1) evaporation of seawater (2) cementation of rounded rock particles (3) compression of fine particles of clay (4) weathering and erosion of a sand dune

40. In which natural environment are most sedimentary rocks formed? (1) molten, liquid magma inside a volcano (2) high mountains covered by snow and ice (3) ocean basins with thick deposits of silt and mud (4) Earth's center where heat and pressure are extreme

41. Which rock is composed of particles of feldspar, quartz, and other minerals that are approximately 0.001 cm in diameter? (1) siltstone (2) breccia (3) dolostone (4) coal

42. How are conglomerate and breccia different? (1) mineral composition (2) shape of fragments (3) density (4) grain size

Part B

43. A hill of gravel (sedimentary particles) is composed primarily of rounded pebbles.

What series of events would most likely transform this material into a sedimentary rock? (1) heating and melting followed by slow crystallization (2) water washing in clay followed by compression (3) weathering by exposure to rain, snow and wind (4) transportation by water to a new gravel location

44. One kind of sedimentary rock is sometimes called "claystone" because of the smooth feel of the rock. What is the most likely more formal name of this kind of rock?
(1) sandstone (2) breccia (3) shale (4) conglomerate

Metamorphic Rocks

Metamorphic rocks form when igneous, sedimentary, or pre-existing metamorphic rocks are changed by heat and/or pressure. Unlike sedimentary rocks, most metamorphic rocks have been greatly changed by intense heat and/or pressure. This change sometimes takes place deep within Earth. However, unlike igneous rocks, metamorphic rocks never quite melted completely. Metamorphic rocks are the only kind of rock that forms directly from another rock.

During metamorphism, structures, such as layering, may become distorted or disappear; new minerals may form, and/or crystals may grow. Metamorphism often results in the alignment of crystals, called **foliation**, which is shown in Figure 2-13. Foliation often causes metamorphic rocks to break apart according to crystal growth and not along the original layering. Schist is an example of a foliated rock.

Metamorphism sometimes causes minerals to separate into light and dark layers, a property called **banding**. Gneiss, Figure 2-14, is an example of a rock that has light and dark bands. Figure 2-15 shows how these properties are used to classify and identify metamorphic rocks. The gray vertical bars on the left side of the chart show the mineral composition of the most common foliated metamorphic rocks. For example, only mica is common in all four foliated rocks shown in this chart.

Most metamorphic rocks formed deep (up to 20 kilometers) within Earth. How do these rocks reach the surface where we can see them? These rocks are pushed to the surface when mountains form. The core of the mountains may be metamorphic rock. In time, erosion wears down the mountains. Erosion may remove many kilometers of rock material, exposing the deep metamorphic core of the mountains that once lay deep below Earth's surface. The process by which rocks are changed by pressure and heat deep within Earth is **regional metamorphism**.

Another process by which metamorphic rocks are formed is **contact metamorphism**. In contact metamorphism, rocks are changed at or near Earth's surface as the result of nearby magma or lava. For example, when molten magma cuts through a layer of siltstone, the

Figure 2-13. The growth of mica crystals in this sample of schist has given the rock a strong foliation (alignment of crystals). The direction of foliation, parallel to the white arrow, causes the schist to break most easily in this direction. The pencil is for scale.

Figure 2-14. This exposure of gneiss shows two common characteristics of metamorphic rocks. It shows that the light and dark colored minerals have separated into roughly parallel bands and these bands are folded.

Scheme for Metamorphic Rock Identification

TEXTURE		GRAIN SIZE	COMPOSITION	TYPE OF METAMORPHISM	COMMENTS	ROCK NAME	MAP SYMBOL
FOLIATED	MINERAL ALIGNMENT	Fine	MICA	Regional (Heat and pressure increases)	Low-grade metamorphism of shale	**Slate**	
		Fine to medium	MICA, QUARTZ, FELDSPAR, AMPHIBOLE, GARNET		Foliation surfaces shiny from microscopic mica crystals	**Phyllite**	
			MICA, QUARTZ, FELDSPAR, AMPHIBOLE, GARNET, PYROXENE		Platy mica crystals visible from metamorphism of clay or feldspars	**Schist**	
	BAND-ING	Medium to coarse	MICA, QUARTZ, FELDSPAR, AMPHIBOLE, GARNET, PYROXENE		High-grade metamorphism; mineral types segregated into bands	**Gneiss**	
NONFOLIATED		Fine	Carbon	Regional	Metamorphism of bituminous coal	**Anthracite coal**	
		Fine	Various minerals	Contact (heat)	Various rocks changed by heat from nearby magma/lava	**Hornfels**	
		Fine to coarse	Quartz	Regional or contact	Metamorphism of quartz sandstone	**Quartzite**	
			Calcite and/or dolomite		Metamorphism of limestone or dolostone	**Marble**	
		Coarse	Various minerals		Pebbles may be distorted or stretched	**Metaconglomerate**	

Figure 2-15. This table is also found on page 7 of the *Earth Science Reference Tables.*

sand and clay near the hot, molten rock is baked. This forms a low-grade metamorphic rock called hornfels. Because contact metamorphism does not involve the high pressures found deep inside Earth, the rocks are usually not changed as much as they would be during regional metamorphism. For example, contact metamorphic rocks do not show foliation.

Figure 2-16 shows the origins of the most common metamorphic rocks. Notice in Figures 2-16 and 2-17 (page 54) how the type of metamorphic rock may change as a result of the intensity of the heat and the pressure. Also refer to the vertical mineral bars in Figure 2-15 that indicate the typical mineral compositions of foliated metamorphic rocks.

Classifying Rocks Figure 2-18 on page 54 provides a simplified scheme to classify rocks as igneous, sedimentary, or metamorphic. Their texture (the appearance and feel of the rock surfaces) gives us clues that help identify these three kinds of rocks. Unfortunately, some rocks are difficult to classify because they do not display these characteristics clearly. For example,

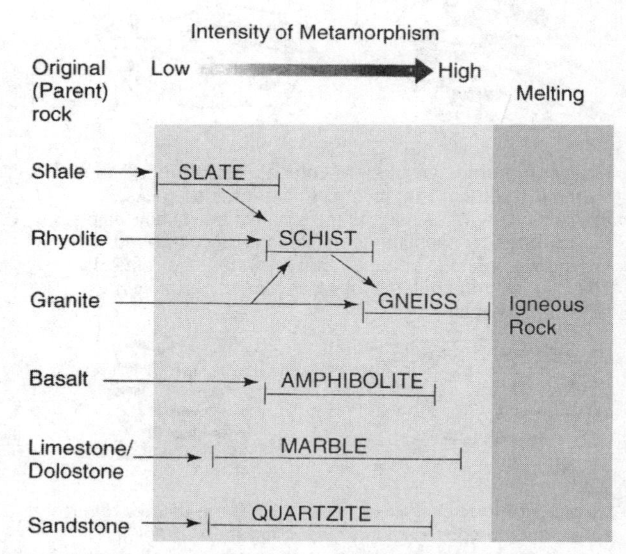

Figure 2-16. This diagram shows the most common varieties of metamorphic rocks, as well as the original rock type before metamorphism. The "I-beam" below the rock names indicates the range of metamorphic change that forms a particular rock type. Please note that the conditions shown on the far right side of the diagram would melt rock, resulting in magma, which would crystallize (solidify) to become igneous rock.

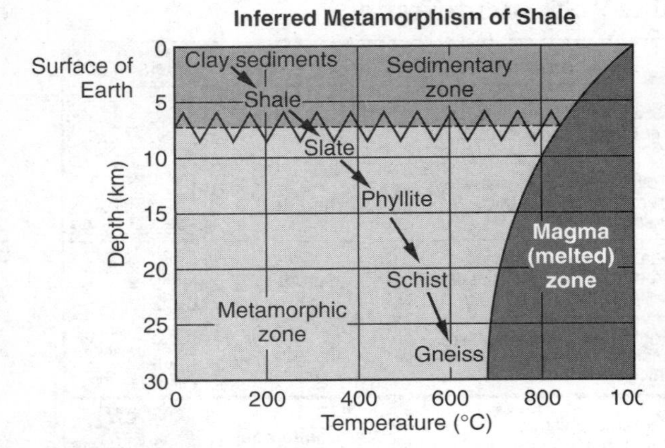

Figure 2-17. Sediments become sedimentary rock. The sedimentary rock becomes metamorphic rock when it is exposed to conditions of increasing pressure and temperature. Temperature and pressure increase as the rock moves deeper within Earth. However, if complete melting occurs, recrystallization will form an igneous rock.

Because **igneous rocks** have formed from molten magma or lava, they are composed of intergrown crystals. Rapid cooling, however, can make the crystals too small to be visible. Igneous rocks are usually quite hard and dense, and layering is rare. Gas bubbles may give igneous rocks a frothy texture.

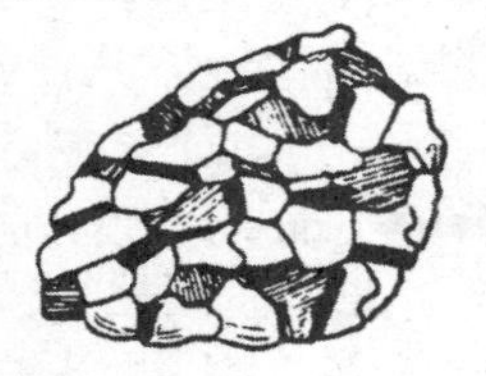

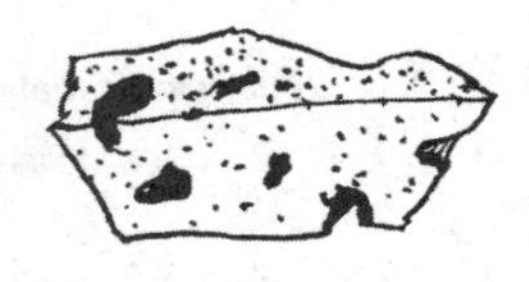

Most **sedimentary rocks** are composed of rounded fragments cemented in layers. In fine-grained rocks, the individual grains may be too small to be readily visible. Rocks made by chemical precipitation are composed of intergrown crystals, although these crystals are relatively soft. A rock that contains fossils is almost certainly a sedimentary rock.

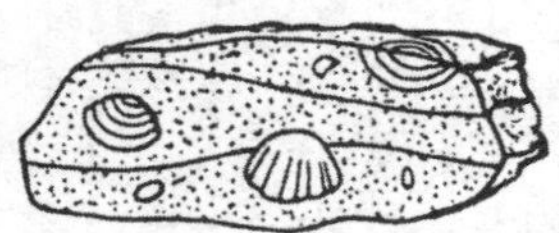

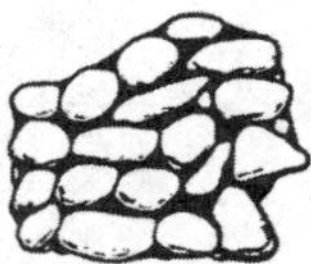

Metamorphic rocks, like igneous rocks, are usually relatively dense and composed of intergrown crystals. But, like sedimentary rocks, they often show layering, foliation (crystal alignment) and banding are common. However, the layers may be bent, or distorted.

Figure 2-18. A simplified way to classify rocks by their appearance.

you might find it difficult to decide if a finely speckled gray rock is sandstone (sedimentary), an impure quartzite (metamorphic), or andesite (igneous). Still, this scheme is often useful in classifying rocks.

QUESTIONS

Part A

45. Which mineral is very common in gneiss, but not common in phyllite? (1) pyrite (2) pyroxene (3) mica (4) garnet

46. Which characteristic of metamorphic rocks is *not* found in sedimentary or igneous rocks? (1) foliation (2) visible crystals (3) layering (4) conchoidal fracture

47. Unlike regional metamorphic rocks, contact metamorphic rocks (1) do not change from the original rock material (2) extend over a greater area of Earth's crust (3) form in relatively narrow bands near an intrusion (4) are heated so much that they melt and crystallize

48. The photograph below is of a metamorphic rock exposure in the floor of a canyon. A credit card shows the scale.

The curved dark layers show the original sedimentary bedding. But the rock splits across these curved layers in the direction of the arrows. What property of this rock causes the rock to split across the bedding? (1) contact metamorphism (2) grain size (3) foliation (4) quartz-rich composition

49. Which rock in the diagram above, right is most likely to be hornfels?

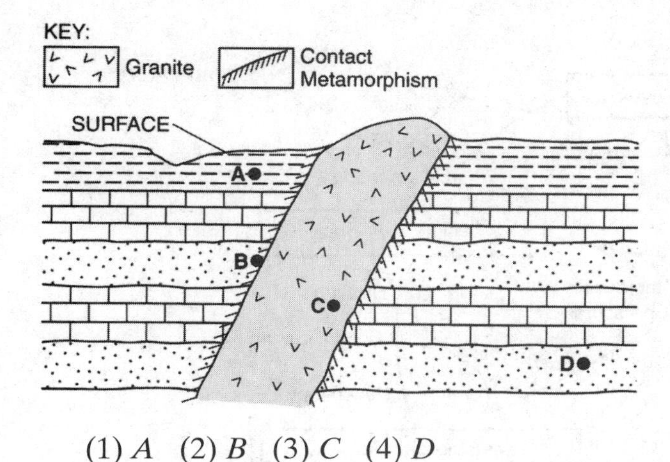

(1) *A* (2) *B* (3) *C* (4) *D*

50. Which mineral family is found in the greatest range of foliated metamorphic rocks? (1) mica (2) feldspar (3) amphibole (4) pyroxene

Part B

Base your answers to questions 51 and 52 on the image below, which shows a series of five rocks labeled A, B, C, D, and E. These rocks represent a progression of changes from a sedimentary rock to a high-grade metamorphic rock.

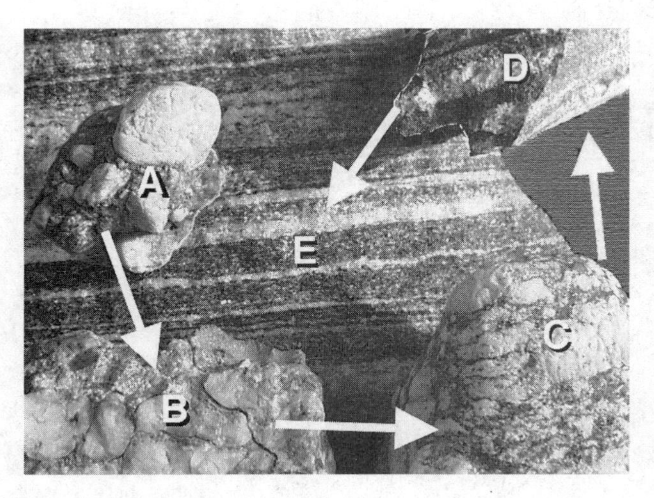

51. The changes shown by the arrows can be summed up as a progression from (1) shale to slate (2) basalt to amphibolite (3) sandstone to quartzite (4) conglomerate to gneiss

52. The change represented by the arrows in the image indicate that the original rock (*A*) was (1) weathered, eroded, and then deposited as sediment (*E*) (2) buried progressively deeper within Earth to become rock *E* (3) changed from a crystalline rock (*A*) to a clastic (fragmental) rock (*E*) (4) heated, melted, and then recrystallized (solidified) to become rock *E*

53. Which process is most like the formation of a metamorphic rock? (1) constructing a house from bricks, wood, nails and other building materials (2) baking bread from flour dough, water and spices (3) melting scrap metal to make molten iron (4) evaporating water into water vapor

ROCK CYCLE

Although Earth receives a small amount of matter in the form of meteorites and a very small amount of our atmosphere is lost to outer space, our planet is basically a closed system. That is, the material of Earth can cycle through various forms. However, Earth materials remain a part of the planet as they change from one kind of rock to another. The **rock cycle** is a model of the natural changes that occur in rocks and rock material.

The diagram in Figure 2-19 on page 56 distinguishes between those processes that take place primarily at Earth's surface (external processes) and those that take place within Earth (internal processes). Internal processes, such as compression and heating, require the conditions of extreme temperature and pressure that occur deep underground.

The rock cycle illustrates several important principles of geology. First, nearly all rocks are made from the remains of other rocks. (Coal and other organic sedimentary rocks are important exceptions.) Second, rocks are classified on the basis of their origin. Finally, there are a variety of ways that rocks can change in response to changing conditions at the surface or within Earth.

If you follow the outside of the rock cycle diagram, you will see that sediments can be compressed and cemented to form sedimentary rocks. These rocks can then be changed by intense heat and/or pressure to become metamorphic rocks. They can then be melted and

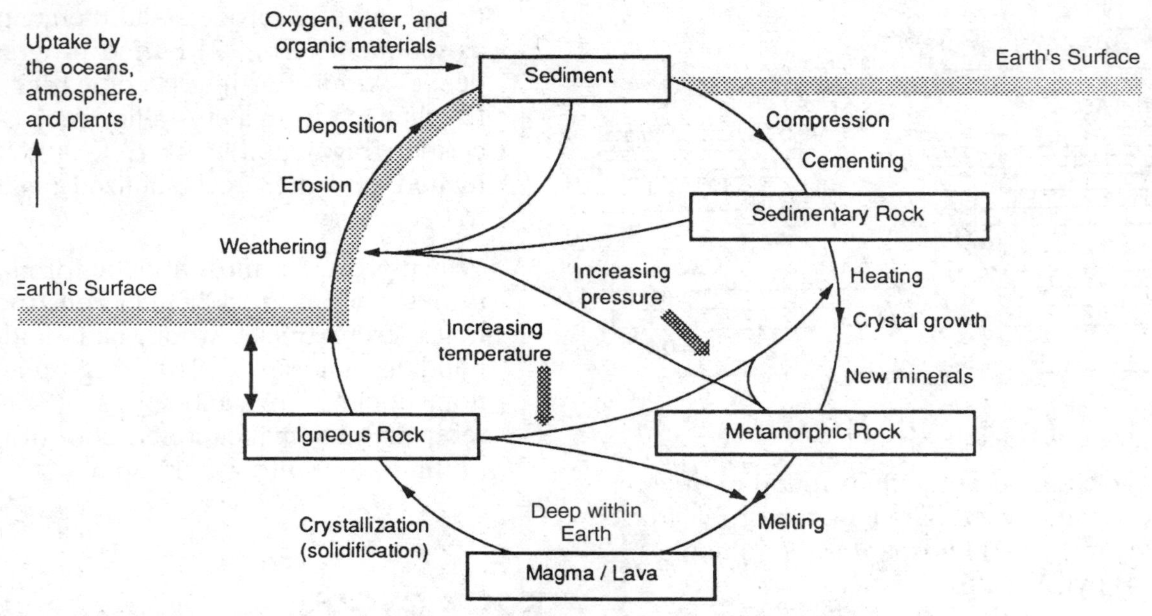

Figure 2-19. A similar diagram is found o page 6 of the Earth Science Reference Tables.

recrystallized to make igneous rock. Weathering and erosion create sediments, and the cycle continues.

There is more than one path through the rock cycle. Each rock type can remain relatively stable for long periods of time. Changes from one rock type to another may require thousands or even millions of years. Sometimes a rock may show evidence of more than one process or origin. For example, conglomerate may be composed of cemented fragments of granite (igneous rock) and gneiss (metamorphic rock). Although conglomerate is a sedimentary rock, some of the rock fragments within it may not be sedimentary.

QUESTIONS

Part A

54. How can granite change directly into magma? (1) weathering (2) compression (3) crystal growth (4) melting

55. How does sedimentary rock change into sediments? (1) cementation (2) weathering (3) compression (4) melting

Part B

Base your answers to questions 56 through 60 on the diagram below, which illustrates various process that occur on the surface and within Earth.

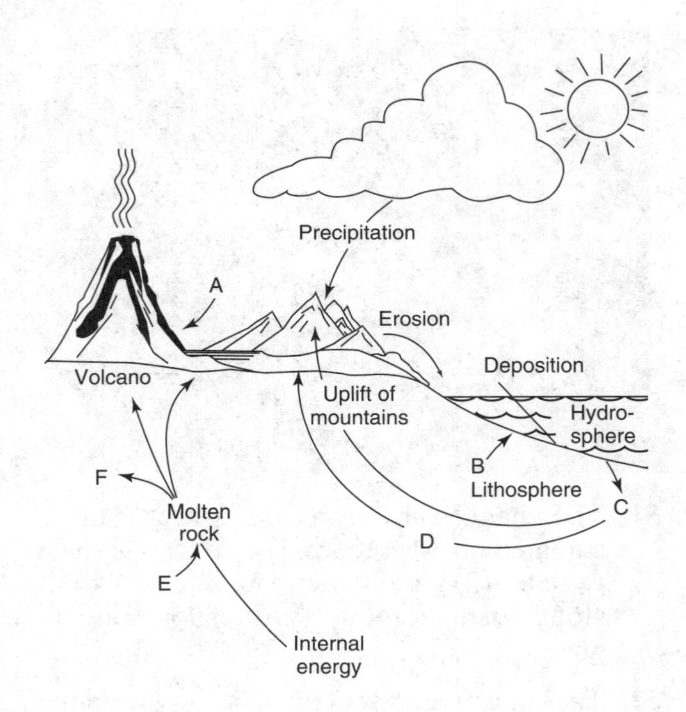

56. As the rocks formed at *C* move toward *D*, how are the conditions likely to change?

(1) heat and pressure will both decrease
(2) heat and pressure will both increase
(3) heat will increase, pressure will decrease
(4) heat will decrease, pressure will increase

57. At which location on the diagram is an igneous rock with small crystals or a glassy texture likely to form? (1) *A* (2) *B* (3) *C* (4) *F*

58. Which location is the most likely to produce a rock with a variety of hard, intergrown crystals and layers (foliation)? (1) *A* (2) *B* (3) *E* (4) *D*

59. Where are rocks composed of rounded particles cemented together likely to form? (1) *A* (2) *F* (3) *C* (4) *E*

60. Fossils are most likely to be found in the rocks at (1) *A* (2) *E* (3) *C* (4) *F*

NATURAL RESOURCES

Natural resources are useful substances that come from Earth. Table 2-2 lists some examples of economically important geological resources. As the need for these resources grows, there may come a time when they will be in short supply. Limited supplies can cause the prices of these resources to increase, or it may influence us to find other substances to meet our needs. Recycling is also a way to help meet our needs and to save resources. In addition, recycling reduces the amount of trash that must be disposed of. There are two types of natural resources: renewable and nonrenewable.

Figure 2-20.

New York State has a variety of natural resources of economic value. While the mineral industry is not among the major businesses of New York State, it does have a significant impact. The most important resources are listed in Figure 2-20.

Renewable

Renewable resources can be replaced after they have been used. For example, trees are a renewable resource. After one crop of trees has been

Table 2-2. Important Geological Resources

Group	Minerals/Products	Common Uses
Metals	Gold	Jewelry, coins, dental fillings
	Copper	Electrical wiring, plumbing, coins
	Iron ore	Construction, transportation, machinery
	Aluminum	Aircraft, food packaging
Nonmetals	Halite (rock salt)	Food, chemicals, melting ice
	Graphite	Lubricants, pencil "lead"
	Garnet	Abrasives, jewelry
	Feldspar	Porcelain, glass, ceramics
Fossil Fuels	Coal	Heating, power plants, synthetics
	Petroleum	Fuels, heating, medicines, plastics

cut, another is planted. Freshwater is generally a renewable resource because rain refills the lakes and rivers. However, in the American southwest, surface and groundwater are being used more quickly than natural processes can replace them. The future of these dwindling water supplies is an important political issue in these areas.

Nonrenewable

Nonrenewable resources, after they are used, cannot be replenished for millions of years, if at all. The native metals, such as gold and copper, as well as the ores of metals, such as iron and aluminum, are considered nonrenewable resources. Although they are being formed in some locations, such as steam vents in the deep oceans, the rate of renewal is far slower than the rate at which they are taken from the ground.

The **fossil fuels**—coal, oil, and natural gas—were formed millions of years ago from the remains of ancient organisms. Fossil fuels are nonrenewable resources. During the past two centuries, these resources have been seriously depleted. In addition, our use of these fuels has accelerated in recent decades. Figure 2-21 shows the major producers of petroleum, our planet's most widely used energy resource.

Although coal is the most abundant fossil fuel, scientists estimate that worldwide coal reserves will last only about 200 years. Supplies of crude oil, or petroleum, are being used up faster than new resources are being found. The United States imports much of the petroleum it uses. If enough oil cannot be imported, shortages develop. A small number of nations around the Persian Gulf control more than half the world's estimated reserves.

In addition to being used for fuel, petroleum is the source of the petrochemicals used to make plastics, synthetic fabrics, medicines, insecticides, fertilizers, and detergents, which are also important to the global economy.

Alternative Energy Sources Scientists are looking for other sources of energy to replace fossil fuels. One possible source of energy is the sun. Solar energy is plentiful and renewable. Solar energy can be used to provide heat, hot water, and to generate electricity. However, the amount of solar energy that reaches Earth varies with the time of day and the seasons. Today, using solar energy to produce electricity is possible, but it is generally more expensive to produce electricity in this way than by using fossil fuels.

Geothermal energy, from heat within Earth can be used to generate electricity, and produce heat and hot water. However, it is available only in certain areas, such as northern California and Iceland.

Wind-driven generators are now used in many areas to provide electricity. These generators are practical only in areas where there are strong, steady winds. Hydroelectric power is

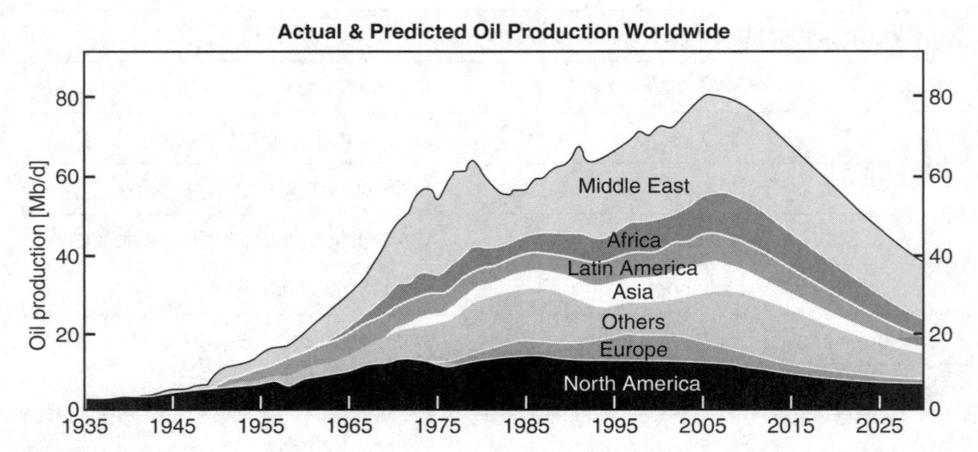

Figure 2-21. As this graph shows, some geologists predict that the output of world energy resources, primarily from fossil fuels, will peak in the near future. Other scientists believe that the pressure of economic forces will promote the discovery of new energy sources and technologies. As the largest user of fossil fuels, the United States depends on the global market for its supply of oil.

available in some locations. However, large areas are flooded when dams are built to harness a river and generate electricity. Nuclear energy is another possibility. It can be used to generate electricity without causing air pollution. The drawback to nuclear energy is the danger of a nuclear accident and the need for long-term storage of radioactive wastes. The potential that terrorists might target nuclear plants or use radioactive materials in weapons is also an important concern.

Problem Solving

Recently, there has been renewed interest in energy reserves found in New York State. Of particular interest is the Marcellus Shale, which is thought to contain trillions of cubic feet of natural gas. This is not a newly discovered resource. For many years, people have tapped the Marcellus Shale. What economic, political and technological factors have caused renewed interest in this fuel source? Why does this rock unit contain so much natural gas? If it is developed, what are the likely effects on people and on the environment? How should Marcellus Shale gas extraction be managed by local and regional agencies?

QUESTIONS

Part A

61. Which of the following is a renewable resource? (1) petroleum (2) solar energy (3) iron ore (4) natural gas

62. Plastics, insecticides, and medicines are made from (1) petrochemicals (2) solar power (3) renewable resources (4) nuclear energy

63. Which of the following natural resources will be conserved if people use public transportation, fuel-efficient cars, and lower their thermostats in the winter? (1) minerals (2) metal ores (3) fossil fuels (4) trees

64. What natural resource is most often used for melting ice and making food taste better? (1) copper (2) petroleum (3) feldspar (4) halite

Part B

65. The map below shows the strength of sunlight throughout most of the United States.

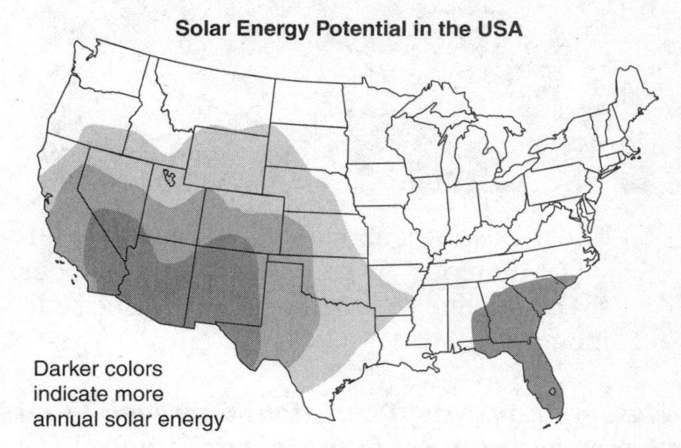

According to the map, why is New York State less likely to use solar energy than some other parts of the United States? (1) The sun does not shine in New York State. (2) Other parts of the country have more sunlight. (3) Other sources of electrical energy are relatively inexpensive in New York. (4) New York State does not offer tax incentives for solar energy production.

Chapter Review Questions

PART A

1. What mineral is often transparent, does not beak along flat surfaces, and has a hardness of 7 on Mohs scale? (1) halite (2) quartz (3) calcite (4) plagioclase feldspar

2. The most common sedimentary rocks are fragmental, or clastic, rocks. Which of the following is a sedimentary rock that is not clastic? (1) basalt (2) dolostone (3) conglomerate (4) shale

3. Classifying rocks as igneous, sedimentary, or metamorphic is based on differences in their (1) origin (2) particle size (3) hardness (4) geologic age

4. When a geologist found a beautiful mineral sample, the only equipment he had with him was a sharp knife blade. This knife blade would be *least* useful in helping identify a mineral's (1) hardness (2) cleavage (3) streak (4) luster

5. The rock shown above is most likely (1) felsic igneous (2) mafic igneous (3) crystalline sedimentary (4) clastic sedimentary

Base your answers to questions 6 through 11 on the images below of four geologic samples. The penny in each image provides scale. One sample is a mineral. The other three are rocks, one igneous, one sedimentary, and one metamorphic rock.

6. Which of the rocks is most likely to contain fossils? (1) *A* (2) *B* (3) *C* (4) *D*

7. Which sample shows prominent banding? (1) *A* (2) *B* (3) *C* (4) *D*

8. Which rock sample probably formed from magma deep underground? (1) *A* (2) *B* (3) *C* (4) *D*

9. Which sample can be represented by the chemical formula SiO_2? (1) *A* (2) *B* (3) *C* (4) *D*

10. Of the four samples, which formed directly by the compression of mud and silt sediments? (1) *A* (2) *B* (3) *C* (4) *D*

11. Which sample does not contain visible crystals? (1) *A* (2) *B* (3) *C* (4) *D*

12. What is the most common mineral in rock salt? (1) silicon dioxide (2) plagioclase feldspar (3) quartz (4) halite

13. How are igneous rocks changed to sediments? (1) crystallization followed by cementation (2) weathering followed by erosion (3) compression followed by melting (4) heating followed by crystal growth

14. Which is generally considered to be a renewable resource? (1) hydroelectric energy (2) iron ore (3) natural gas (4) limestone for cement

15. Which mineral in the table is most dense?

Mineral Sample	Mass (g)	Volume (cm³)
Quartz	10	4
Feldspar	20	8
Galena	15	2
Biotite mica	18	6

(1) quartz (2) feldspar (3) galena (4) biotite mica

16. Silicate minerals are the most common minerals found in Earth's crust. What two elements are most abundant in rocks at Earth's surface? (1) iron and aluminum (2) oxygen and silicon (3) iron and oxygen (4) oxygen and aluminum

17. Which two rock types are likely to be separated by a zone of marble? (1) basalt and limestone (2) granite and rhyolite (3) gneiss and schist (4) sandstone and rock salt

18. Compared with basalt, gabbro (1) contains more quartz (2) is mostly felsic minerals (3) is composed of larger crystals (4) is more likely formed by compression and cementatiom

19. Which sedimentary rock is composed of the smallest particles? (1) sandstone (2) conglomerate (3) siltstone (4) shale

20. Which metamorphic rock is nonfoliated? (1) quartzite (2) gneiss (3) schist (4) phyllite

21. What plutonic rock is composed of primarily of olivine? (1) andesite (2) pegmatite (3) pumice (4) dunite

22. Why is garnet often used as an abrasive? (1) garnet has a dark red color (2) garnet has a hardness of 7 (3) garnet contains silicon (4) garnet has a nonmetallic luster

23. Hydrochloric acid is most useful in identifying (1) amphibole and gneiss (2) quartz and sandstone (3) calcite and marble (4) pyrite and conglomerate

24. What device is most useful in collecting a renewable resource? (1) a windmill (2) a streak plate (3) a blowtorch (4) a stick of dynamite

25. Rhyolite and pumice have vesicular texture. What common object also has a vesicular texture? (1) lead weight (2) kitchen sponge (3) sheet of paper (4) granite countertop

26. Which rock often shows sedimentary layering as well as foliation? (1) slate (2) sandstone (3) basalt (4) rock salt

PART B

Base your answers to questions 27 through 32 on the information below.

Wollastonite Mining in New York State

Wollastonite is a calcium silicate mineral with the chemical formula $CaSiO_3$. It forms when impure limestone or other precipitate rocks are subjected to high temperature and pressure, often in zones of contact metamorphism. The streak of wollastonite is white and its hardness on Mohs scale ranges between 4.5 and 5. Its density is approximately 5 g/cm^3. Wollastonite has several cleavage planes.

Some of the properties that make wollastonite useful are its high brightness and whiteness, low moisture, and low oil absorption. Wollastonite is used primarily in ceramics, automotive products, paint filler, and plastics.

Wollastonite deposits in New York are in the outer parts of the Adirondack Mountain area. It has been mined for more than 50 years at Willsboro, New York (*W* in the figure below). The ore is processed near the mine at the Willsboro quarry. Garnet was removed from the ore. Wollastonite is also mined near Governeur, New York (Location *G*).

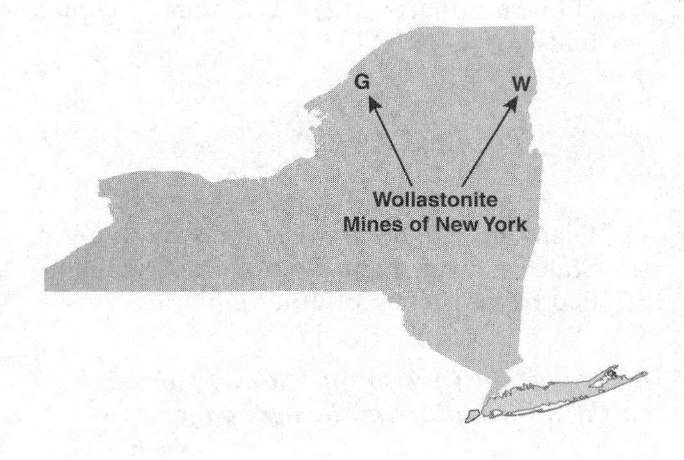

27. Wollastonite is formed by the metamorphism of (1) an intrusive igneous rock (2) an extrusive igneous rock (3) a clastic sedimentary rock (4) a chemical sedimentary rock

28. Pure wollastonite could be used to scratch (1) quartz (2) garnet (3) calcite (4) amphibole

29. What is the approximate mass of a 2-cm^3 sample of wollastonite? (1) 2 grams (3) 5 grams (3) 10 grams (4) 50 grams

30. Wollastonite is commonly used in products made from (1) potassium feldspar and plagioclase feldspar (2) quartz and olivine (3) graphite and galena (4) halite and sulfur

31. Why is wollastonite mined near Willsboro and Goveneur in New York State? (1) The mineral is found only within igneous rocks. (2) In those locations, the mineral is relatively concentrated. (3) Land is inexpensive in those two towns. (4) The wollastonite is pure in those deposits.

32. Wollastonite is most similar in chemical composition to (1) calcite (2) dolomite (3) graphite (4) halite

Base your answers to questions 33 through 39 on the diagram below, which is a cross section of a volcano. Zones of different kinds of rock are labeled along with arrows pointing to positions A through H.

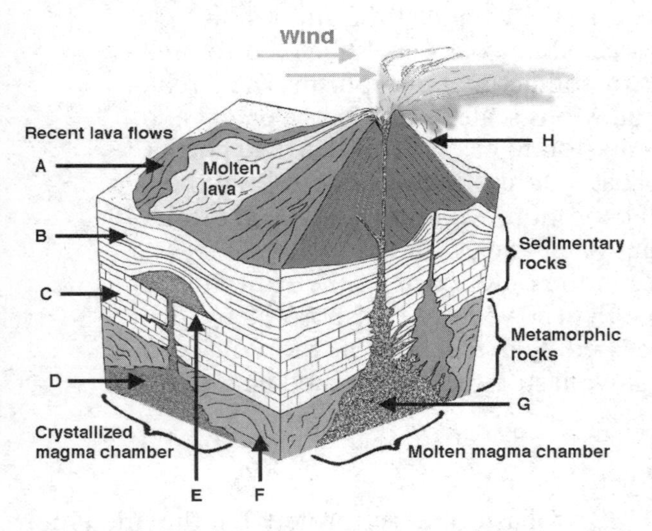

33. What is the most likely place to find basalt? (1) *A* (2) *D* (3) *F* (4) *G*

34. What kind of rock is found at *C*? (1) shale (2) granite (3) schist (4) limestone

35. The rock at *D* is most likely (1) basalt or rhyolite (2) gabbro or granite (3) schist or gneiss (4) quartzite or slate

36. Marble is most likely found at (1) *B* (2) *D* (3) *E* (4) *G*

37. Pumice is probably being deposited at (1) *B* (2) *D* (3) *F* (4) *H*

38. The rock at position *A* and the rock at position *D* have the same mineral content. If you were to compare these rocks, the rock at *D* would probably have (1) more quartz (2) less quartz (3) smaller crystals (4) larger crystals

Base your answers to questions 39 through 41 on the table below, which lists the elements found in certain minerals.

	O	Si	Al	Fe	Ca	Na	C
Quartz	X	X					
Feldspar	X	X	X		X	X	
Olivine	X	X		X	X		
Diamond							X

(X indicates that the element is present.)

39. According to the table, which mineral does *not* contain the two most abundant elements in Earth's crust? (1) quartz (2) diamond (3) olivine (4) feldspar

40. Which mineral contains the greatest variety of elements? (1) feldspar (2) diamond (3) olivine (4) quartz

41. Which of these minerals is the least common in rocks around New York State? (1) quartz (2) mica (3) feldspar (4) diamond

42. Compared with felsic igneous rocks, mafic igneous rocks contain greater amounts of (1) white quartz (2) aluminum (3) pink feldspar (4) iron

PART C

43. Diamond is the hardest known mineral. State one way that a diamond can be identified from a group of other minerals.

Base your answers to questions 44 through 47 on the information on the next page.

Is the Moon Made of Green Cheese?

Sometimes people joke that the moon is made of green cheese. In fact, before we sent rockets to the moon, scientists were not really sure what it was made of. In 1969, humans first set foot on the moon. Over the next 10 years, more than 385 kilograms (845 pounds) of moon rocks were brought back to Earth. Some of the rock was stored for future study. Some of it, such as the small sample in the lobby of the United Nations in New York City, went on public display. The rest went to university and government laboratories for study.

Scientists discovered that moon rocks are similar to rocks found right here on Earth. Most of the moon rock from the lunar maria ("seas") is basalt. In the lunar highlands, breccia and anorthosite (an igneous rock that is rich in plagioclase feldspar) are common.

There are two big differences between moon rocks and rocks of similar composition found here on Earth. Moon rocks have no water trapped in their rock crystals, and they show no chemical weathering. Because there is no air or water to cause chemical weathering, moon rocks are not weathered. However, moon rocks do show evidence of abrasion caused by collisions with meteorites. Mineral grains of lunar "soil" and larger rocks are as fresh as when the rock first formed about 4 billion years ago.

44. Why do the rocks that we find on Earth's surface usually look so different from moon rocks?

45. According to the *Earth Science Reference Tables,* what four minerals are probably most common in igneous rocks from the lunar maria?

46. What can we tell about the formation of the rocks from the lunar maria?

47. How is breccia different from conglomerate?

Base your answers to questions 48 and 49 on the information and the pie graph below.

The United States and Uranium

Like many other mineral resources, the United States imports most of the uranium it uses. Only about 8 percent of the uranium we use comes from our country. About 44 percent comes from Canada and Australia. Russia is the third major source, supplying 35 percent. All other nations provide the remaining 13 percent.

48. Copy the pie graph and label it to show the sources of uranium used in the United States. (You may also make a pie graph of your own to label.)

49. If Russia were to cut off all exports of uranium to the United States, what would be the most logical foreign source to make up the shortage?

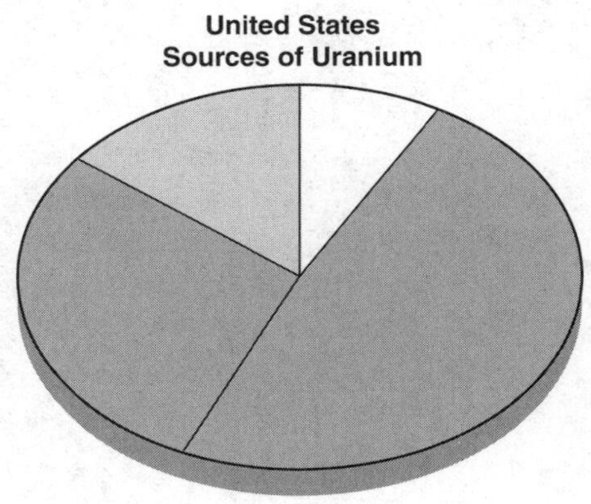

50. What two measurements do you need to make to calculate the density of a mineral sample?

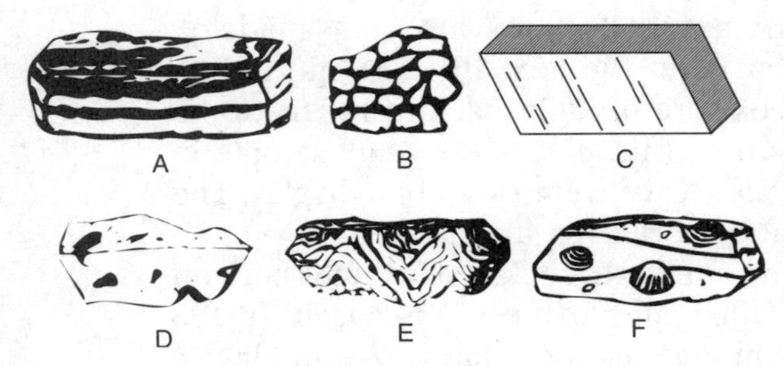

51. The six diagrams above represent rocks. Two represent metamorphic rocks. List the letters of the metamorphic rock samples, and tell what feature of each identifies it as a metamorphic rock.

52. What property is clearly visible in the mineral crystal below that helps you to identify it as calcite rather than halite?

53. Why is copper considered a nonrenewable resource?

54. How is the origin of both granite and gabbro different from that of basalt and rhyolite?

55. What geologic material consists of a certain number of iron atoms and exactly twice the number of sulfur atoms?

56. What piece of geological equipment is usually used to find the color of the powder of a specific mineral?

57. Metallic, glassy, waxy, and earthy are all varieties of what mineral property?

58. The United States imports most of its fossil fuels from other nations. In what two ways could we reduce our dependence on foreign oil?

59. The image below shows four different samples of granite. A pen provides scale. State two properties of these rocks that would allow you to identify these four samples as granite rather than another igneous rock type.

CHAPTER

The Dynamic Crust

Vocabulary at a Glance

asthenosphere	hot spot	reversed polarity
basaltic	inner core	rift zone
compression waves	intensity	Ring of Fire
conduction	island arc	sea-floor spreading
continental crust	magnitude	seismic wave
continental drift	mantle	seismograph
convection	mid-ocean ridge	seismologist
convection cell	Moho	seismometer
convergent boundary	oceanic crust	subduction zone
core	ocean trench	subsidence
crust	origin time	S-waves
divergent boundary	outer core	tectonics
earthquake	Pangaea	transform boundary
epicenter	plate boundary	trench
fault	plate tectonics	tsunami
focus	P-waves	
granitic	radiation	

WHAT CAUSES EARTHQUAKES?

An **earthquake** is any shaking or rapid movement of Earth's solid outer layers. Most earthquakes occur when stress builds along a zone of weakness or a crack where previous movement has occurred. When the stress on the crust is greater than it can resist, the crust shifts and breaks, suddenly along **faults** releasing energy. (See Figure 3-1 on page 66.) The energy radiates in all directions in vibrations called **seismic waves**.

The place underground where the break first occurs is the **focus** (or hypocenter) of the earthquake. The **epicenter** is the location at Earth's surface just above the focus. When the vibrations reach the surface, we feel them as an earthquake, first at the epicenter and then at greater distances from the epicenter. These features of an earthquake are shown in Figure 3-2 on page 66.

Measuring Earthquakes

Earthquakes are measured using intensity and magnitude scales. Scales of **intensity**, such as the Mercalli scale, are based on the reports of people who experienced the earthquake or observed the amount of destruction it caused. A simplified scale of Mercalli intensities is shown in Table 3-1 on page 67. Intensity scales do not measure the energy level of an earthquake, but they do indicate its effects on people at different locations where it can be felt. The intensity at which an earthquake is felt depends on the observer's distance from the epicenter, the nature

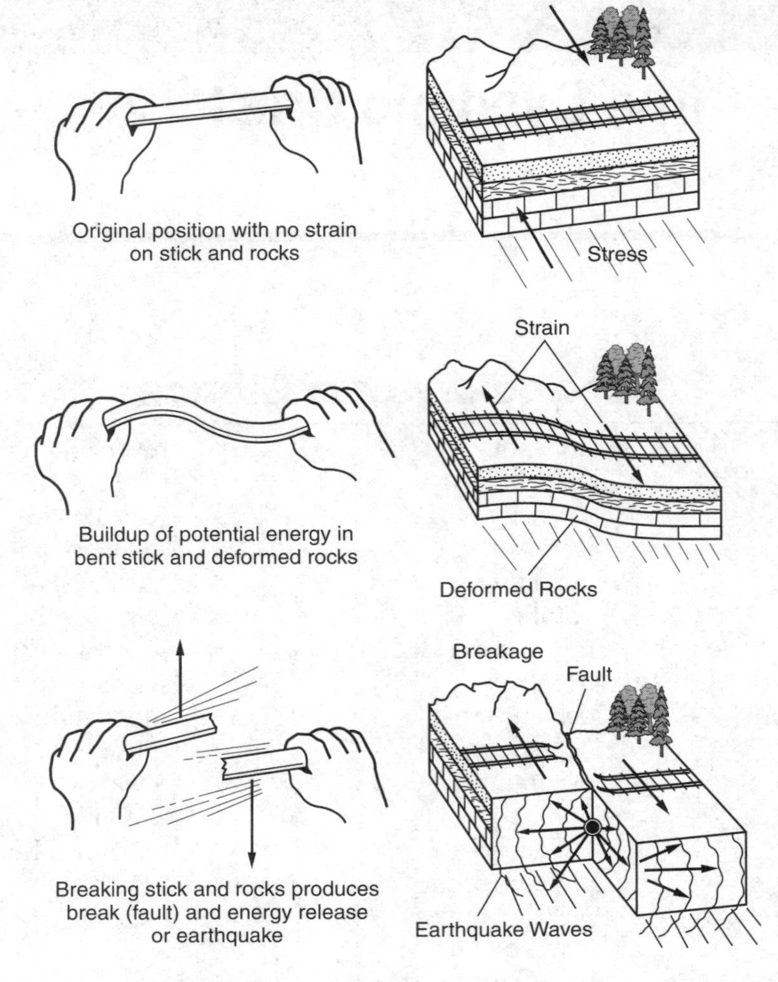

Figure 3-1. When stress exceeds the strength of Earth's crust, rocks break along a zone of weakness. Sudden shifting along the fault releases the energy that you can feel as an earthquake.

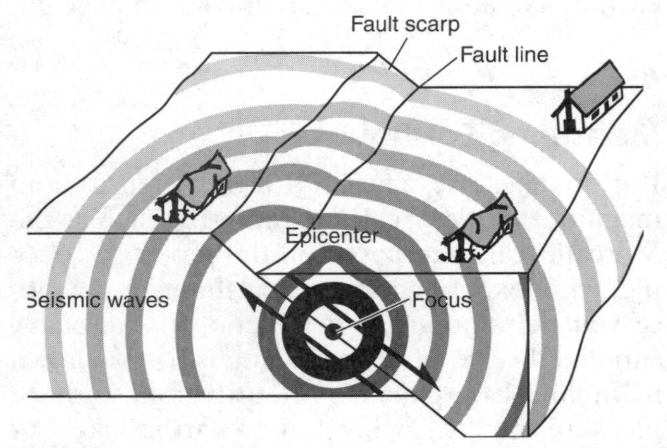

Figure 3-2. Ground motion radiates from the focus of an earthquake in energy waves. The focus, sometimes called the hypocenter, is located underground beneath the epicenter.

of the ground (loose sediments shake more) and the ability of the ground to transmit this energy.

With the widespread use of **seismographs** (instruments that record ground vibrations), you might think that intensity scales would become outdated. However, **seismologists** (scientists who study earthquakes) do not have enough **seismometers** to collect all the data they need. Therefore, they get additional data from the observations of witnesses. From these observations, scientists know that the shaking caused by an earthquake is more intense in buildings constructed on loose sediments and landfill. Structures built on solid bedrock are more likely to survive an earthquake. It is not unusual for the greatest damage (Mercalli in-

Table 3-1. A Simplified Scheme for Measuring Earthquakes

Mercalli Intensity	Observed Effects and Experiences	Number per Year Worldwide	Estimated Magnitude
I to II	May not be felt, but can be recorded by sensitive seismometers.	~1 million	2
III to IV	Slightly felt indoors, dishes rattle, shaking similar to a large truck passing	~Hundreds of thousands	3
V to VI	Felt by most people. Some dishes and windows broken.	~Tens of thousands	4
VII to IX	Difficult to stand up. Collapse of structures that are not reinforced.	~100	6
X to XII	Waves visible on ground. Damage to buildings may be total.	~One every 5 years	8

tensity) to occur many kilometers from an earthquake's epicenter.

The most dependable measures of earthquakes are made with *seismographs*. As shown in Figure 3-3, these instruments operate on the principle that a heavy weight suspended on a spring tends to remain steady while the ground shakes. During an earthquake, the suspended weight and its attached pen remain still, while the ground and the paper shake. The pen makes tracings on the moving paper. Technological advances have led to electronic instruments that can record a wider range of frequencies and magnitudes.

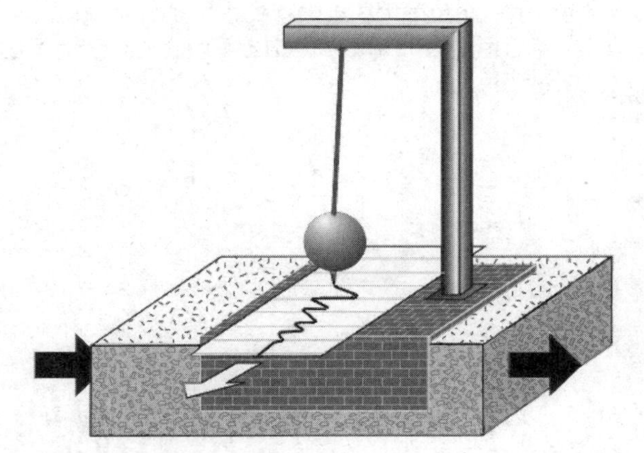

Figure 3-3. Mechanical seismographs record shaking of the ground using a large mass suspended like this pendulum. The mass stays still as the ground and the paper vibrate. A pen records the shaking. However, modern instruments use electronic devices to generate a digital output that can record even very small and very large earthquakes.

Seismographic records of earthquakes have made it possible for geologists to establish **magnitude** scales. One such scale is named for Dr. Charles Richter, its inventor. His original scale has been replaced by improved magnitude scales. However, newscasters still will report a "Richter magnitude" when talking about an earthquake. It is important to note that seismic magnitude scales are logarithmic. That is, each increase of one unit means a 10-fold increase in shaking. Therefore, a magnitude 7 earthquake has 10 times more shaking than a magnitude 6 event and 100 times more shaking than a magnitude 5 event.

EARTHQUAKE WAVES

When an earthquake occurs, its energy radiates in waves away from the focus as shown in Figure 3-2. These waves can be grouped into three categories. **P-waves** travel the fastest. In addition to standing for Primary, "P" also stands for push-pull, which describes the ground's motion as these waves pass. The vibration of a P-wave is like the motion of a spring when it is alternately pulled tighter and then released. It also is similar to the transmission of energy along a line of closely spaced automobiles when they are hit from behind. P-waves are **compression waves**.

The second category is **S-waves**, "S" stands for Secondary or side-to-side. (Actually, S-waves vibrate in all directions perpendicular to the direction of travel.) S-wave motion is like the wave that travels down a rope or spring when one end is moved quickly from side-to-side. S-waves travel more slowly than P-waves. S- and P-waves are shown in Figure 3-4.

The third category of seismic waves is the surface waves. These waves include both push-pull and side-to-side motion. They are similar to waves on the surface of water. Surface waves cause the most damage as they travel along Earth's surface.

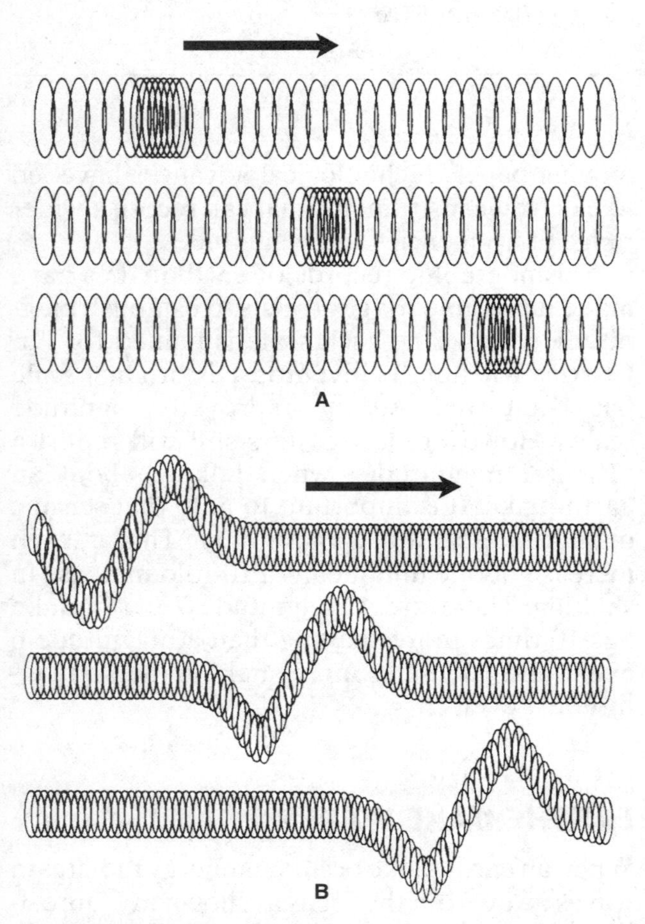

Figure 3-4. As P-waves travel outward from the earthquake focus, they cause the rock to vibrate forward and back, parallel to the direction of motion, like sound waves. (*A*) S-waves vibrate at an angle of 90° to the direction of motion, like light waves. (*B*)

Locating the Epicenter

A seismograph records the magnitude of an earthquake and the time the seismic waves arrive at the recording station. Seismologists use the difference in the speeds of P- and S-waves to locate the epicenter of an earthquake. The graph "Earthquake P-wave and S-wave Travel Time" found on page 11 of the *Earth Science Reference Tables* shows how long it takes P- and S-waves to travel different distances through Earth. Scientists know that as the seismic waves travel outward from the focus, the time delay between the P-waves and S-waves increases.

Finding Your Distance From an Earthquake Epicenter To determine your distance from an earthquake epicenter, follows these directions.

1. Subtract the arrival time of the P-waves from the arrival time of the S-waves. (This can be expressed in hours, minutes, and seconds, such as 08:06:40, or just hours and minutes, such as 15:25 (3:25 P.M.) See Figure 3-5.
2. On the clean edge of a sheet of paper, make two marks to show the observed interval along the travel-time scale on the vertical axis of the "Earthquake P-wave and S-wave Time Travel" graph. See Figure 3-6.
3. Keeping the edge of the paper vertical, slide the marks on the edge of your paper along the P- and S-curves of the travel-time scale until the marks match up with both curves.
4. Carefully follow the marked edge of your paper straight down to the horizontal axis to find the distance to the epicenter. In this example, the distance is 1800 km.

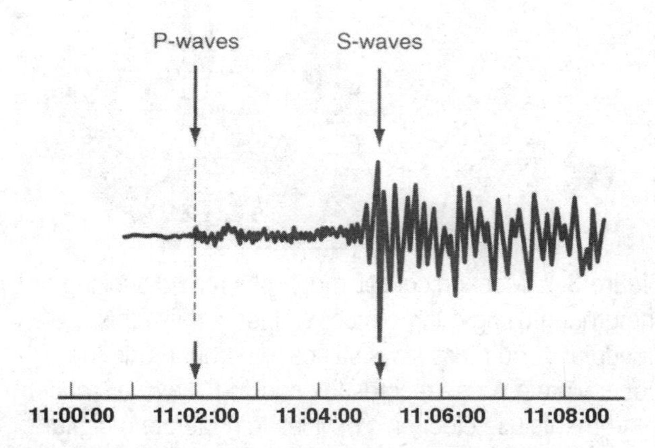

Figure 3-5. Notice that in this seismogram the P-wave arrives 3 minutes before the S-wave.

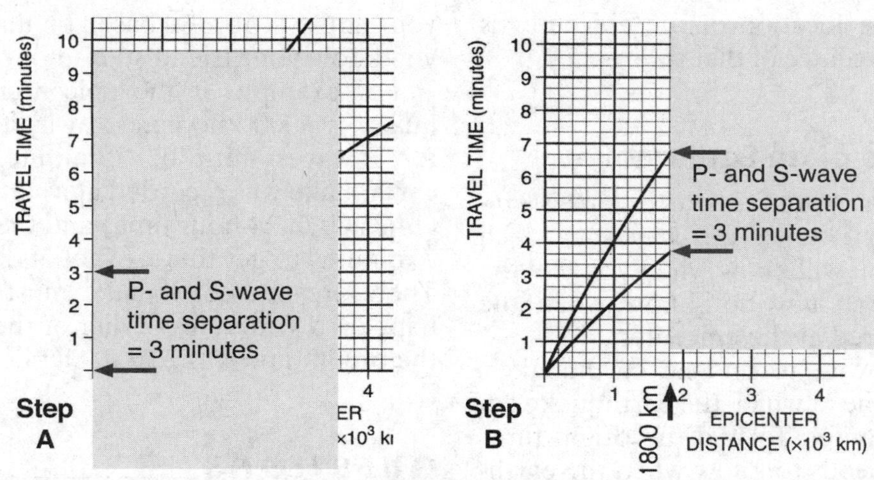

Figure 3-6. The distance to the epicenter of an earthquake can be determined if we know the time difference between the arrival of P-waves and S-waves.

This procedure has not yet located the epicenter, but it has established the distance from the recording station to the epicenter. If you draw a circle around the recording station with a radius equal to that distance, the epicenter should be located somewhere along that circle. To find the exact location, follow the procedure above to find the distance from the epicenter to at least three seismic recording stations. Use a compass to draw a circle around each station at the proper distance for that station. The epicenter is the point at which the circles intersect, or meet. See Figure 3-7. (In practice, the circles seldom intersect at a single point. More often, they make a

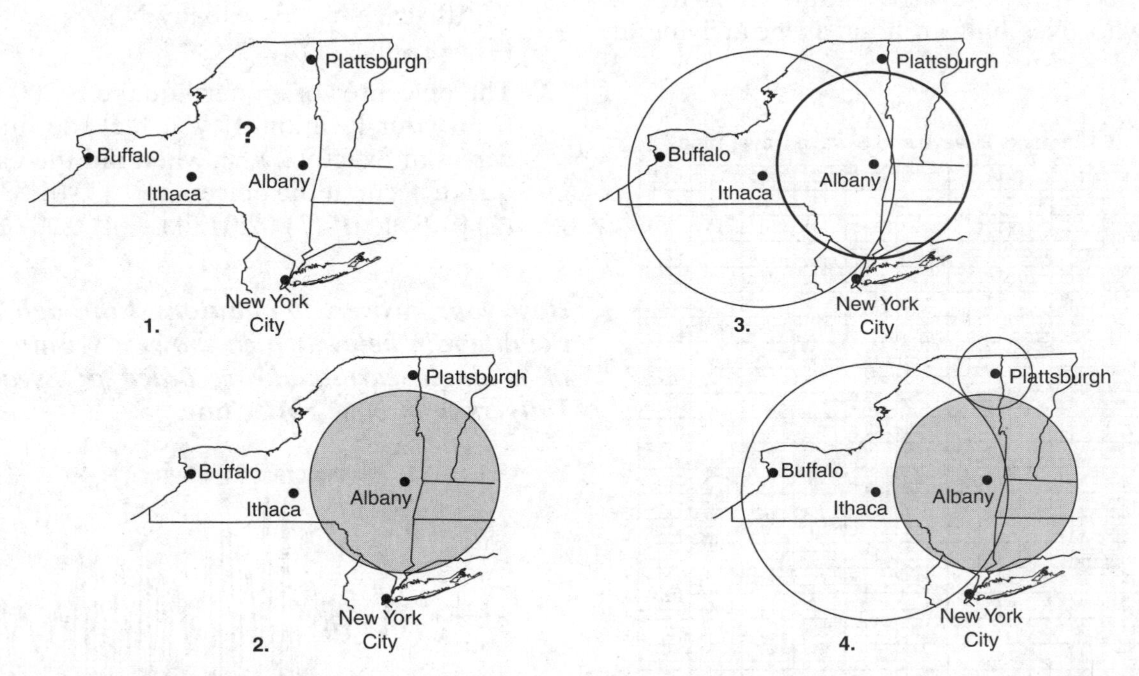

Figure 3-7. This diagram shows four steps to locating an earthquake epicenter. At least three seismic stations are needed to locate an epicenter. Step 1 shows that an earthquake occurred somewhere in New York State. Step 2 shows that the distance from Albany, NY has been plotted. The epicenter could be anywhere on this first circle. In Step 3, we add the distance circle around Ithaca, NY. This narrows the epicenter to the two widely spaced locations where the circles intersect. In Step 4, a third distance circle drawn from Plattsburgh makes the location of the epicenter clear, the only place where all three circles meet.

small triangle. The location of the epicenter is probably near the center of that triangle.)

The Origin Time of an Earthquake

The farther you are from an earthquake's epicenter, the longer it takes the waves to reach you. Therefore, you will know when you feel the vibrations, but you may not know when the earthquake occurred at the epicenter.

Seismologists want to find an earthquake's **origin time**, the time at which the earthquake occurred at its epicenter. To find the origin time, the seismologist needs to know when the earthquake was first recorded at any recording station—that is, the arrival time of the P-waves. The seismologist also needs to know the amount of time that the P-waves took to get there (the travel time). If the distance to the epicenter (as determined in the section above) is known, the travel time can be read from the "Earthquake P-wave and S-wave Travel Time" graph, Figure 3-8. A larger version of the graph is found in the *Earth Science Reference Tables* section of this book. To determine the origin time subtract the travel time from the arrival time. It is like a friend who lives half an hour away, arriving at your house at 10 A.M. based on that information, you know your friend started out at 9:30 A.M.

For example, if the epicenter of an earthquake is 4000 kilometers away, the travel time for P-waves must be 7 minutes. Suppose an earthquake was recorded at a station at 2:25 P.M. (14:25:00 in 24-hour time) and the P-waves took 7 minutes to get there. (That is the travel time.) Therefore, the earthquake must have actually happened 7 minutes earlier at the epicenter. So the origin time was 2:18 P.M. (14:18:00).

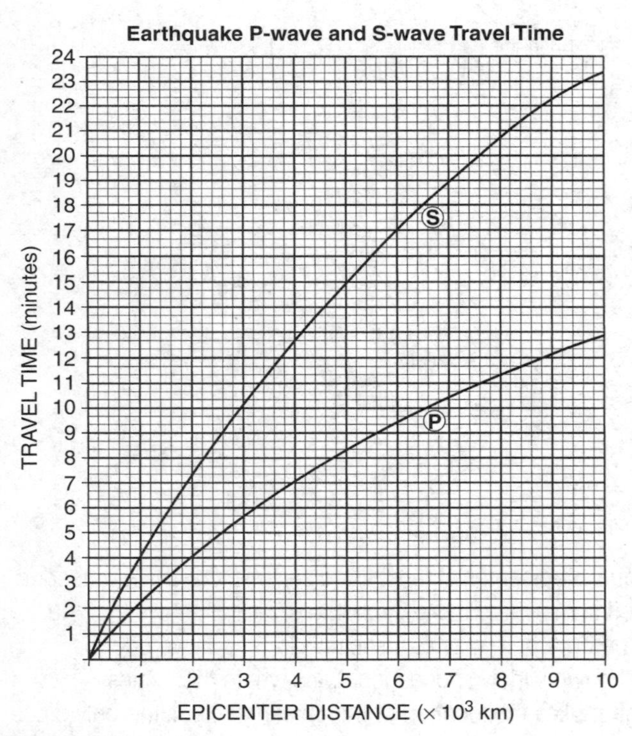

Figure 3-8.

QUESTIONS

Part A

1. A major earthquake occurs in California. It is recorded by a seismograph in Buffalo, NY. Which kind of seismic waves arrive first in Buffalo, NY? (1) light waves (2) surface waves (3) P-waves (4) S-waves

2. If an earthquake were to occur at Binghamton, NY, which location would experience seismic waves first (1) Buffalo, NY (2) Ithaca, NY (3) Albany, NY (4) Massena, NY

3. The epicenter of an earthquake is 2000 km from your location. If you feel the first P-waves at exactly 2 P.M., when did the earthquake occur at the epicenter? (1) 01:53 P.M. (2) 01:56 P.M. (3) 02:04 P.M. (4) 02:07 P.M.

Base your answers to questions 4 through 7 on the diagram below, which shows a seismogram of a single earthquake recorded at Syracuse University in New York State.

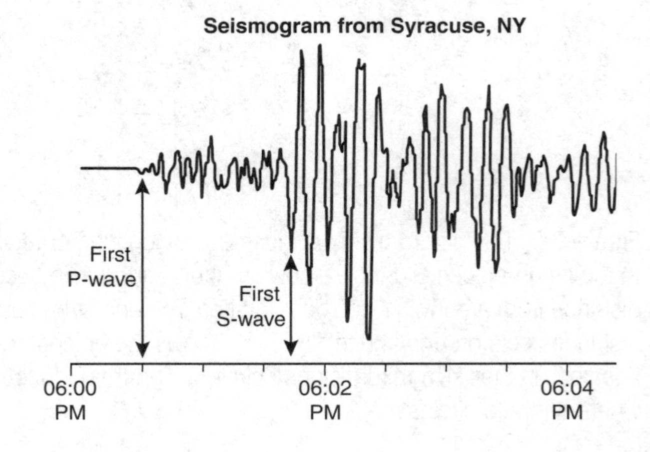

4. When did the first P-wave arrive at Syracuse? (1) 6 P.M. (2) 30 seconds past 6 P.M. (3) 6:01 P.M. (4) 1.5 minutes past 6 P.M.

5. How long after the arrival of P-waves did the S-waves arrive? (1) 30 seconds (2) 50 seconds (3) 1 minute, 15 seconds (4) 1 minute, 30 seconds

6. According to this seismogram, how did the S-waves differ from the P-waves? (1) The S-waves arrive first. (2) The S-waves traveled faster. (3) The S-waves did not transmit energy. (4) The S-waves made the ground vibrate more.

7. Approximately how far from Syracuse was the epicenter? (1) 300 km (2) 600 km (3) 1000 km (4) 1300 km

PART B

Base your answers to questions 8 through 12 on the diagrams in the next column. The top diagram shows three seismograms from the same earthquake recorded at locations X, Y, and Z. The map below the seismograms shows the plotted distance from each of these stations to the epicenter. A coordinate system but no scale is provided.

8. Approximately how far away from station *Y* is the epicenter? (1) 1300 km (2) 2600 km (3) 3900 km (4) 5200 km

9. Station *Z* is 1700 km from the epicenter. How long did it take the P-waves to travel to station *Z*? (1) 1 min, 50 s (2) 2 min, 50 s (3) 3 min, 30 s (4) 6 min, 30 s

10. What are the map coordinates of the epicenter? (1) E-5 (2) G-1 (3) H-3 (4) H-8

11. What was the approximate arrival time of the S-waves at station *X*? (1) 12:01:00 (2) 12:01:01 (3) 12:01:50 (4) 12:02:30

12. A fourth seismic station is located 5000 kilometers from this epicenter. How long after the P-waves did the S-waves arrive at the fourth station? (1) 5 min, 20 s (2) 6 min, 40 s (3) 15 min, 00 s (4) 24 min, 00 s

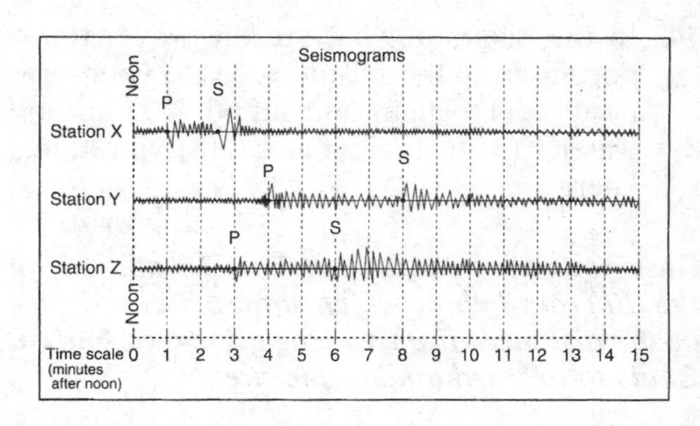

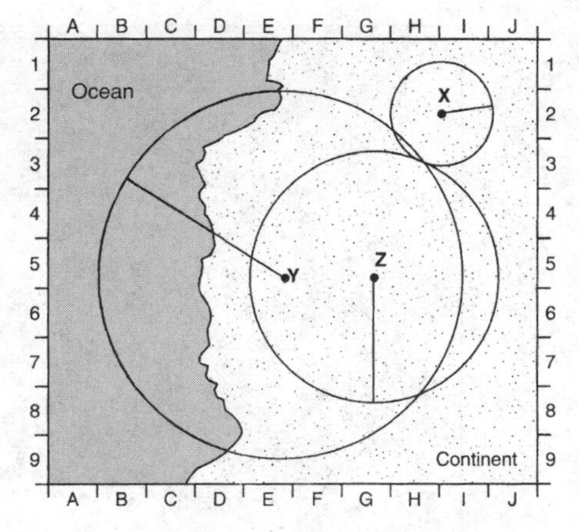

Base your answers to questions 13 and 14 on the diagram below, which shows seismic wave moving through the ground away from an earthquake. The arrows shows the direction the ground moves as the seismic waves pass.

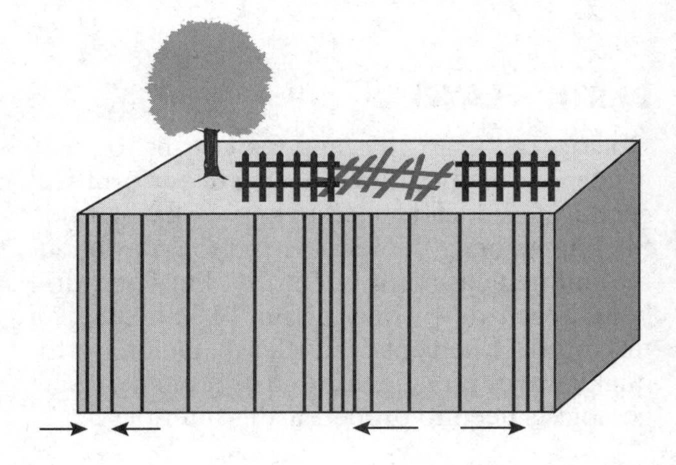

13. What kind of wave is illustrated? (1) P-waves traveling horizontally. (2) P-waves traveling vertically. (3) S-waves traveling horizontally. (4) S-waves traveling vertically.

14. In the diagram, what are the waves transporting from the seismic focus to your location? (1) light and sound (2) air and water (3) tectonic plates (4) vibrational energy

Base your answers to questions 15 and 16 on the diagram below, which shows circles drawn at the distances from Chicago, Denver, and St. Louis to an earthquake epicenter.

15. Where is the epicenter of this earthquake? (1) east of Dallas (2) south of Buffalo (3) west of St. Louis (4) northwest of Chicago

16. At which seismic recording station is the difference in the arrival times of the P- and S-waves the greatest? (1) Chicago (2) St. Louis (3) Denver (4) all three locations have the same S-wave delay

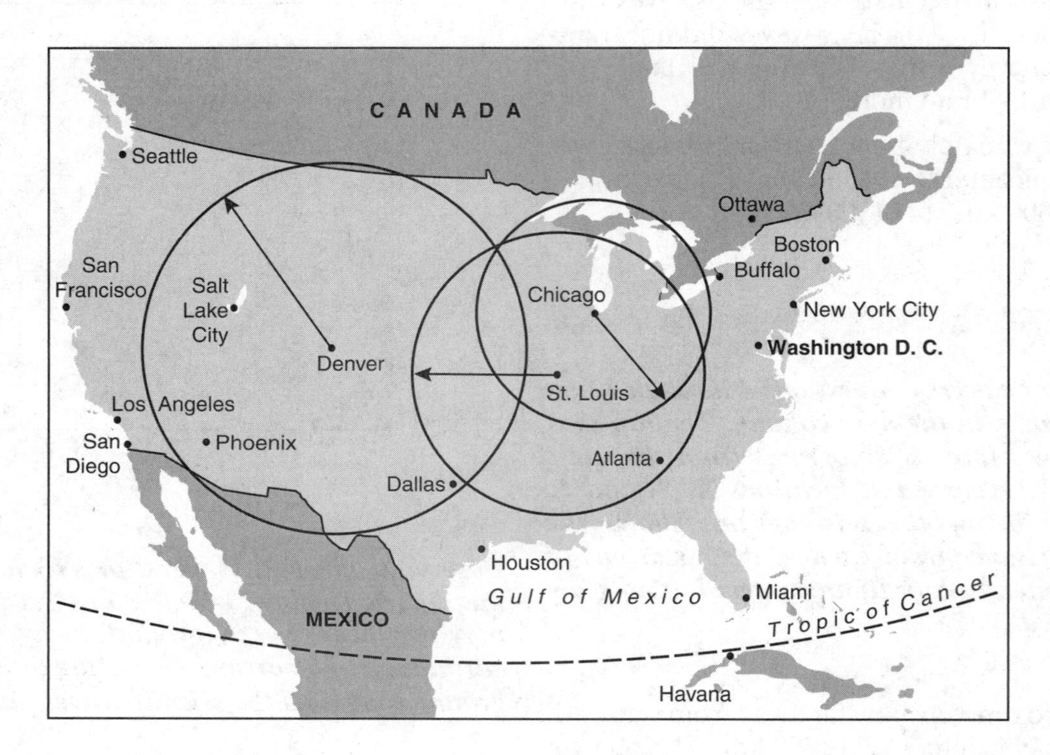

EARTH'S LAYERS

Doctors can use x-rays and CAT scans to "see" inside your body without having to perform surgery. Just as doctors need to know what is inside the human body, geologists need to know about the internal structure of Earth. Mines and drill holes reach down only about 10 kilometers, a tiny 1/10 of 1 percent of Earth's diameter. Earthquake waves provide the kind of remote sensing geologists need to probe Earth's interior.

The Crust

The **crust**, the outermost layer of our planet, varies in thickness from about 5 kilometers under the oceans to about 60 kilometers under the tallest mountain ranges on the continents. The composition of the crust is well known from observations of surface rocks as well as from mines and bore holes. In most places, a thin layer of sedimentary rocks covers the granitelike (**granitic**) rocks of the **continental crust**. The deep continental crust as well as the **oceanic crust**, which is under the layers of marine sediments, is composed mostly of darker and denser mafic rocks similar to basalt (**basaltic**). (See Figure 3-9.)

The Mantle

As earthquake waves travel toward Earth's center, they reach a layer in which their speed suddenly increases. This boundary, or interface, is called the **Moho**, or the Mohorovicíc discontinuity, named after the scientist who discovered it. The layer below the Moho is Earth's mantle.

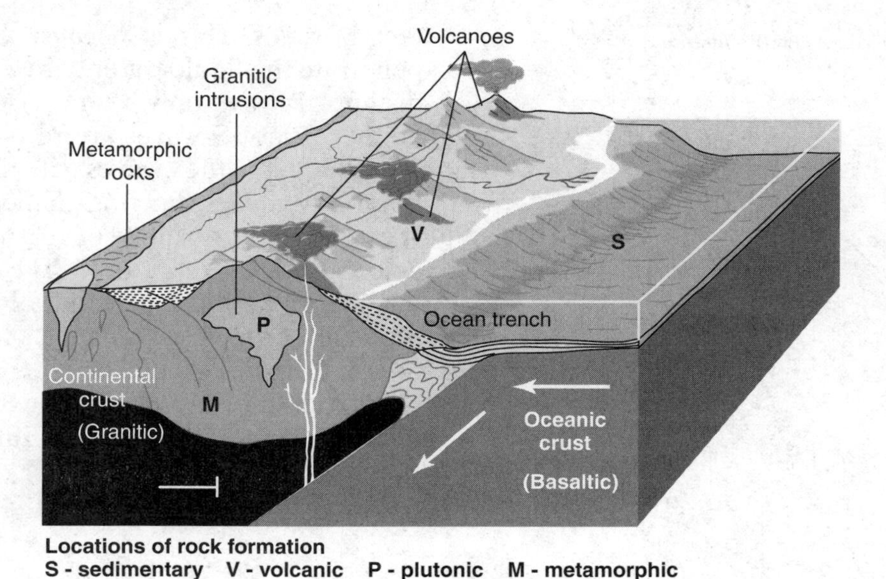

Figure 3-9. The complex continental crust floats high because it is composed mostly of relatively light granitic rocks. The basaltic rocks beneath the oceans are more mafic and more dense. Note the arrows showing movements of the crust and the letters showing places where different classes of rocks may be forming.

The **mantle** extends to a depth of about 2900 kilometers, so it includes most of Earth's volume. The mantle is divided into the plastic mantle or **asthenosphere** and the stiffer mantle. The asthenosphere, between the crust and the stiffer mantle, is partially melted and able to bend and flow slowly. Scientists have inferred the composition of Earth's mantle from a variety of observations:

1. Based on measurements of Earth's curvature, scientists can calculate the volume of our planet. The mass of Earth has been determined from measurements of gravitational attraction. Scientists have calculated the average density of Earth to be about 5.5 times the density of water. That is about twice the density of most rocks found at Earth's surface.
2. Earthquake waves travel faster in the mantle than in the crust.
3. The composition of magmas from deep within Earth includes a high proportion of the dense, mafic minerals.
4. A large portion of the *meteorites* that fall to Earth are composed of the mafic minerals. Scientists believe that these meteorites are the remains of material from which Earth formed billions of years ago. If the meteorites are those remains, they should show an overall composition similar to the greater part of Earth.

These observations suggest that the mantle is composed mostly of the dense, dark, mafic minerals olivine and pyroxene.

The Core

The deepest layers of Earth are the outer and inner **cores**. Geologists think that these layers are made of mostly iron mixed with a smaller amount of nickel. These elements, also found in nickel-iron meteorites, are relatively dense. It is logical that these densest materials would have moved into Earth's center.

Geologists think Earth's **outer core** is a liquid because S-waves cannot pass through the outer core. The **inner core** seems to be a solid, because P- and S-waves can pass through it. (Some S-wave energy is converted into weak P-waves at the core-mantle boundary.) Although the outer core and the inner core are both composed of iron and nickel, the higher pressures at Earth's

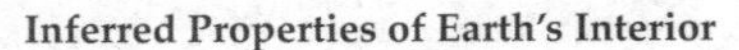

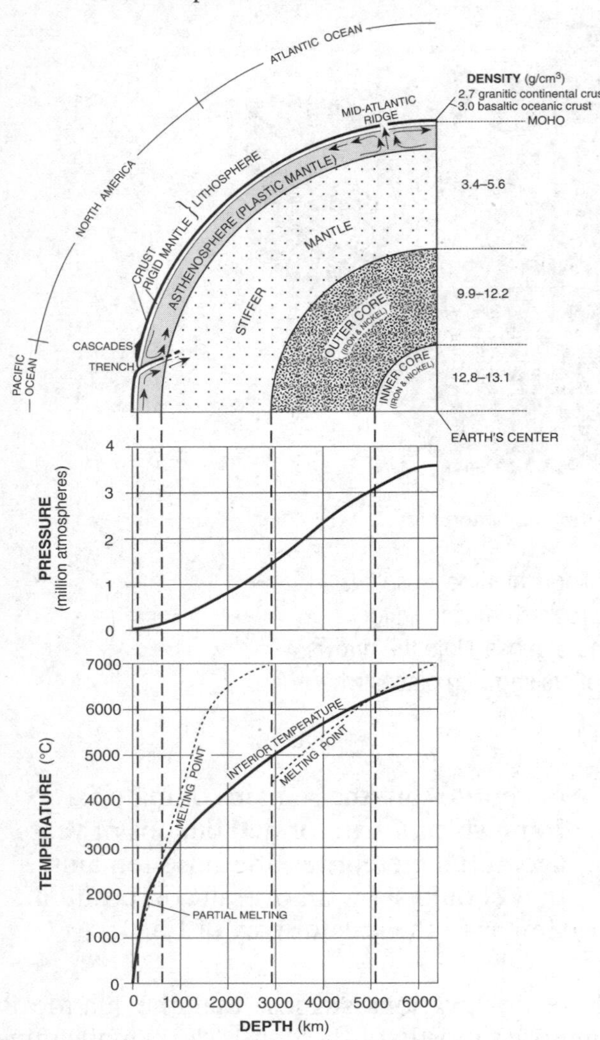

Figure 3-10. The scale diagram and two related graphs in this diagram provide a variety of information about Earth's interior. The graphs show how pressure and temperature change within Earth. The inferred density of each layer is also shown.

center cause the inner core to be solid. The "Inferred Properties of the Earth's Interior" shown in Figure 3-10 and also found on page 10 of the *Earth Science Reference Tables* summarizes scientists' ideas about Earth's interior.

Earthquake Shadow Zones

When a major earthquake occurs, both P-waves and S-waves are received over most of Earth. The part of Earth opposite the side where the earthquake occurs receives P-waves, but no direct S-waves. This is because S-waves cannot penetrate the liquid outer core. Surrounding this distant "P-wave-only" zone is a region where neither P- nor S-waves are received. Refraction (bending) of the waves at the mantle-core boundary causes this ring-shaped shadow zone. It extends from an angle of 102° to 143° from the epicenter. See Figure 3-11. The curved path of the energy waves is the result of the way Earth's physical properties change with increasing depth. In general, the deeper rocks are more rigid, and this causes seismic waves to travel faster and bend down within Earth.

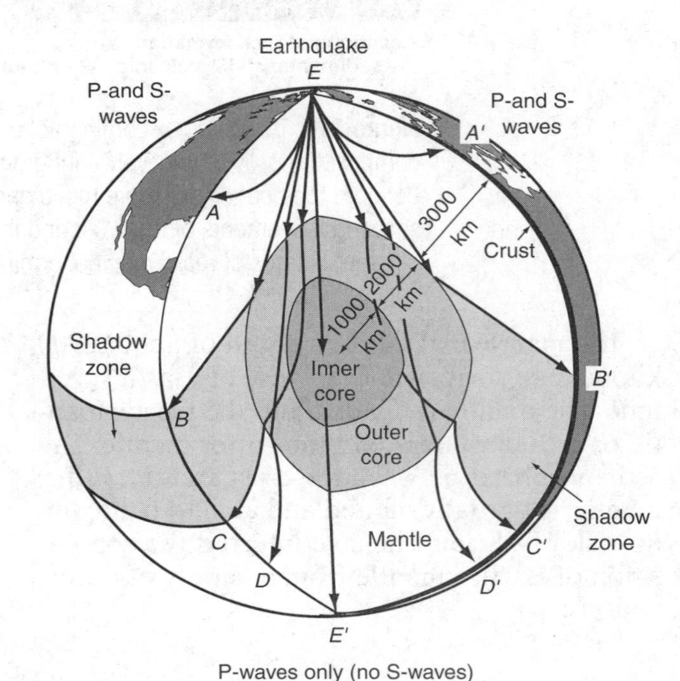

Figure 3-11. Layers of Earth have been defined based on the passage of earthquake energy waves. The crust and the mantle allow both P- and S-waves to pass. Both waves travel faster in the more rigid rocks of the mantle. The liquid outer core will not let S-waves pass. Refraction (bending) of the waves at the mantle-core boundary causes a shadow zone that encircles Earth. In this zone, neither direct P- nor S-waves are received.

QUESTIONS

Part A

17. How do geologists know that the outer core is liquid? (1) Deep drilling has penetrated to the outer core. (2) Temperatures in the outer core are below the melting temperature. (3) P-waves do not reach the far side of Earth. (4) S-waves do not reach the far side of Earth.

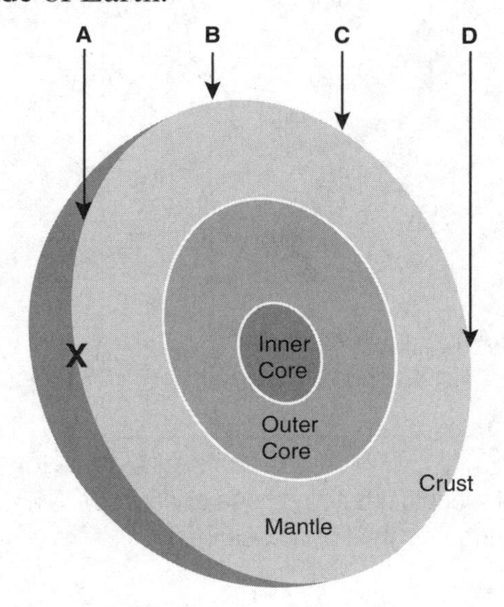

18. The diagram above shows a cross section of Earth. If position *X* is the focus of a major earthquake, at which seismic station would P-waves, but not S-waves, be recorded. (1) *A* (2) *B* (3) *C* (4) *D*

Base your answers to questions 19 through 24 on Figure 3-10 on page 74.

19. Inside Earth, how does pressure change with increasing depth? (1) It increases. (2) It decreases. (3) It increases and then decreases. (4) It decreases and then increases.

20. The asthenosphere is labeled "plastic mantle." Why is the word "plastic" used to describe this part of the mantle? (1) The asthenosphere is not made of rocks. (2) The asthenosphere is lower in density than any other layer of the solid Earth. (3) The asthenosphere is more able to bend and flow than other layers of Earth's interior. (4) The asthenosphere is a liquid.

21. In what part of Earth's interior does the temperature change fastest with depth? (1) the crust (2) the mantle (3) the outer core (4) the inner core

22. According to scientists' best estimates, what is the approximate pressure at the center of Earth's outer core? (1) 2.3 atmospheres (2) 2.3 million atmospheres (3) 3.4 atmospheres (4) 3.4 million atmospheres

23. What is the approximate depth to Earth's center? (1) 1200 km (2) 6400 km (3) 12,700 km (4) 55,000 km

24. Why is it difficult to see the crust layer on this diagram? (1) The crust is actually a part of Earth's mantle. (2) The crust is very thin compared with other layers. (3) The crust does not occur everywhere on Earth's surface. (4) The crust is the most dense layer of Earth.

Part B

25. When a major earthquake occurs, there is a ring around Earth's surface between 102° and 143° from the epicenter where P-waves are absent. Why are P-waves waves not detected in this area? (1) P-waves are bent and reflected away from this region. (2) P-waves can pass through a solid, but not a liquid. (3) This zone is too far from the epicenter to record seismic waves. (4) Seismic waves cannot travel through Earth's interior.

26. What kind of scientist is best trained to investigate the physical and chemical properties of Earth's cores? (1) an expert in meteorology and weather hazards (2) a scientist who studies natural objects that fall from the sky (3) a geologist who supervises drilling for petroleum (4) a researcher who uses satellite images to draw precise maps

27. If a seismic event is recorded by a modern seismograph, how is the strength of the earthquake specified? (1) by the direction of shaking (2) by a logarithmic magnitude scale (3) by how many landslides occur (4) by how far below the surface it occurs

EARTHQUAKES AND VOLCANOES

Earthquakes and volcanoes are not distributed randomly over Earth. There are geologic zones where volcanoes, earthquakes, and mountain building are especially active. For example, a surprisingly large number of the world's volcanoes

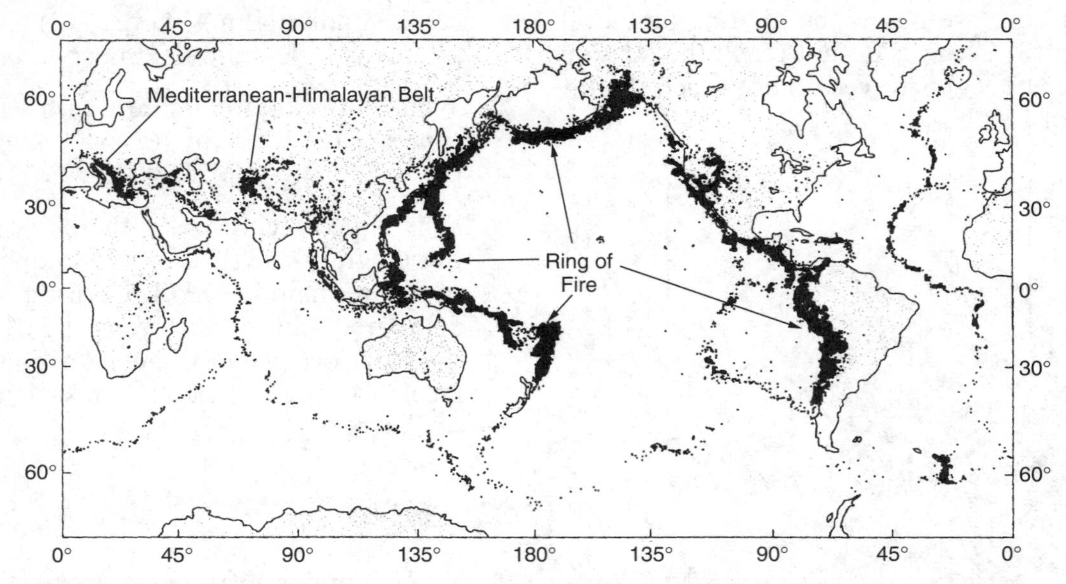

Figure 3-12. Earthquakes, active volcanoes, and regions of mountain building generally occupy distinct geographic regions. One is the "Ring of Fire" that surrounds the Pacific Ocean.

and earthquakes occur around the rim of the Pacific Ocean, in a region called the "**Ring of Fire**." See Figure 3-12. Countries, such as Japan, and states on the western coast of the United States, such as California and Alaska, are on the Ring of Fire. Earthquakes and volcanoes frequently cause damage in these areas. In fact, two of the greatest natural disasters in human history have been earthquakes in China, each of which is thought to have caused more than half a million deaths.

Seismic Hazards

Earthquakes cause damage in several ways. Because of the shaking, buildings and bridges may collapse, or become unsafe. **Subsidence** (sinking land) and landslides damage buildings, roads, and utilities, such as gas, water, electric and phone service. Earthquakes in or near the oceans may cause large waves, called **tsunamis**. (The term "tidal wave" is sometimes wrongly applied to these waves. Tsunamis have nothing to do with the tides.) When tsunamis reach land, they increase in size and cause devastation in coastal locations. Fire is also a danger in earthquakes. Gas lines can rupture, and sparks from damaged power lines may ignite the gas. In addition, water pipes may be broken, which stops the flow of water. This makes fighting fires difficult.

Buildings and other structures constructed on landfill or thick sediments face a triple threat. Thick sediments usually intensify the shaking. In addition, without a firm bedrock foundation, these structures have reduced support. Furthermore, sediments saturated with groundwater can actually become like a fluid when they are shaken. This process is called liquefaction.

In areas that experience earthquakes frequently, governments have developed building codes that call for the reinforcement of bridges and buildings. In addition to making these structures able to withstand moderate earthquakes, these precautions make it less likely that people will be injured by falling debris. However, the safest place to be is away from buildings, such as in a car parked in an open field. Figure 3-13 is a seismic risk map for the United States. Table 3-2 lists selected seismic events

Volcanic Hazards

When volcanoes erupt they often spew hot lava, ash, and/or toxic gases. Lava and ash can bury buildings and people, and toxic fumes can flow down the volcano's slopes to suffocate living things. Other dangers can result from pyroclastic flow, landslides, mudflows, and floods from melted snow pack, especially from high volcanic mountains. Figure 3-14 on page 78 shows some of these hazards.

Table 3-2. Selected Historic Seismic Events

Location	Date	Magnitude	Notes
Shensi, China	1556	Unknown	Worst natural seismic disaster known; 830,000 deaths
New Madrid, MO	1811–1812	~8 (est.)	Flow of Mississippi River briefly reversed.
San Francisco, CA	1906	~8.3 (est.)	Most damage was caused by uncontrolled fires.
Massena, NY	1944	~6 (est.)	Largest in NY state. Chimneys destroyed, water lines broken.
Lebu, Chile	1960	9.6	Largest event measured with seismographs. Occurred in the Pacific Ocean.
Anchorage, AK	1964	9.5	Extensive tsunami; 131 deaths.
Haicheng, China	1975	7.5	Predicted by scientists. City evacuated; only ~130 deaths.
Tangshan, China	1976	7.6	Prediction failed. Worst modern natural disaster; ~650,000 deaths.
Northridge, CA	1994	6.7	$10 billion in damages; 61 deaths.
Au Sable Forks, NY	2002	5.1	Felt throughout NY state. Caused damage to local roads.
Indian Ocean, Southeast Asia	2004	9.0	The tsunami waves caused by this ocean-bottom event killed more than 200,000 people.
Sichuan, China	2008	7.9	With 70,000 deaths and 5 million people left homeless, this was the most destructive land-based earthquake in 30 years.

In the long run, volcanoes can also be helpful. Some of Earth's most fertile soil is composed of weathered volcanic material. Volcanoes built the beautiful Hawaiian Islands. Some volcanoes are used as sources of geothermal energy.

We need to understand that forces of nature, including earthquakes, tornadoes, and other natural events, although disastrous in the short term, are important parts of Earth's cycles. For example, the people in California may treasure

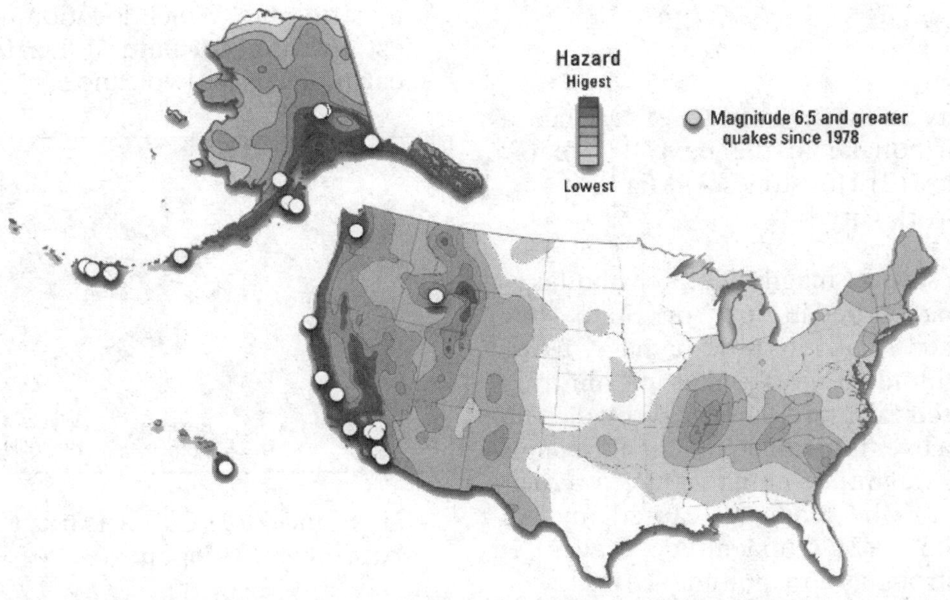

Figure 3-13. Earthquake predictions in any region are generally based on historic events. The white dots show recent major seismic events. Shading on the map indicates where earthquakes are most likely in the future. Note that the most active areas are along the Pacific Coast of the United States.

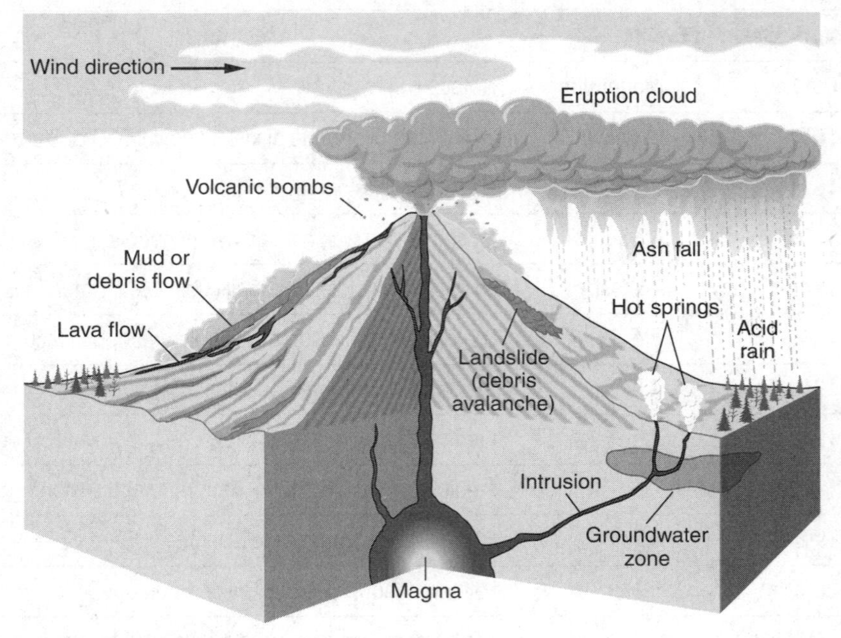

Figure 3-14. The most serious hazards to life and property due to a volcano.

their climate and scenery, but not the threat of major earthquakes. However, they should remember that seismic events helped to shape the land and thereby influence the climate they enjoy.

QUESTIONS

Part A

28. Which city is most likely to experience a major earthquake in the next 100 years? (1) Seattle (2) Houston (3) Miami (4) New York City

29. Recently, several magnitude 3 earthquakes have been recorded in New York State. They have caused relatively little damage. However, magnitude 6 earthquakes in California have caused far more damage. Why did the California events cause so much more damage? (1) Magnitude numbers have no relationship to the damage caused by an earthquake. (2) A magnitude 6 event is twice as strong as a magnitude 3 event. (3) As the magnitude number increases, the damage usually is less. (4) A magnitude 3 earthquake has only 1/1000 the shaking as a magnitude 6 event.

30. Most of the fatalities in the Indonesia earthquake of 2004 resulted from a huge tsunami. How did most of the tsunami victims die? (1) They were shaken severely. (2) They died when large buildings collapsed. (3) They drowned. (4) They fell into large cracks in the ground.

31. The map below identifies four locations on a world map. Which location has the greatest potential for natural hazards related to earthquakes and volcanoes?

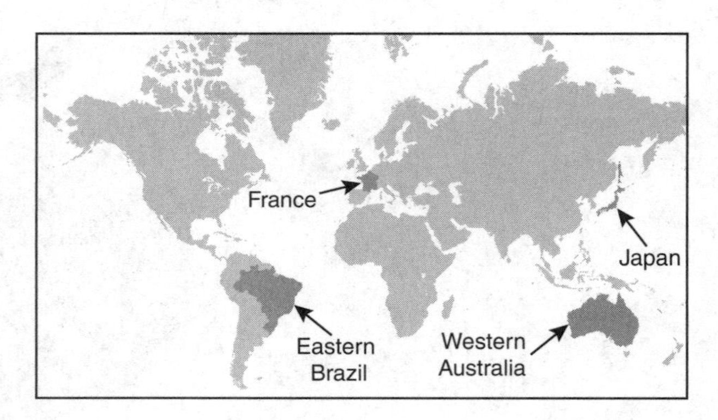

(1) France (2) Eastern Brazil (3) Western Australia (4) Japan

32. Which hazard is *least* likely to be associated with earthquakes or volcanoes? (1) tsunamis (2) tornadoes (3) poisonous gas (4) landslides

33. The geologic structure shown in the diagram below is probably evidence of what past event?

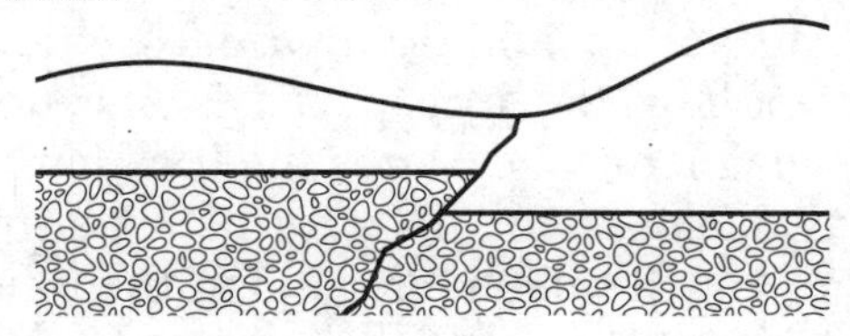

(1) volcano (2) flood (3) tsunami (4) earthquake

Part B

34. Which action would be most useful to prevent earthquake damage inside your home? (1) Glue down objects located on high shelves. (2) Install more insulation within the walls. (3) Replace small windows with larger windows. (4) Replace carpeting with tile floors.

35. In 2004, 10-year-old Tilly Smith was vacationing with her family at a beach resort in Thailand. She is credited with saving approximately 100 lives shortly after the magnitude 9 Indian Ocean earthquake. How did she save so many people? (1) She told her family to spend an extra week at the beach. (2) She recognized the signs of an approaching tsunami and warned the people. (3) She predicted exactly when the earthquake would occur. (4) She convinced the people to avoid eating before going swimming.

36. Which graph below best shows the relationship between the likelihood of volcanic events and earthquakes for different locations on Earth?

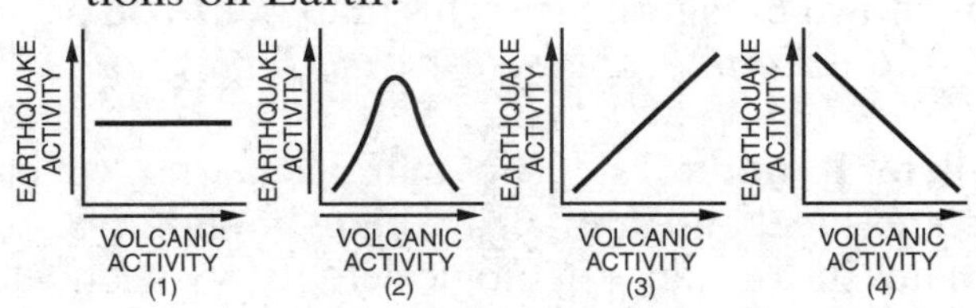

37. Where is the safest place to be if a large earthquake occurs? (1) inside a car parked in an open parking lot (2) inside a car parked under a highway bridge (3) at the top of a bluff along the ocean (4) in the school cafeteria

CONTINENTAL DRIFT

Ever since maps correctly showed the eastern and western shores of the Atlantic Ocean, people noticed that the shoreline of the Americas would fit remarkably well with the shores of Europe and Africa. In 1912, Alfred Wegener proposed that this was more than a coincidence. He suggested that this jigsaw puzzle fit was the result of the opening of the Atlantic Ocean. The opening of the Atlantic Ocean broke apart an ancient supercontinent Wegener named **Pangaea**. Figure 3-15 shows a modern interpretation of the break-up of Pangaea.

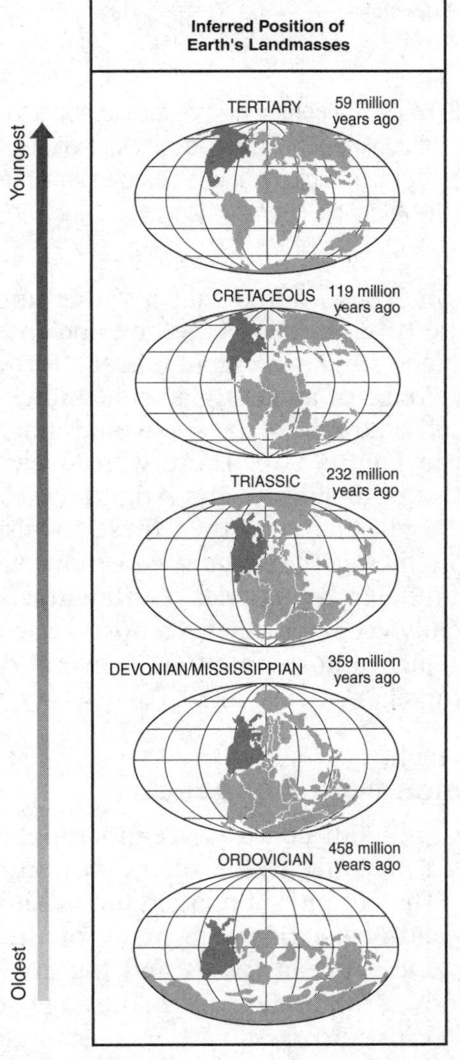

Figure 3-15. The Atlantic Ocean formed as the continent of Pangaea began to split apart approximately 200 million years ago. While this was happening, Australia and India were moving north from the South Polar region to their present positions.

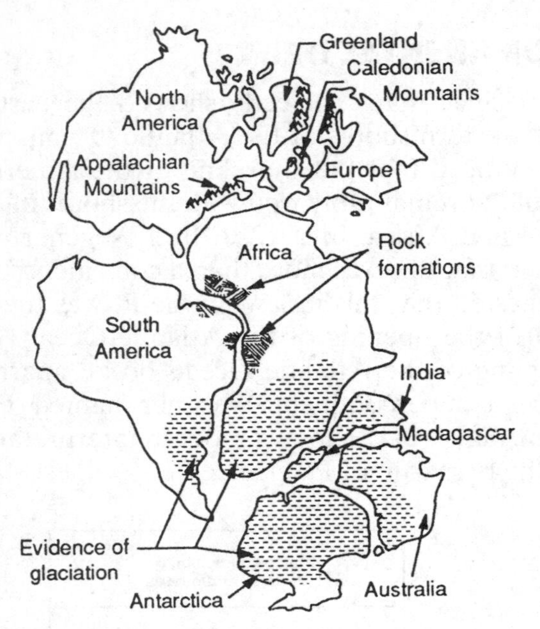

Figure 3-16. If the continents are reassembled to form the ancient continent of Pangaea, glacial features, rock formations, and mountain ranges come together from both sides of the Atlantic Ocean.

Wegener also noted that if these land areas could be brought back together, ancient mountain ranges, similar continental rock formations, and evidence of ancient glaciers (scratched and polished bedrock surfaces) would line up as shown in Figure 3-16. There were even similar fossils on both sides of the Atlantic. If Pangaea could be put back together, these fossils would be brought together again. Wegener called his idea of moving landmasses "**continental drift**." It took many years for other scientists to accept his ideas, which unfortunately was several decades after Wegener's death in 1930.

Evidence from the Oceans

In the mid-20th century, oceanographers used sonar to map the shape of the bottom of the oceans. They found submerged mountain ranges called **mid-ocean ridges** in many of the major oceans. Deep-ocean fossils and the analysis of radioactive materials show the age of the oceanic crust increases with increasing distance from the mid-ocean ridges. It appears that some oceans are growing wider from the middle out.

Scientists using measurements of Earth's magnetic field discovered additional evidence. The scientists found stripes of rock with normal magnetic polarity alternating with stripes that had **reversed polarity**. These patterns of magnetic polarity resulted from molten rock coming to the surface at mid-ocean ridges. While the rock is molten, the iron particles line up with Earth's magnetic field. When the rock hardens, it keeps that alignment. This forms a permanent record of Earth's magnetic field at the time that the oceanic crust formed. Scientists also learned that the north and south magnetic poles reverse. The most recent reversal occurred about 780,000 years ago. This fact led them to understand that they were looking at diverging and slow-moving sections of ocean crust. Parallel stripes of magnetism on both sides of the ridges preserve a record of many reversals of Earth's magnetic field in the geologic past. Scientists recognized that oceanic crust is being created at the mid-ocean ridges and destroyed at the ocean trenches. Tectonic plates slide down into Earth's interior at the **ocean trenches**. This established a new paradigm (theoretical framework). Figure 3-17 shows how the ocean floor is created by divergence at the ocean ridges. This is called **seafloor spreading**. The evidence convinced most scientists to accept Wegener's ideas.

Plate Tectonics

We now call this new paradigm the unified theory of **plate tectonics**. (**Tectonics** refers to large-scale deformation of Earth's crust.) According to this theory, Earth's surface is composed of about a dozen major rigid, moving lithospheric plates and several smaller plates. These plates contain areas of low-density granitic continental crust as well as dense basaltic oceanic crust. For example, the western Atlantic Ocean and the North American continent occupy a single plate as shown in Figure 3-18. This map is also on page 5 of the *Earth Science Reference Tables.*

Plate Boundaries The margins along which plates meet and interact are **plate boundaries**. As plates move, major geologic activity—volcanoes, earthquakes, and mountain building—occurs mostly along these boundaries. At the western edge of South America, the South American Plate is colliding with the Nazca Plate. As a result of this collision, part of the crust of the Pacific Ocean is diving under the South American Plate. The volcanoes of the nearby Andes Moun-

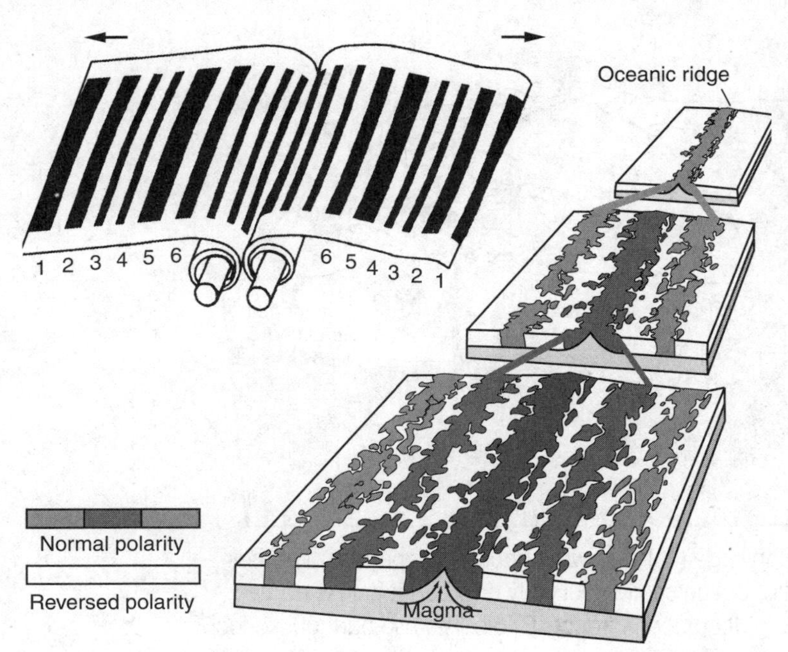

Figure 3-17. As the ocean crust spreads away from the ocean ridges, new crust is created. Changes in Earth's magnetic field create the "zebra stripes" of normal and reversed directions of magnetism. It is as if new crust were rolling away from the ocean ridges like two conveyor belts with magnetic stripes for time markers.

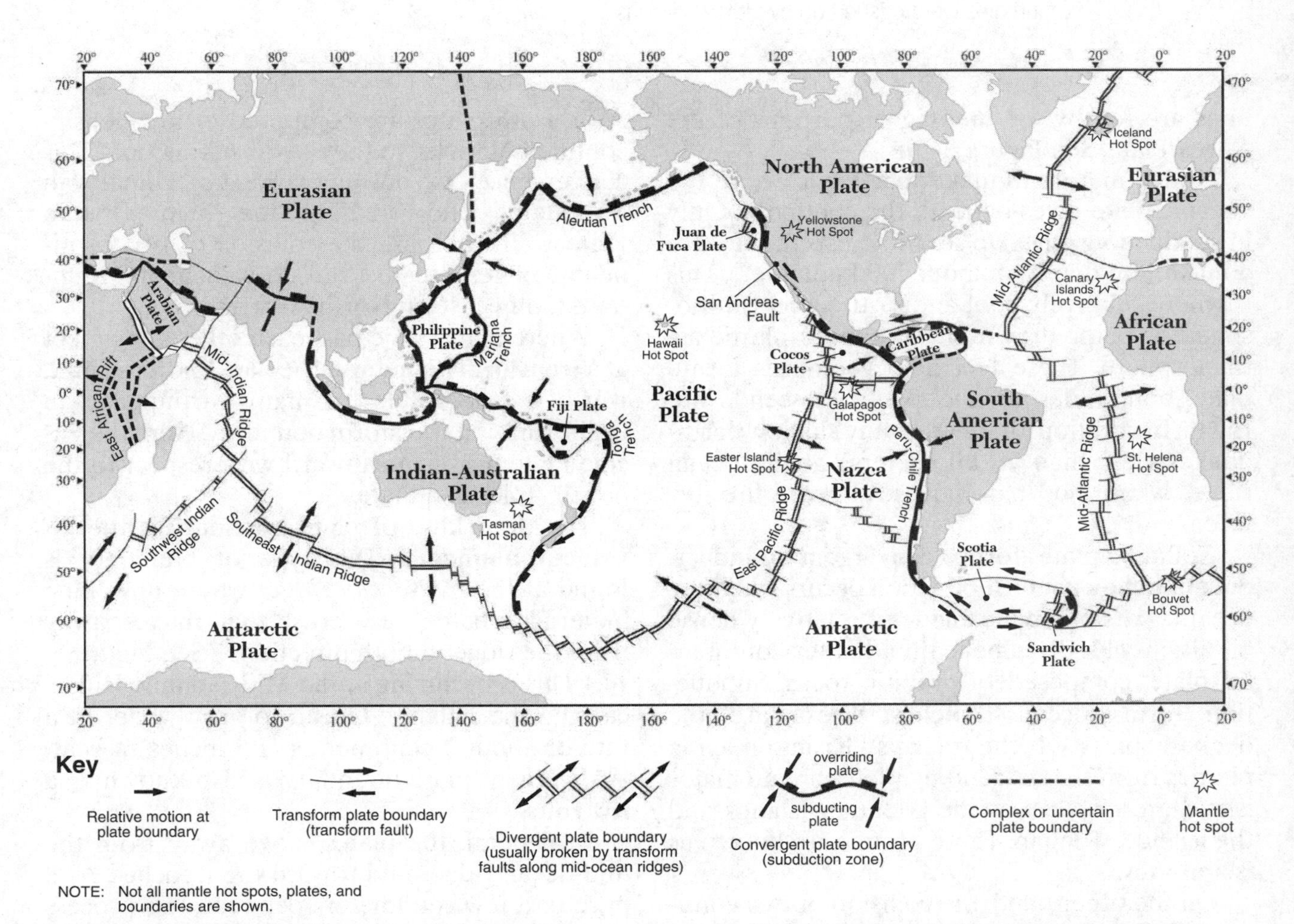

Figure 3-18. Earth's crust is composed of about a dozen major plates and many smaller plates. These plates move toward each other at convergent boundaries, move away from each other at divergent boundaries, and shift horizontally past each other at transform boundaries.

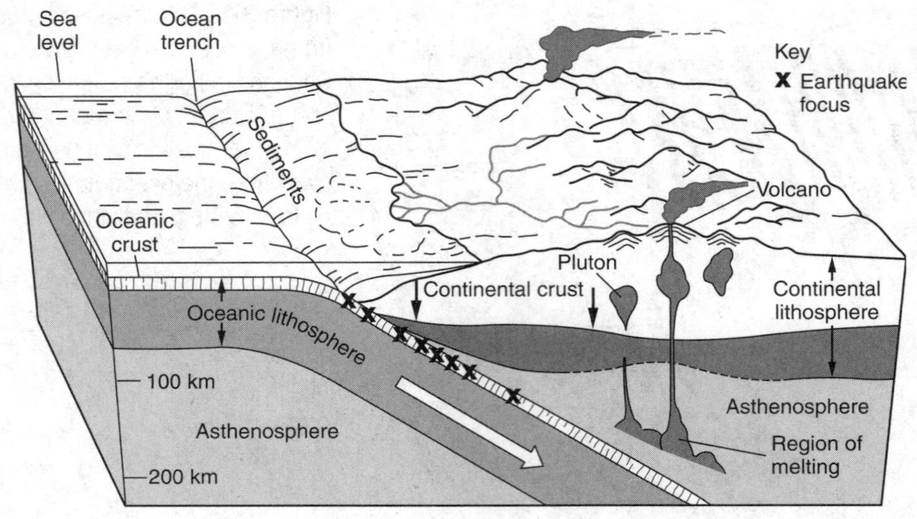

Figure 3-19. When two plates converge (collide), what happens depends upon their relative densities. Dense (basaltic) ocean sections usually dive beneath less dense (granitic) plates. This may create a deep ocean trench. Geologists use earthquake foci to find the boundaries of the descending oceanic crust. The world's highest mountains were pushed up by plate collisions.

tains are fed by the melting of portions of the Nazca Plate. See Figure 3-19.

Notice that earthquake foci occur where the oceanic plate dives beneath the continent. Only in subduction zones do seismologists find foci to a maximum depth of about 700 kilometers. This is where relatively cool and brittle (rigid) lithosphere is being drawn into the more plastic asthenosphere. These foci help geologists locate plate boundaries as the plates descend into Earth. In addition, there are many shallow earthquakes (less than 70 kilometers) at the ocean ridges where hot material rises toward the surface.

Colliding plates form a **convergent boundary**, or **subduction zone**. Subduction occurs when the oceanic part of a plate, made of relatively dense basaltic rocks, dives beneath a lighter continental plate, composed of granitic rocks. Subduction forms ocean **trenches**. These are the deepest parts of the oceans. Rising magma plumes from the descending plates create major island groups such as the Aleutian Islands and the islands of Japan. These groups are known as **island arcs**.

On the other hand, the collision of two continental plates leads to folding, faulting, and mountain building. For example, the landmass of India is moving northward and crashing into Asia. Both plates are composed of low-density continental rocks, so they resist subduction. This convergence is pushing up the Himalayan Mountains and creating the high Tibetan Plateau. In the past, the collision of two continental plates created the Appalachian Mountains along eastern North America.

Where plates slide past each other, they meet at a **transform boundary**. The San Andreas Fault, which is responsible for many earthquakes in California, is a transform boundary. Here the Pacific Plate moves northward with respect to the North American Plate.

The third kind of plate boundary is the **divergent boundary**. This type of boundary is found at the *mid-ocean ridges* where upwelling material creates new crust that moves away from the ridge in both directions. (See Figure 3-17.) This is occurring at the Mid-Atlantic Ridge, causing the Atlantic Ocean to grow wider at a rate of about 3 centimeters (1.5 inches) a year. A divergent plate boundary is also known as a **rift zone**.

In general, the plates move away from the mid-ocean ridges and toward the trenches. At a pace of a few centimeters per year, this process may seem very slow. But when we look at the millions of years of Earth's history, the continents have moved great distances. (See Figure 3-15.)

QUESTIONS

Part A

38. Why does California have so many large earthquakes? (1) California is located at a mid-ocean ridge. (2) California is located where two plates are converging. (3) California is on a transform plate boundary. (4) California is near the center of a tectonic plate.

39. The Ordovician Period was approximately 460 million years ago. Where do most geologists infer that the rocks that underlie the United States were located at that time? (1) south of Africa (2) south of the equator (3) next to Australia (4) at the North Pole

Part B

40. The Atlantic Ocean divided into the North Atlantic and the South Atlantic approximately at the equator. What feature runs roughly north-south near the middle of both oceans? (1) a transform boundary (2) a divergent boundary (3) a convergent boundary (4) a subduction boundary

41. Compared with the position of Africa, tectonic motion is shifting South America toward the (1) north (2) south (3) east (4) west

42. If you wanted to trace the creation of tectonic plates using magnetic bands of normal and reversed polarity, your best place to measure the magnetic stripes would be (1) in the southeastern Pacific Ocean (2) in the northern Pacific Ocean (3) the west coast of North America (4) the east coast of South America

Base your answers to questions 43 through 48 on the diagram below that is a model of Earth's interior and some surface features.

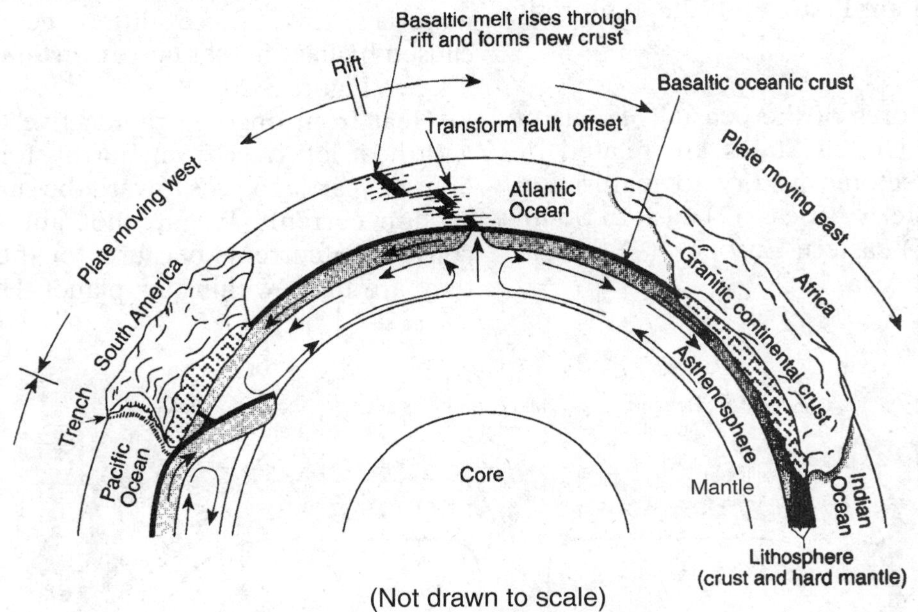

(Not drawn to scale)

43. Mid-ocean ridges (rifts) such as the Mid-Atlantic Ridge normally form where tectonic plates are (1) converging (2) diverging (3) stationary 4) sliding past each other in a transform motion

44. According to the diagram, the deep trench along the west coast of South America is caused by movement of the ocean crust that is (1) colliding with the Atlantic oceanic crust (2) uplifting over the continental crust (3) sinking at the mid-ocean ridge (4) sinking beneath the continental crust

45. At which location are you most likely to find very old igneous rocks that contain mostly quartz and feldspar? (1) Mid-Atlantic

Ridge (2) South America (3) the asthenosphere (4) Earth's outer core

46. Where are deep- focus earthquakes (deeper than 500 kilometers) most likely to occur? (1) Mid-Atlantic Ridge (2) beneath central Africa (3) along the east coast of South America (4) along the west coast of South America

47. Which feature has been located by plotting the epicenters of many shallow earthquakes with no deep seismic foci nearby? (1) Mid-Atlantic Ridge (2) Mohorovicic discontinuity (3) mantle-core interface (4) edge of the continental shelf

48. Which plate boundary is similar to the rift plate boundary at the center (top) of this diagram? (1) where the Caribbean Plate meets the North American Plate (2) at the San Andreas Fault along the edge of North America (3) along the west coast of South America (4) between the Indian-Australian Plate and the Antarctic Plate

49. Bedrock features of the coastal regions of the Eastern United States are related by their plate tectonic history to features of (1) northwestern Africa (2) eastern South America (3) eastern Europe (4) the Pacific Coast of Asia

WHAT DRIVES THE TECTONIC PLATES?

The magnetic pattern of the ocean floor helped scientists understand what drives ocean-floor spreading. Within planet Earth, there is a great deal of heat. People who work in the gold mines of South Africa know that the deeper you go in the mines, the hotter it gets. Very deep wells also confirm that Earth's temperature increases with depth. Heat always flows from places of high heat to areas of lower heat. In the case of Earth, this means heat flows from the center toward the surface.

Heat flows in three ways. Earth receives solar energy by **radiation**, but radiation can travel only through the vacuum of space and through transparent materials, such as air and glass. Since our planet is not transparent, heat cannot travel through it by radiation. Energy flows through solids by **conduction**, but it is a very slow process. On the other hand, **convection** distributes heat through fluids. Most materials expand when they are heated, which makes them less dense. Materials that are less dense tend to rise, while denser materials sink. These differences in density, caused by heating, set up **convection cells** in fluids. See Figure 3-20.

Heat from ongoing radioactive decay along with heat left over from Earth's formation billions of years ago rises toward the surface in convection currents. We are not able to observe convection currents because, for the most part, they are deep within our planet. However, we

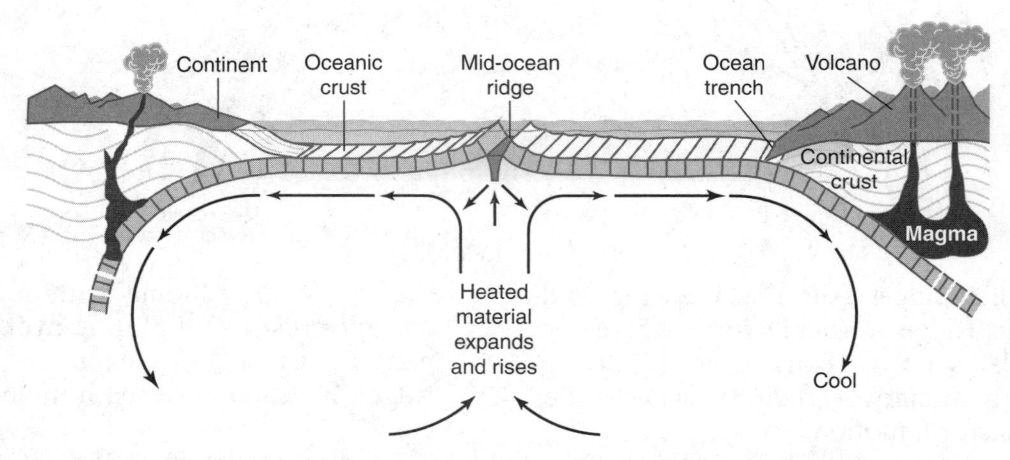

Figure 3-20. Convection currents within Earth enable heat to escape from Earth's interior. These currents create and expand the ocean bottoms, and they carry the continents as passive "rafts" of lighter rock.

can observe changes at the surface. These events are part of a global system of energy flow. So convection currents drive plate motions.

Most of Earth's mantle is brittle, so it would shatter if you hit it hard with a hammer. However, under the conditions of extreme heat and pressure found within Earth, even this rock behaves as a fluid, and it is able to flow very slowly. The plates move because they are driven by convection within the mantle. The rigid portion of the plates includes Earth's crust as well as a small part of the upper mantle. This rigid layer is the *lithosphere*. The plates float on a more fluid and partially molten portion of the mantle called the *asthenosphere*. Convection cells bring material to the surface at the ocean ridges and pull it back into Earth at the trenches. Figure 3-20 shows how heated material rises toward the surface and diverges from the ocean ridge, forming new ocean floor.

In some places, a stationary hot plume of magma breaks through the crust. As a lithospheric plate moves over this source of magma, volcanoes form at the **hot spot**. This movement of a plate over a hot spot can lead to the formation of a chain of volcanoes of decreasing age. The most dramatic example of this is probably the Hawaiian Islands. Kauai, the westernmost island, is the oldest island in the chain. It formed about 4 to 5 million years ago. As you travel eastward, the age of the islands decreases. The Big Island, Hawaii, has several active volcanoes. It is the youngest island in the chain that rises above the ocean. See Figure 3-21. What looks like a progression of volcanic activity through time toward the southeast actually shows the slow movement of the Pacific plate toward the northwest. This plate movement created a chain of submerged volcanoes and volcanic islands thousands of kilometers long. Figure 3-18 on page 81 shows the locations of many hot spots both within and at the boundaries of the tectonic plates.

QUESTIONS

Part A

50. The motion of the convection currents in Earth's mantle beneath the Atlantic Ocean appears to be making this ocean basin
(1) deeper (2) shallower (3) wider
(4) narrower

51. Unlike rocks of the continents, rocks of the ocean bottoms are generally (1) lighter in color and less dense (2) lighter in color and more dense (3) darker in color and less dense (4) darker in color and more dense

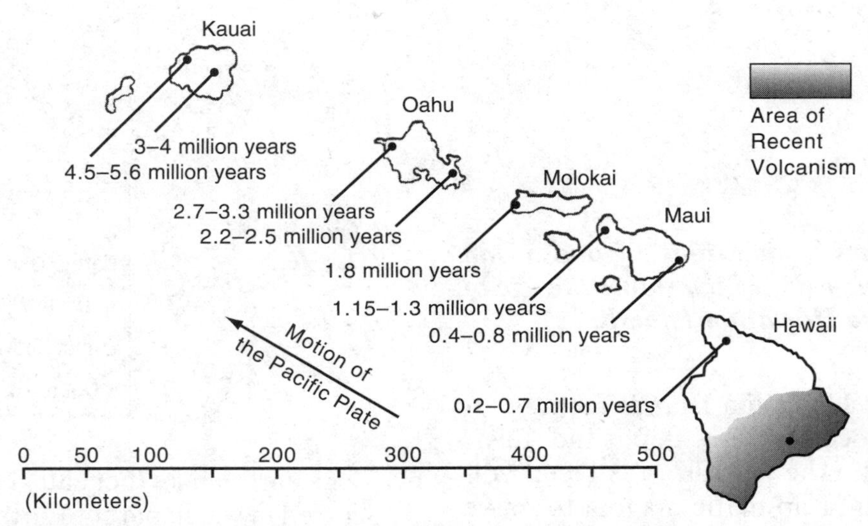

Figure 3-21. As you move northwest from the current zone of volcanic activity on Hawaii, the age of the islands increases. When you reach Kauai, the rocks are 3 to 6 million years old. This is a record of the motion of the Pacific Plate toward the northwest over a stationary source of magma (hot spot) deep below the surface.

52. What energy source causes the North American Plate to move away from Africa and Europe? (1) solar energy (2) electromagnetic stellar radiation (3) heat escaping from Earth's interior (4) global warming due to carbon dioxide

53. Which diagram below best shows the relative motion of tectonic plates along the San Andreas Fault?

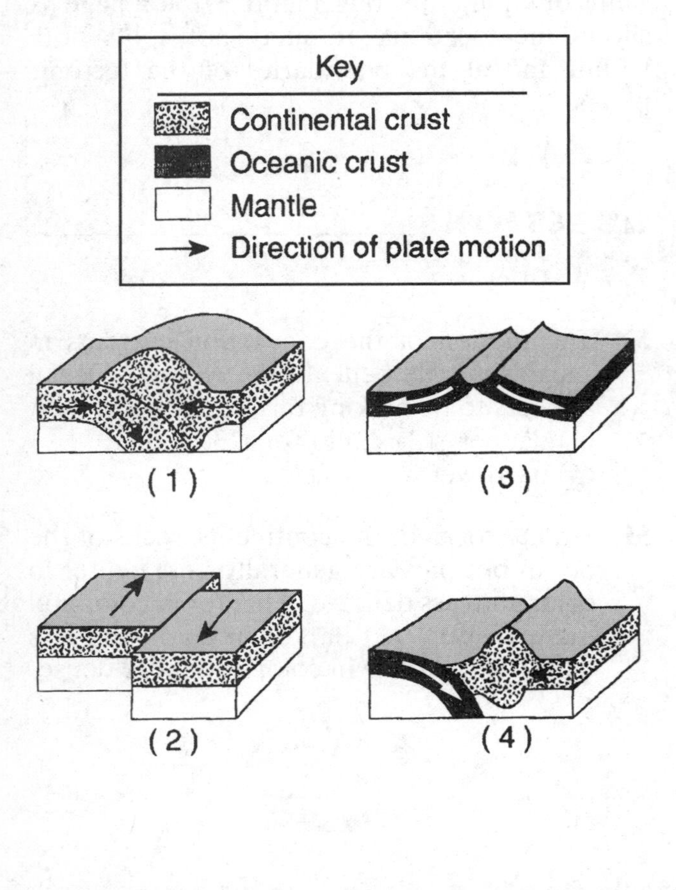

Part B

Base your answers to questions 54 and 55 on Figure 3-21 on page 85, which shows the age of the bedrock in the Hawaiian Islands.

54. How did the Hawaiian Islands form? (1) A mid-ocean ridge deposited felsic rocks to make the islands. (2) Deep volcanism brought up mafic magma to make the islands. (3) The islands are the result of two colliding continental land masses. (4) The bedrock of the islands is made of sedimentary rock deposited by ocean currents.

55. Figure 3-21 indicates that (1) these islands become younger to the west (2) the islands are composed of granitic rocks (3) the Pacific Plate is moving northwest (4) the Pacific Plate is moving southeast

56. The form of energy transfer that drives plate tectonics is most like the energy flow that (1) distributes energy throughout the air in a room (2) transfers energy from the sun to Earth (3) brings sound from its source to your ears (4) moves heat energy within a spoon from one end the other

Base your answers to questions 57 and 58 on the seismic map of South America below, which shows the depths of earthquake foci. Letters* A *through* D *show epicenter locations along a west to east line.

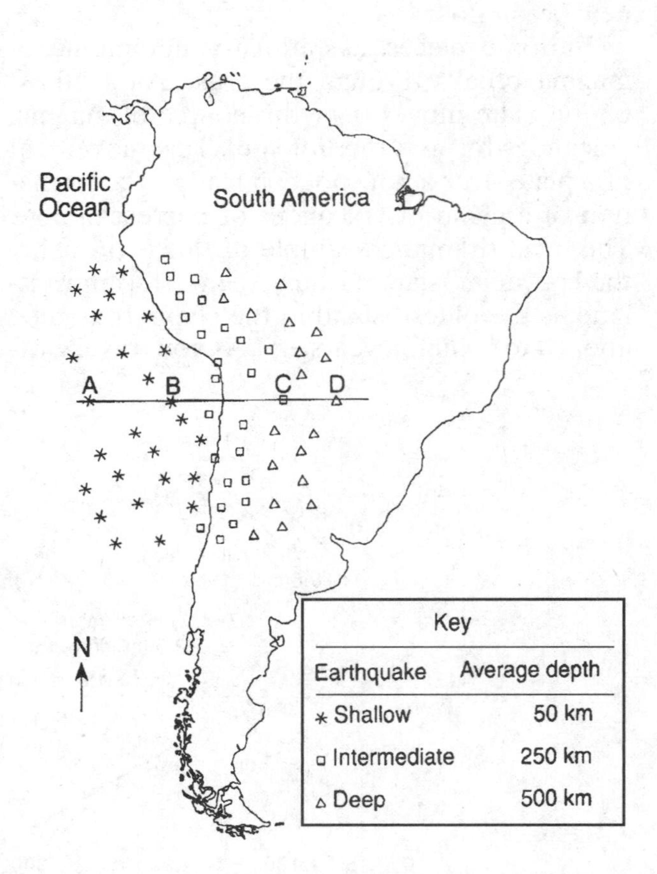

57. The earthquake beneath epicenter *D* occurred in which part of Earth's interior? (1) crust (2) mantle (3) outer core (4) inner core

58. Which graph above right best shows the depths of earthquake foci along line *A–D*?

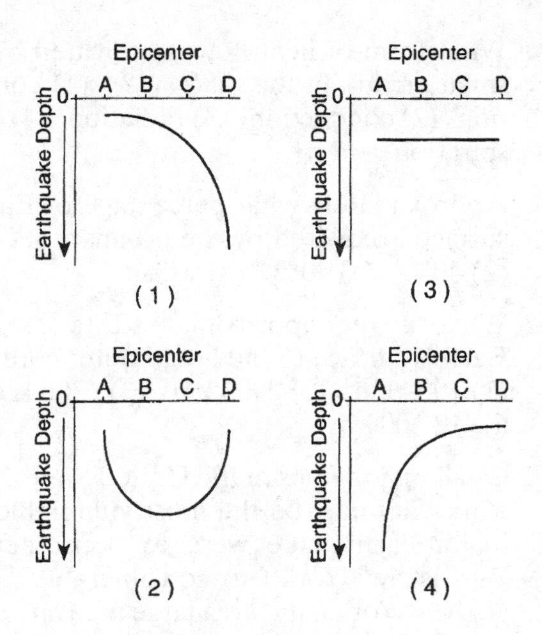

Problem Solving

Working in cooperative groups, produce a report that identifies places in and around your school building that could be damaged in the event that an earthquake occurred in your area. Include a recommendation of ways to reduce those hazards. Your recommendations should include an emergency preparedness plan for people to follow in case of a major seismic event. Include plans for changes, including structural modifications, in the school building and school grounds. Keep in mind that your objective is to save lives and property.

Chapter Review Questions

PART A

1. There is a plate-tectonic hot spot located in Iceland. What does this indicate? (1) There are many volcanic eruptions in Iceland. (2) Iceland has a tropical climate. (3) Earthquakes in Iceland occur deep in Earth's mantle. (4) Iceland is the most popular tourist destination for Europeans.

2. The following map shows the positions of Greenland and North America just before most scientists infer that they moved apart.

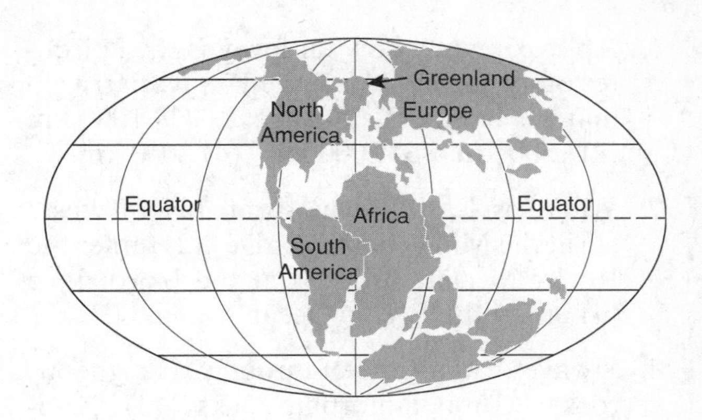

Approximately how long ago were Greenland and the nearby continents in these positions? (1) 59 million years ago (2) 119 million years ago (3) 362 million years ago (4) 458 million years ago

3. The *X* on this map indicates a place where a rare *Mesasaurus* fossil was found in South America.

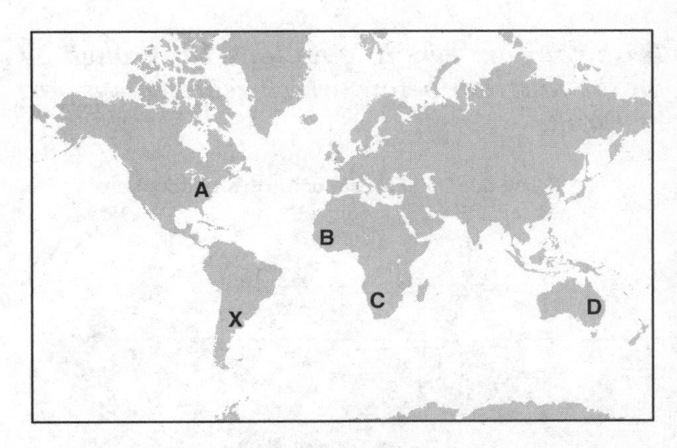

Which other letter on this map indicates the most likely place to find more *Mesasaurus* fossils? (1) *A* (2) *B* (3) *C* (4) *D*

4. What type of plate boundary is most likely at the center of parallel bands of rock that are magnetized alternately positive and negative in response to changes in Earth's magnetic field? (1) divergent plate boundary (2) convergent plate boundary (3) transform plate boundary (4) inactive plate boundary

5. Which tectonic plate is being subducted under the North American Plate? (1) the Nazca Plate (2) the South American Plate (3) the Juan de Fuca Plate (4) the Eurasian Plate

6. Approximately how far away is the epicenter of an earthquake if the P-waves arrive 3 min 20s before the S-waves? (1) 1000 km (2) 1500 km (3) 2000 km (4) 3000 km

7. Where is Earth's crust probably thickest? (1) at the Mid-Atlantic Ridge (2) under the Rocky Mountains (3) at the North Pole (4) under the Pacific Ocean

8. S-waves cannot travel through (1) igneous rocks (2) metamorphic rocks (3) sedimentary rocks (4) molten rock

9. How do geologists know that Earth has a molten outer core? (1) Deep drill holes bring magma to the surface. (2) Magma from many volcanoes is felsic in composition. (3) Some seismic waves do not pass through the center of Earth. (4) Below Earth's surface, the temperature decreases with depth.

Base your answers to questions 10 through 14 on the diagram below, which represents a part of Earth.

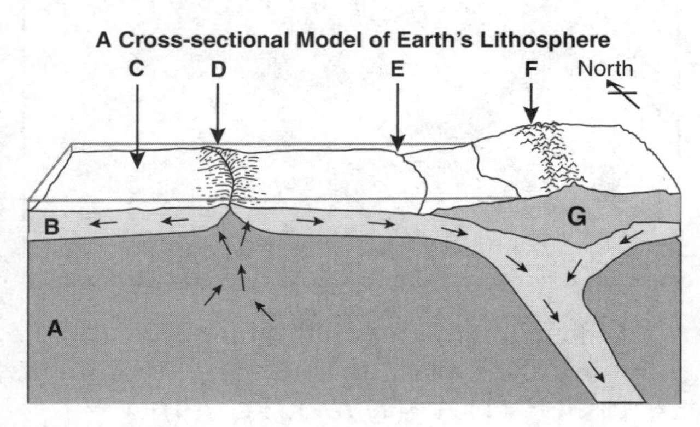

10. Where along the top of the lithosphere would heat flow from Earth's interior probably be greatest? (1) *C* (2) *D* (3) *E* (4) *G*

11. Which arrow best identifies a granitic part of Earth's crust? (1) *A* (2) *B* (3) *D* (4) *G*

12. Which letter best identifies a convergent plate boundary? (1) *C* (2) *D* (3) *E* (4) *F*

13. The diagram best represents (1) the east coast of North America (2) the west coast of South America (3) the north coast of Australia (4) the south coast of Africa

14. What form of heat flow is indicated by the small arrows in the diagram? (1) convection (2) conduction (3) radiation (4) transpiration

15. Approximately what percentage of Earth's surface is covered by the oceans? (1) 25% (2) 40% (3) 70% (4) 99%

16. What is the approximate distance from Earth's surface to the liquid outer core? (1) 1000 km (2) 2900 km (3) 6300 km (4) 12,600 km

17. Of all major cities in the United States, New York City may be the most vulnerable if a major earthquake were to occur nearby. Why is New York City so vulnerable? (1) New York City has a large area and a relatively low population density. (2) New York City is located far from water sources that could be used to put out fires. (3) New York City was constructed below sea level and it is protected by dikes. (4) New York City has many building that were not built to resist shaking.

18. Below is a map of the contiguous United States. Which location is most likely to have a volcanic eruption due to heated magma relatively close to Earth's surface?

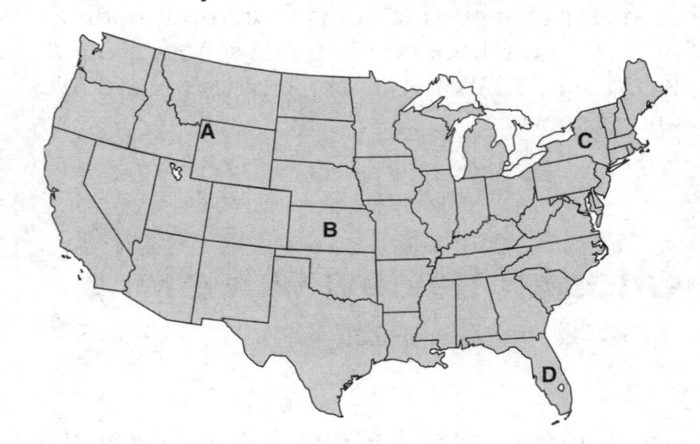

(1) *A* (2) *B* (3) *C* (4) *D*

PART B

19. Seismometers at the State University at Buffalo recorded an earthquake. The first P-waves arrived at 03:45:15 (hours: minutes: seconds) and the first S-waves arrived at 03:45:55. Which of the following locations was the most probable epicenter of this seismic event? (1) Jamestown, NY (2) Albany, NY (3) central South America (4) Australia

Base your answers to questions 20 through 25 on the diagram below, which presents a cross section of Earth showing seismic waves radiating from the position labeled "Earth's Surface."

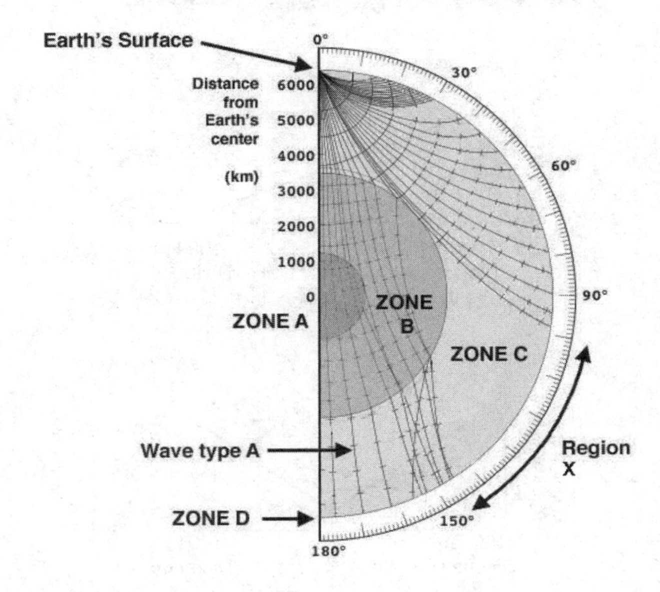

20. What is Earth's radius? (1) 180 km (2) 360 km (3) 5000 km (4) 6400 km

21. Which zone is primarily in the liquid state? (1) zone *A* (2) zone *B* (3) zone *C* (4) zone *D*

22. Region *X* along the surface is a place where (1) direct P-waves arrive, but no direct S-waves arrive (2) direct S-waves occur, but no direct P-waves arrive (3) neither direct P-waves nor direct S-waves arrive (4) both direct P-waves and direct S-waves arrive

23. Why is zone *D* difficult to see on this diagram? (1) Zone *D* is very thin compared with the other zones. (2) Zone *D* is very thick compared with the other zones. (3) Zone *D* is composed of only felsic rocks. (4) Zone *D* is composed of only mafic rocks.

24. What is the inferred temperature at the interface between zones *B* and *C*? (1) about 2000°C (2) about 3000°C (3) about 4000°C (4) about 5000°C

25. What is represented by the line labeled "Wave type *A*?" (1) P-waves (2) S-waves (3) light waves (4) sound waves

26. To find the distance from a seismic recording station to an epicenter, what information is needed? (1) the arrival time of P-waves (2) the arrival time of S-waves (3) the arrival time of P-waves added to the arrival time of S-waves (4) the arrival time of P-waves subtracted from the arrival time of S-waves

Base your answers to questions 27 through 31 on the diagram below, which shows the seismic recordings of a single earthquake made at stations* A, B, C, *and* D. *The initial arrivals of P- and S-waves are shown by the labeled arrows. All times were recorded after noon.

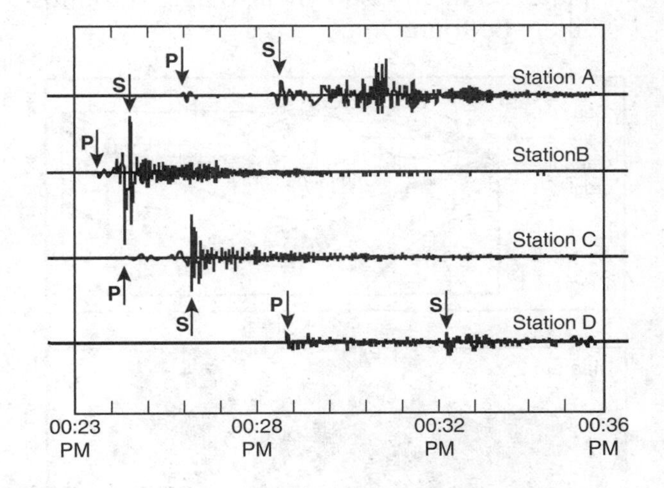

27. Which seismic recordings must be used to locate the epicenter? (1) Any one of them will be enough to find the epicenter. (2) Stations *A* and *B* will allow a scientist to find the epicenter. (3) Stations *B*, *C*, and *D* can be used to find the epicenter. (4) There are not enough recordings here to find the epicenter.

28. At which seismic station location would damage be most likely? (1) *A* (2) *B* (3) *C* (4) *D*

29. When did the first S-wave arrive at Station *B*? (1) 23 minutes past noon (2) 24 minutes and 30 seconds past noon (3) 29 minutes past noon (4) 32 minutes and 10 seconds past noon

30. At Station *A*, how did the arrival time of P- and S-waves compare? (1) The P-waves arrived 2 minutes after the S-waves. (2) The P-waves arrived 2 minutes and 40 seconds after the S-waves. (3) The S-waves arrived 2 minutes after the P-waves. (4) The S-waves arrived 2 minutes and 40 seconds after the P-waves.

31. At station *D*, the S-wave was 4 min:20 s behind the P-wave. How far was the epicenter from Station *D*? (1) 1000 km (2) 2000 km (3) 3000 km (4) 4000 km

32. Landing seismographs on the moon allowed scientists to investigate whether the moon has (1) liquid water (2) a liquid core (3) radioactivity (4) an atmosphere

33. At which location in diagram 2 would the plate boundary shown in diagram 1 most likely be found?

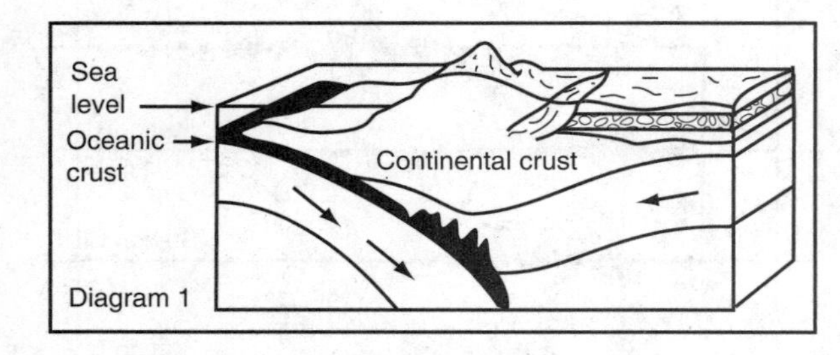

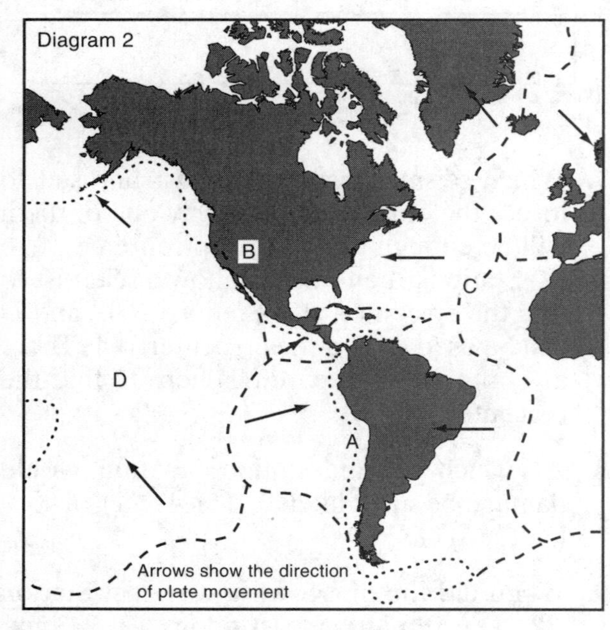

(1) *A* (2) *B* (3) *C* (4) *D*

34. The four graphs below represent the age of Earth's crust near the Mid-Atlantic Ridge. Which graph best shows how the age of the crust changes on either side of the ridge?

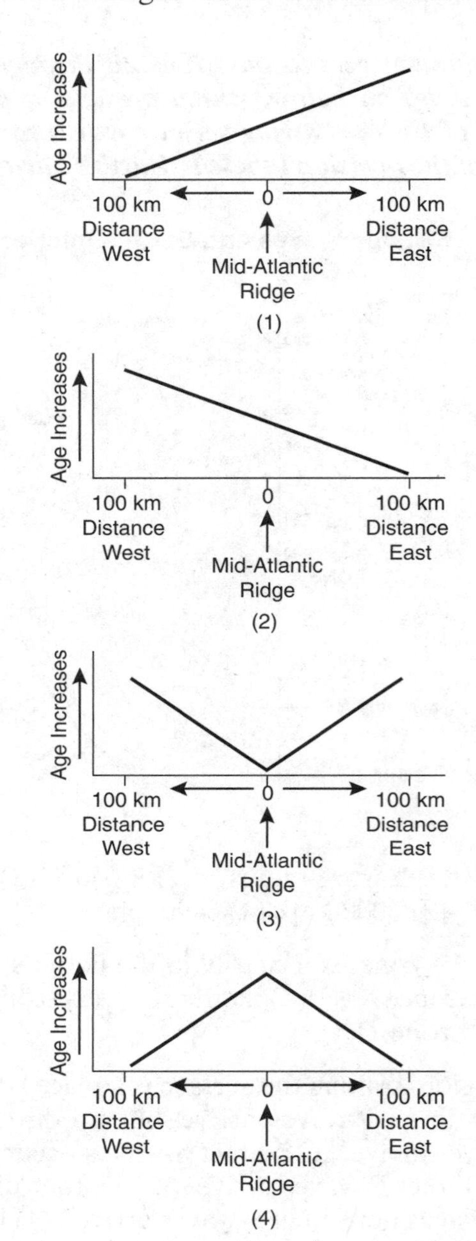

35. An earthquake of magnitude 5 had an aftershock of magnitude 3. How did these two seismic events compare? (1) The second earthquake caused the ground to vibrate 1/100 as much. (2) The second earthquake caused the ground to vibrate 1/10 as much. (3) The second earthquake caused the ground to vibrate 1/2 as much. (4) The second earthquake caused the ground to vibrate twice as much.

36. An earthquake was felt at the epicenter at 3:41 in the afternoon (41 minutes past 3 P.M.). At what time did the first P-waves arrive at a location 4000 km away? (1) 3:42 (2) 3:48 (3) 4:00 (4) 4:41

37. Which observation about the Mid-Atlantic Ridge region provides the best evidence that the seafloor has been spreading for millions of years? (1) The bedrock of the ridge and nearby seafloor is igneous rock. (2) The ridge is the location of irregular volcanic eruptions. (3) Several faults cut across the ridge and nearby seafloor. (4) Seafloor bedrock is younger near the ridge and older farther away.

38. Which diagram at the right best represents the boundary between the Arabian Plate and the Eurasian Plate?

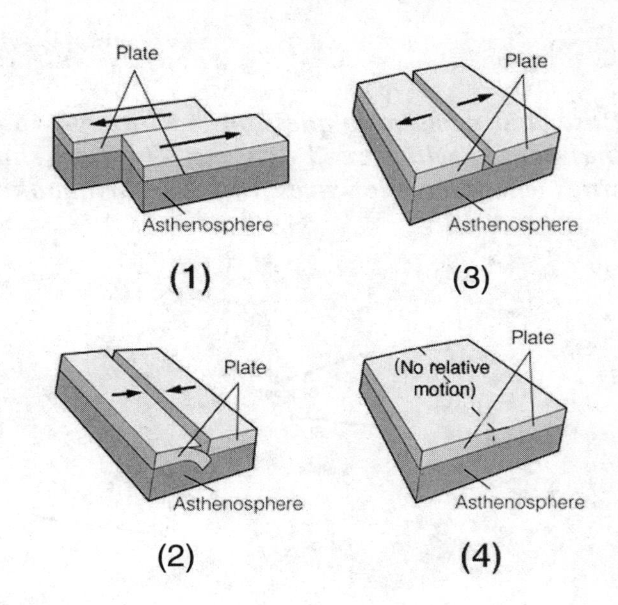

***Base your answers to questions 39 through 42 on the concept map below, which represents disasters that cause large numbers of human deaths and massive destruction. Four positions have been left empty and labeled* A, B, C, D.**

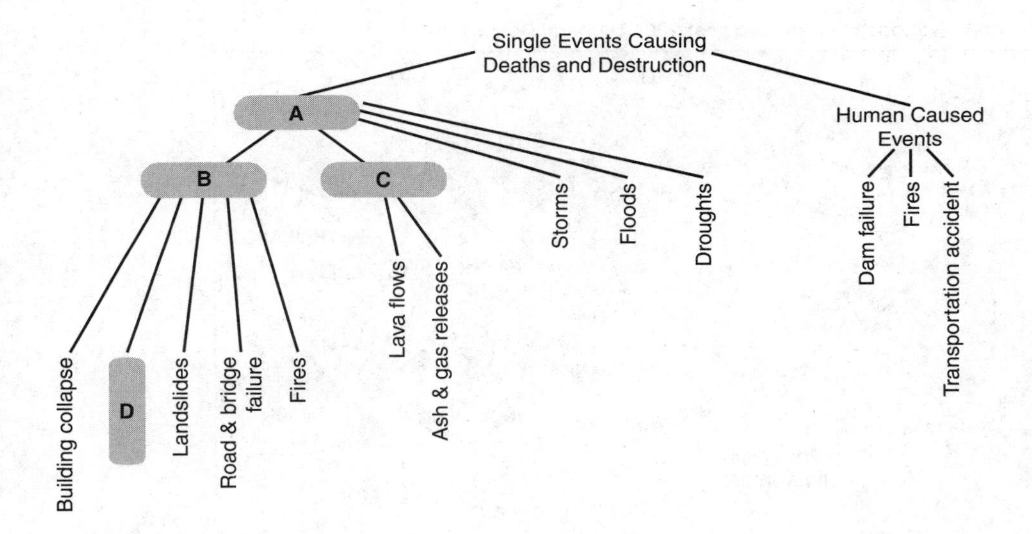

39. Which of the following choices best fits in position *A*? (1) earthquakes (2) natural disasters (3) tsunamis (4) volcanic eruptions

40. Which of the following choices best fits in position *B*? (1) earthquakes (2) natural disasters (3) tsunamis (4) volcanic eruptions

41. Which of the following choices best fits in position *C*? (1) earthquakes (2) natural disasters (3) tsunamis (4) volcanic eruptions

42. What word best fits in position *D*?
(1) earthquakes (3) tsunamis
(2) natural disasters (4) volcanic eruptions

PART C

Base your answers to questions 43 through 46 on the map below, which represents an earthquake that caused widespread damage. The map shows epicenter distance circles drawn around three cities where seismic waves from the earthquake were recorded.

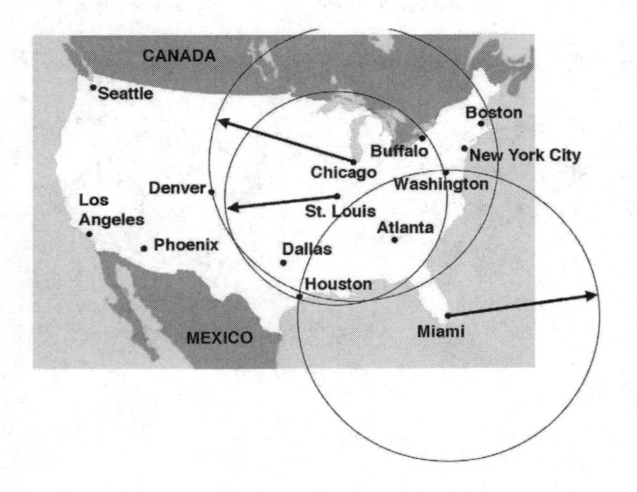

43. What is the name of the city closest to the epicenter?

44. Why would the people in Seattle be less likely to feel this earthquake than people in Phoenix?

45. What would be the advantage of using more than the three seismic stations (Miami, Chicago, and St. Louis) to locate this epicenter?

46. What city on this map is most likely to experience a major earthquake in the next 10 years?

Base your answers to questions 47 through 51 on the map and paragraphs that follow.

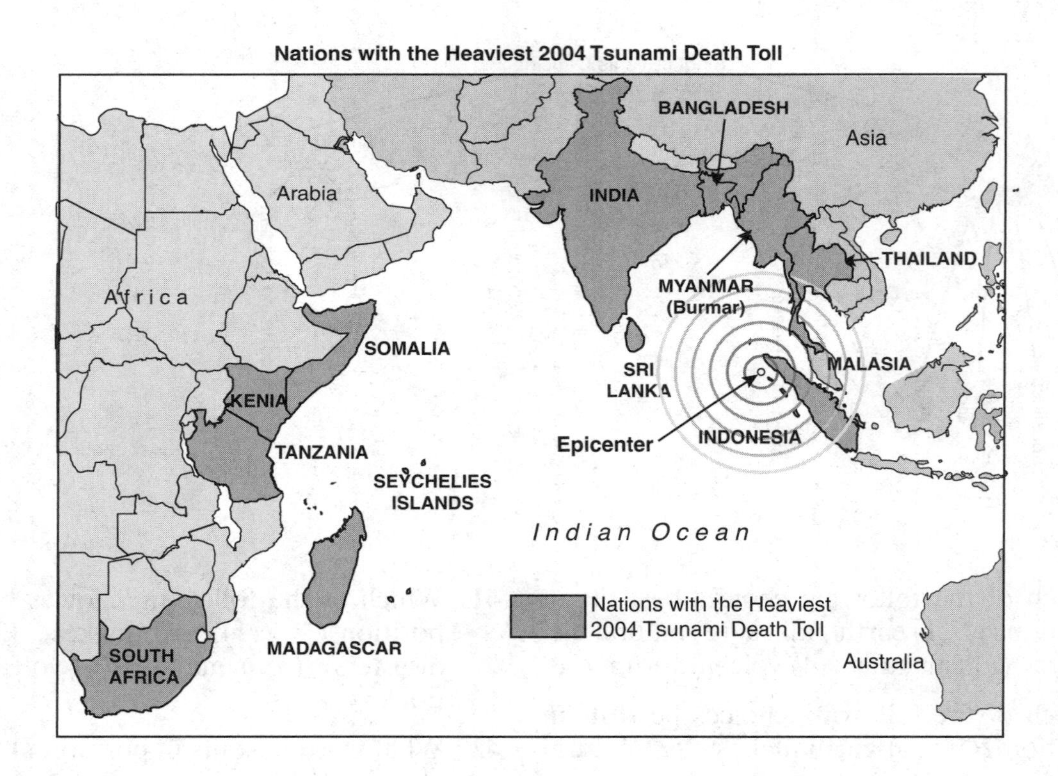

The Great Indian Ocean-Indonesian Earthquake of 2004

On December 26, 2004, an earthquake occurred undersea off Indonesia at about 9 A.M. local time. The epicenter was in the Indian Ocean about 200 kilometers from the west coast of Sumatra, Indonesia, at 3°N, 96°E.

The earthquake was caused by subduction. It triggered a series of devastating tsunamis along the coasts of most landmasses that border the Indian Ocean. Scientists estimate that the tsunamis killed more than 225,000 people in eleven countries. Coastal communities were covered with waves up to 30 meters (100 feet) high. It was one of the deadliest natural disasters in recorded history. Indonesia, Sri Lanka, India, and Thailand were hardest hit.

With a magnitude of between 9.1 and 9.3, it is the second largest earthquake ever recorded on a seismograph. This earthquake had the longest duration of faulting ever observed, between 8.3 and 10 minutes. It caused the entire planet to vibrate as much as 1 centimeter (0.4 inch) and triggered other earthquakes as far away as Alaska.

In deep ocean water, tsunami waves form only a small hump, barely noticeable and harmless. These waves generally travel at a very high speed of 500 to 1000 kilometers per hour (310 to 620 miles per hour). In shallow water near coastlines, tsunamis slow and form large destructive waves. Scientists investigating the damage in the city of Aceh found evidence that the wave reached a height of 24 meters (80 feet) when coming ashore along large stretches of the coastline, rising to 30 meters (100 feet) in some inland areas. The worldwide community donated more than $7 billion (2004 U.S. dollars) in humanitarian aid.

47. What two tectonic plates come together near the epicenter of this earthquake?

48. The diagram below represents the plate boundary where this earthquake occurred. Draw an arrow to show how Plate 1 moved with respect to Plate 2 during this earthquake.

(Be sure the vertical component of this motion is clear from your arrow.)

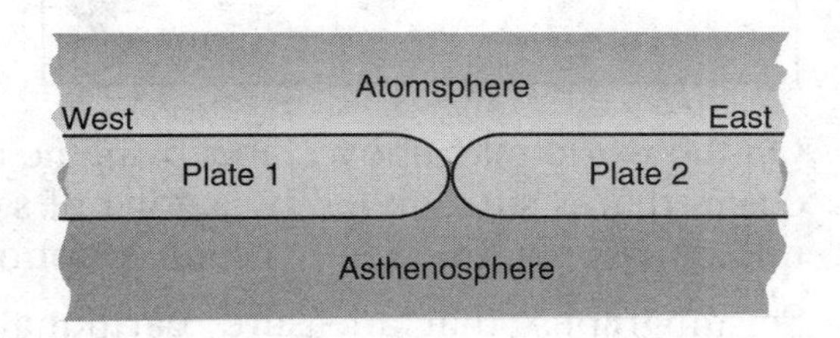

49. Most people in Sri Lanka did not feel this earthquake. If you were vacationing at a beach resort on the east coast of Sri Lanka, how could you best protect yourself from personal harm from an earthquake?

50. Name two common causes of death other than tsunamis for people who experience major earthquakes.

51. Along what kind of geologic feature do *all* earthquakes occur?

Base your answers to questions 52 and 53 on the map of New York State below.

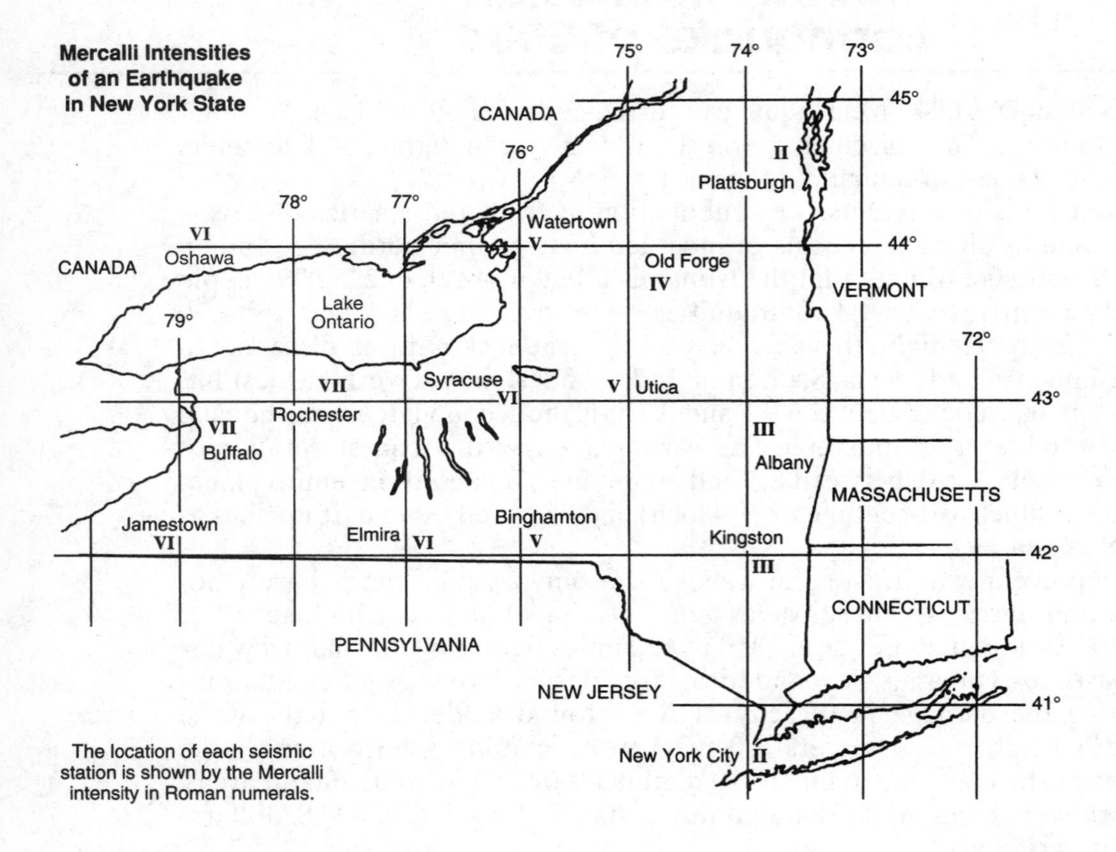

52. Use the map to construct six isoseismals (isolines of equal seismic intensity) that cover New York State. (Use an interval of one Mercalli intensity unit.) End your isolines just past the boundaries of New York State.

53. Draw an X on the map at the most likely location of the epicenter of this earthquake.

54. What makes the Richter scale different from the scales we use to measure mass or distance?

55. Base your answer to this question on your knowledge of Earth science and on the *Earth Science Reference Tables*. At a distance of 5000 kilometers from the epicenter, how far ahead of the S-wave is the P-wave?

56. Why does California have more earthquakes than most other parts of the continental United States?

57. What could your community do to prepare for a major seismic event? Be sure to tell how your suggested procedure would help in the event of a major earthquake.

58. Notice that the lines on the "Earthquake P-wave and S-wave Travel Time" graph in the *Earth Science Reference Tables* curve. Why are both of these lines concave downward? (There are two answers. Please note that this is an especially difficult question.)

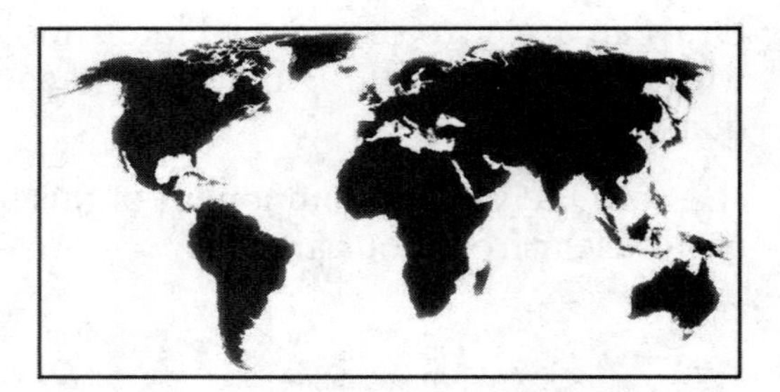

59. On the world map above, label by name the ocean that is surrounded by a zone of seismic activity and frequent volcanic eruptions.

60. Seismographs that measure earthquakes use a Richter-type (magnitude) scale. Based on what observations can a scientist establish the Mercalli intensity of an earthquake?

CHAPTER

Weathering, Erosion, and Deposition

Vocabulary at a Glance

abrasion	horizontal sorting	soil
agent of erosion	humus	soil horizon
bedrock	ice age	solution
chemical weathering	leaching	sorted
deposition	mass movement	stream discharge
discharge	meander	stream velocity
erosion	mass wasting	striation
flotation	outcrop	suspension
frost action	physical weathering	till
frost wedging	precipitate	transported soil
glacial erratic	precipitation	unsorted deposit
glacier	residual soil	vertical sorting
graded bedding	sediment	weathering

WHY DO ROCKS WEATHER?

An **outcrop** is bedrock exposed at Earth's surface. When rocks that formed within Earth's crust are uplifted and exposed to wind, water, and biological processes, their new environment changes them. Rocks change in observable properties and mineral composition. The breakdown of rock due to physical or chemical changes is called **weathering**. See Figure 4-1.

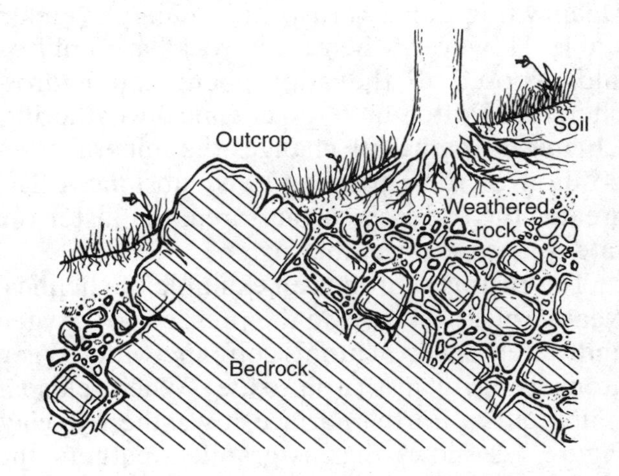

Figure 4-1. Most weathering occurs when rock is exposed to the atmosphere or groundwater. As solid rock is broken apart, weathering usually increases.

Physical Weathering

Physical weathering changes the size and/or shape of a rock without changing the rock's chemical composition. For example, the rock may be broken into smaller pieces. **Frost action** (**frost wedging**) is an important agent of physical weathering in climates that have seasonal temperature changes. In these areas the temperature is alternately above and below the freezing temperature of water, 0°C. In weathering by frost action, water seeps into cracks in rocks. As the water freezes, its volume increases. The increase in volume makes the cracks in the rock a little larger each time the water freezes. When the ice melts and the resulting liquid water evaporates, the rock is left more open than before. Over time, the freezing and melting of water in the cracks causes the rock to break apart.

Plant roots that grow in the cracks of rocks and animals that burrow beneath the ground are also agents of physical weathering. As a plant grows, its roots invade the crevices of a rock and hold them open. When animals burrow, they expose new rock surfaces to weathering.

When flowing water in a stream carries rock particles, the particles bump and rub against one another and the streambed. These collisions wear down the particles by **abrasion**, a form of physical weathering. Wind also causes particles of sediment to abrade exposed rock surfaces. Moving ice in glaciers drags, scrapes, and breaks rocks apart. Wave action constantly attacks rocks and sediment along shorelines. Even gravity alone can cause rocks to fall and break. There are five major natural agents of abrasion: flowing water, moving ice, waves, wind, and gravity.

Some rock-forming minerals are harder than others. Quartz (silica) and feldspars resist physical weathering. However, softer minerals, such as mica and clay, are quickly broken by physical weathering.

Chemical Weathering

Deep within Earth's crust, most minerals remain stable. However, when these rocks are uplifted and exposed to the atmosphere and hydrosphere, they often undergo chemical weathering. **Chemical weathering** changes the mineral composition of rock, forming new substances. The greater the exposed surface area, the faster the rate of chemical weathering.

The rusting of iron is an example of chemical weathering. Iron rusts in the presence of oxygen and water. The iron atoms combine with oxygen atoms to form rust (iron oxide). When feldspar is uplifted to Earth's surface, it weathers to clay. Figure 4-2 shows that as granite weathers, the percentage of feldspar in a sample eventually decreases dramatically. The original minerals are replaced mostly by clay and iron oxide.

Chemical weathering usually requires water to bring about mineral changes. Heat also speeds most chemical weathering. Thus, chemical weathering takes place most rapidly in warm, moist climates. See Figure 4-3.

Some minerals resist chemical weathering better than others. Quartz, a common mineral in beach sand, is relatively stable and resistant to chemical weathering. However, olivine, a mineral that is common deep within Earth, quickly weathers to clay when it is exposed to the atmosphere and hydrosphere. For this reason, quartz is very common in sediments but olivine is rarely found in exposed sedimentary bedrock. Limestone is a fairly hard rock that resists physical weathering; but its calcite is decomposed by

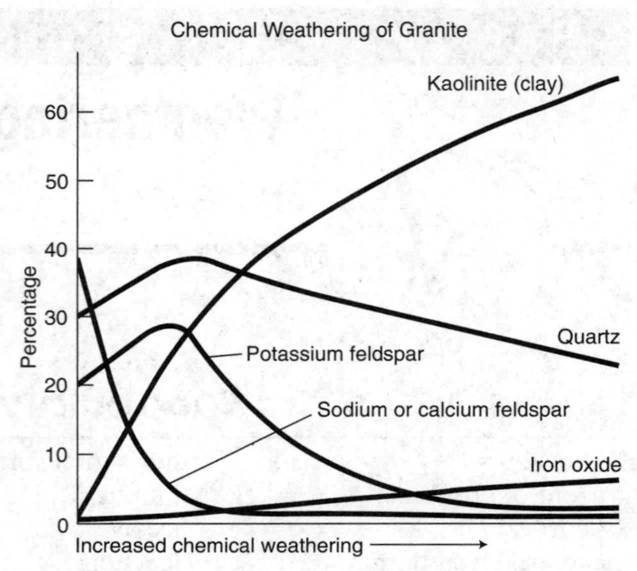

Figure 4-2. As granite weathers, the amount of the feldspars decreases as the amount of clay minerals and iron oxide increases.

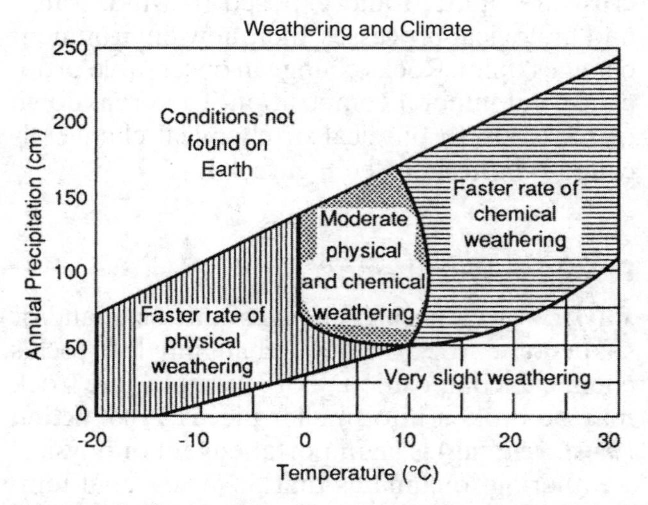

Figure 4-3. The type of rock weathering largely depends on the local climate. In general, chemical weathering dominates in warm moist climates, while cooler climates favor physical weathering (especially frost action).

exposure to acids. Rainwater absorbs carbon dioxide from the atmosphere and organic acids from soil to become slightly acidic. Atmospheric pollutants, such as the oxides of sulfur and nitrogen, can make rainwater unnaturally acidic. When limestone is exposed to these acids, the acids react with the calcite in the limestone. The product of the reaction is soluble in water and is carried away by water.

QUESTIONS

Part A

1. Chemical weathering is most active is a climate that is (1) warm and dry (2) warm and moist (3) cold and dry (4) cold and moist

2. If you leave a bottle of water outside on a cold winter night, the water may freeze, causing the bottle to break. This is an example of (1) physical weathering (2) chemical weathering (3) water erosion (4) ice erosion

3. The rock below was exposed at Earth's surface for a long time. What caused layers *A* and *B* to be different from layers *X* and *Y*?

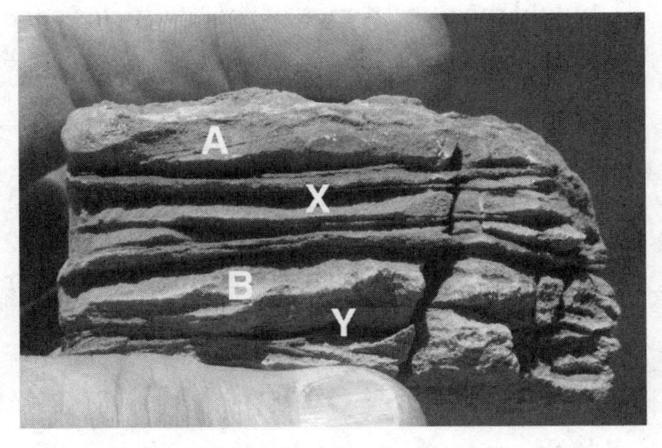

(1) Layers *A* and *B* were more exposed to the elements than layers *X* and *Y* (2) Layers *A* and *B* are composed of minerals with lower numbers on Mohs scale. (3) Layers *A* and *B* have less mineral cement than and are softer than layers *X* and *Y*. (4) Layers *A* and *B* contain minerals more resistant to weathering than layers X and *Y*.

4. Which property probably has the *least* effect on the rate at which a rock weathers? (1) how long it is exposed to the atmosphere (2) the average density of its minerals (3) the hardness of the minerals (4) how much surface is exposed

5. What is the most common form of weathering at high-latitude and high-altitude locations? (1) frost action (2) chemical weathering (3) mineral changes (4) solution in water

6. An iron nail will rust when it is left outside. What type of weathering is this? (1) frost wedging (2) rock abrasion (3) physical weathering (4) chemical weathering

7. What natural process is best represented by the diagram below?

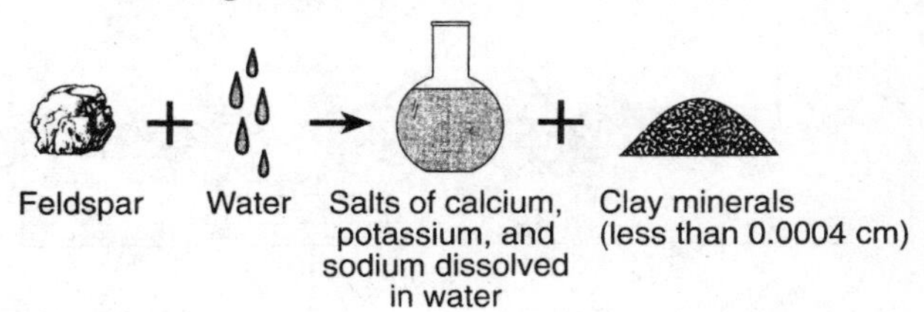

(1) frost wedging (2) rock abrasion (3) physical weathering (4) chemical weathering

8. A student broke the piece of rock shown below from a much larger rock. The arrow points from what was the outer surface of the large rock toward the former inside. The change in color by the arrow shows a form of weathering similar to

(1) water expanding when it freezes to break a metal pipe (2) a rock being rolled along the bottom of a fast-moving stream (3) the body of a car decomposing due to exposure to salt and water (4) very small par-

ticles of sediment blown around due to strong winds

9. In moist climates where temperatures alternate between warm days well above 0°C and cold nights well below 0°C, why do rocks break apart? (1) Cooling causes rocks to expand. (2) Frost forms when ice melts. (3) water expands when it freezes (4) water boils and contracts at 100° C

Part B

10. The diagram below represents a cross-sectional profile of a particular location.

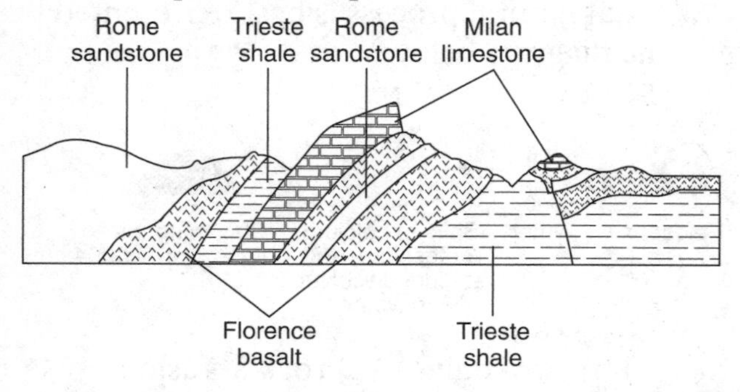

Which rock type appears to best resist weathering and erosion? (1) Rome sandstone (2) Florence basalt (3) Milan limestone (4) Trieste shale

Base your answers to questions 11 through 16 on the following data table. The data are the results of an experiment in which a student shook, one type at a time, 200 grams of four types of different rocks mixed with water. The rocks were shaken with equal energy for a total of 30 minutes.

Mass of Rock Sample Remaining (g)

Rock-Shaking Time (min)	Shale	Marble	Rock Salt	Limestone
0	200	200	200	200
5	160	200	120	200
10	125	200	60	195
15	100	190	20	170
20	75	180	0	150
25	55	175	0	135
30	50	175	0	125

11. Which best explains why the mass of each rock sample remaining after 30 minutes of shaking was different? (1) Each rock was in a different container. (2) Some rocks had more initial mass than others. (3) Some rocks were shaken longer than others (4) Each rock had a different composition.

12. How did the amount of shale change through time? (1) The mass increased, but it increased most quickly at the beginning. (2) The mass increased, but it increased most quickly at the end. (3) The mass decreased, but it decreased most quickly at the beginning. (4) The mass decreased, but it decreased most quickly at the end.

13. Which rock type was most easily weathered by abrasion? (1) shale (2) marble (3) rock salt (4) limestone

14. What percent of shale remained after 30 minutes? (1) 10% (2) 25% (3) 50% (4) 100%

15. This experiment was an investigation of rock (1) weathering (2) transportation (3) erosion (4) foliation

16. Compared to the original samples, how would the appearance of rock particles have changed? The rock particles remaining would be (1) more angular (2) larger (3) more rounded (4) softer

17. In which month is physical weathering likely to be especially active and chemical weathering relatively inactive in New York State? (1) January (2) April (3) July (4) October

HOW DO SOILS FORM?

Soil is the mixture of weathered rock, microorganisms, and organic remains that usually covers bedrock. The texture of a soil depends on the size of the particles it contains. Clay-rich soils tend to feel smooth. Soils rich in sand are more likely to feel gritty. The composition of a soil depends on the rocks from which it weathered and the local climate. Under natural conditions, biological, physical, and chemical weathering processes are usually involved in the development of soils.

Physical weathering breaks solid rock into small particles. Chemical weathering changes the minerals, often increasing the clay content. Plants and animals add organic materials in the form of waste products and the remains of dead organisms. The decay of organic remains produces organic acids, which speed chemical weathering. Burrowing animals, such as earthworms, insects, and rodents, help air and water circulate through the soil and mix mineral and organic matter. Figure 4-4 shows how soil can be classified by texture. In addition to their mineral content, soils contain air, water, and organic matter in highly variable percentages. The local climate is an important factor in determining this. For example, a bog soil in a moist climate can be nearly all organic matter and water.

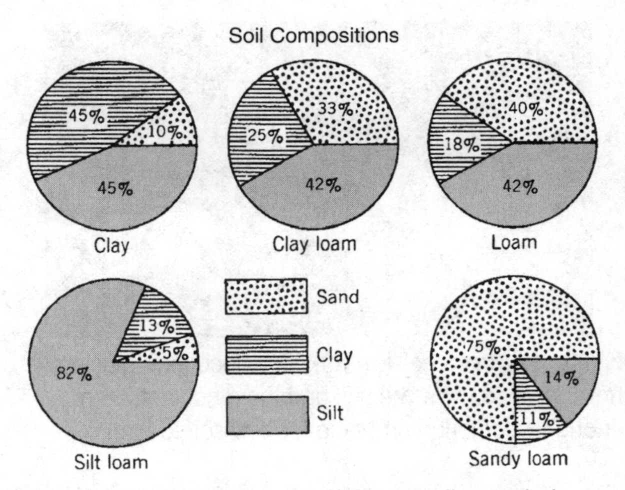

Figure 4-4. Soils can be classified according to their content of sand, silt, and clay-size particles. Each pie graph is labeled by soil type.

The slow formation of soil in place produces layers known as **soil horizons**. When soil remains where it formed, it is a **residual soil**. See Figure 4-5. The top layer is usually the best layer for growing crops because it is rich in dark-colored organic matter called **humus**. However, some important minerals may have been transported deeper into the soil by groundwater infiltration (**leaching**). The lowest layer of the soil is usually composed of broken bedrock, which may merge into solid bedrock. **Bedrock** is the solid layer of rock beneath the soil horizons.

Most of the soils of New York State do not show the complete development of horizons. Continental glaciers from the north repeatedly covered the area that became New York. These glaciers stripped the soils from where they originally formed and moved them southward to become **transported soils**. The most recent withdrawal of continental ice sheets was only about 10,000 to 20,000 years ago. As a result, the lowest soil horizon (broken bedrock) is generally missing, and weathered soil often sits directly on top of a hard, smoothed bedrock surface as you can see in Figure 4-6 on page 100.

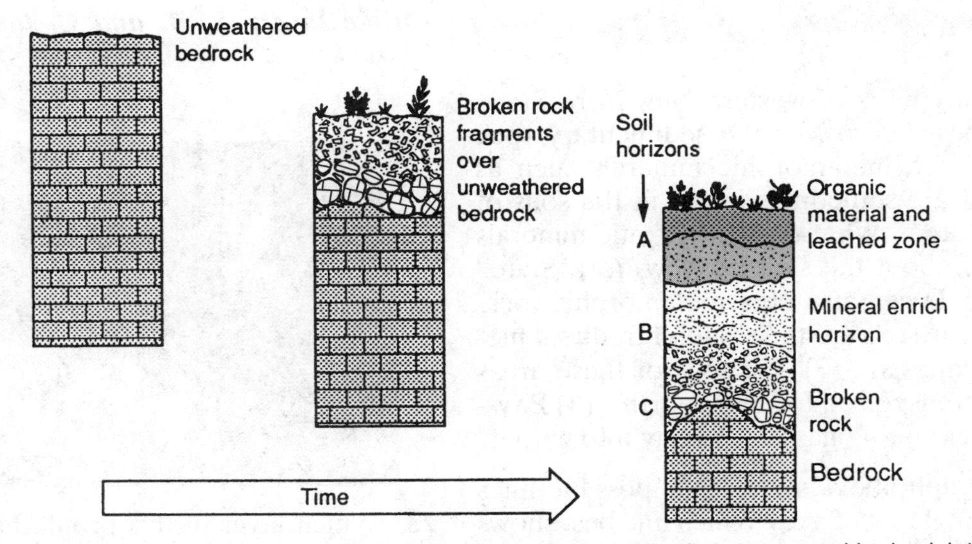

Figure 4-5. Weathering, leaching, and organic processes gradually change exposed bedrock into mature soil. These processes produce layers known as soil horizons.

Figure 4-6. In some New York State locations, glaciers have scraped away the soil and there has not been enough weathering and time for new soil to form.

Protecting the Soil

Soil is a resource that must be protected. It may take hundreds of years for just 1 centimeter of topsoil to form. Human technology has contributed to the loss of soil. For example, construction and mining projects have moved great amounts of rock and soil from their original locations. Destruction of plant cover and poor farming and forestry practices have left soil exposed and unprotected. Running water and wind quickly carry away the exposed soil.

Salt used to remove ice from roads in winter is washed into the soil at the side of the road. If the concentration of salt in the soil is high enough, plants will not grow there. Without plants to hold the sediments in place, erosion can rapidly carry away the soil.

QUESTIONS

Part A

18. In many parts of western New York State, the local bedrock is the sedimentary rock shale. Yet, metamorphic minerals such as garnet are sometimes found in the soils of these areas. What do these exotic minerals tell us about the soils of New York State? (1) Shale can also be a metamorphic rock. (2) Some soils are the result of frequent meteor impacts. (3) The soils of these areas were transported from the north. (4) Physical weathering has turned clay into garnet.

19. The graph above shows four possible lines on a single set of axes. Which line best shows how the depth of a soil depends upon how long it has been forming?

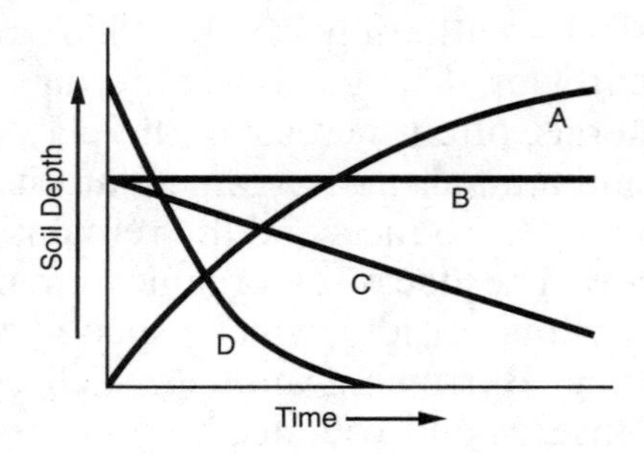

(1) *A* (2) *B* (3) *C* (4) *D*

20. The composition of a soil that formed in place is primarily a function of (1) the elevation above sea level (2) the average annual temperature (3) the minerals in the bedrock (4) the age of the bedrock

21. Most plants grow best in a soil that is composed of (1) solid bedrock (2) only organic material (3) broken up fragments of bedrock (4) weathered rock and organic material

22. Which of the following changes would most likely cause the soils in New York State to become thicker? (1) frequent floods (2) dryer conditions (3) increased biological activity (4) longer winters with little snowfall

PART B

Base your answers to questions 23 and 24 on the photo below, which shows a soil profile exposed by beach erosion along a rocky shoreline. Unlike layers A, B, and C, layer D is solid bedrock.

23. Which layer in this profile has the greatest percentage of organic content? (1) *A* (2) *B* (3) *C* (4) *D*

24. How did this soil originate? (1) erosion of organic materials (2) weathering and biological activity (3) melting followed by crystallization (4) changed caused by heat and pressure deep within Earth

Figure 4-7. This pile of rocks is the result of mass movement of the weathered rock of this cliff.

HOW ARE WEATHERED MATERIALS TRANSPORTED?

Rocks that have been broken into fragments, regardless of their size, are called **sediments**. The mineral composition or other characteristics of sediments may be unrelated to the properties of the underlying bedrock. In such a case, the sediments must have formed elsewhere and then been transported from their place of origin. **Erosion** is the transportation of sediments most often by water, wind, or glaciers—the **agents of erosion**. These sediments are deposited in a different location.

The force of gravity drives most forms of erosion. For example, weathering weakens a rock on a cliff. Gravity causes the weakened rock to fall to the bottom of the cliff. Piles of talus (broken rock) accumulate at the bottom of a cliff. Continued erosion, perhaps helped by wind or water, moves the sediment downslope, away from the cliff.

Erosion by Gravity Acting Alone

Have you ever seen rock debris that has fallen onto a road or to the bottom of a cliff as in Figure 4-7? The downhill movement of rock or sediment without being carried by water, wind, or ice is known as **mass movement**, or **mass wasting**. Material can slide, flow, or fall to its resting place. Although the sediment is not really carried by water, water in the sediment can act as a lubricating agent that makes mass movement more likely. Alternate freezing and thawing also accelerates this process. Mass movement includes *slow creep*, *slumping*, *landslides*, and even the falling of individual rocks.

Erosion by Water

Each year, the streams and rivers of the world carry millions of tons of sediment downstream and into the oceans. Running water, the main agent of erosion in moist areas, transports sediments in several ways. The smallest particles (ions) are carried in solution. **Solution** particles are so small they cannot be filtered out of the water. Sediments in suspension can be filtered out of the water. However, particles in **suspension** are too small to settle on their own. The flowing water rolls or bounces the largest and most dense particles along the streambed. Particles of low density, especially organic matter, are carried along the surface by **flotation**.

The relationship of the size of the transported particle to water velocity is shown in Figure 4-8. The graph shows that the water velocity needed

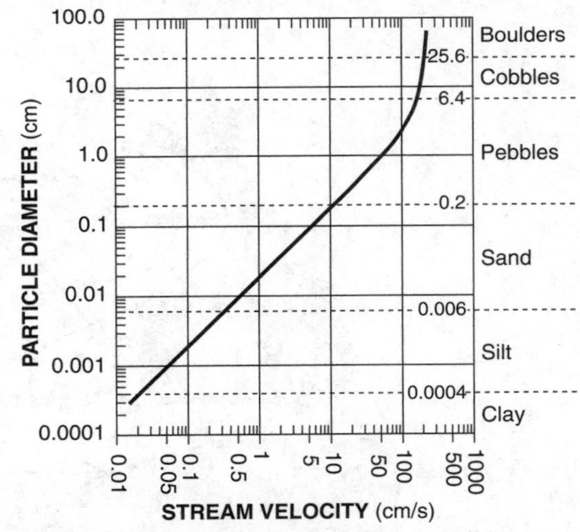

Figure 4-8. There is a direct relationship between the velocity of a stream and the size of the particles it can transport. This generalized graph shows the water velocity needed to maintain, but not start, movement of particles. This graph can be found on page 6 of the *Earth Science Reference Tables*.

to transport particles of sediment depends on the size of the mineral and rock particles. Particles in solution and suspension can be carried by slow-moving water. However, particles rolled along the bottom of the stream require faster stream velocities to move them. A convenient method to estimate the velocity of a stream is to observe the size of the sediment particles that have been carried along the bottom of the stream. Faster streams contain larger particles of sediment. Slow-moving streams can transport only the smaller particles of sediment.

Velocity of Streams Although the term "velocity" often includes both speed and direction, "**stream velocity**" will be used here as a synonym of stream speed. The slope (*gradient*) and the amount of water flowing in the stream (or **stream discharge**) control the stream's velocity. As the stream gradient increases, so does the velocity of the water flowing in the stream. Velocity is also increased by an increase in the amount of water flowing in the stream. Most of the erosion caused by running water takes place when streams are in flood stage. Flooding causes an increase in both the amount of water and the velocity of the water. Therefore, it usually takes a flood to move large boulders.

The speed of a stream is a balance between the force of gravity pulling the water downhill and the frictional forces slowing the stream.

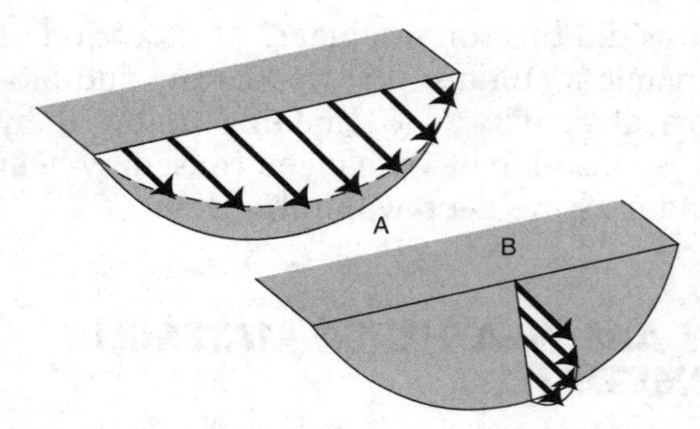

Figure 4-9. Gravity pulling water downhill and friction with the banks of the stream, the streambed, and even with the air determine the velocity of a stream. Therefore streams have the greatest velocity near the center of the stream, just below the surface.

Water usually flows fastest near the center of the stream away from the stream banks and the streambed. There is even a small amount of friction with the air above the water. Therefore, the fastest flow is commonly found at midstream just below the water's surface as shown in Figure 4-9.

Streams with broad, flat valleys often develop S-shaped curves called **meanders**. At a bend in a stream, the fastest-flowing water swings to the outside of the bend, causing erosion along the outer bank of the meander. The slowest-moving water stays to the inside of the bend, causing deposition along the inner bank of the meander. Figure 4-10 shows where streams generally flow fastest in meanders.

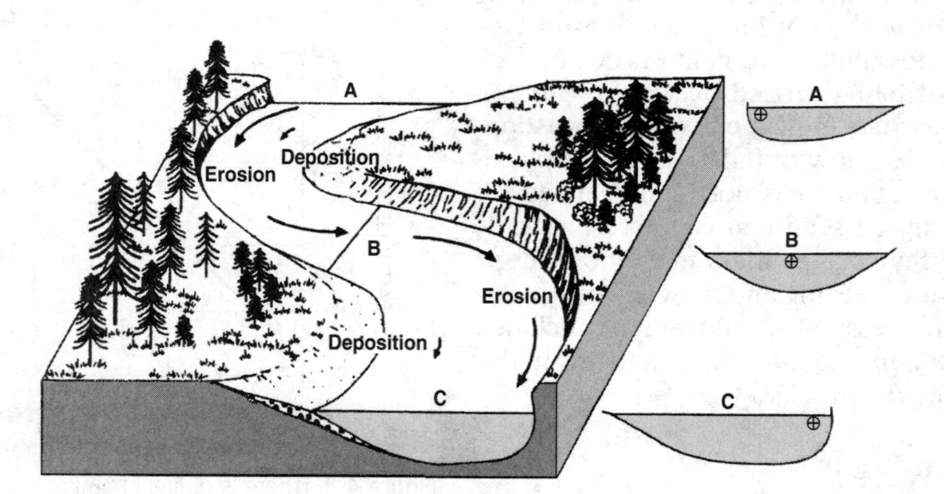

Figure 4-10. When a stream flows around a meander, or bend, the fastest current swings to the outside of the bend as shown by the longer arrows. The circles in the cross sections at A, B, and C also show there the current is fastest. Note that the stream is deepest where the current flows fastest. Therefore, erosion usually occurs on the outside banks of the meanders, while deposition takes place where the water moves slower on the inside banks. This is the way streams change their courses.

Erosion by Wind

Another agent of erosion is wind. It can pick up loose rock materials, such as sand, silt, and clay, and carry them away. Wind erosion occurs mainly in areas, such as deserts and beaches, where there is little plant life to hold soil in place.

You might think that wind is the dominant agent of erosion in desert regions such as the American Southwest. In fact, isolated areas where wind erosion is active can be found throughout the United States. But the infrequent thunderstorms that occur in most desert areas actually cause far more erosion than wind.

Erosion by Ice

During the winter, most of North America receives precipitation as snow. In most places, the snow melts long before the following winter. However, if more snow were to accumulate in the winter than melted in the summer, the snow on the bottom would turn to ice. This is most likely to occur in the Polar Regions and at high altitude. When the ice became thick enough, its weight would cause it to move under the pull of gravity.

A **glacier** is a large mass of slowly flowing ice. As a glacier moves, it carries, pushes, and drags loose rock material. The glacier, with pieces of rock embedded in its ice, acts like a huge abrasion system. It smoothes, striates (scratches), and grooves bedrock. When the ice melts, unsorted rocks and boulders are left scattered around on hilltops and the sides of valleys.

A continental glacier deepens and widens valleys parallel to its movement. It grinds down the hills, leaving them polished and rounded. As a valley glacier moves, it scours away the rock to make a U-shaped glacial valley in which the valley walls may be nearly vertical. Valleys eroded by streams are more often V-shaped with narrow valley floors. Figure 4-11 shows the erosion caused by valley glaciers.

In some ways, valley glaciers are like streams and rivers, but glaciers move much more slowly. The ice may move forward less than 1 meter (3 feet) a day. However, like the water in streams, the ice in a valley glacier moves the fastest near the center of the flowing ice.

Figure 4-11. The top photo shows a typical stream-eroded landscape with a V-shaped valley. If such a landscape were covered by a glacier, such shown in the center photo, the glacier would carve the valley into broader, U-shape. The bottom image is a glacier-eroded U-shaped valley in the western Finger Lakes region of New York State.

Characteristic Changes Caused by Agents of Erosion

Each agent of erosion causes characteristic changes in particles of sediment that can give us clues to how the material was transported. Rough and angular particles deposited at the base of a cliff usually indicate that gravity alone is responsible for transporting the rock a short distance. In this case, the rock particles often have freshly exposed surfaces, as well as older surfaces that are more weathered. Sedimentary particles carried by a stream are usually rounded and polished because the current tumbled them about in the stream. A rock exposed to the wind will develop a smooth flat surface where the wind and wind-blown sand hit it. As wind direction changes and different surfaces are exposed to wind erosion, the rock develops smooth, flat surfaces, or facets, with distinct edges. These angular rocks are called *ventifacts*. Wind-worn rocks are often pitted where softer minerals have been scoured by the wind. Rocks transported by a glacier are usually partially rounded by abrasion and are often scratched (striated) on some faces as a result of being dragged along the bottom of the glacier. See Figure 4-12.

Figure 4-12. Each agent of erosion produces a characteristic shape and texture in rocks. Rocks tumbled in running water are round and smooth. Glacial rocks are often partly rounded with scratches (striations). Wind-eroded rocks (ventifacts) are smooth and angular with pitted surfaces. Talus rocks (from rock falls) are rough and angular and may show some fresh and some weathered surfaces.

QUESTIONS

Part A

25. What is the minimum rate of flow at which water can usually keep pebbles 1.0 centimeter in diameter moving downstream? (1) 50 cm/s (2) 100 cm/s (3) 150 cm/s (4) 200 cm/s

26. Which movement of Earth materials is *not* driven by the force of gravity? (1) rocks pushed and carried by a glacier (2) dissolved salt carried by a large river (3) broken rock that falls to the bottom of a cliff (4) water evaporating into the atmosphere

27. The photo below shows five natural rocks, including igneous, sedimentary, and metamorphic rocks. What do all five rocks appear to have in common?

(1) They have the same mineral composition. (2) They formed by the same geological processes. (3) They were eroded by abrasion in a stream. (4) They have the same hardness and density.

28. Which term below means almost the same as erosion? (1) weathering (2) transportation (3) chemical change (4) gravitational force

Base your answers to questions 29 through 32 on the diagram below, which represents a meandering river. Neither the gradient nor the stream volume changes within this part of the river. Arrows show the general direction in which the water is flowing.

29. At which letters is deposition most likely taking place? (1) *X*, only (2) *Y*, only (3) *Z*, only (4) *X* and *Y*, but not *Z*

30. Where is the stream most likely flowing the fastest? (1) *X*, only (2) *Y*, only (3) *Z* only (4) *X* and *Y*, but not *Z*.

31. Which figure below best represents a cross section of the river along line *a–b*?

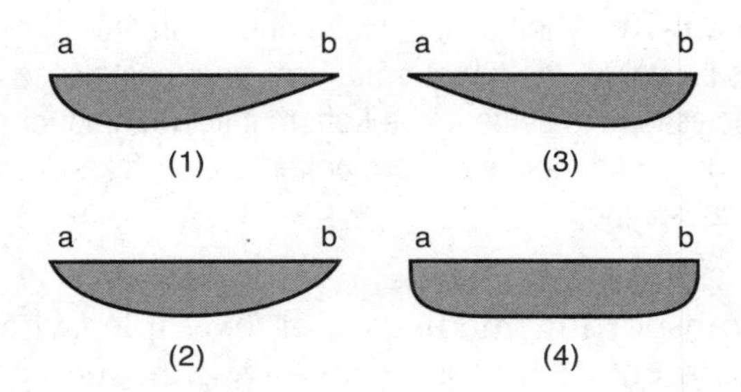

32. The photo below shows a bend in the Susquehanna River near Oneonta in central New York State. What letter on the stream diagram on page 104 best represents the position that the arrow is pointing to?

(1) *X*, only (2) *Y*, only (3) *Z*, only (4) *X* and *Y*, but not *Z*.

33. An observer notes that in a particular part of a stream, the water transports clay, silt, sand, pebbles and cobbles, but not boulders. What is the approximate stream velocity? (1) 50 cm/sc (2) 100 cm/s (3) 250 cm/s (4) 400 cm/s

34. The photo above shows two rocks. Each coin in this photo is approximately 1 cm across. How are these rocks best classified?

(1) silt (2) sand (3) pebbles (4) cobbles

35. What change is most likely to cause a river to flow faster? (1) an increase in discharge (2) a decrease in gradient (3) temperatures well below freezing (4) an increase in sediment load

Part B

36. The rock particles in the diagram below were nearly the same shape when they were dropped into a fast-flowing stream. They are all the same kind of rock. They were later taken from the stream at different places downstream. Which rock was recovered from a location closest to where they were dropped into the stream?

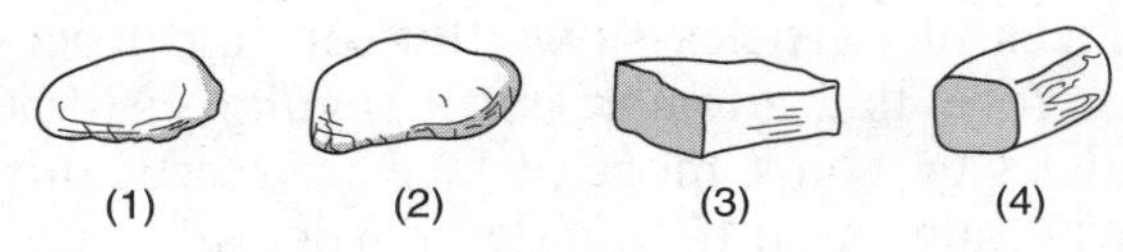

37. If a river is flowing at a speed of 50 centimeters per second, what kind of sediments can it transport? (1) silt and sand (2) silt, sand, and pebbles (3) silt, sand, pebbles, and cobbles (4) silt, sand, pebbles, cobbles, and boulders

38. How are particles of sand and pebbles that are less dense than water carried by a stream? (1) by bouncing along the bottom in traction (2) by solution in water (3) by chemical weathering (4) by floating on the surface

WHAT IS DEPOSITION?

When an agent of erosion deposits, or stops transporting, particles of earth materials (sediments), the process is called **deposition**. Deposition is also called *sedimentation*. Agents of erosion, such as gravity, water, ice, and wind, are also agents of deposition.

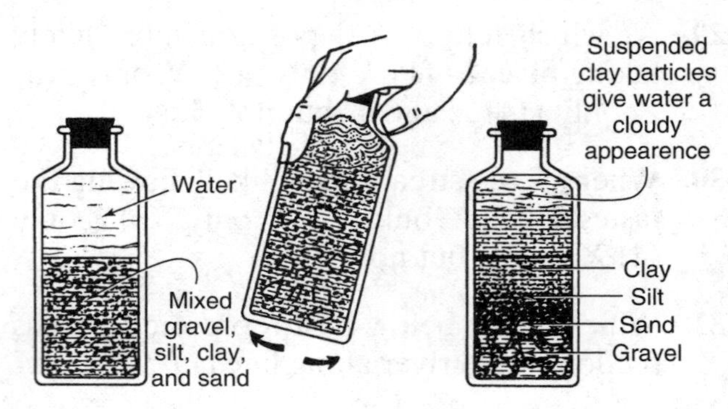

Figure 4-13. When a jar of water and sediments of mixed sizes is shaken, the largest, roundest, and most-dense particles tend to settle to the bottom first. The gradual change in sediment size from bottom to top is known as vertical sorting.

Factors That Affect Deposition

The rate at which sediments are deposited by water and wind depends on the size, shape, and density of the sediment particles and the speed of the transporting medium.

Particle Size Smaller particles, such as clay and silt, settle more slowly than cobbles and boulders. Very small particles (less than 0.001 millimeter in size) may be held in solution or suspension indefinitely. The smallest particles do not settle out of a water solution unless the solution becomes saturated. For example, in some parts of the Persian Gulf, evaporation leaves the remaining water so rich in salt that the water becomes saturated. This means that it can hold no more salt in solution. If more water evaporates, salt crystallizes and settles to the bottom. The salt crystals that form and settle out of solution are called **precipitates**, and the process is called **precipitation**.

Particle Shape Friction between water and the surfaces of particles slows the settling process. Therefore, flat, angular, and irregularly shaped particles that have more surface area settle more slowly than smooth, rounded particles.

Particle Density Among particles of the same average size and shape, denser particles settle faster, while less dense particles take longer to settle.

Settling Rate and Settling Time There is an inverse relationship between the rate of settling and the settling time. Sediments that settle at a faster rate require less settling time. Thus, as the rate of settling increases, the settling time decreases.

Sorting of Sediments

The velocity of a transporting medium plays a major role in determining when the deposition of particles will occur. Deposition is usually the result of a reduction in the velocity (slowing) of a transporting medium. For example, when a stream enters a large body of water, such as an ocean, the stream's velocity decreases as it mixes with the still water. Particles of sediment begin to settle. The largest, roundest, and densest particles are deposited first near the ocean's shoreline. The smallest, flattest, and least-dense particles are carried farthest from shore. The separation of particles, in this case, is called **horizontal sorting**. Figure 4-14 shows a similar pattern of deposition at the end of a steep canyon where floodwater slows.

When particles settle in calm water, the roundest, largest, and densest particles quickly settle at the bottom of a layer, while the flattest, smallest, and least-dense particles settle later at the top of the same layer. This kind of **vertical sorting** may occur when a landslide suddenly dumps particles of many sizes into still water. Figure 4-13 shows a procedure you can use to watch this process.

A series of depositional events, such as a succession of underwater landslides in deep water, can cause a type of vertical sorting known as **graded bedding**. An abrupt change in the particle size—from coarse sediments upward to very fine sediments—separates one graded bed from another. Each graded bed represents a period of deposition or a single, distinct depositional event, such as one landslide as shown in Figure 4-15.

Deposition by Gravity

When gravity acts alone as the agent of erosion, the sediments deposited are not sorted. Where

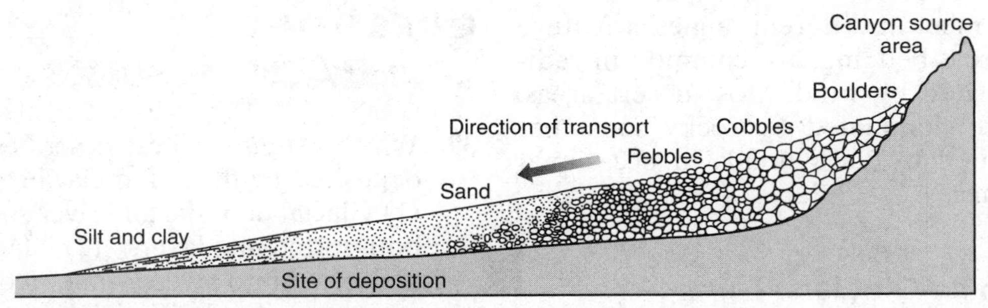

Figure 4-14. Horizontal sorting occurs at the end of a steep canyon. Larger particles are deposited first. The smaller sediments can be carried a greater distance from the end of the canyon. Also notice that the thickness of the sedimentary layer decreases downhill.

pieces of weathered rock have fallen, you will find a talus pile of angular rocks of many different sizes mixed together. Figure 4-12 on page 104 shows characteristics that can help you decide what agent of erosion and deposition has shaped particles of sediment.

Deposition in Water

The size of sediment particles carried by a stream indicates the stream's velocity. Large boulders and cobbles are generally found in the fastest parts of streams. Finer sediments indicate a slower current. As a river slows, it drops some of the sediment it carries. To keep the slower sections of a river deep enough for boats to pass, sediments that accumulate in those sections must be dredged out.

Where a river or stream enters a lake or ocean, the water slows and drops its load of sediment, often in a fan-shaped deposit. This deposit is called a delta because it resembles the Greek letter delta (Δ), an ancestor of our letter D. The land around New Orleans, Louisiana, is a delta deposited by the Mississippi River as it slows upon entering the Gulf of Mexico.

Deposition by Wind

Wind-blown sediments tend to be finer than those deposited by other agents of erosion. Air is less dense than water, so the wind usually does not have enough force to move the largest particles, such as pebbles. You can see hills of wind-blown sand, called dunes, at some beaches as well as in certain desert valleys. Sand is blown up the windward side of the dune and deposited along a steeper slope on the leeward (downwind) side as you can see in Figure 4-16.

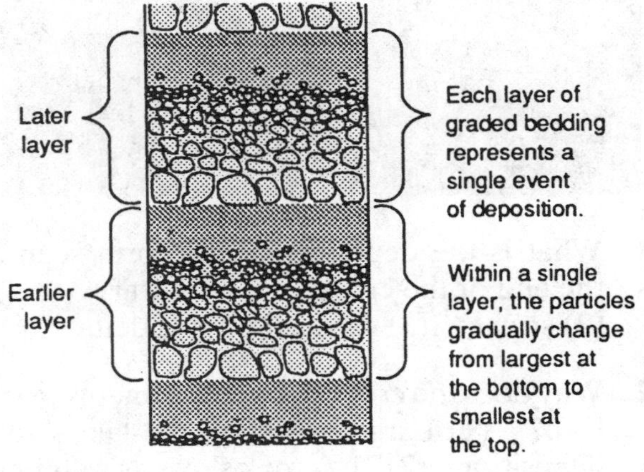

Figure 4-15. Layers of graded bedding represent a series of depositional events each of which shows vertical sorting.

Figure 4-16. In the formation of a sand dune, wind usually carries the sand up the gentler windward slope and drops it on the steeper down-wind slope that is more shielded from the wind. This dune is being formed by winds blowing left to right.

Layers that meet at different angles, a feature known as cross-bedding, are common in sediments deposited by wind. Most desert areas, however, are dominated by rocky soils that were deposited by flash floods in occasional thunderstorms.

Deposition by Glaciers

Throughout Earth's history there have been **ice ages**, long periods when glacial ice covered large parts of the continents. Ice has advanced over Eastern North America at least four times over just the past 2 million years. Geologists have learned about this from the sediments the ice transported southward and the features it left on the land. In New York, transported soils are more common than residual soils. The Pleistocene (10,000 to 2 million years ago) ice sheets that covered mountains as well as valleys in New York State have left rounded mountaintops and polished, grooved, and striated bedrock.

Glaciers produce two kinds of deposits. Material deposited directly by moving ice contains a wide range of particle sizes that are not sorted or layered. Sediments deposited by streams of meltwater, on the other hand, usually contain layers in which sediments are **sorted** by particle size. When we see ice-age sediments, we can usually tell whether they were deposited by the advancing ice or deposited by meltwater.

Deposition directly by glacial ice occurs when a glacier melts and sediments are released. Unlike water deposits, sediment left by melting glacial ice usually contains clay, sand, cobbles, and boulders mixed together. These **unsorted deposits (till)** left by glacial ice are often found in ridges and mounds. **Glacial erratics** are large rocks that were transported by glacial ice without being broken into small particles. They often rest high above stream valleys, which shows that they could not have been deposited by running water. Partial rounding and **striations** (scratches) on smooth surfaces also indicate transport by glaciers. Erratics, as with other glacial deposits, commonly differ in composition from the bedrock on which they rest.

QUESTIONS

Part A

39. Which statement best describes sediments deposited by the ice in glaciers and rivers? (1) Glacial deposits and river sediments are both sorted into layers. (2) Glacial deposits are sorted into layers, while river sediments are unsorted. (3) Glacial deposits by ice are unsorted and river sediments are sorted into layers. (4) Glacial deposits and river sediments are both unsorted and not layered

40. As a stream slows, which sediments are deposited first? (1) the largest and most-dense particles (2) the largest and least-dense particles (3) the smallest and most-dense particles (4) the smallest and least-dense particles

Base your answers to questions 41 through 44 on the image below, which is a satellite photograph of Taughannock Falls State Park near Ithaca, New York. In this image, Taughannock Creek flows toward the northeast. Use this image to answer the following four questions.

41. What is the depositional landform seen at the end of the creek? (1) tributary (2) stream (3) deposition (4) delta

42. Why does the creek deposit sediments in the lake? (1) Large waves on the lake cause deposition. (2) The creek slows as water enters the lake. (3) The sediments become waterlogged. (4) Water stirs up pebbles on the bottom of the lake

43. Where would you find particles of clay with no sand or pebbles? (1) in the bottom of the creek bed (2) in the middle of the delta (3) at the bottom of the high stream banks (4) in the deepest parts of the lake

44. Taughannock Creek transports most of its sediment load in occasional floods. Which diagram below best represents sediments deposited in the lake by flood events.

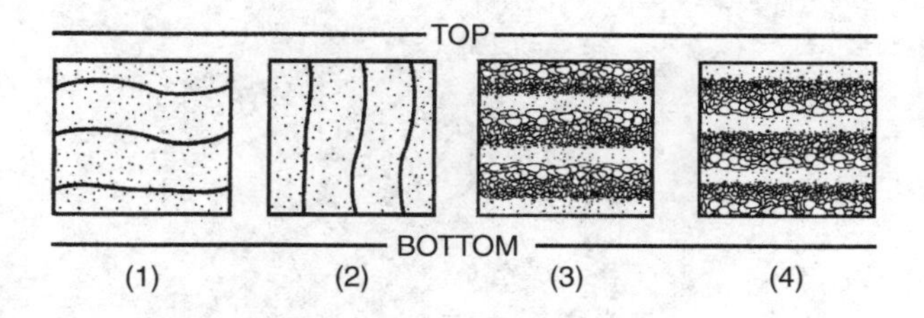

45. How many floods are represented by the sediments in choice (4) above? (1) 1 (2) 2 (3) 3 (4) 4

Part B

46. The photograph below is a view of Amargosa Dunes in California. The camera was pointing west.

Most likely, the winds have been blowing from the (1) north (2) south (3) east (4) west

Base your answers to questions 47 through 49 on the diagram below, which is a cross-sectional profile of a river entering an ocean. Arrows show the direction of the current. The data table describes the sediments in zones A–D.

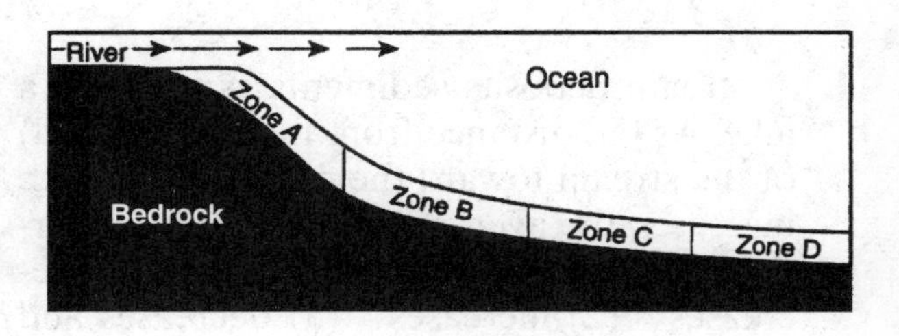

Data Table

Zone	Major Sediment Sizes
A	0.04 cm to 6 cm
B	0.006 cm to 0.1 cm
C	0.0004 cm to 0.006 cm
D	Less than 0.0004 cm

47. How is this pattern of horizontal sorting produced? (1) Larger particles are generally deposited first. (2) Rounded particles generally settle more slowly. (3) Dissolved minerals are generally deposited first. (4) High-density sediments generally settle more slowly.

48. The sedimentary rock siltstone is most likely to form for sediments in which zone? (1) *A* (2) *B* (3) *C* (4) *D*

49. Which graph below best shows the relationship between the density of particles and their positions on the diagram above?

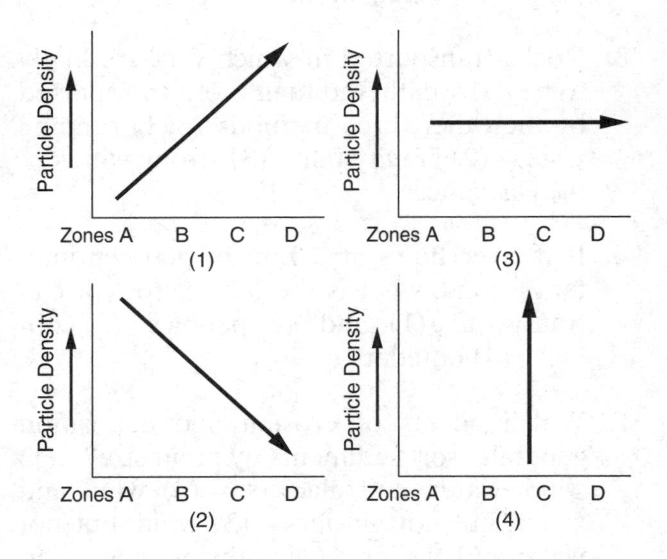

Chapter Review Questions

PART A

1. A stream deposits sediments as it enters a lake. As the distance from the mouth (end) of the stream toward the center of the lake increases, the average diameters of the particles along the bottom of the lake (1) decreases (2) increases (3) decreases and then increases (4) increases and then decreases

2. What agent most likely transported the rock shown in the image below? (A coin is shown for scale)

(1) a river (2) the wind (3) a glacier (4) gravity acting alone

3. Rocks transported in which way are likely to be *less* weathered than those transported by the other three methods? (1) running water (2) gravity falls (3) strong winds (4) glacial ice

4. If a river flows at a rate of 500 centimeters/second, what is the largest rock it can transport? (1) sand (2) pebbles (3) cobbles (4) boulders

5. Which agents of erosion and deposition generally sort sediments by grain size? (1) wind, water, and glaciers (2) wind and water, but not glaciers (3) wind, but not water and glaciers (4) not wind, not water, and not glaciers

6. The photo below shows a marble gravestone in a remote part of the Adirondack Mountains of New York State.

What has been the major cause of changes in this gravestone over the past 50 years? (1) wind abrasion (2) chemical weathering (3) water erosion (4) metamorphism

7. Which two factors are most likely to affect the amount of erosion done by a stream? (1) direction of flow and water temperature (2) elevation and the wind direction (3) longitude and mineral content of sediments (4) gradient and discharge volume

Base your answers to questions 8 and 9 on the photo below, which was taken in Southern California.

8. What was the major event that formed this sand dune? (1) wind erosion (2) water erosion (3) wind deposition (4) water deposition

9. What is the most likely direction of the prevailing winds in this location? (1) left to right (2) right to left (3) foreground to background (4) background to foreground

10. When is physical weathering most active in New York State? (1) winter (2) spring (3) summer (4) autumn

PART B

Base your answers to questions 11 through 13 on the graph above, which shows how the mineral composition of a sample of rock that was exposed to the atmosphere changed over time.

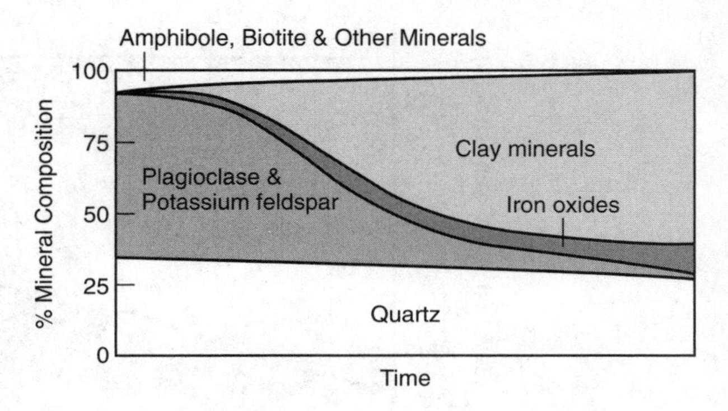

11. What process does this graph best illustrate? (1) erosion (2) weathering (3) recrystallization (4) deposition

12. Which mineral seems to be the most resistant to change? (1) quartz (2) the feldspars (3) amphibole (4) biotite

13. It appears that over time plagioclase and potassium feldspar can change mostly to (1) quartz (2) iron oxides (3) amphibole and biotite (4) clay

Base your answers to questions 14 through 24 on the map, lake profile, photograph, and the text below.

Unintended Consequences at Lake Powell

We want carbon-free sources of energy. Hydroelectric power is an attractive alternative. However, every form of energy comes at an environmental price. Sedimentation, or silting, is the environmental price of dams built to provide hydroelectric power.

The Glen Canyon Dam on the Colorado River was completed in 1964. Lake Powell formed behind the dam. Sedimentation occurs in Lake Powell because the Colorado River is very rich in sediments. Since 1964, the Colorado's nearly 100-million-ton average annual sediment load has collected at the upper end of Lake Powell. The sediments sinking to the lake's bottom will gradually fill the submerged canyon. Scientists estimate that sediment accumulation will force the closing of the dam within 200 years.

Sediment fill at the upper end of Lake Powell—2008

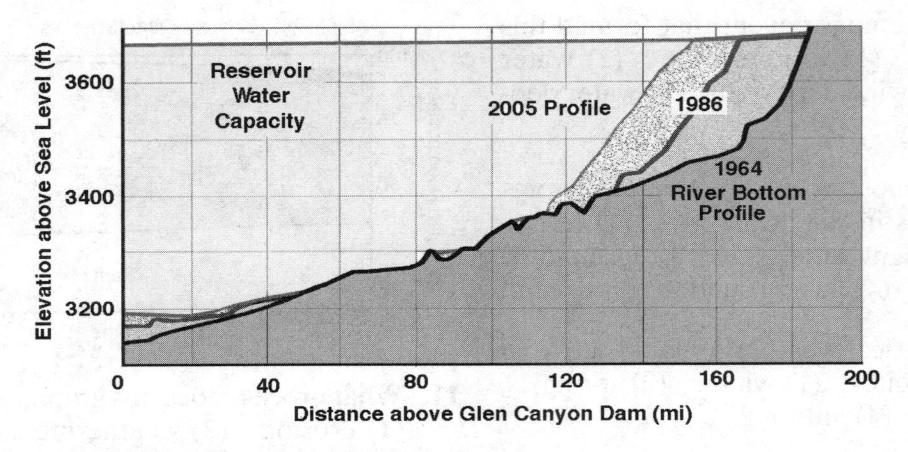

Sedimentation of Lake Powell 1964–2005

Another problem is that the water released from the dam is sediment-free. This causes beaches to wash away in the Grand Canyon. Before the dam was built, sand that washed away from the beaches was replaced by sediment carried by the river. This kept the sandy banks in equilibrium. The loss of the beaches will adversely affect rafting and camping down river as shores erode away to bare rock.

14. What geological process is filling Lake Powell? (1) erosion (2) weathering (3) deposition (4) crystallization

15. What is the major source of most of the sediments flowing into Lake Powell?

(1) Sea of Cortez (3) State of Arizona
(2) Rocky Mountains (4) Mexico

16. In what general direction does the Colorado River flow? (1) southwest (2) southeast (3) northwest (4) northeast

17. What is the straight-line distance from the most northerly source of the Colorado River to the place where it ends at the Sea of Cortez? (1) 300 miles (2) 600 miles (3) 850 miles (4) 950 miles

18. The Glen Canyon Dam is contributing to what problem downstream in the Grand Canyon? (1) flooding (2) warm water (3) deposition (4) erosion

19. What is the greatest depth of sediments deposited in Lake Powell in the period between 1964 and 2005? (1) 100 feet (2) 200 feet (3) 300 feet (4) 400 feet

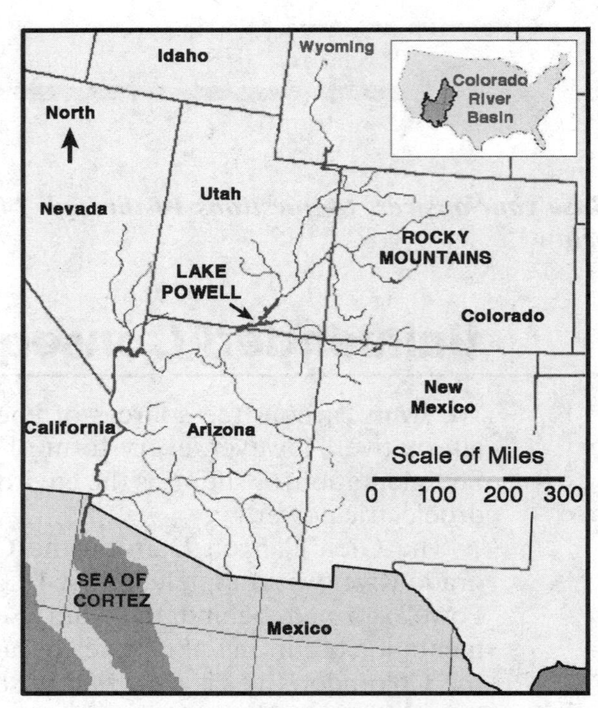

The Colorado River and Its Tributaries

20. How many US states contain tributaries of the Colorado River or border the river? (1) 1 (2) 3 (3) 5 (4) 7

21. What is the approximate distance from the Glen Canyon Dam to the place where the photo of sediment fill was taken? (1) 40 miles (2) 80 miles (3) 120 miles (4) 160 miles

22. The photo below shows two samples of granite from the same part of the Rocky Mountains. One of them was collected along the Colorado River in the state of Colorado. The second sample was collected about 500 miles downstream in Arizona.

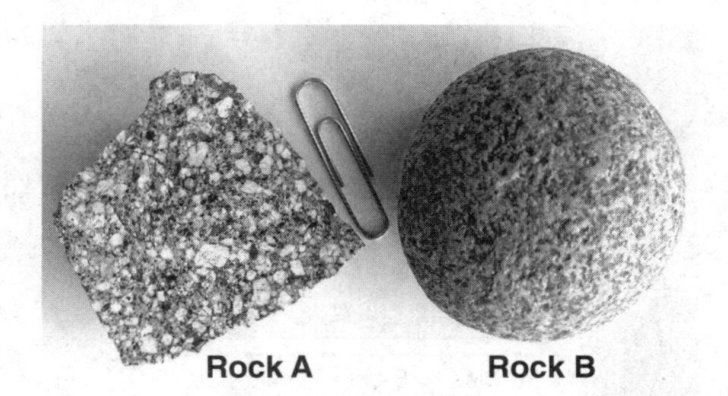

Where were these two rocks found?
(1) Rock *A* was recovered from Colorado and *B* was from Arizona. (2) Rock *A* was recovered from Arizona and *B* was from Colorado. (3) Rock *A* was found at the center of Lake Powell and *B* was at its source. (4) Rock *A* was found in a riverbank and *B* was found in a sand dune.

23. If the original shape of rock *B* was similar to rock *A*, what process changed rock *B*?
(1) wind abrasion (3) metamorphism
(2) abrasion in water (4) crystallization

24. What major river-deposited landform is shown in the photo of the upper end of Lake Powell? (1) a glacial deposit (2) a sand dune (3) a delta (4) a soil profile

25. Paved roads in the Adirondack Mountains of New York State need more frequent repairs than roads built to the same standards on Long Island, New York. Why do the Adirondack roads break down more quickly? (1) Long Island roads get more traffic than the roads in the Adirondacks. (2) Winter snows last longer on Long Island. (3) Long Island is north of the Adirondacks. (4) The Adirondacks have more freeze-thaw cycles.

26. A student quickly dumped the contents of a beaker filled with particles of mixed sizes into a tall transparent column filled with water. Which diagram below best shows how the sediments settled?

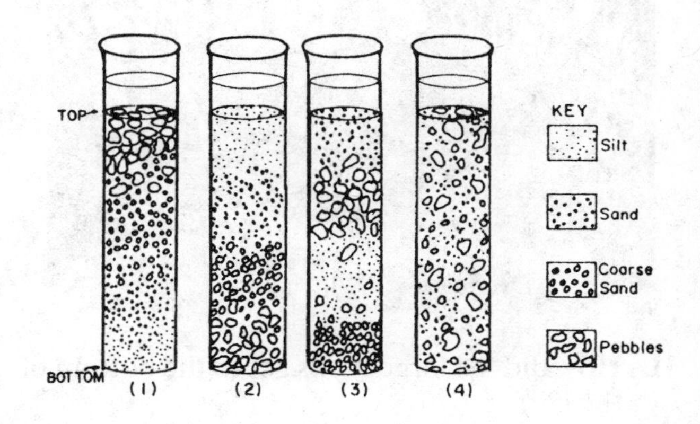

PART C

Base your answers to questions 27 through 30 on the diagram below, which shows changes in the bedrock of a location in New York State.

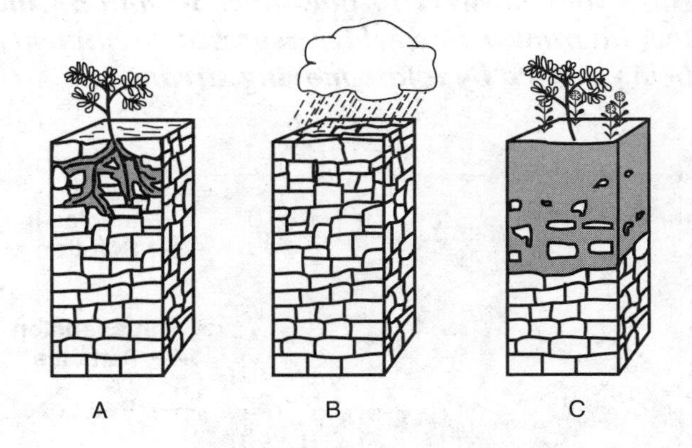

27. What is the correct sequence of the changes shown in these three diagrams?

28. According to the symbols in the *Reference Tables*, what is the mineral composition of this rock?

29. What kind of bedrock weathering is probably dominant in this location?

30. What is the end product of the weathering and biological activity shown in this diagram?

Base your answers to questions 31 and 32 on the image below, which was taken along an ocean shoreline.

31. How did the large rocks get to the bottom of the cliff?

32. State two ways in which the particles of sediment along the beach have changed since they were eroded from the steep area above the beach.

33. Draw one layer of graded bedding.

Base your answers to questions 34 and 35 on the diagram below, which represents sediments being carried by a fast-moving stream.

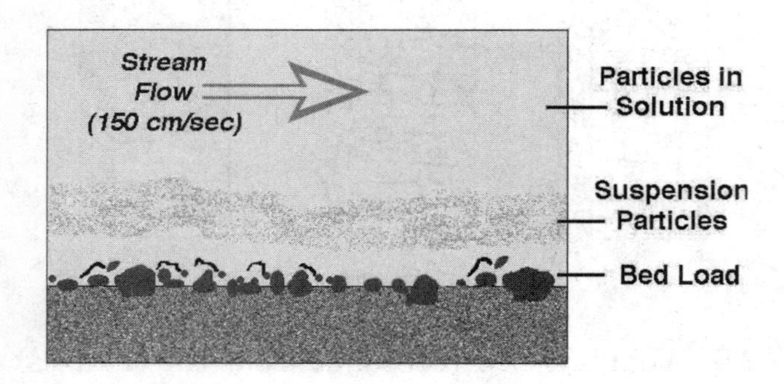

34. According to the information in this diagram, what is the largest type of particle that the stream can transport?

35. Specify three characteristics of the dark particles bouncing along the bottom of the stream that make them different from the particles carried at midstream.

36. Give two characteristics of steam deposits that are not usually found in sediments deposited by glaciers.

Base your answers to questions 37 through 40 on the three graphs below, which show the particle sizes of sediments found in three locations in New York State. Each was deposited by a different agent of erosion and deposition. One location is a sand dune near the shoreline on eastern Long Island. One is a small hill composed entirely of sediments. The third is from the bottom of a fast-moving stream in the Catskill Mountains of New York State.

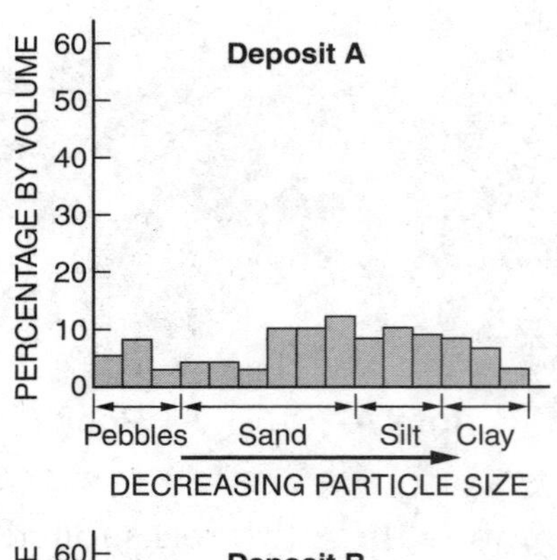

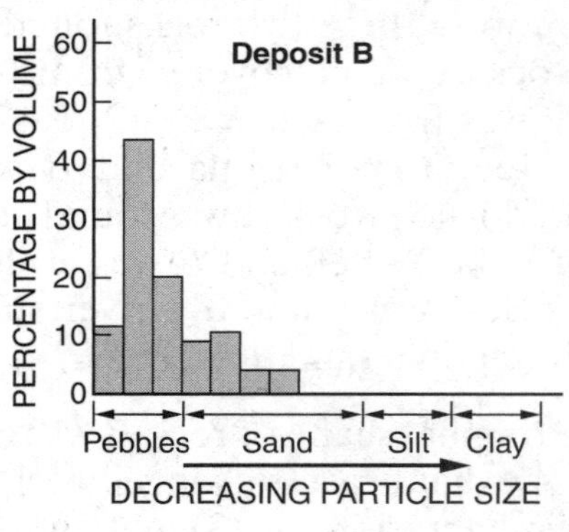

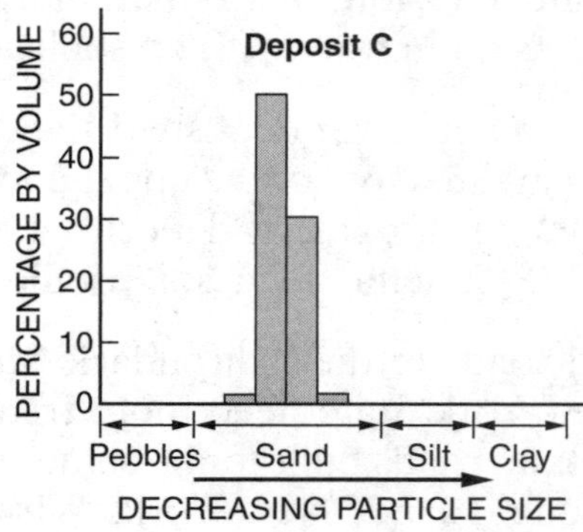

37. Which sediment sample is probably from the Catskill stream? How can you tell?

38. What is the most likely agent of erosion and deposition for Deposit *A*?

39. Name two characteristics of Deposit *C* that help you identify it as a sand dune deposit.

40. What is the approximate average diameter of the sediments in Deposit *C*? (Express your answer in quantity and units.)

Base your answers to questions 41 through 43 on the passage below.

Muck Soil

Muck, or bog soil, is a soil made up mostly of humus from drained swampland. It is used in the United States for growing specialty crops such as onions, carrots, celery, and potatoes. These crops can tolerate the poor drainage and lack of minerals. Farming on drained bogs is an important part of agriculture in southeastern New York State, primarily in Orange County. The soil is deep, dark-colored, and crumbles easily. It is often underlain by clay.

Muck has been hyped as miracle soil, capable of supporting vast yields. However, this does not hold up in reality. It is prone to problems, such as the fact that it is very light and usually windbreaks must be provided to keep it from blowing away when dry. It also can catch fire and burn underground for months. Oxidation also removes a portion of the soil each year, so it becomes more and more shallow.

41. What is humus?

42. The graph below represents a typical composition of muck soil. (Air is not included.)

The portion of the graph labeled Material *X* is a substance very common in most soils but mostly absent in muck. What is substance *X*?

Principal Solid & Liquid Components of Muck Soil

Material "X"

Organic materials (mostly carbon)

Water

43. The figure below shows the soil horizons of a typical soil. Muck has a much thicker *A* horizon than typical soils. What is the nature of the material below the bottom of the *C* horizon in the figure?

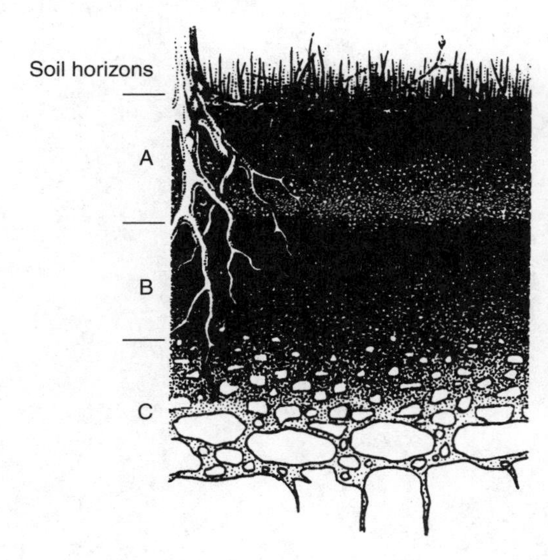

Base your answers to questions 44 through 47 in the photo below of three girls walking in a nature preserve.

44. What is the agent of erosion that deposited the sediment the girls are walking near?

45. Why does the size of the sedimentary particles decrease from boulders to clay in the direction of the arrow?

46. What name is given to the kind of change in the size of the sediments that you see in the photo?

47. You can see boulders in the distance. If the smallest boulder has a diameter of 25.6 centimeters, how fast was the stream moving that carried it?

CHAPTER

Glaciers, Oceans, and Landscapes

Vocabulary at a Glance

alpine glacier	esker	mountain
arid	Finger Lakes	outwash plain
annular	glacial polish	plain
continental glacier	kame	plateau
dendritic	kettle	radial
drainage pattern	kettle lake	relief
drumlin	landform	topography
erratic	landscape	trellis
escarpment	moraine	

NEW YORK AND THE ICE AGES

Glaciers are accumulations of ice large enough to survive the summer melt. In present-day New York State, the climate is too warm to support glaciers, even in the highest mountains. However, as recently as 20,000 years ago, a great continental ice sheet covered most of New York State. The ice sheet grew when the snow that accumulated in Eastern Canada did not completely melt in the summer. Eventually, it formed a gigantic sheet of ice a mile or more thick. Figure 5-1 shows how far south the ice reached. By studying past global ocean temperatures, scientists have learned that ice ages have been occurring for much of Earth's history.

When thick layers of snow build up, the weight of the snow causes ice to form at the bottom of the pile. The ice will flow downhill, or away from its thickest accumulation. The rate at which the ice flows ranges from a few centimeters to several meters per day. **Alpine glaciers**, also known as valley glaciers, build in mountain regions, and often carve out U-shaped valleys during their steady advance. **Continental glaciers**, also known as ice sheets, were responsible for most of the glacial features found throughout

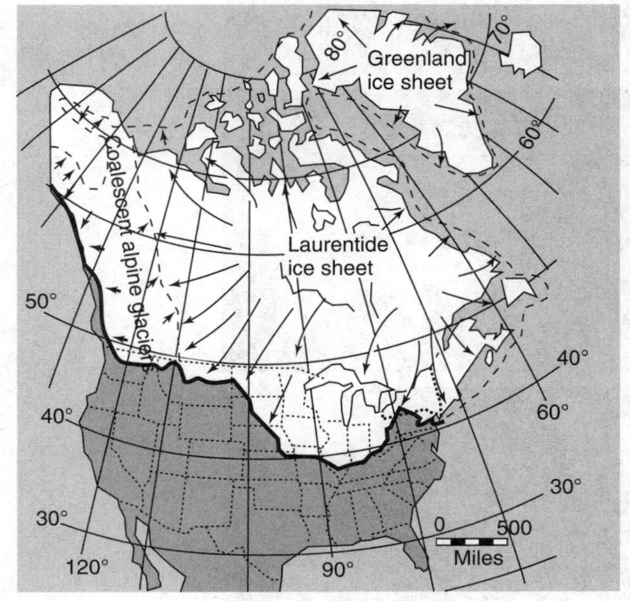

Figure 5-1. In the past, continental ice sheets have repeatedly moved southward out of Canada to cover much of the northern part of the United States, including nearly all of New York State.

New York State. The erosion caused by a glacier is not the result of ice rubbing against rock. Rather, it is the abrasion by rocks carried within the glacier or dragged under its flowing ice that causes most glacial erosion.

As the ice sheets advanced over New York during the ice ages, the rock-encrusted ice acted like sandpaper. It smoothed the jagged edges from the mountains and polished the hard bedrock surfaces (**glacial polish**). Figure 5-2 shows six glacial features in New York State. These features provide evidence that glaciers once covered nearly all of New York State. Rocks within the glacier also left *striations*, parallel grooves and scratches in the bedrock. As the advancing ice deeply scoured former north-south river valleys it carved out the **Finger Lakes** in western New York State.

In addition to producing features through erosion, ice sheets also produced features by deposition of unsorted glacial sediment (till). **Drumlins** are teardrop-shaped hills of glacial sediment. They formed as ice moved up and over rock and soil that had built up in front of the glacier. The round end of the hill faces the direction from which the glacier advanced. Where the ice front stopped its southward advance, it left piles of unsorted soil and rock called terminal **moraines**. The twin forks of Long Island were formed by terminal moraines (see Figure 5-3).

Terminal moraines also dammed valleys, creating the Finger Lakes, and forcing the water to flow north toward Lake Ontario as shown in Figure 5-4. Generally, when two streams join, they make a V that points downstream. However, some streams flowing into the Finger Lakes area reverse direction due to the dams of glacial debris that made the lakes. Low spots in the glacial deposits and places where large, buried ice blocks melted left dry depressions called **kettles** and ponds called **kettle lakes**. Figure 5-5 on page 120 shows the formation of the variety of glacial features that can be seen throughout New York State. These features include **eskers** (ridges of sand deposited in tunnels under the ice), **kames**

Figure 5-2. Evidence of past glacial activity is found throughout New York State. Unsorted glacial soils (A) contain no sorting or layering. Closed depressions called kettles (B) sometimes contain ponds. Large boulders known as erratics (C) are often very different from the bedrock they sit on. Polish and striations (D) can produce a sheen on bedrock. Grooves (E) show the direction of ice movement. Drumlins (F) are hills of unsorted sediment pushed into place by the moving ice.

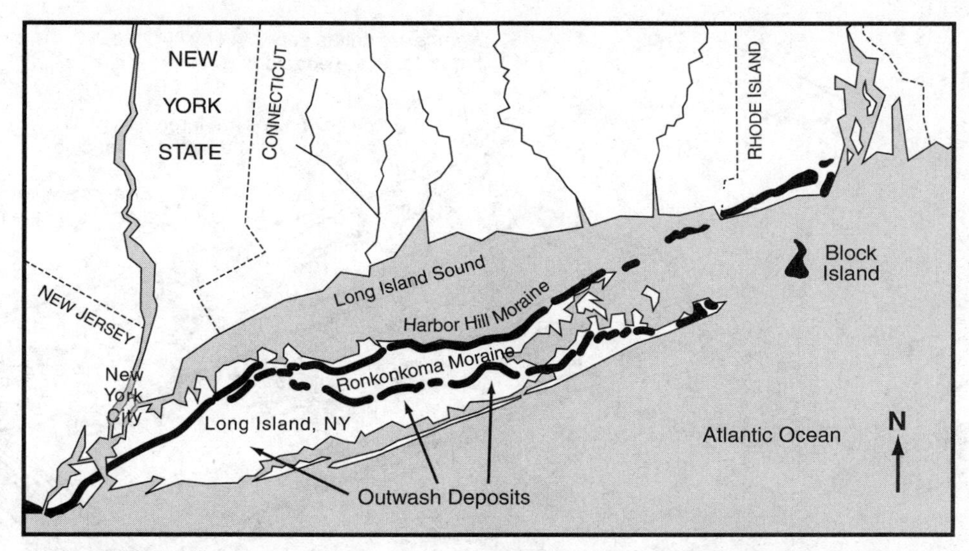

Figure 5-3. The "backbone" of Long Island is made of two terminal moraines of unsorted sediment pushed into place by advancing ice. South of the moraines the soils are layered and sorted outwash deposits.

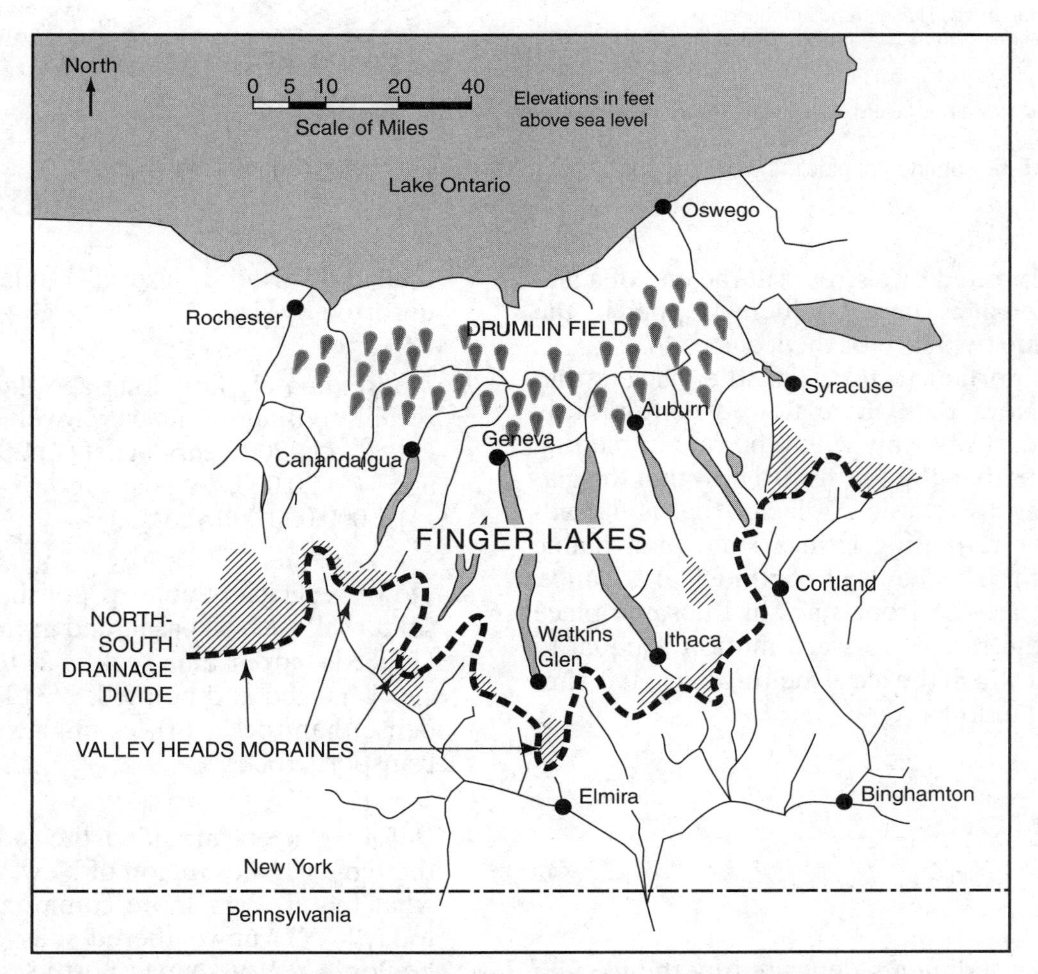

Figure 5-4. The western part of New York State has a variety of glacial landforms. These include the Finger Lakes, which were widened and dammed by glacial sediments of the Valley Heads Moraine. Drumlins were deposited north of the Finger Lakes.

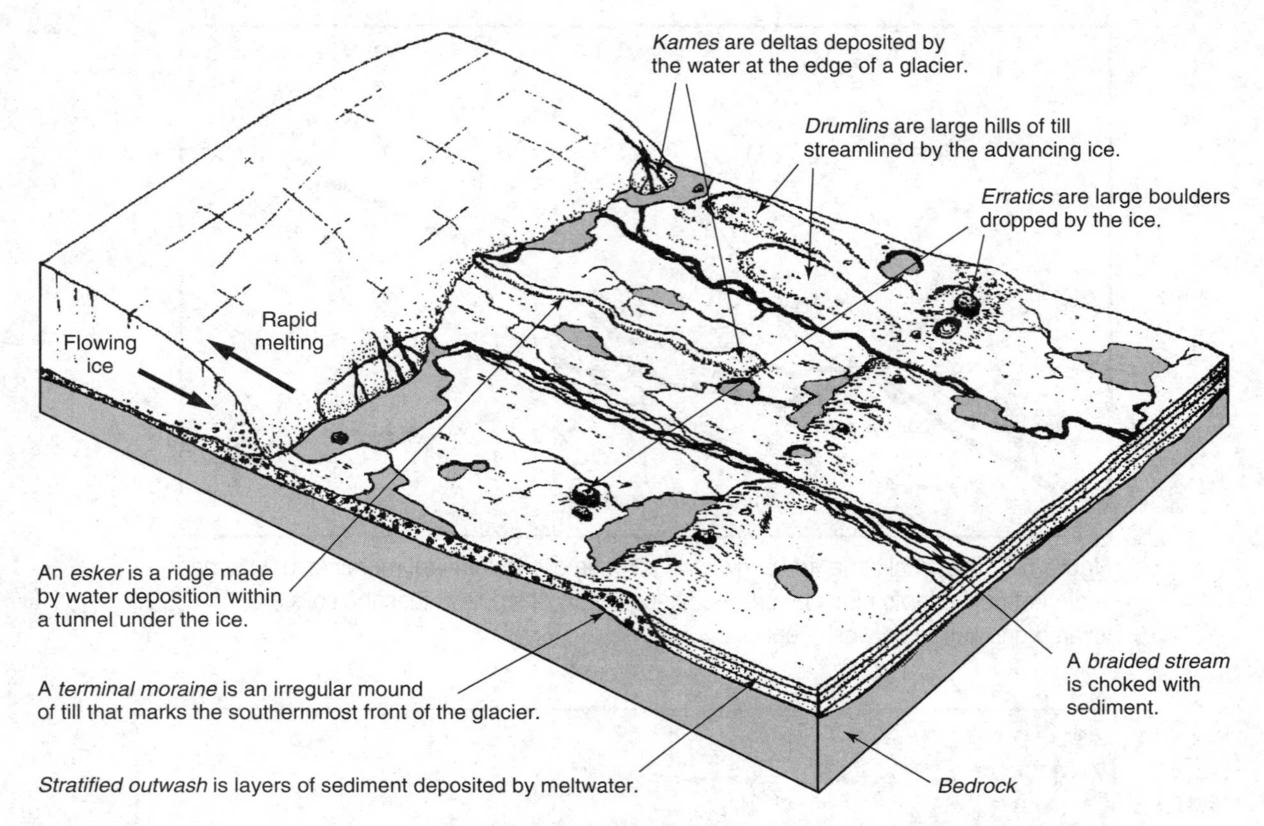

Figure 5-5. Continental glaciers were responsible for the formation of many depositional features.

(deltas deposited by a stream at the end of a glacier) and **erratics** (large boulders dropped by the ice) provide evidence of the ice ages.

It is important to note that the glaciers that covered New York State flowed continuously southward. It was only when the rate of melting was greater than the southward flow that the glaciers began to retreat. A glacier that is not advancing or retreating is in a state of dynamic equilibrium. The ice itself continues to advance. However, the ice front stays in the same place because the rate of forward motion is equal to the rate at which the ice is melting back. It is similar to a standoff.

QUESTIONS

PART A

1. Unlike sediments deposited by the ice in a glacier, sediments deposited by its meltwater tend to be (1) layered and sorted by size (2) layered and unsorted by size (3) unlayered and sorted by size (4) unlayered and unsorted by size

2. Approximately how long ago did the most recent continental glacier cover New York City? (1) 200 years ago (2) 20,000 years ago (3) 20,000,000 years ago (4) 2,000,000,000 years ago

3. How are glaciers able to polish and carve solid rock such as basalt and granite? (1) Ice is harder than rock. (2) Ice in a glacier is jagged and pointed. (3) Ice is more dense than rock. (4) Ice in glaciers usually transports rocks.

4. Before glaciers modified the landscape of the Finger Lakes region of New York State, what feature was more common than it is today? (1) unweathered soils with many boulders (2) V-shaped north-south stream valleys (3) small ponds with poorly developed drainage (4) lakes aligned north-south

5. The graph below shows Earth's average temperatures over the past 250,000 years.

Comparison of Past and Present Temperatures

A B C D

Difference (°C)

4 2 0 −2 −4 −6 −8

0 50 100 150 200 250

Thousands of Years Ago

At which time did glaciers probably cover the greatest part of Earth's land surface? (1) *A* (2) *B* (3) *C* (4) *D*

6. The diagrams below show sediments deposited by several agents of erosion and deposition.

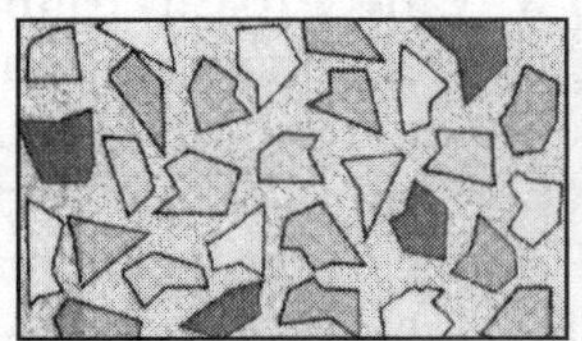
Sample A

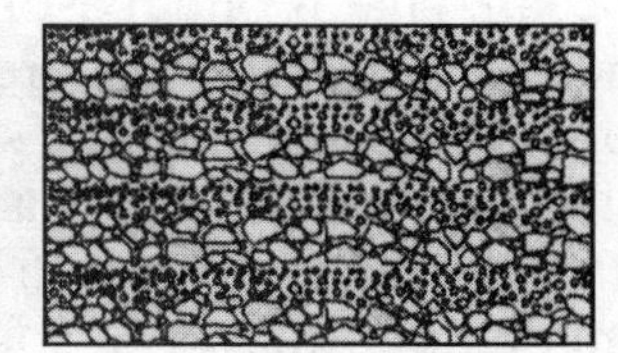
Sample B

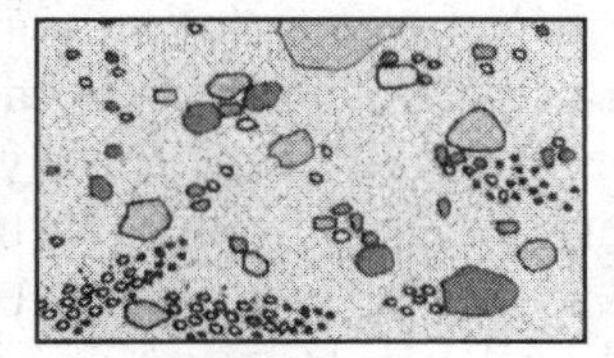
Sample C

What is the best explanation for these sediment textures? (1) Sample *A* was deposited by glacial ice, *B* by a gravity fall, and *C* by meltwater. (2) Sample *A* was deposited by meltwater, *B* by a gravity fall, and *C* by glacial ice. (3) Sample *A* was deposited by a gravity fall, *B* by meltwater, and *C* by glacial ice. (4) Sample *A* was deposited by glacial ice, *B* by meltwater, and *C* by a gravity fall.

Part B

Base your answers to questions 7 through 11 on the following diagrams. Diagram I shows a typical boulder found at various deposit locations shown in Diagram II. In the second diagram, letters A through E mark places where rocks were picked up and later deposited by an agent of erosion and deposition.

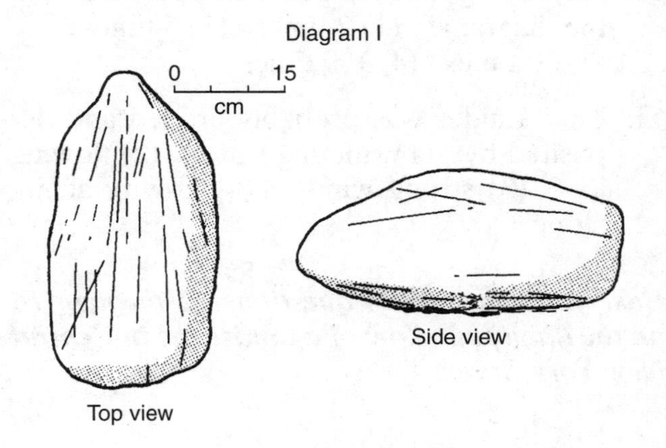

7. What characteristic was probably common to the sedimentary deposits that contained the boulders? (1) They are composed of a single type of rock. (2) They are found at the ends of large rivers. (3) They are found in piles of unsorted sediments. (4) They were eroded from source region *A*.

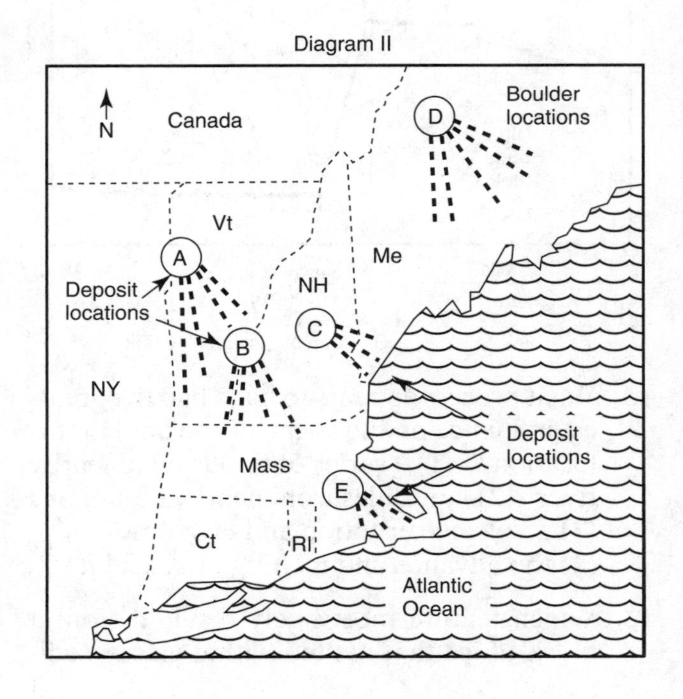

8. The agent that eroded and then deposited these rocks probably came from which direction? (1) northwest (2) northeast (3) southwest (4) southeast

9. What probably caused the scratches in the rock? (1) erosion by running water (2) pitting by windblown particles (3) movement of the boulder over bedrock (4) splitting caused by frost

10. How many times larger is the real rock than the diagram? (1) 2 times (2) 5 times (3) 10 times (4) 100 times

11. The boulder was probably eroded and deposited by (1) running water (2) flowing ice (3) strong winds (4) gravity acting alone

Base your answers to questions 12 through 16 on the diagram below of a landscape in Central New York State.

A Landscape in Central New York State

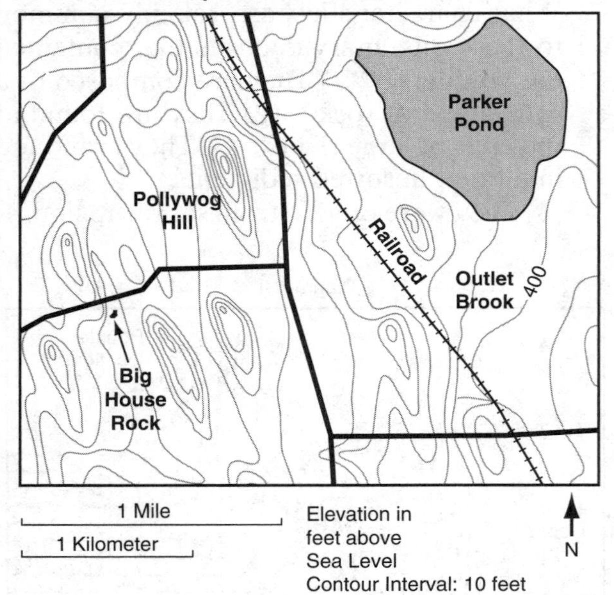

12. What process appears to have had the greatest influence on the shape of the land in this location? (1) a series of floods on a nearby river (2) landslides from nearby mountains (3) a volcanic eruption and lava flows (4) the advance and melting of glacial ice

13. Which feature most likely resulted from a block of ice that melted within glacial sediments? (1) Pollywog hill (2) Big House Rock (3) Parker Pond (4) Outlet Brook

14. What feature is most likely a glacial erratic? (1) Pollywog hill (2) Big House Rock (3) Parker Pond (4) Outlet Brook

15. From what direction did the glaciers move into this location? (1) northwest (2) northeast (3) southwest (4) southeast

16. A hiker is standing on top of Pollywog Hill. If she wants to walk down the hill, avoiding a steep slope, in what direction should she walk? (1) north (2) south (3) east (4) west

OCEANS AND COASTAL PROCESSES

As you can see in Figure 5-6, seawater covers almost 71 percent of our planet. The oceans are also very deep, with an average depth of about 4 kilometers (2.5 miles). If the world's tallest mountain (Mount Everest) could be moved to the deepest part of the Pacific Ocean, the water would cover the mountain and there would still be another mile of ocean above its peak.

Most of the sediments deposited on land or in freshwater will be eroded and deposited in the ocean. That is one reason why most fossils found in sedimentary rocks are the remains of marine organisms. Some of the sediment carried in solution to the oceans consists of simple compounds called salts. As Figure 5-7 shows, these salts are made up primarily of sodium chloride (common table salt) with a variety of other elements and compounds present in smaller concentrations. Evaporation of ocean water may cause the salts to become too concentrated to stay in solution. At that point, they precipitate forming, for example halite. Other sediment may become sedimentary rock.

Ocean currents help distribute heat energy over the whole Earth. Solar energy is strongest in the tropics and weakest near the poles. The oceans help distribute this energy and make Earth's climates more moderate, especially along coastlines and for islands in the oceans. The *Earth Science Reference Tables* has a full-page map of world ocean currents. This map shows the path of the currents and which currents are warm or cold. For example, the Gulf Stream and North

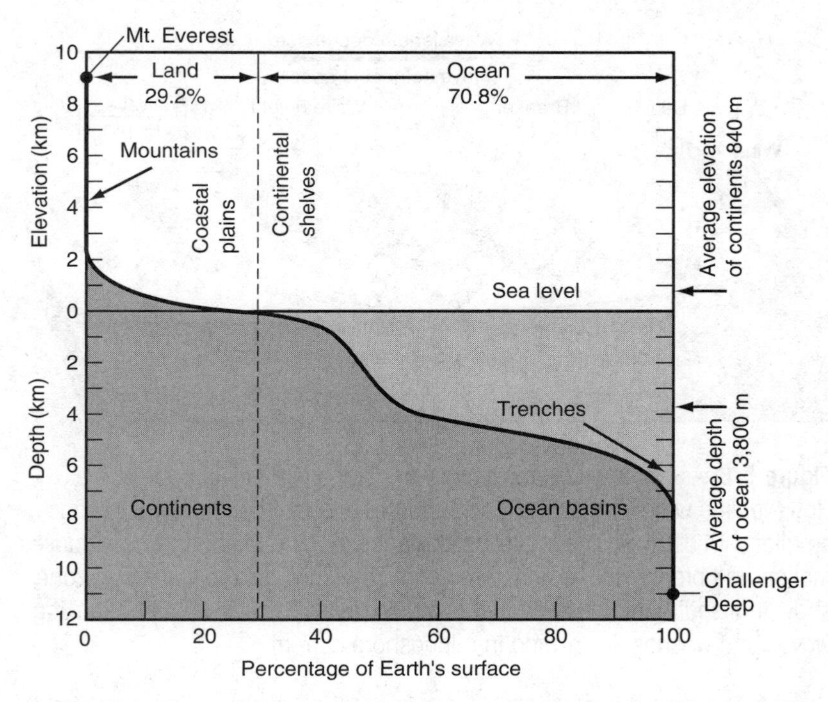

Figure 5-6. This diagram shows that most of Earth's surface (70.8 percent) is covered by oceans. Therefore the most common elevation of the geosphere is below sea level.

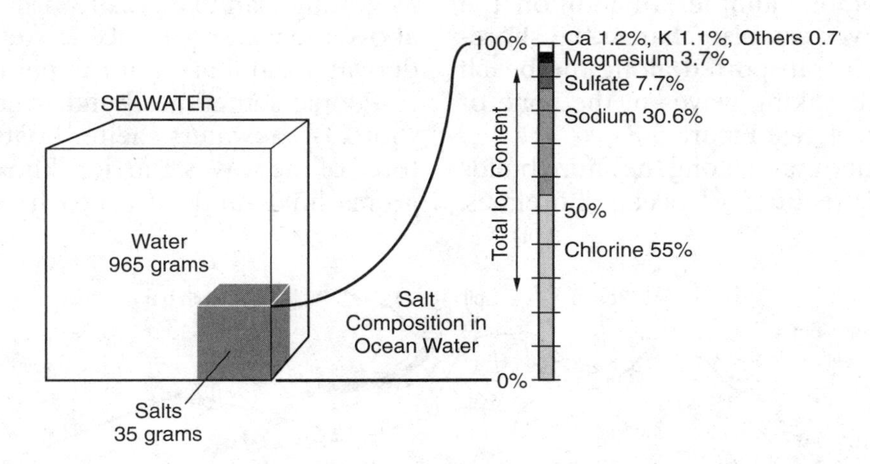

Figure 5-7. One liter of seawater contains about 3.5 percent dissolved solids. Sodium chloride (table salt) is the most abundant compound, but calcium and potassium compounds are also common.

Atlantic Current are warm and the California Current and Canary Current are cold. You will learn more about how oceans affect the climate in Chapter 9. Because of Earth's rotation, all major ocean currents curve to the right (clockwise) in the Northern Hemisphere and to the left (counterclockwise) in the Southern Hemisphere. This is due to the *Coriolis effect*, which you will read about in Chapter 7.

Waves and longshore currents cause rapid change along the shorelines of the oceans. Although ocean waves may seem to carry water, water does not actually move with the waves. Instead, the surface water usually moves in little circles or ellipses (ovals) as the wave passes. The waves carry energy. The energy is released as the waves slow and break (tumble forward) when they reach shallow water. This energy can

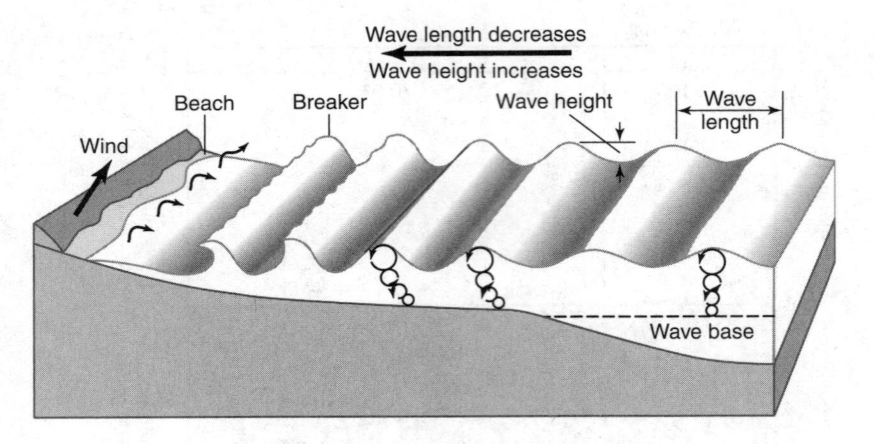

Figure 5-8. As water waves move, they transfer energy toward the shore. However, the water itself usually moves in circles that become smaller and smaller with depth. At the shoreline, the energy of the breaking waves causes friction and breaks up the beach material. The small arrows in the surf zone show that sediment is carried in the surf zone as it moves with the breaking waves and washes down-wind in a longshore current.

change the sediments along the beach. Beach sediments are rounded and reduced in size by abrasion as the breaking waves cause the particles to rub against one another. In addition, the energy in the waves can also change the shoreline. Sand is often transported along the beach just outside the breaking waves in the zone of longshore transport. See Figure 5-8.

This movement of sand along the shore builds the characteristic features of ocean shorelines, including sandbars (ridges, sometimes submerged, that run parallel to the shore), barrier islands, and sand spits, as shown in Figure 5-9. Migrating sandbars can form these features above the water's surface as well as forming underwater sandbars that run parallel to shore.

People sometimes build structures along the shore. Breakwaters shelter boats from the main force of the waves. Barriers known as jetties and groins hold sand on a beach (see Figure 5-9).

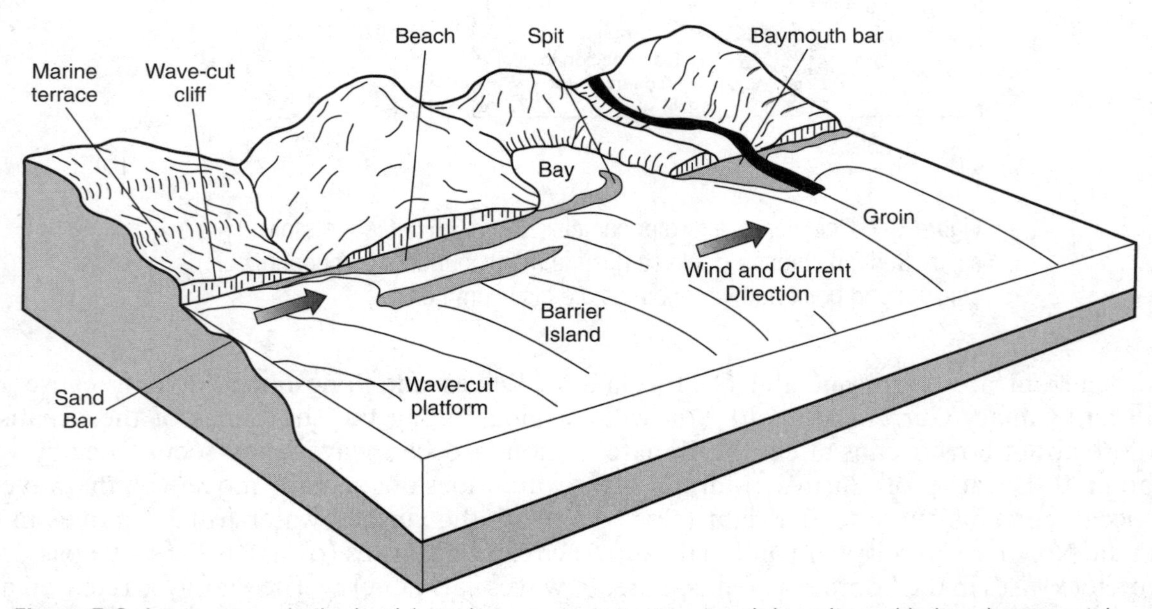

Figure 5-9. As waves erode the land, longshore currents transport and deposit sand in beaches, coastal sandbars, and barrier islands.

Breakwaters usually allow sand to accumulate in the relatively calm water behind the structure. Sand tends to accumulate on the upwind (windward) side of walls built perpendicular to the shore. However, the result of these walls is a narrower beach on the downwind (lee) side.

Such efforts to control the force of the waves often create other problems, such as accelerating beach erosion somewhere else along the shore, or causing unwanted deposition in a harbor. Many oceanfront communities are beginning to understand that beaches come and go in natural cycles. They are learning to plan for these changes. Some environmentally conscious towns discourage the building of homes or other permanent structures along the shore or even in the sand dunes behind the beach.

QUESTIONS

Part A

17. What portion of Earth's surface is covered by oceans? (1) about 25% (2) just under 50% (3) nearly 75% (4) nearly 90%

18. What kind of sediment is deposited in the ocean bottoms far from land? (1) mostly clay and dissolved materials (2) mostly sand evaporating water (3) mostly pebbles and glacial deposits (4) mostly granite and metamorphic rocks

19. What mineral is most likely to be deposited from evaporating ocean water? (1) quartz (2) mica (3) amphibole (4) halite

Base your answers to questions 20 and 21 on the photo below, which was taken in England (at approximately 51°North, 3°West). Note the palm tree near the right side of the picture.

20. This location is hundreds of miles farther north than the New York-Canadian border. What could cause climatic conditions here with temperatures mild enough to allow palm trees to grow? (1) winds from the arctic regions (2) unusually cool summer conditions (3) ocean currents that circulate south of the equator (4) an ocean current that originates to the south

21. What ocean currents cause these mild temperatures at such high latitude? (1) the North Pacific Current and the West Wind Drift (2) the Falkland Current and the North Equatorial Current (3) the Gulf Stream and the North Atlantic Current (4) the South Equatorial Current and the West Wind Drift

Part B

Base your answers to questions 22 through 25 on the graph below, which shows how latitude affects salinity and the difference between evaporation and precipitation.

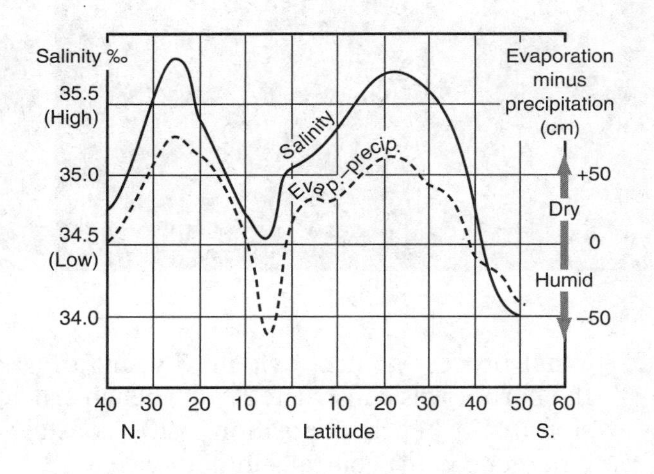

22. Based on the graph, what process causes ocean water to become saltier? (1) inflow from rivers (2) wind-driven currents (3) precipitation (4) evaporation

23. In what direction do most major ocean currents circulate? (1) clockwise in the Northern Hemisphere and counterclockwise in the Southern Hemisphere (2) Counterclockwise in the Northern Hemisphere and clockwise in the Southern Hemisphere

(3) clockwise in both hemispheres (4) counterclockwise in both hemispheres

24. Based on this graph, the salinity of the oceans at 40 N latitude and 50 S latitude is (1) relatively high (2) relatively low (3) variable with the seasons (4) completely unknown

25. At what latitude does precipitation appear to be most abundant? (1) 5°North (2) 25°North (3) 15°South (4) 30°South

26. What is the gradient between the following two points located on the continental slope? The first point is 200 meters below sea level and the second point is at a depth of 1850 meters. The distance between the two points is 2.0 kilometers. (1) 9.25 m/km (2) 12 m/km (3) 825 m/km (4) 925 km/m

Base your answers to questions 27 through 30 on the image below, which shows the harbor at Santa Barbara, California. The image on the left was taken when the harbor structures were new. The image on the right was taken 5 years after construction.

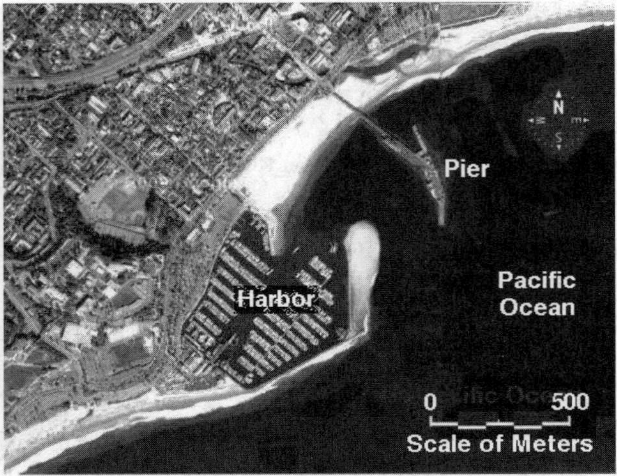

27. What process is most evident 5 years after the harbor was constructed? (1) beach erosion (2) beach deposition (3) coastal emergence (4) coastal submergence

28. From this image it is clear that most of the sediments on the beach are being transported from the (1) northeast (2) northwest (3) southeast (4) southwest

29. To keep the harbor open and yet protect the boats from storm waves, what corrective measure should the City of Santa Barbara take? (1) Just let natural processes continue as they have. (2) Destroy the groin and breakwater structures. (3) Remove beach sediments from the harbor entrance. (4) Completely fill the harbor with concrete.

30. In the image on the right, how wide is the widest part of the sand spit that has formed between the harbor and the pier? (1) 60 meters (2) 120 meters (3) 240 meters (4) 500 meters

31. How does the Peru Current affect the climate of South America? (1) It makes the west coast climate warmer. (2) It makes the west coast climate cooler. (3) It makes the east coast climate warmer. (4) It makes the east coast climate cooler.

WHAT IS A LANDSCAPE?

A **landscape** is a region on Earth's surface in which various **landforms**, such as hills, valleys, and streams, are related by a common origin. The climate, local bedrock, geologic structures, and human activities determine the shape, or **topography**, and composition of the landscape. Landscapes often reveal the interaction between the natural history and human history of a region.

Landscape Regions

Most landscape regions can be classified as *mountains*, *plateaus*, *ridges and valleys,* or *plains*. Each has its own geologic structures and topographic relief. Topographic **relief** is the change in elevation between the highest and the lowest places. Figure 5-10 shows three major types of landscapes.

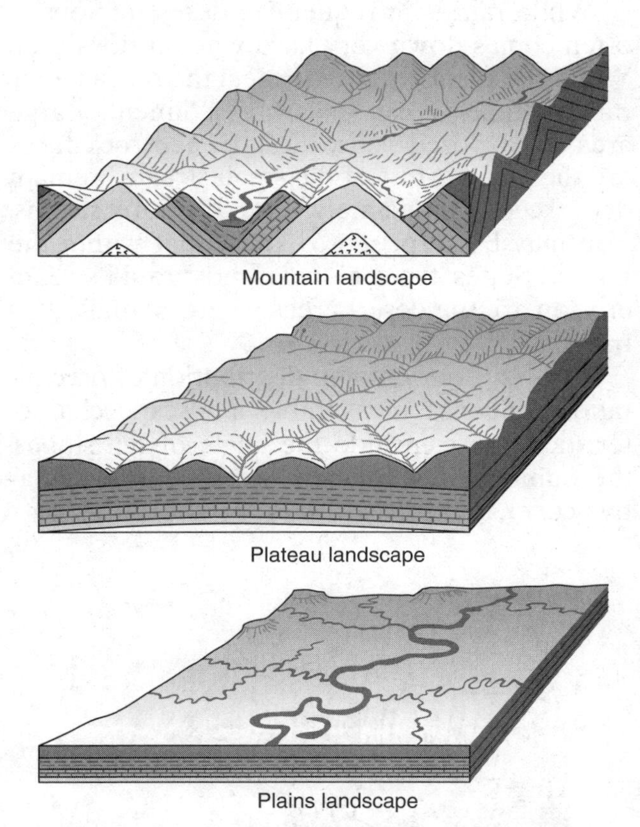

Figure 5-10. Landscapes are generally classified by their topographic relief. Mountain landscapes are the most complex. They show a variety of structures and variable rock types with large changes in elevation. The lower elevations of plateaus and plains are usually a result of their uniform or horizontal rock layers as well as a longer period of erosion than mountain landscapes.

Mountains A **mountain** landscape has the greatest relief between the highest peaks and the deepest valleys. A great variety of rock types are common in mountain landscapes. These often include igneous and metamorphic rocks, as well as folded and faulted sedimentary rock. Tectonic forces within Earth push up mountains. This explains why marine fossils can be found high on mountains. Mountains are especially common where converging tectonic plates collide. In mountainous regions, stream gradients are high, and the fast-moving streams quickly erode deep V-shaped valleys between the mountain peaks. In landscapes such as the Rockies, the Alps, Andes, and the Himalayas, much of the land is steep mountain slopes.

Plateaus A **plateau** landscape is relatively flat or rolling uplands. Streams often cut deep valleys into plateaus.. Flat layers of sedimentary rock usually form the base of plateaus. The Colorado Plateau near the Grand Canyon of northern Arizona is a good example. Plateaus usually have smaller topographic relief than mountains; however, they have more relief than plains.

Ridges and Valleys Areas classified as ridges and valleys have moderate relief, rounded peaks, and wide valleys. The bedrock under these areas is folded sedimentary rock.

Plains The least topographic relief appears in a plain. Although plains may contain a few small hills, they are generally flat and at a low elevation. Flat layers of sedimentary rock form the base of **plains**. Most of Florida and the agricultural areas of the Midwest exhibit plains topography.

Large areas, such as North America, and even smaller areas, such as New York State, can be divided into various landscape regions. Page 2 of the *Earth Science Reference Tables* shows the landscape regions of New York State, and Figure 5-11 on page 128 is a map of landscape regions of the continental United States. Each landscape region has its own characteristic hill slopes, drainage patterns, and soil associations.

The Influence of Climate

In moist climates, landscapes are generally rounded. Those in arid climates are character-

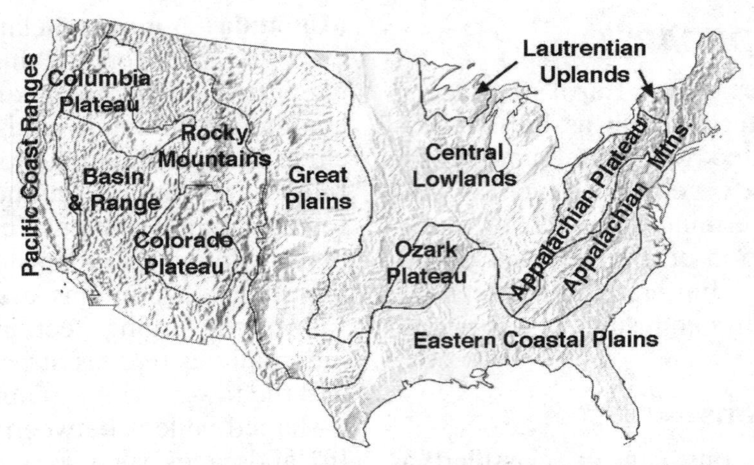

Figure 5-11. Major landscape regions of the contiguous Unites States.

ized by sharp angles and steeper slopes. This is shown in Figure 5-12. Moisture plays an important part in chemical weathering. Humid climates support the development of mature soils with a good balance of minerals and humus (organic remains of plants and animals). You might think that the streams in a moist climate would quickly erode these soils. However, this does not happen. Abundant rainfall throughout the year supports the growth of plant cover. Plant cover protects the soil from rapid runoff and erosion and produces the rounded slopes characteristic of humid climates.

Arid, or dry, climates generally produce thin soils with little humus. Physical weathering, particularly frost action in the colder months, is a dominant form of weathering. This kind of weathering tends to form soils with large and angular grains. There is little fine-grained material to fill the pores and slow the rapid and deep infiltration of precipitation. Only specialized desert plants can survive the long periods between rainstorms.

While rain is infrequent in desert regions, it often comes down very hard when it does rain. With little plant cover to protect the soil, a heavy rainfall quickly carries away sediment. Large areas of exposed bedrock and steep rock faces are the result. Most of the small streams remain dry, except immediately after the rare storms. You may be surprised to know that within the United States, the regions of most rapid stream erosion are the desert areas where rainfall is infrequent.

Thus, the amount and distribution of precipitation influence measurable landscape characteristics. These include the angle of hill slopes, the number of permanent streams, the vegetative cover, and the nature of the soil.

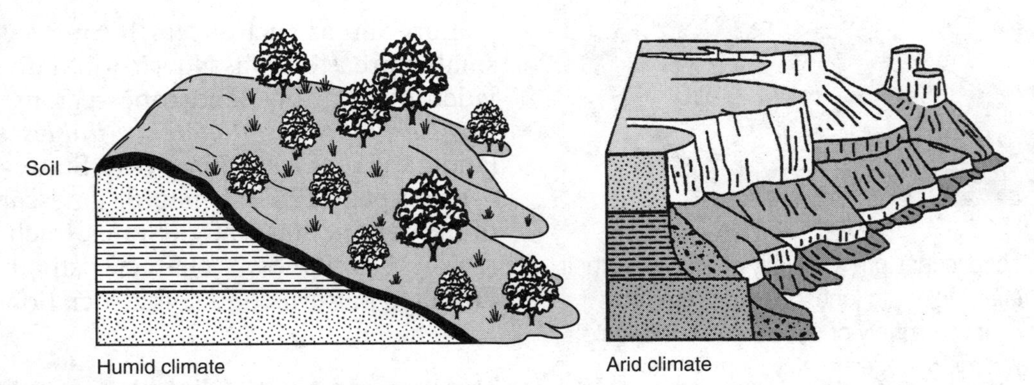

Figure 5-12. The general shape of a landscape is a result the local climate. Landscapes in moist regions are often rounded with enough soil cover to protect the underlying rocks. However, desert climates often have steep slopes of bare rock.

The interfaces (boundaries) between landscape regions are often easy to identify. East of Denver, Colorado, are the gently rolling Great Plains. The plains landscape ends suddenly just west of Denver, where the Rocky Mountains begin. This sudden change in landscape is a result of a change from the horizontal sedimentary rocks of the plains to the complex folds and faults of the mountain landscape. In New York State, the Catskill Escarpment is a steep slope that marks the eastern edge of the Appalachian plateau.

HOW DO GEOLOGIC FACTORS INFLUENCE THE LANDSCAPE?

Even in the same climate, locations with different rock types or different structures can develop very different landscape features. Rocks that are durable resist weathering and erosion. Durable, or competent, rocks form the higher portions of a landscape—the plateaus, mountains, and **escarpments**, or cliffs. The softer rocks and rocks that have been fractured (broken) usually lie beneath valleys and other low areas.

Softer rock erodes faster than harder, durable rock. If all the exposed bedrock on a hill slope has about the same degree of durability, the slopes will be worn down fairly evenly. If the layers of the hill slope differ in durability, the softer rocks will wear away faster, giving the hill slope an uneven or stepped appearance. The more durable layers will form the steepest slopes and cliffs.

Landscape Affects Drainage Patterns

Drainage patterns are determined by the way small streams, or *tributaries*, join to form larger streams. If two very different rock types occur in the same area, the harder rock will form the hills and ridges. Erosion of the weaker rock will form the major valleys. Therefore, streams tend to follow zones of weaker rock. If the bedrock is uniform or made of flat-lying layers, a branching drainage pattern forms, which is called **dendritic**. On a round volcano, a radial, like the spokes of a wheel, drainage pattern forms. Figure 5-13

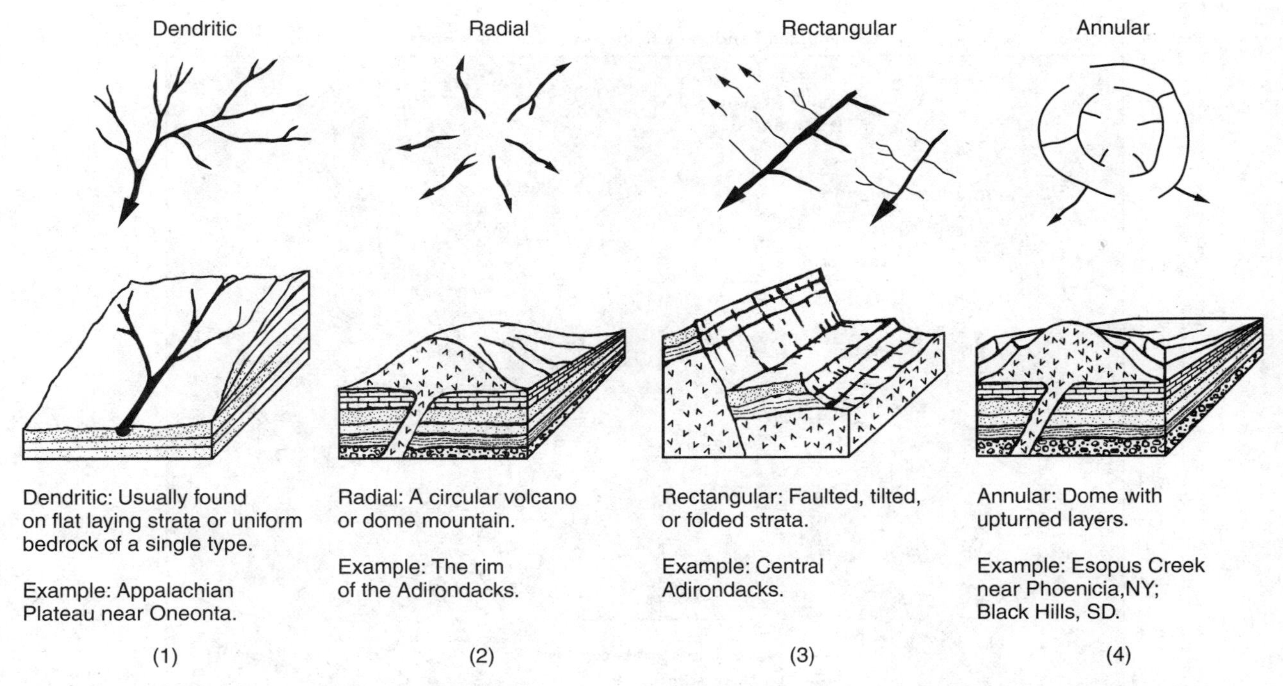

Figure 5-13. Stream drainage patterns are associated with geological structures. (1) On flat-lying layers of uniform composition without zones of weakness, a branching, dendritic pattern usually forms. (2) On a volcano or rounded hill (dome) with little difference in rock durability, a **radial** pattern forms. (3) Joints or faults produce a rectangular, or **trellis**, pattern. Rectangular drainage is often found in areas where rock layers with differing competence have been folded or tilted and eroded to form ridges. (4) If a dome has great differences in rock competence, an **annular** pattern of concentric circles is a likely result.

shows four common drainage patterns that develop on different bedrock structures.

Landscapes of New York State

New York State can be divided into several distinct landscape areas as shown in Figure 5-14.

The *St. Lawrence/Champlain Lowlands* is New York's northernmost region. It is a low-lying plain along the St. Lawrence River and Lake Champlain. This lowland region partially encircles the Adirondack Mountains. The bedrock is predominantly layers of sedimentary rock that gently slope away from the Adirondacks.

The *Adirondack Mountains* in the northeast are New York's only true mountain landscape. The ancient metamorphic and igneous rocks of this region were pushed up in the middle, forming a dome. Most of these rocks are very hard and resistant to erosion. Major valleys run northeast to southwest along joint and fault structures. New York's highest point, Mt. Marcy (1628 meters, or 5344 feet, above sea level), is within this region. The lower areas of these highlands are filled with lakes, ponds, and swamps. Many of these features were formed by glacial deposits.

The Appalachian Plateau (Uplands) is New York's largest landscape region. Within New York State, the Appalachian Plateau has been divided into the Catskills and the lower hills of the Allegheny Plateau to the west. The base of this region is mostly flat layers of sedimentary rocks. Hundreds of millions of years ago, these layers were laid down in a shallow, sinking ocean basin. The land was later pushed up more than 1000 meters to form a plateau. Many streams cut into the plateau.

The Catskills, near the eastern end of this region, rise to 1280 meters (4200 feet) above sea level. The *Finger Lakes*, near the center of this region, were formed when continental glacier moved into existing north-south valleys. The glacier widened and deepened the valleys, and reversed the direction in which they drained. As

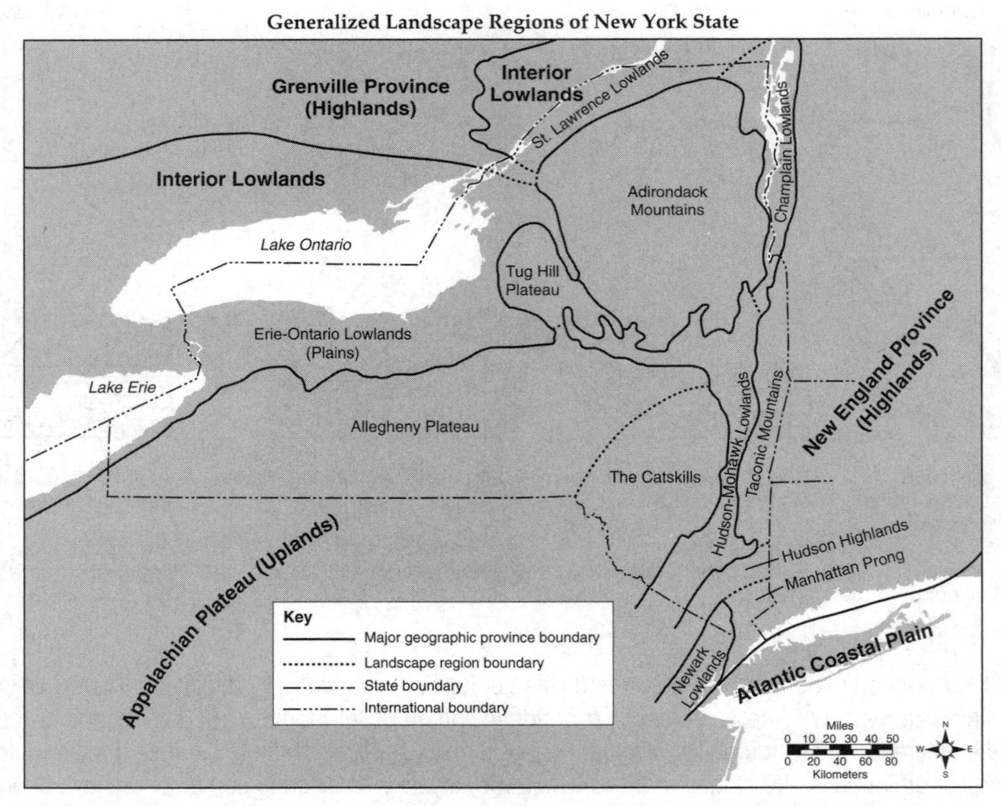

Figure 5-14. New York State has a variety of landscape regions including mountains, plateaus, and plains. (This map is on page 2 of the *Earth Science Reference Tables.*)

the most recent glacier melted back, it deposited till that blocked natural outlets to the south. Long north-south lakes formed in the blocked valleys.

The *Erie-Ontario Lowlands* lie south of two of the Great Lakes and north of the Appalachian Uplands. Although it is a plains landscape, these lowlands have many north-south oriented hills (drumlins) composed of unsorted glacial till. Layered sediments left by meltwater from the continental glaciers are also common. Good soils, which were deposited and mixed by glaciers, and a climate moderated by the Great Lakes make this an important agricultural area.

The *Tug Hill Plateau* is a small region of elevated sedimentary layers near the eastern end of Lake Ontario. The Tug Hill region is an isolated part of the Appalachian Plateau. Poor drainage and abundant winter snowfall make this one of the least-inhabited and least-used areas of the state.

The *Hudson-Mohawk Lowlands* (plains) follow a zone of easily eroded limestones and shales. The Mohawk River valley provided water-level access to the interior of America. This trade route was especially important in the development of New York in the 1700s and 1800s. Consequently, many towns and cities were established in this region.

The *New England Province (Highlands)* is a region of intensely folded and faulted metamorphic rocks. It includes the Taconic Mountains along the eastern boundary of the state. The last continental glacier covered the entire area. The passing of the glacier left rounded hills and a variety of glacial and meltwater deposits. In its southern portion, the highlands split into two prongs. One extends westward toward New Jersey as the Hudson Highlands. The other section extends to the south through Westchester and into the Bronx and Manhattan.

The *Newark Lowlands*, also known as the *Triassic Lowlands*, are a region of sandstones and shales deposited in a fault basin between the lower portions of the New England Highlands and the Hudson River. The region is geologically younger than the surrounding highlands. An intrusion of basaltic magma (sill) cooled, solidified, and later was eroded by the Hudson River. This still forms the high cliffs, called the Palisades, seen on the New Jersey side of the river.

The *Atlantic Coastal Plain*, made up of Staten Island and Long Island, is a coastal plain largely composed of glacial sediments. Along the north shore and the center of Long Island are two terminal moraines left by the last great glacier that covered the state during the Pleistocene Epoch. The land south of these moraines is composed of sorted material washed out of glaciers. This relatively flat area south of the terminal moraines is an **outwash plain**. It is made up of layers of sandy sediment deposited by streams flowing out of the glaciers. Landforms of glacial origin, such as *kettles* (where buried ice blocks may have melted) are common on Long Island.

Human Activities Affect Landscapes

Agricultural and construction projects can increase erosion and affect landscape development. In planning their projects, farmers and engineers must be guided by appropriate conservation practices. For example, contour plowing is practiced on farms that have hilly areas. Contour plowing curves around a hill, rather than up and down the hillside. The furrows, which run level across the slopes, slow water runoff and erosion. This practice helps keep the soil from being washed away. Strip mining removes layers of soil to dig up the minerals found in the bedrock. Rather than leaving a scar on the landscape when mining is finished, engineers can even out the land, replace the soil, and add plants to hold the soil in place. It is estimated that humans currently move more rock and soil than do all rivers combined.

QUESTIONS

Part A

32. In what New York State landscape region is 44°North latitude, 74°West longitude?
(1) the Catskills (2) the Adirondack Mountains (3) the Erie-Ontario Lowlands
(4) the Manhattan Prong

Base your answers to questions 33 through 36 on the diagram below, which shows four drainage patterns.

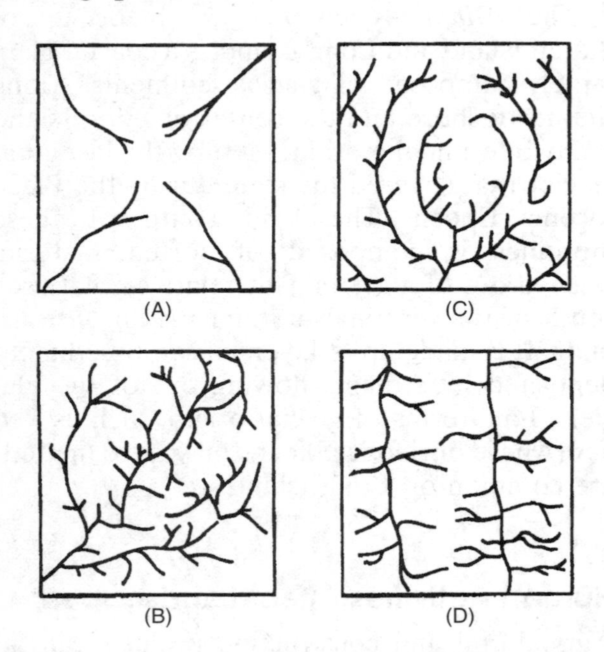

33. Which drainage pattern is likely to develop on fault-block mountains? (1) *A* (2) *B* (3) *C* (4) *D*

34. Which drainage pattern is likely to develop on bedrock of uniform resistance to erosion? (1) *A* (2) *B* (3) *C* (4) *D*

35. The most likely drainage pattern on the landform below is

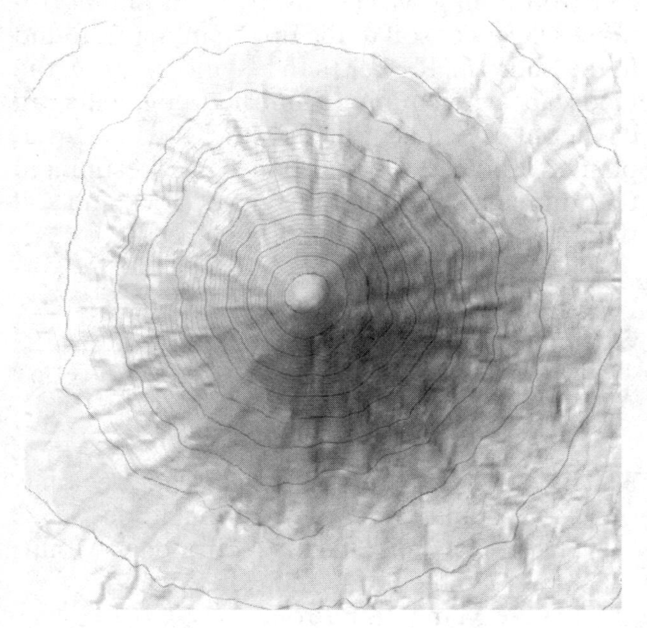

(1) *A* (2) *B* (3) *C* (4) *D*

36. Which of the drainage patterns above is likely to form on the landscape shown in the profile below?

(1) *A* (2) *B* (3) *C* (4) *D*

37. In which New York State landscape region is Niagara Falls located? (1) Erie-Ontario Lowlands (2) Allegheny Plateau (3) Adirondacks (4) Hudson-Mohawk Lowlands

38. What climate is likely in the landscape shown below?

(1) tundra (2) arid (3) tropical (4) moist

39. What landscape regions are found along the entire New York State–Canadian border? (1) lowlands (2) plateaus (3) mountains (4) highlands

40. Which region of New York State has the greatest topographic relief? (1) Hudson-Mohawk Lowlands (2) Catskills (3) Allegheny Plateau (4) Adirondacks

41. The photo below was taken from an airplane looking down on a meandering river, roads, and a forest.

This photo was most likely taken in a nature preserve in what part of New York State? (1) Catskills (2) Taconic Mountains (3) Hudson Highlands (4) Atlantic Coastal Plain

42. For regions that have the same bedrock types and structures, in which landscape has the processes of weathering and erosion most likely taken place for the longest time? (1) plains (2) plateaus (3) highlands (4) mountains

Part B

Base your answers to questions 43 through 45 on the diagram and table below. The diagram is a cross section of Earth's surface. The table describes the landscape regions indicated on the diagram.

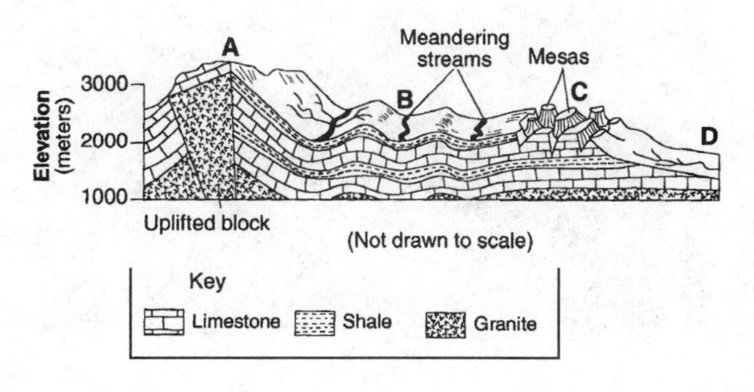

Landscape Region	Relief	Bedrock
A	Great relief, high peaks, deep valleys	Faulted and tilted structure; many bedrock types, including igneous
B	Moderate relief, rounded peaks, wide valleys	Folded sedimentary bedrock
C	Moderate to high relief	Horizontal sedimentary bedrock layers
D	Very little relief	Horizontal sedimentary bedrock layers

43. Which terms best describe the landscapes *A–D*? (1) *A* = mountains, *B* = ridges and valleys, *C* = plateau, *D* = plain (2) *A* = plateau, *B* = plain, *C* = mountains, *D* = ridges and valleys (3) *A* = plain, *B* = mountains, *C* = plateau, *D* = plain (4) *A* = ridges and valleys, *B* = plateau, *C* = plains, *D* = mountains

44. What landscape has the greatest variety of rock types and geologic structures? (1) plains (2) plateaus (3) mountains (4) lowlands

45. In what part of New York State was the photo below most likely taken?

(1) Adirondack Mountains (2) Allegheny Plateau (3) St. Lawrence Lowlands (4) Atlantic Coastal Plains

Problem Solving

Write a set of guidelines to help local residents and builders avoid environmental damage when they plan construction and landscaping projects. The guidelines should be in two parts. Part 1 is the specific guidelines; Part 2 is the justification, or reason, for each of the guidelines. Make your suggestions and justifications specific to the area in which you live. How do your guidelines compare with local building codes?

Chapter Review Questions

PART A

1. What characteristic of a stream is most important in determining how much erosion it produces? (1) water temperature (2) width (3) depth (4) gradient

2. Which part of New York State has the greatest topographic relief? (1) Long Island (2) Catskills (3) Adirondacks (4) Tug Hill Plateau

3. Why do some flat areas of New York State have soils that are very deep down to bedrock while other relatively flat areas have no soil at all? (1) There are great differences in climate within New York State. (2) Thick soils form over the hardest bedrock regions of New York State. (3) Glaciation left sediments thick in some places and thin in others. (4) Farming has caused extensive erosion of some thick soils.

4. What is the largest landscape region of New York State? (1) Allegheny Plateau (2) St. Lawrence Lowlands (3) Erie-Ontario Lowlands (4) Tug Hill Plateau

5. A continental ice sheet covered most of New York State 20,000 years ago. However, the ice is gone now. Why? (1) The ice moved back to the north. (2) The glaciers fell into the Atlantic Ocean. (3) Ice melts at a lower temperature now that it used to. (4) The glaciers melted back faster than the ice moved south.

6. Which stream pattern below would form in a region with low slopes and uniform bedrock?

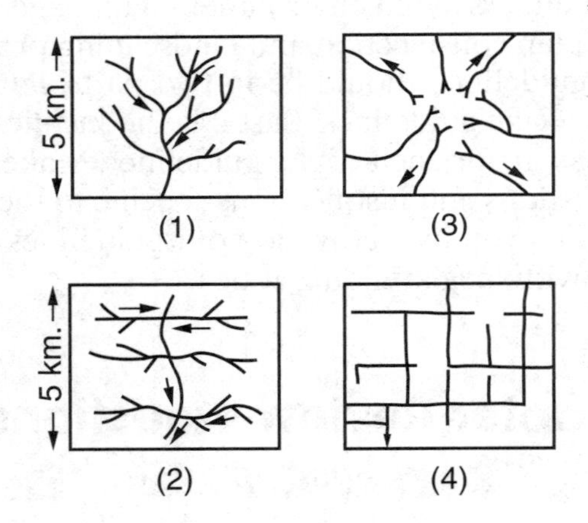

7. Central Park in New York City has many rocky outcrops that are rounded with grooves that run roughly north-south. What agent of erosion formed these features? (1) running water (2) wind (3) glaciers (4) ice carried in a river

8. In parts of New York State, there are many giant boulders that sit on ridges well above nearby stream valleys. How could the boulders have been moved there? (1) They were left by floods. (2) They were pushed by flowing ice. (3) They were moved by strong winds. (4) They bounced into place during earthquakes.

9. What features have the most influence on the pattern of streams that drain a landscape? (1) rock types, folds and faults (2) changes in climatic conditions (3) the growth of trees and shrubs (4) latitude and longitude

10. The photo below shows a part of the Rocky Mountains as seen in the winter from an airplane.

What evidence shows that this location was covered by major glaciers in the past? (1) There is snow in the mountains. (2) No trees grow on the highest mountaintops. (3) The valleys are broad and rounded. (4) Summers here are cooler than at lower elevations.

11. The diagram below shows the end of an advancing glacier.

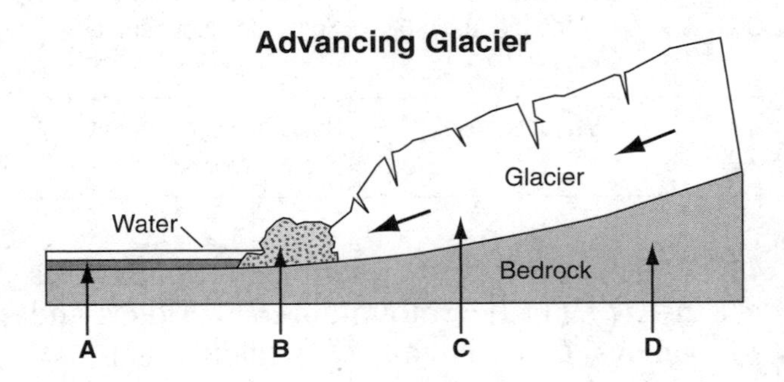

At which position would an observer be likely to find sand mixed with relatively few particles of silt and no pebbles, cobbles or boulders?
(1) *A* (2) *B* (3) *C* (4) *D*

12. Which feature of New York State was *not* a result of the continental glaciers that covered about half of North America? (1) The "fish-with a tail" shape of Long Island. (2) The roughly parallel north-south lakes of the Finger Lakes region. (3) Hills south of Lake Ontario that trail off gently to the south. (4) The height of Mount Marcy, New York State's highest point.

PART B

Base your answers to questions 13 through 15 on the diagram below, which represents an overhead view of concrete groins built along a beach on Long Island. The groins were built to make the beach wider along this section.

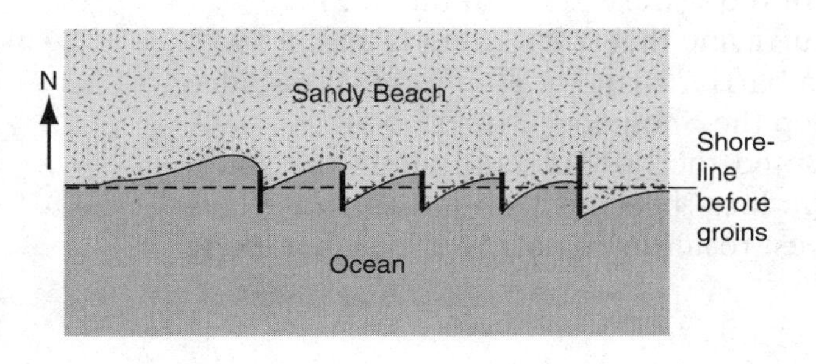

13. What processes have occurred along this shoreline since the groins were constructed? (1) only erosion (2) only deposition (3) erosion and deposition (4) neither erosion nor deposition

14. What is the primary direction of sand transport along this part of the shoreline?

(1) from the north (3) from the east
(2) from the south (4) from the west

15. The construction of these groins illustrates that (1) beach structures are always successful at making beaches wider (2) when we alter natural systems, unintended consequences may occur (3) natural systems are easy to change with the desired results (4) the sand on the beach stays where it is for a very long time

Base your answers to questions 16 through 19 on the map below. The map shows the major water bodies in the Finger Lakes Region of New York State.

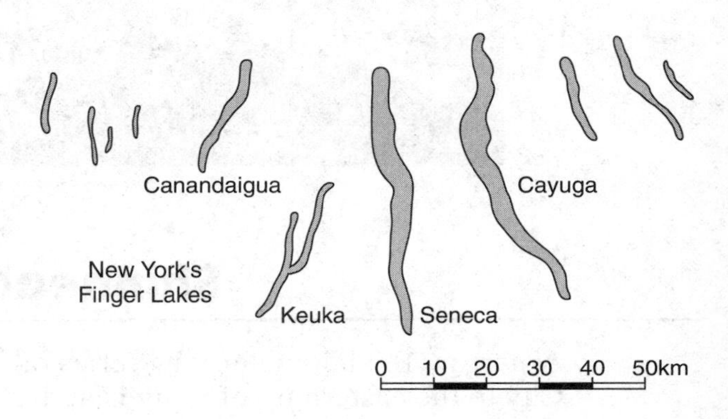

16. What agent of erosion and deposition is most responsible for the shapes of these lakes? (1) running water (2) ocean waves (3) wind (4) flowing ice

17. The agent of erosion that shaped these lakes came into New York State (1) from the north (2) from the south (3) from the east (4) from the west

18. What is the approximate length of Seneca Lake? (1) 12 km (2) 27 km (3) 38 km (4) 52 km

19. Why is Keuka Lake shaped like a *Y*? (1) Keuka Lake is aligned east-west. (2) Keuka Lake was not influenced by flowing ice. (3) Keuka Lake was created when the ice flowed back to the north. (4) Keuka Lake used to be two streams that joined to flow south.

Base your answers to questions 20 through 24 on the photo and text below. The photo shows a portion of the South Shore of Long Island.

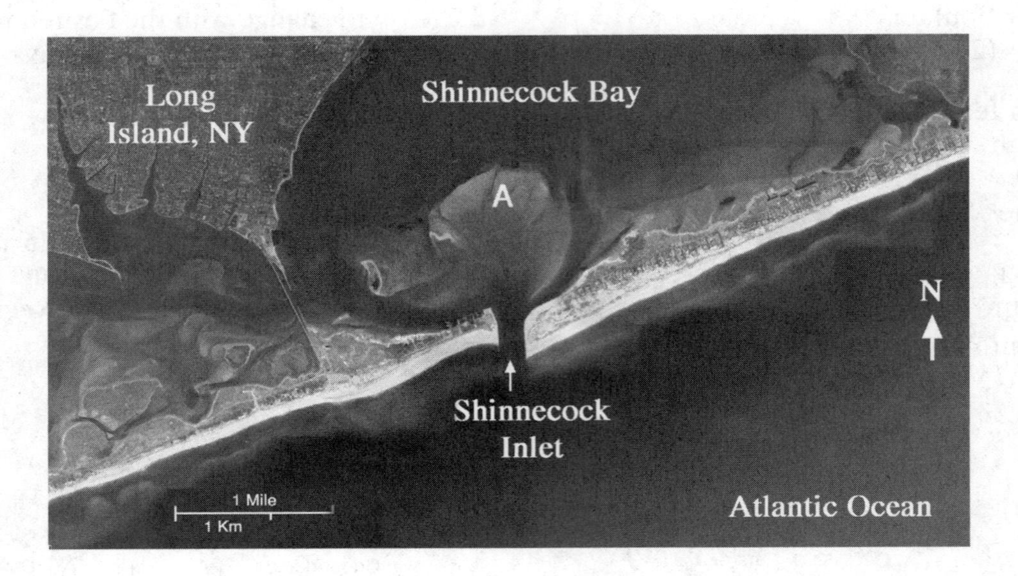

Shinnecock Inlet

A narrow 160-kilometer-long series of barrier islands stretches from New York City to the eastern tip of Long Island at Montauk Point. These sand islands protect the South Shore from the full force of storms. However, the barrier islands also prevent uninterrupted access to the Atlantic Ocean. The Shinnecock Inlet connects Shinnecock Bay and other sheltered waterways with the ocean.

The inlet was formed by the Great Hurricane that stuck Long Island in 1938. This storm permanently broke through a barrier island at Hampton Bays. Since then, dredging has been necessary to keep the Shinnecock Inlet open.

Maintenance of the inlet has interrupted the flow of sand along the south shore of Long Island. Efforts to preserve local beaches with jetties and groins have pushed erosion problems farther west resulting in narrower beaches along Long Island's south shore.

20. What process threatens to fill in Shinnecock Inlet? (1) weathering (2) erosion (3) deposition (4) recrystallization

21. Where is the source area of most of the sand along this part of Long Island? (1) to the northeast (2) to the northwest (3) to the southeast (4) to the southwest

22. For millions of years, what has caused the changing shorelines on this part of Long Island? (1) dredging and construction of shoreline structures (2) abrasion and chemical weathering (3) winds and ocean currents (4) map makers and historians

23. What would be the most likely result if no further construction or dredging was performed at the Shinnecock Inlet? (1) The inlet would become wider. (2) The inlet would fill in with sand. (3) The position of the inlet would migrate to the east. (4) The inlet would become deeper.

24. In what landscape province of New York State is the Shinnecock Inlet? (1) Atlantic Coastal Plain (2) Manhattan Prong (3) Erie-Ontario Lowlands (4) Grenville Province

Base your answers to questions 25 through 31 on the diagram below, which represents a landscape that has been changed by a continental glacier.

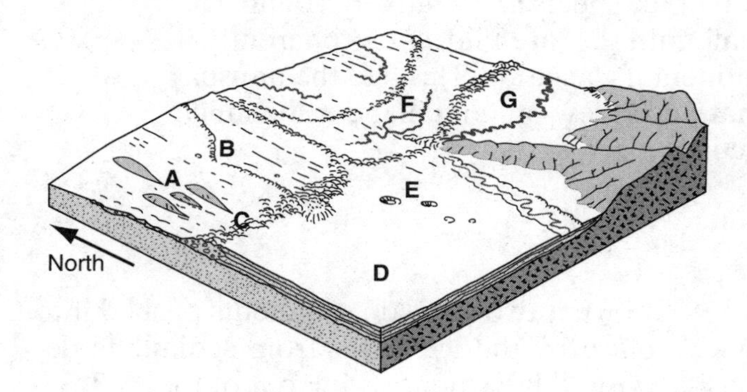

25. Sediments collected at location *D* are most likely (1) sorted by size and layered (2) sorted by size and not layered (3) unsorted by size and layered (4) unsorted by size and not layered

26. Which letter represents a terminal moraine indicating the farthest southern advance of the front of the glacier? (1) *A* (2) *C* (3) *F* (4) *E*

27. In some locations, ice blocks were buried within the sediments. These ice blocks melted to make landforms known as kettles. Which letters represent the most likely locations of kettles? (1) *A* and *B* (2) *C* and *D* (3) *E* and *F* (4) *A* and *G*

28. Where did the advancing ice move up and over the debris that it was pushing southward? (1) *A* (2) *D* (3) *F* (4) *G*

29. At the time these features were formed by an ice sheet, the glacier was most likely the thickest 100 miles to the (1) north (2) south (3) east (4) west

30. Scratched (striated) and polished bedrock is most likely found under the sediments at (1) *E* (2) *B* (3) *D* (4) *G*

31. A meltwater stream flowing in a tunnel under the glacier deposits a ridge called an esker. The esker often forms a small delta, called a kame, where it emerges from the ice. The diagram above shows an esker at (1) *A* (2) *B* (3) *F* (4) *G*

PART C

32. The glaciers that once covered New York State moved southward from Canada. What caused them to flow?

Base your answers to questions 33 through 38 on the diagram and passage that follow. The diagram shows features of the Greenland Ice Sheet; the passage describes it.

The Greenland Ice Sheet

At Last Glacial Maximum (LGM), the Laurentide ice sheet covered much of Canada and North America. An ice sheet is a massive glacier that covers an area of more than 50,000 square kilometers (20,000 square miles). The only current ice sheets are in Antarctica and Greenland.

The Greenland ice sheet covers 81 percent of the island and the mean altitude of the ice is more than 2000 meters (6500 feet). The weight of the massive Greenlandic ice cap has depressed the central land area, forming a basin that is more than 300 meters (1000 feet) below sea level. The ice generally flows to the coast from the center of the island.

Although the surface is cold, the base of an ice sheet is usually warmer due to heat from Earth's center. In places, melting occurs and the melt-water lubricates the ice sheet so that it flows more rapidly. This process produces fast-flowing channels within the ice sheet: these are ice streams.

In geological terms, the present-day polar ice sheets are relatively young. The Greenland ice sheet did not develop at all until the late Pliocene. Apparently it developed very rapidly with the first continental glaciation. This had the unusual effect of allowing the fossils of plants that once grew on Greenland to be much better preserved than in those in Antarctica.

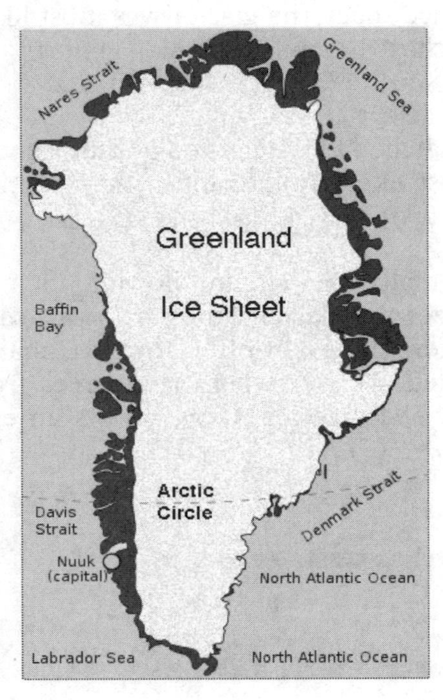

33. What force causes the Greenland ice sheet to flow toward the edges of the continent?

34. What is the ice thickness in central Greenland?

35. Why is Greenland cold enough to have an ice sheet, while North America is not?

36. How are the valleys where the Greenland ice sheet flows toward the sea different from the valleys in places that were never glaciated?

37. In places where the Greenland ice cap has melted back, exposed bedrock surfaces tend to display different characteristics from the bedrock surfaces in most other places. Name two ways in which these Greenland rock surfaces would appear different.

38. In what two ways do sediments pushed into place by the ice differ from sediments deposited by water flowing out of the ice?

39. The map below shows inferred ice depths at the maximum of the last major glacial episode in North America. This was the Wisconsin Era and the major glacier was the Laurentide Ice Sheet. Construct isolines of ice depth at an interval of 2 kilometers depth.

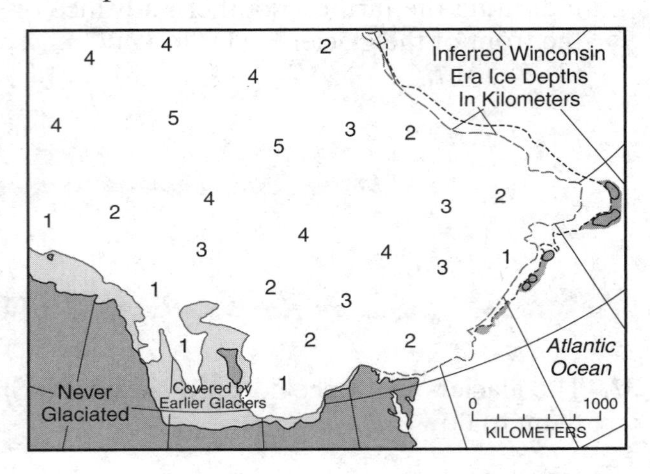

Base your answers to questions 40 through 42 on the photo below. The photographer was looking to the west when he took this image of a drumlin in western New York State.

40. If the photographer dug a deep hole in the drumlin, what would he notice about the degree of sorting?

41. How can you tell from this picture that the glacier advanced from the north?

42. What caused continental glaciers build up and advance southward into New York State?

Base your answers to questions 43 through 45 on the photo below, which shows two of the largest lakes in New York State. The camera was pointing generally toward the north.

43. What agent of erosion and deposition is most responsible for these lakes?

44. If you could drain the water from these lakes and clean out the sediments down to bedrock, what would be the east-west cross-sectional shapes of the valleys that the lakes now occupy?

45. These two lakes are primarily in what landscape region of New York State?

Base your answers to questions 46 through 48 on text and the image below. Two major hotels, Hotel A and Hotel B, were built along the beach as shown in the photo below. The owners of both hotels worried that the sandy beach was becoming narrower each year. As a result, the owner of one of the hotels considered building a groin out into the water to stop the sand flow at the boundary between the two hotels.

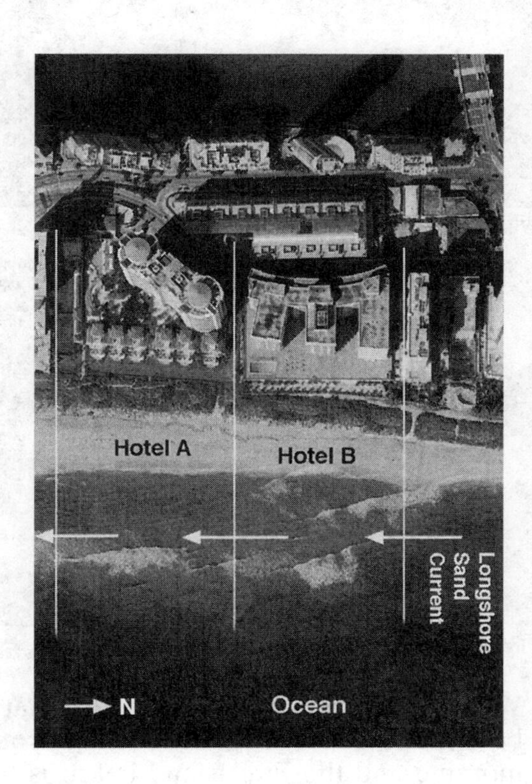

46. Why would the owner of Hotel *A* object to the groin at that location?

47. Where could a groin be constructed that might benefit both hotels, but not benefit some other nearby owners?

48. What is the source of energy that ultimately drives the longshore sand currents?

49. The image below shows one beach on the north shore of Long Island and another on the south shore.

Wildwood State Park, Long Island North Shore

Fire Island, Long Island South Shore

What observation supports the idea that the beach at Wildwood State Park is from a moraine and the Fire Island beach is from outwash?

50. Sketch the drainage pattern that would occur on the landform shown in the photo below.

CHAPTER

Interpreting Earth's History

Vocabulary at a Glance

absolute age	fold	orogeny
absolute time scale	geologic time scale	period
carbon-14	half-life	radioactive
correlation	index fossil	relative age
decay product	intrusion	relative time scale
decay-product ratio	isotope	superposition
era	law of superposition	tilted sedimentary layers
evolution	mass extinction	unconformity
extrusion	original horizontality	uniformitarianism

THE GEOLOGIST AS DETECTIVE

Earth formed about 4.6 billion years ago by the accumulation of rock, dust, and gases drawn together by gravity. The rocks of Earth's crust preserve clues that help us unravel the mystery of our changing planet, its environments, and the development of life on Earth. Geologists must be like detectives hunting for clues and evidence. The clues and evidence detectives find allow them to determine how a crime occurred. In a similar way, geologists locate, observe, and interpret the evidence and clues recorded in Earth's rocks. A primary role of geologists is to use the evidence to reconstruct a sequence of geologic events. There are several fundamental principles that guide geologists in interpreting Earth's history.

Uniformitarianism

Geologists infer that the processes they observe today are similar to processes that have occurred throughout Earth's history. However, these Earth-shaping processes do not necessarily occur at a steady, or uniform, rate. The idea that the processes that shape Earth's surface today, such as volcanoes, earthquakes, deposition, and erosion, are the same processes that occurred in the geologic past is called **uniformitarianism**. Another way to state the principle of uniformitarianism is "The present is the key to the past."

Superposition

The second fundamental principle is the **law of superposition**. This states that the rocks at the bottom of an undisturbed exposure are usually the oldest. After all, the lower layers must be in place before the younger layers can be deposited on top of them. Therefore, geologists generally assign relative ages to rock layers starting from the bottom (oldest) and moving to the top (youngest). There are occasional exceptions to the law of superposition, such as the effects of folding and faulting, which will be explained in the sections that follow.

Original Horizontality

The third principle is **original horizontality** and it states that sediments, such as those at the bottom of an ocean or lake, are usually deposited in flat layers. Even if the basin in which the sediments were deposited was not flat, the layers of sediment usually are laid down flat. When geologists see **tilted sedimentary layers**, they usually infer that these layers were deposited horizontally, and that the layers were tilted after they had turned into sedimentary rock. Figure 6-1 shows a place where rock layers that were originally horizontal have been tilted and folded by movements of Earth's crust.

Igneous Intrusions, Extrusions, and Inclusions

The fourth fundamental principle is that a unit of rock is always older than the processes that changed it. For example, when molten magma comes into contact with older rocks, heat changes the older rocks. Clearly, the changed rocks must have been there before they were changed.

An **extrusion** occurs when molten rock flows onto Earth's surface. There it cools and crystallizes to form igneous rock. An extrusion is therefore younger than the rock below it. The rock beneath an extrusion will include a zone of *contact metamorphism* where the hot lava baked the older rock. Rock that formed above an extrusion after it cooled will not show contact metamorphism. Sedimentary rocks above an igneous layer were probably deposited after the lava cooled and solidified.

Figure 6-1. These rocks were laid down originally as horizontal layers. Later, forces within Earth tilted and bent the layers.

Figure 6-2. A dark, basaltic magma intruded into a crack in the light-colored granite, and turned to solid rock. Erosion has exposed the evidence of this event in an outcropping.

An **intrusion** occurs when magma squeezes into a pre-existing rock. Figure 6-2 shows an intrusion of basaltic magma into granite. When an intrusion occurs, hot, molten rock changes the surrounding rock directly above, below, and next to it by contact metamorphism. Therefore, an intrusion can be identified by the narrow zones of metamorphism that surrounds it on all sides. Figure 6-3 illustrates a sequence of events. Magma intruded into layers of sedimentary rock. The magma also extruded at the surface. Notice that a later layer of sedimentary rock on top does not have a zone of contact metamorphism.

Sometimes molten magma surrounds pieces of older rock. The magma does not melt the older rock fragments. These fragments of older rock (inclusions) must be older than the igneous rock in which they are found. Figure 6-4 shows an inclusion of relatively old, layered, metamorphic rock that was moved and surrounded by magma before the magma became solid granite.

Crosscutting Relationships

Intrusions of magma are always younger than the rocks they invade. In Figure 6-5 on page 144, the granite intrusion must be younger than the layers of sedimentary rock it has invaded. However, the gabbro intrusion is younger than the granite because it cuts through the granite intrusion. The basalt intrusion is the most recent

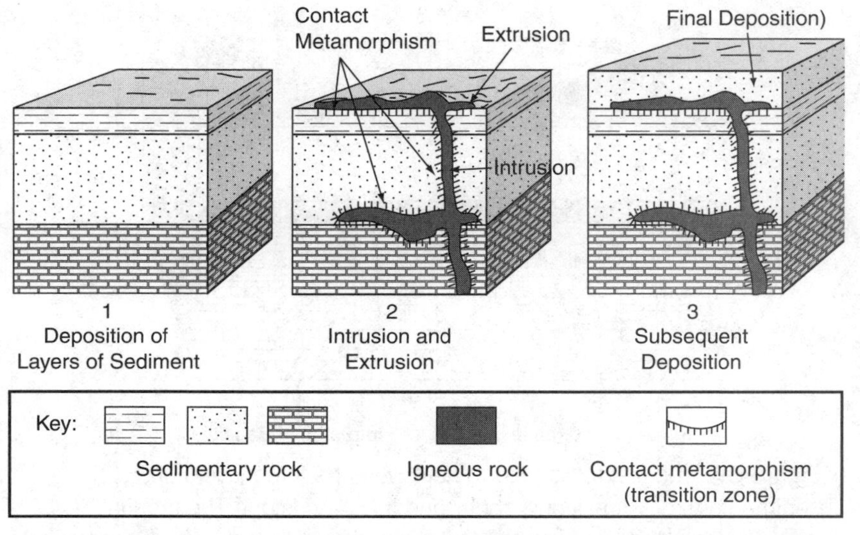

Figure 6-3. This series of three diagrams illustrates a sequence of events. Part 1 shows three layers of sedimentary rock produced by deposition and compaction. Part 2 adds a magma intrusion (underground) as well as an extrusion (a lava flow on the surface). Note the baked zone of contact metamorphism that surrounds the intrusion and under the lava flow. In part 3, the top sandstone does not show contact metamorphism because it was deposited after the lava cooled.

(youngest of all) because it extends through all the other rock units, including the sedimentary rocks and both other intrusions.

Folds and Faults

Tectonic processes within our restless planet can warp or distort rock layers. **Folds** are bends in rock layers produced by crustal plate movements. *Faults*, as you may recall from Chapter 3, are breaks in the rock where movement has occurred. Faults are often associated with earthquakes. Offset layers, such as those shown in Figure 6-6, indicate faulting. By the principle of original horizontality, folds and faults occur after the rock has formed.

Figure 6-4. This boulder is mostly dark-colored basalt. The light-colored part in the center is a piece of solid rock that was brought to the surface within the basaltic magma. Therefore, the white inclusion is older than the basalt around it.

Folding or faulting can lead to exceptions to the law of superposition. As shown in Figure 6-6 on page 144, older rocks can be found on top of younger rocks in places where folding or faulting has occurred. Note that layer 1 is the oldest, layer 2 formed next, and layer 3 is the youngest. In general, layer 3 is above layers 1 and 2, but not always.

Fossils

Fossils are the preserved remains or traces of ancient life. Fossils tell scientists a great deal about the life-forms and environments that existed in Earth's past. Fossils can also provide geologists with clues about the past geological events and **relative age** of rock layers. Relative age tells whether an event occurred before or after other events without specifying the actual age in units,

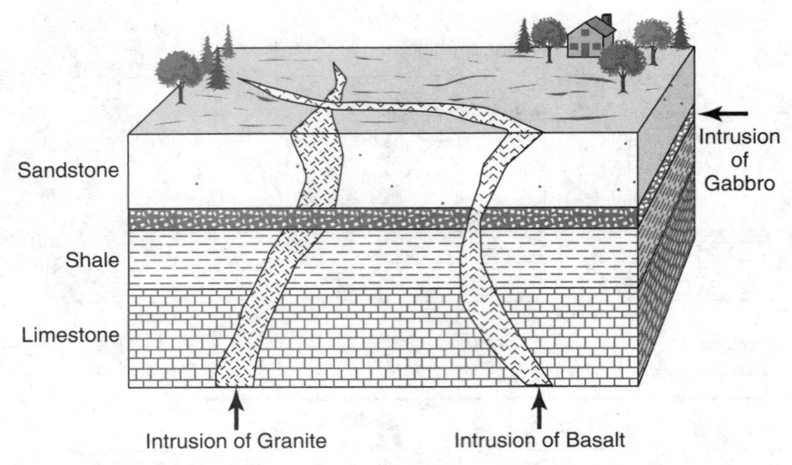

Figure 6-5. The first rocks to form here were the three layers of sedimentary rock: limestone, shale, and then sandstone. The granite intrusion was next because it cuts through the sedimentary rocks, but not through the other two intrusions. The gabbro must have been the next intrusion. Finally, the basalt cuts through the sedimentary rocks and both earlier intrusions.

such as years. The processes of metamorphism (extreme heat and/or pressure) and melting (to form igneous rocks) usually destroy fossils. Therefore, with few exceptions, fossils are found in sedimentary rock. Most fossils do not look like modern life-forms. Most of the organisms (living things) that turned into fossils became extinct millions of years ago. A good example is the Eurypterid, which is the official fossil of New York State. (See Figure 6-7.)

There are many types of fossils. Body fossils are parts of an organism that have been preserved. Fossils of hard parts such as shells, teeth, and bones are common. These body parts decay relatively slowly and therefore are more likely to be preserved than flesh and other soft parts. In

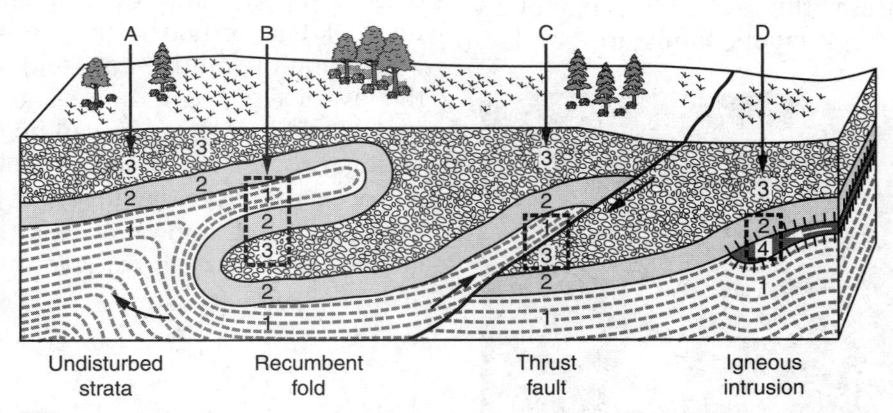

Figure 6-6. In some situations, the law of superposition does not apply. Section A shows three layers in their original sequence with the oldest layer (1) on the bottom. The tight fold in section B has created a reverse order of layers within the dashed box where the oldest layer is actually on top. (The sequences above and below the box are normal.) Section C shows a thrust fault that has pushed an older layer (1) above a younger layer (3). In the final section, D, an intrusion of magma (4) has caused a younger igneous rock to form below layer 2, which is older.

Figure 6-7. The long-extinct Eurypterid, sometimes called a "sea scorpion" is the state fossil of New York..

fact, some limestone layers are composed almost entirely of fossil shells. Sometimes, minerals in groundwater replace the original compounds that made up the preserved body part. The remains are then said to be petrified. Figure 6-8 shows ancient logs in the Petrified Forest of Arizona. The wood in these logs has been completely replaced by colorful quartz minerals.

Occasionally, whole animals are preserved as fossils. These fossils include insects encased in amber (fossilized tree sap) and mammoths (ice age elephants) frozen in Arctic tundra. The La Brea tar pits of Los Angeles preserved the complete skeletons of many animals unfortunate enough to have wandered into these death traps.

Some fossils do not include any actual remains of the organisms that produced them. These trace fossils include the impressions of shells, dinosaur footprints, oddly shaped formations resulting from sediments filling an ani-

Figure 6-8. The Petrified Forest National Park in Arizona has thousands of fossil trees. After the tree trunks were buried, minerals (mostly quartz) gradually replaced the wood. This process often preserves such features as surface texture and growth rings.

mal's burrow, and even petrified animal droppings. Usually, trace fossils do not reveal what a prehistoric organism looked like. However, they can reveal much about its behavior and relationship to its living and nonliving environment.

Fossils can provide important clues about environmental conditions in the geologic past. For example, New York State has fossils of coral and warm-water marine organisms, indicating that the area was once a tropical sea. It also has more recent fossils of wooly mammoths and mastodonts, revealing the cold, snowy climate of the last ice age. These fossils and others found in New York State are illustrated on pages 8 and 9 of the *Earth Science Reference Tables*, which you will find following page 336 of this book.

ESTABLISHING A GEOLOGICAL SEQUENCE

No single location shows a complete record of the geologic past. If an area was above sea level for a time, it is likely that sediments were not deposited and that older sediments or rocks were eroded. Thus, erosion causes gaps in the geologic record. A new layer of rock laid down on a surface left by erosion forms a buried erosion surface, or **unconformity**.

When a rock outcrop shows an unconformity, it indicates that the area, at some time in the past, was uplifted above water level and then eroded. Later the area subsided below water level and new layers of sediment were deposited on top of the eroded surface. The gap in the geologic record is sometimes indicated by an uneven interface or by gaps in the fossil record in the strata. Figure 6-9 shows an unconformity in southern Utah.

Many rock outcrops show evidence of many geologic events. The original rock layers were created by deposition of sediments or the solidification of molten magma, or lava. Folding may bend the layers, and faulting may cause them to be offset. Intrusion of magma can result in veins of igneous rock and metamorphism. In turn, metamorphism can form new minerals and distort or destroy layering and other structures. *Weathering* and erosion can destroy rock strata and expose underlying layers. Figure 6-10 shows

Figure 6-9. This unconformity (buried erosion surface) shows folded layers of sedimentary rock that were eroded to a flat surface. Then, the thick layer of conglomerate was deposited on top of the unconformity. The sequence of events revealed here is deposition, folding, and erosion, followed by new deposition.

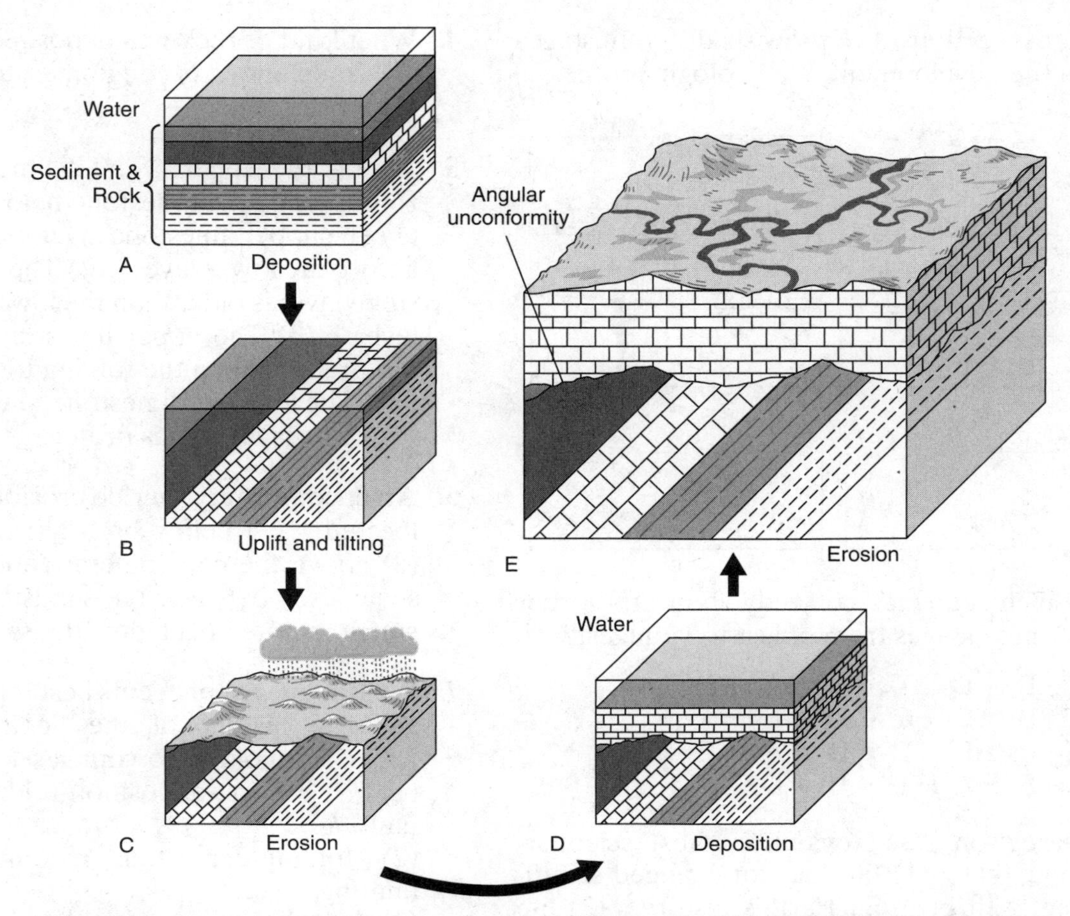

Figure 6-10. This figure shows a sequence of events that produced a final geologic profile. Block A shows deposition under water. Block B adds uplift and tilting. Next, block C shows erosion at Earth's surface. Block D adds a new phase of deposition, which is followed by more erosion in E. However, this whole history can be inferred by examination of just the block E.

a sequence of events that result in a final geological profile.

A geologist can study a rock outcrop to determine its geologic history, that is, the sequence of events that made the rock outcrop what it is today. If fossils are present, they will help the geologist to date the layers of rock and the events that occurred in them.

QUESTIONS

Part A

1. Bedrock outcrops *A* and *B* are located along the Genesee River in western New York State. Rock layers 1, 2, and 4 are the same at both outcrops.

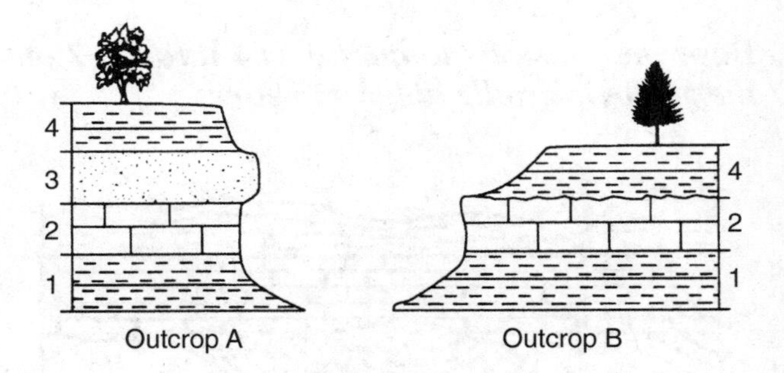

Why is layer 3 missing in outcrop *B*? (1) There is a fault between outcrops *A* and *B*. (2) Erosion occurred at location *B* after the deposition of layer 3. (3) Rock layer 3 was destroyed by a volcanic eruption near location *B*. (4) Layer 3 at outcrop *A* is a metamorphic rock.

2. Cross sections *A–F* show six different stages in the development of a geologic profile.

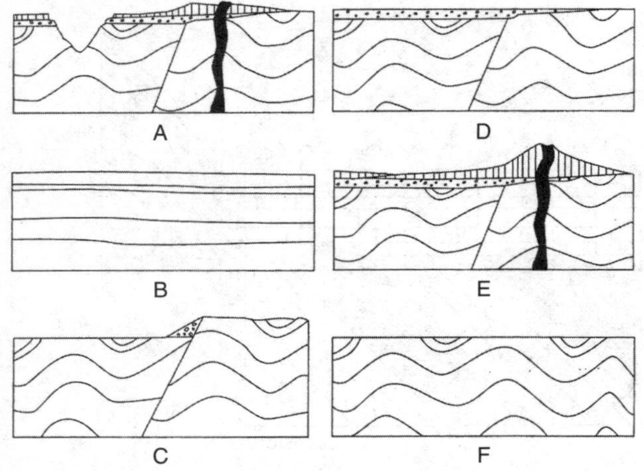

Which sequence correctly shows the order of these stages from oldest to youngest?

(1) B → D → C → F → A → E
(2) B → F → C → D → E → A
(3) B → F → C → D → A → E
(4) B → A → F → E → D → C

3. Based on fossil evidence, most scientists infer that (1) life has not changed significantly throughout Earth's history (2) life has evolved from complex to simple forms (3) many organisms that lived on Earth have become extinct (4) mammals developed before any other organisms

Part B

Base your answers to questions 4 through 7 on the geologic profile diagram below.

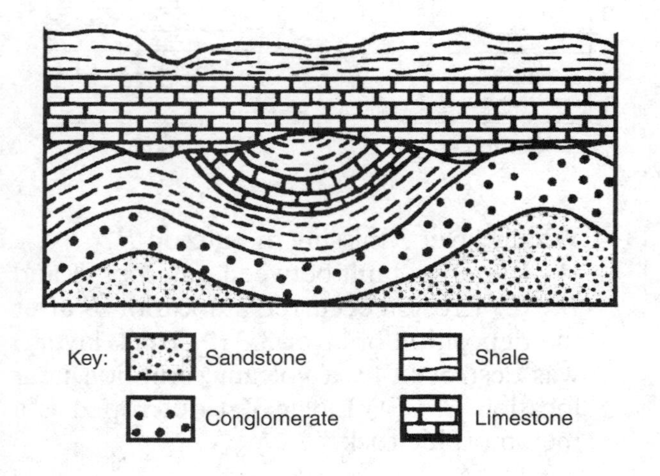

4. What kind of rock was deposited first? (1) sandstone (2) conglomerate (3) shale (4) limestone

5. Why is the upper layer of limestone horizontal, while the limestone below is folded? (1) The upper limestone layer was deposited before the lower layer. (2) The upper limestone layer is older than the lower limestone layer. (3) The upper limestone layer was not present when the folding took place. (4) The upper limestone layer is the youngest layer in this profile.

6. An unconformity (buried erosion surface) is located (1) below both limestone layers (2) above the conglomerate and the lower shale (3) between the sandstone and the conglomerate (4) below the sandstone

7. What sequence of events best represents the geologic history of the geologic profile above from oldest to youngest?
(1) erosion-deposition-folding-erosion-faulting
(2) folding-deposition-erosion-deposition-faulting
(3) deposition-erosion-folding-erosion-deposition
(4) deposition-folding-erosion-deposition-erosion

Base your answers to questions 8 through 11 on the geologic cross section below.

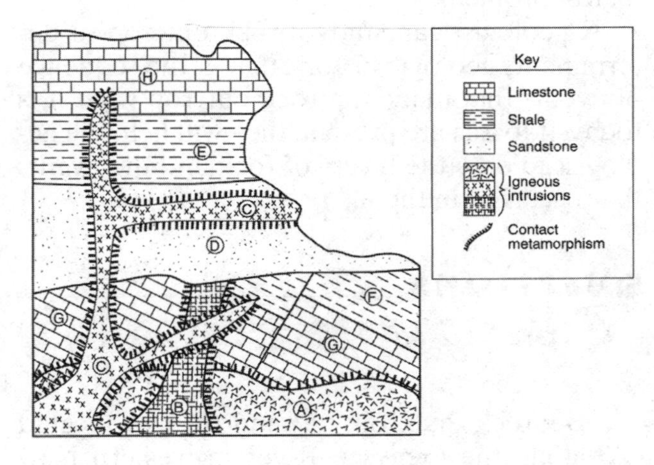

8. Which statement is most accurate about the bedrock in this outcrop? (1) Intrusion *C* is

younger than intrusion *B*. (2) Intrusion *A* is younger than intrusion C. (3) Intrusion *B* is older than intrusion *A*. (4) Intrusion *C* is older than layer *E*.

9. What event occurred after the intrusion of *B*, but before the deposition of layer *D*? (1) tilting of *G* (2) erosion of *B* (3) intrusion of *C* (4) deposition of *F*

10. Which rock type seems to be the softest and most easily eroded? (1) limestone (2) shale (3) sandstone (4) igneous rock

11. What kind of rock can be found at the boundary between *C* and *H*? (1) granite (2) sandstone (3) gypsum (4) marble

Base your answers to questions 12 through 16 on the geologic cross section below, which shows rock units* A *through* G. Q *is a fault.* W, X, Y, *and* Z *are unconformities.

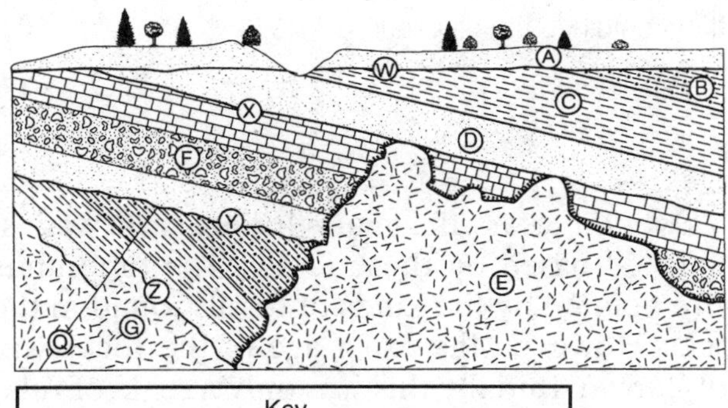

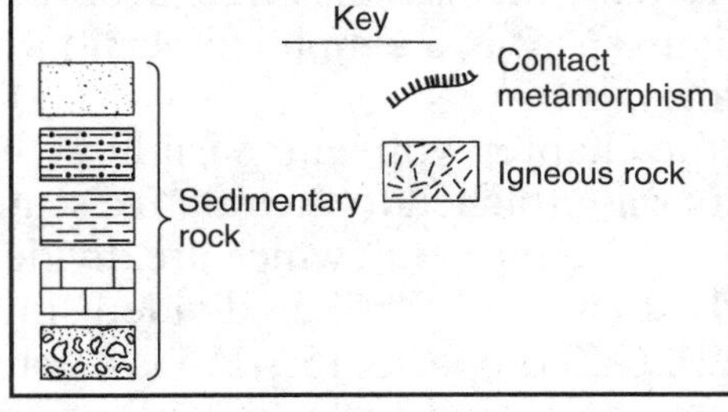

12. Which is the oldest rock or feature in the profile above? (1) rock layer *A* (2) rock layer *G* (3) fault *Q* (4) unconformity *Z*

13. The wavy line labeled *Y* shows that (1) part of the geologic record is missing (2) an earthquake took place in the past (3) folding took place after layer *F* was deposited (4) the rocks below are all metamorphic rocks

14. Rock unit *E* is granite. What rock type most probably is found at the boundary between rock units *E* and *F*? (1) shale (2) metaconglomerate (3) obsidian (4) gabbro

15. Rock units *A* and *D* are (1) limestone (2) breccia (3) rock salt (4) sandstone

16. What is the last event for which there is evidence in this bedrock profile? (1) faulting (2) folding (3) deposition (4) erosion

HOW CAN ROCKS BE CORRELATED?

Geologists try to match similar rock strata in different locations to infer if they formed at the same time or under similar conditions. This is **correlation**. There are several ways to correlate rock formations.

Scientists can match the rock strata in one location with the strata in different location by comparing the properties of the rocks, such as color, texture, or composition. However, the same rock types are not necessarily the same age. Using the sequence of layers, such as limestone, sandstone, and then shale, is better than matching by looking at a single layer. However, even if the sequence of rock layers in two locations is the same, the layers may not have formed at the same time.

Time correlation requires other methods. One way to do this is to compare index fossils contained in the strata. The best **index fossils** are the remains of organisms that existed for a relatively brief geologic time but can be found over a wide area. Thus, in a rock outcrop, an index fossil would *not* be found in many layers vertically, but would be widespread horizontally from one location to another.

It is likely that, millions of years from now, humans will be an excellent index fossil. Humans have existed for only about 2 million years, yet our remains and signs of our existence can be found worldwide. Figure 6-11 on page 150 shows some of the techniques and limitations of geological correlation.

If a layer does not contain a specific index fossil, it can sometimes be tentatively correlated by the kinds of fossils it contains. For example, a layer of rock that contains dinosaur fossils could be correlated with another layer that contains

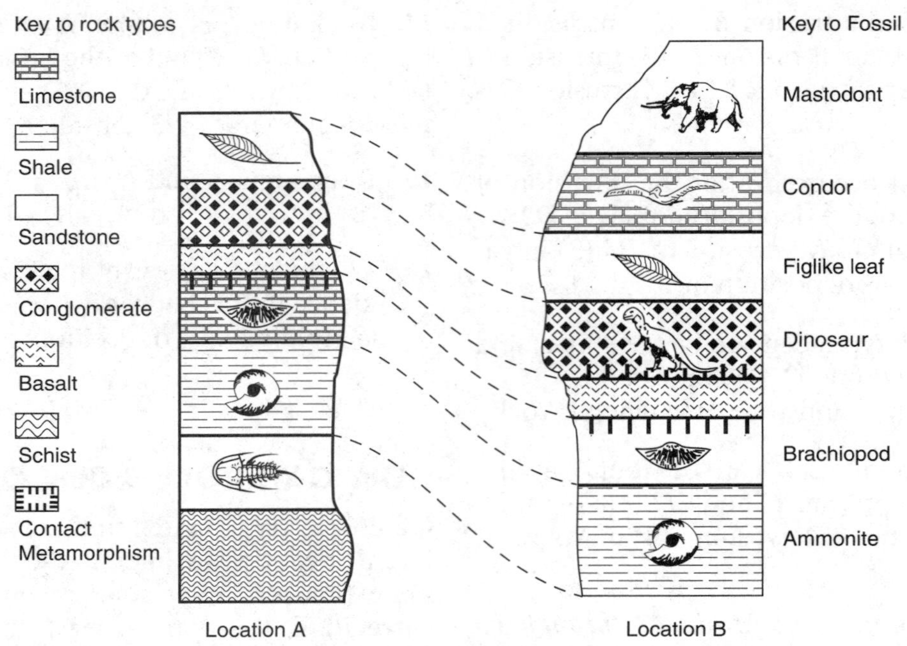

Figure 6-11. Rock layers can be correlated by rock type or by geologic age. Fossil correlation is used to match rocks by age, regardless of rock type. In the diagram, although both locations have sandstone at the top, the fossils show that the top layers cannot be the same age. The absence of dinosaurs at location A may indicate that dinosaurs did not live in that area, that the dinosaurs did not have any recognizable fossils, or the layer that contained the dinosaur fossils was eroded.

fossils of marine organisms that existed at the same time as the dinosaurs.

GEOLOGIC TIME SCALE

In the late 18th and early 19th centuries, geologists in Europe noticed that rock formations could often be identified by the fossils they contained. The geologists also found that certain formations were consistently located above or below other formations. Eventually, a worldwide pattern of a progress of life-forms became evident. From these observations of a sequence of fossil groups from oldest to youngest, the scientists established a **relative time scale**. A relative time scale indicates whether the object in question is older or younger than something else. Each of these groups of fossils was named for a location where its characteristic fossils could be easily observed in the rocks. For example, fossils characteristic of the Devon region, in the south of England, were named Devonian.

Over the years, the **geologic time scale** was established based on these fossils. The time units were defined by the kinds of fossils found in the rocks. However, the exact age of the rocks was not known. Initially this was only a scale of relative time. Figure 6-12 is a simplified version of this time scale.

Later in the chapter, you will learn how the scale of years was attached to the relative scale. The scale is divided into **eras**, which are divided into **periods** that are further divided into epochs. Figure 6-13 on pages 152–153 is a comprehensive geologic time scale that is specific to New York State. This table is also printed in the center of the *Earth Science Reference Tables*. You should use the geologic time scale when studying major events in New York State's geologic history.

Paleozoic means "ancient life," Mesozoic means "middle life," and Cenozoic means "recent life." The Precambrian was the dawn of life. For the most part, Precambrian organisms had no hard parts that were easily fossilized. As a re-

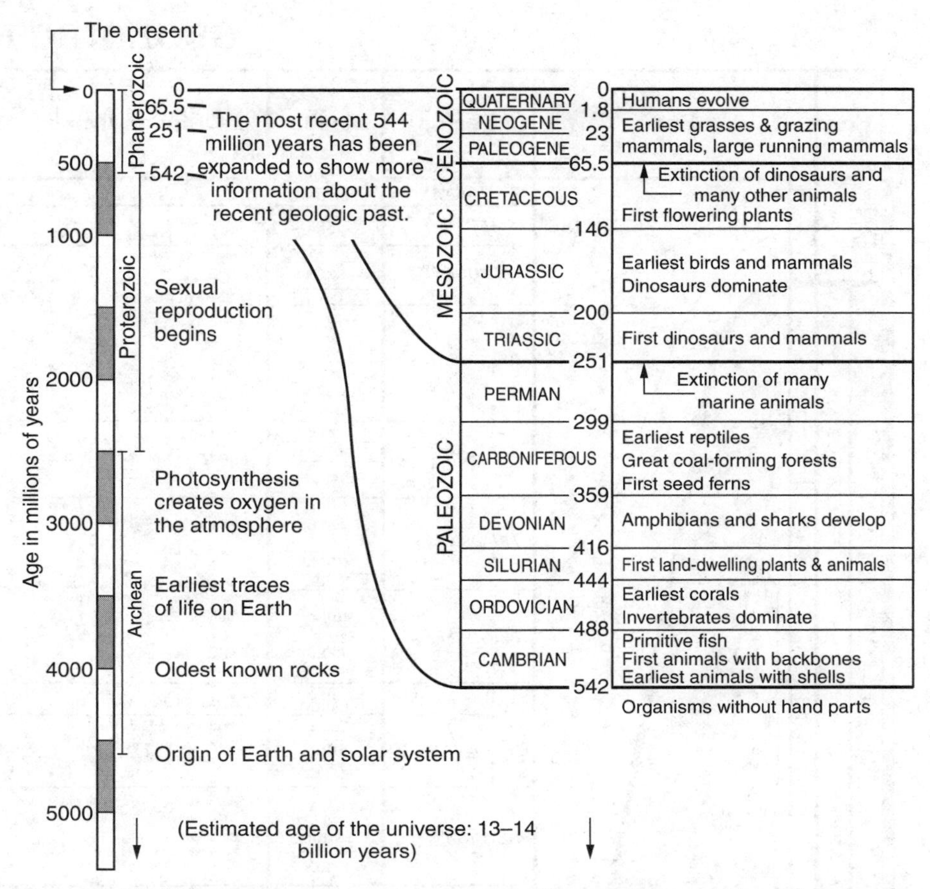

Figure 6-12. The geologic time scale was originally a relative scale based on fossils. However, by using radioactive materials, scientists made it an absolute (numerical) scale. Note that the top 544 million years have been greatly expanded on the right.

sult, there are few Precambrian fossils. At the beginning of the Paleozoic Era during the Cambrian Period, there was "explosion of life." At this time, a wide variety of animals with hard parts developed. These animals included corals, shellfish, trilobites, and fishes. Reptiles, including land-dwelling dinosaurs and flying pterosaurs, dominated the Mesozoic Era. We live at the end of the Cenozoic Era, which is sometimes called the Age of Mammals.

Note in Figure 6-12 and Figure 6-13 on pages 152–153, which is in the *Earth Science Reference Tables*, that the most recent portion of the geologic time scale has been expanded on the right side of the diagram. Some terms on the reference table time scale may not be familiar to you. **Orogeny** is the process of mountain building. The center of the diagram shows the four major orogenies in New York's geologic history. For example, the Taconic Orogeny occurred during the Ordovician Period.

Geologic Maps

Geologists use the techniques explained in this chapter to study the local bedrock, to relate it to other regions, and to construct geologic maps. The Generalized Bedrock Geology of New York State (Figure 6-14 on page 154) and the Generalized Landscape Regions of New York State (Figure 5-14 on page 130) are in this book and in the *Earth Science Reference Tables*. Use these maps along with the geologic time scale to determine the age of bedrock throughout New York State. They also provide general information about local rock types, the age of the bedrock, types of fossils likely to be found in the rocks, and the geologic history of your area. More detailed maps as well as textbooks and field guides are available from the United States Geological Survey, the New York State Geologic Survey, universities, and other professional organizations. These resources can help you to understand the unique geologic setting in which you live.

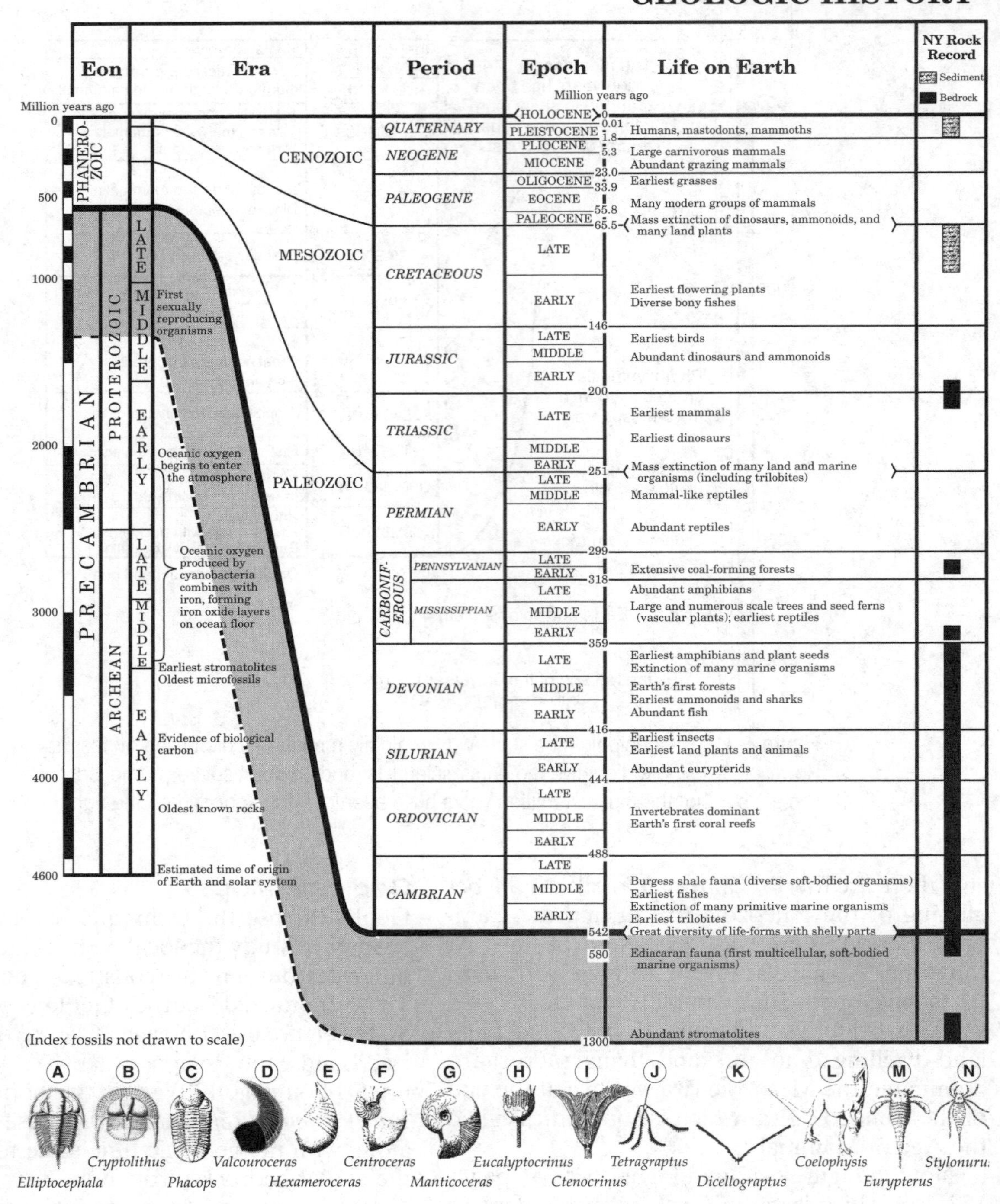

Figure 6-13.

EVOLUTION OF LIFE

Early geologists noticed that the oldest rocks found in Europe did not seem to contain fossils. Those scientists inferred that life began with the life-forms whose fossils characterized the early Cambrian Period, about 540 million years ago.

Scientists are not sure how life began on Earth. They know that life on Earth began before the Cambrian Period, perhaps as long as 4 billion years ago in the Precambrian. Fossils of simple marine organisms, such as algae that left calcite reefs, have been found in rocks more than 3 billion years old. Geologists believe that many

OF NEW YORK STATE

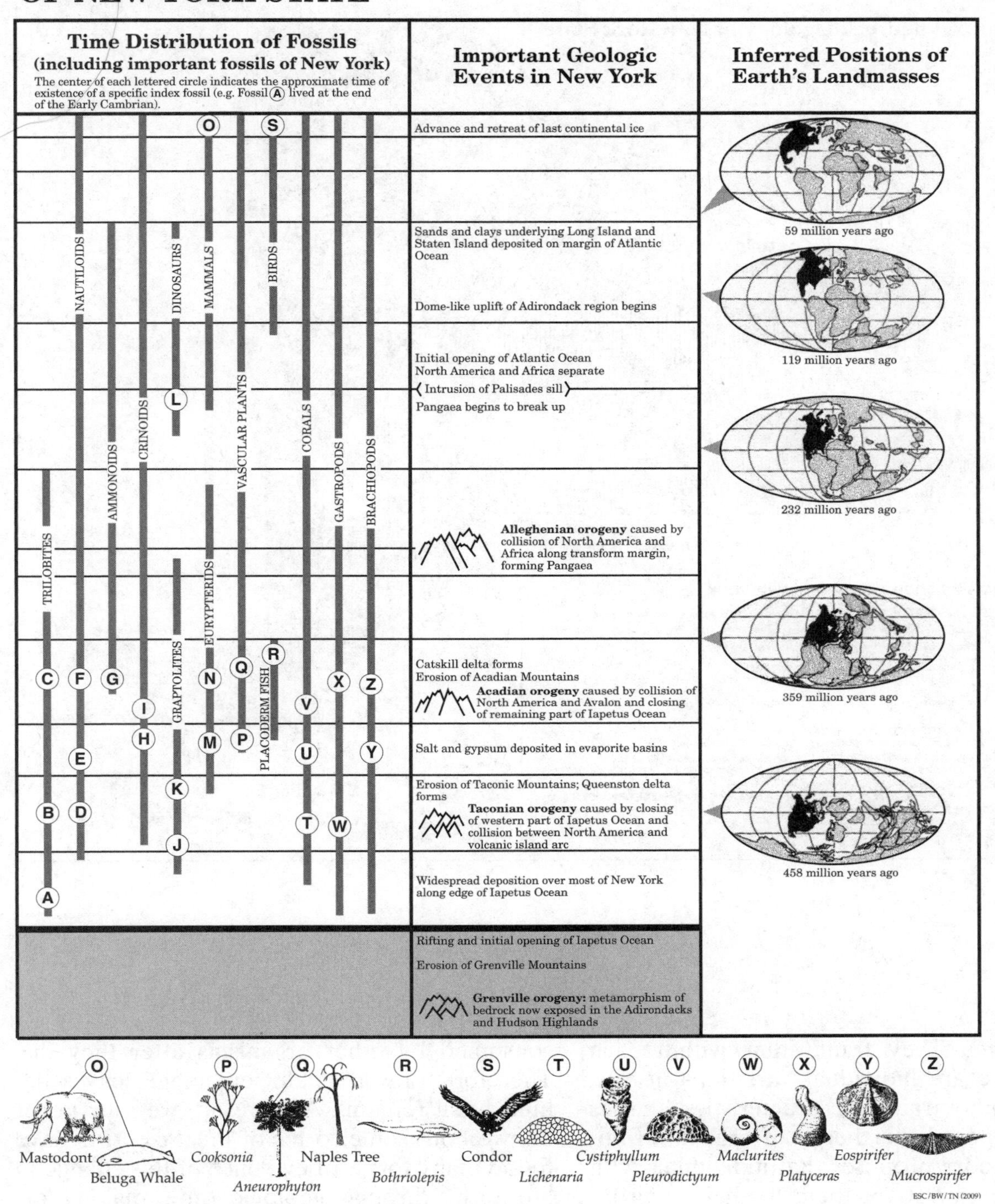

other forms of life may have existed in the Precambrian. However these life-forms had no hard parts and, therefore, left poor fossils. In addition, these oldest rock formations are more likely to have been changed by erosion and metamorphism, which would have destroyed the fossils. In the Cambrian Period, a variety of more complex life-forms developed, many with skeletons and shells that left a more detailed fossil record.

As geologists studied the fossil record, they found that more complex organisms developed as time went on. Some organisms disappeared from the fossil record. That is, they became extinct. Scientists provided an explanation for

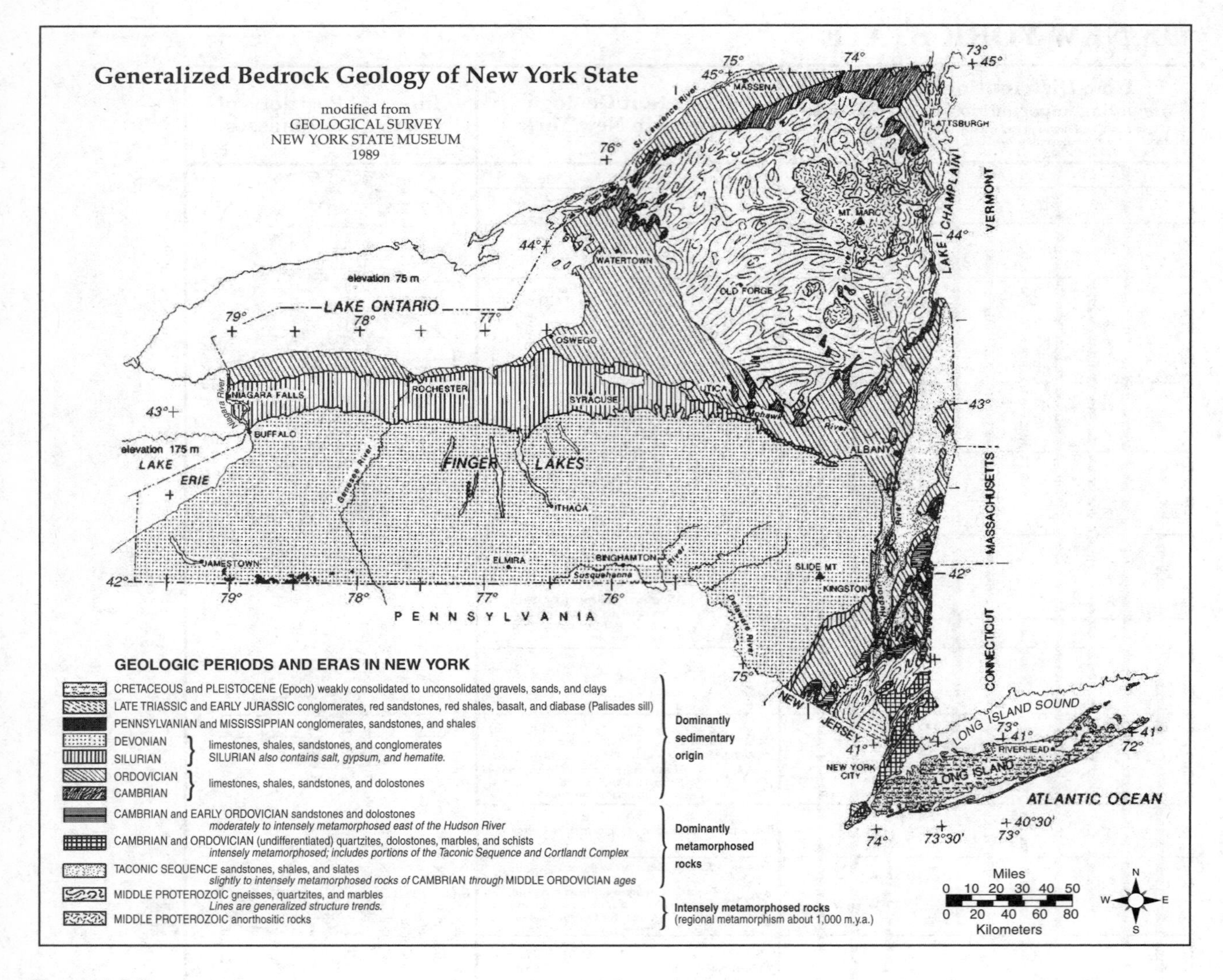

Figure 6-14.

these changes. They found that within each species there are individual variations in size, shape, and other traits, or characteristics. Charles Darwin proposed the theory of organic **evolution**. This theory proposes that individuals with traits that better adapt them to their environment are more likely to survive and have offspring. The offspring inherit these desirable traits. Eventually, an increasing number of individuals in a species possess the desirable traits. By natural selection, the number of those who do not have the desirable traits would decrease.

The process of natural selection leads to the extinction of some species and the formation of new species. Most organisms decompose or are consumed by other organisms after they die. Therefore, only a small percentage leaves behind fossil remains. As a result, we may never know about some forms of life. Scientists have found many forms of evidence of this change in life-forms through geologic time, making organic evolution one of the most important and well-documented principles of science. See the column "Life on Earth" in Figure 6-13.

Humans are probably the most complex life-form to have evolved. Fossils found in Africa indicate that over the past 4 million years humans evolved from apelike animals. This is much less than 1 percent of Earth's age. Thus, humankind is a very recent form of life.

Mass Extinctions

The study of fossils has shown several points in geologic time where a large number of species became extinct. This is called a **mass extinction**. Although geologists do not completely agree on the cause or causes, research has linked some of these mass extinctions to asteroid, comet, or meteoroid impact events. A prime example is a recently investigated and largely buried crater in southern Mexico. This crater may have been caused by the impact of a massive object from outer space at the end of the Mesozoic Era about 65 million years ago. This impact occurred at the time the dinosaurs and many other species became extinct.

The mass extinctions were not caused directly by the impacts. It is more likely that global climatic change resulted from debris thrown into the atmosphere by the impact. The debris prevented sunlight from reaching Earth's surface. The dramatic climatic changes may have interrupted the whole food chain, including climate sensitive plants.

Life and the Atmosphere

When life first developed on Earth, about 4 billion years ago, the atmosphere probably consisted of a mixture of carbon monoxide, carbon dioxide, hydrogen, nitrogen, ammonia, and methane. Today, the mixture of gases in the air is very different: 78 percent nitrogen and 21 percent oxygen. Scientists now understand that microscopic organisms that developed photosynthesis more than 3 billion years ago caused this dramatic change in the atmosphere.

The very first forms of life that developed on Earth were bacteria that did not need oxygen to get energy from food. In time, organisms evolved that could make their own food using an energy source, such as sunlight, and simple raw materials, such as carbon dioxide and water. The organisms that made food from carbon dioxide and water released oxygen as a waste product. Over time, the oxygen built up in the atmosphere to the point that the earliest, anaerobic organisms could no longer survive.

Oxygen is essential for the survival of modern organisms, such as humans. Modern organisms use oxygen to get energy from food. As the oxygen in the air was increasing, modern organisms evolved and came to dominate the planet.

Now, human activities are affecting the composition of the atmosphere. Are we again changing the atmosphere with our pollutants in a way that will harm current forms of life and may favor some forms of life yet to evolve? Will the waste products we vent into the atmosphere cause the extinction of our familiar life-forms? Only time will tell.

QUESTIONS

Part A

17. What is the most reliable method to determine whether two rock layers 100 miles apart are the same age? (1) Both layers are the same kind of sedimentary rock. (2) Both layers crystallized from molten magma. (3) Both layers contain the same group of fossils. (4) Both layers are the same thickness.

18. The three columns below represent three widely separated rock outcrops and the fossils found in each location. The rock layers at each outcrop span the same large range of geologic time.

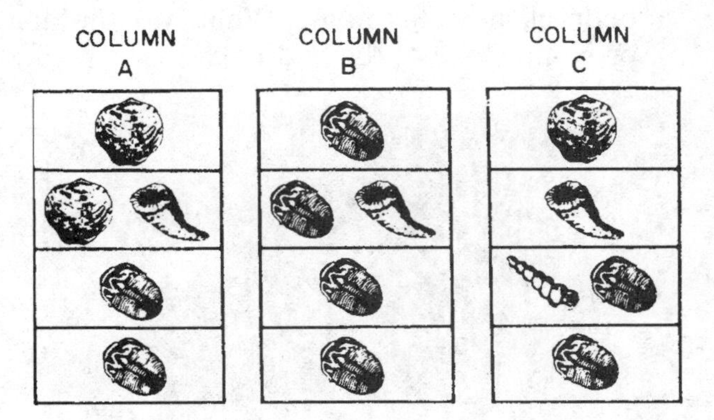

Which fossil appears to be the best index fossil?

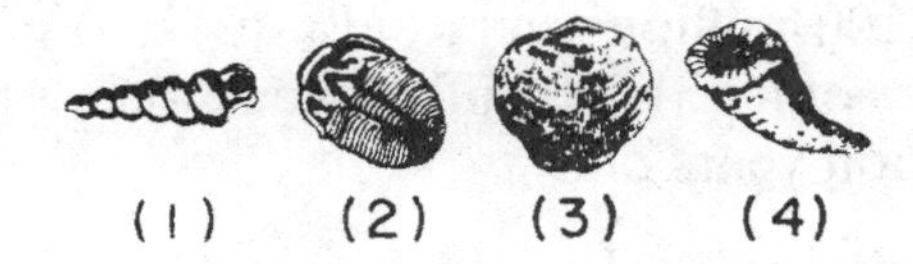

19. There is evidence that a bolide (meteorite or comet) impacted southern Mexico at the end of the Mesozoic Era. Geologists infer that this caused the extinction of (1) trilobites (2) ammonoids (3) corals (4) mammoths

20. Which inference is best supported by the rock and fossil record in New York State? (1) *Eurypterids* lived in shallow seas near present-day Syracuse. (2) *Coelophysis* hunted smaller animals near present-day Albany. (3) The earliest coral reefs formed along the shores of present-day Long Island. (4) Condors flew over the present-day Adirondacks during the Grenville Orogeny.

21. Which New York State geologic event occurred first? (1) sediments and rocks deposited on Long Island (2) first sexually reproducing life-forms (3) intrusion of Palisades sill (4) Taconian Orogeny

22. Scientists have located a large asteroid or comet impact structure in southern Mexico that is approximately 65 million years old. The creation of this impact structure may be the cause of (1) the first appearance of trilobites (2) the advance of the Pleistocene glaciers (3) the extinction of dinosaurs (4) the initial opening of the Atlantic Ocean

23. A student found this index fossil in the bedrock near his home. What was the approximate age of the bedrock?

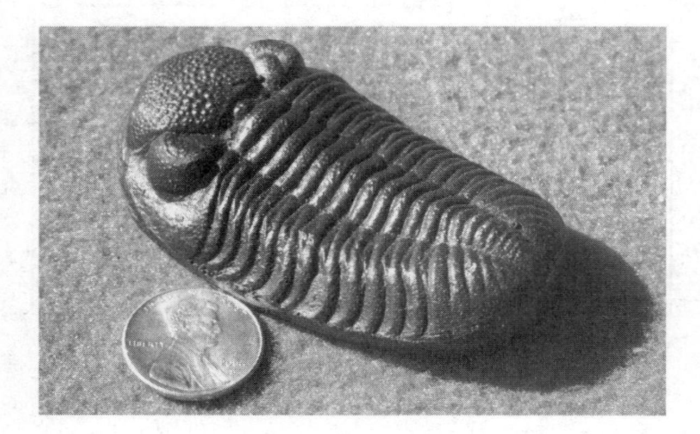

(1) 100 million years old (2) 200 million years old (3) 300 million years old (4) 400 million years old

24. The diagram above is a part of a time line representing Earth's history. According to geologists, in which time period did birds first evolve?

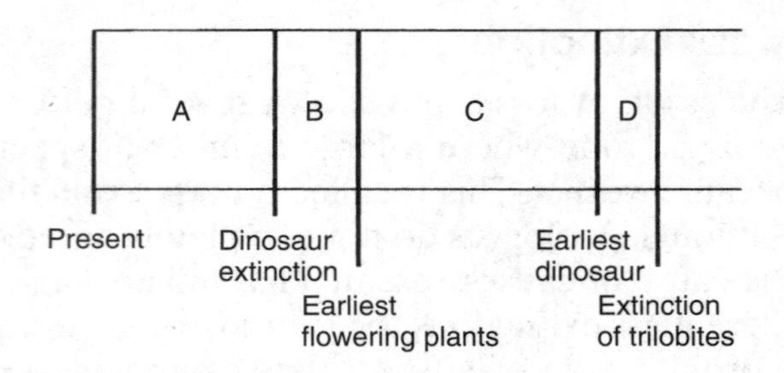

(1) *A* (2) *B* (3) *C* (4) *D*

25. What is the age of the surface bedrock throughout most of the Finger Lakes region of New York State? (1) Cambrian (2) Devonian (3) Pennsylvanian (4) Permian

Part B

Base your answers to questions 26 and 27 on the drawing below, which shows the fossil nautiloid Centroceras.

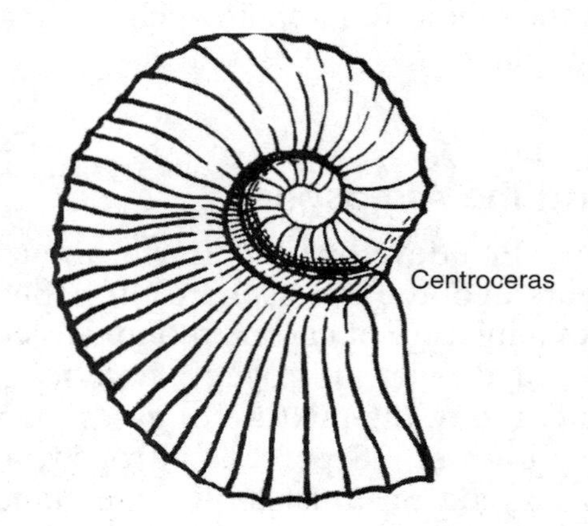

26. What other fossil is likely to be found in the same New York State rock layers with *Centroceras*? (1) the condor (2) the nautiloid *Valcouroceras* (3) the brachiopod *Mucrospirifer* (4) the dinosaur *Coelophysis*

27. If you wanted to collect *Centroceras*, where would be the best place to look in the local bedrock? (1) Massena, NY (2) Albany, NY (3) New York City, NY (4) Ithaca, NY

Base your answers to questions 28 through 32 on the diagram below, which shows rock layers along a part of Earth's surface. The dashed lines connect points of the same geologic age. Drawings of fossils show in which layers they are found.

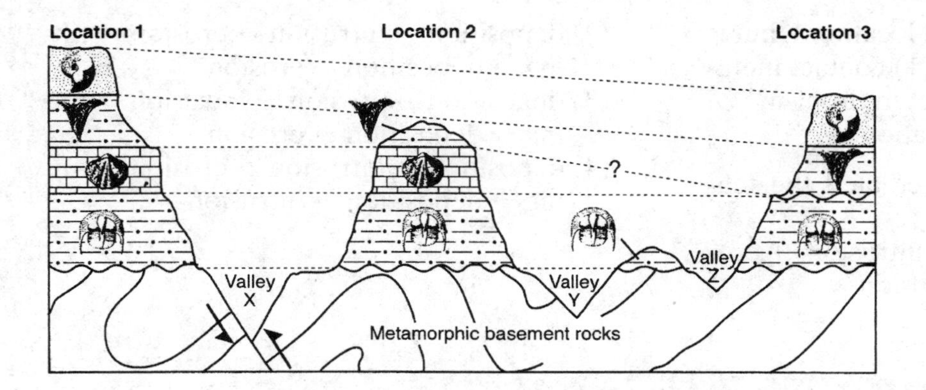

28. The fossils show that at the time the layers containing these fossils were deposited (1) glaciers did not exist on Earth (2) this was a mountainous area (3) rivers were cutting deep valleys at *X*, *Y*, and *Z* (4) an ocean covered this region

29. Which fossil organism below most likely lived when the species of trilobite shown in the diagram was also alive?

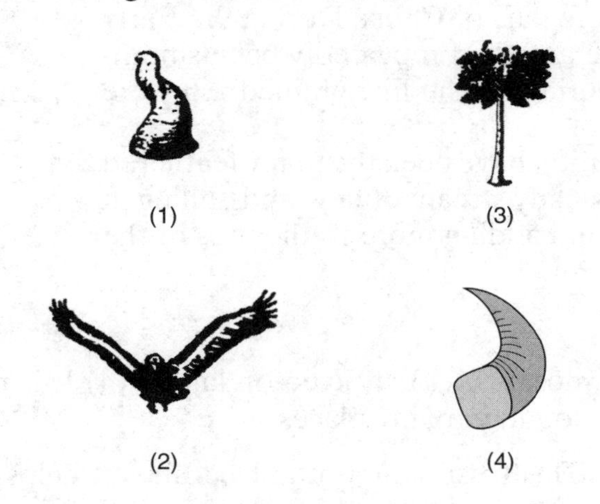

30. The fossil in the top layer at locations 1 and 3 is an ammonoid. What is the approximate numerical age of that rock layer? (1) 362 years (2) 390,000 years (3) 418,000 years (4) 390,000,000 years

31. Why is the limestone layer containing *Eospirifer* missing at location 3? (1) The rocks are folded. (2) There is an unconformity. (3) An earthquake took place. (4) *Eospirifer* became extinct at location 3.

32. Of the three valleys, Valley *X* is eroded the deepest. Why has this occurred? (1) There is a geological fault in Valley *X*. (2) Valleys *Y* and *Z* contain a rock that is softer than in Valley *X*. (3) Trilobites have weakened the rock at Valley *X*. (4) The river in Valley *X* is eroding its way into basalt bedrock.

Base your answers to questions 33 through 35 on the two outcrops below, which are 10 kilometers apart.

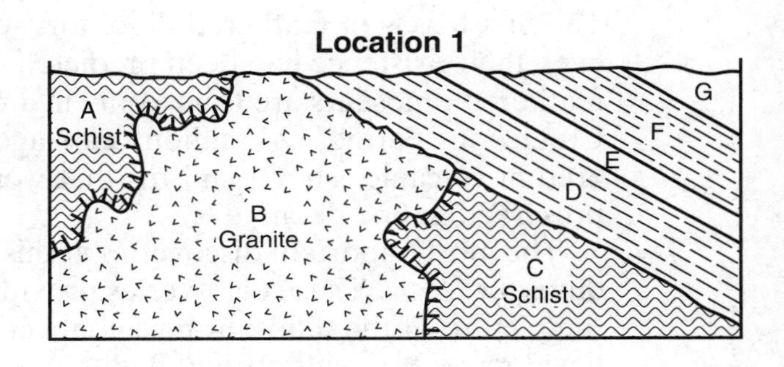

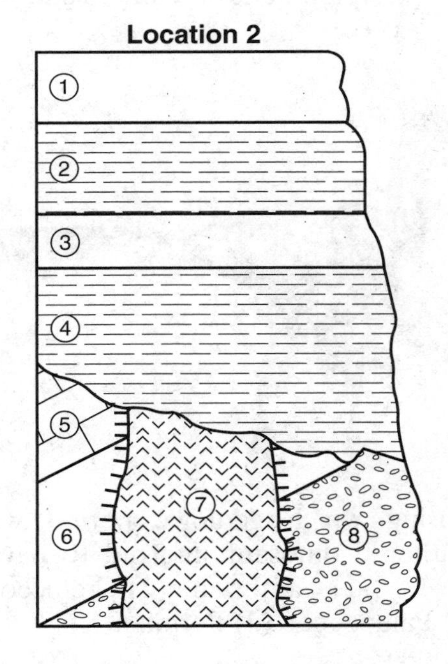

33. Based on the sequence of rock types, layer *E* is most likely the same age as layer (1) 1 (2) 2 (3) 3 (4) 6

34. What event has occurred at location 1, but is not shown at location 2? (1) contact metamorphism (2) regional metamorphism (3) intrusion (4) crystallization

35. What sequence of events created the features in location 2?
(1) deposition → tilting → intrusion → erosion → deposition → erosion
(2) deposition → intrusion → erosion → tilting → deposition → erosion
(3) deposition → erosion → intrusion → tilting → deposition → erosion
(4) deposition → intrusion → erosion → tilting → deposition → intrusion

Base your answers to questions 36 through 39 on the illustration and passage below. The illustration is an artist's conception of a feathered dinosaur that is based on fossil evidence.

Linking Dinosaurs With Modern Birds

The first fossils of feathered dinosaurs were discovered during the 1990s. However, their existence had been predicted for more than a century. These fossils of feathered dinosaurs are found only in a remote part of China. During the Early Cretaceous Period, 124 million years ago, the area had repeatedly been smothered in volcanic ash. Organisms that were buried by the fine-grained ash were preserved in fine detail.

The most important discoveries at this location have been the many feathered dinosaur fossils. These discoveries provide a steady stream of new finds, filling in the picture of the dinosaur-bird connection and adding more to theories of the development of feathers and flight.

36. During which geologic period was this feathered dinosaur inferred to have been alive? (1) Cambrian (2) Cretaceous (3) Paleogene (4) Permian

37. Why is the fossil dinosaur found in China not a good index fossil? (1) It lived too long ago. (2) It was too similar to other dinosaurs. (3) It lived on land. (4) It is not found in many places.

38. This fossil is important because it helps us understand (1) biological evolution (2) motions of the tectonic plates (3) recrystallization of rocks (4) the formation of unconformities

39. In what part of New York State would a geologist be most likely to find fossils of feathered dinosaurs? (1) Tug Hill Plateau (2) Catskill Mountains (3) Erie-Ontario Lowlands (4) Newark Lowlands

Base your answers to questions 40 through 50 on the following figure, which shows two widely separated rock profiles. The rocks have not been overturned in either location. Three specific fossils were found where the arrows indicate them.

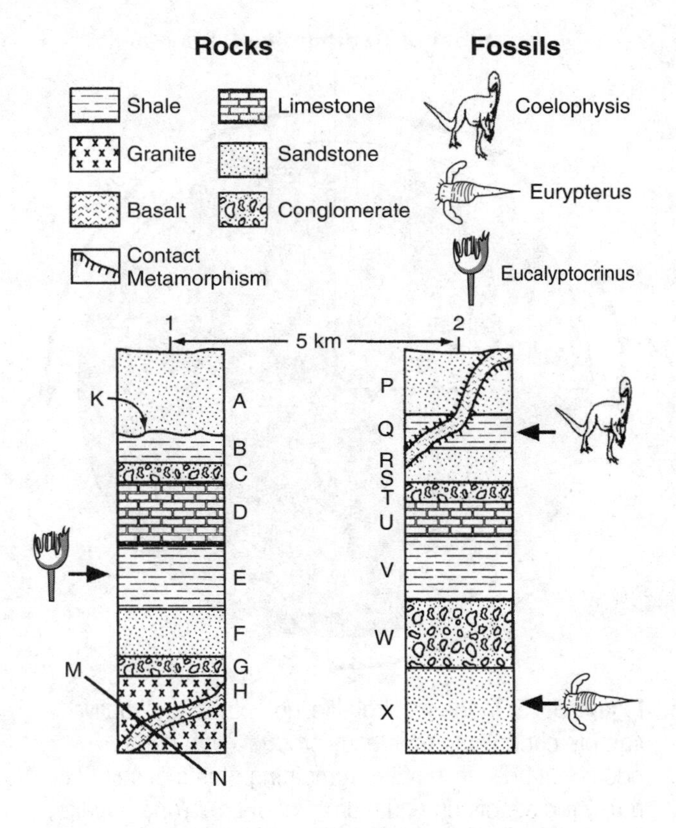

40. In location 2, which rock layer is probably the youngest? (1) *P* (2) *R* (3) *W* (4) *X*

41. In which rock layer will you most likely find quartzite? (1) *A* (2) *E* (3) *P* (4) *W*

42. In what geologic period did rock layer *E* form? (1) Cambrian (2) Silurian (3) Permian (4) Triassic

43. What rock layer is the same age as layer *E*? (1) *U* (2) *V* (3) *W* (4) *X*

44. Which graph below best shows the relative ages of the rock layers along line M–N?

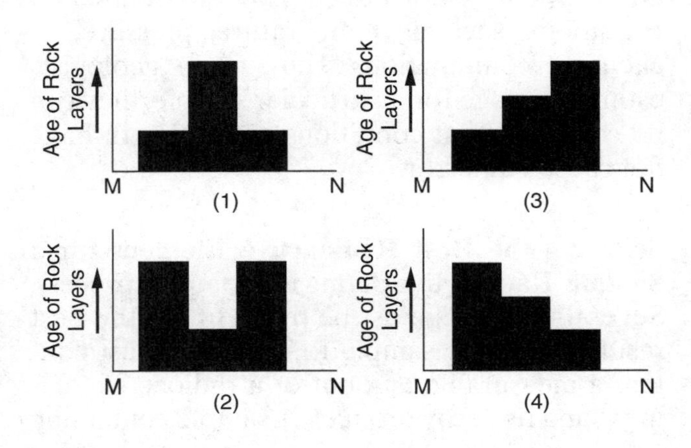

45. *K* is an unconformity. What can we conclude about the rock layers adjacent to *K*? (1) Layer *A* was deposited before layer *B*. (2) Layer *B* was not deposited flat and level. (3) Part of layer *A* is missing. (4) Part of layer *B* is missing.

46. All the sedimentary layers were deposited in water. How fast was the water moving when it deposited layer *C*? (1) 0.1 cm/s (2) 1 cm/s (3) 10 cm/s (4) 100 cm/s

47. At location 2, during which geologic period did layer *Q* form? (1) Cambrian (2) Devonian. (3) Triassic (4) Quaternary

48. Which rock layer is least likely to contain fossils? (1) *E* (2) *F* (3) *G* (4) *I*

49. What is the most likely age of layer *P*? (1) 100 million years (2) 300 million years (3) 500 million years (4) 900 million years

50. Which rock layer is composed of particles of sediment smaller than 0.0004 cm in diameter?

(1) *A* (2) *B* (3) *C* (4) *I*

WHAT IS RADIOMETRIC DATING?

Through fossil evidence, geologists established the geologic time scale shown in Figure 6-13 on pages 152–153. At first, this was only a relative time scale. It simply showed an inferred order, or sequence, of events. Fossils could not tell geologists how many years ago the organisms lived or just when the rock strata formed. The geologic time scale became an **absolute** (numerical) **time scale**, using specific times, when numbers and units of time (years) were determined. An absolute time scale is not absolutely correct. There is always a margin of error in scientific measurements. However, radiometric dating techniques can make the error remarkably small.

Chemical elements often have several forms, called **isotopes**, which differ in the number of neutrons in their atomic nuclei. For example, carbon-12 has 6 protons and 6 neutrons in its nucleus, and **carbon-14** has 6 protons and 8 neutrons in its nucleus. If the nucleus of an isotope has more or fewer than the number of neutrons in its most stable form, the isotope may be unstable, or **radioactive**. A radioactive isotope will break down naturally into a different element called a **decay product**. In the process, it gives off radiation and/or particles. For example, the most common form of carbon, carbon-12, is not radioactive; but carbon-14, with two extra neutrons in its nucleus, is unstable. Carbon-14 will change into its stable decay product, nitrogen-14.

Since atoms decay randomly, scientists cannot predict when a particular atom will decay. However, even a small sample of a radioactive element contains millions of atoms, from which scientists can predict a reliable rate of decay.

Half-life

The decay of a radioactive element is measured by its half-life. Different radioactive elements have different half-lives. A **half-life** is the time required for half of the atoms in a sample of a radioactive element to change to the decay product. At the end of one half-life, a sample contains equal amounts of the radioactive element and its decay product. In each succeeding half-life, half of the remaining atoms decay (no matter how large the sample). Figure 6-15 is a model of radioactive decay. Each dark arrow represents an equal period of time (one half-life). Through time, fewer radioactive atoms remain in the sample while more decay product accumulates.

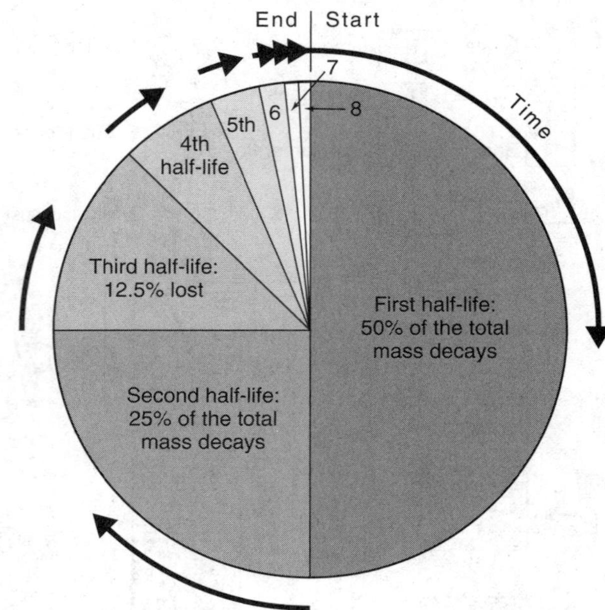

Figure 6-15. In the first half-life, half of the radioactive sample changes to the decay product. With each additional half-life, half the remaining atoms decay. Each arrow represents an equal amount of time (one half-life). So each smaller and smaller section of the circle still represents the same amount of time. In theory, the radioisotope never gets to zero mass. It just gets too small to measure.

The Decay-Product Ratio The ratio between the mass of a radioactive element and its decay product in a sample is the **decay-product ratio**. After scientists determine this ratio, they can calculate how many half-life periods have occurred since the sample was formed, and, in turn, determine its **absolute age**, its age in years. Figure 6-16 shows this progression toward the decay product.

For example, if a sample contains equal amounts of carbon-14 and its decay product, nitrogen-14, the sample must have gone through one half-life. From laboratory work, scientists know that the half-life of carbon-14 is 5.7×10^3 years. The half-life period of commonly used radioisotopes are listed in the *Earth Science Reference Tables* and in Figure 6-17.

If the ratio in a sample is three-quarters nitrogen-14 and one-quarter carbon-14, then the sample has gone through two half-life periods. (Therefore it is 11.4×10^3 years old.) After three half-life periods, the amount of carbon-14 would be cut in half again, so that the ratio in the sample would be one-eighth carbon-14 and seven-eighths nitrogen-14. See the top boxes in Figure 6-17.

Laboratory studies show that the half-life for any element is not affected by environmental conditions, such as temperature, pressure, or chemical combinations. Thus, when geologists estimate an age for a particular sample, they can be confident that conditions within Earth have not caused an error.

Selecting the Best Radioactive Element for a Sample Radioactive dating is a complex process. Several decisions must be made to get the best results. First, the sample to be dated must contain a measurable amount of a radioactive element and its decay product. A sample containing

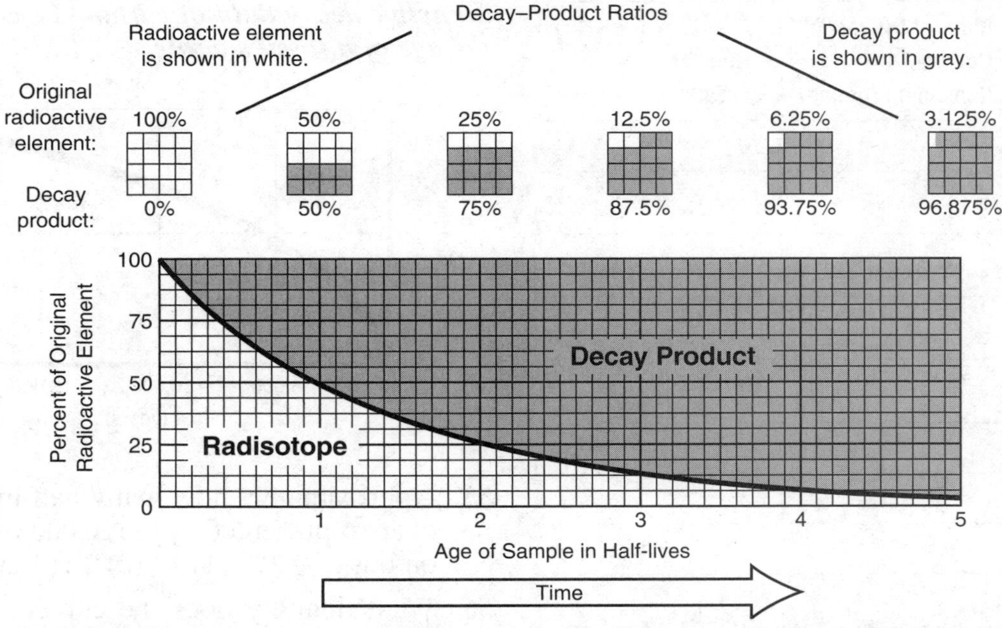

Figure 6-16. This diagram is another way to illustrate the half-life disintegration of a radioactive substance. It presents the same information as Figure 6-15, only in a different format. Through time, the parent (radioactive) substance is replaced by decay product. However, as the original radioactive substance decreases in mass, fewer atoms disintegrate with each half-life time period.

the remains of living organisms is likely to contain radioactive carbon-14.

The next factor to consider is the sample's estimated age. With a half-life of only 5700 years (5.7×10^3 years), carbon-14 can date samples no older than about 50,000 years. If the sample is older than 50,000 years, so many half-life periods have passed that too little of the original carbon-14 remains to be measured accurately. On the geologic time scale, 50,000 years is very recent, covering only the very recent part of the Cenozoic Era.

However, uranium-238, with a half-life of 4.5×10^9 years, can measure samples of the oldest rocks on our planet. For very recent samples of rocks, however, too little uranium-238 would have decayed to lead-206 to make the decay product measurable. Thus, a geologist must select the radioactive isotope with a half-life that best measures the age of a sample.

Radioactive Decay Data

RADIOACTIVE ISOTOPE	DISINTEGRATION	HALF-LIFE (years)
Carbon-14	$^{14}C \rightarrow {}^{14}N$	5.7×10^3
Potassium-40	$^{40}K \rightarrow {}^{40}Ar$, $^{40}K \rightarrow {}^{40}Ca$	1.3×10^9
Uranium-238	$^{238}U \rightarrow {}^{206}Pb$	4.5×10^9
Rubidium-87	$^{87}Rb \rightarrow {}^{87}Sr$	4.9×10^{10}

Figure 6-17.

QUESTIONS

Part A

51. At the origin of Earth, a radioactive sample was 100% rubidium-87. Approximately how much of the sample would be changed to strontium-87 by now? (1) 10% (2) 50% (3) 90% (4) 99%

52. The following table gives information about the radioactive decay of carbon-14. Part of the table has been left blank for you to complete.

Half-life	Mass of Original Carbon-14 Remaining (grams)	Number of Years
0	1	0
1	$\frac{1}{2}$	5700
2	$\frac{1}{4}$	11,400
3	$\frac{1}{8}$	17,100
4	$\frac{1}{16}$	
5		
6		
7		

After how many years will $\frac{1}{128}$ of a gram of the original carbon-14 remain? (1) 22,800 yr (2) 28,500 yr (3) 34,200 yr (4) 39,900 yr

53. If a radioactive sample of uranium-238 were allowed to decay for billions of years, what element in the sample would increase over that time? (1) carbon-14 (2) strontium-87 (3) lead-206 (4) uranium-235

54. Carbon-14 is not used to determine the age of geological objects and events because of its (1) inorganic origin (2) relatively short half-life (3) abundance in igneous rocks (4) resistance to weathering

Part B

Base your answers to questions 55 and 56 on the following graph, which represents the radioactive decay data of carbon-14 used to find the age of a fossil sample.

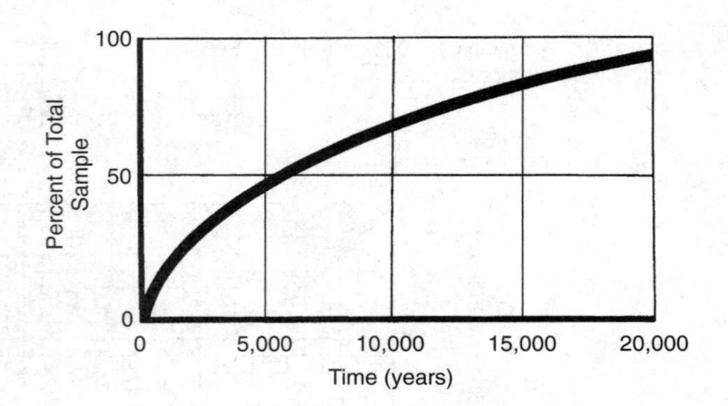

55. Approximately how many half-live periods of are represented by the 20,000 years shown on this graph? (1) 1 (2) 2 (3) 3 (4) 4

56. What quantity does the curved graph line best represent? (1) the change in the amount of uranium-238 in a radioactive sample (2) the change in the amount of lead-206 in a radioactive sample (3) the change in the amount of carbon-14 in a radioactive sample (4) the change in the amount of nitrogen-14 in a radioactive sample

57. A student filled a 1-liter graduated cylinder with water to use as a model for radioactive decay. She selected 30 seconds as the half-life. After 30 seconds, she poured out 500 milliliters, and after another 30 seconds she poured out another 250 milliliters. How much water did the student need to pour out when she reached the 2-minute mark? (1) 12.5 mL (2) 25.0 mL (3) 62.5 mL (4) 125 mL

Base your answers to questions 58 through 65 on the text and the diagram below.

Potassium-Argon Dating

Potassium-argon (K-Ar) dating is used by geologists to determine the age of geologic samples. Potassium is an element commonly found in clay, evaporites, and minerals such as mica.

Quickly cooled lavas are excellent candidates for K-Ar dating. These rocks trap the argon gas decay product and minimize the possibility of contamination with atmospheric argon. Great care is needed in collecting a sample for K-Ar dating. Many potential samples have been contaminated.

Due to the long half-life of potassium-40, the technique is most useful for dating minerals and rocks more than 100,000 years old. K-Ar dating was used to determine the ages of magnetic reversals at the ocean ridges where new crust preserves Earth's magnetic polarity at the time the magma cools.

The graph shows the decay of a sample of potassium-40. The line labeled "Radioisotope *Y*" indicates a very different rate of radioactive decay.

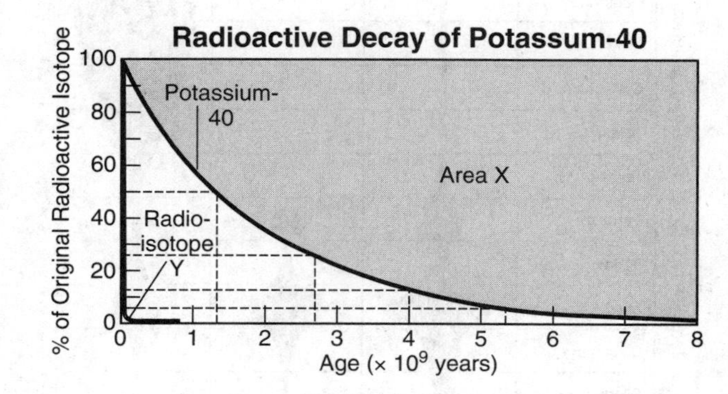

58. What is the approximate half-life of potassium-40? (1) 1.3 thousand years (2) 1.3 million years (3) 1.3 billion years (4) 1.3 trillion years

59. The gray part of the graph labeled "Area *X*" best represents (1) the amount of the original radioactive substance (2) open space within the radioactive sample (3) the amount of the decay product (4) a different radioactive substance

60. Potassium-40 dating has been especially important in measuring (1) the rate at which the Atlantic Ocean formed (2) the amount of heat flow at the ocean ridges (3) the rate at which the Rocky Mountains are being uplifted (4) when the universe began as the big bang

61. The following question has only three responses. The steep line labeled "Radioisotope *Y*" best represents the decay of (1) carbon-14 (2) uranium-238 (3) rubidium-87

62. A 10-gram sample of pure potassium-40 decays for 5.2×10^9 years. How much of the sample will still be potassium-40? (1) 5 grams (2) 2. 5 grams (3) 1.25 grams (4) 0.625 grams

63. Radiometric dating with substances such as potassium-40 has helped geologists find (1) the order in which different species came about in organic evolution (2) a numerical time scale for Earth's geologic history (3) the energy output of the sun (4) the luster of quartz

64. Potassium-40 decays into argon-40 as well as into (1) uranium-40 (2) sodium-40 (3) calcium-40 (4) oxygen-40

65. What problem can occur in measuring potassium-40 and its decay product that is much less likely when a scientists measures uranium-238 and its decay product? (1) The potassium-40 decay rate depends upon the temperature of the sample. (2) Only potassium-40 decreases in mass with time. (3) Only uranium-238 is found in Earth's crust. (4) The decay product of potassium-40 often escapes.

Chapter Review Questions

PART A

1. Three fossils found in New York State are shown in the diagram below.

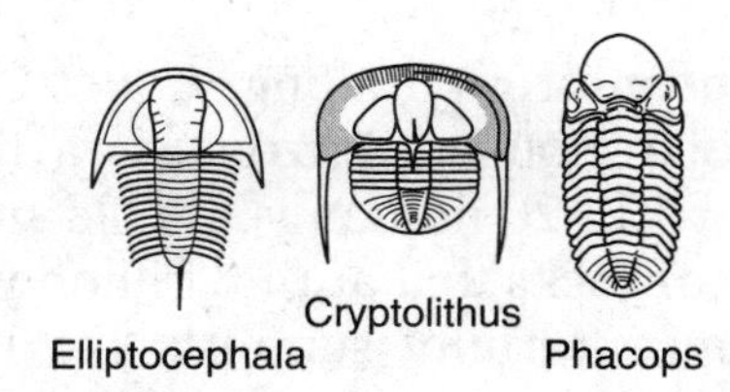

Which statement about these fossils is most likely true? (1) These three animals were all alive at the same time. (2) These three animals are all species of trilobites. (3) These three animals lived in forests and grasslands. (4) These three animals can be found alive today.

2. The diagram below represents rock layers found in an outcrop. The three index fossils shown are found within these layers.

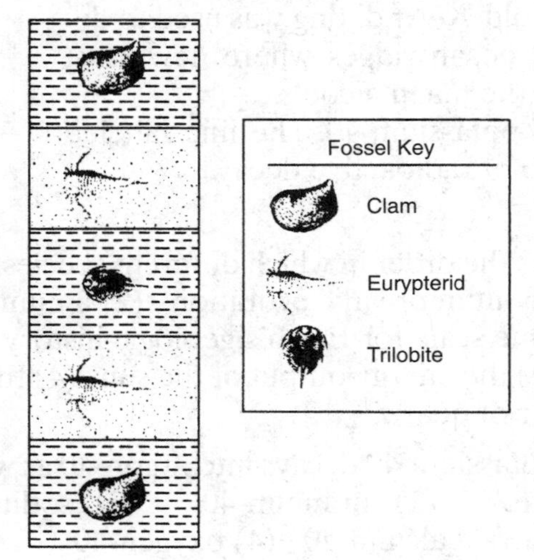

What event could most easily result in the sequence of layers shown above? (1) faulting (2) folding (3) uplift (4) erosion

3. What radioisotope would be most useful to tell when humans first migrated to North America based on the excavation of their campsites? (1) uranium-238 (2) potassium-40 (3) carbon-14 (4) rubidium-87

4. The Paleozoic Era includes what approximate portion of the entire history of our planet? (1) 5% (2) 25% (3) 50% (4) 75%

5. A fossil shell contains 25% of its original amount of carbon-14. Approximately when did the animal that made the shell live? (1) 5700 years ago (2) 11,400 years ago (3) 17,100 years ago (4) 22,800 years ago

6. An unconformity along Cave Creek, in Maricopa County, Arizona, is a boundary where rocks 20 million years old sit directly on top of rocks that are 1.8 billion years old. This unconformity supports the inference that (1) no rocks were deposited in this location between 1.8 billion and 20 million years ago (2) if fossils are found in this location, they are most likely beneath the unconformity (3) Cave Creek has caused deposition but no erosion in the past million years (4) a long period erosion at this location ended 20 million years ago

Base your answers to questions 7 through 10 on the diagram below, which gives information about three different groups of fossil organisms. The number of species of each—crinoids, blastoids, and echinoids—is represented by the width of the black figures above the names.

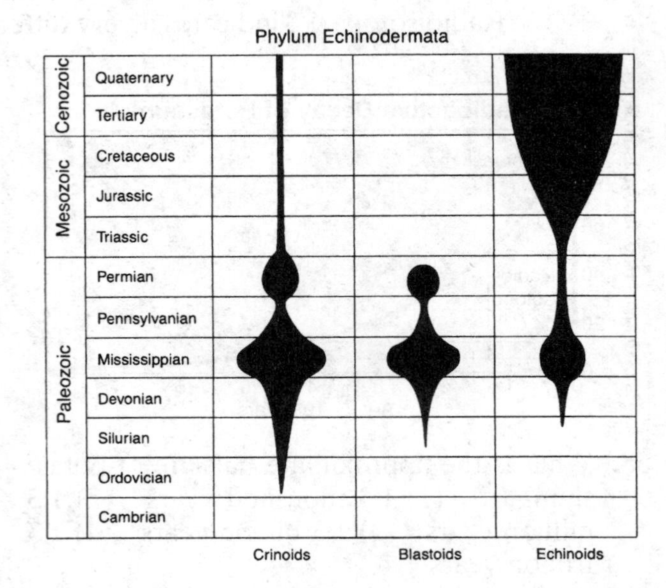

7. Which statement about these fossil organisms is best supported by the diagram? (1) All three groups of organisms are now extinct. (2) All three groups of organisms came into existence in the same geologic period. (3) All three groups of organisms existed during the Devonian Period. (4) All three groups of organisms constantly increased in abundance since they first appeared.

8. When were crinoid species most abundant? (1) 250 million years ago (2) 350 million years ago (3) 450 million years ago (4) 550 million years ago

9. Crinoids were very common in New York State during the Paleozoic Era. Which of the following were species of crinoids? (1) *Eucalyptocrinus* and *Ctenocrinus* (2) *Cryptolithus* and *Phocops* (3) *Cystiphyllum* and *Pleurodictyum* (4) *Eospirifer* and *Mucrospirifer*

10. Of the geologic periods shown in the figure above, which was the longest? (1) Permian (2) Triassic (3) Jurassic (4) Cretaceous

11. Which of the following graphs best shows the radioactive decay of carbon-14?

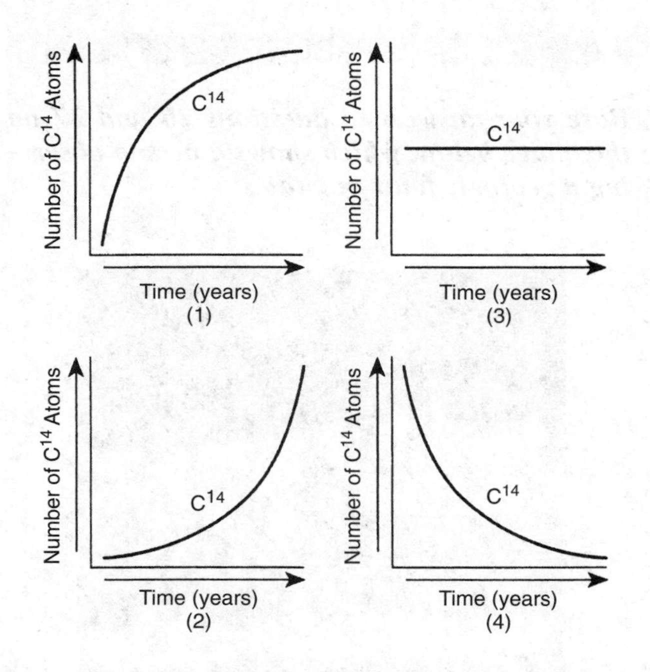

12. Which of the following landscape regions would be *least* likely to have abundant fossils? (1) Atlantic Coastal Plain (2) Erie-Ontario Lowlands (3) Adirondack Mountains (4) Allegheny Plateau

PART B

Base your answers to questions 13 through 18 on the diagram below, which represents a portion of Earth's crust.

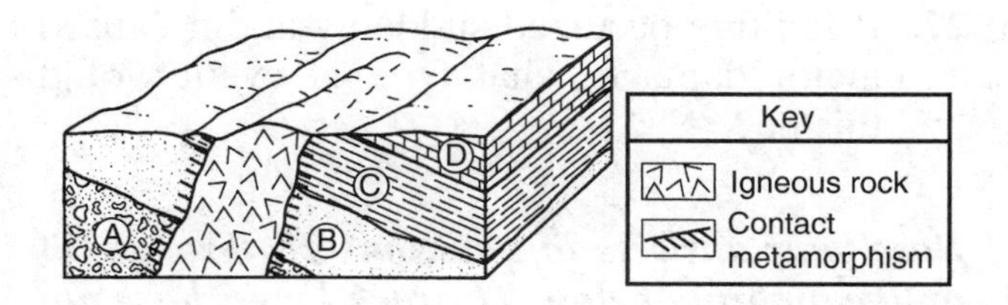

13. If the igneous rock is not tilted, which of these events occurred most recently? (1) formation of layer *A* (2) formation of layer *D* (3) tilting of layers *A*, *B*, *C*, and *D* (4) contact metamorphism of layer *A*

14. An igneous rock is light in color and composed primarily of crystals of quartz and feldspar. This rock cooled quickly so the crystals are too small to be easily seen. What kind of rock is it? (1) granite (2) basalt (3) rhyolite (4) pegmatite

15. The symbols in layer *D* show that the rock type is (1) limestone (2) sandstone (3) marble (4) gneiss

16. What process created layer *C*? (1) recrystallization (2) metamorphism (3) deposition (4) erosion

17. What sequence of events led to the final outcrop?
(1) intrusion → crystallization → erosion → deposition → tilting
(2) deposition → tilting → intrusion → crystallization → erosion
(3) deposition → crystallization → intrusion → erosion → tilting
(4) contact metamorphism → deposition → tilting → intrusion → erosion

18. A similar group of rocks is found in an outcrop 50 kilometers away. What would be the best evidence that the rocks in both places are similar in age? (1) Both locations include an intrusion with contact metamorphism. (2) Both locations contain sedimentary rocks that are tilted. (3) Both locations contain many of the same fossil species. (4) Both locations have conglomerate on the bottom next to a layer of sandstone.

Base your answers to questions 19 and 20 on the diagram below, which represents a location in Montana.

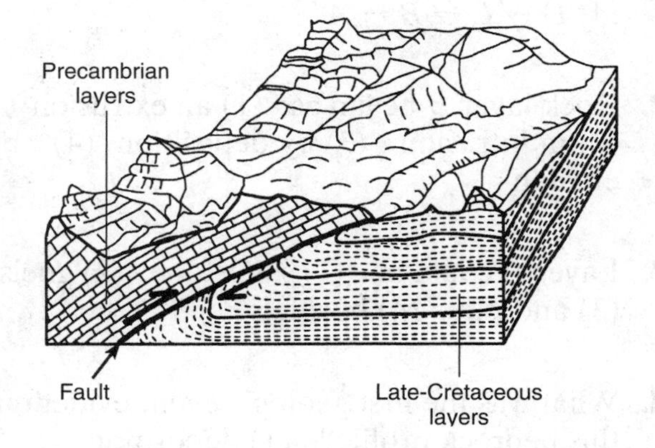

19. The fault could have occurred (1) 1 billion years ago (2) 544 million years ago (3) 142 million years ago (4) 1 million years ago

20. How is the rock structure in this location different from rocks found in most other places? (1) Erosion is presently occurring at the surface. (2) The rocks have been moved horizontally by plate tectonic forces. (3) Older rocks are positioned on top of younger rocks. (4) This mountain region has major folds and faults.

Base your answers to questions 21 through 25 on the geologic cross section below.

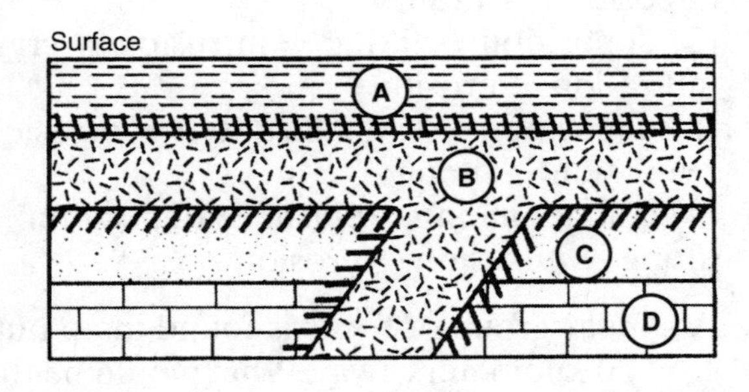

21. What is the correct sequence in which these rock layers formed from the oldest to the youngest?
(1) $A \rightarrow B \rightarrow C \rightarrow D$
(2) $D \rightarrow C \rightarrow A \rightarrow B$
(3) $A \rightarrow C \rightarrow D \rightarrow B$
(4) $D \rightarrow C \rightarrow B \rightarrow A$

22. Rock layer *B* began as (1) an extrusion (2) in intrusion (3) by deposition (4) by erosion

23. Layer *B* could be (1) sandstone (2) gneiss (3) andesite (4) hornfels

24. What was the first geologic event evident in this bedrock profile? (1) deposition (2) erosion (3) intrusion (4) extrusion

25. The contact metamorphism occurred at the same time that (1) layer *A* was deposited (2) layer *B* crystallized (3) layer *C* was eroded (4) layer *D* was weathered

PART C

Base your answers to questions 26 and 27 on the image below, which shows a person observing a geologic fault in Utah.

26. Make a sketch of this image and draw arrows along both sides of the fault to show the relative directions of movement that occurred here. Clearly indicate which side went up and which side went down.

27. If faulting occurred suddenly and it caused major damage, what type of event would this be?

Base your answers to questions 28 through 33 on the diagram below. The rock layers have not been overturned. The age of the granite intrusion is 279 million years, and of the vesicular basalt 260 million years.

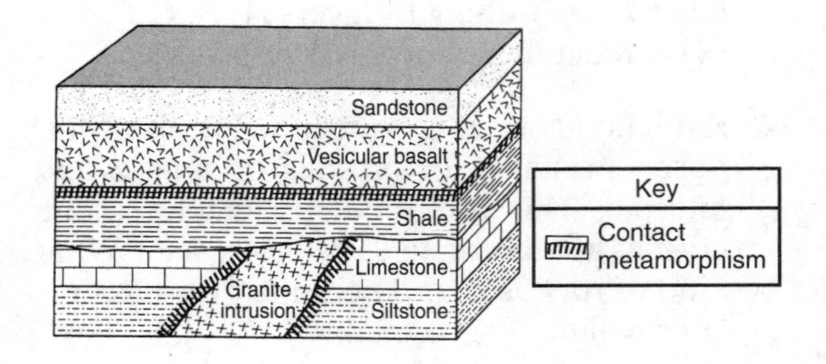

28. List the six rock units in order from oldest to youngest.

29. During which geologic time period did the shale form?

30. What rock type does the zone of contact metamorphism between the shale and the vesicular basalt indicate?

31. If you were to look carefully at these rocks, how could you tell that the vesicular basalt crystallized more quickly than the granite?

32. What event occurred after the intrusion of granite and before the deposition of shale?

33. How can you tell that the vesicular basalt is not an intrusion?

Base your answers to questions 34 through 37 on the text passage and the diagram below.

The World's Biggest Trilobite

While examining a rock unit along the shore of Hudson Bay in northern Manitoba, Canada, a team of Canadian paleontologists found the largest known complete fossil of a trilobite. This animal is thought to have lived during the Ordovician Period. The fossil animal was 60 centimeters long and classified in the genus Isotelus.

Map of North America

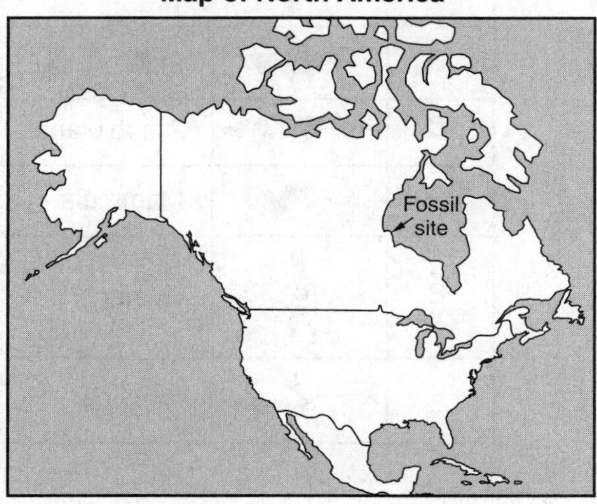

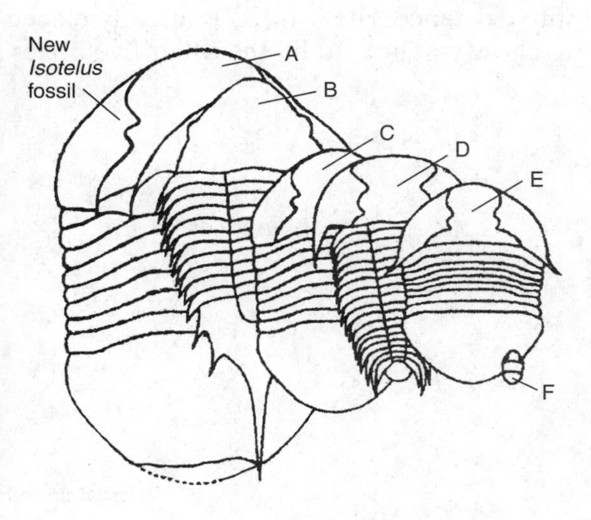

34. In what type of rock was the fossil most likely found?

35. According to plate tectonic theory, what was the approximate latitude of the fossil site during the Ordovician Period?

36. The actual new fossil *Isotelus* is approximately how many times as large as the scale drawing above?

37. What are the present coordinates of latitude and longitude where these fossils were found?

38. A student observed the exposed vertical rock surface shown at the right and made the following four inferences. State one form of evidence that is shown in the diagram to support each inference.

Inference 1: The shale is older than the basaltic intrusion.
Inference 2: The shale is older than the sandstone.
Inference 3: The shale was deposited long after the limestone and the granite formed.
Inference 4: The limestone is older than the granite.

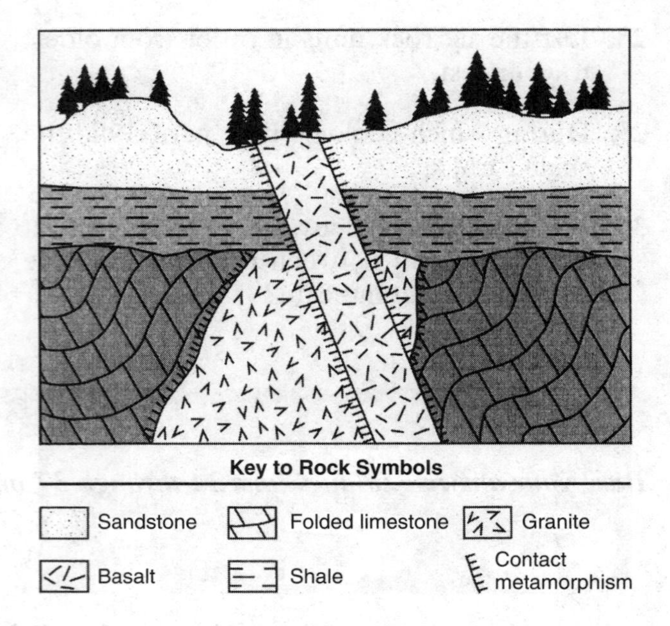

39. The table at the right shows several animal groups that have lived on Earth. Below the table is a time line to represent part of Earth's history. Indicate when each of the animal types shown on the table first came into existence. The fish (*B*) is already placed to show you how to fill the other five boxes.

Animal Key

Letter	Picture	Animal Group
A		Birds
B		Fish
C		Amphibians
D		Mammals
E		Humans
F		Reptiles

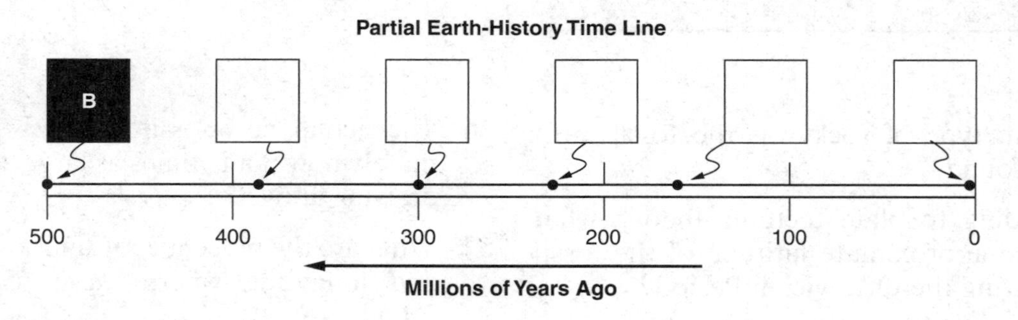

Base your answers to questions 40 and 41 on the diagram below, which represents an outcrop in New York State.

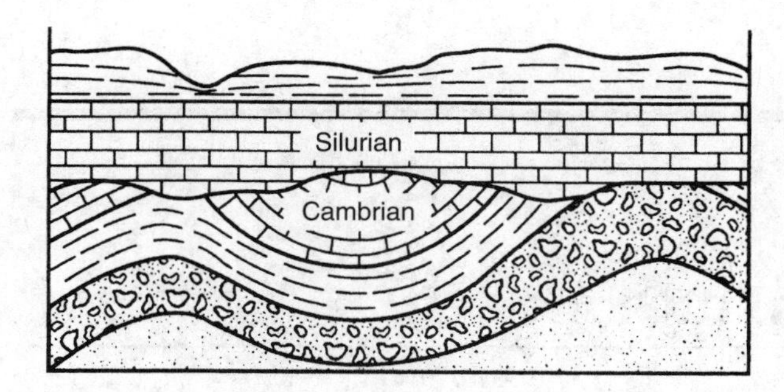

40. What geologic period is missing in the rocks at this location?

41. This profile was produced by three geologic processes: folding, deposition, and erosion. One or more of these processes occurred more than once. Make a list of the five events that produced this profile in sequence from the first to the last process.

Base your answers to questions 42 through 47 on the figure below, which is a time line showing part of Earth's history. Letters a through g are specific points on this timeline.

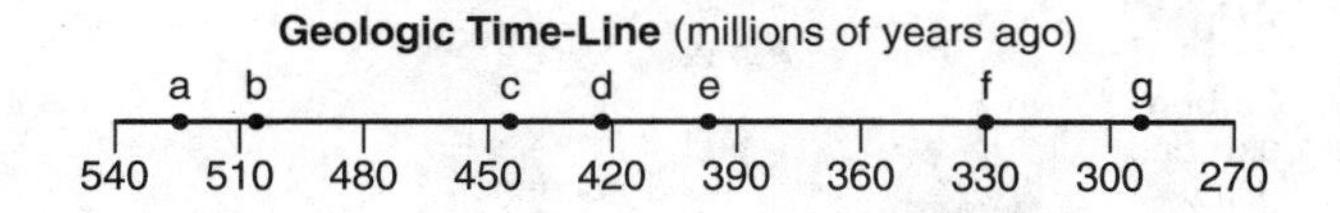

42. Letter *e* represents a time how long before the present?

43. What species of trilobite inhabited the ancient oceans at the time represented by letter *a*?

44. The time line above represents a time span completely within what geologic era?

45. What mountain-building event occurred at the time of letter *g*?

46. Identify one letter that represents a time that is not represented within the rock record in New York State.

47. What kind of marine animal first appeared at the time of letter *b*?

48. What is the approximate numerical (absolute) age of the bedrock found at the surface in most of the Tug Hill Plateau?

49. The only known dinosaur fossils in New York State are footprints of Coelophysis. Draw an *X* on the map of New York State below to show where a geologist would be most likely to find Coelophysis footprints.

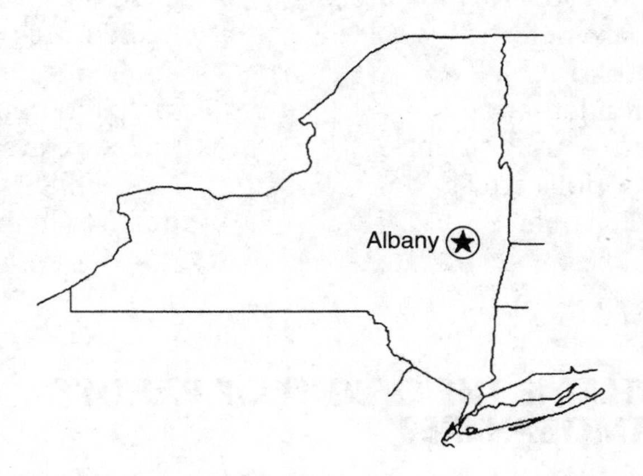

50. The time line below represents the whole 4.6 billion years of Earth's history. It is drawn to a uniform scale. The label "First Marine Animals with Shells" is written above the line. Draw an arrow from this label to the time it occurred on the time line.

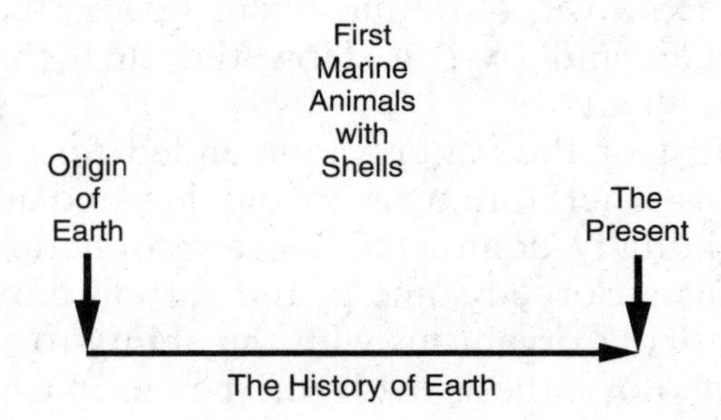

CHAPTER

Properties of the Atmosphere

Vocabulary at a Glance

air pressure	humidity	saturated
anemometer	jet stream	sea breeze
atmospheric pressure	land breeze	sling psychrometer
barometer	meteorologist	specific heat
cloud	millibars	temperature
condensation	precipitation	water vapor
condensation nuclei	pressure gradient	weather
Coriolis effect	prevailing winds	wind
dew point	relative humidity	

WHAT IS THE ORIGIN OF EARTH'S ATMOSPHERE?

Earth's early atmosphere was probably very different from the atmosphere of today. The most likely origin of our atmosphere is the gases vented by erupting volcanoes as shown in Figure 7-1. This process is *outgassing*. The volcanoes vented mostly **water vapor** (water in the form of a gas), carbon dioxide, and nitrogen. However, the atmosphere now is mostly nitrogen and oxygen. How did that change come about?

Most of the water vapor ended up in the oceans. Therefore, carbon dioxide and nitrogen initially dominated the atmosphere. The oceans absorbed some of the carbon dioxide. Eventually, organisms with the ability to carry out photosynthesis evolved. They used carbon dioxide to make living tissue. As a result of that process they gave off oxygen. In addition to the carbon dioxide used by plants, much of it is now tied up in limestone deposits that can be traced back to an organic origin. Thus, the modern atmosphere is a direct result of billions of years of volcanic and biologic processes.

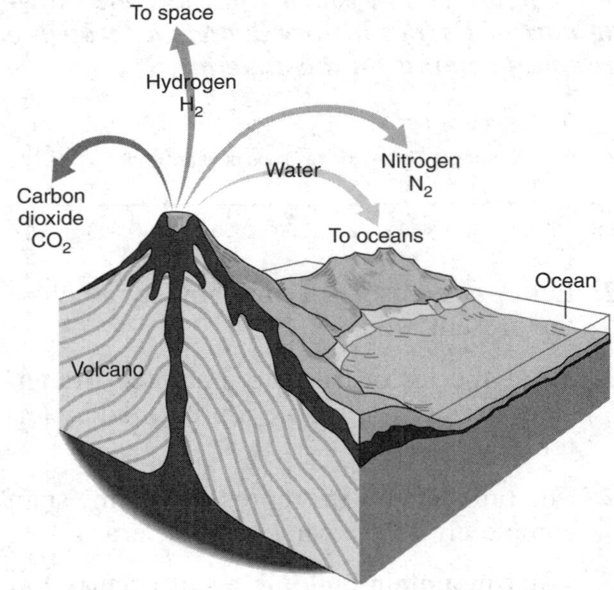

Figure 7-1. Earth's atmosphere came from gases vented by erupting volcanoes. Most of the hydrogen, because it is very light, escaped into outer space. Therefore, carbon dioxide was abundant in the atmosphere of early Earth. Eventually most of the carbon dioxide was used by living organisms that gave off oxygen. The carbon dioxide was also absorbed by the oceans. Nitrogen, a relatively unreactive gas, gradually became the principal component of our atmosphere.

WHAT IS THE STRUCTURE OF THE ATMOSPHERE?

In Chapter 1, you learned that the atmosphere is divided into four layers: *troposphere* (where most weather occurs), *stratosphere*, *mesosphere*, and *thermosphere*. The troposphere is the lowest layer. Although the troposphere is relatively thin, it is the most important layer because it contains most of the mass of the atmosphere.

The upper boundaries of the atmospheric layers have names that end with the suffix *-pause*, such as tropopause or stratopause. A changing trend in temperature identifies each boundary. For example, as you move higher within the troposphere, the temperature grows cooler. When you cross the tropopause and enter the stratosphere, however, the temperature of the air increases with increasing altitude. The table "Selected Properties of Earth's Atmosphere" in the *Earth Science Reference Tables* and Figure 1-3 on page 19 show the temperature changes that define the layers of the atmosphere.

CHANGES IN AIR TEMPERATURE

At Earth's surface, changes in air temperature tend to be cyclic. Two cycles control our weather: the daily weather cycle and the seasons. The temperature is usually lowest in the early morning and warmest at mid-afternoon. Similarly, the annual temperature in New York State is lowest in January, a month or so after the December solstice, and warmest in July, about a month after the June solstice, (You will learn about the cycles of sunlight in Chapter 10.) In addition to these cycles, short-term factors such as cloud cover and regional weather systems affect temperature. Clouds reduce daytime temperature by reflecting sunlight (solar energy) back into space. At night the clouds help hold heat energy to Earth.

Measuring Temperature

Temperature is a measure of the average vibrational energy of molecules. The higher the temperature, the more kinetic (vibrational) energy the atmosphere contains. Molecules are too small for us to see them vibrating, but we can feel and measure this energy.

We measure temperature with a thermometer. The most common kind of thermometer has a bulb that contains a liquid, such as colored alcohol. The liquid expands when it is heated, moving up the narrow, calibrated neck of the thermometer. When the temperature decreases, the liquid contracts and moves down the neck. You can make your own thermometer with a laboratory flask, a one-hole stopper, and a thin glass tube. When meteorologists record official air temperature readings, the thermometer is kept in the shade as shown in Figure 7-2.

Figure 7-2. A thermometer exposed to direct sunlight is heated by solar radiation. To measure air temperature accurately, a thermometer should be kept in the shade.

There are three temperature scales used to measure temperature: Fahrenheit, Celsius, and Kelvin. Figure 7-3 on page 172, taken from the *Earth Science Reference Tables,* allows you to convert between the three scales. Fahrenheit and Celsius temperatures are measured in units of degrees (°C and °F). However, Kelvin scale temperatures (K) have no degree sign. For example, 0°C = 32°F = 273K. The following Sample Problems are based on Figure 7-3.

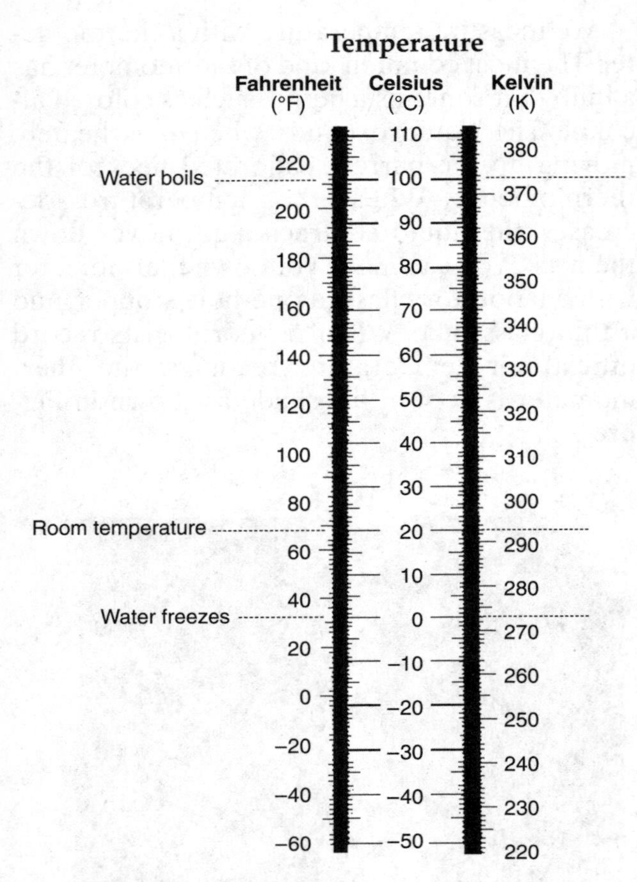

Figure 7-3. You are probably most familiar with the Fahrenheit scale of temperature. The Celsius scale divides the difference between the freezing and boiling points of water on Earth at sea level into 100 divisions. The Kelvin scale uses the same interval as the Celsius scale. However, kelvin temperatures start at absolute zero, which is considered to be the lowest possible temperature.

Sample Problems

1. What property of the atmosphere does a thermometer measure? (1) energy (2) volume (3) mass (4) density

Solution

A thermometer measures energy. (1)

2. What Fahrenheit temperature is equal to 20°C? (1) –2°F (2) 10°F (3) 68°F (4) 100°F

Solution

On page 13 of the *Earth Science Reference Tables*, go to the temperature chart. Find 20° on the Celsius scale, read across to the Fahrenheit scale, and you will find 68° (3).

3. Water vapor condenses when it is cooled to what temperature? (1) 32°C (2) 32°F (3) 100K (4) 373K

Solution

As the temperature drops, water condenses at 212°F, 100°C, or 373K as shown on the temperature chart from the *Earth Science Reference Tables*, (4).

WHAT CAUSES AIR PRESSURE?

Air pressure (atmospheric pressure) is caused by the weight of the atmosphere. You can prove that air has weight by using a sensitive scale to first weigh a deflated soccer ball, then inflating the ball and weighing it again. The inflated ball will weigh more than the empty ball.

Although air is relatively light, the atmosphere extends many kilometers above Earth's surface and therefore air exerts a pressure of nearly 15 pounds per square inch, or 1013.2 millibars. The **millibar** is the metric unit of atmospheric pressure. This is shown in Figure 7-4. Air pressure is greatest at Earth's surface and decreases with altitude.

You may have noticed that in this chapter not all measurements are given in metric, or SI, units. Most weather maps in newspapers and on television use the more familiar measurements of Fahrenheit temperature and wind speeds in miles per hour. Therefore these traditional units have been used in this portion of this book.

Because air pressure is exerted in all directions, we are seldom aware of it. However, air pressure allows you to drink liquids through a straw. You may think that when you drink through a straw, your lungs pull water up through a straw. However, you really cannot pull on water. In fact, as you begin to drink through a straw, you lower the air pressure in the straw. The air pressure on the surface of the liquid outside the straw is now higher than the pressure in the straw. The pressure on the liquid outside the straw pushes the liquid up the straw and into your mouth. Normal atmospheric pressure is so great that in a tube, such as a water pipe, it can

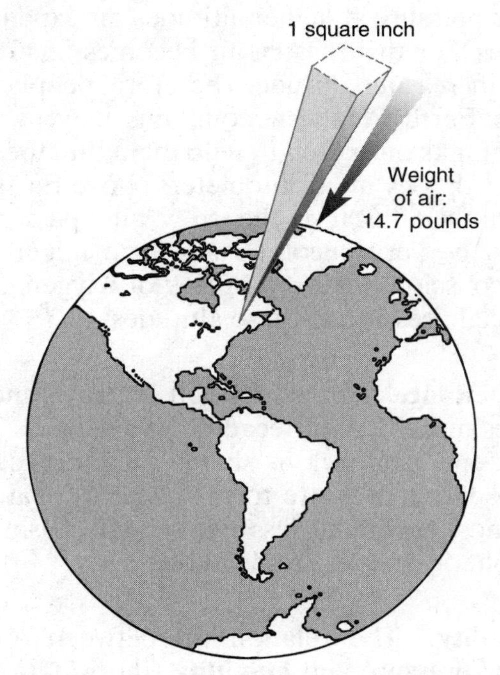

Figure 7-4. Air pressure is caused by the weight of the atmosphere. At sea level above each square inch of Earth's surface, there is a column of air that weighs 14.7 pounds. Most of the mass of the atmosphere is in the troposphere. The upper atmosphere becomes very thin with increasing altitude.

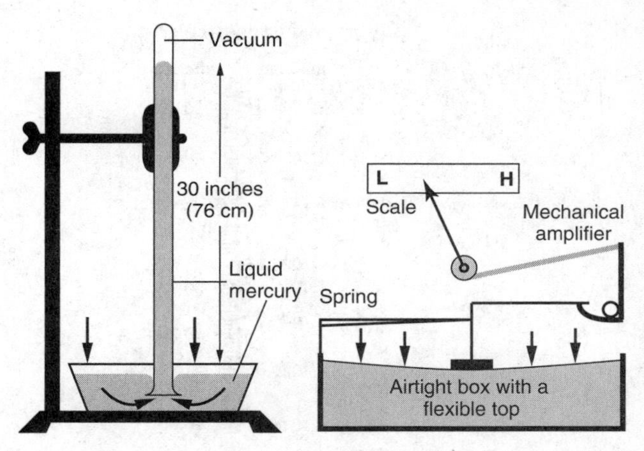

Figure 7.5. Barometers are used to measure air pressure. In a simple mercury barometer, the force of atmospheric pressure balances the weight of a column of liquid mercury. An aneroid barometer contains a sealed box that expands and contracts with changes in air pressure. This motion is linked to a pointer on a scale.

push water to a height of nearly 10 meters (33 feet). The water would not move above that height, even if you could create a total vacuum.

Measuring Air Pressure

The fact that air pressure can support a column of liquid is the principle behind the **barometer**, an instrument used to measure air pressure. However, instead of using water, which would require a very long tube, scientists more often use the dense, liquid-metal mercury. Atmospheric pressure can support a column of mercury about 76 centimeters (30 inches) high. This is the principle behind the first barometers. Most modern barometers do not use mercury, water, or any other liquid. They are called aneroid barometers. They use the pressure air exerts on an airtight box to rotate a pointer across a calibrated dial. Figure 7-5 shows both types of barometers.

All barometers measure the effect of the weight of the atmosphere. Meteorologists measure air pressure in millibars (mb). The average sea level pressure is about 1013.2 millibars, which is also known as 1 atmosphere. Figure 7-6, also found in the *Earth Science Reference Tables*, links millibars and inches of mercury. The following Sample Problems are based on Figure 7-6 on page 174.

Sample Problems

4. What units of measure are commonly used to indicate air pressure? (1) millibars and kelvins (2) millibars and inches of mercury (3) inches and meters (4) inches of mercury and kelvins

Solution

On page 13 of the *Earth Science Reference Tables*, use the pressure chart, which shows that pressure is measured in inches and millibars (2).

5. Which measurement is the equivalent of 1000 millibars? (1) 29.5 inches of mercury (2) 29.53 inches of mercury (3) 29.56 inches of mercury (4) 1013.2 inches of mercury

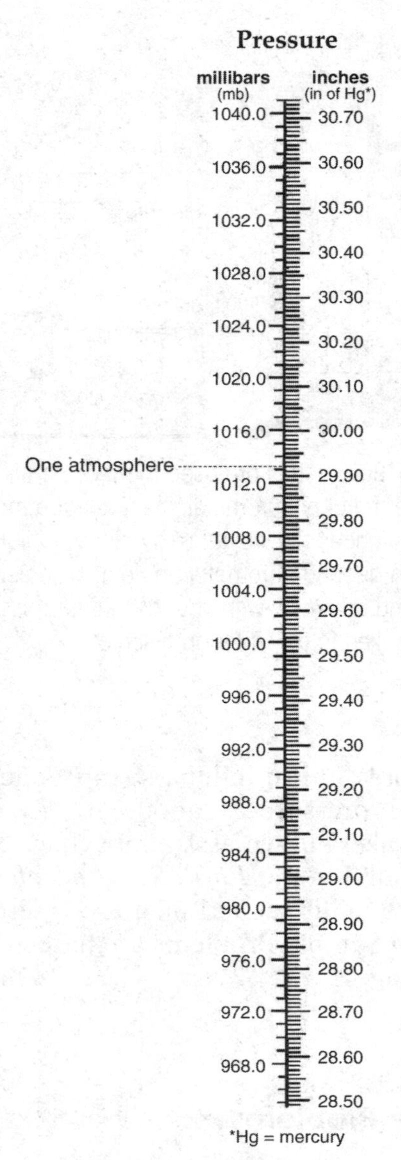

Figure 7-6. You can use this *Reference Tables* chart to convert air pressure readings from millibars to inches of mercury or from inches to millibars.

Solution

As shown on the pressure chart of the *Earth Science Reference Tables*, 1000 millibars equals 29.53 inches of mercury (2).

Factors That Affect Atmospheric Pressure

Altitude Atmospheric pressure decreases with altitude because as air rises, there is less of the atmosphere above it pushing down. Because there is less pressure at higher altitudes, air expands as it rises. For this reason, air becomes less dense with increasing altitude. Therefore, people who climb Earth's highest mountains may need to carry tanks of oxygen to help them breathe. Airplanes that fly many kilometers above Earth are pressurized because the crew and passengers would be very uncomfortable and might even lose consciousness from a lack of oxygen at the reduced pressure at these altitudes.

Temperature If air is cooled, it contracts and becomes more dense. Because the air exerts pressure on whatever is below it, this causes atmospheric pressure to rise. If air is heated, it expands, becoming less dense and causing atmospheric pressure to decrease.

Humidity The relationship between atmospheric pressure and **humidity** (the water vapor content of the air) is less obvious. It is helpful to know that at a constant temperature and pressure, a given volume of any gas contains the same number of molecules. As a result, when water evaporates into the atmosphere and becomes water vapor, the air must expand to take in the added molecules. Since air is free to move in the atmosphere, the air can expand horizontally and vertically.

Humid air is actually less dense than dry air. This is because the mass of water molecules (18 amu) is less than the mass of nitrogen (28 amu) and oxygen (32 amu) molecules. As you can see in Figure 7-7, when water molecules enter the atmosphere, they displace heavier air molecules (mostly nitrogen). When heavy molecules are replaced by lighter molecules the density of the air becomes lower. The lower air density results in lower atmospheric pressure. Therefore, atmospheric pressure usually decreases as humidity increases.

MOISTURE IN THE ATMOSPHERE

Of all the gases in the atmosphere, the amount of water vapor is the most variable. Each gas exists independently of the other gases. A sponge can soak up water. Within the atmosphere, nothing "soaks up" water vapor. Water vapor must displace other gas molecules.

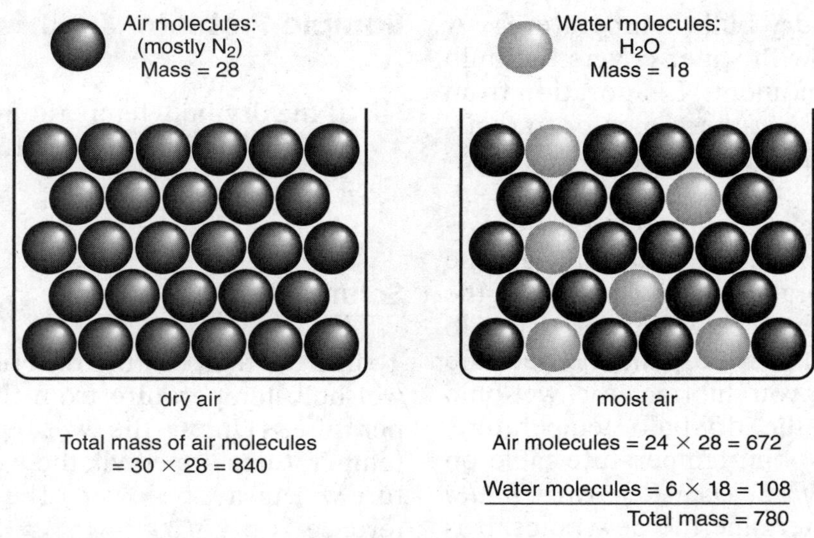

Figure 7-7. When moisture is added to air, the air becomes less dense. Water vapor is less dense than air (mostly nitrogen). Therefore, when water vapor is added to air, the light water molecules replace the heavier nitrogen molecules. (The units of mass in these diagrams are atomic mass units, amu.)

When the air includes as much water vapor as it can, the air is **saturated.** The air's ability to contain water vapor depends on the temperature. The higher the temperature, the more moisture the atmosphere can contain before it becomes saturated. For every 10°C increase in temperature, air can contain approximately twice as much water vapor. Compared with warm air, cooler air will become saturated when it contains less water vapor. It follows that, if the temperature of moist air is lowered enough, the air will become saturated. The **dew point** is defined as the temperature to which the air must be cooled to become saturated. If the temperature falls below the dew point, **condensation** usually occurs as water vapor changes to liquid water.

Measuring Moisture in the Atmosphere

Meteorologists use several methods to determine the moisture content of the atmosphere. They use a sling psychrometer (sigh-CRAH´-met-er) and a dew-point-temperature table to determine the dew point. A **sling psychrometer**, as shown in Figure 7-8, has two thermometers mounted so that they can be swung through the air. One thermometer records the

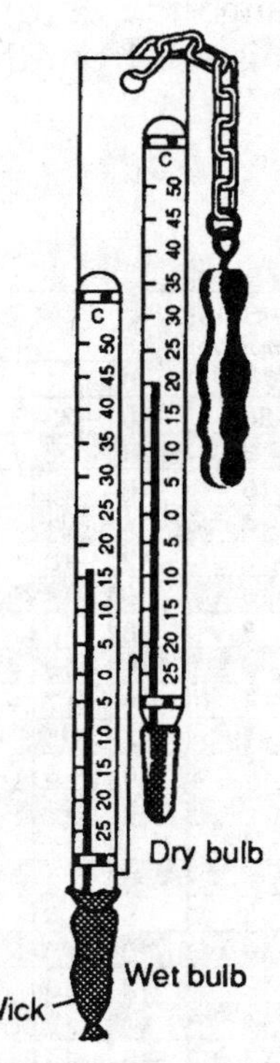

Figure 7-8. The sling psychrometer has two thermometers mounted side by side on a narrow frame. One thermometer measures the air temperature. A wet cloth covers the bulb of the other thermometer. As the thermometers are swung through the air, evaporational cooling causes the wet-bulb thermometer to register a lower temperature. The drier the air, the greater the evaporational cooling.

air temperature (dry-bulb temperature). A cloth wick soaked with water covers the bulb of the second thermometer. Evaporation from the wick causes the wet thermometer to register a lower temperature (wet-bulb temperature).

The cooling effect of evaporation is the key to measuring the humidity. The drier the air, the greater the evaporational cooling, and the greater the difference between the dry-bulb temperature (air temperature) and the wet-bulb temperature. After you subtract the wet-bulb temperature from the dry-bulb temperature, you can use a dew-point-temperature table on page 12 of the *Earth Science Reference Tables* or Figure 7-9 to determine the dew point. It is important to remember that although the dew point is expressed in degrees Celsius (or Fahrenheit), we use dew point to indicate the water vapor content of the atmosphere. Therefore, the higher the dewpoint, the more moisture is present in the atmosphere. You should also note that although the dew point is a temperature, the only way to change the dew point is by adding or removing moisture from the air.

Sample Problem

6. If the dry-bulb temperature is 20°C and the wet-bulb temperature is 15°C, find the dew point.

Solution

To find the temperature difference, subtract the wet-bulb temperature from the dry-bulb temperature. (Unless the wet-bulb and dry-bulb temperatures are equal, the wet-bulb temperature will always be lower.) The temperature difference is 5°C (20° − 15° = 5°). This step of subtracting is the one most often forgotten by students. Now, turn to the dew-point temperature table. At the left side of the table locate 20°C; follow that row over to where it meets the column coming down from 5°C (the temperature difference). They meet at 12°C. Therefore, the dew point is 12°C, and the air would become saturated if it were cooled to that temperature.

Dewpoint (°C)

Dry-Bulb Temperature (°C)	Difference Between Wet-Bulb and Dry-Bulb Temperatures (C°)															
	0	1	2	3	4	5	6	7	8	9	10	11	12	13	14	15
−20	−20	−33														
−18	−18	−28														
−16	−16	−24														
−14	−14	−21	−36													
−12	−12	−18	−28													
−10	−10	−14	−22													
−8	−8	−12	−18	−29												
−6	−6	−10	−14	−22												
−4	−4	−7	−12	−17	−29											
−2	−2	−5	−8	−13	−20											
0	0	−3	−6	−9	−15	−24										
2	2	−1	−3	−6	−11	−17										
4	4	1	−1	−4	−7	−11	−19									
6	6	4	1	−1	−4	−7	−13	−21								
8	8	6	3	1	−2	−5	−9	−14								
10	10	8	6	4	1	−2	−5	−9	−14	−28						
12	12	10	8	6	4	1	−2	−5	−9	−16						
14	14	12	11	9	6	4	1	−2	−5	−10	−17					
16	16	14	13	11	9	7	4	1	−1	−6	−10	−17				
18	18	16	15	13	11	9	7	4	2	−2	−5	−10	−19			
20	20	19	17	15	14	12	10	7	4	2	−2	−5	−10	−19		
22	22	21	19	17	16	14	12	10	8	5	3	−1	−5	−10	−19	
24	24	23	21	20	18	16	14	12	10	8	6	2	−1	−5	−10	−18
26	26	25	23	22	20	18	17	15	13	11	9	6	3	0	−4	−9
28	28	27	25	24	22	21	19	17	16	14	11	9	7	4	1	−3
30	30	29	27	26	24	23	21	19	18	16	14	12	10	8	5	1

Figure 7-9.

Relative Humidity (%)

Dry-Bulb Temperature (°C)	Difference Between Wet-Bulb and Dry-Bulb Temperatures (C°)															
	0	1	2	3	4	5	6	7	8	9	10	11	12	13	14	15
−20	100	28														
−18	100	40														
−16	100	48														
−14	100	55	11													
−12	100	61	23													
−10	100	66	33													
−8	100	71	41	13												
−6	100	73	48	20												
−4	100	77	54	32	11											
−2	100	79	58	37	20	1										
0	100	81	63	45	28	11										
2	100	83	67	51	36	20	6									
4	100	85	70	56	42	27	14									
6	100	86	72	59	46	35	22	10								
8	100	87	74	62	51	39	28	17	6							
10	100	88	76	65	54	43	33	24	13	4						
12	100	88	78	67	57	48	38	28	19	10	2					
14	100	89	79	69	60	50	41	33	25	16	8	1				
16	100	90	80	71	62	54	45	37	29	21	14	7	1			
18	100	91	81	72	64	56	48	40	33	26	19	12	6			
20	100	91	82	74	66	58	51	44	36	30	23	17	11	5		
22	100	92	83	75	68	60	53	46	40	33	27	21	15	10	4	
24	100	92	84	76	69	62	55	49	42	36	30	25	20	14	9	4
26	100	92	85	77	70	64	57	51	45	39	34	28	23	18	13	9
28	100	93	86	78	71	65	59	53	47	42	36	31	26	21	17	12
30	100	93	86	79	72	66	61	55	49	44	39	34	29	25	20	16

Figure 7-10.

Finding Relative Humidity

Relative humidity compares how much moisture is actually in the air with how much moisture could be there if the air were saturated. Air is saturated if it contains all the moisture it can at its present temperature. Relative humidity is expressed as a percent of saturation. For air that is *saturated*, the relative humidity is 100 percent. To determine relative humidity you need a sling psychrometer and a relative humidity table such as Figure 7-10 from the *Earth Science Reference Tables*.

Sample Problem

7. Find the relative humidity when the dry-bulb temperature is 14°C and the wet-bulb temperature is 9°C.

Solution

First, find the temperature difference. It is 5°C (14° − 9° = 5°). On the relative humidity table, follow the row across from 14° and the column down from 5°. As you can see, they meet at 50. Therefore, under these conditions, the relative humidity is 50 percent.

When the relative humidity is 50 percent, it means that the air could actually contain twice as much water vapor as it actually has. If the relative humidity were 20 percent, then the air would contain only one-fifth the water vapor it could have before it became saturated. Figure 7-11 on page 178 illustrates how changes in air temperature and changes in the amount of water vapor in the air affect the relative humidity. As temperature of the air approaches the dew point, the relative humidity approaches 100 percent.

QUESTIONS

Part A

1. Scientists infer that Earth's atmosphere came primarily from (1) erosion of Earth's

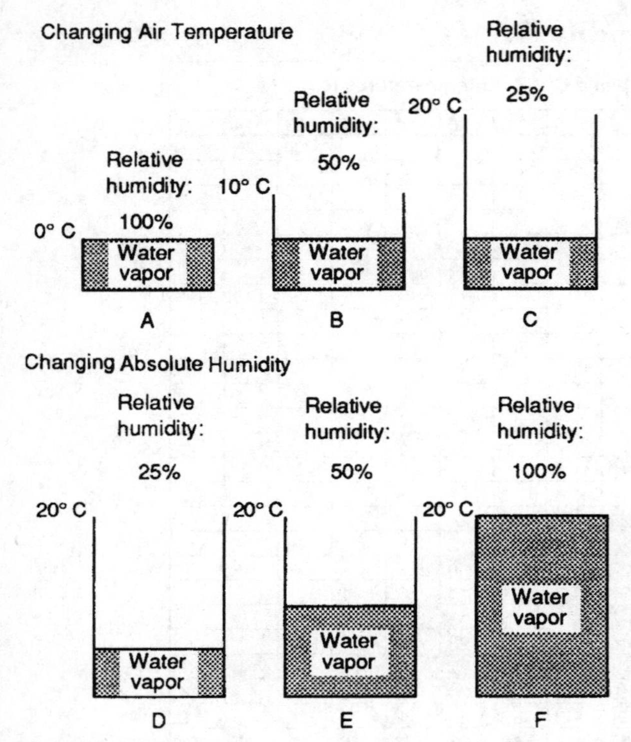

Figure 7-11. Changing air temperature and changing the moisture content of the air also change the relative humidity. In diagrams *A–C,* you can see that the air is able to contain more moisture as it gets warmer. If moisture is not added to the atmosphere, the relative humidity decreases as the temperature increases. In diagrams *D–F,* the temperature remains constant. As moisture is added, the relative humidity increases. Air does not actually "hold" water vapor as suggested here. In fact, the water vapor can be present whether or not other gases are present.

surface (2) decay of marine microorganisms (3) magma in volcanic eruptions (4) radioactive decay of unstable atoms

2. A weather balloon was launched into the atmosphere to record the air temperature at various altitudes. The instrument broke when it reached a temperature of 95°C. How high above sea level was the balloon when the instrument malfunctioned? (1) 70 km (2) 75 km (3) 100 km (4) 120 km

3. A student used a sling psychrometer to find the relative humidity. If the wet-bulb temperature and the dry-bulb temperature were both 20°C, what was the relative humidity? (1) 20°C (2) 20% (3) 100°C (4) 100%

4. A student measured the air temperature and the dew point in the morning and in the afternoon on the same day. The difference between the two was 12°C in the morning and 4°C in the afternoon. How was the weather changing on that day? (1) The relative humidity was decreasing and the precipitation was becoming more likely. (2) The relative humidity was decreasing and the precipitation was becoming less likely. (3) The relative humidity was increasing and the precipitation was becoming more likely. (4) The relative humidity was increasing and the precipitation was becoming less likely.

5. A student used a sling psychrometer to measure the humidity of the air. If the relative humidity was 65% and the dry-bulb temperature was 10°C, what was the wet-bulb temperature? (1) 5°C (2) 7°C (3) 3°C (4) 10°C

6. A student used a sling psychrometer to find the wet-bulb and dry-bulb temperatures. Both thermometers registered 15°C. Which statement below best describes the atmospheric conditions? (1) The dew point is 0°C. (2) The atmosphere is very dry. (3) The air is saturated. (4) The air pressure is relatively high.

7. Which graph below best shows how air pressure is affected by changes in humidity?

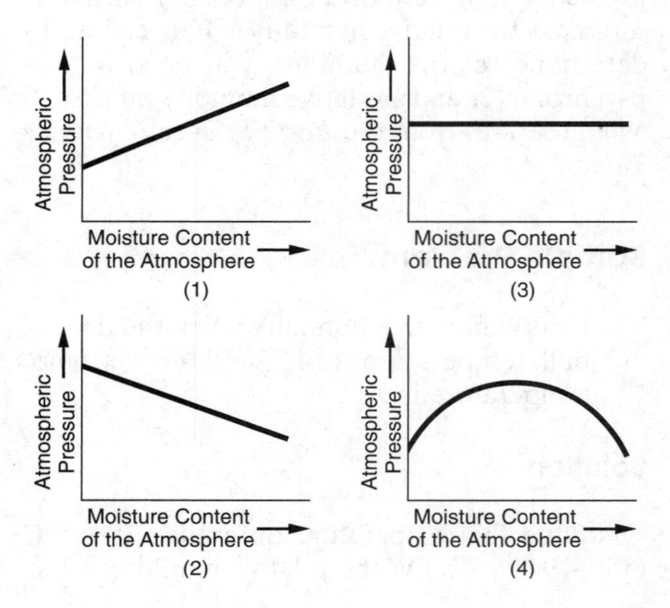

Part B

8. A parcel of air has a dry-bulb temperature of 24°C and a relative humidity of 55%. What is the dew point of this parcel of air? (1) 6°C (2) 14°C (3) 24°C (4) 30°C

9. The data table below shows winter weather observations at a location in New York State.

Air temperature (dry-bulb temperature)	0°C
Relative humidity	81%
Present weather	Snow

What was the dew point at this time? (1) 1°C (2) 2°C (3) –3°C (4) –5°C

Base your answers to questions 10 through 12 on the image below, which shows a climber near the top of Earth's highest summit, Mt. Everest in Asia between Nepal and Tibet.

10. Mt. Everest is nearly 9000 meters above sea level. In what part of the atmosphere is the summit of Everest? (1) troposphere (2) stratosphere (3) mesosphere (4) thermosphere

11. The climber in the photo is wearing heavy clothing because of the winds and cold temperatures, even in the summer. Why is it so cold at the top of the highest mountains? (1) Mt. Everest is near the South Pole. (2) Earth's climate is becoming cooler. (3) As air is compressed, its temperature falls. (4) As air expands, its temperature decreases.

12. What is the approximate air pressure at the top of the stratosphere? (1) 5×10^{-4} atmosphere (2) 5×10^{-2} atmosphere (3) 5×10 atmospheres (4) 5×10^{2} atmospheres

WHAT CAUSES THE WIND?

Wind is the natural movement of air along, or parallel to, Earth's surface. In general, winds are the result of uneven heating of Earth's surface. This uneven heating causes differences in air pressure to develop. Cool, dry air is generally more dense than warm moist air. Therefore, cool places usually have relatively high air pressure and warmer places have lower air pressure. Winds always blow from areas of high pressure to areas of low pressure.

Furthermore, winds blow fastest where the gradient in air pressure is greatest. On a map, that is where the *isobars* (lines of equal atmospheric pressure) are close together. Figure 7-12 on page 180 shows how winds blow from high pressure to low pressure. The wind speed depends on the change in air pressure (the **pressure gradient**). Although Earth's rotation does not cause winds, it does influence the directions in which they travel. The Coriolis effect will be explained shortly.

Land and Sea Breezes

As shown in Figure 7-13 on page 180, on a hot summer day the land heats more quickly than the ocean. Rock and soil have a relatively low **specific heat**. This means that when land and water absorb the same amount of energy, the land areas heat up more quickly. As the land warms, low pressure develops over the land as the air warms and rises. The result is a cool **sea breeze** that blows off the water. At night, the land cools quickly, which cools the air above it. Now the air over the water is warmer than the air over the land. Therefore, low pressure develops over the water. The result is a **land breeze** that blows at night from the land to the sea.

The Coriolis Effect

Earth's rotation influences the direction of the winds. If Earth were not spinning, the pattern of wind circulation would be relatively simple. Winds would blow from the poles straight to the

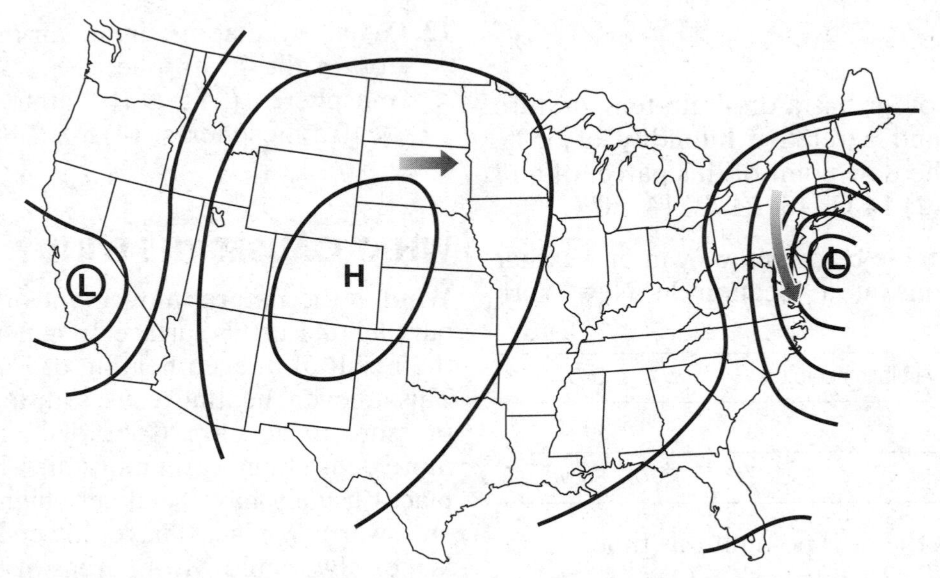

Figure 7-12. The isolines on this map are called isobars. Isobars show changes in atmospheric pressure. Notice three things. First, winds, as shown by the two arrows, blow from higher toward lower pressure. Second, the length of the arrows shows that winds are stronger where the isobars are closer together. (A strong pressure gradient.) Finally, over long distances the winds do not blow directly from high to low. This is due to the Coriolis effect.

equator. However, Earth rotates, which causes the path of the winds to curve, as shown in Figure 7-14. Rather than blowing straight from areas of high pressure, such as the North Pole, to places of low pressure, such as the equator, the winds in the Northern Hemisphere curve to the right of their original path. This curvature is called the **Coriolis effect**. However, in the Southern Hemisphere, the Coriolis effect causes winds to appear to curve to the left of their original path.

It can be hard to distinguish between a right curve and a left curve. It is essential that you think in terms of the air moving in the direction of the arrow. Therefore, if the wind is blowing toward you, as the wind turns to its right, you will see it curve to your left. Figure 7-15 shows four lines curving to their right. This is why the black

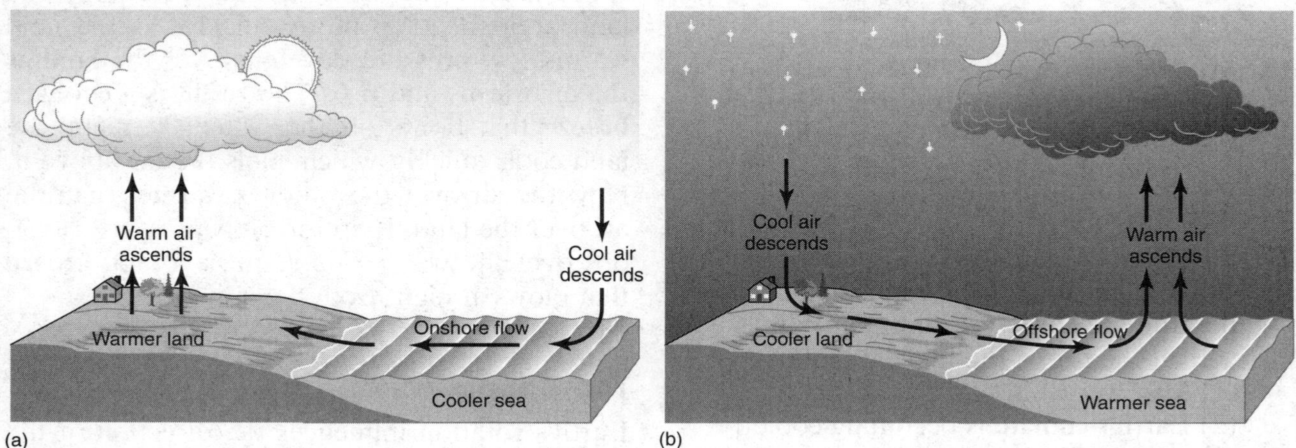

Figure 7-13. If regional weather systems do not dominate, heated, low-density air over the land rises. This draws in a sea breeze toward the shore during the day (*a*). At night, the land cools more than the water. The cool, denser air blows in the opposite direction out to sea (*b*).

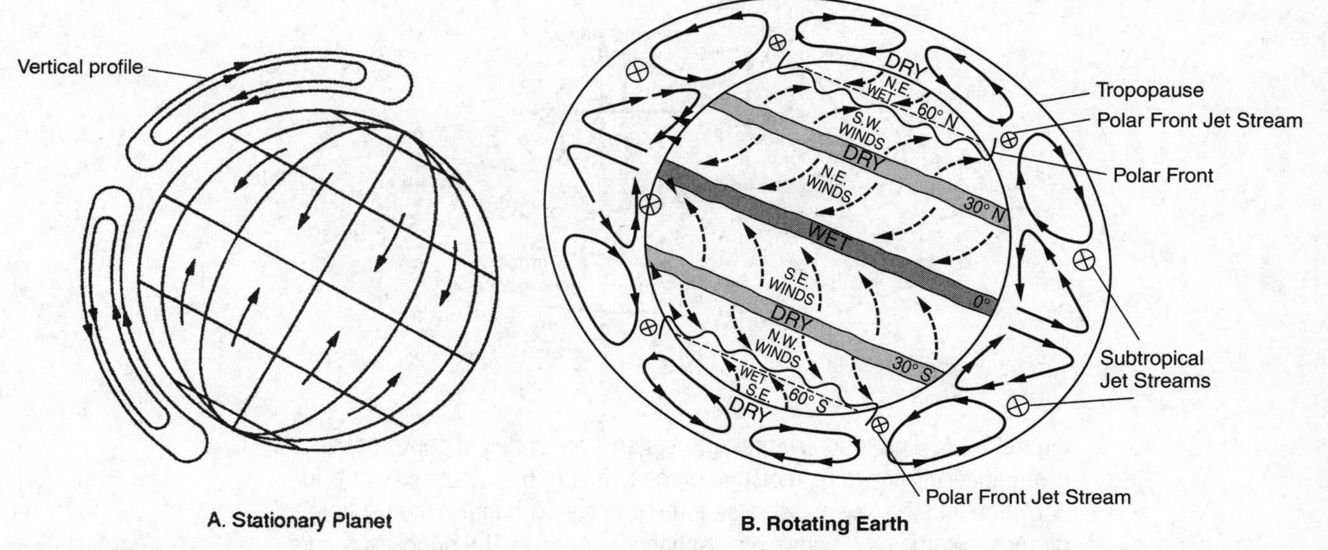

Figure 7-14. If Earth did not spin, wind patterns would be relatively simple. However, because of Earth's rotation, winds curve to the right in the Northern Hemisphere and to the left in the Southern Hemisphere. This causes the pattern to break into three giant convection cells in each hemisphere.

wind arrows in Figure 7-12 on page 180 do not blow straight from the high-pressure center toward the low. The Coriolis effect turns them to the right.

In the Northern Hemisphere, the winds blowing out of a high-pressure area turn clockwise. The winds blowing into a low-pressure area turn counterclockwise. Figure 7-16 on page 182 illustrates how surface winds are part of atmospheric convection currents that include a return flow high above Earth. Winds that converge into a low at the surface and rise into the atmosphere must produce winds that diverge aloft. Areas of low pressure are called cyclones, and areas of high pressure are called anticyclones.

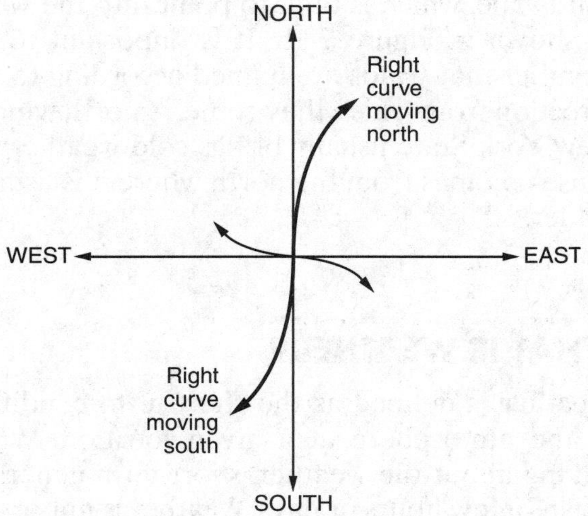

However, when regional winds blow into a low-pressure region, they actually begin to curve in the opposite direction. If they did not, the winds would not be able to converge toward the center of the low-pressure area. Instead of continuing their curve to the right, which would drive them against the pressure gradient, they curve counterclockwise (toward the left) and into the center of the low pressure. So winds curve to the right (clockwise) out of a high-pressure area and reverse their direction of curvature, turning left (counterclockwise) into the low pressure center as you can see in Figure 7-17 on page 182.

Figure 7-15. This diagram shows the four compass directions and four lines curving to their right. Left or right is defined according to the direction of movement.

Jet Stream Winds

Jet streams are very fast wind currents in the upper atmosphere. These upper-level winds seldom follow surface winds. They usually blow from west to east in mid-latitudes where cold, polar air meets warmer, tropical air. Jet stream winds are important because they influence the development and movement of storm systems. Looking back at Figure 7-14, you will see the

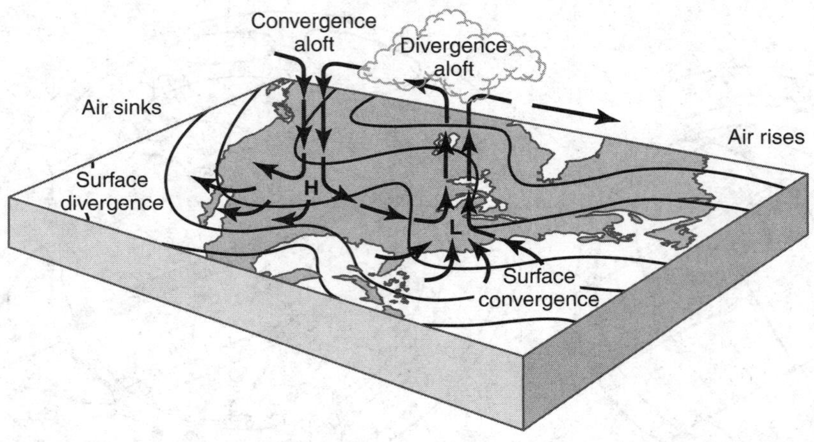

Figure 7-16. High- and low-pressure systems are zones of convergence and divergence. Rising warm, moist air at the center of a low causes winds to converge into the low-pressure system. However, high in the atmosphere this air must diverge as a part of atmospheric convection. The opposite occurs where the atmospheric pressure is high at the surface. Divergence at the surface is supported by convergence high in the atmosphere.

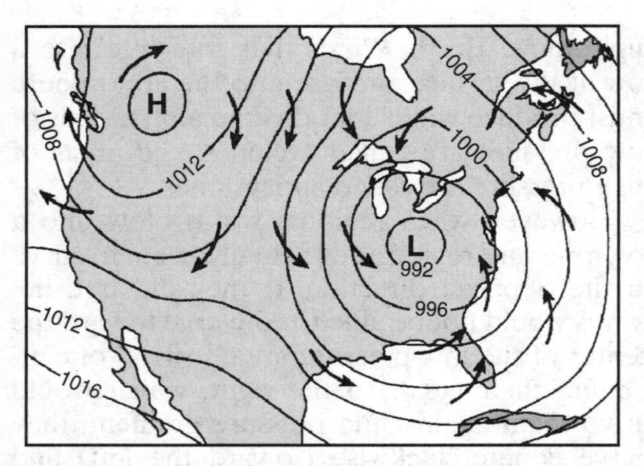

Figure 7-17. Winds around high- and low-pressure centers.

typical positions of Earth's jet stream winds shown by circles with *X*s in them.

Prevailing Winds and Ocean Currents

Prevailing winds blow more often from one direction than from any other. If you compare Earth's prevailing winds, shown in Figure 7-14, with the surface ocean currents shown on page 4 of the *Earth Science Reference Tables*, you will find that the winds and ocean currents generally move in the same direction. This is not just a coincidence. Wind is one factor that moves the surface waters of Earth's oceans. Like the winds, ocean currents also curve in response to the Coriolis effect. Usually, ocean currents curve to the right of their original direction in the Northern Hemisphere and to the left of their original direction in the Southern Hemisphere.

Measuring the Wind

To fully describe winds, you need to determine the wind speed and the wind direction. Wind speed is measured with an **anemometer**. The cups on an anemometer catch the wind, causing the anemometer to spin. An indicator shows the wind speed. The wind direction is shown by a wind vane, which is built to point into the wind, as shown in Figure 7-18. It is important to remember that winds are named according to the direction from which they come. A north wind in New York State usually brings cold weather because it comes from the north where it is usually colder.

WHAT IS WEATHER?

Weather is defined as the short-term condition of the atmosphere at a given location. When talking about the weather, short-term generally means a few hours or days. Weather is important to everyone. No matter what your outdoor plans, it is likely that weather will be an important issue. The weather conditions that concern us

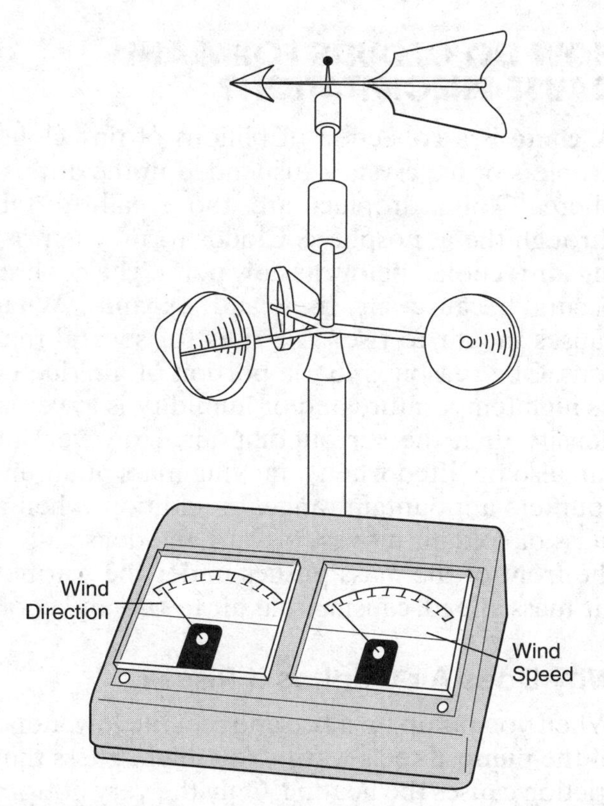

Figure 7-18. An anemometer, which measures wind speed with rotating cups, is often combined with a wind vane, which tells wind direction.

most are temperature, sky conditions, precipitation, atmospheric pressure, humidity, wind speed, and wind direction. Scientists who study and predict the weather are called **meteorologists**.

It may help you to understand weather if you think of the workings of the atmosphere in terms of energy distribution. Earth gets most of its energy in the form of radiation (heat and light) from the sun. However, the distribution of solar energy is not the same all over the planet. The tropical regions, near the equator, receive the most solar energy. Toward the poles, which receive the least solar energy, the strength of the sunlight decreases. Weather helps distribute solar energy over Earth's surface. If it were not for its atmosphere and weather, Earth would be a much less comfortable place. We would face extreme temperature changes over Earth's surface.

Figure 7-19 shows that there is extra heat energy in the tropics (around the equator) where absorbed solar radiation is greater than the energy Earth emits as infrared energy. However, at

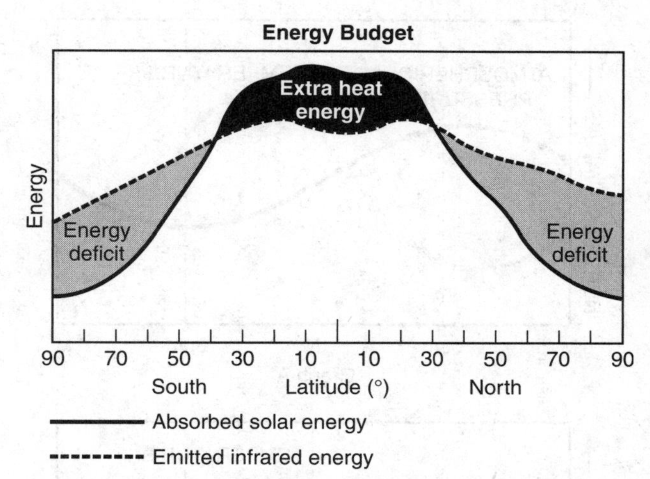

Figure 7-19. Earth absorbs more energy in the tropics than near the poles. The solid line shows the solar energy absorbed by Earth at various latitudes. The atmosphere and oceans move the excess energy away from the equator and toward the poles. The result is the dotted line that shows energy has been moved by air and water to moderate surface temperatures.

the poles Earth emits more infrared energy than it absorbs from solar radiation, causing a deficit. Air and water generally carry excess heat energy from the region around the equator to places where it is usually cold. Earth would be hotter in the tropics and colder than it is now in the polar regions if it were not for the energy transported by the atmosphere and the oceans.

HOW ARE WEATHER VARIABLES RELATED?

The daily temperature cycle of the atmosphere affects air pressure, wind speed, and relative humidity. Figure 7-20 on page 184 shows the relationship between temperature and other weather variables. Air pressure is indirectly related to temperature. During the afternoon, when the air temperature is usually highest, the air expands and becomes lighter. This causes atmospheric pressure to decrease. Then, at night when the air cools, air pressure increases as shown in Figure 7-20, graph *A*.

Winds are primarily the result of uneven heating of the atmosphere, most of which occurs during the day. The wind speed is usually highest when the temperature is highest. See Figure 7-20, graph *B*.

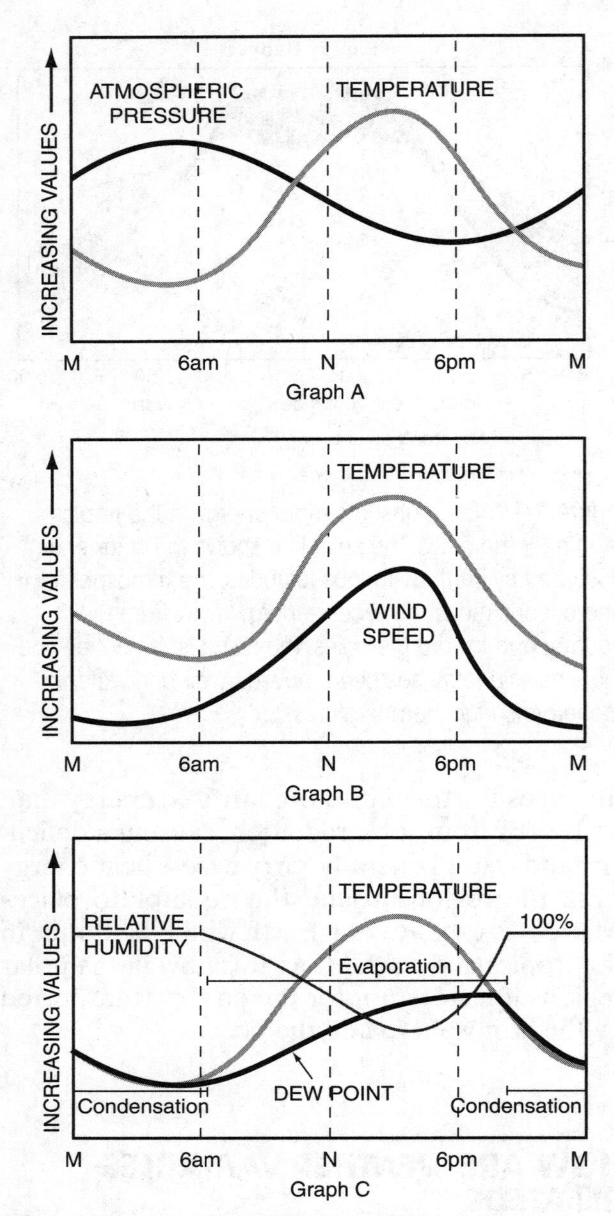

Figure 7-20. The daily cycles of atmospheric change are caused by changes in temperature, air pressure, and humidity. Increasing temperature generally causes air pressure to decrease, wind speeds to increase, evaporation to increase, and relative humidity to decrease.

The dew point and relative humidity often follow a daily cycle of change. Water vapor leaves the atmosphere by condensation at night and enters the air by evaporation during the day. The dew point rises until the air temperature drops to the dew point in the evening. At this point, the relative humidity is 100 percent, where it remains, and condensation occurs until the air temperature rises above the dew point in the morning. These changes are shown in Figure 7-20, graph *C*.

HOW DO CLOUDS FORM AND CAUSE PRECIPITATION?

A **cloud** is a collection of billions of tiny cloud droplets or ice crystals suspended in the atmosphere. Cloud droplets are too small to fall through the atmosphere. Clouds form when rising air is cooled below its dew point. The cooling occurs because air rises and expands. What causes the air to rise? Air rises for several reasons. One reason is that a portion of air, due to its high temperature and/or humidity, is lower in density than the surrounding air. However, air can also be lifted when a moving mass of air encounters a mountain range. In addition, when a mass of cold air moves forward, the dense air at the front of the mass pushes under the warmer air mass ahead, causing that air to rise and cool.

Why Does Air Cool as It Rises?

When you pump up a tire and feel the lower end of the pump, it feels warm. You might guess that friction causes the heating. Only the very bottom of the pump, where the air is compressed the most, becomes very warm. It takes mechanical energy to compress air, and that energy changes to heat, which makes the pump warm. The opposite occurs as air rises into the atmosphere. The rising air expands due to decreasing pressure. Just as compression warms a gas, this expansion of rising air causes it to become cooler.

Condensation nuclei are tiny particles in the atmosphere on which water can condense. If the air is cooled below the dew point and condensation nuclei are present, water vapor will condense on them and a cloud will form. If the air is too clean, the temperature can actually fall below the dew point without condensation occurring. A variety of processes add these particles to the air: salt spray from the oceans, dust storms, fires, and burning of fossil fuels.

Figure 7-21 shows that as air rises, the temperature drops relatively quickly. But the dew point does not decrease as quickly. At the altitude where the air temperature reaches the dew point, a cloud starts to form. This is why clouds often have a flat cloud base.

What is Precipitation?

Precipitation is water that falls from the sky. It may be in the form of rain, snow, sleet, hail, or

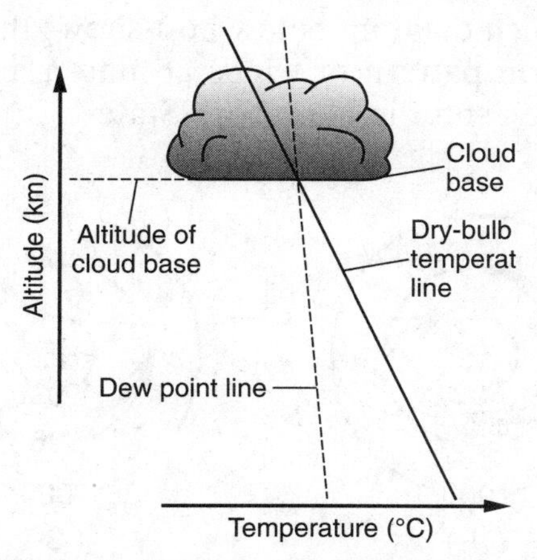

Figure 7-21. As air rises, the air temperature decreases more quickly than the dew point. When the air temperature reaches the dew-point temperature, a cloud begins to form.

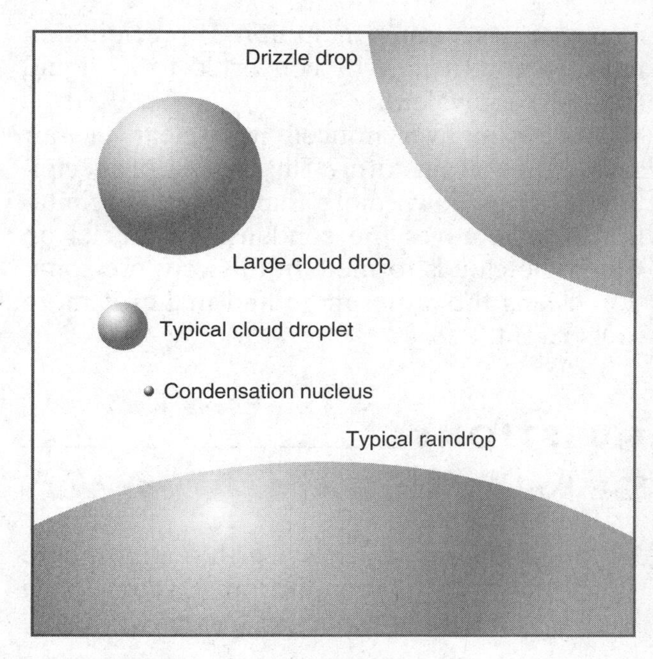

Figure 7-22. Cloud droplets can remain suspended in the atmosphere because they are so much smaller than droplets that fall as precipitation. The upper size limit for cloud droplets is approximately 0.2 mm in diameter, but most are much smaller.

other less common forms of precipitation. What causes precipitation? Scientists know that the cloud droplets (or ice crystals) must combine and become heavy enough to fall through the atmosphere. However, scientists do not know exactly what causes them to combine. Since a raindrop may be a million times the size of a cloud droplet, it is clear that many cloud droplets must unite to make each drop of rain. Figure 7-22 shows the great difference in the size of cloud droplets and raindrops.

Actually, most precipitation starts as snow. High in the atmosphere, the temperature is usually well below the freezing temperature of 0°C. As the snowflakes fall to lower elevations, the air temperature is above the freezing point, and the snow melts, changing to rain.

Although rain and snow are the most common forms of precipitation, there are several others. Drizzle is made up of small raindrops that fall slowly. Sleet is a partially frozen mixture of rain and snow that occurs when the surface temperature is just above freezing. Hail is precipitation in the form of ice balls, which usually occurs in violent thunderstorms. Hailstones begin as snowflakes that start to melt and gather more moisture as they fall. Updrafts in the storm cloud repeatedly push the partially melted snow back up into cooler air high in the cloud. Here the coating of water freezes. This process of falling and rising continues until the hailstones are too heavy to be carried by the updrafts. Hailstones larger than baseballs sometimes fall in the American Midwest.

Precipitation is measured with a rain gauge shown in Figure 7-23. This instrument can be as simple as a cylindrical container open at the top. The amount of rain is reported as a depth in

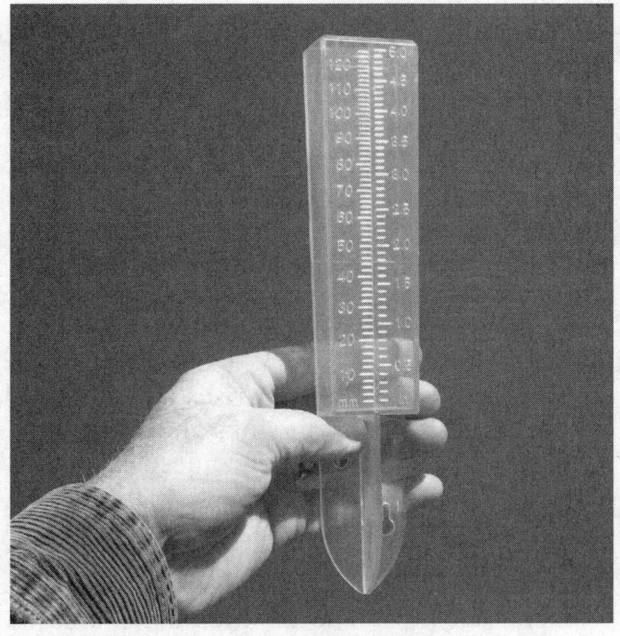

Figure 7-23. Precipitation is usually measured with a relatively simple graduated container called a rain gauge.

inches or other convenient unit. If precipitation falls as snow, it may be converted to its liquid (rainfall) equivalent.

Have you ever noticed how clear the air seems after a rainstorm? The process of precipitation brings down more than water. Precipitation also removes the condensation nuclei on which the clouds formed. In this way, precipitation cleans the atmosphere and makes it more transparent.

QUESTIONS

Part A

13. Solar energy reaches Earth's atmosphere primarily by (1) conduction (2) convection (3) radiation (4) nuclear fusion

14. The winds that blow snow and clouds over the tops of mountains are a part of Earth's energy distribution by (1) conduction (2) radiation (3) convection (4) deposition

15. The four diagrams below show different patterns of vertical air movement above cool and warm places on Earth's surface. Which pattern is most likely above places with these temperature conditions?

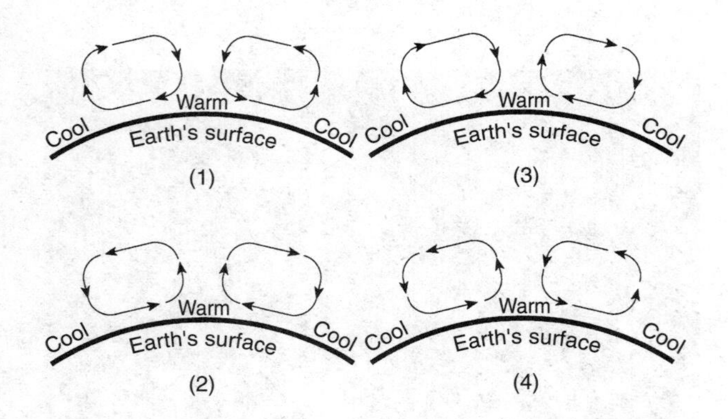

16. Most clouds form when (1) moist air rises and cools (2) moist air rises and becomes warmer (3) moist air sinks and cools (4) moist air sinks and becomes warmer

17. When a mass of air passes over Lake Erie and picks up moisture, which change in the air is most likely? (1) The air temperature increases. (2) The air pressure increases. (3) The density of the air decreases. (4) The relative humidity decreases.

18. Which diagram below best shows the circulation pattern of winds around a low-pressure center in New York State?

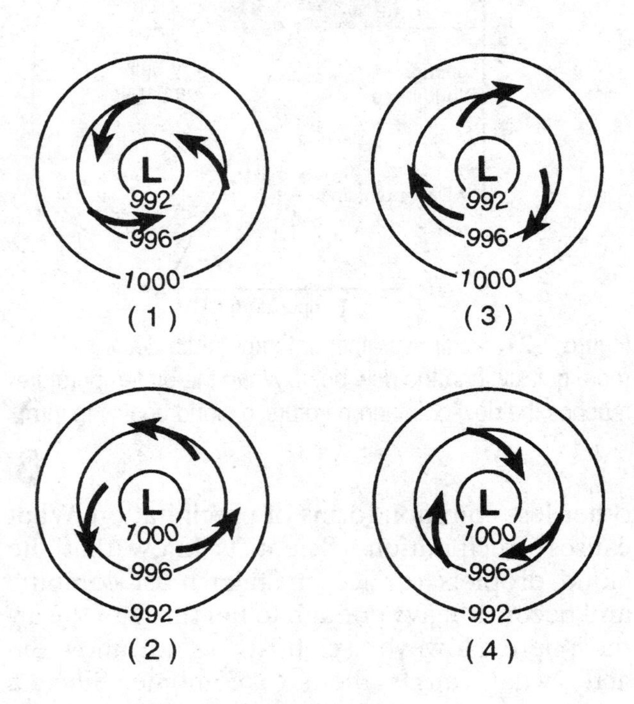

19. Which diagram below best shows how surface winds are deflected (curved) in the Northern and Southern Hemispheres?

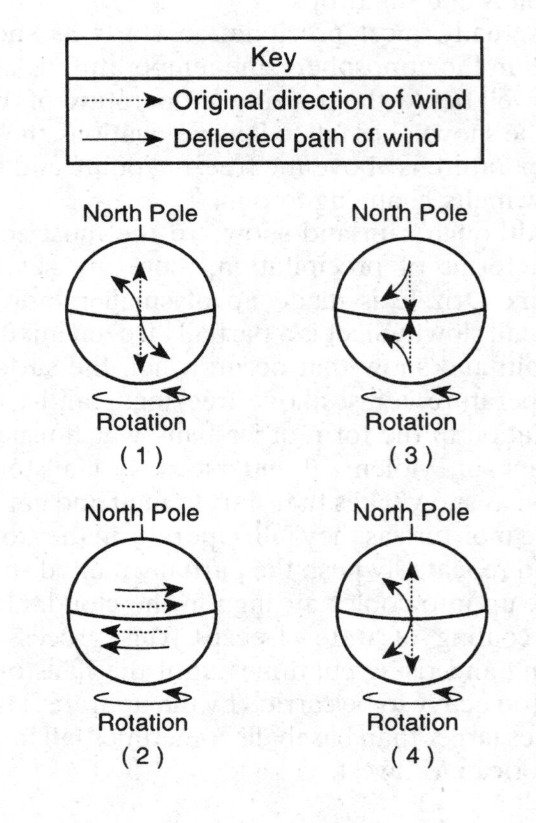

20. Of the four weather maps below, which one best shows the direction of surface winds within this regional pressure field.

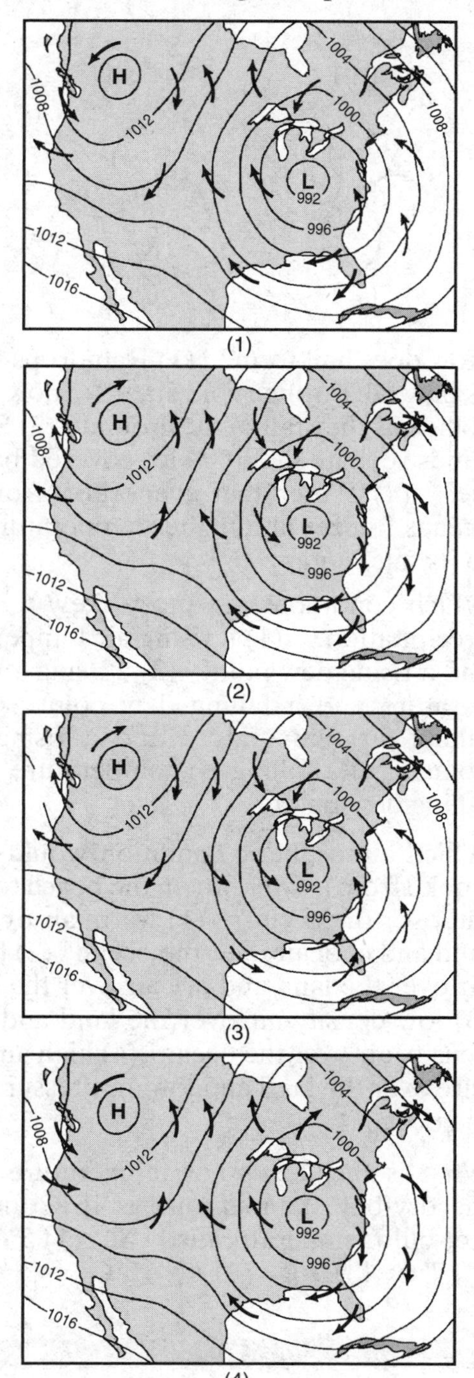

Part B

Base your answers to questions 21 through 26 on the two maps on the right, which show relatively stable high- and low-pressure areas near New York State. The only difference between map A and B is the pressure difference between the high- and low-pressure centers. A dot on each map shows the location of the city of Binghamton, NY.

21. What is the wind direction at Binghamton, NY on both maps? (1) north wind (2) south wind (3) east wind (4) west wind

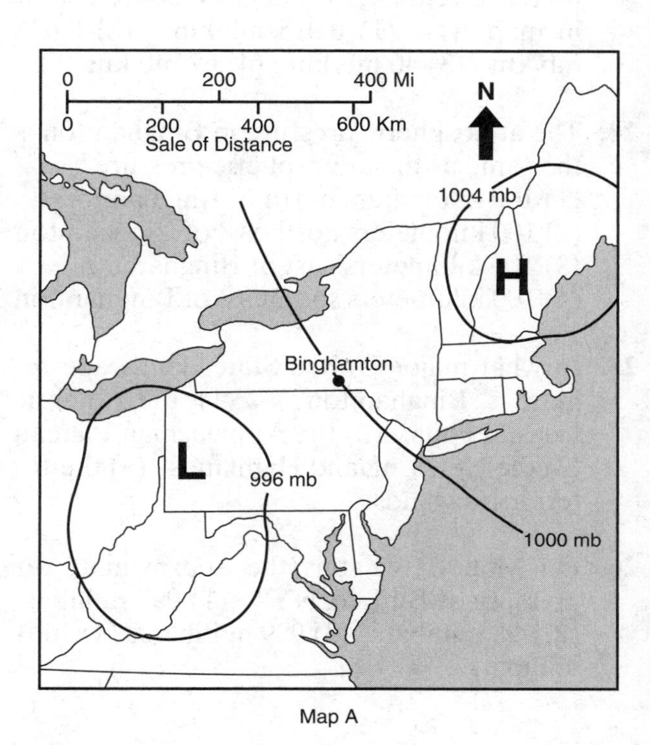

Map A

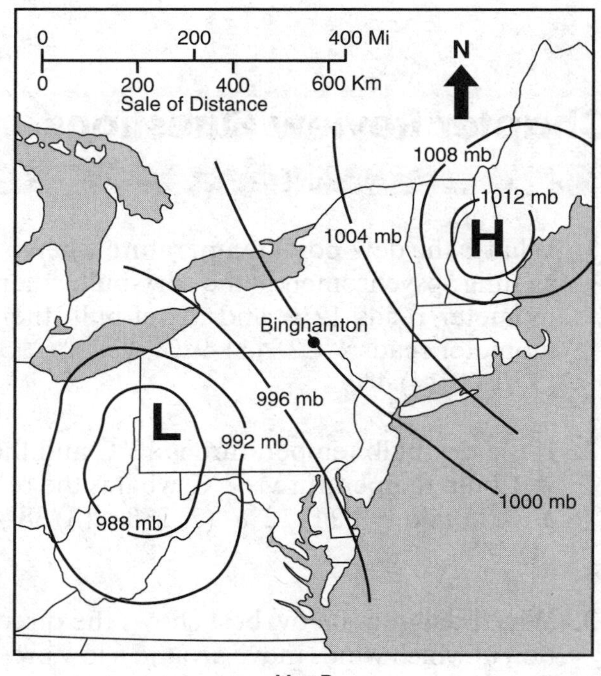

Map B

22. If map *B* were correct, rather than map *A*, how would this be apparent to the people in Binghamton? (1) The winds would be calmer. (2) The winds would be faster. (3) The winds would not be as steady. (4) There would be no difference.

23. What is the pressure gradient from the low-pressure center to the high-pressure center in map *A*? (1) 0.015 mb/km (2) 0.025 mb/km (3) 40 mb/km (4) 67 mb/km

24. The atmospheric pressure in Binghamton is the same as the atmospheric pressure (1) 100 kilometers north of Binghamton (2) 100 kilometers northeast of Binghamton (3) 100 kilometers east of Binghamton (4) 100 kilometers southeast of Binghamton

25. In what major United States landscape region is Binghamton? (1) the Atlantic Coastal Plain (2) the Appalachian Plateau (3) the New England Highlands (4) the Interior Lowlands

26. On Map *A*, what is the approximate air pressure at Buffalo, NY? (1) 997 millibars (2) 998 millibars (3) 999 millibars (4) 1001 millibars

Chapter Review Questions

PART A

1. What is the dew-point temperature when on a sling psychrometer the dry-bulb thermometer reads 12°C and a wet-bulb thermometer reads 4°C? (1)–9°C (2)–4°C (3) 0°C (4) 4°C

2. If the dry-bulb temperature is 12°C and the wet-bulb temperature is 7°C, what is the relative humidity? (1) 12% (2) 28% (3) 38% (4) 48%

3. Which diagram below best shows the direction in which winds move around a low pressure system?

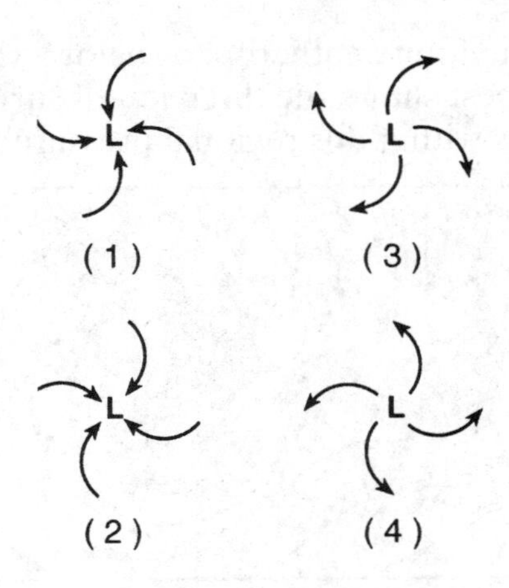

4. How does hail form? (1) Raindrops freeze as they fall through cold air. (2) Snow flakes collide as they fall to the ground. (3) Strong winds repeatedly lift water-covered balls of ice. (4) Water that evaporates from the oceans freezes at a higher temperature due to its salt content.

5. Which conditions are most likely to cause precipitation? (1) a rising air temperature and a rising dew point (2) a rising air temperature and a falling dew point (3) a falling air temperature and a rising dew point (4) a falling air temperature and a falling dew point

6. Which atmospheric condition would cause smoke from a campfire on the beach to blow out over the ocean? (1) warm air over the land and cool air over the ocean (2) humid air over the land and dry air over the ocean (3) low-density air over the land and high-density air over the ocean (4) high air pressure over the land and low air pressure over the ocean

7. What is the dew-point temperature when the dry-bulb temperature is 16°C and the wet-bulb temperature is 11°C (1) 5°C (2) 7°C (3) 9°C (4) 17°C

PART B

8. A student measured the wet-bulb temperature and air temperature to determine the dew point. If the relative humidity is 51% and the temperature is 20°C, what is the order of those three temperatures from lowest to highest? (1) wet-bulb temperature, air temperature, dew point (2) wet-bulb

temperature, dew point, air temperature (3) dew point, air temperature, wet bulb temperature (4) dew point, wet-bulb temperature, air temperature

Base your answers to questions 9 through 12 on the graph below, which shows atmospheric conditions at a weather station at Kennedy Airport in New York City

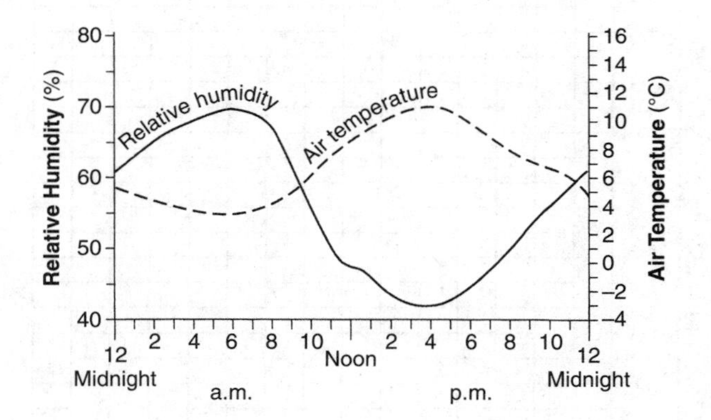

9. What was the air temperature at noon? (1) 0°C (2) 8°C (3) 45°C (4) 70°C

10. What was the range of the relative humidity from its highest value in the morning to its minimum in the afternoon? (1) 10% (2) 28% (3) 42% (4) 70%

11. What is the relationship between relative humidity and air temperature on this day? (There are only three choices for this question.) (1) direct (2) indirect (inverse) (3) a change in one has no affect on the other

12. Why does the relative humidity drop so low in the late afternoon? (1) More moisture can exist in the atmosphere when the air is warm. (2) Evaporation causes the relative humidity to decrease. (3) Humid air moved into this location in the afternoon. (4) The dew point reached a maximum temperature in the afternoon.

PART C

13. The diagram above right shows a beaker of water being heated by a flame. A dye pellet has been added to show the pattern of circulation. Draw arrows to show the directions in which the water is moving in response to the uneven heating.

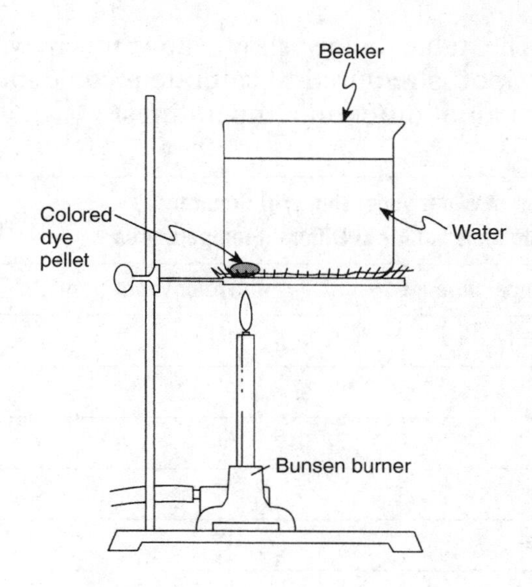

14. The diagram below shows the way local daytime and night breezes commonly blow during the summer at a location along the south shore of Long Island, NY. On a copy of the diagrams, write the letter H to show where the surface atmospheric pressure is relatively high and L to show where it is relatively low on each of the two diagrams.

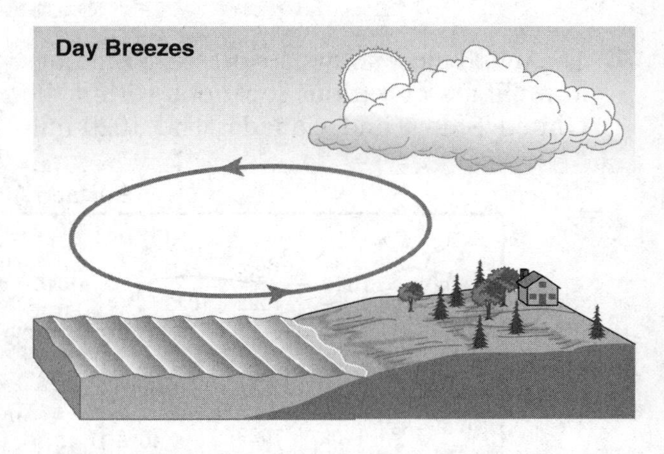

15. The table below shows how much water vapor is required to saturate a cubic meter of air at different temperatures.

Amount of Water Vapor That Will Saturate 1 Cubic Meter of Air at Different Temperatures

Air Temperature (°C)	Water Vapor (g/m³)
–20	1
–10	2
0	5
10	9
20	17
30	29
40	50

On a copy of the grid at the right, draw a line graph of this data. Be sure to:

a. Place the name of the correct variable along the *y*-axis and include the correct units.

b. Mark an appropriate numerical scale showing equal intervals along the *y*-axis

c. Plot each value in the table as a point and connect the points as a smooth curve.

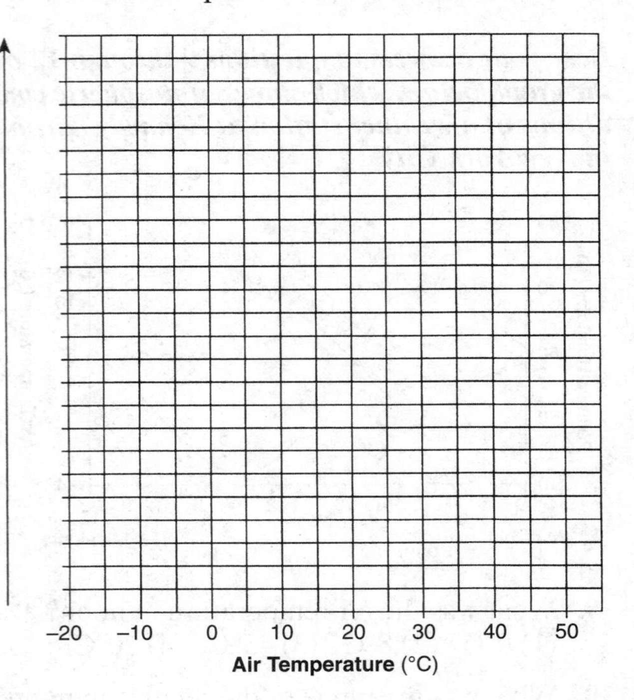

16. The map below shows air-pressure readings in millibars at various locations within the United States and Canada. The 1020-millibar isobar has been drawn and labeled already. Copy the map and draw and label the 1024- and 1028-millibar isobars.

Surface Air Pressure

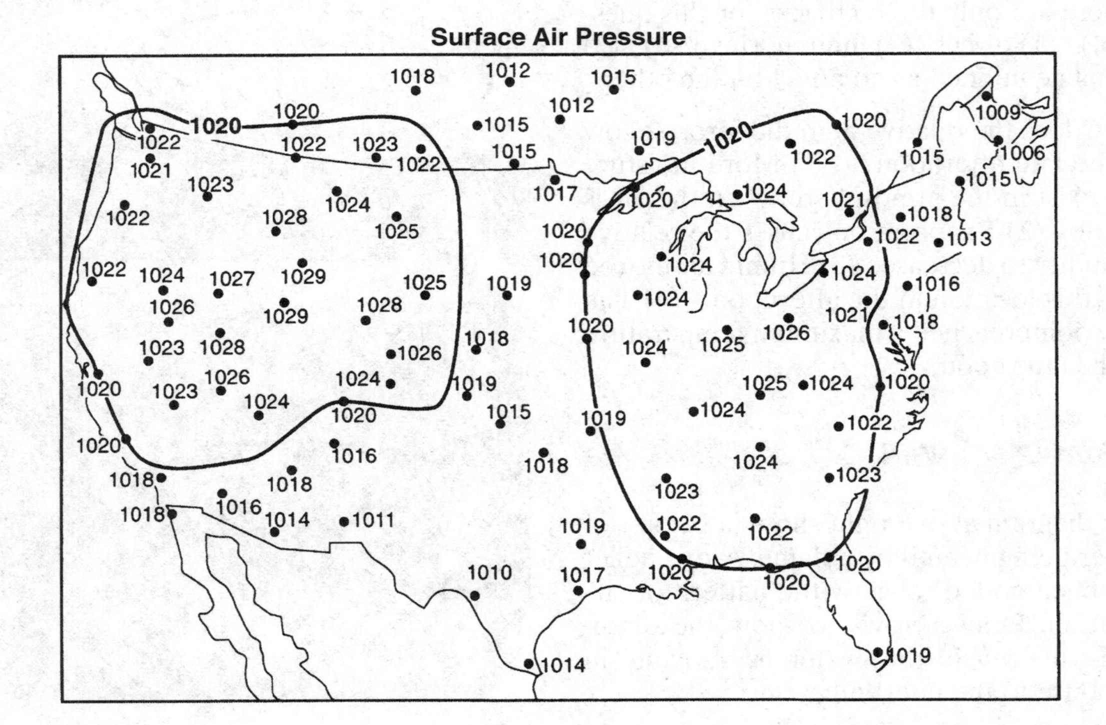

17. What instrument is usually used to measure atmospheric pressure?

18. The isoline map below shows air-pressure values in a large geographic region.

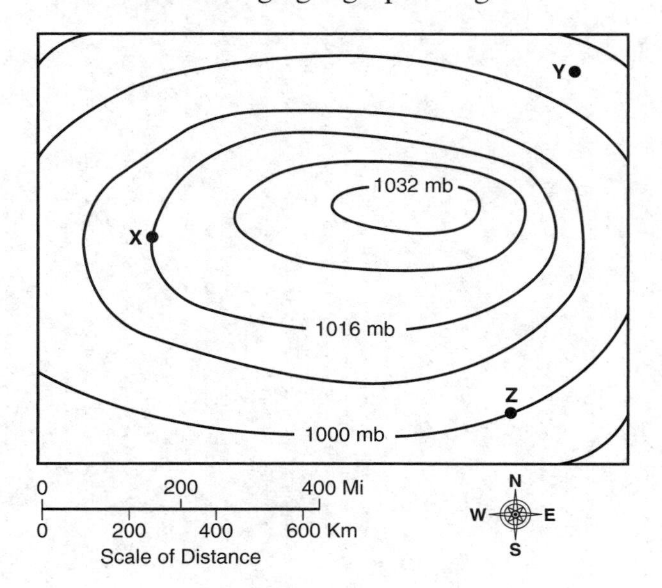

(*a*) On a copy of the diagram, draw an arrow through the point on the isobar labeled *X* to show the most likely wind direction at that position.

(*b*) What is the atmospheric pressure at point *Z in inches of mercury*?

(*c*) What is the atmospheric pressure at the point labeled *Y*.

19. (*a*) Describe how clouds form when warm, humid air rises. Your answer must include the terms *dew point* and *expansion* or *expands*.
(*b*) State the phase change that occurs when the air reaches the dew point.

20. The data table below shows the temperatures and air pressures recorded as instruments attached to a weather balloon rose over the airport at Buffalo, NY.

On a copy of the grid at the right plot each data point using an *X* and connect the points with a smooth curve.

Altitude Above Sea Level (m)	Air Temperature (°C)	Air Pressure (mb)
300	16.0	973
600	16.5	937
900	15.5	904
1200	13.0	871
1500	12.0	842
1800	10.0	809
2100	7.5	778
2400	5.0	750
2700	2.5	721

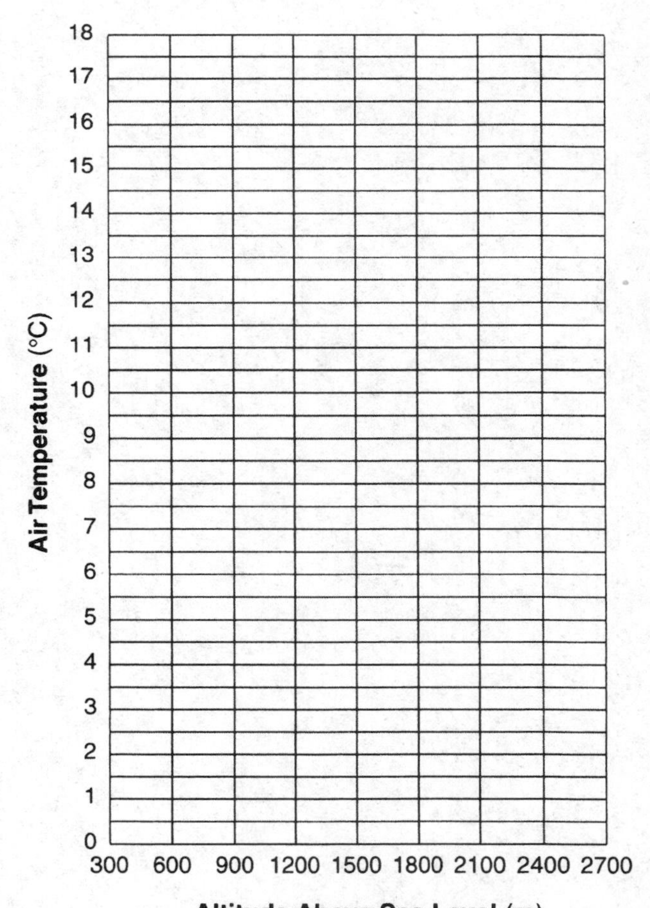

21. State the relationship between the altitude above sea level and the air pressure.

22. The image below shows a cloud in the sky. On a sketch, draw an arrow to show the direction in which air moves in order to form a cloud.

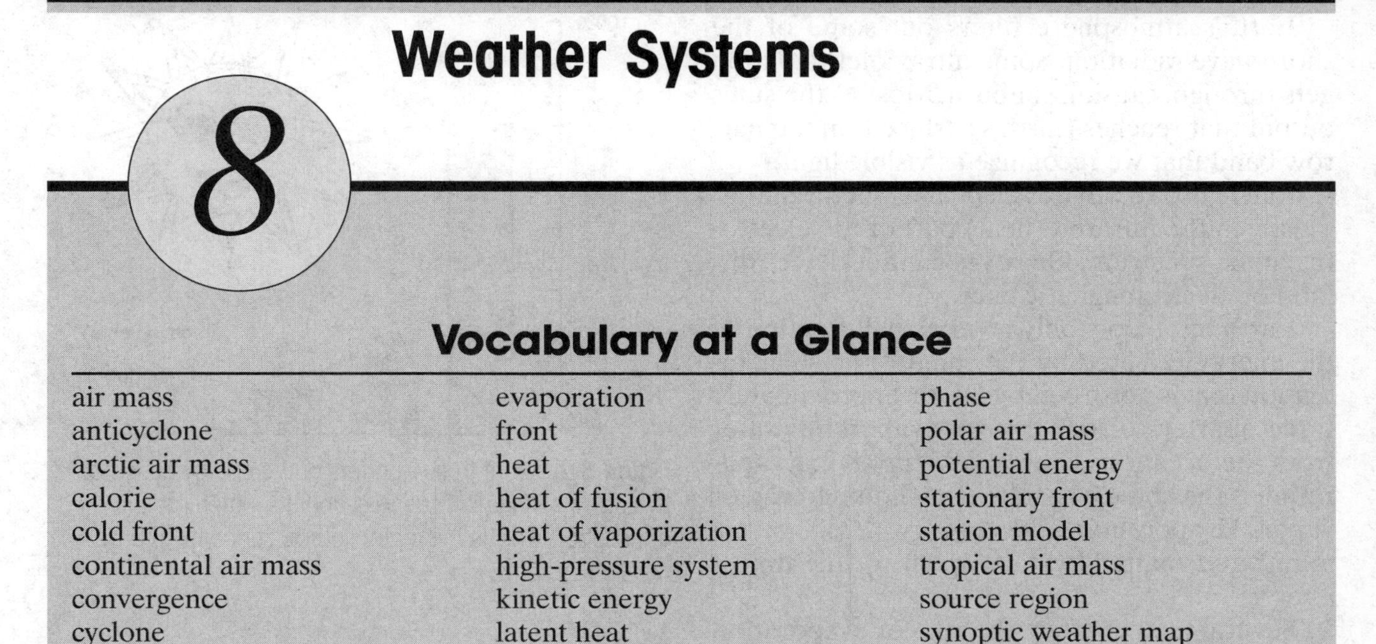

CHAPTER

8 Weather Systems

Vocabulary at a Glance

air mass
anticyclone
arctic air mass
calorie
cold front
continental air mass
convergence
cyclone
divergence
electromagnetic energy
electromagnetic spectrum
evaporation
front
heat
heat of fusion
heat of vaporization
high-pressure system
kinetic energy
latent heat
low-pressure system
maritime air mass
occluded front
phase
polar air mass
potential energy
stationary front
station model
tropical air mass
source region
synoptic weather map
warm front

HOW DOES ENERGY ENTER THE ATMOSPHERE?

It takes a great deal of energy to drive Earth's weather. Where does this energy come from? The sun is the major source of energy for Earth. Stars, including our sun, give off **electromagnetic energy** over the wide range of wavelengths that make up the **electromagnetic spectrum**. Figure 8-1 shows the range of electromagnetic energy from short-wave radiation, such as x-rays and ultraviolet, to long-wave radiation, that includes

Electromagnetic Spectrum

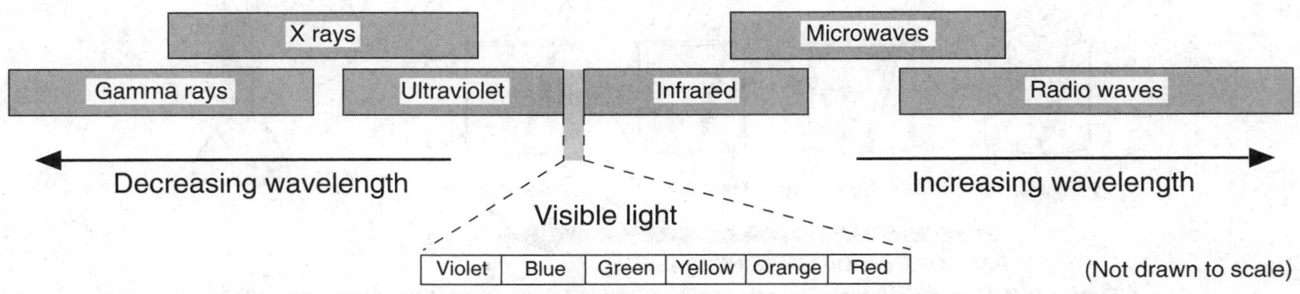

Figure 8-1. Visible light is a small part of the total electromagnetic energy spectrum. We cannot see most forms of electromagnetic energy; however, they are very important in natural processes and scientific applications.

infrared and radio waves. This figure is also found on page 14 of the *Earth Science Reference Tables*.

Earth's atmosphere filters out some of the short-wave radiation. Some ultraviolet radiation gets through, causing sunburn. Most of the sun's output that reaches Earth's surface is in the narrow band that we recognize as visible light.

Earth also reradiates electromagnetic energy, mostly in the infrared (heat) part of the electromagnetic spectrum. Our eyes cannot detect this kind of electromagnetic energy.

Earth intercepts only a very small fraction of the energy radiated by the sun. Of the solar radiation that is absorbed by Earth, approximately three-quarters of this energy evaporates water from the oceans as shown in Figure 8-2. **Evaporation** is the change in state from liquid to gas, or vapor. Evaporation adds energy (2260 joules/gram) and matter (water vapor) to the atmosphere.

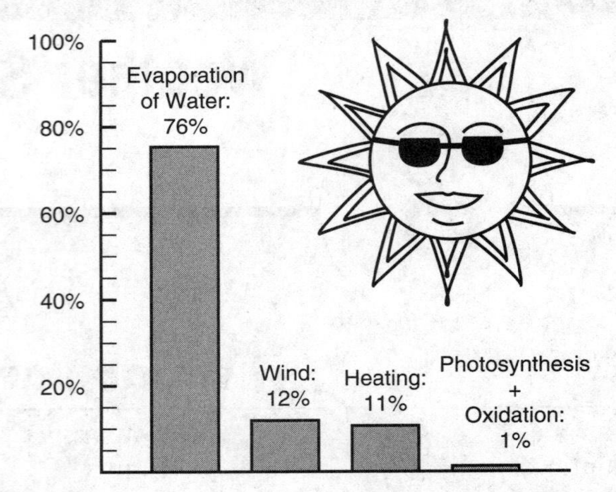

Figure 8-2. About three-quarters of the energy absorbed by Earth's surface is used to evaporate water into the atmosphere. This is the major source of energy to drive weather systems.

Several factors affect the rate of evaporation of water. Evaporation decreases when water is covered or cooled. The rate of evaporation increases when water is uncovered, heated, and when wind blows over it. Figure 8-3 illustrates these factors.

Three States of Water

You are probably familiar with the three states of water: ice, liquid water, and water vapor. Figure 8-4 shows how the temperature of a 1-gram sample of ice changes as it is heated. The sloping parts of the graph show changes in temperature. The flat parts of the graph, where the temperature does not change, show that energy is used for a change in state (sometimes called a change in **phase**). Energy that is absorbed or released during a change in state is called **latent** (hidden) **heat**, a form of **potential energy**. It is latent energy because it does not cause a change in temperature.

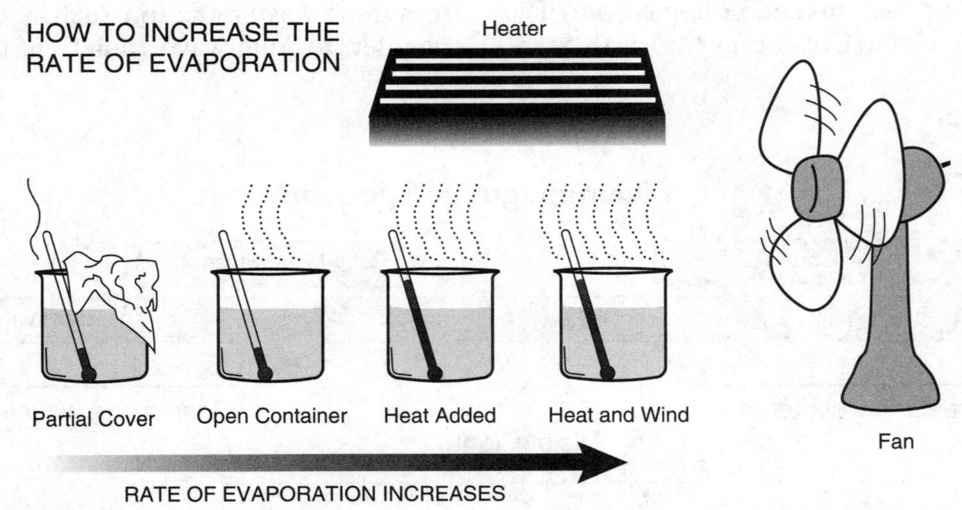

Figure 8-3. Water evaporates most quickly if it is (1) uncovered, (2) heated, and (3) exposed by wind to a constant source of dry air.

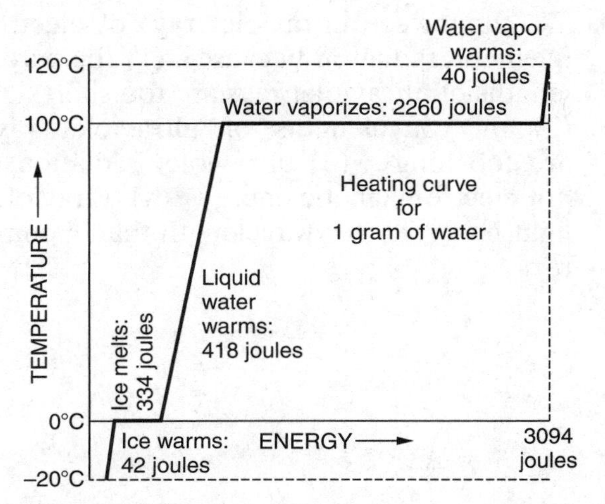

Figure 8-4. This graph shows how the temperature of a 1-gram sample of water changes as it absorbs energy at a constant rate. The horizontal parts of this graph show that water does not change temperature as it changes from solid (ice) to a liquid (water) or from a liquid to a gas (water vapor). These changes of state involve much more energy than changes in temperature.

People often confuse heat and temperature. **Heat** is a form of energy that is the result of the vibrational motion (**kinetic energy**) of atoms and molecules. Energy is measured in **joules**. On the other hand, *temperature* is a measure of the average kinetic energy of the atoms and molecules of a substance. Temperature is measured in degrees (Kelvin, Celsius, or Fahrenheit) with a thermometer.

According to the graph, before the ice was heated its temperature was −20°C. Heating increased the temperature of the ice to 0°C, at which point it began to melt. The temperature of the ice does not change until all the ice has melted. As the ice changed to liquid water, it absorbed 334 joules/gram (latent heat). The energy needed to melt 1 gram of a substance at its melting point is the **heat of fusion**. In the *Earth Science Reference Tables* this value is called energy gained during melting and energy released during freezing. The heat of fusion of ice is 334 joules/ gram (334 J/g).

Heating continued and the temperature of the liquid water increased until it reached 100°C, the boiling point of water. At the boiling point, liquid water changed to water vapor. The energy necessary to change 1 gram of a substance from the liquid to the vapor state is the **heat of vaporization**. The heat of vaporization of water is 2260 J/g. The

Properties of Water

Heat energy gained during melting	334 J/g
Heat energy released during freezing	334 J/g
Heat energy gained during vaporization	2260 J/g
Heat energy released during condensation	2260 J/g
Density at 3.98°C	1.0 g/mL

Figure 8-5 Page 1, the *Earth Science Reference Tables* lists these properties of water.

gram of water absorbed 2260 joules of latent heat as it changed to vapor. After all the liquid water became vapor, the temperature of the vapor began to rise. The specific heat of water vapor is 2.00 J/g•°C. Figure 8-5 shows that evaporation and condensation involve a relatively large exchange of energy.

If the water vapor is cooled to 100°C, it will condense and release the 2260 joules of latent heat it absorbed during evaporation. As cooling continues, the water will release the heat it absorbed while it was heated.

Water can evaporate at temperatures lower than its boiling point. You can observe this by leaving a glass of water indoors undisturbed for a few days. You will notice that the amount of water in the glass decreases. Every time 1 gram of water evaporates into the atmosphere it carries with it 2260 joules of latent heat. That energy is released when clouds form by condensation or deposition. (Deposition occurs when water vapor changes directly into ice. The opposite process, sublimation, occurs when ice or snow evaporates without first melting.)

This may help you to understand how much energy that is. The potential energy alone is more than 5 times greater than the energy it takes to heat the same gram of water from the freezing temperature (0°C) to boiling point (100°C). Within the atmosphere, water vapor holds a great amount of energy. Water vapor is the principal reservoir of energy that drives thunderstorms and all other forms of violent weather.

QUESTIONS

Part A

1. What is the principal way energy enters Earth's atmosphere?

(1) Terrestrial radiation is reflected by clouds. (2) Sunlight travels through space by conduction. (3) Solar energy causes evaporation of seawater. (4) Visible light is reflected off the land areas.

2. Which process increases the temperature of the atmosphere? (1) evaporation of water (2) convection of air (3) radiation loss to outer space (4) cloud formation

3. We cannot see ultraviolet rays of electromagnetic radiation because (1) the wavelengths of ultraviolet rays are too short (2) the wavelengths of ultraviolet rays are too long (3) ultraviolet radiation is not electromagnetic energy (4) ultraviolet light has a shorter wavelength than gamma rays

Part B

Base your answers to questions 4 through 8 on the following passage.

What Is Transpiration?

Transpiration is the process by which moisture is carried through plants from the roots to small pores on the underside of leaves. At the leaves, the moisture changes to vapor and is released to the atmosphere. Transpiration is essentially evaporation of water from plant leaves.

Roots in the soil draw in water and nutrients, which are carried up into the stems and leaves. Some of this water is returned to the air by transpiration (when combined with evaporation, the total process is evapotranspiration). Transpiration rates vary widely depending on weather conditions, such as temperature, humidity, and availability and intensity of sunlight. Furthermore, some plants use and transpire more water than others.

Studies have shown that about 10 percent of the moisture found in the atmosphere is released by plants through transpiration. The remaining 90 percent is mainly supplied by evaporation from oceans, seas, and other bodies of water (lakes, rivers, streams).

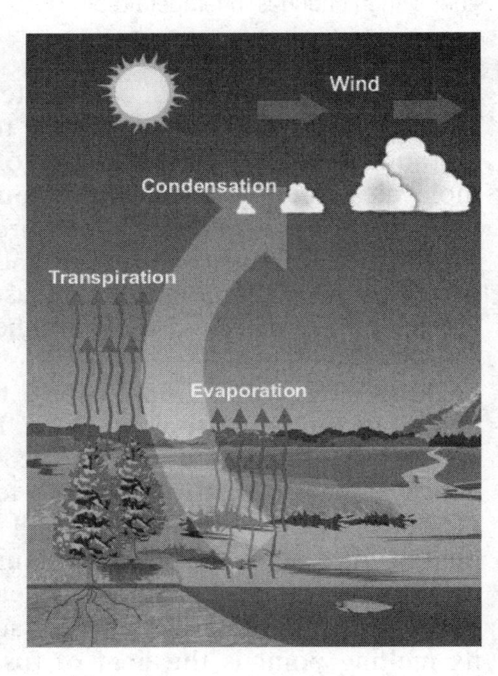

4. How do evaporation and transpiration change Earth's atmosphere? (1) They add water vapor and energy to the atmosphere. (2) They add water vapor but take energy from the atmosphere. (3) They take out water vapor but add energy to the atmosphere. (4) They take both water vapor and energy out of the atmosphere.

5. Both evaporation and transpiration are most active in (1) polar regions (2) tropical regions (3) high mountains (4) locations with calm winds

6. A student planned experiments to observe water lost to the air by a flowering plant through transpiration. In the first part of the procedure, a transparent plastic bag was placed over the plant and droplets of water were observed on the inside of the bag. However, relatively little water was lost by the plant. Why did the plant lose so little water in this part of the procedure? (1) The student gave the plant too much water. (2) The air in the bag became saturated. (3) A breeze was shaking the bag. (4) No light could get to the plant.

7. What is the main source of energy to support both evaporation and transpiration? (1) the sun (2) Earth's internal heat from radioactive decay (3) the soil (4) pollutants added to the atmosphere by humans

8. For every gram of water that enters the atmosphere (1) 334 joules are lost from the atmosphere (2) 334 joules are gained by the atmosphere (3) 2260 joules are lost from the atmosphere (4) 2260 joules are gained by the atmosphere

9. What is the major source of water for Earth's atmosphere? (1) plants (2) rivers (3) lakes (4) oceans

Base your answers to questions 10 through 13 on the data table below, which describes the addition of energy at a constant rate to a 1-gram sample of water, changing its temperature from −20°C to 120°C.

Interval	Temperature Change (°C)	Heat Added (joules)	Total Heat Added (joules)
1	−20 to 0	42	42
2	None	334	376
3	0 to 100	418	794
4	None	2260	3054
5	100 to 120	40	3094

10. The changes from interval 1 to interval 5 represent the changes from (1) ice to liquid water to water vapor (2) water vapor to liquid water to ice (3) liquid water to water vapor to ice (4) ice to water vapor to liquid water

11. The reverse of which interval best represents the formation of a cloud or fog? (1) 1 (2) 2 (3) 3 (4) 4

12. What portion of the 3094 J expended was used in the vaporization change? (1) 10% (2) 20% (3) 54% (4) 73%

13. During interval 2 and 4 there was no change in temperature. What does this tell you about what was happening during these intervals? (1) There was no energy absorbed during these intervals. (2) A change in state is not a change in the total energy. (3) A phase change occurs at a constant temperature. (4) A thermometer can measure both latent and kinetic heat energy.

WHAT IS A SYNOPTIC WEATHER MAP?

Synoptic weather maps show atmospheric field quantities such as temperature, air pressure, precipitation, and other weather conditions at a particular time and over a large geographic area. These maps provide a general summary, or synopsis, of weather patterns. Figure 8-6 on page 198 is a weather map that shows temperatures with isotherms, as well as high- and low-pressure centers, weather fronts, and precipitation patterns. Because most weather systems move across the United States from west to east, synoptic weather maps provide information that can be used to make weather predictions.

Interpreting Information Shown at a Weather Station

Some weather maps show a variety of measurements recorded at each weather station. Because so much information must be recorded at each station on a weather map, the information is presented in a standardized form called a **station model**. Each measurement is at a preset position next to the circle that indicates the geographic location of the reporting station. To interpret the information recorded at weather stations on a synoptic weather map, follow the steps below that interpret the station model in Figure 8-7 on page 199.

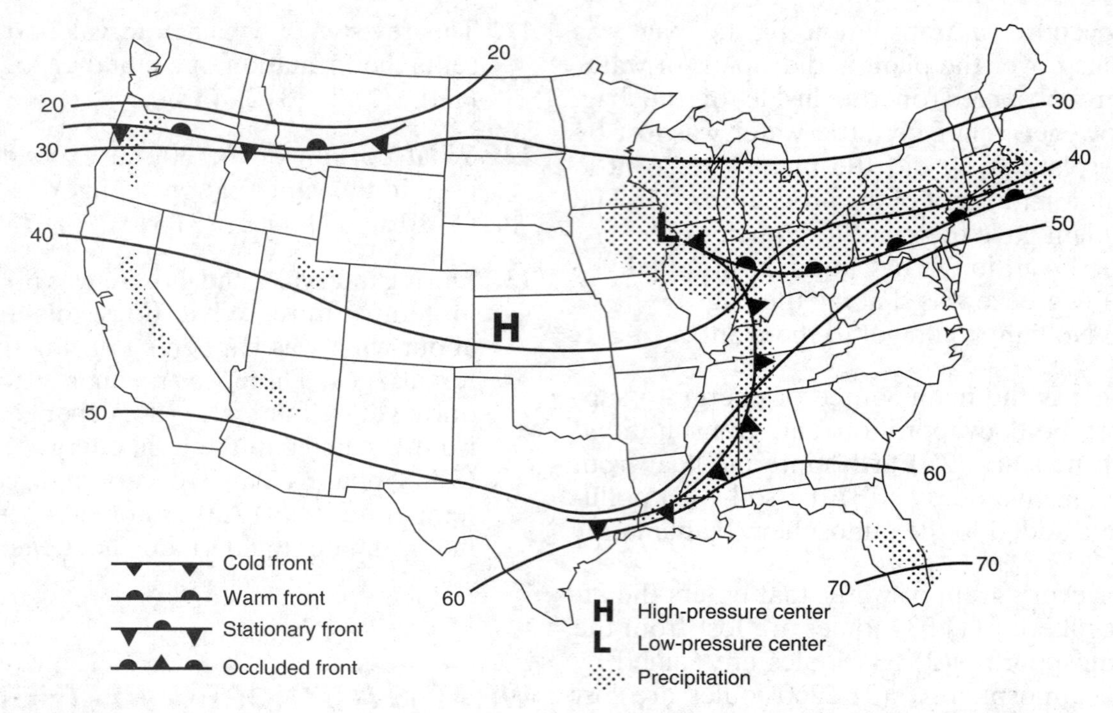

Figure 8-6. Synoptic weather maps often show the centers of **cyclones** (lows marked with an *L*) and anticyclones (highs marked with an *H*) as well as the frontal boundaries along which weather changes and precipitation generally occur. This map shows a typical weather pattern.

1. The circle shows the geographic location of the weather station. The cloud cover is shown by the fraction of the circle that is dark. This example shows a 75 percent cloud cover.
2. The air temperature in degrees Fahrenheit is shown at the top left. The air temperature in this example is 28°F.
3. The symbol at the left side of the circle shows the present weather. The symbol in Figure 8-7 indicates snow.
4. The number next to the symbol for present weather is the atmospheric visibility in miles; here it is $^{1}/_{2}$ mile.
5. The dew point is shown at the bottom left. In this example, the dew point is 27°F.
6. The line connected to the circle shows the direction from which the wind is blowing. This example shows a wind from the southwest. (Winds are always shown and named by the direction from which the wind comes and not by the direction toward which the wind is going.) The feathers indicate the wind speed. Each whole feather represents 10 knots (10 nautical miles per hour; a nautical mile is 1.15 miles), and each half feather is 5 knots. In this example, $1^{1}/_{2}$ feathers mean a wind of approximately 15 knots.
7. At the top right is a three-digit code showing the atmospheric pressure. To decode it, you must add "9" or "10" in front of the code numbers, and a decimal point must be placed before the final digit. When the first digit is less than 5, add a 10. However, if the first digit is 5 or greater, add a 9. For example, "196" would mean 1019.6 millibars. If the heading was "979," this would mean 997.9 millibars. The final recording should result in a number that is close to the normal reading for atmospheric pressure, which is roughly 1000 millibars as indicated on page 13 of the *Earth Science Reference Tables.*
8. The number to the right of the circle shows the change in atmospheric pressure over the past 3 hours. (Again, a decimal point must be added.) The line to the right of the number shows how the atmospheric pressure has changed. This one shows a steady increase of 1.9 millibars. Decreasing atmospheric pressure often indicates that clouds and rain are coming. Rising

Key to Weather Map Symbols

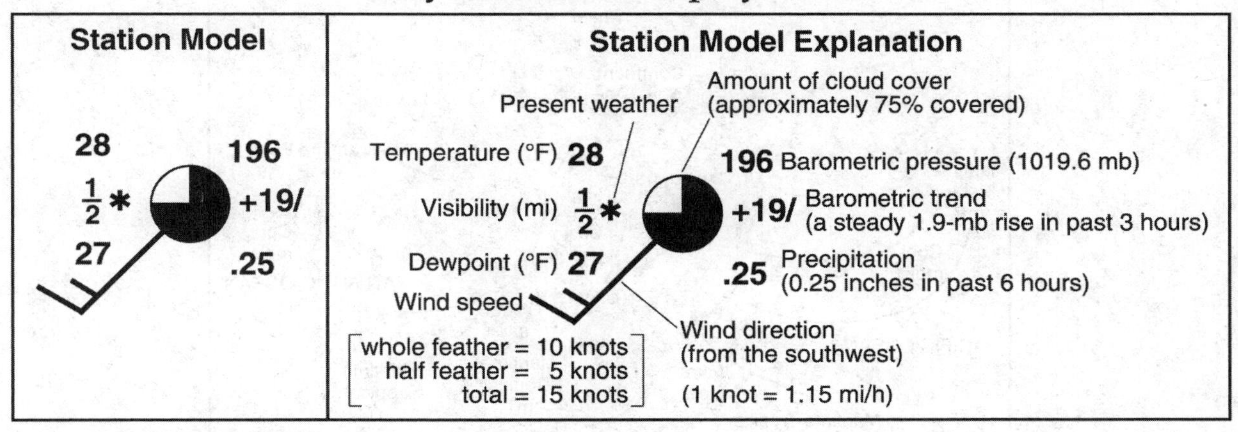

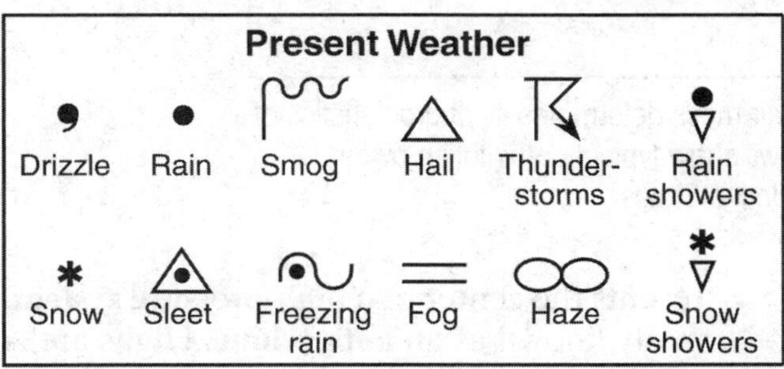

Figure 8-7. On page 13 of the *Earth Science Reference Tables*, you will find these charts showing how to read weather data from a station model.

atmospheric pressure generally means improving weather.

9. The number below the change in atmospheric pressure shows the amount of precipitation in inches over the past 6 hours. In this example, 25 hundredths of an inch of precipitation has fallen in the past 6 hours.

Some weather stations may display more or less information, but it will always be displayed in the format shown here.

Air Masses

Weather maps are especially useful to identify **air masses**, which are large bodies of air with uniform temperature, pressure, and humidity. The movement of air masses brings changes in the weather. The leading edge of an air mass or where two air masses meet is a **front**. When a front passes, precipitation often accompanies changes in temperature, humidity, and atmospheric pressure. These changes mark the arrival of a new air mass. Within the new air mass, conditions generally stabilize.

The character of an air mass depends on its geographic origin or **source region**. For example, an air mass from central Canada that moves southward into the United States tends to be cold and dry. This air mass causes an increase in atmospheric pressure. However, an air mass from the Gulf of Mexico will usually be warm and moist, and produces relatively low atmospheric pressure.

Meteorologists have developed a system of two-letter codes to identify the temperature and humidity characteristics of air masses. Each code has a small (lowercase) letter followed by a capital letter. These codes are listed on page 13 of the *Earth Science Reference Tables*. You should be familiar with the five air mass descriptions and their abbreviation codes.

cA continental arctic: an unusually cold and dry air mass from arctic Canada

cP continental polar: a cold, dry air mass, which may have originated over central Canada

cT continental tropical: a warm, dry air mass, which may have originated over Mexico or

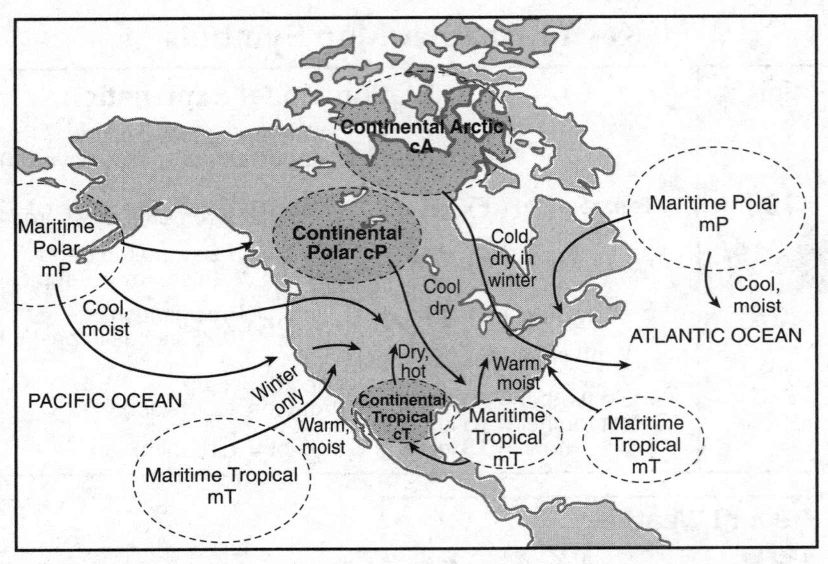

Figure 8-8. The source area of an air mass determines its characteristics of temperature and humidity. The arrows show typical paths taken by air masses that affect weather in the United States.

the desert region of the American Southwest

mT maritime tropical: a warm, moist air mass, which may have originated over the Gulf of Mexico

mP maritime polar: a cold, moist air mass that may have originated over the North Atlantic or North Pacific Ocean

In particular, you should know that a **continental air mass** is dry and a **maritime air mass**, from off the oceans, is moist. A **tropical air mass** is warm and a **polar air mass** is cold. An **arctic air mass** is very cold. Figure 8-8 shows the geographic origins of air masses coming into the United States.

HOW CAN WE PREDICT THE WEATHER?

Weather systems usually move across the United States from west to east. Storm systems that start in the American Midwest usually affect New York State several days later.

Highs and Lows

If you look at the weather map in a newspaper or on television, you will notice that they have a number of features in common. A large *H* represents the center of a **high-pressure system**, formally known as an **anticyclone**. Highs are zones of **divergence** at ground level where sinking air at the center causes the winds to blow outward. Often, high-pressure systems bring cool and dry air with clear skies and stable conditions. (You learned about this in Chapter 7, especially Figure 7-16 on page 182.)

On the other hand a large *L* represents the center of a **low-pressure system**, also known as a mid-latitude cyclone, or a zone of **convergence**. Rising air at the center of the low draws in contrasting air masses with their associated weather fronts. This generally produces unsettled weather conditions. Low-pressure systems are commonly associated with changeable weather, cloudy skies, and precipitation.

Fronts *Fronts* are shown by lines that separate two air masses. Fronts are often found in low-pressure systems because of the different air masses that converge to form a cyclone. Symbols along these fronts indicate the direction in which the fronts are moving and what kind of fronts they are: cold, warm, stationary, or occluded. Figure 8-9 shows the weather map symbol for air masses, fronts, and a hurricane.

Warm and cold fronts are named for the temperature changes they bring. A **warm front** is the boundary between a mass of warm air and the colder mass of air it is replacing. A **cold**

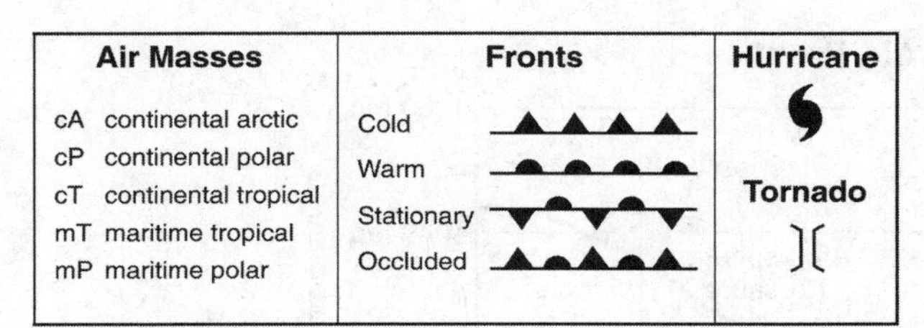

Figure 8-9. Weather map symbols from page 13 of the *Earth Science Reference Tables*.

front is the boundary between a mass of cold air and the warmer air it is replacing. **Stationary fronts** are boundaries between air masses that are not moving (stationary). **Occluded fronts** form when warm air is pushed above Earth's surface by cooler air that is closing in from both sides.

Figure 8-10 shows the interactions of air masses at fronts. A cold front (*A*) passes quickly. An advancing mass of cold air pushes warmer, moister air upward, often causing cooling, rapid cloud formation, and intense precipitation. Summer cold fronts sometimes bring severe thunderstorms, followed by cooler, drier weather and high atmospheric pressure. A cold front may pass in a matter of several hours. Figure 8-11 on page 202 shows how the weather variables of temperature, dew point, and air pressure can change as a strong cold front passes.

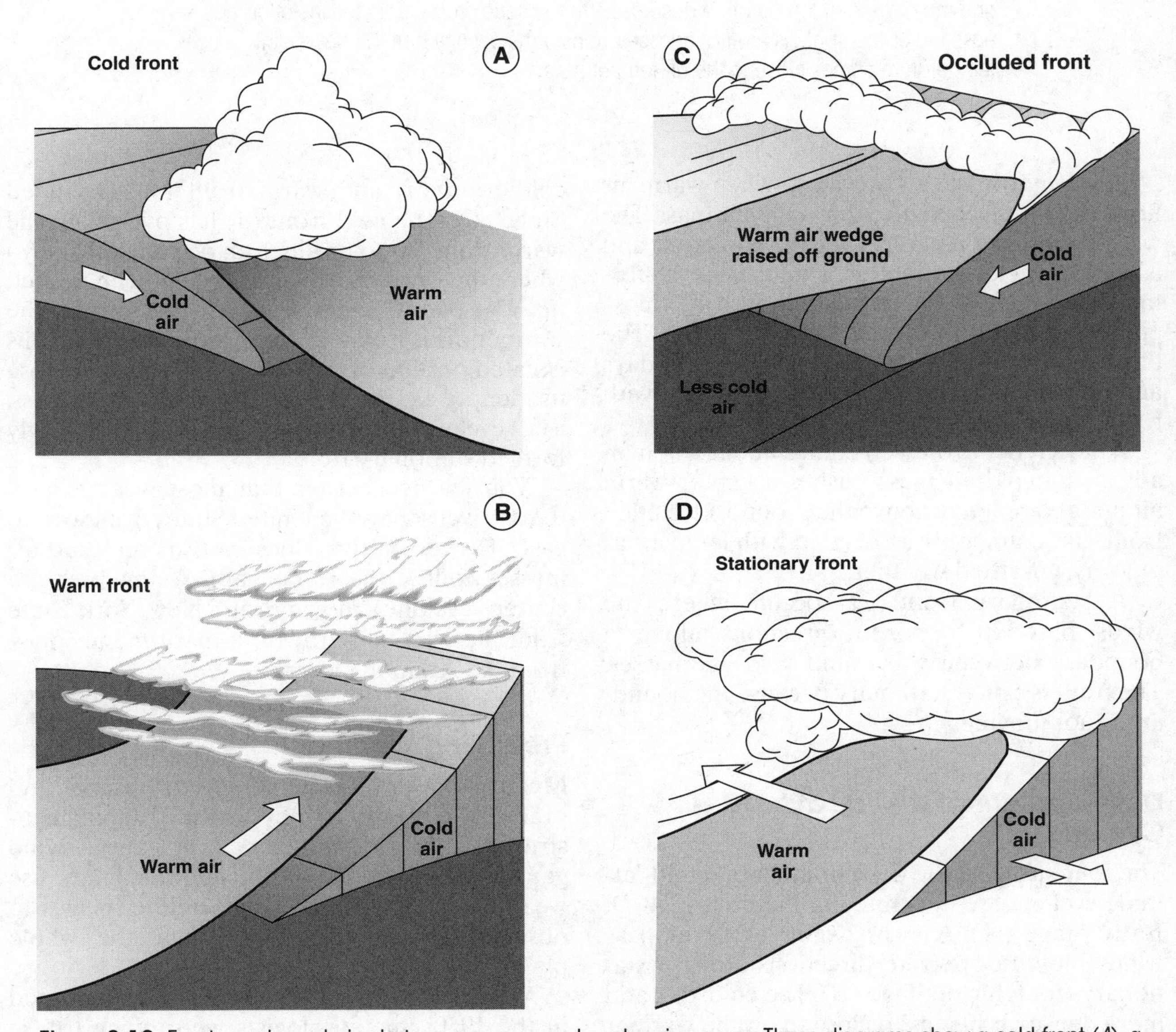

Figure 8-10. Fronts are boundaries between warmer and cooler air masses. These diagrams show a cold front (*A*), a warm front (*B*), an occluded front (*C*), and a stationary front (*D*).

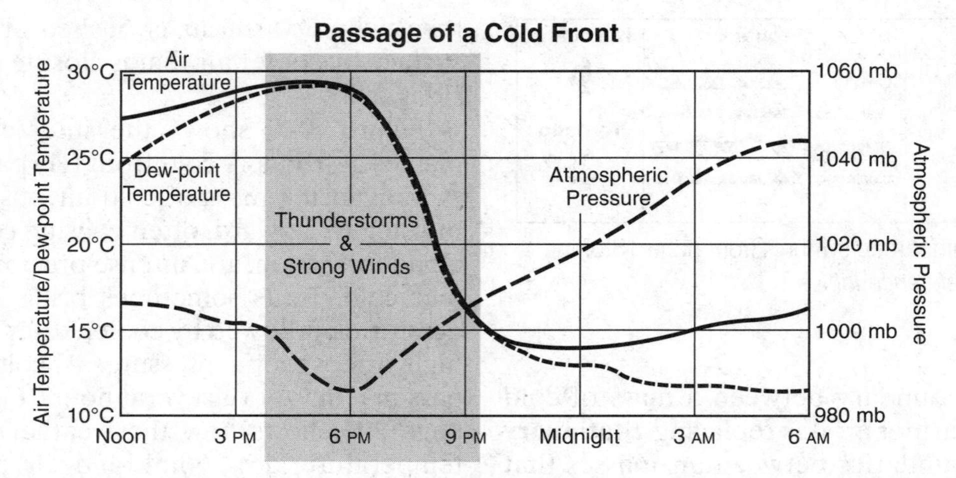

Figure 8-11. This figure shows how weather conditions change during the passage of a strong cold front. As a cold, dry air mass replaces warmer, moist air, air temperature and the dew-point temperature decrease. Atmospheric pressure increases after the passage of the front. As the front passes, rainstorms occur when the dew-point temperature comes close to the air temperature.

A warm front (*B*) is produced when warm air flows in to replace a retreating cold air mass. The warmer, moister air rises over the cooler air and is cooled by expansion. Rising air causes the formation of high, wispy clouds followed by thickening clouds and steady precipitation. The passage of a warm front may take several days and bring in warm, hazy weather conditions with relatively low atmospheric pressure.

An occluded front (*C*) comes about when an advancing cold air mass pushes a lighter warm air mass completely above the ground. Occluded fronts are commonly associated with large areas of rainy, unsettled weather.

A stationary front (*D*) occurs where the winds blow in opposite directions along a boundary between warm and cold air masses. The front is called stationary because the boundary is not moving.

Development of a Mid-Latitude Cyclone

The four stages in the development of a mid-latitude cyclone are illustrated in Figure 8-12*A*–*D*. Early Stage (*A*): A swirl begins to develop as winds blow in opposite directions along a stationary front. Open Stage (*B*): The cold, dry, and more dense air begins to close in on the warmer and moister air mass. Note the appearance of the cold front (*a*) and warm front (*b*). Occluded Stage (*C*): The cold front (*a*) has overtaken the warm front (*b*) to produce an occluded front (*c*) where the whole warm air mass has been pushed up. Dissolving Stage (*D*): A portion of the warmer air is isolated as the swirl closes in. This isolated portion of the warmer air mass will lose its identity as it is absorbed by the cold air mass. The cycle can begin again along the newly formed stationary front.

You learned earlier that most weather systems travel across the United States from west to east. However, this does not mean that air masses follow the same path. A low-pressure center (cyclone) moving into New York State from the west can draw in a maritime air mass from the Atlantic Ocean.

Predicting Weather

Meteorologists use technology to observe and predict the weather. They use such familiar instruments as thermometers, rain gauges, wind gauges, and barometers. In addition, they use weather balloons, radar, and satellite images to observe weather conditions over the whole planet.

When electronic computers were developed in the 1950s, meteorologists were quick to use them. Computers allowed them to analyze more

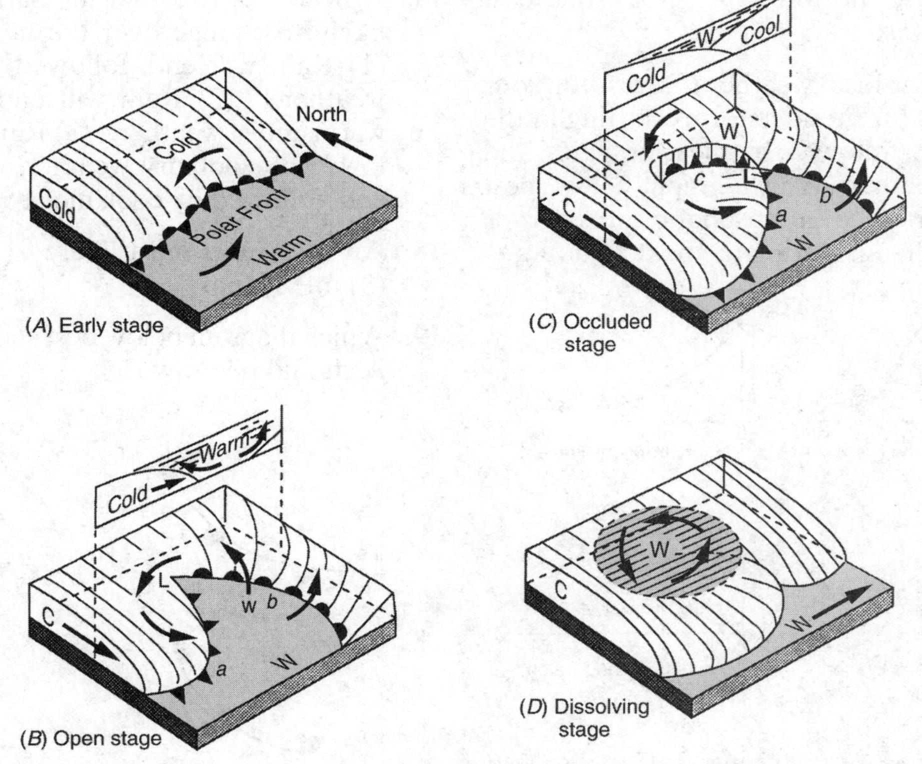

Figure 8-12. Weather systems and their associated fronts commonly follow a sequence of development beginning with a bend or swirl in the Polar Front. The eddy intensifies and then fills as the warm air sector is isolated high in the atmosphere.

data, which increased the reliability of weather forecasts. With the help of computers, meteorologists could determine the kind of weather that resulted from weather patterns similar to those occurring now. Based on this data, meteorologists make predictions that include a probability of occurrence. For example, if a low-pressure system with specific measurements of temperature and dew point is advancing from the west, computers may indicate that in the past these conditions produced rain approximately 80 percent of the time. This means that in four out of five similar events, rainfall followed these conditions.

The more information meteorologists have about present weather conditions, the more reliable their forecasts become. At present, day-to-day forecasts are correct about 90 percent of the time. Meteorologists also want to make accurate long-range forecasts weeks or months in advance. However, forecasts that far into the future are not reliable based on our present understanding of weather development.

Chaos theory, a modern branch of science and mathematics, has shown meteorologists that predictions for more than a week or two into the future are unlikely ever to be reliable. The complex development of global circulation and its extreme sensitivity to even minor changes in conditions probably make accurate long-term weather predictions impossible.

QUESTIONS

Part A

14. Clear skies and mild temperatures in New York City on the afternoon of Monday, May 1 were replaced by cold, rainy weather on Tuesday May 2. Most likely the air mass that came in overnight was from the (1) northeast (2) northwest (3) southeast (4) southwest

15. The 24-hour local weather forecast for your location is for thunderstorms starting in the next hour, followed by clearing skies and cooler weather. Which graph below best shows how the atmospheric pressure is likely to change during the next 24 hours?

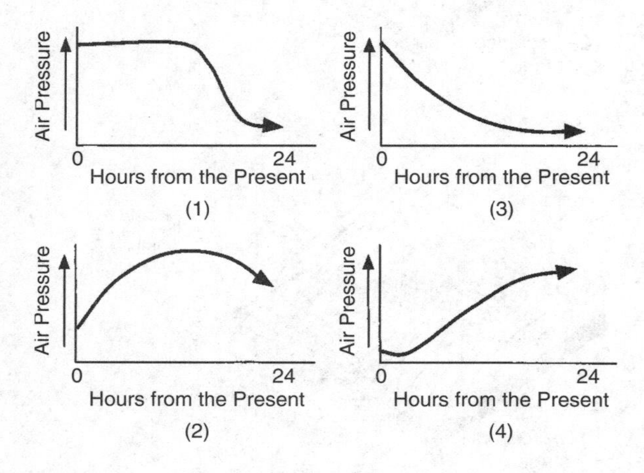

16. An air mass from central Canada that moves into New York State during the month of March would most likely be designated on a weather map as (1) cP (2) cT (3) mT (4) mP

Part B

Base your answers to questions 17 through 21 on the weather map below, which shows a mid-latitude cyclone passing through New York State. The locations of fronts A and B, air masses X and Y, as well as the city of Saratoga Springs, NY, are marked on the map.

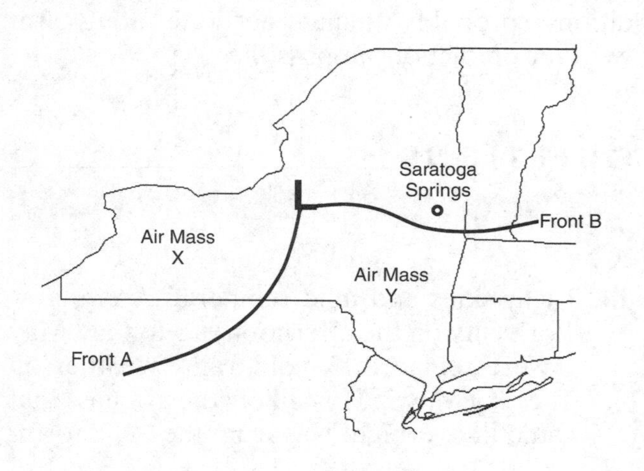

17. How is the weather in Saratoga Springs likely to change over the next few hours? (1) Rain will end, followed by warm, dry weather. (2) Rain will end, followed by warm, moist weather. (3) Rain will end, followed by cool, dry weather. (4) Rain will end, followed by cool, moist weather.

18. Air mass *X* is most likely (1) cP (2) cT (3) mP (4) mT

19. Which diagram below best shows how Front A should be drawn?

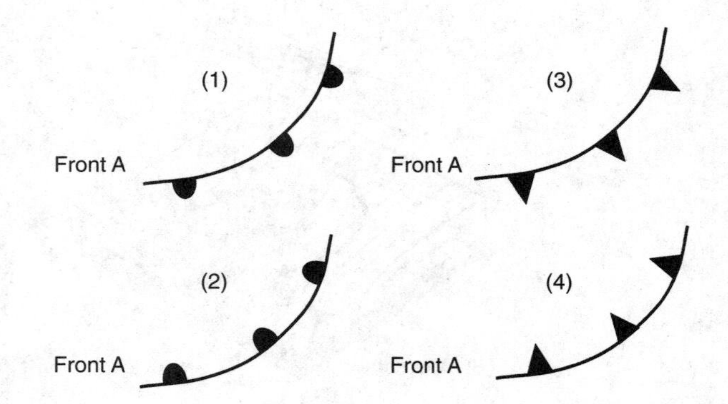

20. Which city in New York State probably has the lowest atmospheric pressure at the time this map represents? (1) Buffalo (2) Rochester (3) Syracuse (4) Plattsburgh

21. Which term best describes the circulation pattern of the winds around this low-pressure center? (1) downward (2) outward (3) clockwise (4) counterclockwise

Base your answers to questions 22 through 26 on the weather station model below.

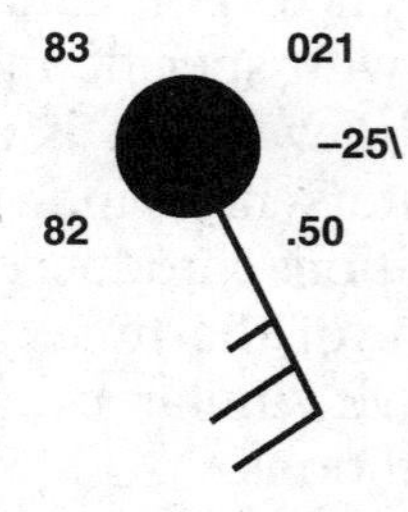

22. What is the wind speed at this location? (1) 15 knots (2) 25 knots (3) 15 miles per hour (4) 25 miles per hour

23. What is the atmospheric pressure at this location? (1) 21 millibars (2) 82 millibars (3) 1002.1 millibars (4) 1021 millibars

24. What is the approximate relative humidity at this location? (1) 95% (2) 83% (3) 50% (4) 21%

25. Over the past few hours the air pressure has been (1) decreasing (2) increasing (3) remaining constant (There are only three choices for this question.)

26. Which of the following instruments was *not* used to gather this weather data? (1) a sling psychrometer (2) an anemometer (3) a seismometer (4) a barometer

CAN WE LEARN TO LIVE WITH NATURAL HAZARDS?

Some weather hazards, such as smog and haze, are the result of pollution. Other hazards such as hurricanes and tornadoes are natural occurrences. Figure 8-13 shows where certain natural hazards are more likely to occur in the contiguous United States. Each year, violent storms cause property damage and loss of life. Among natural hazards, weather events cause the greatest loss of life. Drought is sometimes called the "silent killer" because it has no violent onset, although it has led to the deaths of millions of people, especially in Africa.

Thunderstorms

Most thunderstorms, such as the one shown in Figure 8-14 on page 206, occur in the summer when the air is warm and moist. An advancing cold front pushes the warm air upward, causing the formation of giant cumulus clouds and heavy precipitation. Energy becomes available when clouds form through the process of condensation, which releases latent heat. Because violent storms have a large amount of energy, they can

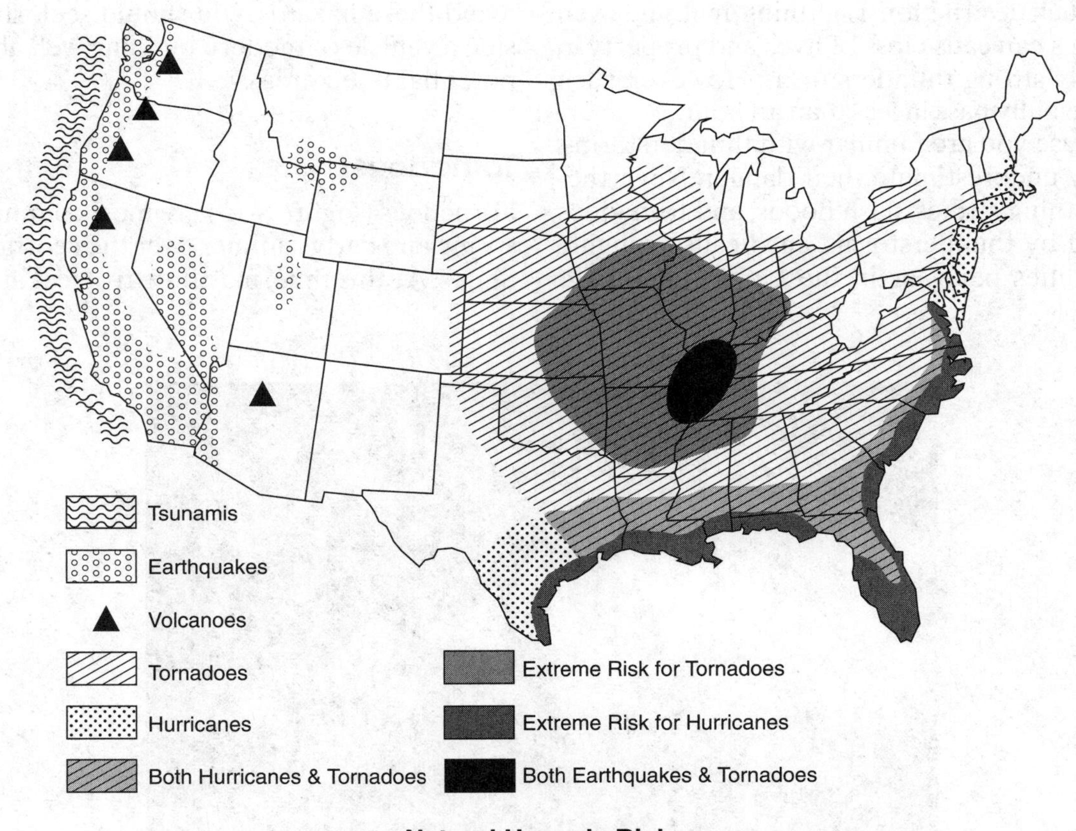

Figure 8-13. Natural hazards are more likely in some areas than in others. For example, the greatest risk of hurricanes is along the coastlines of the Gulf of Mexico and Atlantic Ocean. Tornadoes are most common in the south-central states. Earthquakes and volcanoes are most common in the Pacific States.

Figure 8-14. Most thunderstorms occur in the summer when rising air currents cause rapid condensation and heating of the rising air. The latent heat of vaporization intensifies the energy of the storm.

cause much destruction. Lightning, hail, and even tornadoes can cause loss of lives and property in especially strong thunderstorms. However, such storms usually pass in less than an hour.

Because you are familiar with thunderstorms, you may underestimate their danger. Nevertheless, lightning strikes, flash floods, and tornadoes spawned by thunderstorms, each cause roughly 100 fatalities per year in the United States. To avoid these hazards, you should seek shelter inside a vehicle or modern building well above any potential flood areas.

Tornadoes

Tornadoes (Figure 8-15) are most common in the spring and early summer over the central United States. At this time, maritime tropical air masses

Figure 8-15. Although tornadoes are most common in the American Midwest, they also occur in New York State.

from the Gulf of Mexico collide with cooler and drier air from the north. However, tornadoes can occur in any part of the United States.

Tornadoes are relatively small and short-lived. They are usually less than 0.5 kilometer (0.3 mile) in diameter. Although some have lasted for hours, most tornadoes last 10 minutes or less. As you can imagine, it is very difficult to measure the wind speeds in tornadoes. Instruments in the paths of tornadoes have been destroyed along with the buildings to which they were attached. Winds in some tornadoes have been measured at 450 kilometers/hour (280 miles/ hour) before the instruments broke. However, winds of more than 500 kilometers/hour (300 miles/hour) have been identified based on radar measurements and observations of damage caused by powerful tornadoes.

When tornadoes threaten, the best way to learn that a tornado is approaching is to listen to the radio. Battery-operated radios will work even if the power goes out. Because tornadoes form and disappear so quickly, warnings are usually short-term, allowing people only a few minutes notice of a nearby tornado. This is unlikely to give time to protect property. However, people can move to interior rooms for safety. Basements and special underground storm cellars provide even better protection in case a major tornado passes over.

Figure 8-16. Hurricanes are intense tropical cyclones that produce their energy by rapid condensation (cloud formation) as winds circulate counterclockwise and toward the low-pressure center.

Hurricanes

Tropical depressions are areas of low pressure. They usually develop in the late summer and early autumn in the Atlantic Ocean between South America and Africa. This region has an abundance of solar energy and warm tropical water. Tropical depressions can grow into hurricanes. Figure 8-16 shows how rising air and condensation within a hurricane can supply energy to support and strengthen the storm. These storms gather strength as they drift across the Atlantic Ocean. When sustained winds in the storm are over 120 kilometers/hour (74 miles/hour), the storm's classification is changed from a tropical storm to a hurricane. Figure 8-17

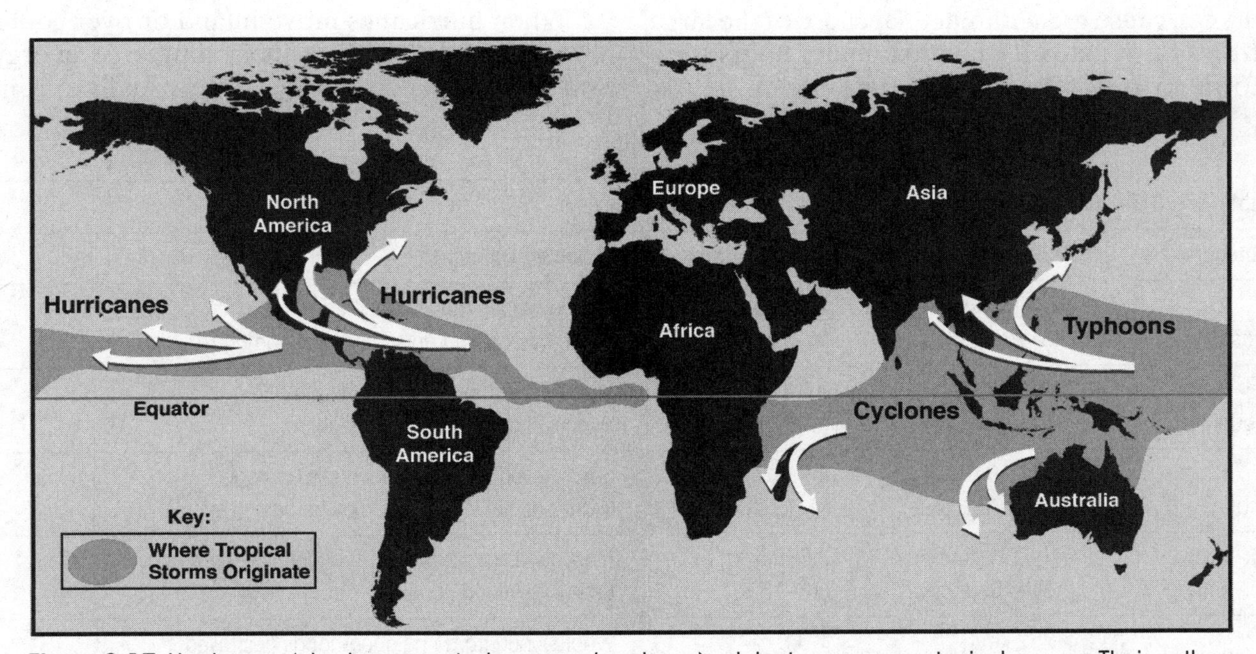

Figure 8-17. Hurricanes (also known as typhoons and cyclones) originate over warm, topical oceans. Their paths curve due to the Coriolis effect and often move into the mid-latitude westerly wind zones.

shows the most common paths of hurricanes worldwide.

Hurricanes that start in the Atlantic basin usually drift westward and a little north until they are influenced by the prevailing west winds over the continental United States. Pacific Ocean hurricanes sometimes cause damage in Hawaii, but they most often move away from the western coast of North America. Figure 8-18 shows the path of Hurricane Katrina August 23–30, 2005. This storm followed a typical track for a major Atlantic hurricane.

Hurricanes are rated on the Saffir-Simpson Scale from Category 1 to Category 5 based mostly on the speed of the winds. Table 8-1 is an example of the Saffir-Simpson scale.

Hurricanes grow into huge rotating systems that average 650 kilometers (400 miles) in diameter. At the center of a hurricane, the air pressure is very low. Therefore, a large pressure gradient develops, which generates the strong winds associated with these storms. The winds in a hurricane may reach speeds up to 240 kilometers/hour (150 miles/hour). Winds whirl around the center of a hurricane, which is called the eye. However, in the eye itself, winds are relatively calm.

Hurricanes produce fierce winds, intense precipitation, and a storm surge, which causes unusually high tides. When a hurricane comes ashore along the southern and eastern coasts of the United States, the combined effects of these factors can cause great damage. Because of the large size of hurricanes, it may take many hours for a storm to pass. During this time, flying debris, flooding, and even tornadoes that form within intense hurricanes can cause great destruction.

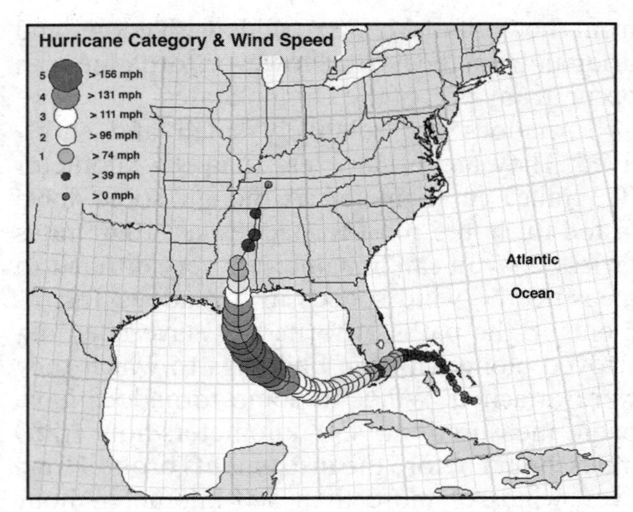

Figure 8-18. Hurricane Katrina started as a tropical depression over the western Atlantic Ocean on August 23, 2005. It strengthened to a Category 5 hurricane and then weakened to a Category 3 as it hit New Orleans. This map shows its progress for 8 days until it died out over the central United States.

There are many similarities between hurricanes and mid-latitude cyclones. Both are low-pressure systems with surface winds that converge counterclockwise. However, hurricanes are more intense and do not include weather fronts.

When hurricanes move inland or over cooler ocean water, they lose their source of energy: warm, tropical ocean water. They usually expend

Table 8-1. Saffir-Simpson Scale

Category	Wind Speed	Storm Surge	Expected Damage
1 Weak	120–153 km/h (75–95 mph)	1.2–1.5 m (4–5 ft).	Mobile homes may be toppled. Some trees blown down and roof damage.
2 Moderate	154–177 km/h (96–110 mph)	1.8–2.4 m (6–8 ft)	Manufactured homes sustain damage. Windows and doors may break.
3 Strong	178–210 km/h (111–130 mph)	2.7–3.7 m (9–12 ft)	Small, wood frame homes badly damaged. Older and light buildings may be destroyed.
4 Very Strong	211–251 km/h (131–155 mph)	3.8–5.5 m (13–18 ft)	Most buildings damaged. Roof collapse common. Extensive inland flooding.
5 Devastating	>252 km/h (>155 mph)	>5.8 m (>19 ft)	Whole wood frame neighborhoods destroyed. Storm surge devastates costal areas.

their most violent winds in a matter of days and become mid-latitude cyclones with a fraction of the power of a hurricane.

Hurricanes cause the most damage in coastal areas where a 1- to 2-day warning is possible. People shutter windows and may move valuables to interior rooms above expected flood levels. Ocean shore communities are evacuated by designated routes as people move to inland shelters such as schools. People who wait too long to evacuate may be caught in traffic or subjected to extreme weather conditions.

Blizzards

Blizzards are winter snowstorms that produce heavy snow and restricted visibility along with wind speeds of 56 kilometers/hour (35 miles/hour) or greater. The near-coastal regions of New York are sometimes hit by storms that transport great quantities of water vapor from the Atlantic Ocean. Locations near Buffalo and Syracuse are downwind from the Great Lakes and are subject to "lake-effect storms." These storms can dump as much as 2 meters (6 feet) of blowing snow in just a day or two.

Blizzards can paralyze cities and suburban areas by making roads impassable, even for emergency vehicles. Public services, such as electric power and telephone may be interrupted. A person stranded in a blizzard, even in an automobile trapped in the blowing snow, can suffer frostbite and a deadly loss of body heat.

The best way to weather a blizzard is to stay home. However, you must be prepared for extended loss of electrical power and telephone service. In winter, it is wise to keep plenty of fuel and food along with battery-powered radios on hand. After a blizzard, roads may be blocked for days; so do not drive until you know the roads are safe. If you must go out, dress in warm layers.

QUESTIONS

Part A

27. In the United States, most tornadoes are classified as (1) large low-pressure centers (2) large high-pressure centers (3) small low-pressure centers (4) small high-pressure centers

28. Hurricanes, cyclones, and typhoons are examples of (1) powerful tropical cyclones (2) powerful polar cyclones (3) powerful tropical anticyclones (4) powerful polar anticyclones

29. Which location in New York State is likely to experience the heaviest winter snowfall when the winds are blowing from the northwest? (1) New York City (2) Binghamton (3) Oswego (4) Plattsburgh

Part B

Base your answers to questions 30 through 34 on the graph below, which shows weather changes in Dallas, Texas over a 24-hour period during the month of April.

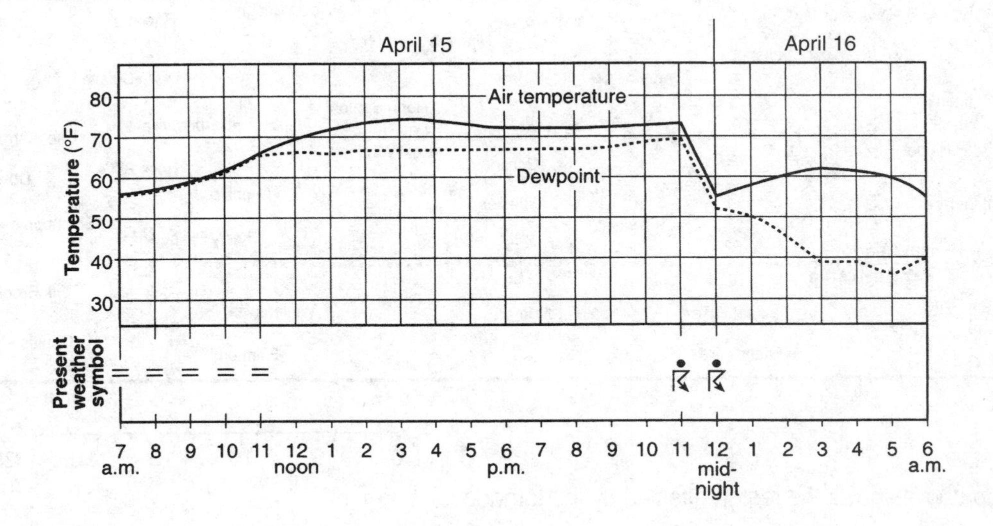

30. What event occurred just before midnight? (1) A warm front passed thorough. (2) A cold front passed through. (3) The atmospheric pressure suddenly dropped. (4) A tropical air mass moved in.

31. What weather hazard might occur in Dallas in the very late evening? (1) a hurricane (2) a blizzard (3) a tornado (4) a drought

32. There were a series of weather-related traffic accidents on the morning of April 15. What was the most likely cause of these accidents? (1) poor visibility (2) high winds (3) freezing rain (4) thunder storms

33. There were thunderstorms late in the evening of April 15. At what other time was precipitation most likely? (1) early in the morning of April 15 (2) mid-afternoon of April 15 (3) early in the evening of April 15 (4) near sunrise on April 16

34. What was the Celsius temperature at noon on April 15? (1) 19°C (2) 21°C (3) 68°C (4) 70°C

Base your answers to questions 35 through 39 on the text below.

A New York Tornado

A small tornado formed near the town of Apalachin, New York at about 5:30 P.M. producing winds as strong as 70 miles per hour. It caused a path of destruction approximately 200 feet wide. By 5:45 P.M. the tornado had strengthened as it traveled thorough Vestal, New York with winds as high as 100 miles per hour. Considerable damage was reported along the path of the tornado.

By 6:10 in the evening the tornado moved to Binghamton with winds of about 130 miles per hour. It pushed over a 1000 foot high television transmission tower near Binghamton. Then the tornado lifted off the ground on its way toward Windsor, New York where only a few tree were damaged. In Sanford, New York a mobile home was completely overturned, although no greater damage was found.

The tornado arrived at Deposit, New York at about 6:30 P.M. and surface winds reached their greatest speed of approximately 180 miles per hour, causing severe damage to homes. However, after passing through Deposit, the tornado soon ended its more than 1-hour rampage and died out.

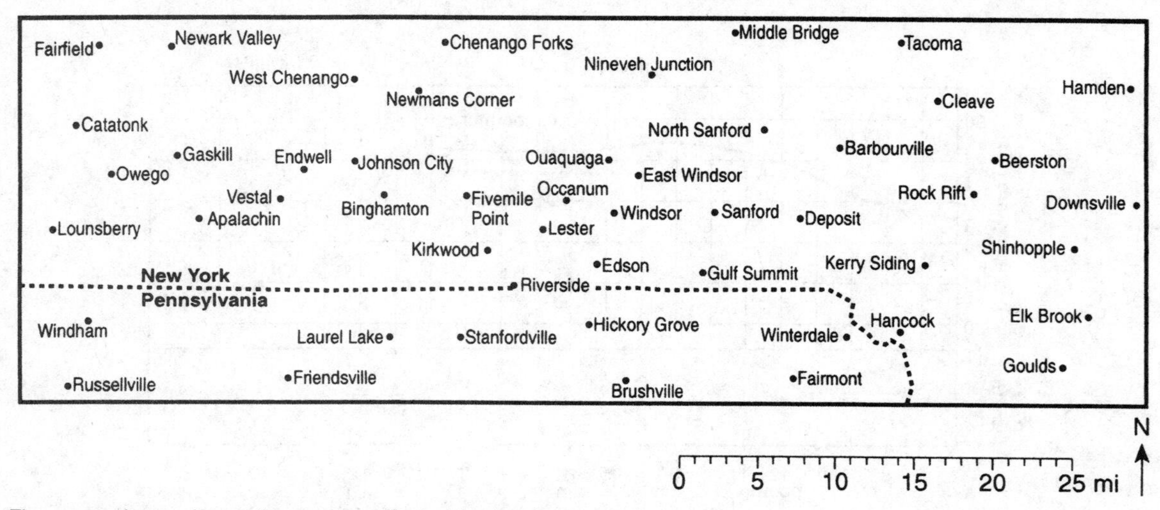

The map above shows the region affected by the tornado.

35. The tornado occurred as a continental-polar air mass replaced a maritime-tropical air mass. Which front symbol should be used on a weather map of this region to show this weather front advancing along with the tornado?

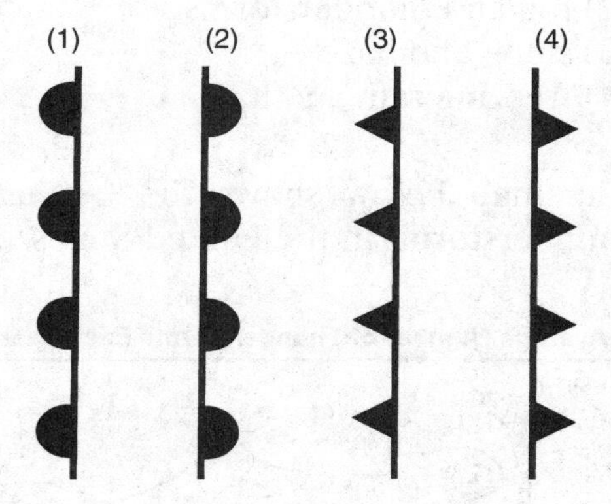

36. What was the average rate of speed at which this tornado advanced during the 1½ hours it was observed? (1) 0.04 mile per hour (2) 25 miles per hour (3) 38 miles per hour (4) 57 miles per hour

Base your answers to questions 37 through 39 on the table below, which shows the Enhanced Fujita Scale that is used internationally to classify tornadoes.

Enhanced Fujita Scale

EF-Scale Number	Wind Speed (mph)	Type of Damage Done
EF-0	65–85	Some damage to chimneys; breaks branches off trees; pushes over shallow-rooted trees; damages sign boards.
EF-1	86–110	Peels surfaces of roofs; mobile homes pushed off foundations or overturned; moving autos pushed off the roads; attached garages may be destroyed.
EF-2	111–135	Considerable damage; roofs torn off frame houses; mobile homes demolished; boxcars pushed over; large trees snapped or uprooted; light-object missiles generated.
EF-3	136–165	Roofs and some walls torn off well-constructed homes; trains overturned; most trees in forest uprooted.
EF-4	166–200	Well-constructed houses leveled; structures with weak foundations blown off some distance; cars thrown and large missiles generated.
EF-5	>200	Strong frame houses lifted off foundations and carried considerable distances to disintegrate; automobile-sized missiles fly through the air in excess of 100 meters; trees debarked; steel-reinforced concrete structures badly damaged.

37. What was the maximum Enhanced Fujita Scale number of this tornado and where did it occur? (1) EF-1 at Apalachin (2) EF-3 at Apalachin (3) EF-4 at Deposit (4) EF-5 at Deposit

38. What was the Enhanced Fujita scale number of this tornado in Sanford?
(1) EF-1 (3) EF-3
(2) EF-2 (4) EF-5

39. The image below shows all that remained of a well-constructed frame house after if was struck by a tornado in the American Midwest. Most of the remains of the house were found approximately 1/4 mile from its original location. What was the approximate wind speed of this tornado?

(1) 50 miles per hour (2) 100 miles per hour (3) 200 miles per hour (4) more than 200 miles per hour

Problem Solving

The National Weather Service reports that massive Hurricane Sally is now located several hundred miles to the south of your location. It is likely to move into your location. A weekly newspaper in your community has asked you to write an article to inform the public about how this storm may affect people in your area. Submit an edited article to your teacher including images of possible hurricane damage.

Chapter Review Questions

PART A

1. Most cyclonic weather systems in the United States (1) contain winds that rotate clockwise and move east to west (2) contain winds that rotate clockwise and move west to east (3) contain winds that rotate counterclockwise and move east to west (4) contain winds that rotate counterclockwise and move west to east

2. To assist pilots, a meteorologist wanted to identify airports that have poor visibility. He used past weather records to identify airports that frequently had low-visibility conditions. Which weather conditions indicate poor visibility due to high pollution levels from traffic and the use of fossil fuels?
(1) rain and drizzle
(2) fog and thunderstorms
(3) smog and haze
(4) freezing rain and hail

3. The map below shows the frequency of thunderstorms in the United States.

Average Number of Thunderstorms Each Year

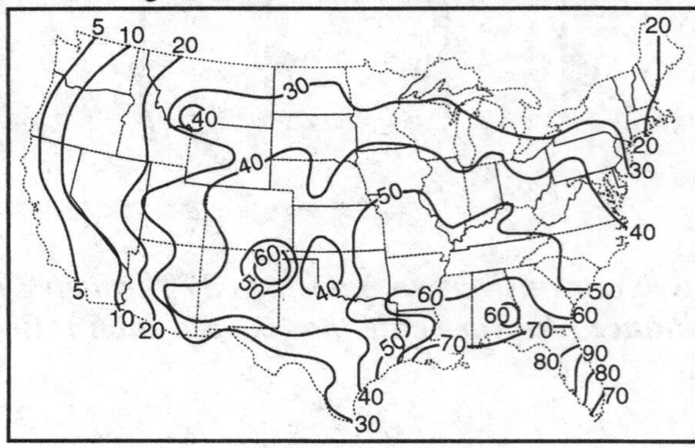

Based on this map, what kind of air mass is most often associated with thunderstorms?
(1) continental arctic
(2) continental polar
(3) continental tropical
(4) maritime tropical

4. Which process is the most important source of energy that heats the atmosphere, causing and sustaining violent weather events?
(1) cloud formation (2) violent winds
(3) release of electromagnetic radiation
(4) conduction from land surfaces

5. Which of the following lists the forms of electromagnetic energy in order of increasing wavelength?
(1) red→yellow→blue
(2) green→blue→violet
(3) blue→green→red
(4) x-rays→visible light→untraviolet

6. Which graph best shows how the surface temperature of a body of water affects the rate of evaporation into the atmosphere?

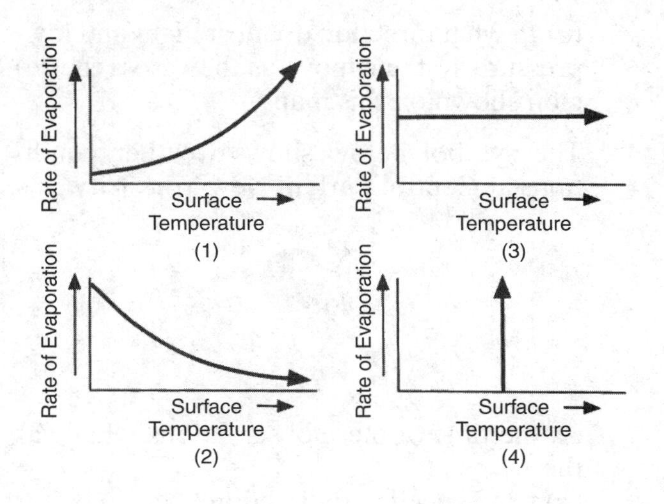

PART B

Base your answers to questions 7 through 11 on the diagram below, which represents a weather system as it moves across the central United States. Shading shows regions of overcast skies (light shading) and precipitation. (darker shading)

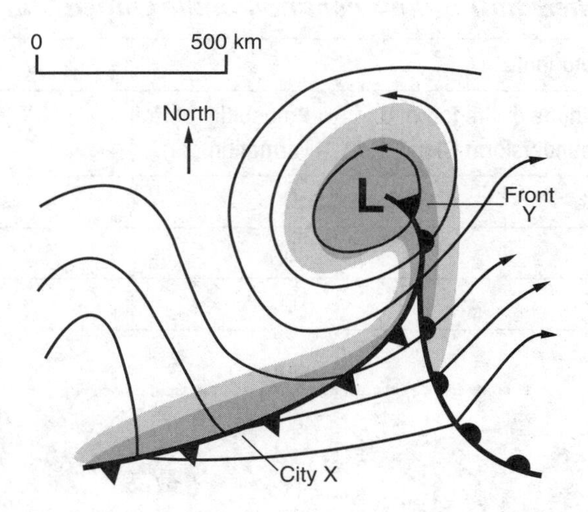

7. Another name for this kind of weather system is (1) high-pressure system (2) tornado (3) mid-latitude cyclone
(4) anticyclone

8. What caused the formation of Front *Y*?
(1) Weather fronts move across the United States from east to west.
(2) Warm fronts stay in the same position until they dissolve.
(3) Cold fronts usually travel faster than warm fronts.
(4) Warm air is more dense that cold air.

9. Which diagram best represents this weather system 24 hours before the diagram above?

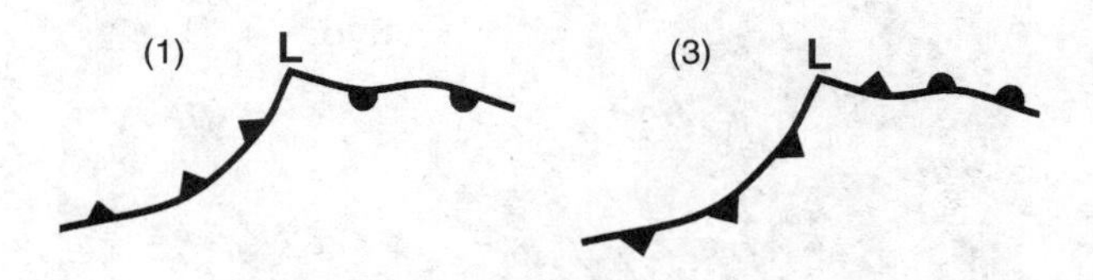

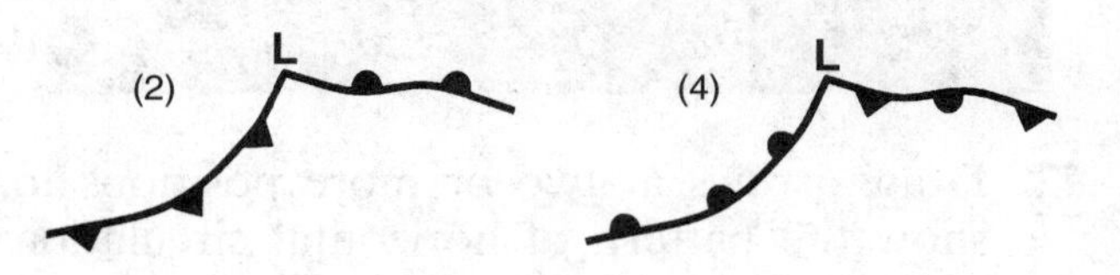

10. Note the position of City *X*. In 24 hours, how is the weather at City *X* likely to be different?
(1) Atmospheric pressure will be higher and the relative humidity will be higher.
(2) Atmospheric pressure will be higher and the relative humidity will be lower.
(3) Atmospheric pressure will be lower and the relative humidity will be higher.
(4) Atmospheric pressure will be lower and the relative humidity will be lower.

11. In 48 hours this weather system will disappear. How will this most likely occur?
(1) The warm, moist air mass will move upward and lose its character. (2) The warm front will wedge under the cooler air and cause its temperature to rise. (3) A continental tropical air mass will move in from the northwest. (4) The temperature of the continental polar air mass will warm the whole region.

PART C

Base your answers to questions 12 and 13 on the following satellite image, which shows a

major tropical storm approaching the United States in the Gulf of Mexico.

12. Draw arrows at two or more positions to show the pattern of horizontal circulation within this storm.

13. This storm is one of the most powerful storms ever observed in the Gulf of Mexico. To which class of tropical storm in the Atlantic Ocean and the Gulf of Mexico does this storm belong?

14. Below is a map of the United States showing a mid-latitude cyclone moving across the United States. High-pressure (*H*) and low-pressure (*L*) centers are shown. The position of a warm front has also been shown.

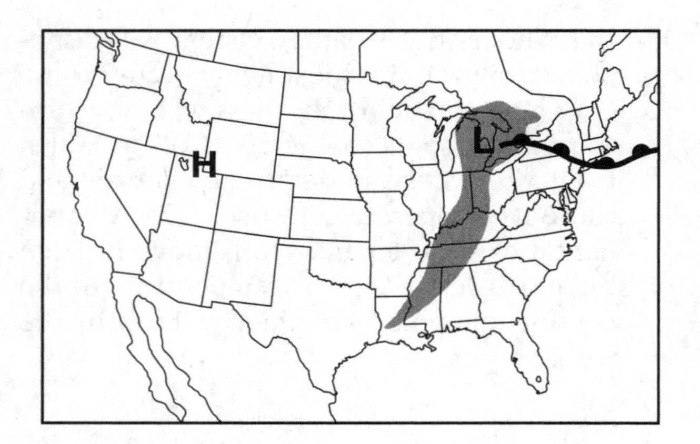

(*a*) Draw the most likely location of the cold front. Be sure the cold front symbols show the direction in which the front is moving.

(*b*) Why are there no fronts near the high-pressure center?

(*c*) In what direction do most high- and low-pressure systems move as they cross the region shown on this map?

15. The symbol below shows weather conditions at Central Park in New York City.

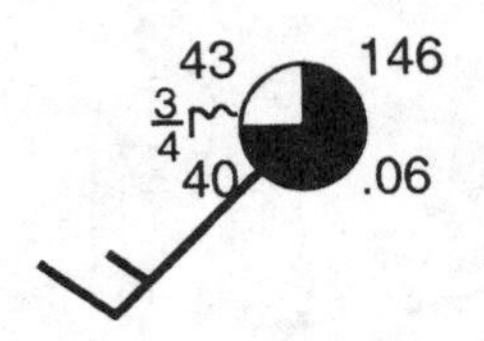

At the time of these observations, what was the

(*a*) wind speed and direction?
(*b*) atmospheric pressure?
(*c*) Celsius temperature
(*d*) Why is the visibility only 3/4 mile?

16. Describe the conditions of temperature and humidity of an air mass that is labeled mP on a regional weather map.

Base your answers to questions 17 through 20 on the data table and map below. The data table shows how cloud height is related to the formation of hail in New York State. The map shows where hail is most common in the United States.

Data Table

Altitude of the Top of a Thunderstorm Cloud (km)	Probability of Hail Formation (%)
13	50
15	75
17	100

17. Describe the relationship between the maximum height of thunderstorm clouds and the likelihood of hail.

18. If a storm cloud has sufficient height to have a 100% probability of hail, in what layer of the atmosphere is the top of the cloud?

19. State the average number of days per year that a person in Albany, New York is likely to experience hail.

20. State one way that you can protect yourself from injury during a hailstorm.

21. The table below shows weather conditions recorded at Albany, New York. Draw a weather station model showing how all this data should be presented on a weather data map.

Wind direction	Northwest
Wind speed	20 knots
Visibility	1/4 mile
Present weather	Hail
Amount of cloud cover	100%
Barometric pressure	990.0 millibars

Base your answers to questions 22 through 26 on the diagram below, which shows a weather front moving over New York State in the month of September.

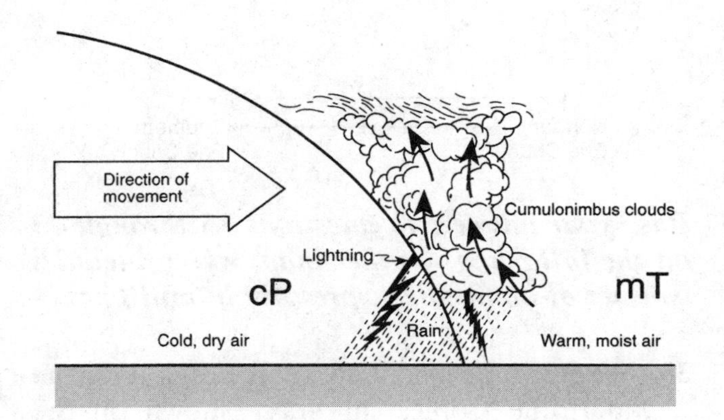

22. What class of front is this?

23. Why is the warm, moist air rising as the front moves forward?

24. What change in state is occurring to make the clouds?

25. What is the most likely source area for the air mass labeled mT?

26. What change in atmospheric pressure would an observer most likely notice as the cP air mass replaces the mT air mass after the passage of the front?

Base your answers to questions 27 through 31 on the information given by the four weather station models below. The data at four New York State locations was collected at the same time.

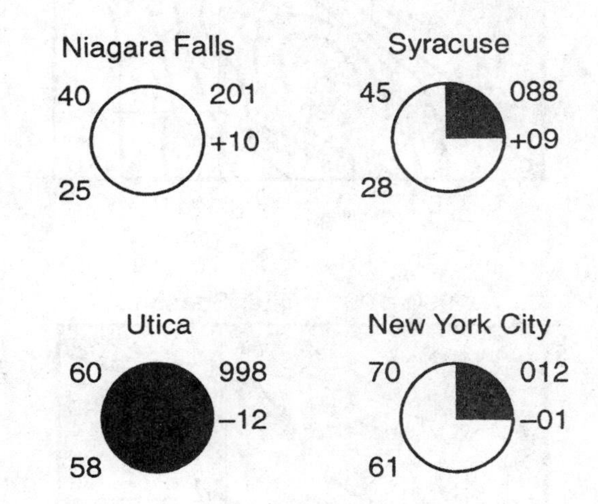

27. What was the air pressure at Niagara Falls?

28. Give two reasons why Utica appears to have the greatest chance of precipitation.

29. New York City was experiencing a south wind at 15 knots with hazy conditions and a visibility of 3/4 of a mile. Complete the station model for New York City to show this additional data.

30. What is the current air pressure in Utica and what was the air pressure three hours before this reading?

31. If these conditions show the passage of a cold front, in what direction was the front moving?

Base your answers to questions 32 through 34 on the three maps below, which show the total snowfall in a region of western New York State

during three winter seasons. The names of some counties appear on the maps.

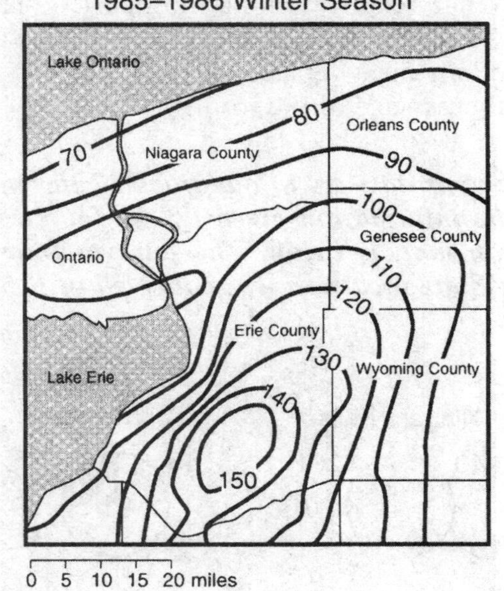

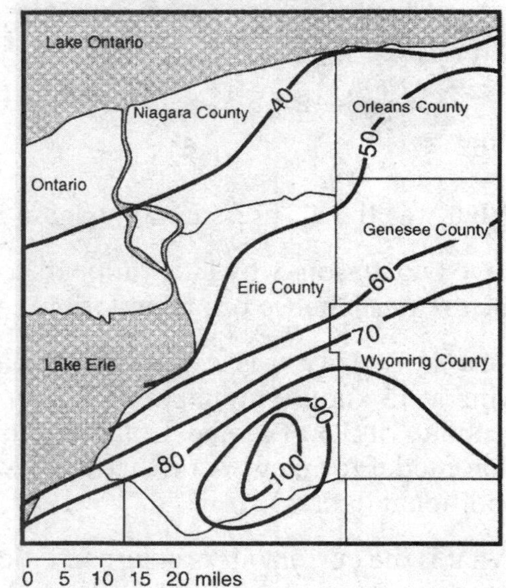

32. Calculate the average snowfall gradient along the dotted line *A–B* on the 1991–92 map.

33. The amount of snowfall in each storm is usually smaller later in the winter after Lake Erie freezes over. Explain why the freezing of Lake Erie reduces snowfall in this area.

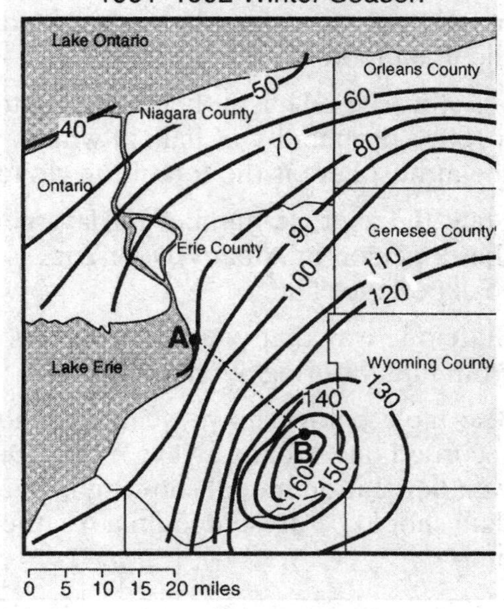

34. Use the diagram below to draw a line graph showing the general relationship between the snowfall in Erie County and the north-south location in Erie County.

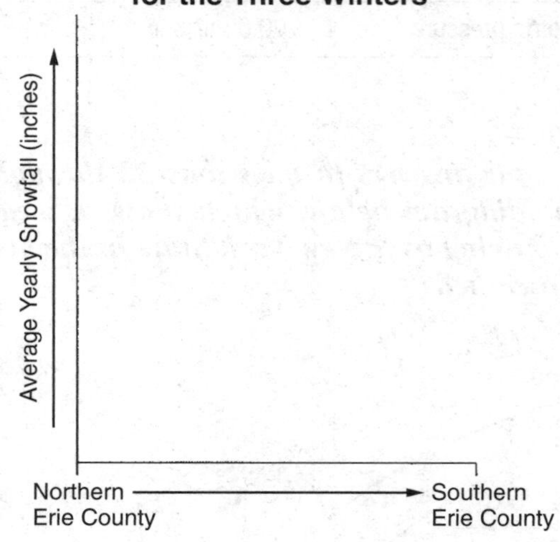

Base your answers to questions 35 through 43 on the following weather map, which includes isolines of atmospheric pressure in millibars.

35. What is the most likely source region for the maritime tropical air mass shown on the map?

36. Calculate the pressure gradient along a straight line from point *A* to point *B*.

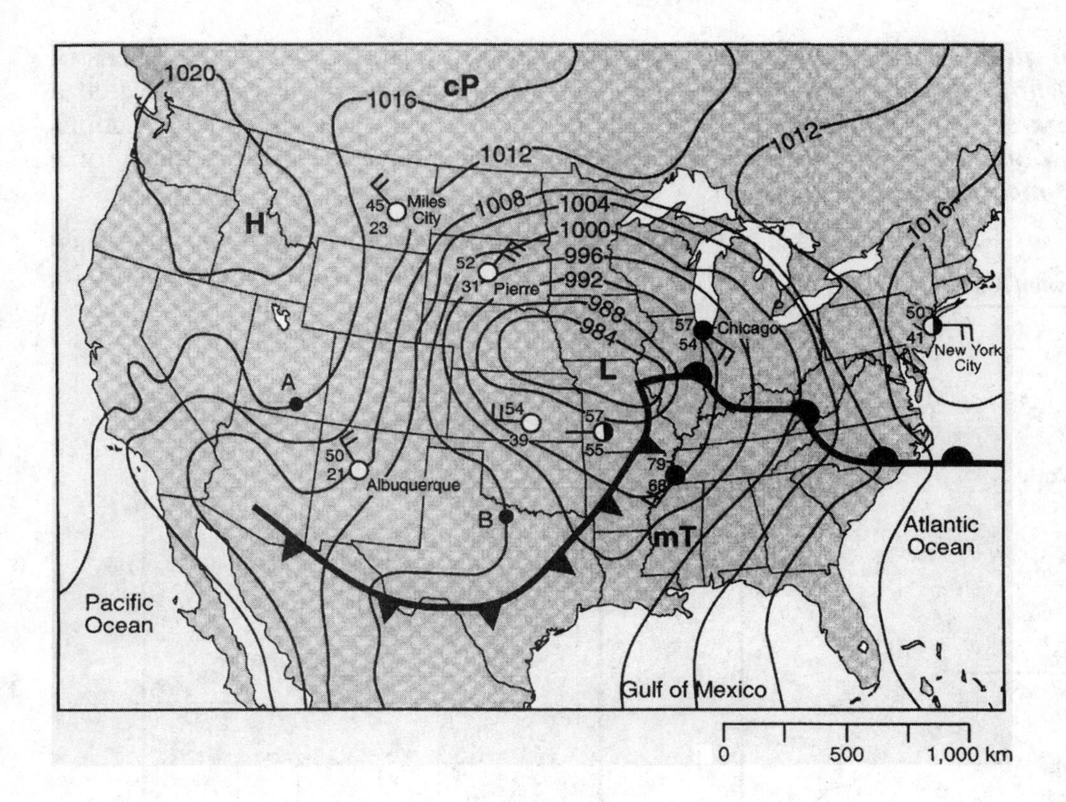

37. What evidence on the map shows that the winds 500 km west of Miles City are relatively light?

38. List the following locations in order from lowest relative humidity to highest relative humidity: Albuquerque, Chicago, and New York City

39. Describe the pattern of the surface winds around the low-pressure center.

40. What is the atmospheric pressure at Miles City?

41. What is the atmospheric pressure at the location marked *H*?

42. What is the wind speed and direction indicated for Pierre?

43. What type of weather front is located about 500 km south of New York City?

(*Questions continue on page 218.*)

Base your answers to questions 44 through 46 on the map below, which shows the depth of snowfall from a blizzard in the month of December. The decimal point of each number is the location at which the measurement was made.

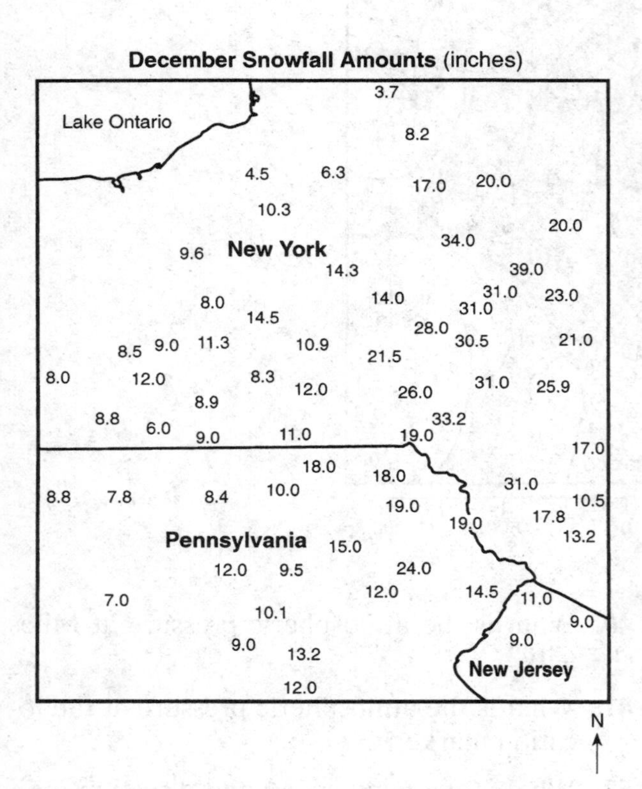

44. Draw the 30-inch snowfall isoline.

45. How did the topography of The Catskills most likely influence the distribution of snowfall shown on the map?

46. The media was able to give the citizens of New York City several days' notice that a major snowstorm was coming. What actions could families have taken to prepare for a snow emergency?

47. The map below shows the snowfall measured in the 1984-85 winter in the Buffalo, New York area. Draw an isoline that represents 120 inches of snowfall.

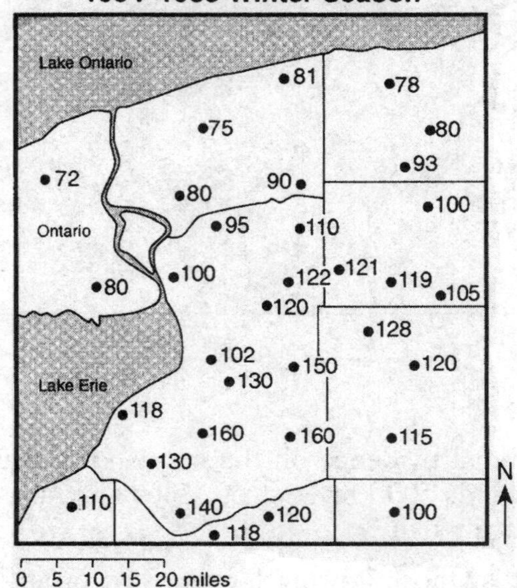

CHAPTER

The Water Cycle and Climate

Vocabulary at a Glance

adiabatic	groundwater	reflected
angle of insolation	infiltrate	runoff
capillarity	insolation	transpiration
capillary water	insolation-temperature lag	vertical ray
climate	leeward	water cycle
continental climate	maritime climate	water table
duration of insolation	permeability	watershed
El Niño	planetary wind belts	windward
greenhouse effect	porosity	zone of aeration
greenhouse gases	radiative balance	zone of saturation

THE HYDROLOGIC CYCLE

Earth has a limited supply of water. As shown in Figure 9-1, most of that water is saltwater in the oceans. The majority of the freshwater is in glacial ice, mostly near the South Pole. Groundwater and fresh surface water form only a small part of Earth's water.

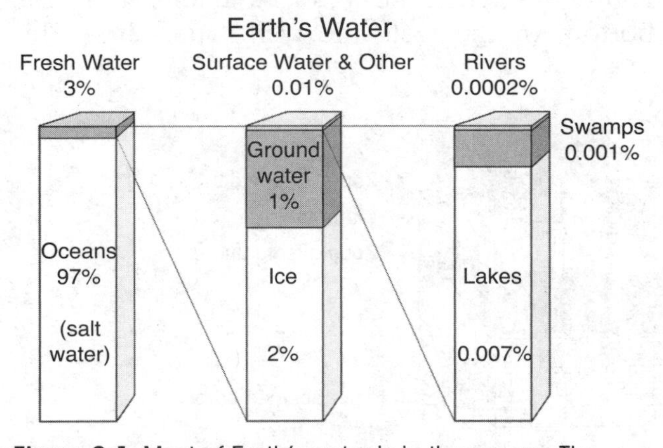

Figure 9-1. Most of Earth's water is in the oceans. The water you see in Earth's rivers and lakes make up only about 1 part in 10,000 of Earth's total water supply.

Earth's water supply is being recycled constantly between the oceans, atmosphere, and land in a process called the **water cycle** (hydrologic cycle). During this recycling process, water enters the atmosphere by *evaporation* and transpiration. **Transpiration** is the process by which living plants release water vapor to the atmosphere. These two processes (sometimes called evapotranspiration) are driven by energy from the sun. Because the oceans cover approximately 71 percent of Earth's surface, most of the water vapor in the atmosphere gets there by evaporation from the oceans. Water vapor in the atmosphere condenses and returns to Earth's surface as precipitation. To complete the cycle, groundwater circulates through Earth, as shown in Figure 9-2 on page 220.

Less than 1 percent of Earth's water **infiltrates**, or sinks into, the ground as **groundwater**. However, the amount of groundwater is still about 50 times the total freshwater on Earth's surface. Unlike the oceans, groundwater is mostly freshwater; unlike the glaciers, it is almost all in the liquid state. Plants need groundwater for growth. Groundwater is also an important source of water for our homes, industry, and agriculture.

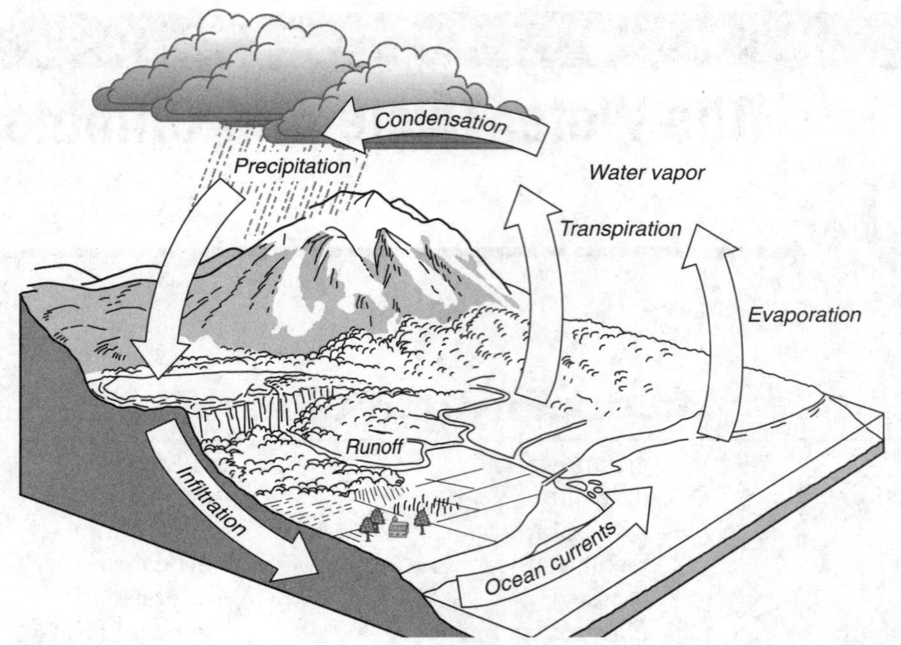

Figure 9-2. Water circulates between the atmosphere, the land, and the oceans in a never-ending series of changes. These include precipitation, infiltration, runoff, evaporation, transpiration, convection (movement via the wind), and condensation.

Groundwater Zones

After infiltrating the soil, groundwater occupies distinct zones. Much of the groundwater is pulled down by gravity until it reaches a layer of solid bedrock. Since the water cannot penetrate the bedrock, it stops sinking. This layer is known as impermeable bedrock. Above the bedrock is the **zone of saturation**. In the zone of saturation, all spaces, cracks, and other openings in soil and rock grains are completely filled with water.

The **zone of aeration** is above the zone of saturation. In the zone of aeration, the spaces between particles are filled with air. The **water table** is the boundary between the zone of saturation and the zone of aeration. It is the upper surface of the zone of saturation. As water passes through the ground, it is filtered by the soil and rock. This makes groundwater a good source of drinking water.

The location of the water table is important to us. Water wells must reach below the water table to provide a reliable supply of groundwater. Pumping water from a well faster than it can be replaced can lower the water table, as shown in Figure 9-3. When the water table falls below the bottom of the well, the well "runs dry." This

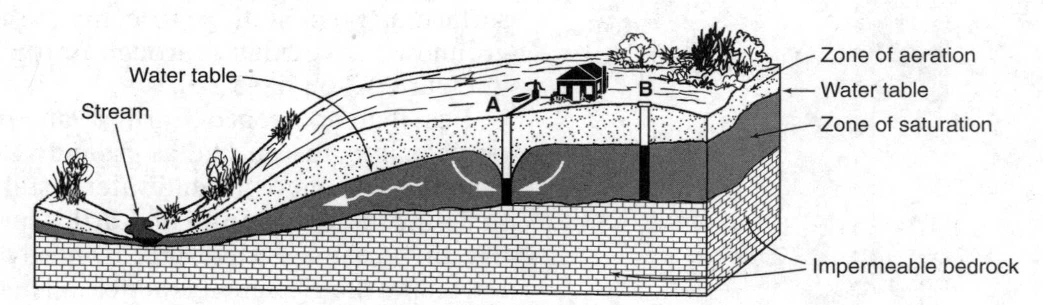

Figure 9-3. Water flows, usually slowly, through the ground under the influence of gravity. Pumping from well *A* has lowered the water table near the well. Well *B* is capped, so there is no pumping, and no lowering of the water table around well *B*.

problem can be solved in two ways. First, you can drill a deeper well, as long as it is not extended into an impermeable layer, such as dense clay or bedrock. Alternatively, you can use less water, allowing groundwater to seep back into the well.

Factors That Affect the Storage and Movement of Groundwater

The characteristics of the soil and rock near the surface control the amount of water in the ground and the movement of water through the ground. Water will infiltrate the ground when there are many openings, or pore spaces, between ground particles for the water to pass through. Almost all materials on Earth's surface are porous. The amount of pore space in a material compared with its volume is called **porosity**. It is porosity, or the percentage of empty space, that determines how much air or water a sample of rock can hold.

The porosity of a loose material, such as soil, largely depends on shape, packing, and the mixture of sizes (sorting) of the soil particles. The most porous soils, as shown in Figure 9-4 (top), are those that contain particles that are all about the same size (*well-sorted*) and particles that are not closely packed. Well-sorted sand can have porosity of greater than 40%. In Figure 9-4 (middle), small particles fill spaces between the larger particles. Therefore, there is less room for water. Mixing particle sizes reduces porosity. Figure 9-4 (bottom) illustrates a soil that has no porosity because a very fine sediment or mineral cement has filled all the open spaces. Like solid rock, this sample has a porosity of 0%

Particle shape and packing also affect porosity. Flattened and angular soil grains, such as clay particles, can pack together closely. Therefore, most clay soils have little pore space and low porosity. Figure 9-5 (left) shows particles packed closely together, so porosity is low. In Figure 9-5 (right) the particles are not closely packed, so this sample is relatively porous.

It is important to note that particle size alone does not affect the porosity of a soil. For example, when two well-sorted samples of different soils are taken—one with large particles and one with small particles—both may have about the same porosity. This is especially true if their soil particles are similarly shaped and packed. Figure 9-6 on page 222 shows that particle size does not affect porosity. The large number of small spaces in sample *A* are equal to the smaller number of larger spaces in samples *B* and *C*.

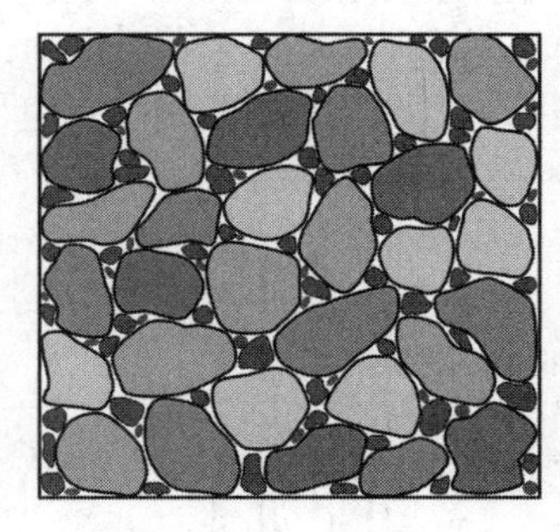

Figure 9-4. The porosity of a sediment or soil is reduced when the sample is poorly sorted by particle size.

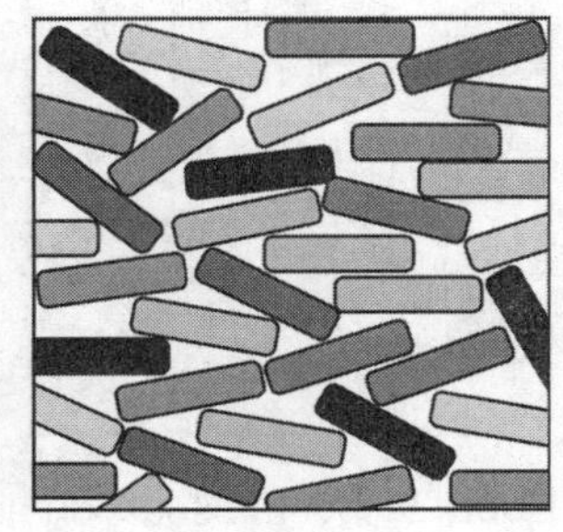

Figure 9-5. Particles that fit closely together reduce porosity as shown in the sample on the left. However, the sample on the right has a much higher porosity because the same particles are loosely packed.

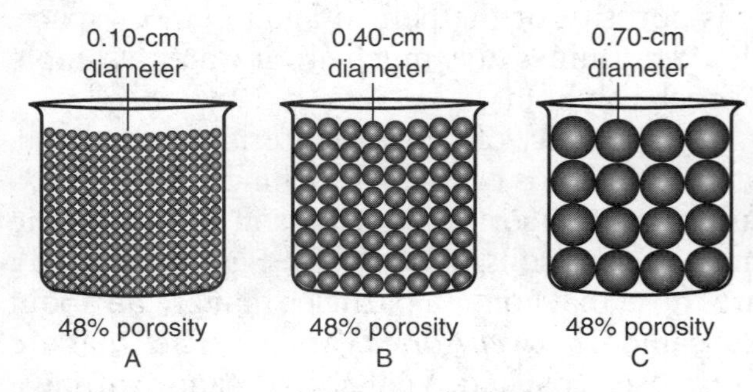

Figure 9-6. Changing the size of the particles does not change the porosity of a sample as long as there is no change in the shape and packing of the particles. Each of these samples of spheres has the same porosity. Although *A* has the smallest pore spaces, it has more spaces than *B* or *C*. These two factors (number of pores and pore size) cancel each other out.

Permeability is the ability of a soil to transmit water. The rate of permeability, or how fast water can pass through a soil, depends on the size of the pores and how the pores are connected. For example, soils with high permeability, such as sand, have large pores that are well connected. Water flows through the pores with little resistance. Soils with low permeability, such as clay, have small pores or pores that are not well connected. Water in very small pores moves slowly and clings to the surfaces of soil particles. Therefore, in Figure 9-6, the permeability is slower in *A* and *B* than it is in *C*.

A soil can have a high porosity and a low permeability. On the other hand, it can have low porosity and high permeability. For example, fractured bedrock or sediments with a few large cracks, such as shown in Figure 9-7 (left) can hold very little water (nonporous), but water may pass quickly through the cracks (permeable). On the other hand, the sample in Figure 9-7 (right) has a high porosity, but a low permeability because the large pores are not well connected.

Figure 9-7. A material can have a high permeability but a low porosity as in the sample on the left. However, in the sample on the right, the large pores can hold a lot of water, but the water cannot flow quickly through the sample because the pores are not well connected.

Surface **runoff** occurs when rainfall exceeds the permeability rate of a soil. (That is, the rain falls faster than it can soak into the ground.) When a soil is saturated, no more water can infiltrate, so it runs off. When the slope (gradient) of a soil's surface is too steep, water runs off before it can soak in. Also at temperatures below 0°C, water within a soil will freeze, thus preventing further infiltration.

Capillarity is the ability of a soil to draw water upward into the tiny spaces between soil grains. Soils composed of very small particles, such as clay, have the most capillarity. Water moves upward against the force of gravity because of the attraction between water molecules and the surfaces of the soil particles. You can observe this when you dip the corner of a paper towel into water and watch the water move upward into the towel. Soils composed of small particles have more surface area per unit volume for the water to cling to. **Capillary water** is water that rises above the water table and becomes available for plants to draw in through their roots. Figure 9-8 illustrates an experiment you can do to observe how capillarity depends upon the size of the space into which water rises. The smaller the opening, the higher water will rise.

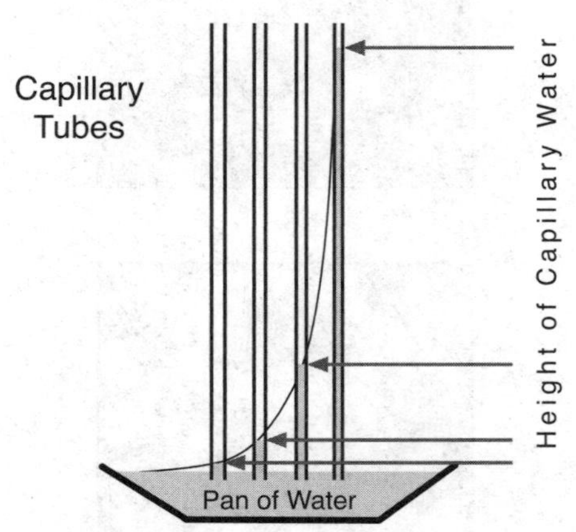

Figure 9-8. Capillarity is the ability of a soil to draw water upward into tiny spaces. The thin line shows how the height to which water rides depends upon the size of the openings.

WATER FOR STREAM FLOW

During and just after precipitation, streams receive water from runoff. Even during dry periods when there is no ground runoff, many streams continue to flow. In these dry periods, stream water actually may flow out of the ground. For example, in regions where groundwater and precipitation are plentiful, water will seep into the streams where the water table comes to the surface in the streambed.

In dry climates, groundwater may be recharged by water flowing from the streams into the ground. Figure 9-9 shows that in a humid region, water usually flows from the ground into streams. But in desert regions, streams may dry up as the stream water seeps into the ground.

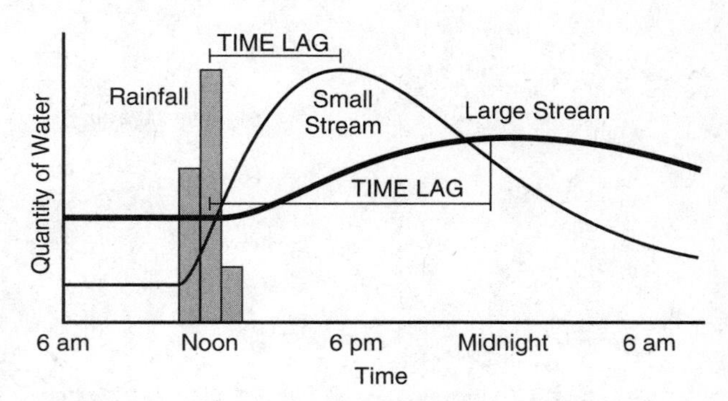

Figure 9-10. The greatest stream flow occurs after the time of maximum precipitation. This delay is called the time lag. In general, small streams respond quickly, with only a brief time lag, and they also return quickly to base flow. On the other hand, the discharge of larger streams increases and decreases slowly, with a longer time lag.

Rainfall and Stream Flow

Streams and rivers do not respond to rainfall immediately. Most precipitation falls on the ground and then must flow over the land as runoff to reach a stream. Therefore, there is a time lag between maximum precipitation and maximum stream discharge. Figure 9-10 compares the time lag for small and large streams. Several other factors determine how quickly and how strongly streams respond to precipitation. Streams will respond more slowly when precipitation falls as snow, when the gradient of the land is low (little slope), or when a great deal of vegetation blocks runoff.

Large rivers respond slowly because most runoff must flow a long distance to reach the rivers. Small streams and streams in mountain areas where the land is steep and rocky respond quickly to rainfall. In addition, runoff is very rapid and very brief in urban areas where the soil is covered by buildings, paved streets, and parking lots.

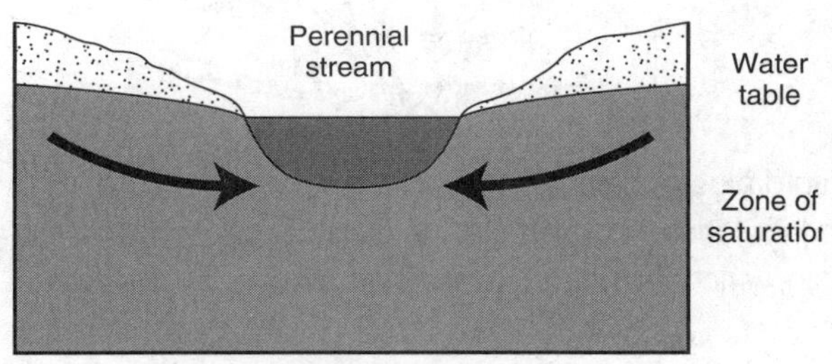

Moist climate: Groundwater replenishes the stream.

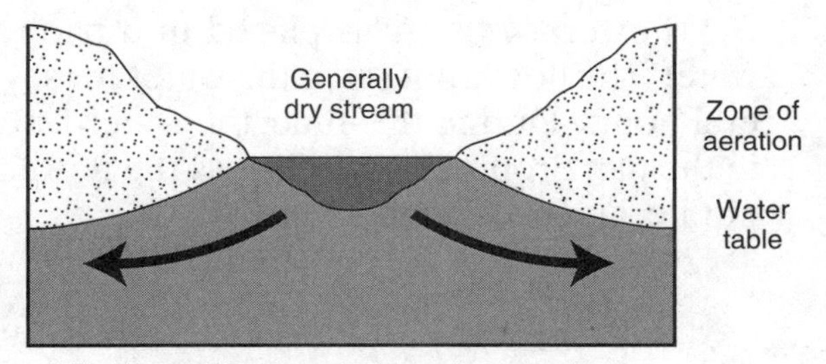

Arid climate: Stream replenishes the groundwater.

Figure 9-9. In moist climates, streams are generally below the surrounding water table; therefore, groundwater flows into the streams to sustain flow between rainstorms. In dry climates, streams generally rest on top of the water table, so the stream drains into the ground.

Watersheds

A **watershed**, or drainage basin, is the geographic area in which water drains into a particular stream or other body of water. A drainage divide, usually a ridge of high land, across which streams do not flow, forms the perimeter of a watershed. Groundwater may be polluted by leaking chemical tanks or a fuel spill that seeps into the ground. Because all the streams in a watershed flow into the same body of water, they will not carry pollution into neighboring watersheds.

You can trace the perimeter of a stream's watershed by carefully drawing a line around the stream and its tributaries. The line should cross the stream only where the main stream exits the drainage area. That line will follow the drainage divides that define the watershed. Figure 9-11 shows the major stream systems in New York

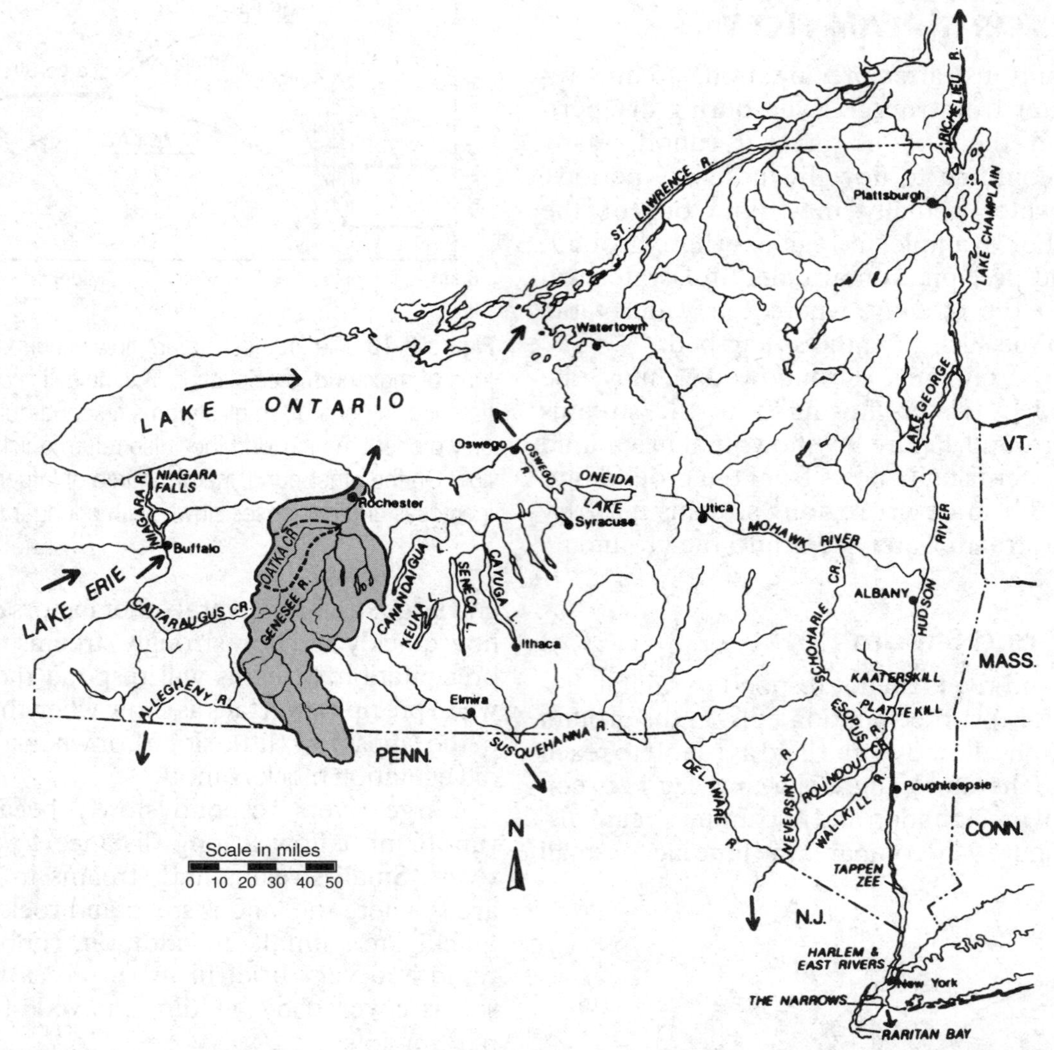

Figure 9-11. Any geographic region can be divided into drainage basins. Large watersheds can usually be divided into the smaller watersheds of their tributaries. For example, the Oatka Creek watershed is a part of the Genesee River drainage basin, which is in turn a part of the Lake Ontario–St. Lawrence River watershed.

State and highlights one watershed. As you can see on the map, Oatka Creek is a tributary of the Genesee River because it flows into the Genesee.

QUESTIONS

Part A

1. Locations *A* and *B* receive the same amount of rain. However, there is less runoff into streams at location *A*. Why does *A* have less runoff? (1) The soil at *A* is composed of smaller, poorly sorted particles. (2) The gradient at *A* is greater. (3) The ground is frozen at *B*. (4) The rain fell more slowly at *B*.

2. The diagram below shows an empty, dry, clay flowerpot that was placed in a pan of water. The flowerpot sat in the water for several hours. During this time, the water level in the pan dropped to level *A*, as the flowerpot became wet up to level *B*.

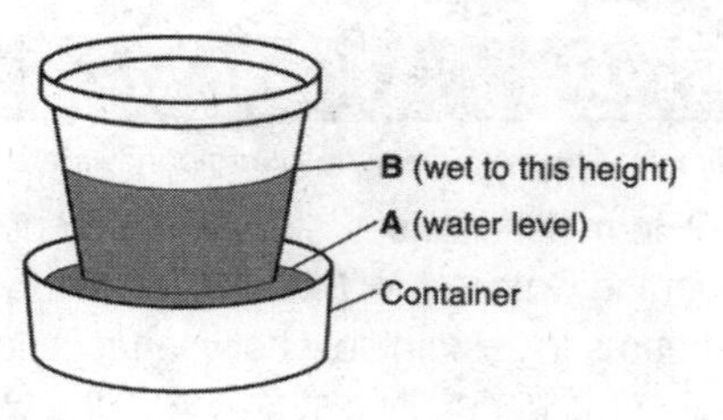

Why has the water moved upward to *B*?

(1) Water is less dense than the clay pot.
(2) Water is more dense than the clay pot.
(3) The clay pot is solid and nonporous.
(4) The clay pot has many tiny pores.

3. An Earth science class decided to investigate capillarity of different sediments. Four 2-meter-tall columns were covered on the bottom with a strong cloth and filled with dry soil. They placed the columns vertically into a large pan of water. Students returned the next day to see how far the wet sediments extended above the water in the pan. In which column was the sediment wet the highest? (1) Column *A*, which was filled with sand (2) Column *B*, which was filled with pebbles (3) Column *C*, which was filled with silt (4) Column *D*, which was filled with cobbles

4. After a wet period in the spring, it did not rain for the whole month of May in the Catskills. Yet, many small streams continued to flow throughout May. At the end of May, what was the principal source of water in the local steams? (1) transpiration from plants (2) infiltration from groundwater (3) precipitation water (4) evaporation from lakes and the ocean

5. Which city is within the St. Lawrence River watershed? (1) Buffalo, New York (2) Binghamton, New York (3) Albany, New York (4) New York City

6. How deep must a well be dug to ensure a supply of water that is dependable for present and future needs? (1) to bedrock (2) to the capillary zone (3) to the water table (4) to below the water table

7. A major chemical spill was found in groundwater at Niagara Falls, New York, north of Buffalo. What other city in New York State may naturally receive polluted water from this spill?
(1) Jamestown (3) Massena
(2) Syracuse (4) Albany

8. The diagrams above represent three identical containers filled with spherical beads.

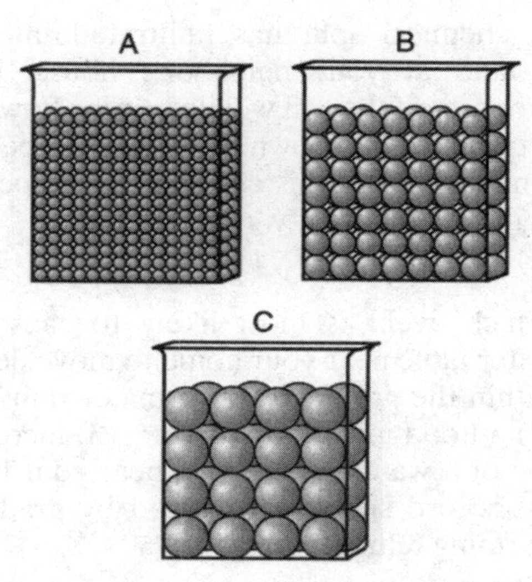

(Not drawn to scale)

Which data table below best represents the properties of these beds?

Beaker	Porosity (%)	Infiltration Time (sec)
A	40	5.2
B	40	2.8
C	40	0.4

(1)

Beaker	Porosity (%)	Infiltration Time (sec)
A	40	0.4
B	40	2.8
C	40	5.2

(2)

Beaker	Porosity (%)	Infiltration Time (sec)
A	20	5.2
B	30	2.8
C	40	0.4

(3)

Beaker	Porosity (%)	Infiltration Time (sec)
A	20	0.4
B	30	2.8
C	40	5.2

(4)

9. A chemical spill has infiltrated into the ground at your neighbor's house. What property of the soil will determine how long it takes for your own water well to be contaminated with this chemical? (1) permeability (2) porosity (3) capillarity (4) density

10. Which event is most likely to cause the water table near your home to move deeper within the ground? (1) a major rainstorm (2) a flood in a nearby river (3) increased use of a water supply well near your home (4) record snowfall followed by gradually warming temperatures

Part B

Base your answers to questions 11 through 18 on the diagram below, which represents the hydrologic cycle. Arrows show the movement of air.

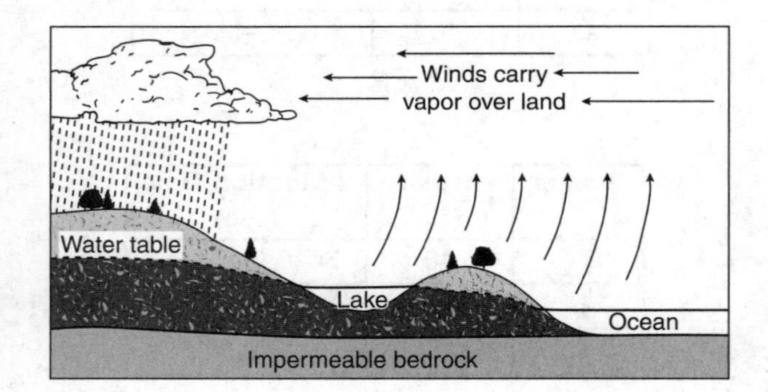

11. What process in the water cycle is dominant at the surface of the ocean? (1) condensation (2) vaporization (3) transpiration (4) convection

12. The horizontal arrows best represent (1) condensation (2) vaporization (3) transpiration (4) convection

13. How will the position of the water table respond to long-lasting precipitation? (1) The water table will fall before precipitation begins. (2) The water table will fall after precipitation begins. (3) The water table will rise before precipitation begins. (4) The water table will rise after precipitation begins.

14. Runoff will occur most rapidly if (1) the soil is permeable (2) the soil is porous (3) the soil is frozen (4) grass covers the ground

15. As the surface of the lake freezes, what energy exchange occurs? (1) The lake gains 334 joules per gram of ice that forms. (2) The lake gains 2260 joules per gram of ice that forms. (3) The lake loses 334 joules per gram of ice that forms. (4) The lake loses 2260 joules per gram of ice that forms.

16. Plants lose moisture to the air by (1) condensation (2) transpiration (3) radiation (4) crystallization

17. What is the major energy source that moves water through the water cycle? (1) gravity (2) radioactive decay (3) solar energy (4) heat of vaporization

18. Groundwater moves through the soil and into the lake and ocean under the influence of gravity. This happens most rapidly when (1) the soil is very porous (2) the soil is very permeable (3) the soil is composed of very small particles (4) the soil supports active capillary movement

Base your answers to questions 19 through 22 on the diagram below, which shows the apparatus used by a student to model properties of different kinds of soil. The experimental data for Tube A is shown below the diagram of the tubes.

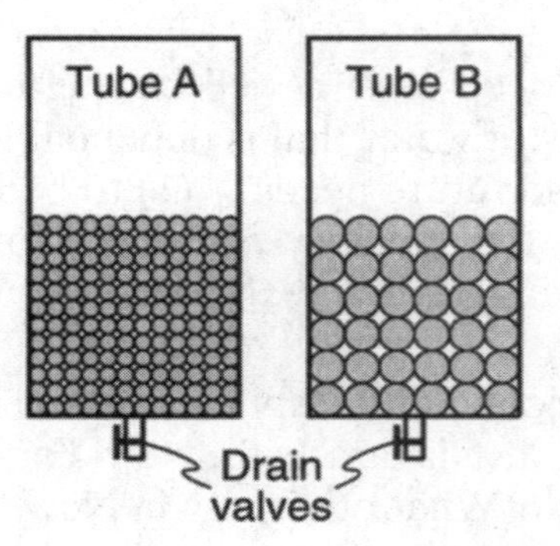

Data Table 1: Tube *A*

Water required to fill pore spaces	124 mL
Time required for draining	2.1 s
Water that remained around the beads after draining	36 mL

19. What properties of the beads were being investigated by the student? (1) capillarity and porosity (2) capillarity and permeability (3) permeability and porosity (4) absorption and density

20. Which data table below best shows the experimental results for Tube *B*?

(1)

Data Table 2: Tube *B*	
Water required to fill pore spaces	124 mL
Time required for draining	1.4 s
Water that remained around the beads after draining	26 mL

(2)

Data Table 2: Tube *B*	
Water required to fill pore spaces	168 mL
Time required for draining	3.2 s
Water that remained around the beads after draining	46 mL

(3)

Data Table 2: Tube *B*	
Water required to fill pore spaces	124 mL
Time required for draining	3.3 s
Water that remained around the beads after draining	36 mL

(4)

Data Table 2: Tube *B*	
Water required to fill pore spaces	168 mL
Time required for draining	1.4 s
Water that remained around the beads after draining	36 mL

21. If the student wanted to select beads that would better show upward movement of water against the force of gravity, what size of beads would draw water the highest?
(1) beads smaller than the beads in tube *A*
(2) beads larger than the beads in tube *A*, but smaller than the beads in Tube *B*
(3) beads larger than the beads in tube *B*
(4) a tube with no beads in it

22. If the beads in tube *B* are the size of pebbles, what is their approximate diameter?
(1) 0.001 cm (2) 0.01 cm (3) 0.1 cm
(4) 1 cm

Base your answers to questions 23 through 29 on the reading passage and the diagram below.

Salt Water Encroachment

New Yorkers use approximately 900 million gallons of groundwater per day. Groundwater aquifers supply water for households and for many small communities. Although bedrock formations are a significant source of groundwater supply, the most productive aquifers in New York State are located in unconsolidated sediments (e.g., sand and/or gravel deposits). Many of these aquifers yield plentiful, high-quality water.

Upstate aquifers in sediments and in bedrock are not as extensive as the Long Island aquifers. However, excessive use of well water in some places on Long Island has caused seawater to intrude into the active aquifer. Freshwater is less dense than saltwater. Therefore, fresh groundwater sits between the land surface and deeper salty groundwater, forming a barrier to saltwater flow. Removal of too much freshwater from the aquifer pulls the saltwater farther inland and closer to the surface, where it can find its way into wells.

Saltwater intrusion is a major problem in coastal locations of central and western Long Island where many deep wells near the shore have become too salty for use. New wells must be sunk farther inland. The New York City boroughs of Brooklyn and Queens on Long Island get their water from reservoirs in upstate New York. As a result, the aquifers in this area are beginning to recharge with fresh water and drive the salty water deeper below the surface.

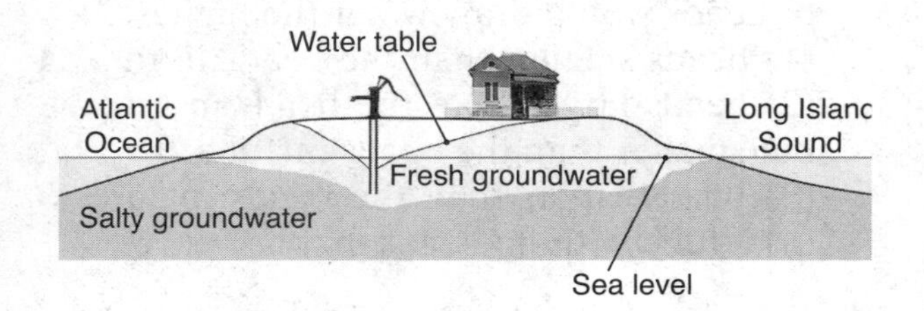

23. What properties of the soil on Long Island make it an excellent aquifer? (1) low porosity and low permeability (2) low porosity and high permeability (3) high porosity and low permeability (4) high porosity and high permeability

24. Why is the aquifer under the part of New York City that is on Long Island increasing in its supply of fresh water? (1) New York City pumps only salt water. (2) New York City uses very little groundwater. (3) The aquifer within New York City has a very low permeability. (4) Rainfall in New York City is diverted into the Atlantic Ocean.

25. Why does the fresh groundwater rest above the salty groundwater under Long Island? (1) Salt water has a density greater than 1 g/cm^3. (2) Salt water has a density less than 1 g/cm^3. (3) Salt water cannot penetrate the Long Island aquifers. (4) Most saltwater infiltrates from deep within Earth's mantle.

26. In what direction is most of the groundwater flowing? (1) toward the Atlantic Ocean (2) toward Long Island sound (3) toward the surface of the ground (4) toward the active water well

27. How can we best describe the underground zone above the water table? (1) This zone contains no water. (2) This zone does not contain air. (3) Capillary water can move up into this zone. (4) Gravity does not affect water in this zone.

28. Why is fractured bedrock less desirable as an aquifer than a deep, sandy soil? (1) Fractured bedrock has a relatively high permeability. (2) Fractured bedrock has a relatively low porosity. (3) Fractured bedrock can become contaminated. (4) Fractured bedrock contains a greater variety of minerals.

29. What is the best way for the resident of the house to avoid salt water entering his well? (1) Install a motorized pump at the wellhead. (2) Build drains to guide runoff into the ocean. (3) Allow purified wastewater to infiltrate the soil. (4) Extend the well pipe to a greater depth.

HOW DOES HEAT ENERGY TRAVEL?

Heat is a form of energy because it can do work. Heat energy can travel in three ways. Figure 9-12 illustrates conduction, convection, and radiation, the three ways that heat travels.

Convection is heat flow caused by density currents within a fluid. The uneven heating of Earth's surface by the sun causes vertical air currents as well as horizontal winds. Wind is a familiar example of heat flow by convection. Warm air is less dense than cold air. The energetic (warm) air molecules carry their vibrational energy as they move from the tropics toward the polar areas. Ocean currents are also caused by differences in density. Like air, water is a fluid and it can carry energy by convection.

If you touch a hot object, you will feel the heat. You may even get burned. *Conduction* is a form of heat flow in solids that occurs when a hot

Figure 9-12. Solar energy reaches Earth by radiation. The boy's feet feel hot because of conduction due to contact with the hot pavement. The wind is heat flow by convection.

substance touches a cooler substance. The vibrational energy of the atoms and molecules in the warmer substance is transferred by contact to the atoms and molecules in the cooler substance. The transfer of energy increases the vibration of the atoms and molecules in the cooler substance, increasing the temperature of that substance. A metal spoon left in hot coffee quickly becomes hot all the way up the spoon because metals are excellent conductors of heat energy.

Earth's primary source of energy is the sun. However, that energy has to travel through millions of kilometers of empty space to reach Earth. Convection and conduction require a substance, or medium, through which the energy can travel, but radiation requires no medium. *Radiation* is the flow of energy as electromagnetic waves, such as visible light. Radiation can travel through space or through transparent materials. For example, air and glass are transparent to radiation in the visible part of the electromagnetic spectrum.

Heat rays, which you may be able to feel even though you cannot see them with your eyes, also carry energy by radiation. In fact, some forms of radiant energy that you cannot feel at all can still cause terrible damage. The sun's ultraviolet rays and radioactivity from natural or artificial sources can cause serious illness or even death.

The more we understand about these forms of radiation, the better we can deal with them. The electromagnetic spectrum in the *Earth Science Reference Tables* shows the broad range of electromagnetic energy, both visible and invisible. The various forms of electromagnetic energy are shown in Figure 8-1 on page 193.

INSOLATION

Earth receives nearly all of its energy from the sun. The sun's electromagnetic energy that reaches the earth is called **insolation** (INcoming SOLar radiATION). The ability of insolation to heat Earth's surface depends upon several factors, such as the angle of insolation, the duration of insolation, and the type of surface the insolation strikes.

Angle of Insolation

The **angle of insolation** depends on how high the sun is in the sky. As the sun rises and sets, the angle of insolation changes. This is a daily cycle. The angle of insolation is measured from the horizon up to the position of the sun. The noon sun has the greatest angle of insolation. Therefore, noon is when we receive the greatest intensity of insolation per unit area. In the morning and in the afternoon, when the sun is lower in the sky, the sunlight is less direct and less intense. Figure 9-13 on page 230 shows that when the sun is high in the sky the rays of sunlight are more concentrated as they reach the ground.

The angle of insolation also changes with the seasons. In the Northern Hemisphere, the lowest noontime angle of insolation is reached at the winter solstice (about December 22), and the highest angle at the summer solstice (about June 21).

The angle of insolation is different at each latitude because Earth is spherical. The **vertical ray** is sunlight that strikes Earth's surface at an angle of 90°. Each day throughout the year the vertical ray hits Earth at noon somewhere in the tropics. At all other latitudes, slanting rays of sunlight strike Earth's surface at acute angles and are weaker in intensity. Slanting rays distribute their energy over a larger area than the vertical ray.

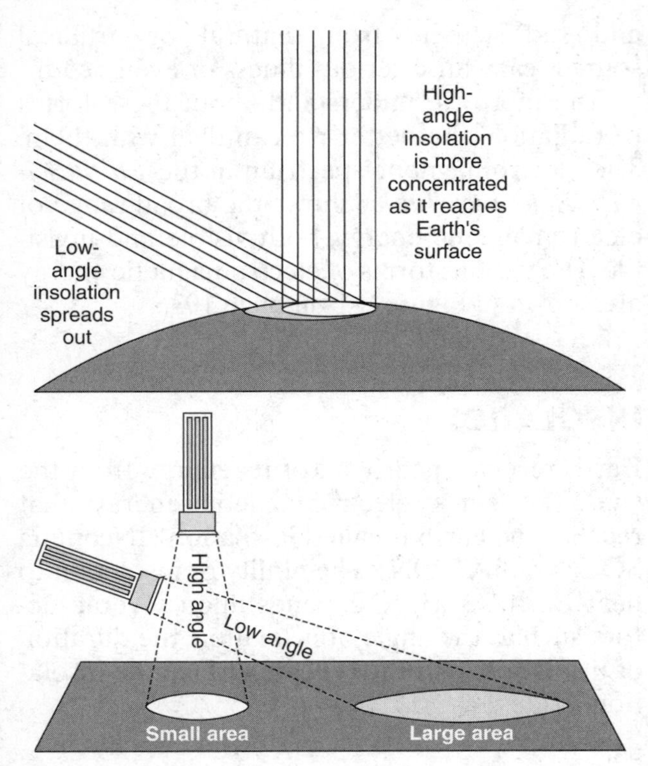

Figure 9-13. When the sun is directly overhead, its rays are concentrated in a small area. When the sun is low in the sky, the same energy covers more area, so the sunlight is not as strong.

As shown in Figure 9-14, for all locations north of the Tropic of Cancer ($23^1/_2$°N), the most direct and most intense insolation occurs at noon on the Northern Hemisphere's summer solstice (about June 21). The low angle (sometimes tangential) rays that strike Earth in the Arctic zone ($66^1/_2$°N to 90°N) and in the Antarctic zone ($66^1/_2$°S to 90°S) account for the low average temperatures in these zones.

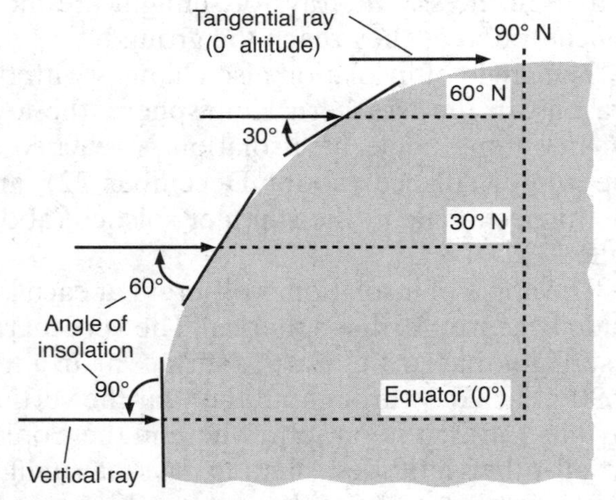

Figure 9-14. As latitude increases, the angle of insolation and the intensity of insolation generally decrease. (This diagram shows the vertical ray at the equator.)

Duration of Insolation

The **duration of insolation** is the length of time from sunrise to sunset. This is the period of daylight during which the sun appears in the sky. A section of Earth's surface receives the most heat energy when the sun is highest in the sky and when the duration of insolation is the greatest. Generally, as the angle of insolation and the duration of insolation increase, temperatures at Earth's surface increase.

The duration of insolation varies greatly with latitude. In the Northern Hemisphere, the greatest duration of insolation occurs at our summer (June) solstice. The further north you go, the longer the daylight period is at this time. For example, at the end of June, the sun is in the sky for the whole day (24 hours) at locations north of the Arctic Circle. At locations south of the Arctic Circle, the duration of sunlight decreases in a regular pattern. At the equator, the duration of insolation is 12 hours every day.

If you add up the number of hours of daylight for any location on Earth, the total is 6 months. At the North and South Poles, this means 6 months of sunlight followed by 6 months of darkness. At the equator, it means 12 hours of sunlight and 12 hours of darkness each day.

Energy Absorption

Some surfaces absorb energy better than others. You are probably aware that dark-colored, rough surfaces absorb more sunlight than light-colored, shiny surfaces. That is why in hot, sunny weather it is best to wear light-colored clothing. You may not realize, however, that dark objects are better at radiating heat energy. If two nearly identical cups of hot water were left outside at night, the darker one would cool faster. Do not make the mistake of thinking that light objects are always cooler. That depends upon whether the objects are primarily absorbing or radiating energy.

Absorption of Insolation by the Atmosphere

When you place a sponge on a spilled liquid, the sponge absorbs the liquid. Earth also absorbs most of the sunlight that falls on it. Most of the energy that penetrates to the surface is in the form of visible light. Ozone and other gases in the upper atmosphere absorb high-energy radiation, such as x-rays and gamma rays. Long-wave radiation, such as infrared, is absorbed by water vapor and carbon dioxide in the atmosphere. Visible-light wavelengths, however, easily penetrate the atmosphere. Figure 9-15 shows the energy absorbed by Earth's atmosphere.

Some of the absorbed energy is changed into infrared heat waves that are reradiated back into the atmosphere. Figure 9-16 shows that most of this terrestrial radiation is absorbed by the atmosphere.

At a given latitude, the average temperature of land and water may be different for the following four reasons. First, water has a higher *specific heat* than soil or rock as you can see in Table 9-1 from page 1 of the *Earth Science Reference Tables*. This means that water must absorb more energy to undergo a given temperature change. Second, water reflects low-angle insolation better than land. Third, because water is transparent, insolation penetrates water more deeply and more quickly than it can be absorbed by land. Fourth, convection currents in water carry heat energy deep into the hydrosphere. In the ground, heat energy can travel only by conduction. Thus, the same amount of insolation travels through a greater volume of water than land. The result of all these factors is that land heats faster than water. Land also loses its limited heat energy more quickly, so land cools faster than water. In other words, good absorbers of energy are good radiators of energy. (Easy come, easy go.)

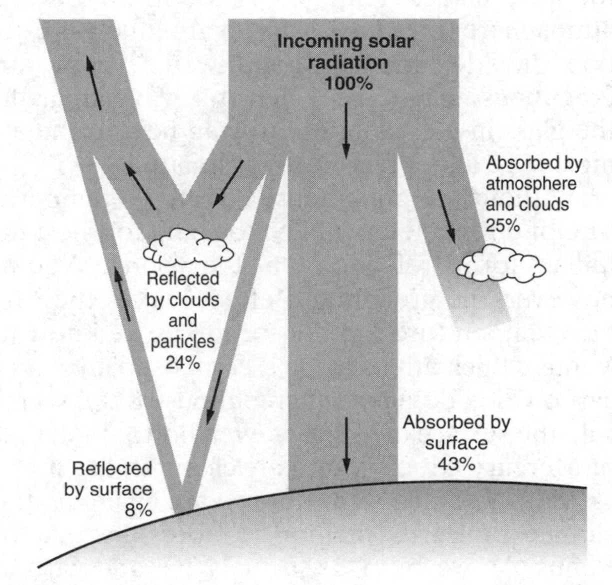

Figure 9-15. The ground absorbs only about half the solar energy that comes to Earth. The rest of the energy is absorbed or reflected by the atmosphere.

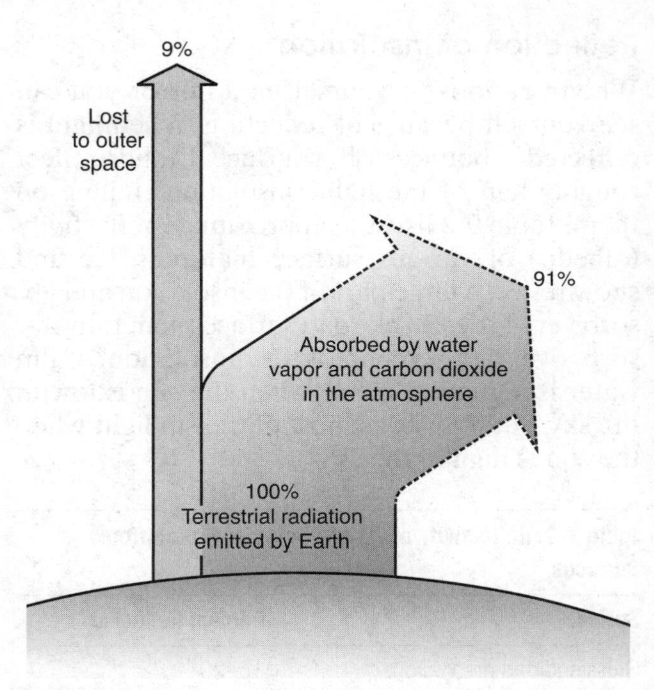

Figure 9-16. We cannot see the infrared (heat) radiation given off by Earth because it is outside the range of energy we can detect with our eyes. The atmosphere traps most of this infrared energy.

Table 9-1.

Specific Heats of Common Materials

MATERIAL	SPECIFIC HEAT (Joules/gram • °C)
Liquid water	4.18
Solid water (ice)	2.11
Water vapor	2.00
Dry air	1.01
Basalt	0.84
Granite	0.79
Iron	0.45
Copper	0.38
Lead	0.13

Reflection of Insolation

When you look at yourself in a mirror you can see yourself because of reflection. When light is **reflected** it bounces off a surface. Clouds reflect roughly half of the light (insolation) falling on them. Table 9-2 lists the approximate reflectivity (albedo) of various surface materials. Ice and snow reflect a large part of the insolation and absorb very little. Black road surfaces generally absorb over 90 percent of the insolation. Calm water is a good reflector when the sun is low in the sky, but it absorbs most of the sunlight when the sun is high in the sky.

Table 9-2. Reflectivity of Visible Solar Radiation from Surfaces

Surface	Percent Reflected
Grassland and green crops	16–20
Broadleaf forests	15–20
Desert	25–30
Blacktop pavement	5–10
Fresh snow	75–95
Water (high solar altitude)	3–10
Water (low solar altitude)	10–100

While the North and South Poles receive the same total duration of insolation as the equator, the temperature is much lower at the poles. There are three reasons for this. First, the snow at the poles reflects much of the insolation back into space. Second, near the poles the sun is always low in the sky, so the sunlight is more spread out and less concentrated than when it strikes near the equator. Third, when the sun is low in the sky, the sunlight must travel a greater distance through Earth's atmosphere. Within the atmosphere, much of the insolation is reflected, refracted, or absorbed. Therefore, the intensity of the radiation that reaches the ground at the poles is reduced.

In the atmosphere, there are tiny particles of airborne solids, such as dust, pollen, pollutants, and water droplets. These particles reflect or scatter insolation and reduce its intensity. Many scientists think that the extinction of the dinosaurs was related to the impact of a large object from space about 65 million years ago. This collision added tons of debris to the atmosphere, which reduced its transparency. This caused less sunlight to reach the surface. Lower temperatures and reduced plant growth may have led to a collapse of the food chain. Violent volcanic eruptions at any time can have a similar effect.

Terrestrial Radiation

Energy waves (infrared) emitted from Earth's surface are longer in wavelength than energy waves in the range of visible light emitted from the sun. The longer infrared heat waves radiated by Earth are absorbed by gases such as carbon dioxide and water vapor and remain trapped in the atmosphere. This process is called the **greenhouse effect** because the glass of a greenhouse traps heat inside just as gases in Earth's atmosphere trap energy.

The Greenhouse Effect In a greenhouse, the short wavelengths of energy from the sun (visible light) pass through the glass and into the greenhouse. When the light strikes objects inside the greenhouse, the energy is absorbed. The energy is then given off, reradiated, as long-wave infrared rays, or heat. As the objects reradiate heat outward, the glass reflects the infrared heat waves back into the greenhouse where they heat the air. Thus, the temperature inside a greenhouse is usually warmer than the temperature of the air outside. Figure 9-17 shows how Earth's atmosphere traps energy like a greenhouse. Carbon dioxide,, methane, and water vapor are **greenhouse gases**. They act in a way similar to the glass in a greenhouse to trap heat in the atmosphere and affect Earth's climate.

Without the greenhouse effect, the temperature of our planet would be too cold for most familiar forms of life. On the planet Venus, however, the greenhouse effect makes the surface temperature too hot for life as we know it. Venus's thick atmosphere is composed mostly of carbon dioxide, a potent greenhouse gas. As a result, the surface of Venus is even hotter than that of Mercury, which is actually closer to the sun.

Many scientists fear that we are changing the climate of Earth, making it more like that of Venus. As we burn fossil fuels, we add carbon dioxide to the atmosphere as shown in Figure 9-18. To make matters worse, we are also cutting down forests all over the world, thereby de-

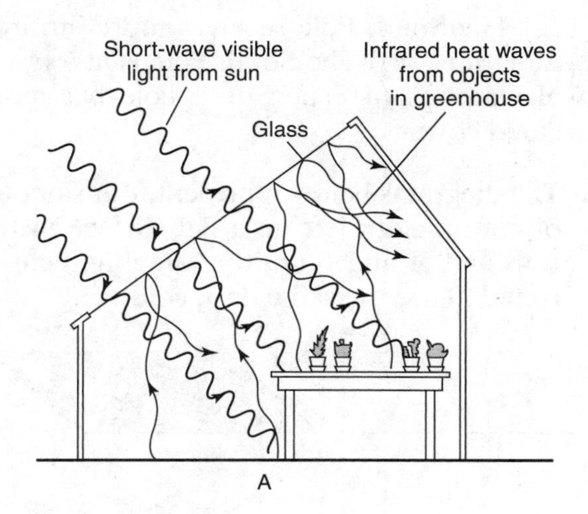

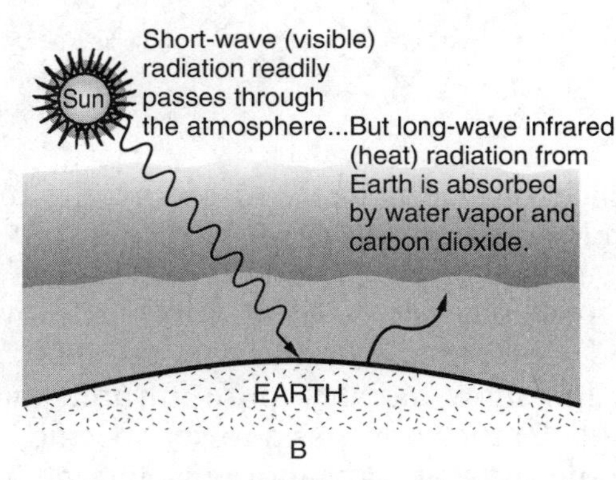

Figure 9-17. Like a greenhouse, Earth's atmosphere traps heat energy. Sunlight penetrates the atmosphere as it does the glass of a greenhouse. Because infrared radiation from the ground cannot penetrate either the glass of a greenhouse or Earth's atmosphere, energy is held in. Thus the temperature on Earth, and in a greenhouse, is higher than it would be without the greenhouse effect.

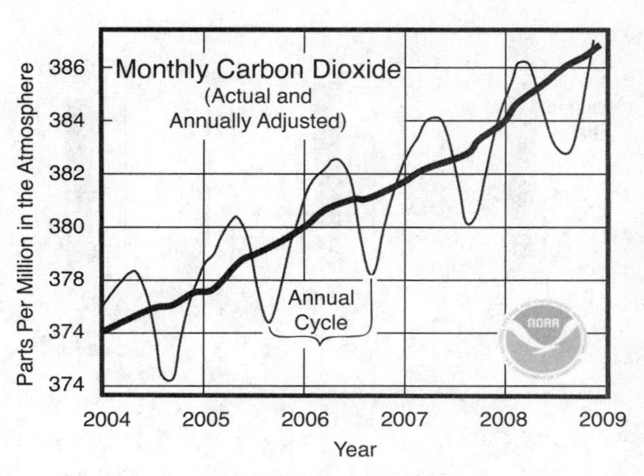

Figure 9-18. Since the Industrial Revolution more than 100 years ago, human use of fossil fuels has increased. This has increased the carbon dioxide content of Earth's atmosphere by about 20%. This graph shows that although there is an annual cycle of carbon dioxide concentration, the upward trend is clear.

stroying a way of removing carbon dioxide from the atmosphere. (Trees and other plants take in carbon dioxide and release oxygen, which is not a greenhouse gas.) Carbon dioxide is the major greenhouse gas. Our use of fossil fuels is the most important human contribution to Earth's greenhouse effect. The quantity of methane being released into the atmosphere is also increasing. Methane is released as a waste product of petroleum refining. In addition, animals (including humans) and decaying organic matter produce methane.

It is not yet clear how severe the effects of the increase in greenhouse gases might be, or how difficult it will be to reverse global warming. To reduce global warming, people could switch from petroleum-based fuels to solar energy, wind energy, and nuclear power. One consequence of global warming is a progressive melting of Earth's ice caps and glaciers. This will cause sea level to rise, flooding coastal cities. In fact, observations of many glaciers worldwide seem to indicate that global warming is already occurring.

The threat of global warming has led to a worldwide campaign to reduce the human production of greenhouse gases, especially carbon dioxide from fossil fuels. The United States has only about 5% of the world's population, but produces nearly a third of the greenhouse emissions. (See Figure 9-19 on page 234.) The developing nations argue that the United States should take the lead to reduce its production of greenhouse gases.

The Insolation-Temperature Lag

A time lag exists between the time of greatest intensity of insolation and the time of highest air temperature. Why? Because insolation energy is first absorbed by Earth's surface and then reradiated as heat energy that warms the air. Let us consider a 24-hour period. At sunrise, the ground is cool; it becomes warmer throughout the morning as it absorbs solar radiation. At noon, the incoming radiation reaches a maximum. For the next 2 or 3 hours, the ground continues to absorb more energy than it radiates. Thus, the surface temperature continues to rise.

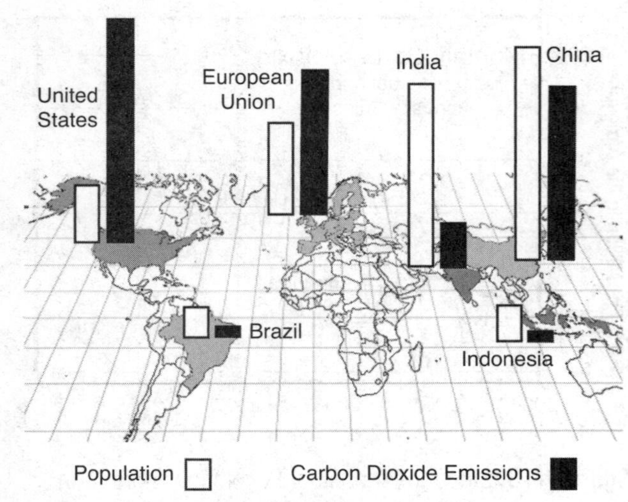

Figure 9-19 The United States produces more carbon dioxide than any other nation. China produces almost as much carbon dioxide, but has more than four times the population of the United States. Most of the developing nations of the world have far smaller emissions in spite of their large populations.

By mid-afternoon, there is equilibrium, or balance, between the incoming energy and the outgoing radiation., The highest air temperature occurs at this point. After this, Earth radiates more energy than it receives from the sun and the net loss of energy causes Earth to cool. This cooling period continues until slightly after sunrise the next morning. The time delay between maximum or minimum insolation and maximum or minimum air temperature is known as the **insolation-temperature lag**.

Seasonal variations in air temperatures also occur. As with the daily cycle, average monthly temperature depends on the balance between insolation and terrestrial radiation. Maximum and minimum annual temperatures occur when there is a **radiative balance** between insolation and terrestrial radiation usually in July and January. In an annual temperature cycle, the insolation-temperature lag is about a month.

QUESTIONS

Part A

30. Why is the climate warmer in the tropics than it is at the South Pole? (1) The South Pole receives less intense insolation. (2) The South Pole receives more infrared radiation. (3) The South Pole is at a lower elevation. (4) The South Pole has more cloud cover.

31. The diagrams below represent four samples of cotton cloth left on a flat surface in the sunshine. *A* and *B* are the same light color; *C* and *D* are the same dark color.

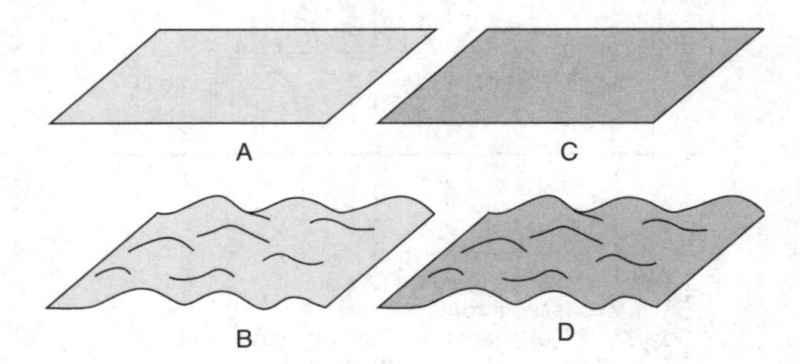

Which surface is likely to absorb the most insolation? (1) *A* (2) *B* (3) *C* (4) *D*

32. A fresh blanket of snow covers a plowed field. Compared with the bare ground (1) the snow radiates more infrared radiation (2) the snow has a lower specific heat (3) the snow absorbs more insolation (4) the snow reflects more radiation

33. A town has two large, identical, above-ground water tanks to hold its municipal water supply. One is painted black and the other is painted white. How do the two tanks adjust to warm day temperatures and cold nights? (1) The black tank heats more quickly in the sunlight and it cools more quickly at night. (2) The black tank heats more quickly in the sunlight and it cools more slowly at night. (3) The black tank heats more slowly in the sunlight and it cools more quickly at night. (4) The black tank heats more slowly in the sunlight and it cools more slowly at night.

34. As the sun moves higher in the sky, the amount of energy that falls on each square meter of the ground (1) decreases (2) increases (3) remains the same (There only 3 choices.)

35. At which location above right is the observer experiencing the greatest intensity of insolation?

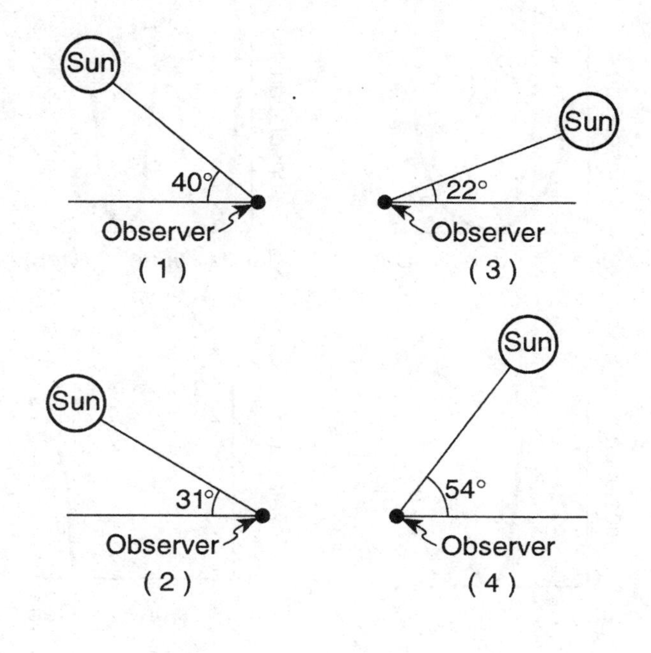

Part B

Base your answers to questions 36 through 43 on the figure below, which illustrates electromagnetic radiation as it enters, leaves, and is internally trapped within a greenhouse. Letters A and B identify rays of two different wavelengths of electromagnetic energy that are very common near Earth's surface.

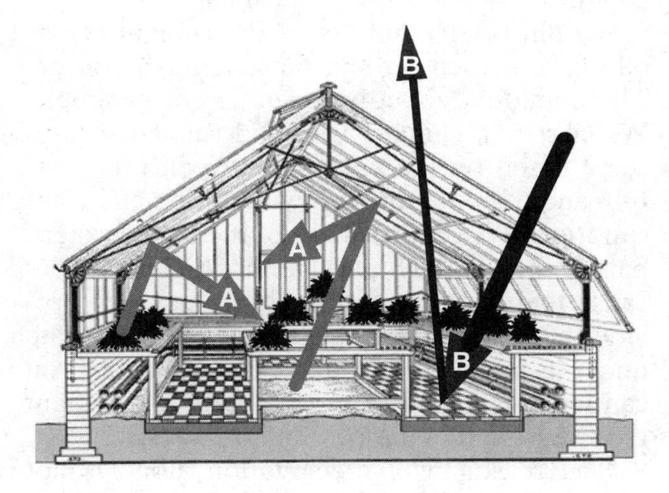

36. What kind of electromagnetic energy is best represented by the letter *A*? (1) infrared radiation (2) visible light (3) ultraviolet energy (4) gamma rays

37. What visible radiation is labeled *B*? (1) ultraviolet (2) visible (3) infrared (4) microwaves

38 Within the greenhouse, what happens to the radiation labeled *B*? (1) All radiation *B* is absorbed. (2) All radiation *B* is all reflected. (3) Radiation *B* is partly absorbed and partly reflected (4) Radiation *B* is reflected as it transforms to radiation *A*.

39. Scientists predict that Earth's average temperature will become warmer in future decades because our atmosphere acts like a greenhouse. What greenhouse gas produced by human activities has dramatically increased in Earth's atmosphere? (1) oxygen (2) nitrogen (3) water vapor (4) carbon dioxide

40. How can humans most effectively slow global warming? (1) Find new technology to discover unknown petroleum sources. (2) Use solar energy, wind energy and nuclear power to replace the use of fossil fuels. (3) Build more highways to allow commuters to live farther from where they work (4) Switch from the use of petroleum products to coal for energy production.

41. How could the greenhouse become more effective in trapping solar energy? (1) cover the greenhouse with a large sheet of opaque (nontransparent) plastic. (2) Plant large trees to surround the greenhouse. (3) Fill the greenhouse with dark colored plants. (4) Remove the glass roof on half the greenhouse.

42. If all windows and doors are tightly closed, how could the greenhouse lose energy? (1) by conduction and radiation (2) by radiation and convection (3) by conduction and convection (4) by radiation only

43. If humans continue to accelerate the addition of greenhouse gases to the atmosphere, what would be one likely result? (1) longer winters in the southern hemisphere (2) the growth of glaciers in Greenland and Antarctica (3) fewer tornadoes, hurricanes and other violent storms (4) displaced people who live in low, coastal areas

Base your answers to questions 44 through 47 on the graph below, which shows the intensity of insolation reaching Earth at four different latitudes.

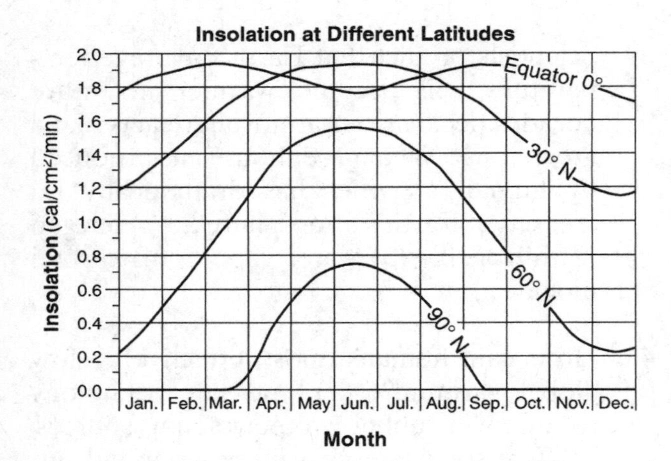

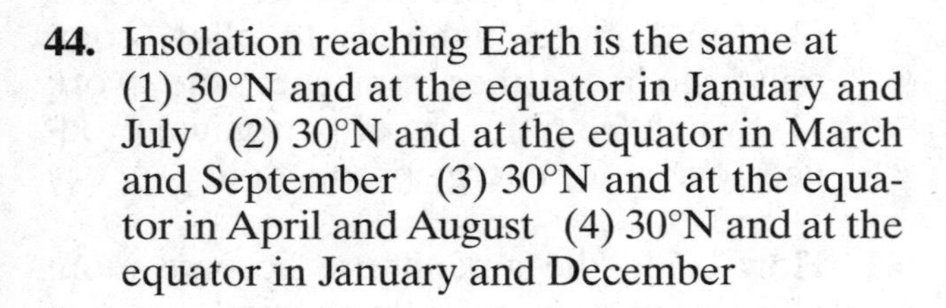
44. Insolation reaching Earth is the same at (1) 30°N and at the equator in January and July (2) 30°N and at the equator in March and September (3) 30°N and at the equator in April and August (4) 30°N and at the equator in January and December

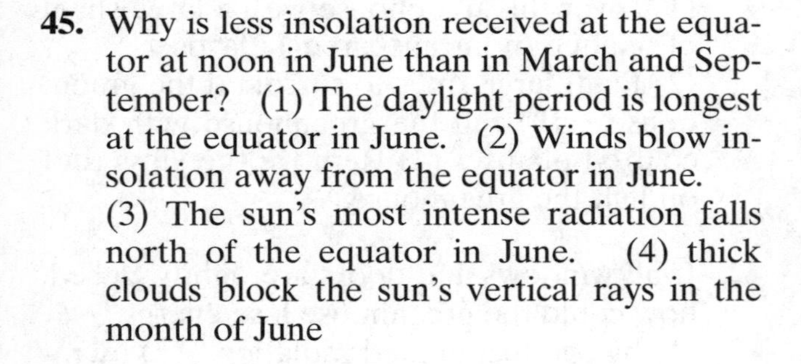
45. Why is less insolation received at the equator at noon in June than in March and September? (1) The daylight period is longest at the equator in June. (2) Winds blow insolation away from the equator in June. (3) The sun's most intense radiation falls north of the equator in June. (4) thick clouds block the sun's vertical rays in the month of June

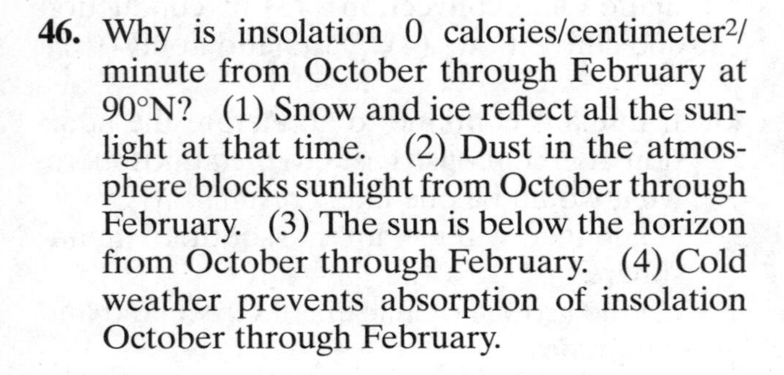
46. Why is insolation 0 calories/centimeter²/minute from October through February at 90°N? (1) Snow and ice reflect all the sunlight at that time. (2) Dust in the atmosphere blocks sunlight from October through February. (3) The sun is below the horizon from October through February. (4) Cold weather prevents absorption of insolation October through February.

47. Which of the following graphs best represents the annual cycle of the intensity of insolation at the South Pole?

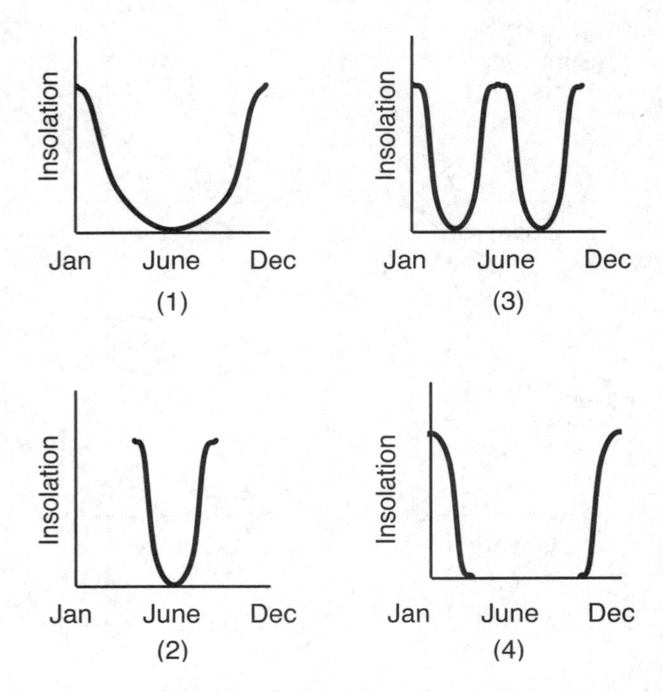

CLIMATE

Unlike weather, the **climate** for a large geographic region is based on atmospheric conditions observed over a long period of time. The average conditions of temperature and precipitation as well as the annual distribution of these conditions define a region's climate.

Within the United States, the climate generally becomes cooler the farther north you go. The humidity factor, however, is not as simple. We classify a climate as humid (moist) or arid (dry) based on the balance between precipitation and the potential for evaporation and transpiration (P/E_p ratio). Most of the southwestern part of the United States has an arid climate because there is not enough precipitation to provide the amount of water that could evaporate under its hot and dry climatic conditions. As you can see in Figure 9-20, precipitation is dominant in the east and in the coastal northwest.

The reason why precipitation alone is not enough to classify a humid or an arid climate may become clear if you consider the Polar regions. Like deserts, the Polar regions get very little precipitation. Both deserts and the poles are regions of prevailing high pressure. However, unlike the deserts, there is no shortage of water at the poles because they are covered by ice and snow. Therefore, in spite of very little precipita-

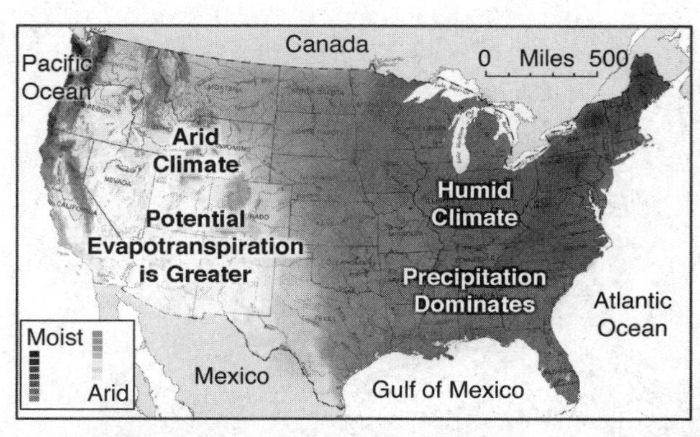

Figure 9-20. Climate is generally classified as humid or arid based upon the ratio of precipitation and potential evapotranspiration. In the United States, the Eastern Seaboard, the Midwest, and Northwest Coast have a humid climate. The Southwest has a dry climate.

tion, polar climates are usually specified without reference to their low precipitation.

Factors That Affect Climate

Major influences on the climate of a region include latitude, altitude, mountain ranges, oceans and large bodies of water, ocean currents, planetary wind belts, and typical storm tracks.

Latitude The latitude of a location is an important factor that determines the average local temperature. As the distance from the equator increases, the average annual temperature generally decreases. Hence, the coldest climate, the polar climate, is found near Earth's poles. Locations near the equator have a tropical climate and are generally warm throughout the year. Between the tropical and polar climate zones, locations have temperate, or middle-latitude, climates. Locations with middle-latitude climates experience large seasonal changes in temperature.

Altitude High-altitude locations have cool climates because of the **adiabatic** cooling of air as it moves to higher elevations. As the air rises, reduced air pressure causes the air to expand and become cooler. Because of this, high plateaus and mountain peaks have lower temperatures than do places at the same latitude at sea level. This is why snow is often found near the top of a high mountain, even in the summer.

Mountain Ranges Mountain ranges can modify precipitation and temperature patterns. For example, warm, moist air from over the Pacific Ocean is often forced to rise along the slope of a mountain barrier, such as the Cascade Mountain range in the Pacific Northwest. Air undergoes adiabatic cooling as it rises along the **windward** side of the mountain. If the air is cooled below the dew point, cloud formation and precipitation will make locations on the windward side of the mountains more moist than surrounding lowlands.

Once the air rises over the mountain range, it will then descend on the **leeward** (downwind) side of the mountains. The air warms by compression as it moves to lower elevations. This causes the climate on the leeward side of the mountains to be warmer and drier than the climate on the windward side. Temperatures on the leeward side will be warmer than temperatures at the same elevation on the windward side. Figure 9-21 on page 238 shows why climate often differs, even at the same elevation, on the two sides of a mountain range.

Oceans and Large Bodies of Water As illustrated in Figure 9-22 on page 238 coastal locations tend to have less temperature change than inland locations for several reasons: (1) Water is a better conductor than land; so the energy moves deeper below the surface. However, the ground acts as a better insulator. (2) Water is a fluid. Surface water and its energy mix with deeper water. (3) Water has a high specific heat; that is, water requires a great deal of energy input to cause a small change in temperature. Furthermore, (4) much of the energy from the sun that falls on water causes evaporation, rather than warming. For these reasons, the climate of locations near the ocean or other large bodies of water (**maritime** or **marine climate**) is more moderate than the inland or continental climate. Areas with a **continental climate** experience large seasonal changes in temperature due to the absence of nearby bodies of water that could moderate the climate

Seasonal temperature changes are greater at locations far from large lakes and oceans. Seasonal temperature changes are smaller for locations along a coastline. Coastal and marine climates are cooler in the summer and warmer in the winter than are inland climates. The temperature moderating effects of large bodies of water

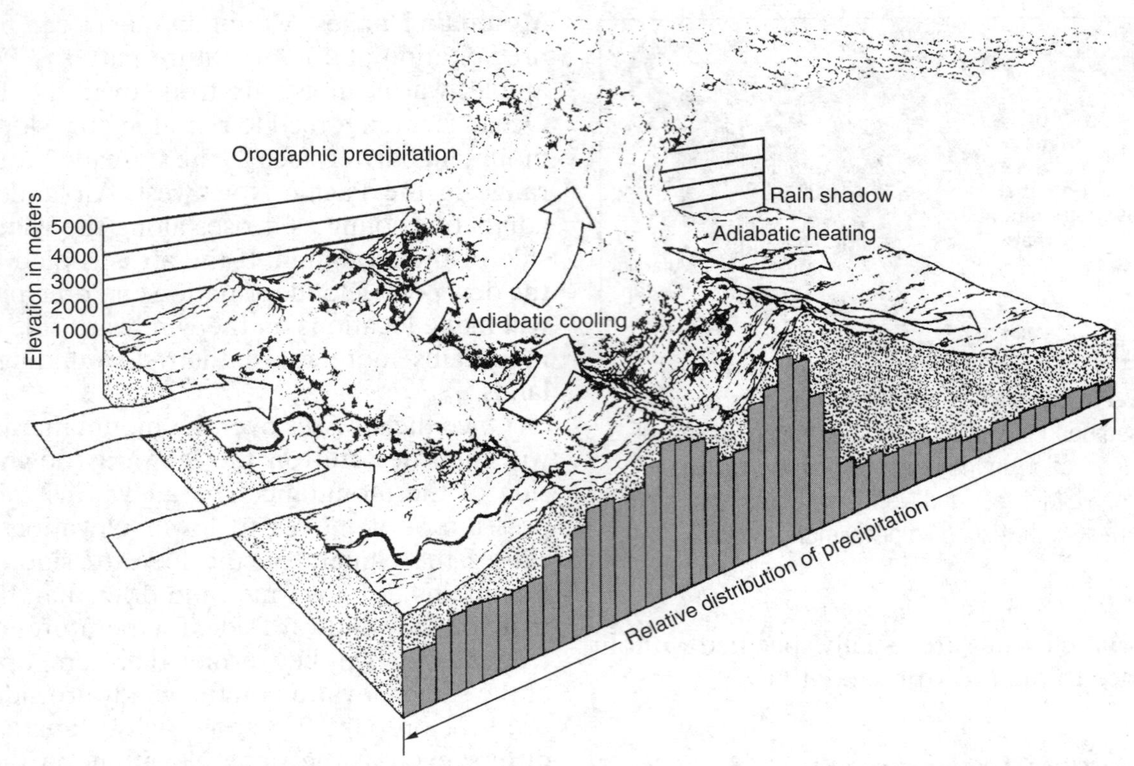

Figure 9-21. Wind rising over mountains cools and drops its moisture on the windward side. Therefore, air on the downwind side is very dry. As a result, the descending air warms quickly. This makes the climate on the downwind (leeward) side of a mountain range relatively warm and dry. Note the bar graph that shows the distribution of precipitation across this mountain range.

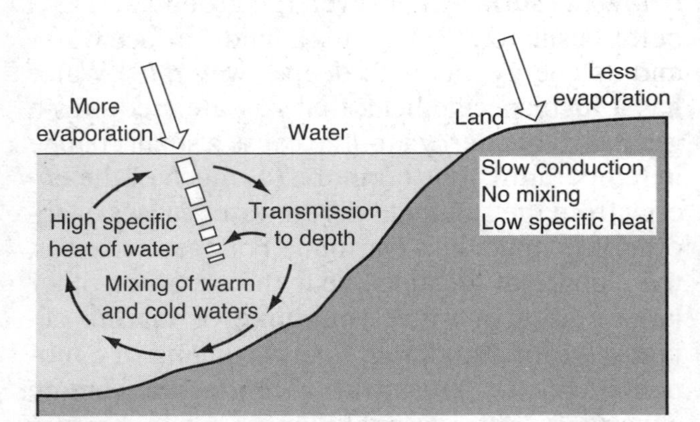

Figure 9-22. Large bodies of water do not change temperature as quickly as land areas. The reasons are sunlight penetrates deeper into water, water conducts energy relatively well, and convection currents carry the energy away from the surface. Water also has a higher specific heat than the materials on land.

are most noticeable where winds usually blow from the water onto land. New York's Long Island and the coastal cities of California, for example, enjoy smaller seasonal changes in temperature than nearby areas that are not so close to the ocean.

Ocean Currents An ocean current is a stream of water that circulates through the ocean basin. The surface temperature of an ocean current affects the air above the water. Cool water will cool the air, and warm water will warm the air. This affects the nearby coastal climate.

Warm tropical waters from the Straits of Florida move northward as the Gulf Stream and the North Atlantic Current. Under the influence of a warm ocean current and the prevailing westerly winds, Great Britain has a remarkably warm and equitable climate for such a high-latitude location. Palm trees can grow in some parts of southern Great Britain. However, these places are much farther north of the equator than colder locations in New England.

Because of the cold California current, the coast of Northern California has a cooler climate than might be expected. Winds blowing off the cold ocean water lower the temperature of the Pacific Northwest. The Surface Ocean Currents in the *Earth Science Reference Tables*, printed near the back of this book, can help you identify coastal areas where the climate is influenced by warm and cold ocean currents. For example, Figure 9-23 shows why prevailing west winds off the Pacific Ocean keep northern California relatively cool, but the Alaska Current makes coastal Alaska warmer than the interior,

El Niño Under normal conditions in the Pacific, ocean water upwells from deep ocean currents along the west coast of South America. This cold water is rich in oxygen and nutrients. The ocean fish have ample food and the fishing is good. However, in some years, there is less upwelling and the warmer surface water is not as productive. This often happens around the Christmas season, hence the name **El Niño** (the Spanish name for the Christ child).

El Niño, or ENSO (El Niño/Southern Oscillation), events do more than cause poor fishing. They can create extra rainfall in the Eastern Pacific and severe droughts in the Western Pacific regions. In fact, they can lead to unusual weather patterns over an even larger area. El Niño events help us understand that the oceans exert a strong influence on Earth's atmosphere.

Surface Ocean Currents

Warm Current
Cold Current
180
80
60
40
20
0
ALASKA C.
N. PACIFIC C.
CALIFORNIA
North America
GULF STREAM

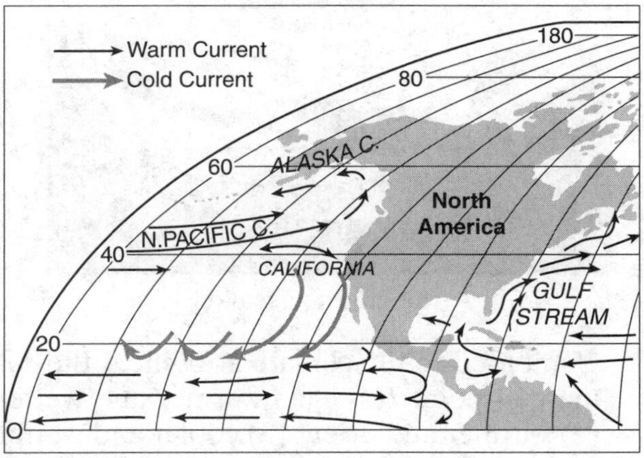

Figure 9-23. Prevailing winds from the west bring cool air over the coast of California and relatively warm air to the west coast of Alaska. This is a portion of the Ocean Currents map in the *Earth Science Reference Tables*, which are printed near the back of this book.

Planetary Wind Belts In Chapter 7, you learned about the influence of the Coriolis effect on the direction of Earth's prevailing winds. In turn, prevailing winds influence local climates. Prevailing winds are important in determining the effect of ocean currents on nearby climates. For example, the eastern coast of North America is not generally influenced by the warm waters of the Gulf Stream because of prevailing winds that come from the west, that is, off the land.

Within the different planetary wind belts, there are regions of rising air currents (low-pressure systems) and sinking air currents (high-pressure systems). When rising air is cooled by adiabatic expansion, moisture in the air may be released by condensation (cloud formation) and precipitation. Along the equator and in the mid-latitudes, zones of low pressure are common. Cyclonic storms often develop, and precipitation is plentiful. The local climate in these regions is humid. Near the poles and in the desert locations, about 20° to 30° on either side of the equator, there are high-pressure zones. The air at these latitudes is sinking within the atmosphere and is warmed by adiabatic compression. Most local climates in these regions are arid. Figure 9-24 on page 240 shows the convection cells and prevailing winds that affect these climate zones.

Monsoons Large continents influence their own weather. For example, winter conditions over Asia cause a mass of very cold, dry air to build up. This high-pressure region results in winds that blow down and off the continent during the winter. The winds become warmer as they descend toward sea level, but they remain relatively dry. Southern Asia is usually dry in the winter months because of these continental winds. However, as the continent warms in the summer, an inland low-pressure region develops. The winds reverse direction. The summer monsoon winds off the ocean bring rainy weather to southern Asia in the summer.

Arizona and the surrounding desert regions also experience monsoon weather. A large portion of the precipitation in this region comes as thunderstorms fed by the inland flow of moist air in the summer. This is a seasonal cycle that is

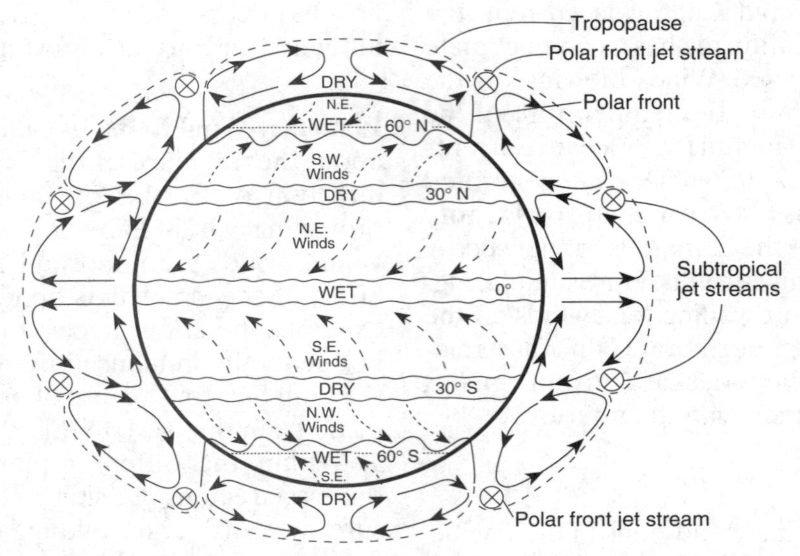

Figure 9-24. Notice the rising air currents at the equator and at 60° north and south latitude. These are generally regions of abundant clouds and precipitation. The down drafts at about 30° and near the poles mark the approximate latitudes of less precipitation and many desert regions.

similar the daily cycle of land and sea breezes shown in Figure 7-13 on page 180.

Typical Storm Tracks Weather systems usually move across the United States from west to east. That's why the weather in Chicago is often the weather that will affect New York several days later. Another characteristic path of weather systems is from the Gulf States northward along the eastern seaboard. Hurricanes, born in the South Atlantic or in the Caribbean Sea, generally move northward and westward toward the Gulf of Mexico and into the United States. The prevailing westerly winds over North America then carry most hurricanes over the eastern states and back toward the Atlantic Ocean. (See Figure 8-17 on page 207.)

QUESTIONS

PART A

48. How do ocean currents affect the coastal regions of South America at 20° south latitude? (1) Locations near both the east and west coasts are warmed. (2) Locations near both the east and west coasts are made cooler. (3) Locations on the east coast are warmed and locations on the west coast are cooled. (4) Locations on the east coast are cooled and locations on the west coast are warmed.

49. Which substance has the highest specific heat? (1) dry air (2) iron (3) liquid water (4) granite

50. The diagram below shows the prevailing winds moving from northern New York State and into Vermont.

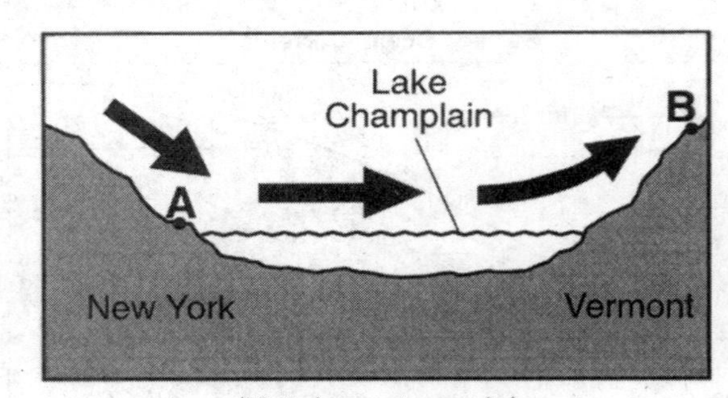

(Not drawn to scale)

Compared to the climate at point *A* the climate at point *B* is (1) warmer and wetter (2) warmer and drier (3) cooler and wetter (4) cooler and drier

51. London, England is approximately 1000 kilometers farther from the equator than New York City. Yet London has warmer winters than New York City. Why are the winters in London warmer than New York City? (1) London gets more insolation in the winter. (2) London is at a higher elevation above sea level than New York City. (3) London is more likely to be influenced by continental polar air masses. (4) London has a climate influenced by the North Atlantic Current.

52. What city in New York State probably has the greatest range of temperatures? (1) Buffalo (2) Albany (3) New York City (4) Oswego

53. The diagram below represents a continent that existed on Earth millions of years ago.

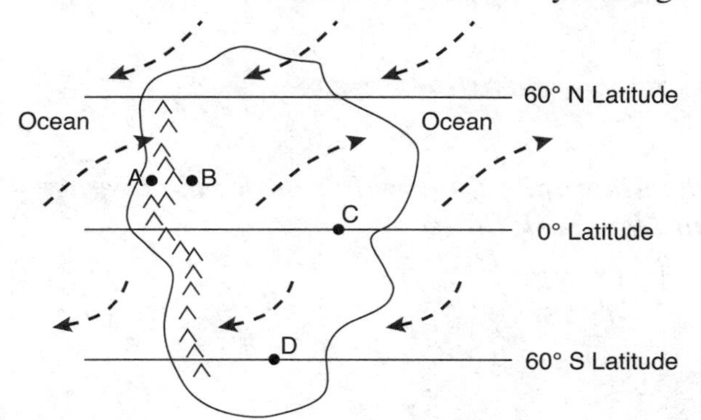

At which location would the warmest temperatures of the year usually occur in January? (1) *A* (2) *B* (3) *C* (4) *D*

Part B

Base your answers to questions 54 through 57 on the diagram below that shows the prevailing wind regions of Earth.

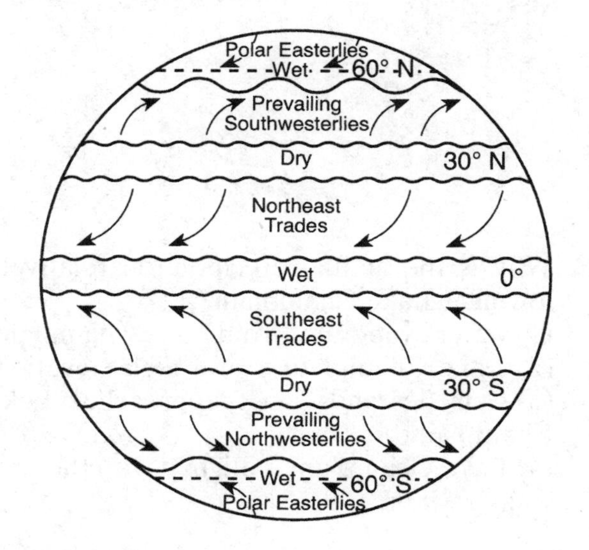

54. Why do the winds curve as shown by the arrows? (1) Earth rotates on its axis. (2) Earth revolves around the sun (3) The heating of Earth's surface is uneven. (4) The moon pulls on Earth's atmosphere.

55. Which wind belt has the greatest affect on the summer climate of New York State? (1) prevailing northwesterlies (2) prevailing northeasterlies (3) prevailing southwesterlies (4) prevailing southeasterlies

56. Why do wet conditions generally occur where the Northeast and Southeast Trade Winds meet? (1) This region has only ocean areas. (2) Rain is common in anticyclones. (3) Convergence heats the air. (4) Clouds form in rising air currents.

57. Why are the Polar Regions generally *not* classified as deserts? (1) There is plentiful rainfall at the poles. (2) There is little potential evapotranspiration at the poles. (3) The Polar Regions are zones of convergent winds. (4) The Polar Regions experience frequent fontal boundaries.

Base your answers to questions 58 through 63 on the diagram below, which shows winds blowing over the Sierra Nevada Mountains of California. These mountains are 3700 meters high. The valley on the leeward side is at an elevation of 305 meters above sea level.

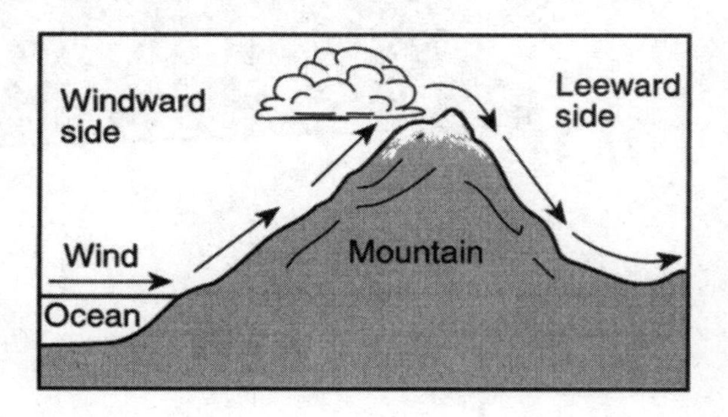

58. Why do clouds form high on the windward side of the mountains? (1) Rising air causes warming and evaporation. (2) Rising air causes warming and condensation. (3) Rising air causes cooling and evaporation. (4) Rising air causes cooling and condensation.

59. What is the air temperature like at the top of the mountains? (1) warmer than the temperature along the ocean (2) warmer than the temperature in the interior valley (3) very close to the dew-point temperature (4) equal to the temperature of the ocean water

60. How does the climate at 305 meters above sea level on the windward side and the leeward side of the mountains differ? (1) The windward side is warmer and drier than the leeward side. (2) The windward side is cooler and wetter than the leeward side. (3) The windward side is warmer and wetter than the leeward side. (4) The windward side is cooler and drier than the leeward side.

61. Where do temperatures probably change the least in both annual and daily cycles? (1) along the ocean shore (2) half way up the mountain on the windward side (3) at the top of the mountains (4) in the interior valley

62. What is the direction of the prevailing wind at this location? (1) north (2) south (3) east (4) west

63. What type of air mass would most likely bring the heaviest precipitation to the windward side of these mountains? (1) mP (2) mT (3) cP (4) cA

Base your answers to questions 64 through 66 on the diagram below, which shows the average yearly precipitation, measured in inches, throughout New York State.

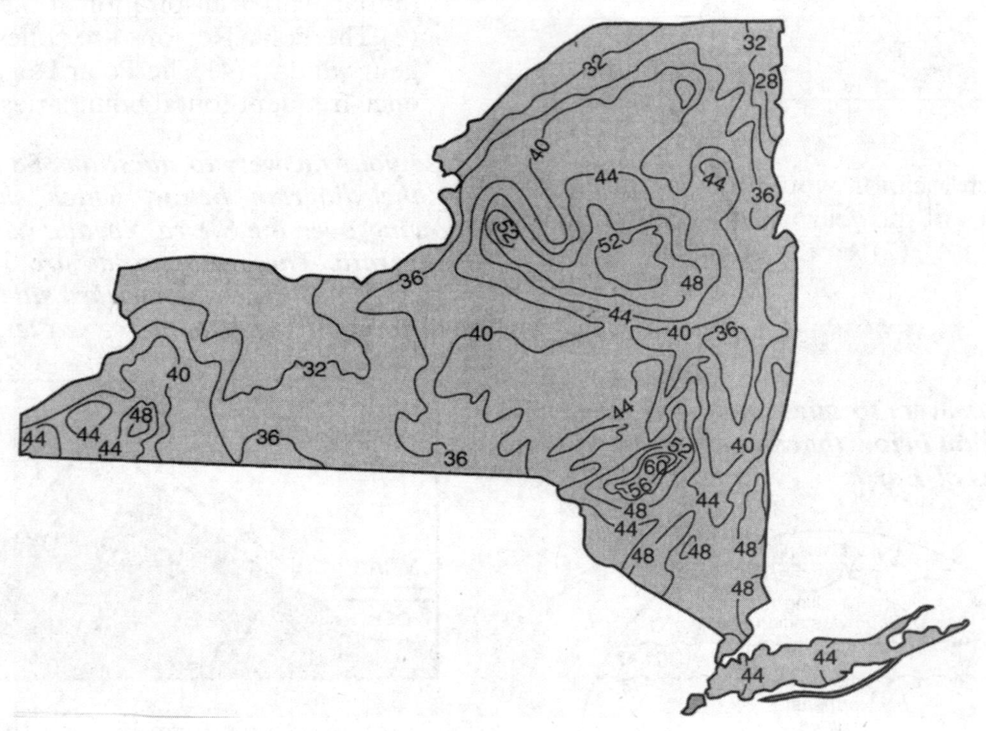

64. The location with the greatest annual precipitation is located in which landscape region? (1) Catskills (2) Tug Hill Plateau (3) Adirondack Mountains (4) Taconic Mountains

65. What is the average yearly precipitation at Rochester, New York? (1) 26 inches (2) 30 inches (3) 34 inches (4) 38 inches

66. Why is the annual precipitation relatively low near Lake Champlain?
(1) water evaporates from Lake Champlain.
(2) Lake Champlain is at a high elevation.
(3) Air descends as it approaches Lake Champlain.
(4) Lake Champlain is aligned with the prevailing winds.

Base your answers to questions 67 and 68 on the information and on the table below. Four samples of solid substances were added to identical insulated cups of warm water at 50°C. After the cold samples were added, each cup was left for 10 minutes until the water and the substance added came to the same temperature. The final temperature of the cup that contains basalt and water has been left blank.

Cup Number	Sample Added	Mass of Sample (g)	Initial Temperature (°C)	Final Temperature (°C)
1	Ice	50	–10	10
2	Lead	50	–10	40
3	Iron	50	–10	38
4	Basalt	50	–10	

67. What was the most likely final temperature of Cup 4? (1) 10°C (2) 30°C (3) 40°C (4) 50°C

68. What two properties of water best explain why the final temperature in Cup 1 was so low? (1) specific heat and energy gained during vaporization (2) energy gained during vaporization and energy gained during melting (3) energy gained during vaporization and transparency (4) specific heat and energy gained during melting

Problem Solving

In 1997, 178 nations met in Kyoto, Japan to negotiate limits on emission of greenhouse gases. An important outcome would be to reduce emissions of carbon dioxide to avoid a global warming. You can find the Kyoto Protocols in many internet sites. Following the Kyoto guidelines has been very controversial in the United States, and we are the only major nation that has not yet ratified the Kyoto Protocols. At the time of this book's publication, a climate summit was scheduled to take place in Copenhagen to draw up a successor to the Kyoto Protocols. Write a paragraph explaining why the United States should follow the guidelines of the Kyoto Protocols and its successor. Then write a second paragraph explaining why the United States should not.

Chapter Review Questions

PART A

1. What ocean current transports energy away from Earth's equatorial region? (1) Brazil Current (2) Guinea Current (3) Falkland Current (4) California Current

2. Equal volumes of the four samples shown below were placed outside and heated by energy from the sun's rays for 30 minutes. The surface temperature of which sample increased at the *slowest* rate?

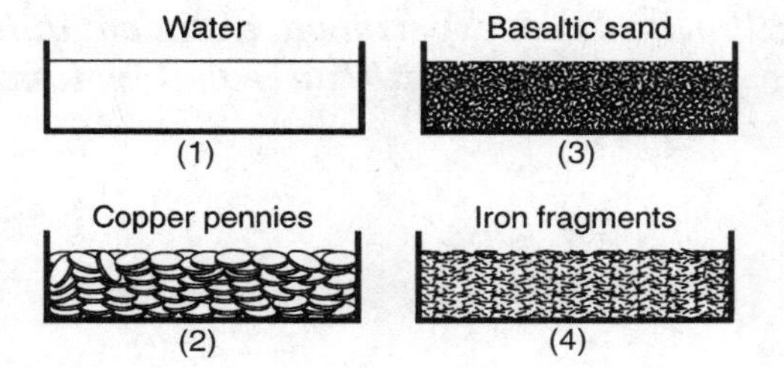

3. How much energy is transferred and where does it go when 1 liter of water evaporates from the surface of a lake? (1) 334 joules of energy enter the lake water. (2) 2260 joules of energy enter the atmosphere. (3) 334,000 joules of energy enter the lake water. (4) 2,600,000 joules of energy enter the atmosphere.

4. Compared to inland locations, coastal areas at the same elevation and latitude generally experience climates that have (1) cooler winters and cooler summers (2) cooler winters and warmer summers (3) warmer winters and cooler summers (4) warmer winters and warmer summers

5. How does the Canaries Current influence the coastal climate of northwest Africa? (1) The Canaries Current makes the climate more humid and warmer. (2) The Canaries Current makes the climate more humid and cooler. (3) The Canaries Current makes the climate less humid and warmer. (4) The Canaries Current makes the climate less humid and cooler.

6. At night, the continents usually become cooler as they radiate which energy into space? (1) ultraviolet (2) infrared (3) visible light (4) microwaves

7. When a lake freezes over, how does the energy content of the lake change? (1) 334 joules per gram are lost. (2) 334 joules per gram are gained. (3) 2260 joules per gram are lost. (4) 2260 joules per gram are gained.

PART B

Base your answers to questions 8 through 10 on the diagram below, which shows four columns of sediment. In an experiment, a student quickly filled each column with an equal volume of water.

Column A

Mixed particles (0.00001 cm to 0.5 cm in size)

Column C

Sorted particles (0.0001 cm to 0.2 cm in size)

Column B

Uniform-sized particles (0.2 cm)

Column D

Dry mud (Smaller than 0.0004 cm in size)

(Not drawn to scale)

8. Which column would allow the water to infiltrate most quickly? (1) column *A* (2) column *B* (3) column *C* (4) column *D*

9. Which column will hold the most water in its pore space? (1) column *A* (2) column *B* (3) column *C* (This question has only three answer choices.)

10. Column *C* was filled by dumping identical samples of poorly sorted sediment into a tall column of water. How many times did the student add sediment using this method? (1) 1 (2) 2 (3) 3 (4) 6

Base your answers to questions 11 through 13 on the map below, which represents an imaginary continent on a planet similar to Earth and on the four graphs, I, II, III, and IV.

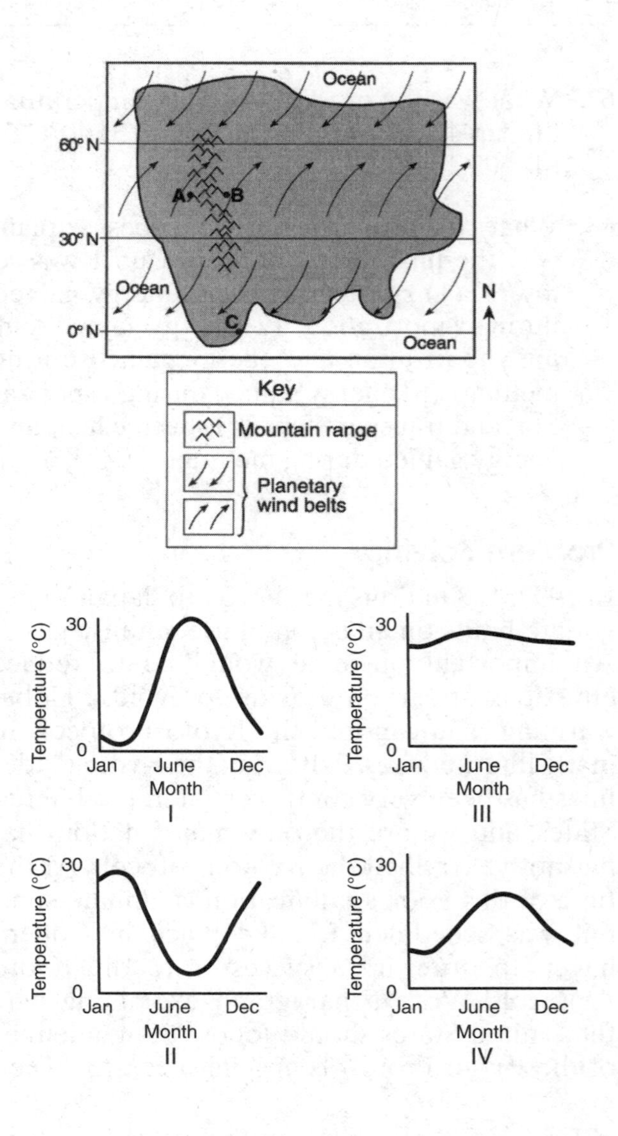

11. Which set of graphs below best represents the climatic conditions at locations *A*, *B*, and *C*? (1) *A*-I, *B*-II, *C*-III (2) *A*-II, *B*-III, *C*-IV (3) *A*-III, *B*-I, *C*-II (4) *A*-IV, *B*-I, *C*-III

12. Which graph best represents a southern hemisphere location? (1) I (2) II (3) III (4) IV

13. Compared with other locations on this map, location *C* probably experiences (1) low air pressure and low precipitation (2) low air pressure and high precipitation (3) high air pressure and low precipitation (4) high air pressure and high precipitation

Base your answers to questions 14 through 19 on the text passage, tables, and graphs below.

What Causes Global Warming?

Earth's atmosphere is a mixture of dozens of different gases. However, when it comes to global warming, not all of them are equal. The two most abundant gases in the atmosphere, nitrogen and oxygen, do not absorb infrared radiation. Water vapor, carbon dioxide, and several other gases absorb infrared terrestrial radiation. When all greenhouse gases are ranked by their contribution to the greenhouse effect, the most important are:

- water vapor, which contributes 36–72%
- carbon dioxide, which contributes 9–26%
- methane, which contributes 4–9%
- ozone, which contributes 3–7%

Humans have little effect on the amount of water vapor in the atmosphere. However, carbon dioxide, methane, nitrous oxide, and chlorofluorocarbons (CFCs) are major greenhouse gases that human activities produce. The contribution to the greenhouse effect by a given gas is affected by two factors. These are the chemical characteristics of the gas and its abundance in the atmosphere. For example, on a molecule-for-molecule basis methane is a much stronger greenhouse gas than carbon dioxide. However, methane is present in much lower concentration, so that its total contribution is smaller.

Humans produce greenhouse gases in a wide range of activities. Burning fossil fuels in transportation, for heating, and for power production add tons of carbon dioxide to the atmosphere each year. Fires used to clear tropical forests and open farmland account for nearly a third of the total carbon dioxide additions. Livestock, rice cultivation, and other farming practices form methane. Chlorofluorocarbons (CFCs) are used in refrigeration, fire suppression, and a variety of manufacturing processes. Any of these can leak CFCs into the atmosphere. Nitrous oxides come mainly from agricultural activities, including the use of fertilizers.

However, as the table and graphs that follow show, carbon dioxide is considered by most experts as the "elephant in the room." Therefore, if we are to manage global warming, controlling emissions of carbon dioxide is the most important factor.

Gas	Increase 1750 to 2009	Human Contribution to Global Warming
Carbon dioxide	37%	76%
Methane	150%	13%
Nitrous oxide	16%	6%
CFCs	None in 1750	5%

Four Greenhouse Gases Increased by Human Activities*

* The greenhouse influence of water vapor is not included here because human activities do not have an important effect on the water vapor content of the atmosphere.

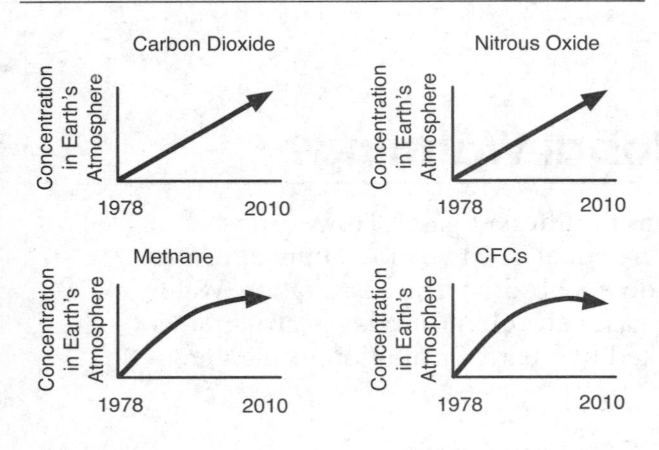

14. Which atmospheric gas absorbs the greatest amount of Earth's heat radiation? (1) ozone (2) water vapor (3) carbon dioxide (4) methane

15. Which greenhouse gases have been increasing most rapidly since the year 2000?
(1) carbon dioxide and nitrous oxide
(2) carbon dioxide and methane
(3) methane and nitrous oxide
(4) methane and CFCs

16. Compared with its abundance in 1750, which gas has increased in Earth's atmosphere by the greatest percentage? (1) carbon dioxide (2) methane (3) nitrous oxide (4) nitrogen

17. Why do environmentalists seldom express concern about water vapor as a greenhouse gas? (1) Water vapor is not a greenhouse gas. (2) Water vapor does not absorb terrestrial radiation. (3) Human activities have little effect on the water vapor content of the atmosphere. (4) Most of the water vapor in the atmosphere comes from condensation of ocean water.

18. Why is methane less important than carbon dioxide in causing global warming? (1) Unlike carbon dioxide, methane does not absorb terrestrial radiation. (2) Methane is not one of the gases found in Earth's atmosphere. (3) Methane gas is not as effective as carbon dioxide in absorbing infrared rays. (4) Methane is not as abundant in the atmosphere as carbon dioxide.

19. Which pie graph best shows the importance of all greenhouse gases in Earth's atmosphere?

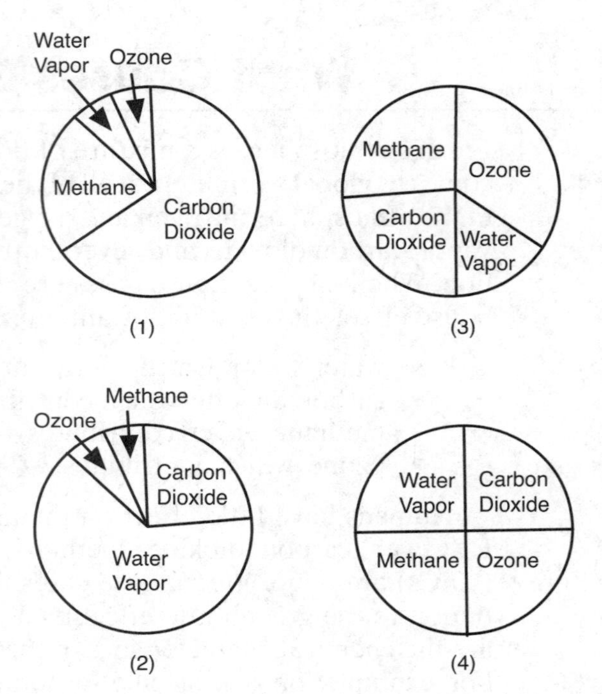

20. Earth absorbs solar energy and radiates energy back into space. How does the energy change in the process? (1) The energy is reradiated at a longer wavelength. (2) The energy is re-radiated at a shorter wavelength. (3) Much more energy is absorbed than reradiated. (4) Much more energy is reradiated than absorbed.

PART C

Base your answers to questions 21 through 23 on the following graph, which shows the aver-

age monthly temperatures for two cities, X and Y. Both are at the same latitude.

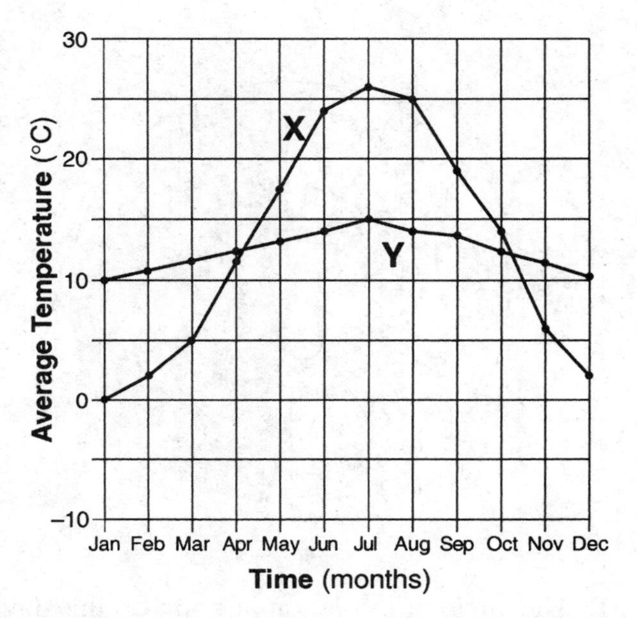

21. What is the range of average monthly temperatures at city *Y*?

22. Why does city *X* have a greater range of temperatures than city *Y*?

23. What evidence does the graph provide that both locations are in the Northern Hemisphere?

24. Seneca Lake, Cayuga Lake, and Canandaigua Lake are the largest of New York's Finger Lakes. These lakes rarely freeze over in spite of the very cold winters. In terms of specific heat, explain why these large bodies of water remain unfrozen, while the land around them is solidly frozen.

Base your answers to questions 25 and 26 on the data table below, which shows the average date on which ice cover on the Tanana River in central Alaska breaks up.

Data Table

Decade	Average Date of First Ice Breakup
1960–1969	May 7
1970–1979	May 5
1980–1989	May 4
1990–1999	April 29

25. Use this data to make a bar graph showing the average date of ice breakup on the grid below.

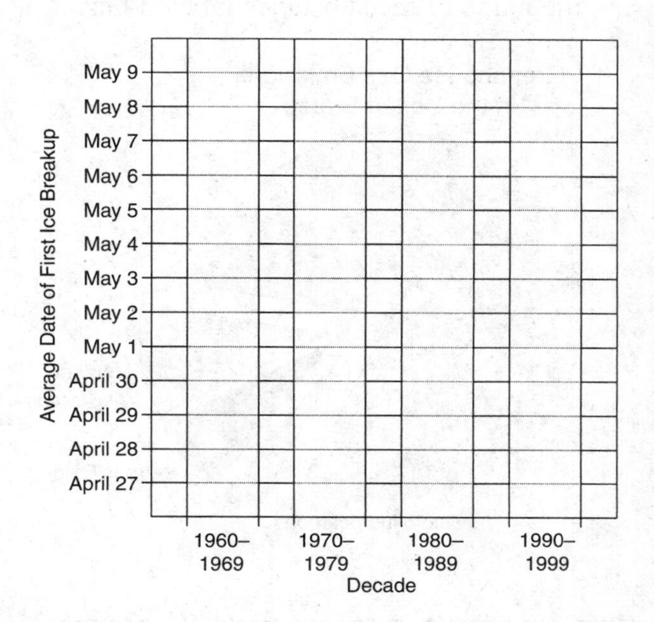

26. What does the data table show about the climate of central Alaska over these four decades?

27. The map below shows a region south of New York State that generally has high pressure due to the locations of global wind belts, and a region to the north that generally has lower pressure. Stating at point X, draw a curved arrow to show how the wind would blow across New York State. Show that the wind initially follows the pressure gradient, but is strongly influenced by the Coriolis effect.

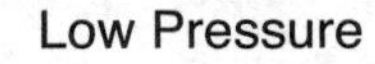

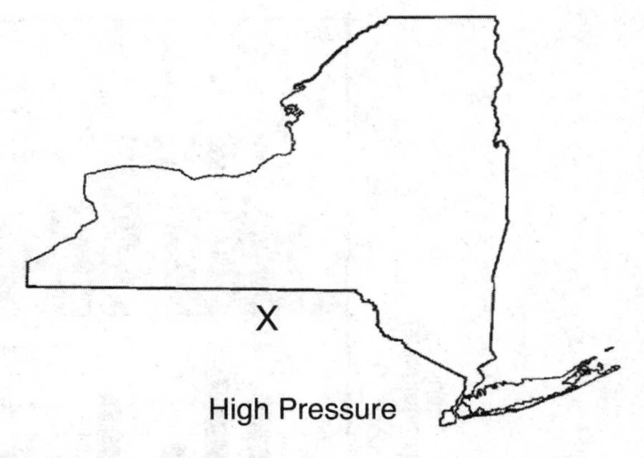

28. What Earth motion is responsible for changing the direction of planetary winds?

29. The pie chart below shows the greenhouse gases added to the atmosphere by human activities within the United States. What is the name of the substance labeled Gas *X*.

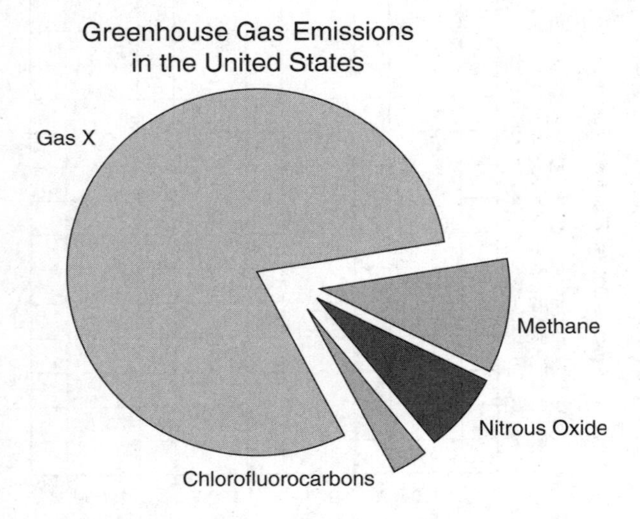

Base your answers to questions 30 through 33 on the map at the right, which represents Earth as seen from a position above the North Pole. The curved arrows show the direction of Earth's motion. The shaded portion represents the nighttime side of Earth. Some of the latitude and longitude lines have been labeled. Points A and B represent locations on Earth's surface.

30. Starting at point *B*, draw a curved arrow to show the direction of the prevailing surface winds.

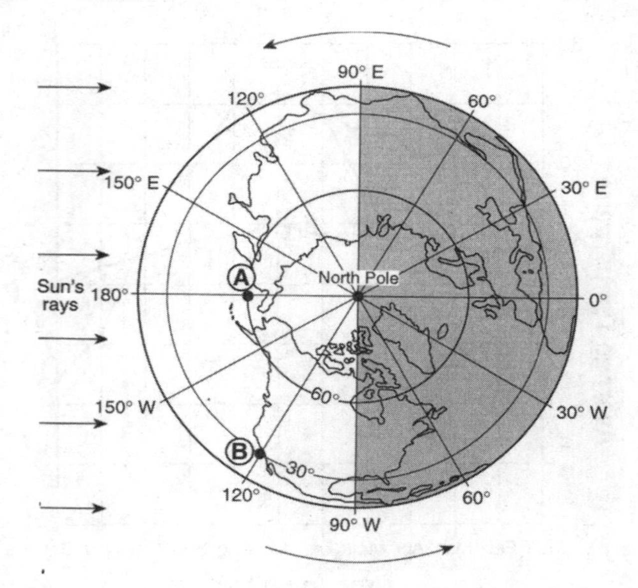

31. The angle of insolation at noon on this day at point *A* is 30° above the horizon. What is the angle of insolation at point *B* on the same day?

32. What ocean current influences the average annual temperature at point *B* and how does it affect the average annual temperature?

33. What is the latitude at point *A*?

Base your answers to questions 34 through 37 on the diagram and bar graph below, which represent climatic conditions experienced on a trip from west to east across California. Arrows show the prevailing winds in this part of the United States.

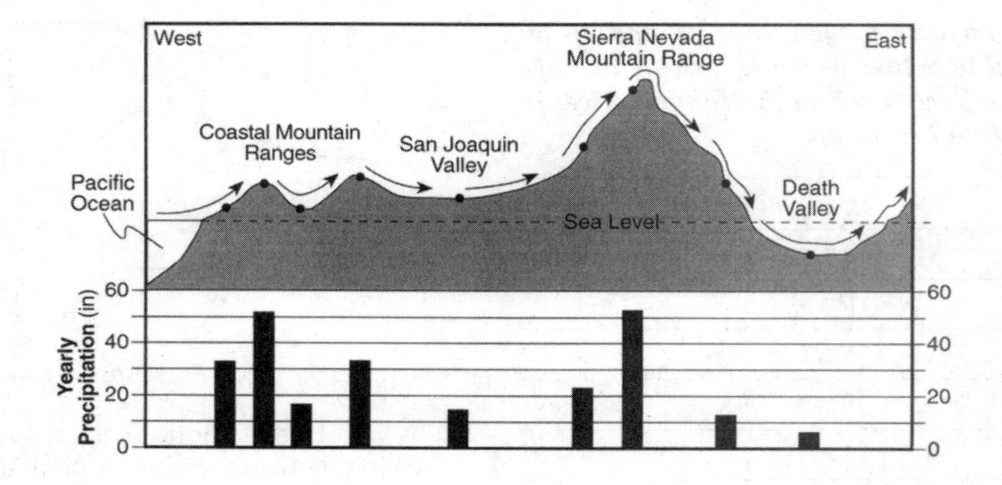

34. Explain why the San Joaquin Valley and Death Valley have less precipitation than nearby mountain locations.

35. What is the range of precipitation for the four locations in the Coastal Mountain Ranges?

36. Give one reason that the climate is colder near the summit of the Sierra Nevada Mountains than it is at nearby lower elevations.

37. What location in this diagram has the warmest average annual temperature?

Base your answers to questions 38 through 42 on the map below, which shows the major streams and rivers of New York State.

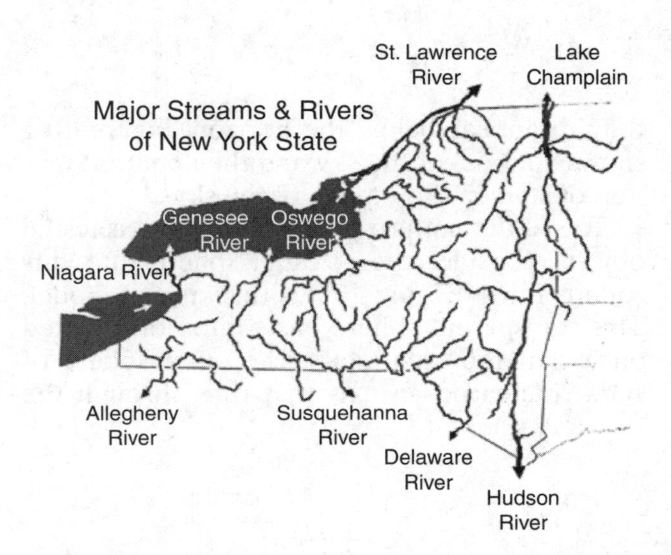

38. Outline the portion of New York State that is drained by the Delaware River.

39. What New York State landscape is part of the watershed of the Susquehanna River?

40. If it rains in Utica, New York, and the water flows into streams and rivers, at what city does the water flow out of New York State?

41. A major chemical spill in the Adirondack Mountains has the potential to contaminate what three rivers or bodies of water that carry water out of New York State.

42. What New York State river drains most of the Finger Lakes region?

43. How does the Labrador Current influence temperatures along the eastern coastline of Canada?

44. Why is there relatively little precipitation at Earth's North and South Poles?

Base your answers to questions 45 through 47 on the following information. A group of campers heated their dinner in a frying pan that they suspended over a campfire. They observed three different ways that the energy of the fire moved away from the flames. The answers to questions 45 through 47 must each involve a different form of heat flow.

45. By what method of heat flow did the handle of the frying pan become almost as hot as the part of the pan directly over the fire?

46. The campers found that their faces became warm only when they looked toward the fire. How did the energy reach their faces?

47. What kind of heat flow caused the smoke to rise?

48. The diagram below shows a stove used to heat a small cabin on a cold night. Draw two convection cells to show how the air in the room circulates when the stove is in use.

CHAPTER

Earth in Space

Vocabulary at a Glance

apparent motion	equinox	rotation
axis	Foucault pendulum	seasons
celestial object	geocentric model	solstice
celestial sphere	heliocentric model	terrestrial
constellation	hour	Tropic of Cancer
cyclic changes	orbit	Tropic of Capricorn
day	revolution	year

MOTIONS OF CELESTIAL OBJECTS

To most observers, the night sky looks like a huge dome, or hemisphere, that extends down to the horizon in every direction. If you were to extend the dome of the sky so that it circles all of Earth, you would produce an imaginary sphere. We call this the **celestial sphere**, on which all objects in the sky appear. A **celestial object**, such as the sun, the moon, a planet, a star, or any distant object visible in the sky, generally appears to rise in an easterly direction and set in a westerly direction. These motions are **cyclic changes**, that is they follow a pattern or cycle. (**Terrestrial** refers to objects that are a part of Earth, such as rocks, oceans, and clouds, not a part of the starry sky.)

Most celestial objects appear to move counterclockwise along circular or curved paths. These circular paths are centered near the North Star, Polaris. The North Star happens to be located very close to the point in the sky directly above Earth's North Pole. Celestial objects appear to move at 15° per hour (360°, or one rotation in 24 hours). In the course of a 24-hour day in the Northern Hemisphere, a star that appears near Polaris will appear to move in a complete circular path around a point near Polaris (the celestial North Pole). Other stars appear to move 15° per hour from east to west along arcs until they disappear below the horizon. Figure 10-1 shows a model of the sky and the apparent motion of stars in three parts of the sky.

To see the pattern of the motions of celestial objects, consider the celestial sphere that surrounds Earth. Figure 10-2 is this kind of model. This transparent sphere has white dots printed on it that represent stars. The outer sphere of stars rotates on an axis that runs through the

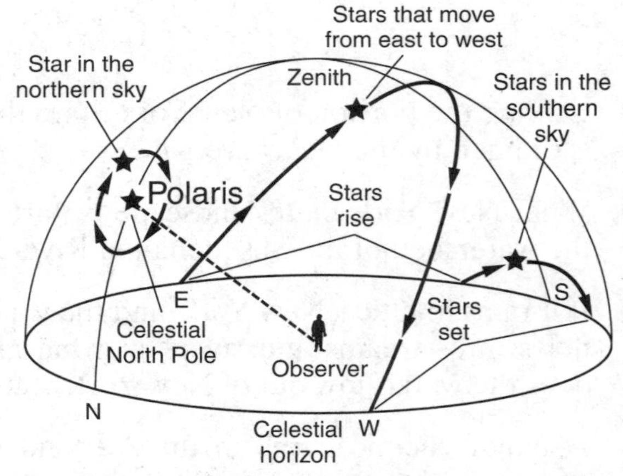

Figure 10-1. The stars appear to rotate counterclockwise around Polaris. Star paths, including the daily path of the sun, can be thought of as circles around the North and South Celestial Poles. Some of these circles are partly or even totally below the horizon.

Figure 10-2. This plastic celestial model shows the stars visible at any location on Earth. The "fixed stars" maintain their relative positions as the external, transparent sphere rotates around Earth.

North and South poles of the Earth model in the center. These stars are sometimes called "the fixed stars" because even as they rotate around Earth, each star keeps its position relative to the other stars. Some celestial objects, such as the sun, moon, and the planets are not marked on the outer sphere. This is because these much closer objects do not hold the same positions among the "fixed stars."

Star Paths

Near the equator, stars that are overhead appear to rise in the east and set in the west. In New York State, the stars high in the southern sky follow nearly the same path as the sun. These stars rise over the eastern horizon and move high into the southern sky. Then they move down to the right and set in the west. In the Northern Hemisphere, we may be able to see Southern Hemisphere stars, but only briefly. These stars usually are below the Celestial Equator. They rise above the horizon briefly in the south as they move from east to west. Only a small part of their path is visible to us. Stars even closer to the South Celestial Pole never rise above our horizon.

Stars over the North Pole move in circles around Polaris. The apparent motion of stars (like that of the sun, moon, and planets) is characterized by a circular motion of 15° per hour around the celestial North Pole. Of course, this is because of Earth's rotation. Although Figure 10-11 on page 258 shows three paths of the sun, it can also be used to understand how other stars move through the night sky.

Paths of the Planets and the Moon

The planets do not remain in the same position among the stars. In addition to their daily 15° per hour apparent circular motion, they change their position from night to night. Each planet has its own characteristic motion. Sometimes a planet appears to drift westward with respect to the stars, and sometimes a planet appears to reverse its direction and moves eastward among the stars. The wandering motions of planets led ancient observers of the sky to recognize that these celestial objects were different from the "other stars." In fact, the word *planet* means wanderer.

We know that the moon revolves around Earth in about 27 days. (See the Solar System Data, Table 11-1 on page 282 and on page 15 of the *Earth Science Reference Tables*.) As a result, you can see the moon rise about 50 minutes later each evening. The moon's path through the sky is higher or lower depending on the season of the year. In short, the moon's path is similar to the sun's as it appears to rise and set.

Star Trails

The apparent motion of the stars becomes clear when you look at a long-exposure photograph of the night sky. Objects on the ground appear still and the stars seem to rotate around the North Celestial Pole. Figure 10-3 on page 252 is a photograph of the night sky. The camera was mounted on a tripod and the shutter was left open for 40 minutes. White arrows show the apparent motion of the stars. Notice that the stars on the left are going down. This is how the sun sets in the west. On the right, the stars are rising as the sun also rises in the east.

Figure 10-3. In this long-exposure photograph, the stars appear to circle around Polaris at a rate of 15° per hour. The white lines that extend from Polaris to the beginning and end of one of the star paths make an angle of 10° (This is true of every star path.) Because 10° is 2/3 of 15°, the shutter must have been open for 2/3 of an hour, or 40 minutes. The direction north is always on the horizon below Polaris.

Figure 10-4. Orion, the hunter, is a prominent constellation of the evening sky in winter. Orion is dominated by the stars Betelgeuse, at the left shoulder, and Rigel, at the right foot of this figure. The three bright stars of the belt are also easy to spot.

WHAT ARE CONSTELLATIONS?

With fire as their only source of light at night, ancient peoples became familiar with the patterns of the stars. They used changes in the night sky to mark the seasons. Stars are randomly distributed throughout the sky. However, ancient peoples imagined that groups of stars formed patterns. These star patterns were often associated with traditions and legends that were part of the culture of early civilizations. Others star patterns were thought to resemble animals, such as Ursa Major, the Great Bear. Still others were thought to be objects, such as Lyra, the harp. Although some of the star patterns do resemble the objects for which they are named, most of them require considerable imagination to look like the person or object they represent.

Orion, the hunter, is shown in Figure 10-4. The drawing of a hunter can be made in any way that somehow uses the positions of the major stars. Because it contains some of the brightest stars, Orion dominates in the southern sky in the evening in our winter.

Astronomers still use the **constellations** to mark the position of the stars. Figure 10-5 shows

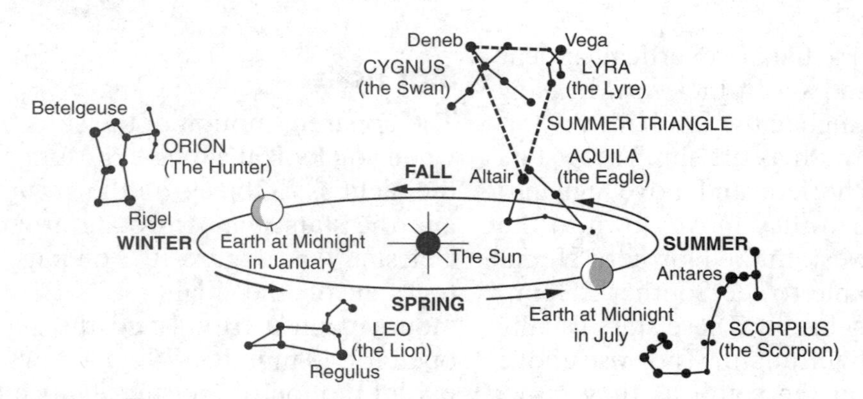

Figure 10-5. As Earth revolves around the sun, the night side faces toward different constellations each season of the year. We can see only those constellations that face Earth's night side.

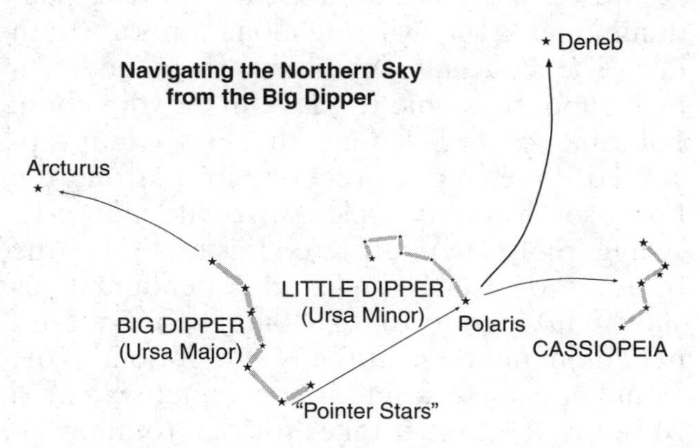

Figure 10-6. The Big Dipper is relatively easy to spot in the northern sky as it rotates around Polaris. The pointer stars will lead you to the North Star (Polaris). This diagram also shows you how to find the Little Dipper and several other nearby features. In different seasons you will need to rotate this chart to align it with the Big Dipper before you navigate to the other stars and constellations.

constellations that are visible in the Northern Hemisphere during different seasons of the year. To the modern astronomer, the constellations have a new meaning. Scientists have divided the night sky into 88 regions, each region associated with a constellation. When astronomers speak of a feature in Cygnus, the swan, they mean that to observe this feature you must look near the stars that form the constellation Cygnus.

You can use familiar constellations or star patterns to locate other features of the night sky. The Big Dipper, which is a part of the constellation Ursa Major (the Great Bear), is especially useful. For example, the two bright stars at the end of the bowl of the big dipper are called "the pointer stars" because they point to Polaris. Figure 10-6 illustrates this method of using the Big Dipper to find other celestial features.

Figure 10-7 is a map of the whole sky that shows the brightest constellations you can see

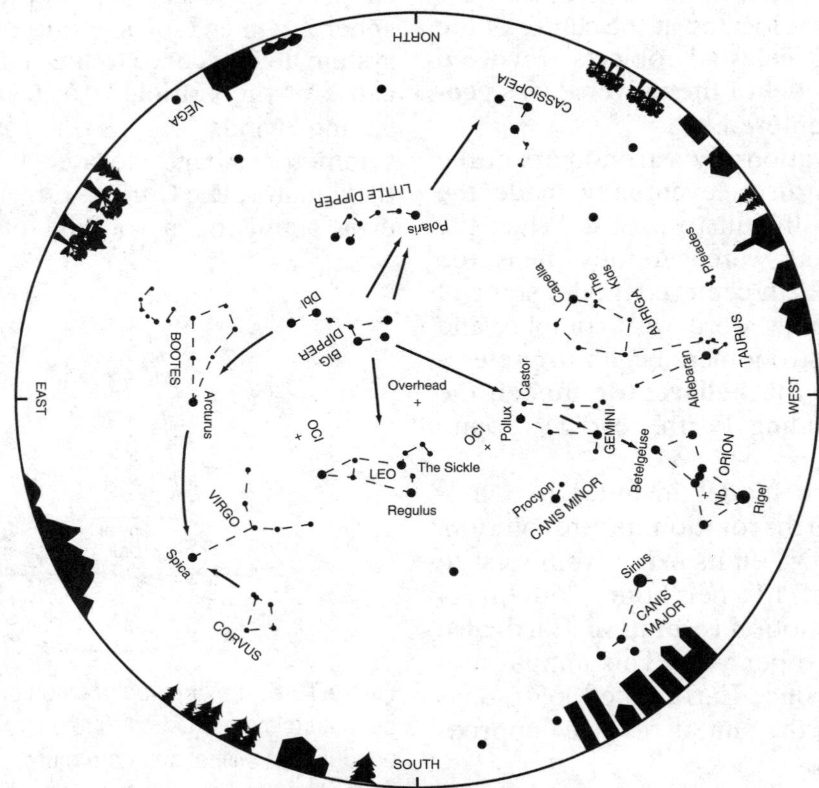

Figure 10-7. This map of the night sky shows the constellations visible at about 40° North latitude after 10 P.M. in March and about 8 to 9 P.M. in April. To use this map, hold it above your head so that the directions (north, south, east, and west) are aligned properly. Arrows show how to navigate to various constellations from the Big Dipper. The objects labeled OCl (open clusters), Dbl (double stars), and Nb (nebulae) are more impressive when viewed through binoculars or a small telescope. Because planets "wander" among the stars and change their position, they are not shown on this chart.

in late spring and early summer. To locate some of the constellations visible at this time of year, take this map of the stars (or an enlarged copy) outside on a clear night. Hold the star chart upside down above your head so that the compass points on the map line up with their true direction. (You may need to use a compass to help you determine exactly where north, south, east, and west are located relative to where you are standing.) Start by finding the Big Dipper (a part of Ursa Major), which is visible throughout the year for observers as far north as New York State. Once you have located this "landmark," you should be able to find the other bright constellations and stars.

MODELS OF THE NIGHT SKY

Early civilizations were aware of how celestial objects, such as the sun and moon, appeared to move through the sky from east to west. Because these people could not feel Earth moving, they believed that Earth did not move. They also believed that Earth was located at the center of the universe, and all celestial objects revolved around it. Their model of the universe was **geocentric**, or Earth-centered.

Detailed observations by astronomers in the 16th and 17th centuries eventually made the geocentric model difficult to accept. When the paths of the planets were carefully measured and plotted in an Earth-centered model, some of the planetary motions were very complex and hard to explain. Astronomers began to prefer a simpler model. In the **heliocentric model**, the eight planets, including Earth, revolve around the sun.

The heliocentric model includes two motions of planet Earth: rotation and revolution. Each day Earth spins on its **axis**—from west to east at the rate of 15° per hour (360° in 24 hours). This daily motion is **rotation**. Earth also **orbits** the sun once per **year**. This annual motion is **revolution**. Since Earth takes $365^1/_4$ days to revolve around the sun, it revolves approximately 1° per day.

Proof for the Heliocentric Model

Experimental proof of the motion of Earth was not found until 1851. At that time, the French scientist Jean Foucault suspended a long pendulum and set it swinging along a north-south line. The **Foucault pendulum**, mounted on a high support, is able to move in any direction. Foucault observed that the pendulum appeared to change direction away from the north-south line in a clockwise direction as it swung freely. He interpreted this motion as the rotation of Earth under the pendulum, as shown in Figure 10-8. Although a Foucault pendulum mounted at the North or South Pole would appear to rotate in a complete circle in 24 hours, the time it takes for one rotation increases when the pendulum is located at lower latitudes. At the equator, the Foucault pendulum would not appear to rotate at all. Many science museums have Foucault pendulums. They are sometimes set up to knock down small posts, as the pendulums appear to rotate slowly to a new swing direction.

The second proof of the Earth's rotation is the Coriolis effect. It was named for another French scientist, Gaspard Coriolis. You may recall from Chapter 7 that in the Northern Hemisphere, winds blowing out of a high-pressure system always curve to the right. In addition, because of the Coriolis effect, the system of prevailing winds on Earth forms a series of symmetrical bands located north and south of the equator. The Coriolis effect is the result of inertia acting on a rotating planet. Figure 10-9

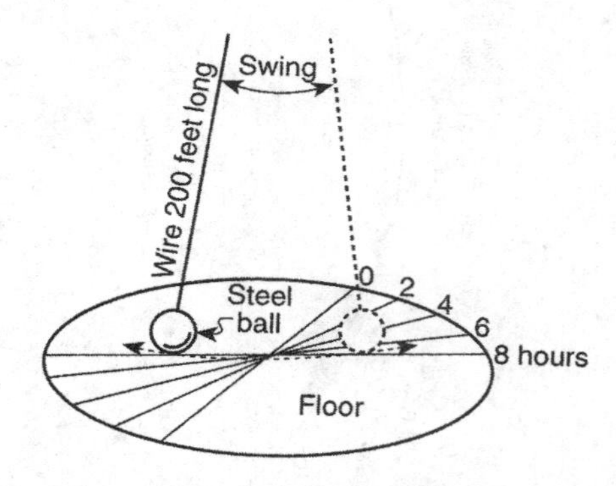

Figure 10-8. The apparent rotation of a ball of a Foucault pendulum is proof of Earth's rotation. This kind of pendulum is suspended from a pivot in the ceiling. The pivot allows the pendulum to swing back and forth, but also to slowly rotate. At the North Pole one complete (apparent) rotation (360°) takes approximately 24 hours. However, in New York State a full rotation takes about 36 hours, as indicated in this diagram.

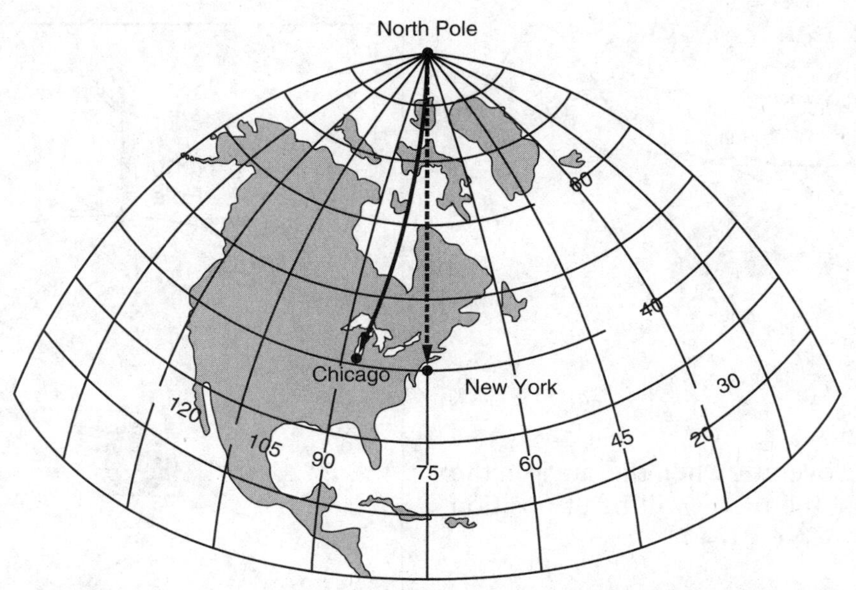

Figure 10-9. If a cannon at the North Pole fired a long-range shell toward New York City, the projectile would appear to curve to its right, toward Chicago. This is an example of the Coriolis effect, which is caused by Earth's rotation. Note that a right curve coming toward you seems to curve to your left.

shows what would occur if a cannon at the North Pole were fired toward the Equator.

QUESTIONS

Part A

1. Understanding the Coriolis effect is most important for (1) clock makers and geologists (2) artillery soldiers and meteorologists (3) map makers and paleontologists (4) construction workers and seismologists

2. The image below is a time-lapse photograph of stars in the night sky over New York State. The camera shutter was left open for 1 hour.

In what direction was the camera pointed? (1) toward the north (2) toward the south (3) toward the east (4) toward the west

3. A student in New York State observing the evening sky in December was able to find the constellation Taurus quite easily. However, when she looked in the evening sky in June, she could see Virgo, but not Taurus. Why were the evening stars different in December and June? (1) The stars revolve around the sun. (2) Earth revolves around the sun. (3) The stars rotate on their axes. (4) Earth rotates on its axis.

4. How can we describe the rate of Earth's rotation and the nightly apparent motion of the stars? (1) The stars rotate faster around Polaris. (2) Earth rotates faster on its axis. (3) Earth rotates at 15°/hour so the stars rotate at 15°/hour. (4) Sometimes the Earth rotates faster, and sometimes the stars rotate faster.

5. The following diagram shows an observer in New York State looking at the moon and the sun as they set in the west.

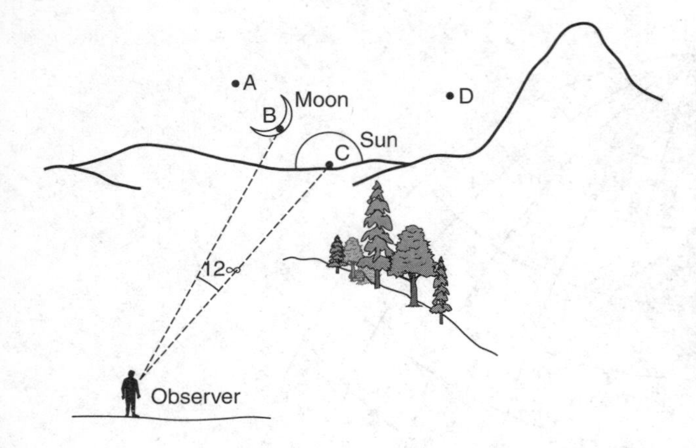

The following evening when the sun is in the same position, the moon will be at position (1) *A* (2) *B* (3) *C* (4) *D*

6. A student in New York State observes the star Regulus in the constellation Leo high in the southern sky. Three hours later, Regulus is still visible, but its position in the sky has changed. In what direction has Regulus moved in those three hours?
(1) up and to the left
(2) up and to the right
(3) down and to the left
(4) down and to the right

7. For thousands of years, most people accepted the geocentric model of the universe. What object did those people think was the center of the universe? (1) the sun (2) Earth (3) Polaris (4) the Milky Way

8. The worldwide pattern of ocean currents is most influenced by
(1) the gravity of the moon
(2) the gravity of the sun
(3) Earth's rotation
(4) Earth's revolution

Part B

Base your answers to questions 9 through 11 on the following diagram, which shows a part of the Northern Hemisphere including the North Pole and North America. A Foucault pendulum is shown at the North Pole. A cannon that can fire a cannon ball hundreds of miles is located in the United States.

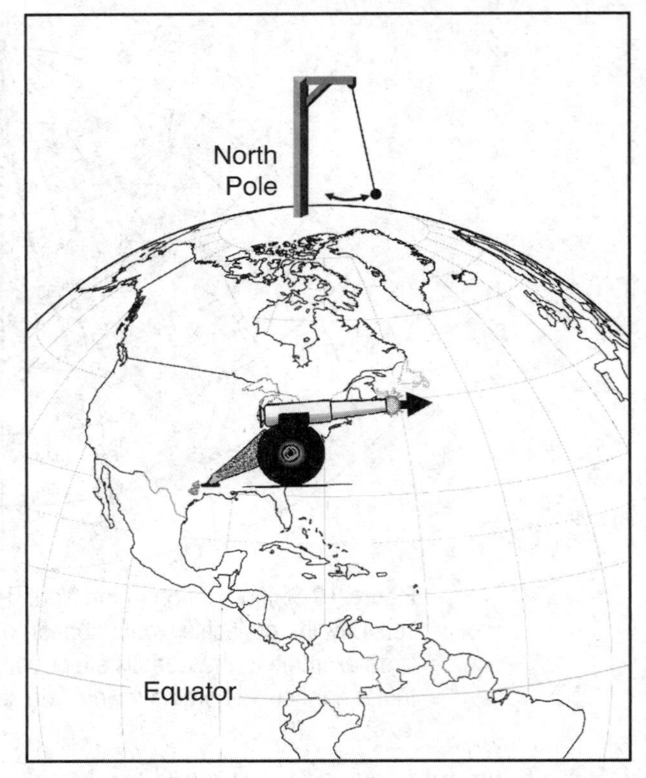

9. The pendulum and the cannon can be used to show that (1) Earth spins on its axis (2) Earth revolves around the sun (3) the moon is a satellite of Earth (4) the moon is approximately 4×10^5 kilometers from Earth

10. An essential feature of the Foucault pendulum is that the (1) swinging weight must be entirely solid (2) cord the weight swings must be flexible (3) pendulum must swing with no loss of energy (4) swing must not be restricted to one direction

11. The cannon fires a long-range cannonball. For an observer standing on the ground behind the cannon and able to see the cannonball fly hundreds of miles, the cannonball would appear to (1) curve to the right (2) curve to the left (3) curve upward (4) fly perfectly straight

Base your answers to questions 12 through 16 on the following diagram, which shows observations of the night sky in New York State. The positions of a star (X) and a group of stars are shown at two different times, A and B. The black arc shows the apparent motion of the group of stars over this period of time.

12. What is the name of the group of stars shown at two different positions? (1) the Big Dipper (2) the Little Dipper (3) Ursa Minor (4) Orion

13. What is the name of the star labeled *X*? (1) Betelgeuse (2) Polaris (3) Rigel (4) Spica

14. In what direction is the observer looking to see these stars? (1) north (2) south (3) east (4) west

15. How long did it take for the star group to appear to move from position *A* to position *B*? (1) 1 hour (2) 3 hours (3) 6 hours (4) 12 hours

16. What caused the apparent motion of the star group? (1) motion of Earth around the sun (2) motion of the sun around the Milky Way Galaxy (3) Earth's revolution (4) Earth's rotation

APPARENT MOTIONS OF THE SUN

Before there were mechanical clocks, people used the movement of the sun through the sky to track the passage of time. Sundials, such as the one shown in Figure 10-10, are among the first timekeeping devices. On a sundial, the passage of the daylight hours is indicated by the movement of the shadow of a pointer (gnomon) across a marked dial or scale. Today, we have replaced sundials with more precise electronic devices, but our measures of time are still based on Earth's motions.

Figure 10-10. As the sun moves across the sky, the shadow of the pointer, or gnomon, moves across the face of the sundial. This shows the passage of the daylight hours.

You can see how a sundial works by performing a simple demonstration with a meterstick. First, find a sunny spot outdoors where the ground is flat. Hold the meterstick vertically (You can use a weight on a string to be sure the meterstick is vertical) Record the length and direction of the shadow of the meterstick. Repeat this procedure at different times of the day. You will discover that the shadow is very long at sunrise. The shadow decreases in length as the morning progresses. It is shortest at noon, and then increases in length throughout the afternoon. Also note that the shadow points directly away from the sun. For example, if the sun is to the south, as it always is in New York State at noon, the shadow points to the north. You can even use this method to create your own simple sundial. Table 10-1 on page 258 is a quick guide to telling the time of day by the sun. Due (north, south, east or west) means in exactly that direction and not just in that general part of the sky.

Our system of time is based on the **apparent motions** of the sun. The motion of a celestial object, such as the sun, the moon, or a planet, through the sky is called apparent motion because the object is not really moving. It only appears to be moving. Part of the fascination of Figure 10-3 on page 252 (and others in this chapter) is understanding that it is not the celestial objects that move. It is Earth that moves and is

Table 10-1. Telling Time by the Sun*

Solar Time	Sun's Position	Shadow Direction	Shadow Length
6 A.M.	Due east	Due west	Long
Noon	Due south	Due north	Shortest
6 P.M.	Due west	Due east	Long

*In autumn and winter, the sun is below the horizon (no shadows) at 6 A.M. and 6 P.M. Solar time precisely agrees with clock time only if you are located at the longitude at which your time zone begins. In most places, time measured by the sun is always a few minutes ahead or behind clock time.

responsible for the apparent daily and annual motions that we see in the heavens. The apparent rising and setting of the sun is actually caused by Earth's rotation. Therefore, the **day** is the average time it takes the sun to appear to move from its highest point on one day to its highest point on the next. The **hour** is, correspondingly, $1/24$ of a day.

Seasons

Because Earth moves in its orbit, the sun's pathway from east to west in the sky also changes in a yearly cycle as shown in Figure 10-11. These changes in the sun's path are the cause of the **seasons.** In New York, the noon sun is high in the southern sky at the beginning of the summer. Through the next 6 months of summer and autumn (Late June until late December), the sun's noon position gets a little lower in the sky each day. At the beginning of summer, the sunrise position is north of due east and moves slowly southward. In addition, each daylight period becomes a little shorter than the day before.

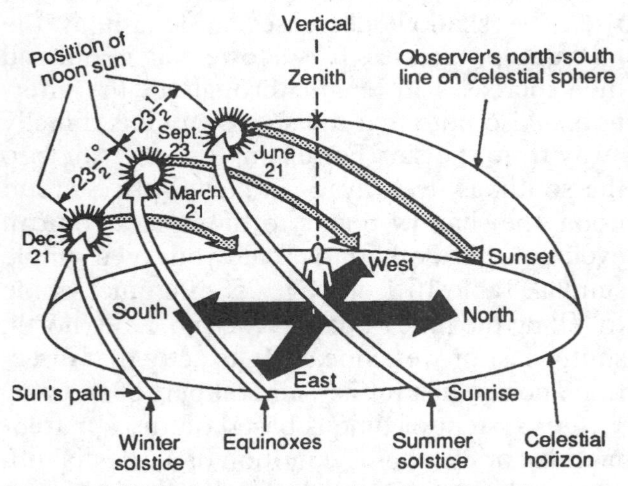

Figure 10-11. To the observer in New York State, the sun's path appears to change in daily and yearly cycles. The positions of sunrise, sunset and the height of the sun's path are all a part of this complex cycle. This diagram also shows that it is noon for the observer when the sun crosses a north-south line through the sky (the observer's meridian).

At the autumnal **equinox** (about September 22), the sun rises due east and sets due west. Sunlight lasts for 12 hours, and this is the beginning of fall, or autumn, in the Northern Hemisphere. (Actually, the sunlight is a little longer than 12 hours for several reasons. For example, the atmosphere bends—refracts—the sun's rays so that it appears to be above the horizon when it is slightly below the horizon.)

At the beginning of winter, the winter **solstice** (about December 22), the sun rises south of due east. The sun's path through the sky is relatively short so this is the shortest daylight period of the year, as measured from sunrise to sunset. Throughout the months of winter and spring (late December until late June), the positions of sunrise and sunset move northward. As the sun's path through the sky becomes longer, the daylight periods become longer.

At the vernal equinox (about March 21), the sun rises exactly in the east (due east) and sets due west. The daylight period is 12 hours long. This is the beginning of spring in the Northern Hemisphere.

When the summer solstice arrives (about June 21), the sun rises north of due east, is high in the sky at noon, and sets north of due west. On the summer solstice the sun's path is as long it ever gets. Therefore, the summer solstice is the longest daylight period of the year, measured from sunrise to sunset. This is also the beginning of summer.

Please keep in mind that the length of daylight changes because the part of the sun's apparent circle of motion that is above the horizon changes. The changing length of daylight hours is not due to changes in the rate of motion of the sun. The sun always seems to move 15° per hour. (This is the constant rate of Earth's rotation.) In spring and summer, when the sun is higher in the sky, more than half of the sun's circular path (more than 180°) is above the horizon. In fall and winter, the days are short because the sun is lower in the sky and most of the path of the sun is below the horizon. Table 10-2 lists the positions of sunrise and sunset and the angle of the noon sun on different dates as seen from central New York State.

Table 10-2. Observations of the Sun's Path for an Observer in Central New York State

Date	Sunrise	Sunset	Angle of Noon Sun (Varies with latitude)
March 21	East	West	47.5°
June 21	Northeast	Northwest	71°
September 21	East	West	47.5°
December 21	Southeast	Southwest	24°

LATITUDE AND THE ANGLE OF THE SUN

The path of the sun through the sky depends on where you are when you observe it. People who live near the equator know that the noon sun is always high in the sky. However, the noon sun can be directly overhead *only* for observers within the tropics. Specifically, the **Tropic of Cancer**, 23½° North, and the **Tropic of Capricorn**, 23½° South, mark the latitudes within which the noon sun can be directly observed overhead at any time. Figure 10-12 illustrates the path of the sun at the equinoxes for observers at five different latitudes.

Observed anywhere within the continental United States, the sun is higher in the sky in the summer than it is in the winter. However, the noon sun is *never* directly overhead. Therefore the position of the noon sun in New York varies. It changes from a low point less than half way up the sky in the winter, to a summer position that

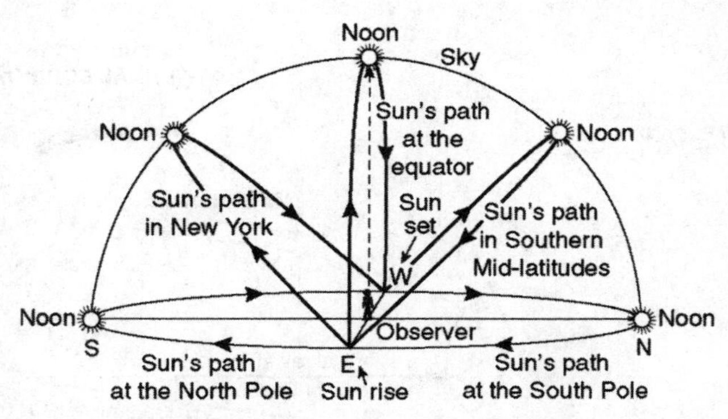

Figure 10-12. The sun's path is shown at the spring and autumn equinoxes for observers at five different latitudes: North Pole, northern New York State, equator, any southern mid-latitude location, and South Pole. On these two days, the sun circles around the horizon at both poles. At other latitudes, the sun rises due east and sets due west. The closer an observer is to the equator, the higher the noon sun will be in the sky.

is much higher. (But, it is never directly overhead.) Figure 10-11 shows three paths of the sun representing the two solstices and the two equinoxes. (The spring and fall equinox paths are the same.)

As you travel northward from any mid-latitude location such as New York State, the noon sun will move lower in the sky. This is a consequence of Earth's spherical shape. At the Arctic Circle (66½°N), the sun can be as high as 47° on the first day of summer, but on the first day of winter the noon sun barely reaches the horizon. At this time, the sun is visible for only a brief period about noon. For observers within the Arctic and Antarctic circles, on the first day of summer, the sun moves around the sky without actually setting. However, the sun does not come up at all on the first day of winter.

On the first day of summer, observers at the poles see the sun move in a big circle 23½° above the horizon. As the autumnal equinox approaches, the sun spirals downward toward the horizon, circling a little lower each day. After the sun sets in the autumn, it does not rise again until the first day of spring. Then it begins to spiral up to its summer solstice position. Thus, observers at the poles see six months of daylight, followed by six months of darkness in a yearly cycle.

Figure 10-13 on page 260 shows that our annual cycle of the seasons is caused by a combination of the 23½° tilt of Earth's axis and its

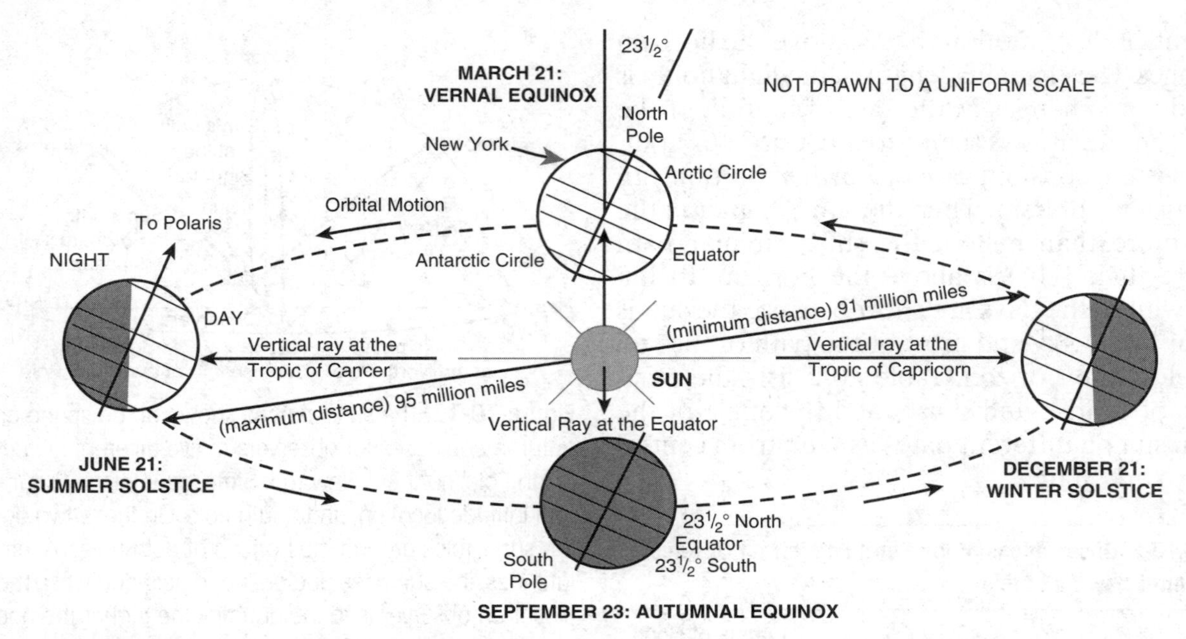

Figure 10-13. The 23½° tilt of Earth's axis and its orbit around the sun cause the seasons. Our summer occurs when the Northern Hemisphere is tilted toward the sun. In the winter, this hemisphere is tilted away from the sun. In the Southern Hemisphere, the seasons are six months earlier or later. Note that our closest approach to the sun occurs in our winter, although this change in distance is very small. In this diagram, Earth may look closer to the sun at the spring and autumn positions. It is not. Also note that throughout Earth's revolution around the sun, Earth's axis always points in the same direction.

revolution around the sun. Although Earth's spin axis always points toward the same direction in space, its journey around the sun causes the North Pole to tilt alternately toward and away from the sun. This is the cause of the seasons.

Hours of Daylight

The length of daylight follows an annual cycle everywhere on Earth. Actual motions of the sun do not cause the changes in the sun's daily path across the sky. The apparent motions of the sun are the result of Earth's spin on its axis. Earth's axis is tilted 23½° from a line perpendicular to the plane of Earth's orbit around the sun. Earth also exhibits parallelism—this means that Earth's axis always points in the same direction. Figure 10-14 is a close-up of Earth at the solstice and equinox positions shown in Figure 10-13.

At the equinox, the sun rises due east and sets due west in all locations except the poles. Therefore, daylight lasts 12 hours. At the equator throughout the year, there are 12 hours of daylight and 12 hours of night. As you approach the poles, the amount of seasonal variation in the length of daylight increases until a maximum of 6 months of daylight and 6 months of darkness is reached at the North Pole or the South Pole, as you can see in Figure 10-15.

QUESTIONS

Part A

17. Which statement below best describes the path of the sun through the sky on December 23 in New York State? (1) The sun rises due east, is high in the sky at noon, and sets due west. (2) The sun rises due east, is low in the sky at noon, and sets due west. (3) The sun rises in the southeast, moves very high across the sky at noon, and sets in the southwest. (4) The sun rises in the southeast, moves low across the sky at noon, and sets in the southwest.

18. The most conclusive evidence that Earth rotates is the motion of (1) the tectonic plates (2) the sun and moon (3) a wind vane (4) a Foucault pendulum

(Questions continue on page 262.)

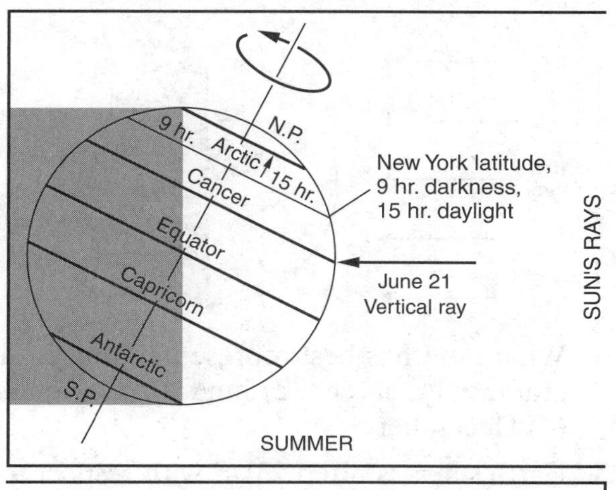

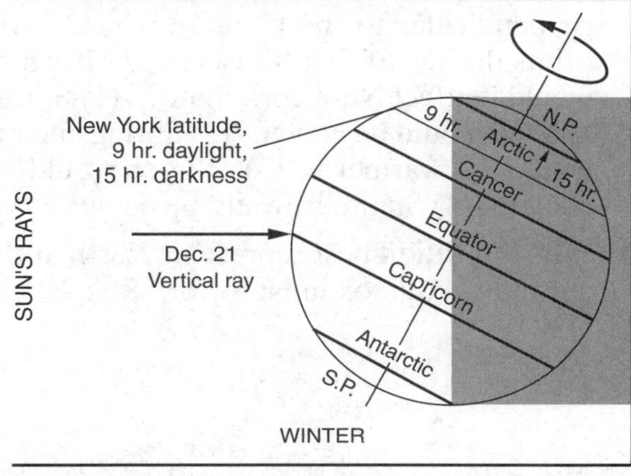

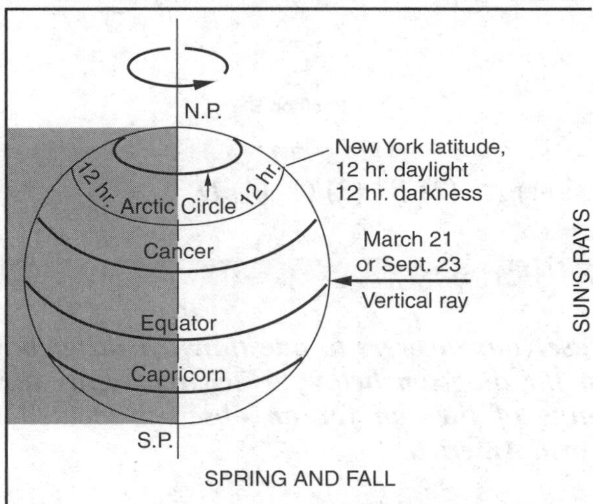

Figure 10-14. These diagrams show how rays of sunlight strike Earth's surface during the course of a year. They also show how day and night are distributed on Earth. The top diagram is our summer solstice when we have about 15 hours of daylight in New York State. The middle diagram is our winter solstice when we have only about 9 hours of daylight. On the equinoxes (bottom diagram) there are 12 hours of daylight and night nearly everywhere on Earth.

One Year at the North Pole

June

September (& March)

December

Figure 10-15. Earth's poles experience the greatest change in daylight hours. This series of images shows that at the North Pole, on the summer solstice, the sun circles around the sky $23^1/_2°$ above the horizon. The sun does not set until September, when it circles down to and below the horizon. The sun does not appear again until it rises in March. The pattern at the South Pole is the same, but 6 months earlier (or later).

19. The diagram below represents observations of the sunrise position of the sun for an observer in New York State.

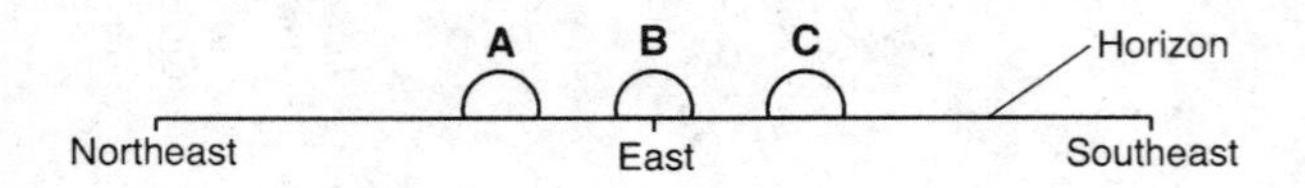

Which list below correctly identifies these sunrise positions?
(1) *A*—June 21, *B*—March 21, *C*—December 21
(2) *A*—December 21, *B*—March 21, *C*—June 21
(3) *A*—March 21, *B*—June 21, *C*—December 21
(4) *A*—June 21, *B*—December 21, *C*—March 21

20. The Coriolis effect causes most prevailing winds in New York State to curve (1) to the right in the direction of travel (2) to the left in the direction of travel (3) upward, away from Earth's surface (4) downward, toward Earth's surface

21. The diagram below represents a swinging pendulum suspended from a pivot point so it is free to rotate as it swings.

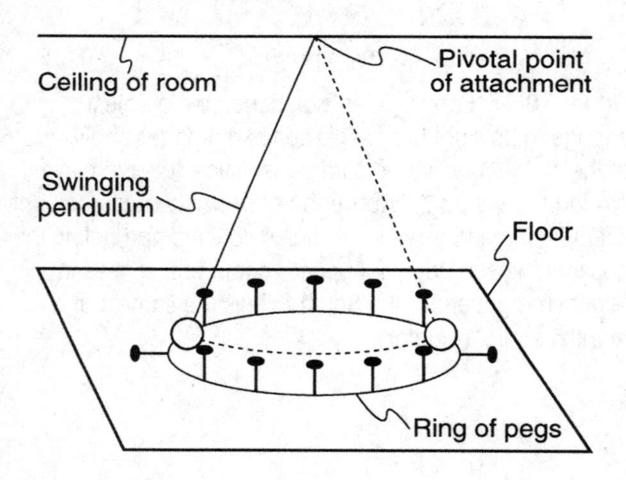

What property of our planet does this apparatus best show? (1) Earth's curved surface (2) Earth's revolution (3) Earth's rotation (4) the tilt of Earth's axis

22. The following diagram represents Earth. The dotted line is Earth's axis.

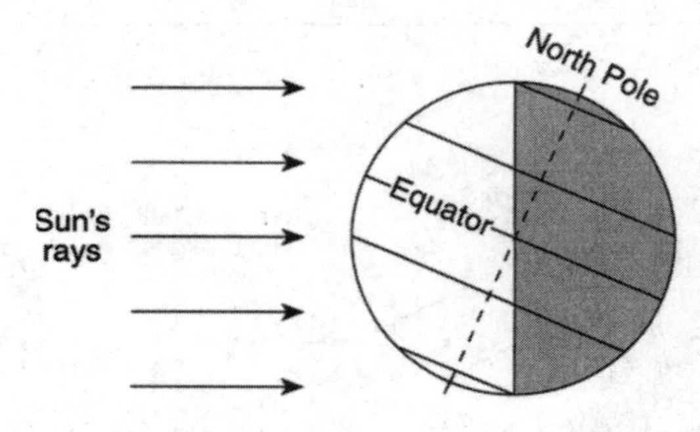

What month is best represented in this diagram? (1) March (2) June (3) September (4) December

23. Earth's axis is tilted $23^1/_2$° with respect to a perpendicular to the plane in which Earth orbits the sun. If that tilt were only 10°, how would it affect New York State? (1) Spring and fall would be cooler. (2) Spring and fall would be warmer. (3) Winter would be cooler. (4) Summer would be cooler.

24. Which position best represents Earth at the autumnal equinox in New York State?

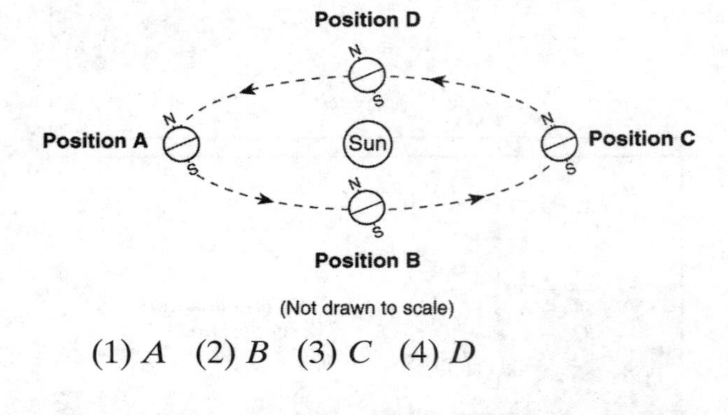

(Not drawn to scale)

(1) *A* (2) *B* (3) *C* (4) *D*

Part B

Base your answers to questions 25 through 31 on the diagram below, which represents three paths of the sun for an observer located in North America.

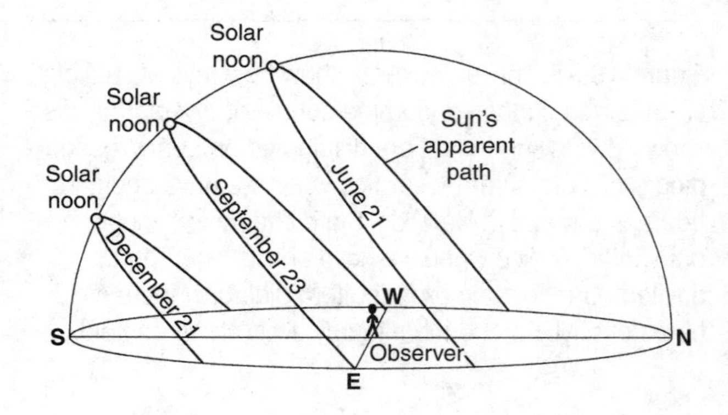

25. What is the direction of sunrise on June 21? (1) northwest (2) northeast (3) southwest (4) southeast

26. How many hours occurred between sunrise and solar noon on June 21? (1) about 5 hours (2) exactly 6 hours (3) about 7 hours (4) exactly 12 hours

27. What is the direction of the sun at solar noon? (1) due east (2) southeast (3) due south (4) southwest

28. Which diagram below best shows the position of Polaris in the sky for this observer?

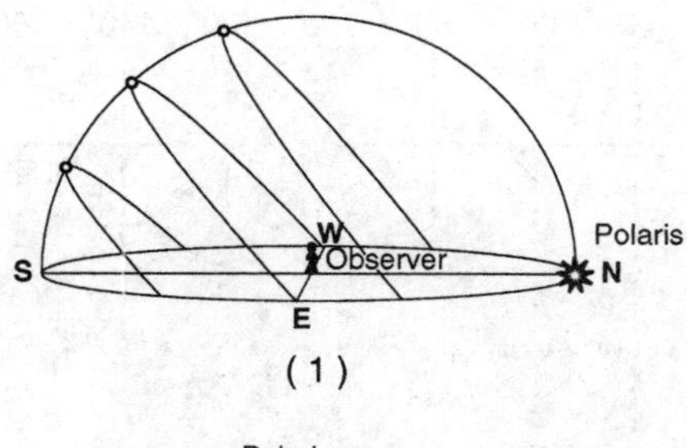

(1)

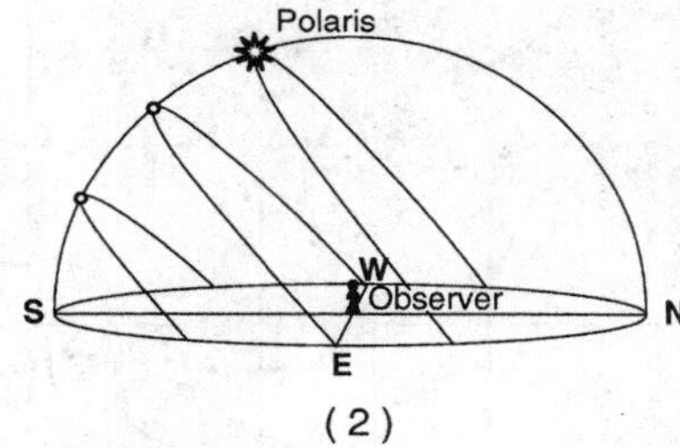

(2)

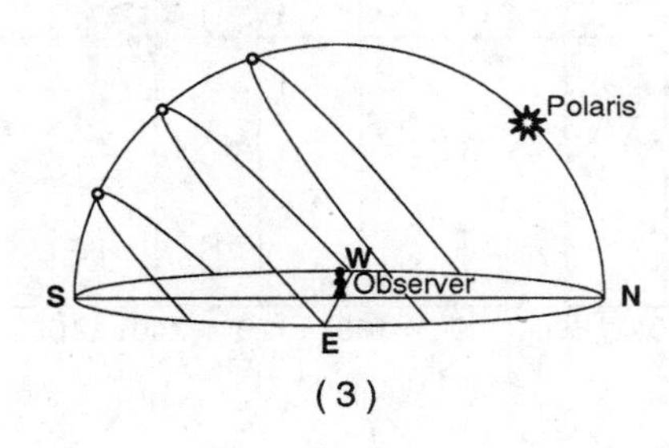

(3)

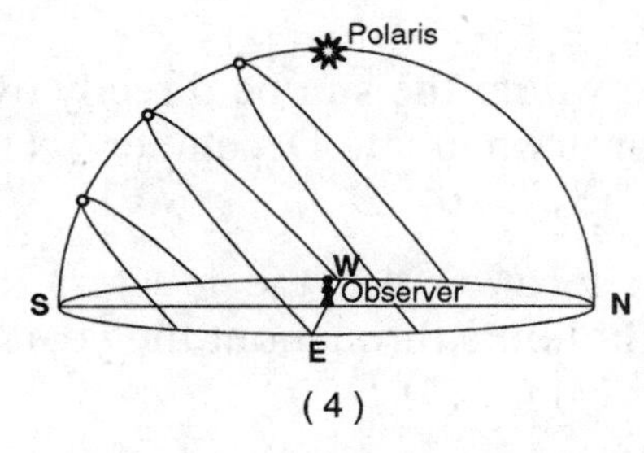

(4)

29. The path of the sun through the sky would always be the same as the September 23 path if (1) Earth's axis did not tilt $23^1/_2°$. (2) Earth did not rotate on its axis. (3) Earth did not revolve around the sun. (4) Earth were farther from the sun than Mars.

30. Why is the path of the sun called an apparent path? (1) The path of the sun changes throughout the year. (2) The sun always moves through the sky from east to west. (3) The sun is sometimes hidden by clouds. (4) The motion of the sun that we see is caused by Earth's rotation.

31. If on December 21 this observer traveled due south 1000 kilometers from New York City, how would the length of the day from sunrise to sunset change? (1) The daylight period would be longer and the noon sun would be higher in the sky. (2) The daylight period would be longer and the noon sun would be lower in the sky. (3) The daylight period would be shorter and the noon sun would be higher in the sky. (4) The daylight period would be shorter and the noon sun would be lower in the sky.

Base your answers to questions 32 through 35 on the diagram below, which represents Earth orbiting the sun. Four Earth positions (A, B, C, and D) are shown as well as Earth's axis of rotation.

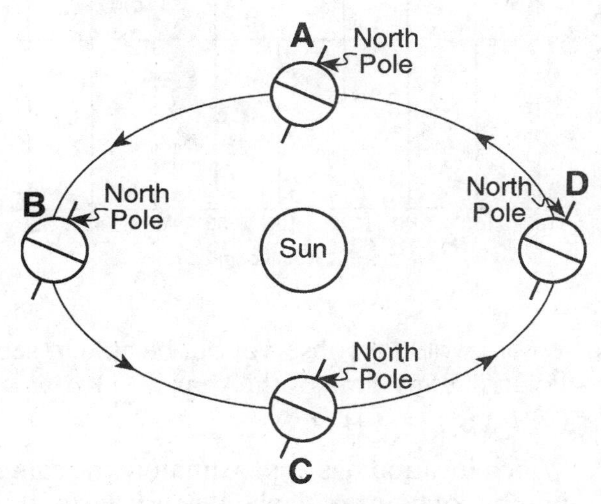

(Not drawn to scale)

32. Which Earth position represents the spring equinox? (1) *A* (2) *B* (3) *C* (4) *D*

33. Approximately how far does Earth move in its orbit over a period of one month? (1) 10° (2) 20° (3) 30° (4) 40°

34. Which Earth position represents the time when the sun is highest in the sky at the South Pole? (1) *A* (2) *B* (3) *C* (4) *D*

35. Where does the vertical ray of sunlight strike Earth's surface at position *A*? (1) $23^1/_2$° S (2) 0° (3) $23^1/_2$°N (4) 90°N

Base your answers to questions 36 through 39 on the map below, on which locations A, B, C, and D are shown.

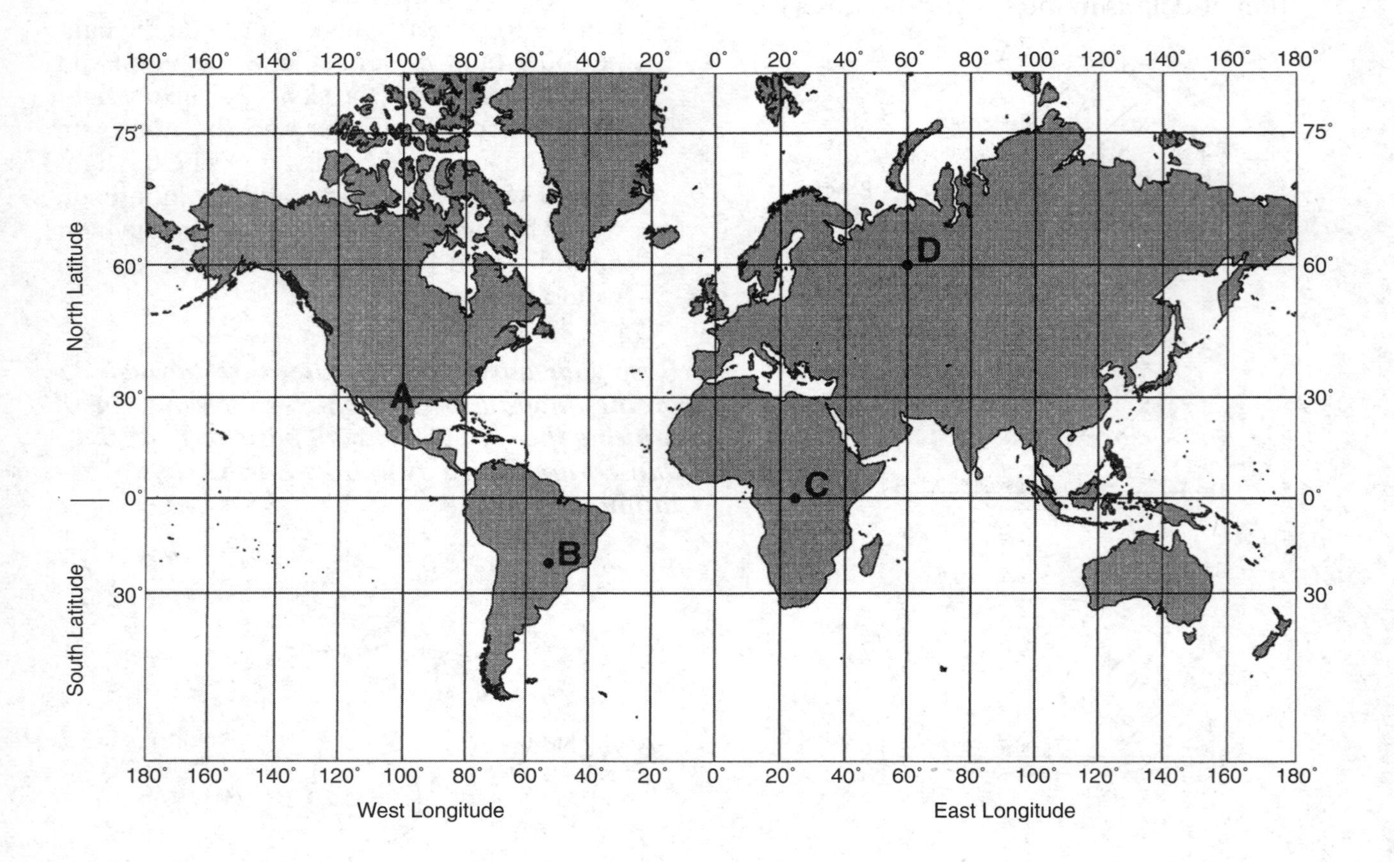

36. Where would an observer *not* be able to see Polaris at any time of the year? (1) *A* (2) *B* (3) *C* (4) *D*

37. Which location has approximately the same number of hours of daylight every day of the year? (1) *A* (2) *B* (3) *C* (4) *D*

38. Where would the sun be directly overhead at solar noon in late December? (1) *A* (2) *B* (3) *C* (4) *D*

39. Which location has the greatest range of daylight hours throughout the year? (1) *A* (2) *B* (3) *C* (4) *D*

Base your answers to questions 40 through 45 on the map below. Shading on the map indicates the locations on Earth that are dark and locations in sunlight at a particular time of the day and the year.

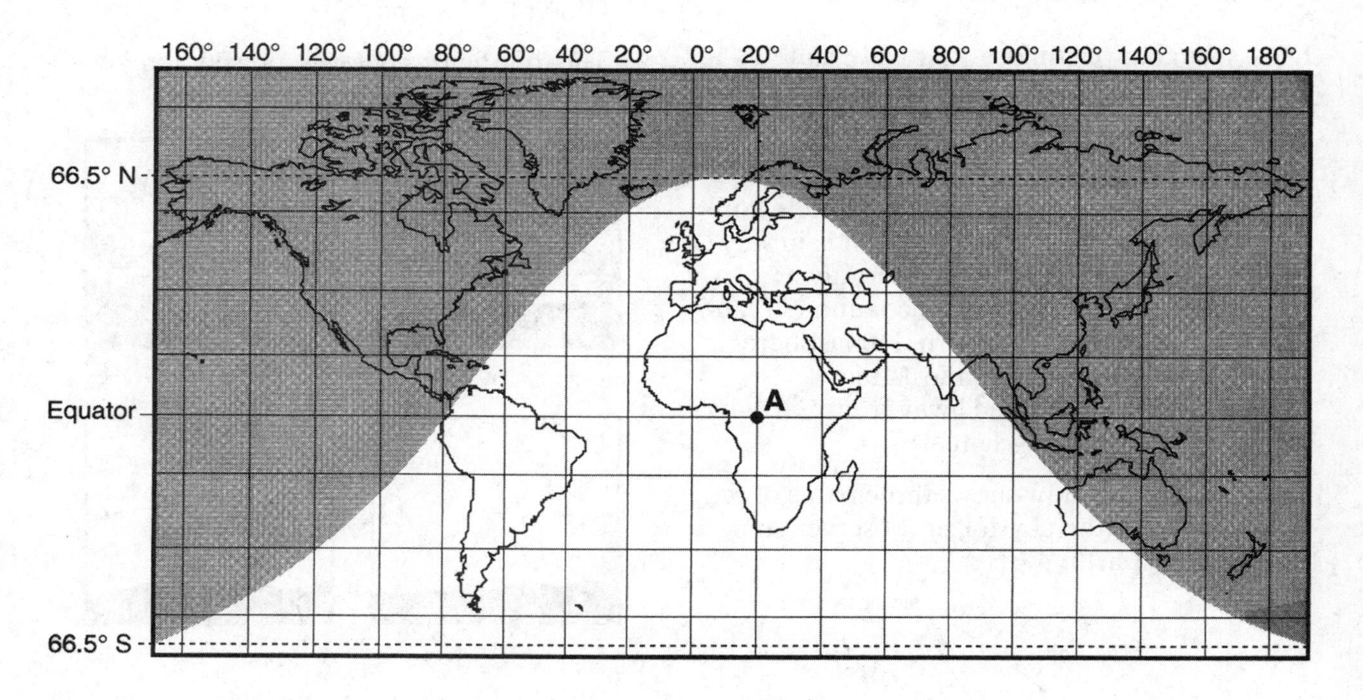

40. Which diagram below best shows the incoming rays of sunlight on the date of the map above?

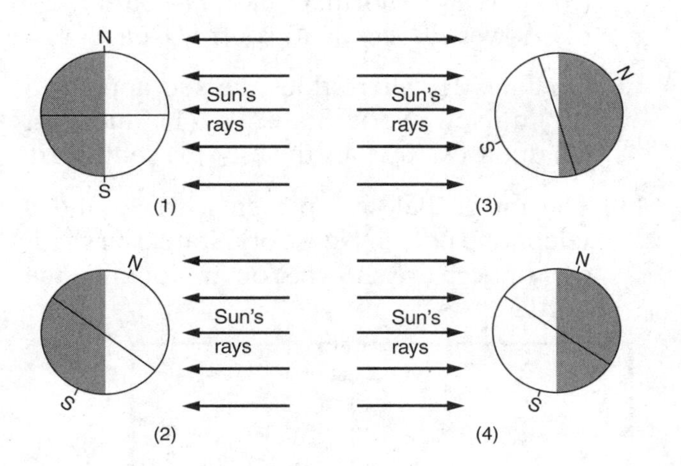

41. Approximately how many hours of daylight occur on this date at position *A*? (1) 6 (2) 9 (3) 12 (4) 15

42. Approximately how many hours of daylight occur on this date at the South Pole? (1) 0 (2) 9 (3) 12 (4) 24

43. What change in seasons is occurring at this date for the residents of New York State? (1) winter to spring (2) spring to summer (3) summer to autumn (4) autumn to winter

44. In the next few hours, daylight will end and the sun will go below the horizon for observers in (1) New York State (2) South America (3) western Australia (4) southern Africa

45. What is the name of the dotted horizontal line that runs near the top of this map? (1) Arctic Circle (2) Antarctic Circle (3) Tropic of Cancer (4) Tropic of Capricorn

Problem Solving

Make a graph of the changing length of the shadow of a vertical meterstick from just after sunrise until just before sundown. As an added challenge, construct a graph to show the direction in which the shadow is cast. (You may elect to use a different kind of graph rather than the standard rectangular grid.)

Chapter Review Questions

PART A

1. A swinging pendulum that can rotate as it swings can be used to show Earth's (1) shape (2) rotation (3) revolution (4) reflectivity

2. Why is the weather in New York State consistently cooler in January than it is in July? (1) Earth is closer to the sun in July. (2) Cloudy weather is more common in January. (3) The Northern Hemisphere is tilted away from the sun in January. (4) Sunlight is reflected away from Earth by ice and snow during January.

3. The diagram below shows three paths of the sun through the sky for an observer somewhere on Earth.

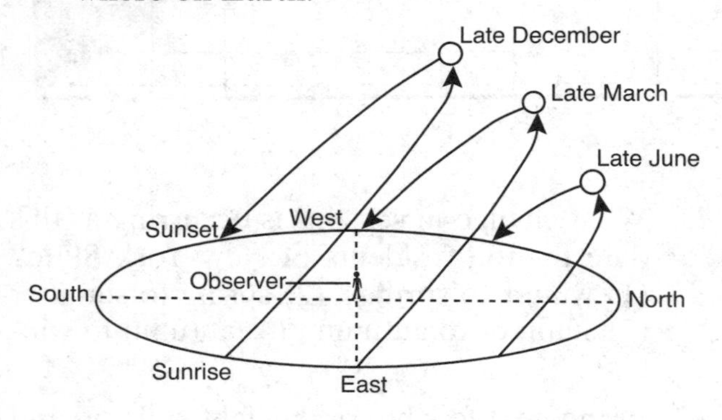

What is the location of this observer? (1) 60°N (2) 30°N (3) 30°S (4) 60°S

4. As Earth orbits the sun, through approximately how many degrees does Earth revolve each day? (1) 1° (2) 15° (3) 24° (4) 30°

5. In what month is the intensity of insolation the greatest in Australia? (1) March (2) June (3) September (4) December

6. From January 1 until December 31 of the same year, how many times does the sun's perpendicular ray cross the equator? (1) 1 (2) 2 (3) 3 (4) 4

7. On the spring equinox, an observer measures the angle from the southern horizon up to the noon sun. That angle was 30°. What is the latitude of the observer? (1) 30°S (2) 0° (3) 30°N (4) 60°N

8. Which New York State location would experience sunrise first on the equinox? (1) Riverhead (2) New York City (3) Buffalo (4) Massena

9. The four figures below represent long-exposure photographs of the night sky.

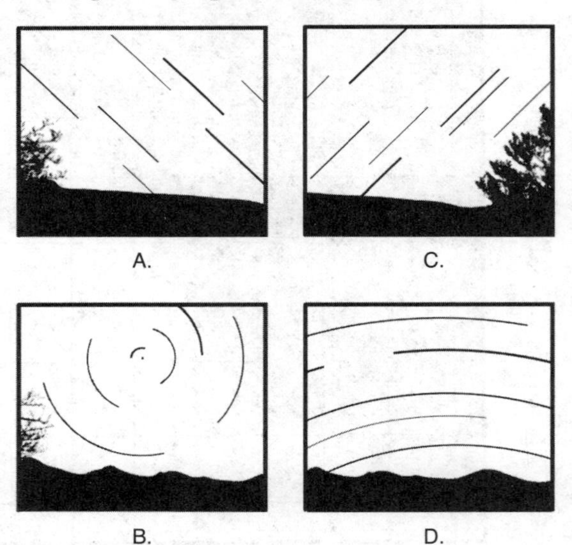

Which choices below correctly match each figure with the direction in which the camera was pointed?
(1) *A*–north, *B*–south, *C*–east, *D*–west
(2) *A*–south, *B*–north, *C*–east, *D*–west
(3) *A*–west, *B*–north, *C*–east, *D*–south
(4) *A*–west, *B*–north, *C*–south, *D*–east

10. On June 21, where does the sun appear to set in New York State? (1) due east (2) due west (3) northwest (4) southwest

11. The image below represents a tree and a telephone pole in New York State. The shadows of both are shown. You are looking due north.

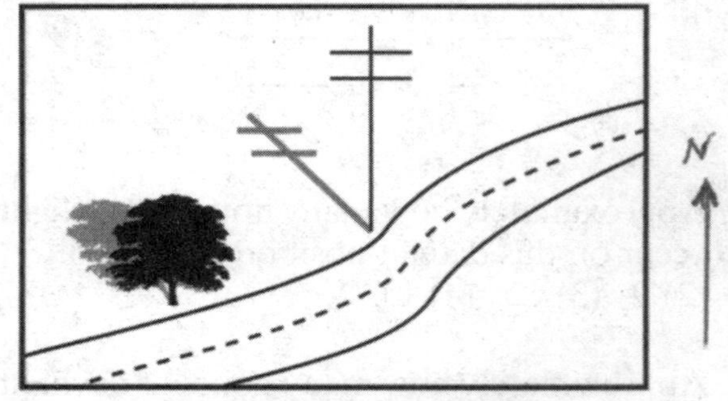

Shadow Cast on March 21

What time does this image best represent? (1) 6 A.M. (2) 10 A.M. (3) 2 P.M. (4) 6 P.M.

PART B

Base your answers to questions 12 through 18 on the diagram below, which shows the position of the noon sun as observed from different latitudes on Earth. Letters A through D represent specific locations on Earth.

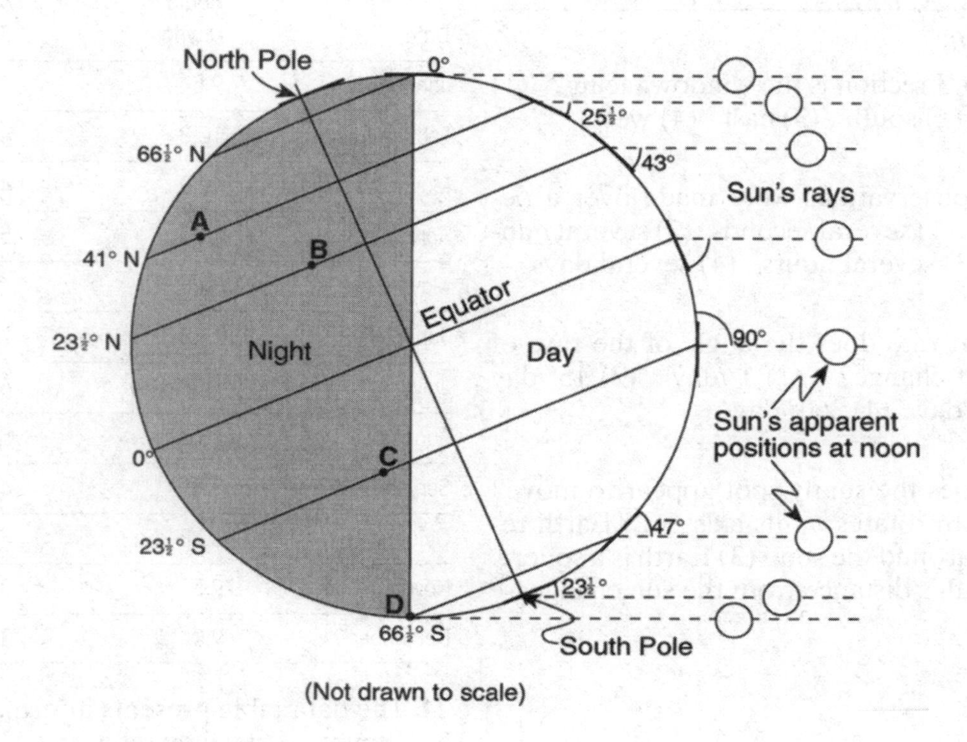

(Not drawn to scale)

12. Which location will experience the shortest daylight period on this date? (1) *A* (2) *B* (3) *C* (4) *D*

13. At the equator, what is the angle from the horizon up to the sun on this date? (1) 23½° (2) 43° (3) 66½° (4) 90°

14. What is the range of altitude of the sun on this date at location *D*? (1) 0° (2) 23½° (3) 47° (4) 66½°

15. How many hours of daylight will a person on the equator experience on this day? (1) 9 hours (2) 12 hours (3) 15 hours (4) 24 hours

16. On this date, what is the angular altitude of the noon sun in New York City? (1) 0° (2) 25½° (3) 47° (4) 66½°

17. Three months after this date, the line between day and night will run from the North Pole straight to the South Pole. Why will this change occur? (1) Earth's axis will point in a different direction in space. (2) Earth rotates on its axis. (3) Earth will move closer to the sun. (4) Earth will move in its orbit around the sun.

18. What is the local time at point *A*? (1) 12 midnight (2) 2 A.M. (3) 6 A.M. (4) 10 A.M.

Base your answers to questions 19 through 22 on the diagrams below, which show light coming through the same window at different times on the same day. These observations were made in New York State.

19. In what direction is the window facing? (1) north (2) south (3) east (4) west

20. These observations were made over a period of (1) several seconds (2) several minutes (3) several hours (4) several days

21. At what rate does the angle of the rays of sunlight change? (1) 1°/day (2) 15°/day (3) 90°/day (4) 360°/day

22. Why does the sunny spot appear to move? (1) Earth rotates on its axis. (2) Earth revolves around the sun. (3) Earth is a sphere. (4) Earth's distance from the sun changes.

PART C

23. The diagram below shows the length and position of a shadow of a flagpole on the nearby level surface.

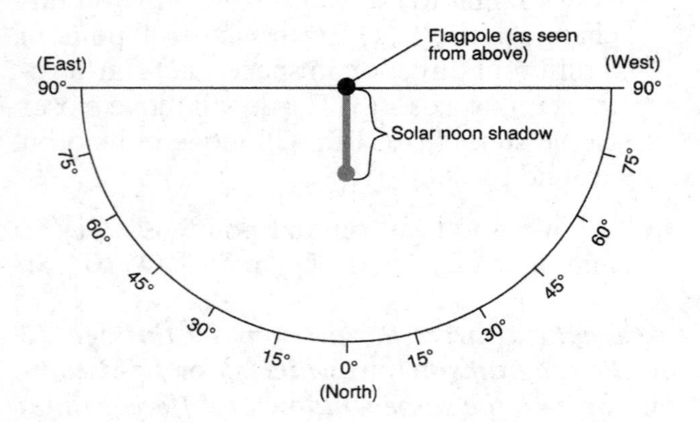

The shadow is observed again 3 hours later. Describe both the angular direction of the shadow and its length compared with the noon shadow.

Base your answers to questions 24 and 25 on the data table and diagram below.

Data Table

Date	Hours of Daylight	Altitude of the Sun at Noon (°)
January 21	9.5	32.3
February 21	10.8	40.1
March 21	12.0	47.3
April 21	13.7	55.1
May 21	14.8	62.5
June 21	15.3	70.4
July 21	14.8	63.3
August 21	13.7	55.5
September 21	12.1	47.7
October 21	10.8	39.9
November 21	9.5	32.1
December 21	9.0	24.4

24. The data table presents information about hours of daylight and the altitude of the noon sun on the 21st of each month near Buffalo, New York. Use the axes below to draw a line graph that shows the general relationship between the altitude of the sun and the total number of hours of daylight throughout the year at Buffalo.

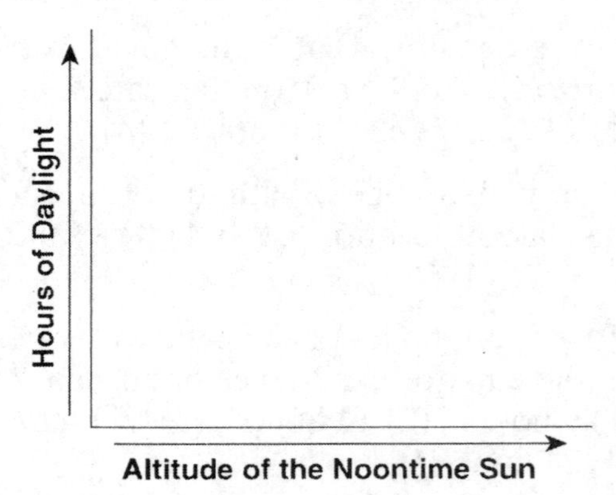

25. The following sky model chart shows one path of the sun through the sky for an observer in Buffalo, New York. Use the diagram to draw a line to show the apparent path of the sun from sunrise to sunset on

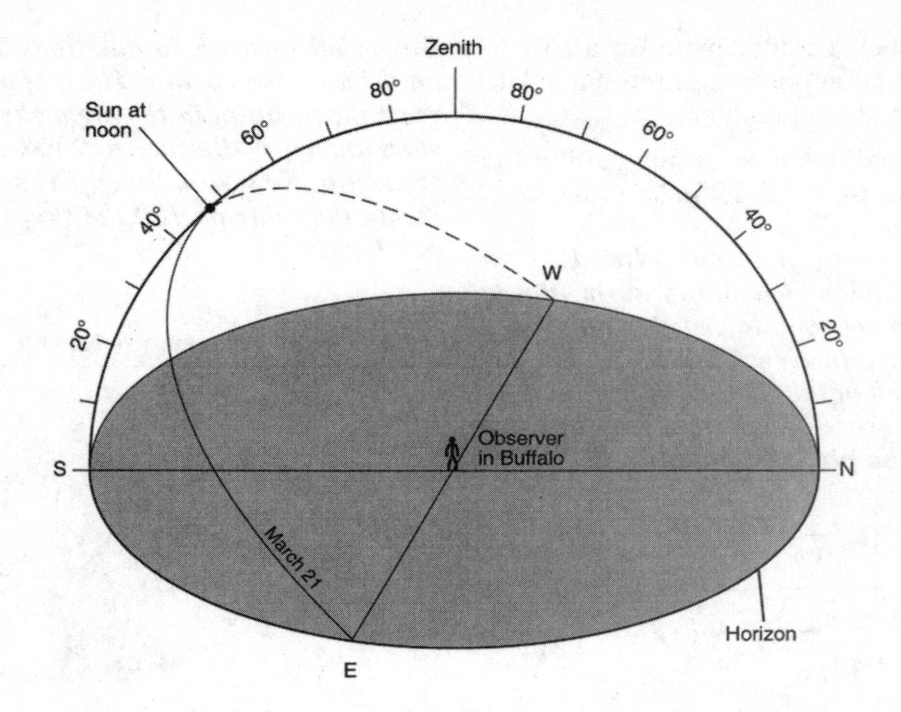

May 21 at this location. Be sure that your line conforms to the data table above.

26. On the same sky model above, place an asterisk (*) to show the position of the North Star for this observer.

27. How does the total number of hours of daylight during a period of one year at the Arctic Circle compare with the total number of hours of daylight during a period of one year at the Equator?

Base your answers to questions 28 through 30 on the map below, which shows three cities in the United States.

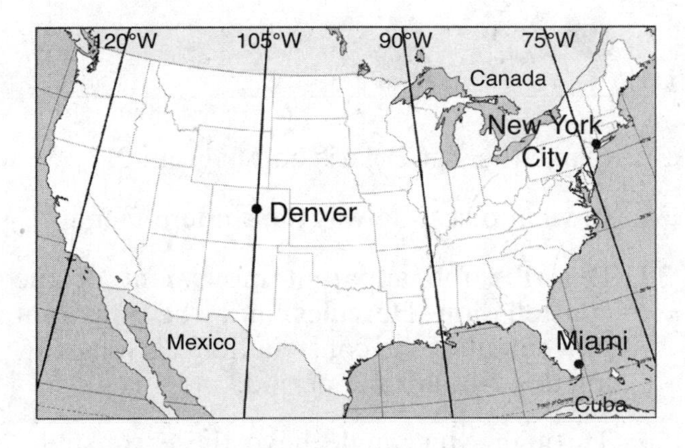

28. If it is solar noon in New York City, what is the approximate solar time in Denver?

29. Near the end of June, how does the number of daylight hours in Miami compare with the number of daylight hours in New York City?

30. Explain why the mid-day insolation is stronger in Miami than it is in New York City.

Base your answers to questions 31 through 33 on the diagram below, which represents an observer in New York State. Two paths of the sun through the sky are labeled "Late December Sun Path" and "Late March and September Sun Paths." A dotted line shows the position in the sky straight overhead from the observer.

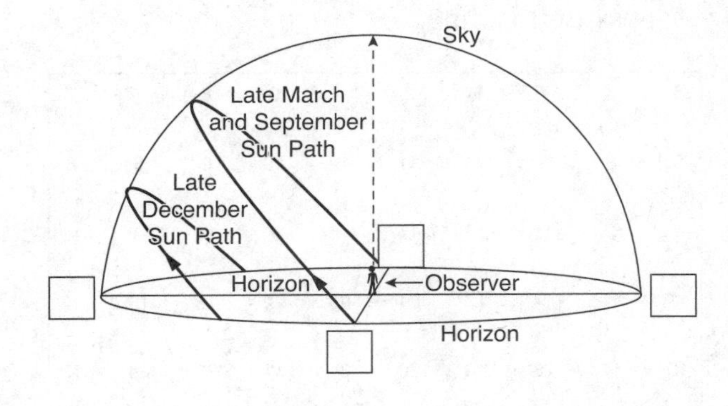

31. Inside each of the square boxes along the horizon on the diagram, write the appropriate letters of direction; *N* for North, *S* for South, *E* for East, and *W* for West.

32. Add the label, 12 noon with an arrow to show the 12-noon position of the sun in late March and Late September.

33. Draw a curved line to show the path of the sun from sunrise to sunset in late June.

Base your answers to questions 34 and 35 on the diagram below, which shows a pin placed vertically into a rectangular card resting on a horizontal surface in New York State. The direction and length of the shadow of the pin have been drawn for three different times on the equinox. The card and pin were not moved.

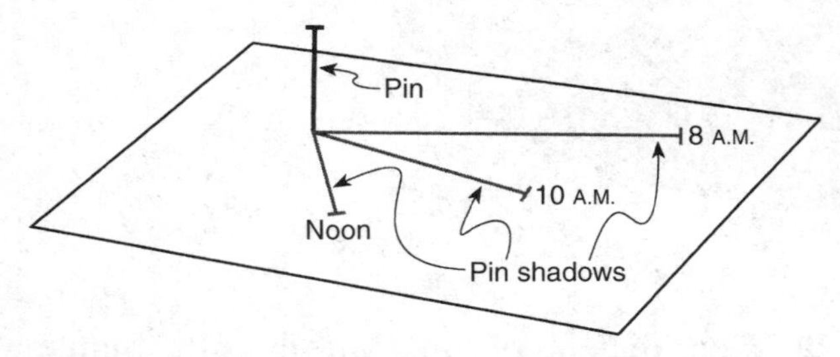

34. On the diagram, draw the shadow that would be cast by the pin at 2 P.M. on the same day.

35. The same procedure was performed at the winter solstice. How would the direction and length of the noon shadow on December 21 compare with the direction and length of the noon shadow shown in the diagram above?

36. The figure below shows information about the relative brightness of the sky just before sunrise. Each number marks the place where the brightness unit data was measured. Use this figure to draw the 25-brightness-unit isoline.

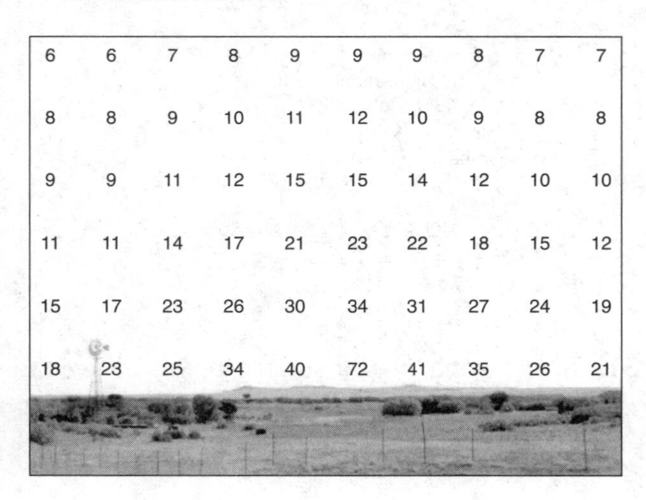

Base your answers to questions 37 through 41 on the diagrams below. The first diagram represents the position in the night sky of prominent stars and constellations at 9 P.M. as viewed by a student in New York State. The second diagram shows the same part of the sky two hours later at 11 P.M.

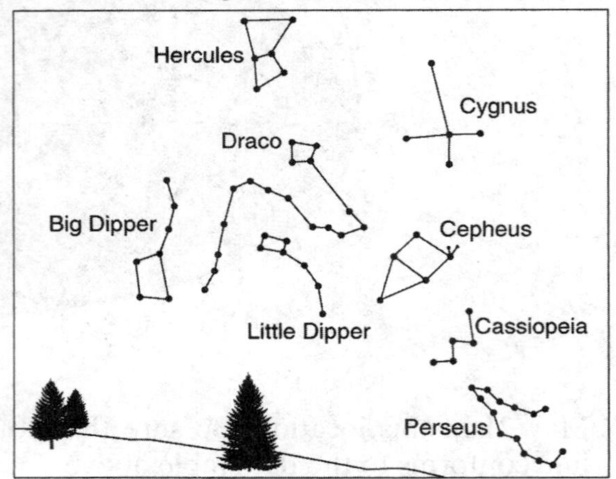

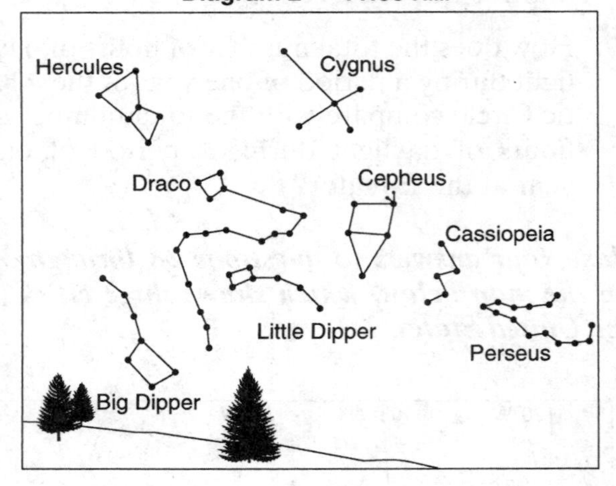

37. Circle the star Polaris on Diagram 2.

38. In what direction was the student facing?

39. Describe the apparent movement of the constellation Hercules and the apparent movement of the constellation Perseus during this 2-hour time period.

40. Through what angle have these constellations apparently rotated between 9 P.M. and 11 P.M.?

41. What causes the apparent motion of these stars and constellations?

Base your answers to questions 42 through 45 on the diagram below, which represents observations of the path of the sun made by an observer somewhere on Earth.

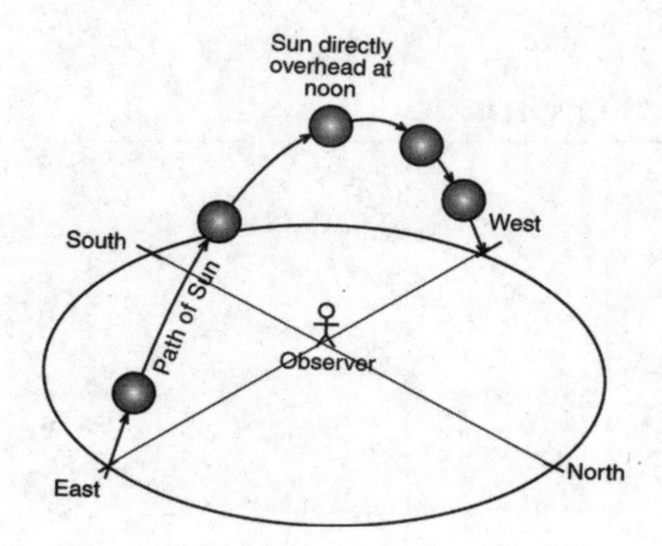

42. Where is the observer located?

43. What is the date of this observation?

44. What is the length of sunlight hours for this observer at this location and date?

45. What is the hourly rate of motion of the sun across the sky?

CHAPTER

Beyond Planet Earth

Vocabulary at a Glance

angular diameter
apparent diameter
asteroid
big bang
big crunch
comet
eccentricity
eclipse
ellipse
focus
galaxy
gravity
inertia
light-year
luminosity
major axis
meteor
meteorite
meteoroid
nuclear fusion
phase
primary
redshift
satellite
solar system
star
sunspot
tides

EARTH'S MOON

No other object in the night sky is as spectacular as Earth's moon. For thousands of years, the moon has inspired myths, songs, poems, and superstitions. Ancient ideas about the moon have been preserved in the English language. For example, the words *lunatic* and *lunacy* come from the illogical belief that moonlight could cause a person to become mentally unstable.

Phases of the Moon

The moon completes one revolution around Earth in a few days less than 1 month. During this trip, the lighted area of the moon appears to change shape. These **phases** (apparent change in shape) of the moon are caused by the changing relative positions of Earth, the sun, and the moon. When the moon and the sun are on the same side of Earth, the moon's dark side faces Earth. At this time, the illuminated side of the moon is not visible from Earth. This is the new moon phase. Figure 11-1 shows the way light and shadow on the orbiting moon appear from Earth. When the moon's position is opposite the sun, we see a fully lighted moon, the full moon. If the position of the moon and the sun are 90° apart, we see half of the moon lighted and half in shadow. At this time we see the quarter moon.

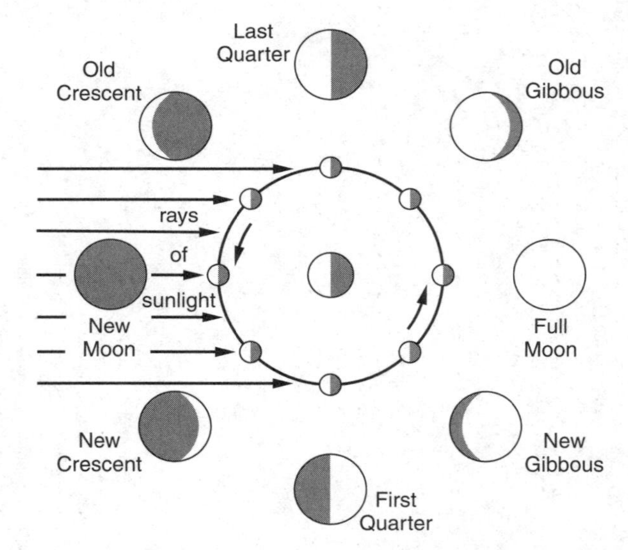

Figure 11-1. As the moon orbits Earth, the sun always illuminates one-half of the moon's surface. Viewed from Earth, the moon's phases are seen as different portions of the illuminated circle. At the new moon phase, the whole moon looks dark because the lighted half of the moon is facing the sun. The proportion of light and shadow on the moon changes in a monthly cycle.

Although one complete orbit of the moon around Earth takes about 27 days, a complete cycle of the moon's phases takes $29^1/_2$ days. This $2^1/_2$-day difference occurs because as the moon orbits Earth, Earth also orbits the sun. When the moon gets back to its original position, it must move through an extra angle of about 30° to compensate for Earth's orbital motion around the sun.

Like other celestial objects, the moon shows two types of motion. *Rotation* is spinning on an internal axis. Earth's rotation is what causes day and night. On the other hand, *revolution* is orbital motion. Earth's revolution (along with the tilt of Earth's axis) causes our annual cycle of the seasons. A good way to remember the difference is to recall that rotation is spelled with two *t*'s, and a *t*op is a *t*oy that rotates.

The moon's periods of rotation and revolution are equal. Because of this, the same side of the moon always faces Earth. Humans did not see the far side of the moon until the 1960s, when a Russian rocket orbited the moon and took pictures. As the first to see the moon's far side, the Russians named craters and other features found there.

Eclipses of the Moon and Sun

When lighted from just one direction, all objects cast a shadow. A shadow has two parts: the umbra and the penumbra. The umbra, the inner part, is the darker part. The penumbra, the outer part, is not as dark as the inner part. The sun shines on Earth and the moon; so they both cast shadows into space, as shown in Figure 11-2. An **eclipse** occurs when one celestial object casts its shadow on another celestial object.

Lunar Eclipse An eclipse of the moon (lunar eclipse) occurs when the full moon moves into Earth's shadow. During a lunar eclipse, the moon turns a coppery red. You can still see the moon because sunlight is bent (refracted) by Earth's atmosphere, which causes a weak illumination of the moon. If the orbits of Earth and the moon were in the same plane, there would be an eclipse of the moon every month. However, we know lunar eclipses are much more rare. This is because the moon's orbit is tilted at an angle of about 5° with respect to Earth's orbit as shown in Figure 11-3 on page 274. As a result, the moon usually passes above or below Earth's shadow.

Solar Eclipse An eclipse of the sun (solar eclipse) occurs when the new moon briefly moves between Earth and the sun. At this time, the moon casts its shadow on Earth as shown in Figures 11-2 and 11-3. Because of the tilt of the moon's orbit, we do not have a solar eclipse at each new moon. Precise observations of the

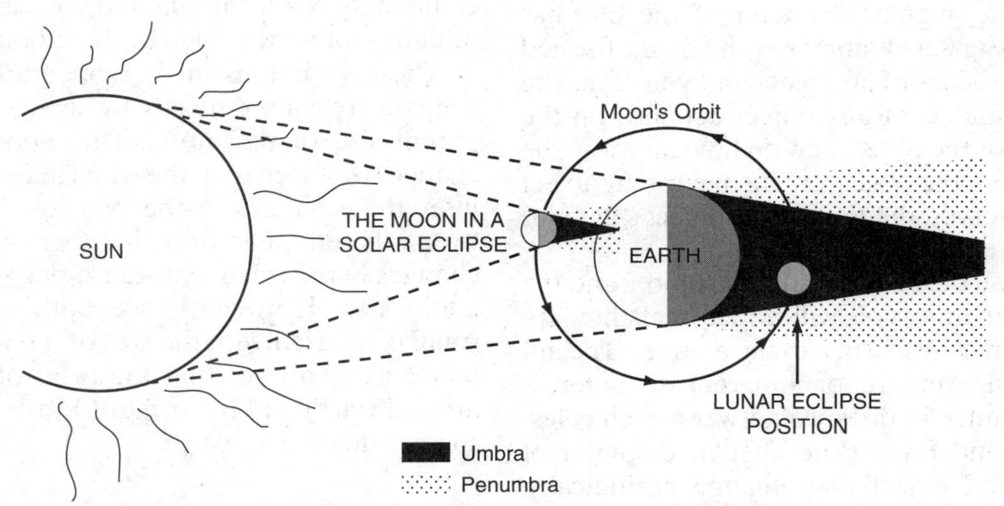

Figure 11-2 An eclipse of the sun and an eclipse of the moon as observed from a position high above Earth's North Pole. Eclipses of the sun are visible over only a small area on Earth's surface. However, eclipses of the moon are visible from all of Earth's night side. Note that this diagram is not to scale. At this Earth scale, the moon would be the same size but about 60 centimeters away, while the sun would be about 200 meters away and 2 meters in diameter.

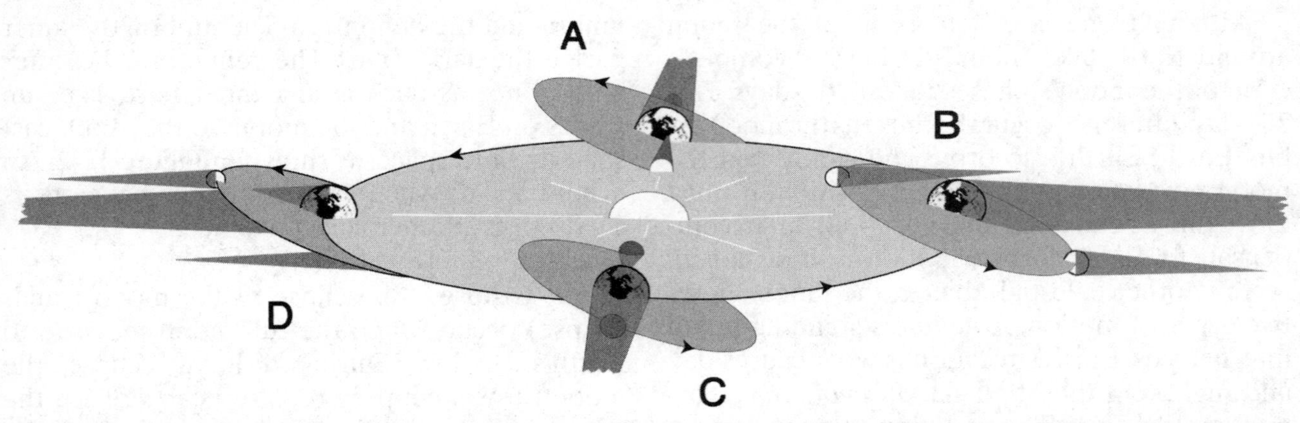

Figure 11-3. The orbit of the moon is tilted about 5° with respect to Earth's orbit. At positions *A* and *C*, the sun, Earth, and moon are in line, so eclipses can occur. However, usually the moon is above or below Earth's orbit during the new and full moon phases, as you see at positions *B* and *D*.

moon made over hundreds of years have enabled astronomers to predict its orbit many years into the future. These predictions include the timing of the eclipses and best places to view eclipses of the sun and the moon.

ANGULAR DIAMETER OF CELESTIAL OBJECTS

If you hold a ping-pong ball at arm's length, it looks rather small. However, if you hold the same ball close to your eye, it appears much larger. Although the size of the ball has not changed, the angular diameter of the ball has changed. **Angular diameter** is the angle formed between the sides of an object and your eye. The angular diameter of any object depends on the actual size of the object and on how far away the object is from the observer. The nearer an object is to an observer, the greater an object's angular diameter.

When you observe the sun, the moon, and the planets over time, you will notice that they appear to change in size in a cyclic manner. The apparent change in size, or **apparent diameter**, is due to a change in distance between each celestial object and Earth. The angular diameter of each celestial object may change periodically, but the real diameter, or size, of each celestial object stays the same.

The sun's angular diameter is slightly larger in winter and a little smaller in summer. These observations of the sun's angular diameter tell us that Earth is actually closest to the sun in January and farthest from the sun in July. This may be opposite what you might expect. Seasonal variations in temperature result from Earth's tilt (Earth's axis always pointing at the same direction in space), and Earth's revolution around the sun—not from changes in Earth-sun distance.

Angular Size and Shape of Orbit

It is a coincidence that as seen from Earth the apparent size of the moon is approximately equal to that of the sun. During many solar eclipses, the moon entirely covers the sun. However, during some eclipses the moon does not completely cover the sun and you can see a ring of light around the edge of the moon.

These variations in the appearance of solar eclipses are caused mostly by changes in the apparent size of the moon. The moon seems to change size more than the sun. Therefore, we can infer that changes in the relative distance between the moon and Earth are greater than the changes in the relative distance between the sun and Earth. However, since both changes are small compared with the size of the average distance, we can infer that the orbit of the moon around Earth and the orbit of Earth around the sun are nearly circular.

THE TIDES

People who live along the seashore are familiar with the twice-daily rise and fall of the oceans we call the **tides**. In most places, the difference be-

Figure 11-4. Most places along the oceans experience two complete cycles of the tides in a little more than 24 hours. A tidal range of about 1 meter is common. The Bay of Fundy in Eastern Canada has the world's greatest range of tides. In the left photo, notice the boy diving into deep water. In the right photo taken several hours later, the water has receded beyond the far end of the pier.

tween high tide and low tide is less than 1 meter (3 feet). The Bay of Fundy in eastern Canada has the world's greatest difference between high and low tide, which can exceed 15 meters (50 feet) as shown in Figure 11-4. The cause of the tides is the gravitational attraction of the moon and the sun. The sun and the moon pull on the water in the oceans and on the solid part of Earth. The water of the oceans is pulled toward the moon, which causes a high tide. On the opposite side of Earth, where the solid Earth is, in effect, pulled away from the oceans, inertia causes another high tide.

The highest high tides and the lowest low tides are called spring tides. They occur about twice a month near the full and new moon phases. In that alignment, the sun and moon pull in the same line. Figure 11-5 shows these positions of the sun, moon, and Earth. When the sun and moon are at right angles at the first and last quarter phases of the moon, the changes in water

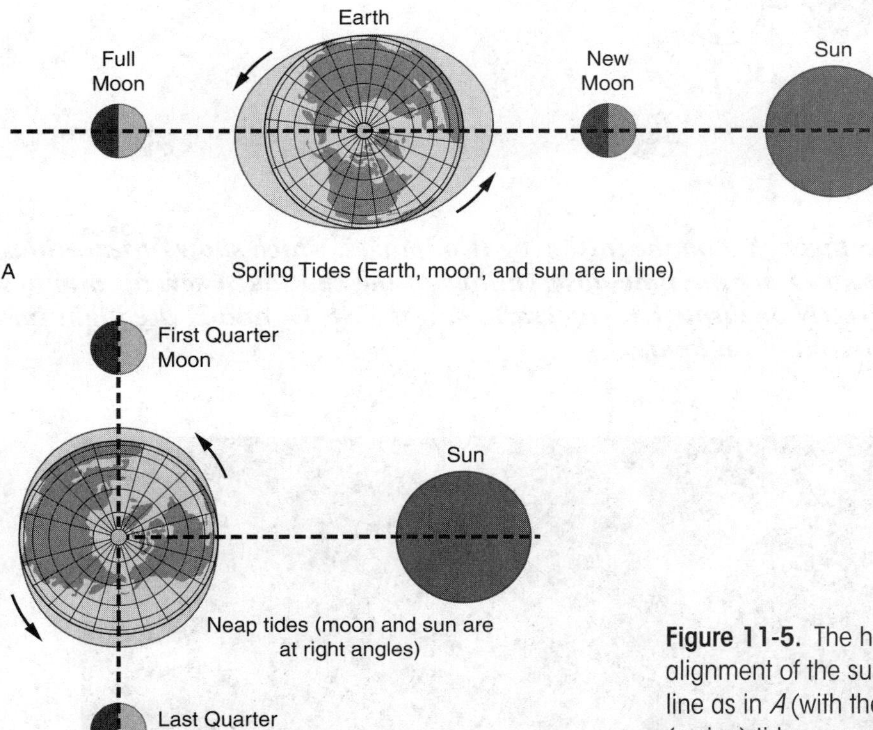

Figure 11-5. The height of the tides depends on the alignment of the sun, moon, and Earth. When they are in line as in *A* (with the moon on either side) the highest (spring) tides occur. When they form a right angle as in *B*, the smallest (neap) range of tides is observed.

level are smaller. These smaller tides are called the neap tides. Although the sun has many times the mass of the moon, the moon has more effect on the tides because it is so much closer to Earth.

QUESTIONS

Part A

1. How far does Earth move in its orbit during one full cycle of the phases of the moon? (1) 1° (2) 15° (3) 30° (4) 360°

2. Which diagram below best shows the monthly sequence of moon phases?

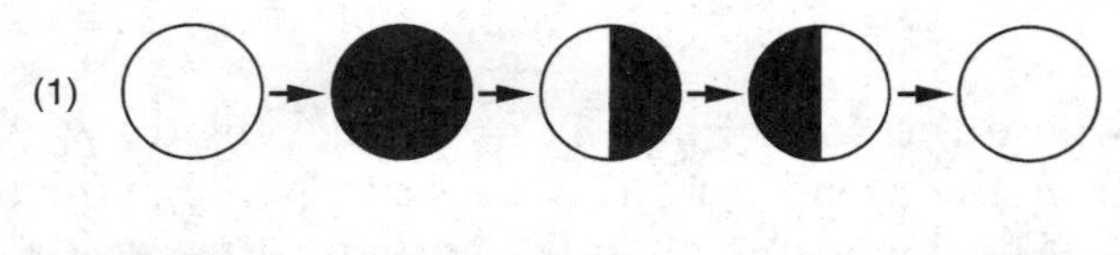

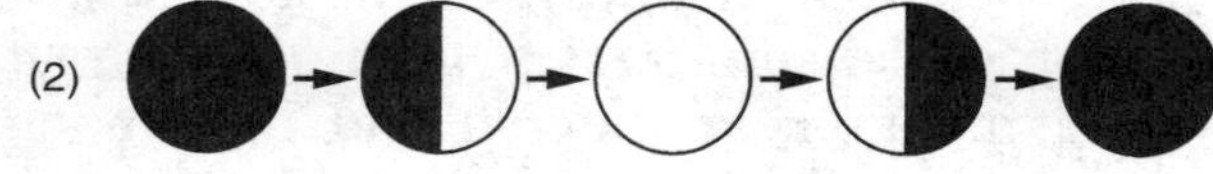

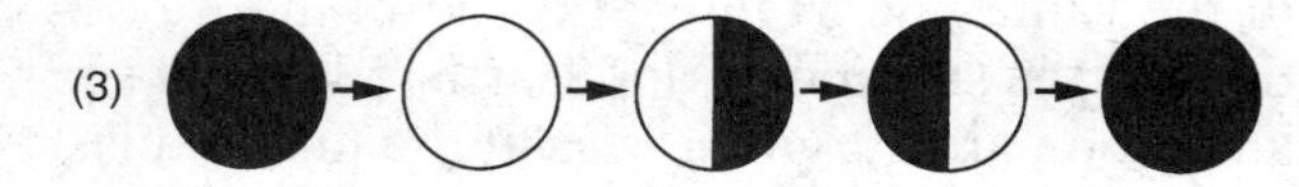

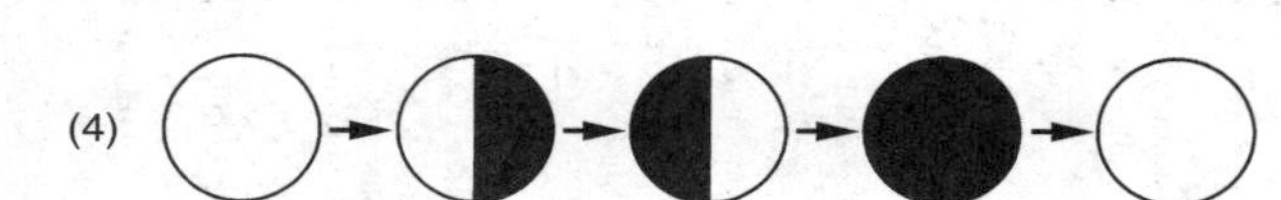

3. In most of Earth's ocean coastal locations, how long does it take to go through one full cycle of the tides? (1) approximately 3 hours (2) approximately 6 hours (3) approximately 12 hours (4) approximately 24 hours

4. Which sequence of moon phases could be observed from Earth over a 2-week period?

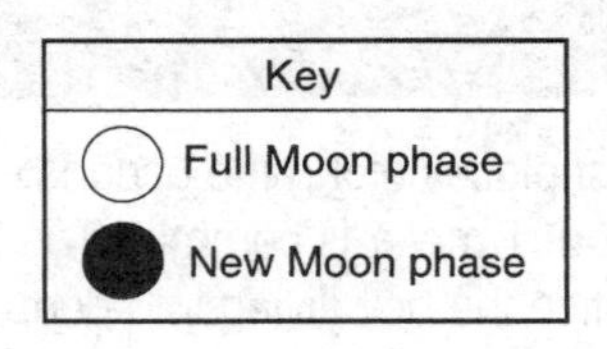

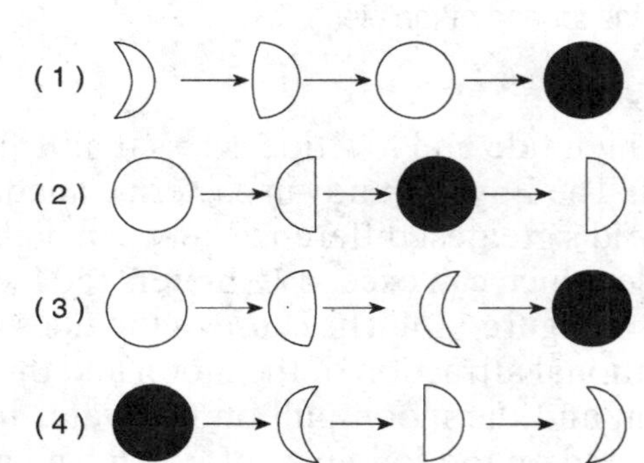

Part B

Base your answers to questions 5 through 9 on the following two images, which show observations of the sun and moon. A, B, C, and D are daytime observations of the sun taken several minutes apart. (Never look at the sun directly or through a camera.) Images E, F, G, and H are night images of the moon taken over a period of an hour.

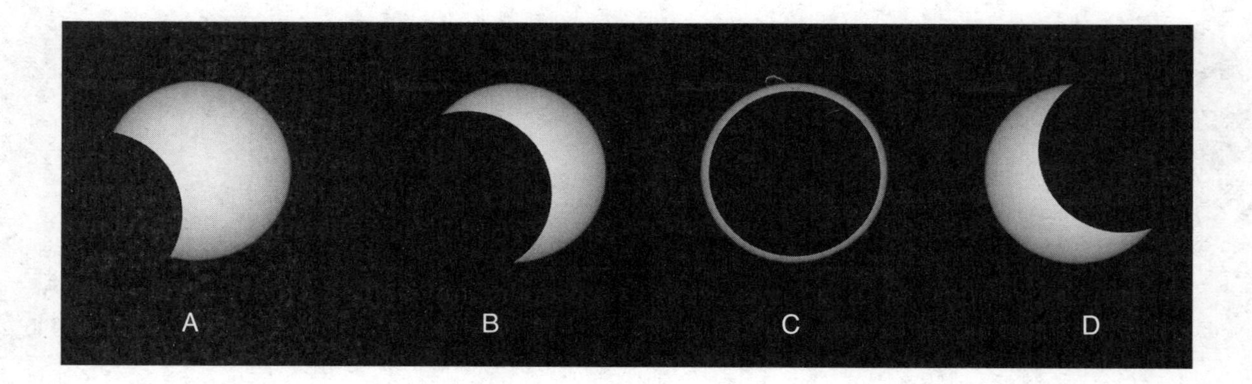

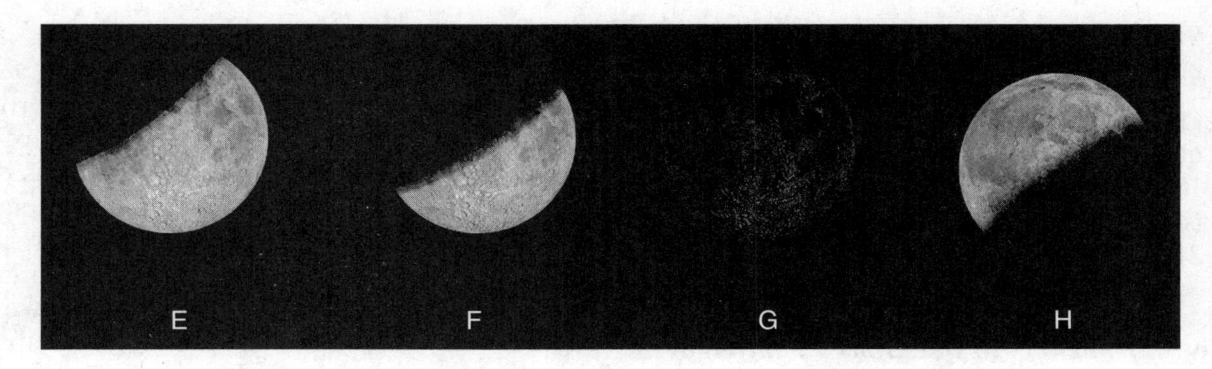

5. What events are shown in both of the previous figures? (1) changes in phase (2) eclipses (3) cloud cover (4) rising and setting of celestial objects

6. What do images *E* through *H* show about Earth and the moon? (1) The moon is hundreds of thousands of miles away from Earth. (2) The moon changes phases on a monthly cycle. (3) The orbit of the moon is nearly circular. (4) The diameter of the moon is smaller than Earth's diameter.

7. What is the moon phase in images *E* through *H*? (1) full moon (2) quarter moon (3) crescent moon (4) new moon

8. Which position of the moon best represents the relative positions of the sun, Earth, and moon for the event shown in *E* through *H*?

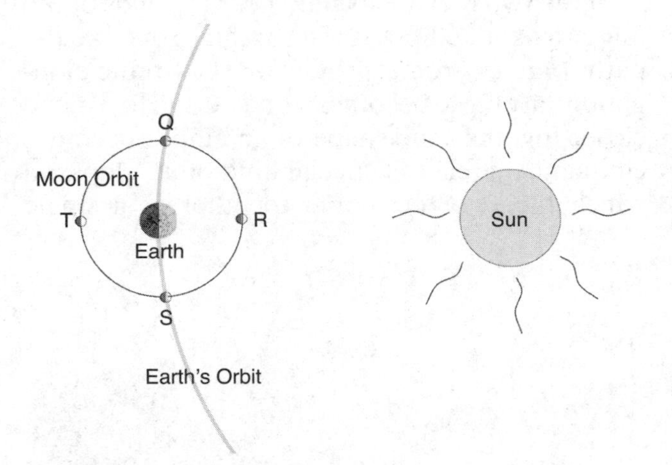

(1) *Q* (2) *R* (3) *S* (4) *T*

9. How does the moon appear in position *S* in the diagram above for observers on Earth? (1) The moon appears all dark. (2) The right half of the moon is lighted. (3) The left half of the moon is lighted. (4) The whole moon is brightly illuminated.

Base your answers to questions 10 through 12 on the diagram below, which shows the relative positions of the sun, Earth, and moon over a period of approximately 1 month.

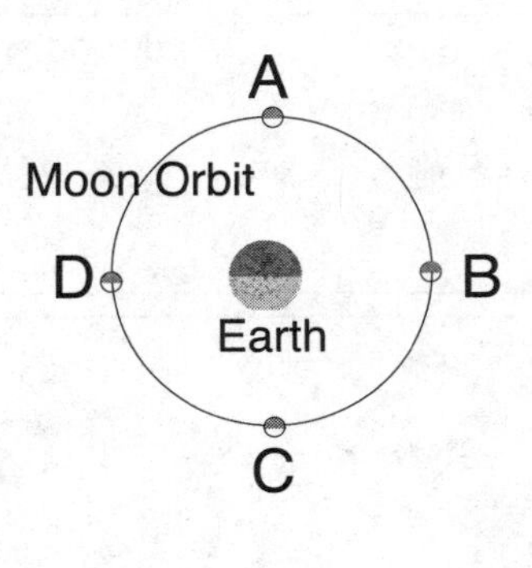

10. In which part of its orbit would the moon appear completely dark to observers on Earth? (1) *A* (2) *B* (3) *C* (4) *D*

11. What causes Earth's cycle of ocean tides? (1) gravity and the winds (2) gravity and inertia (3) inertia and the winds (4) the winds and the Coriolis effect

12. The graph below shows the highest and lowest ocean water levels on the island over the same month.

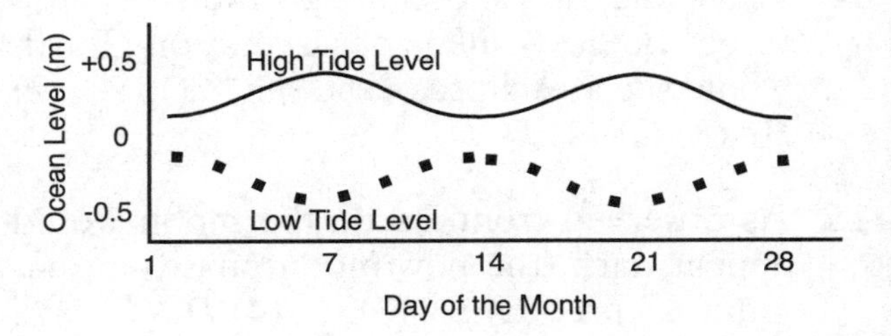

Which choice best represents the moon's positions during this month?

(1) *A* is day 1, *B* is day 7, *C* is day 21, *D* is day 28 (2) *A* is day 4, *B* is day 10, *C* is day 18, *D* is day 24 (3) *A* is day 7, *B* is day 14, *C* is day 21, *D* is day 1 (4) *A* is day 1, *B* is day 21, *C* is day 14, *D* is day 28

Base your answers to questions 13 through 15 on the four diagrams below, which show different relative positions of the sun, Earth, and moon.

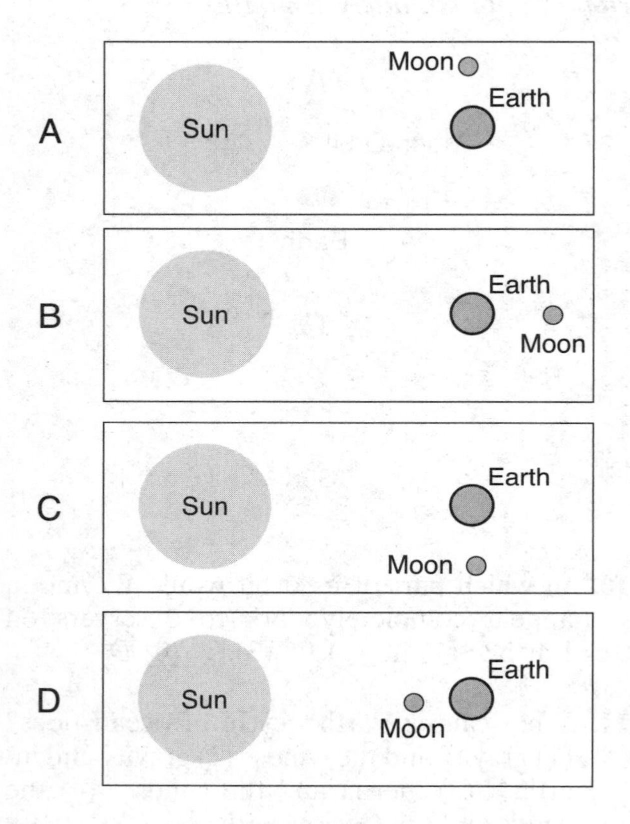

13. Which diagrams show the positions of the sun, Earth, and moon when the highest (spring) tides occur? (1) *A* and *C* (2) *B* and *D* (3) *C* and *D* (4) only diagram D

14. Which diagram shows the positions of these three bodies when observers on Earth would see an eclipse of the sun? (1) *A* (2) *B* (3) *C* (4) *D*

15. As observed from Earth, the moon would appear dark (the new moon phase) in diagram (1) *A* (2) *B* (3) *C* (4) *D*

THE GEOMETRY OF ORBITS

Planets that revolve around the sun have orbits that look like slightly flattened circles. The exact shape of an orbit is an ellipse. An **ellipse** is defined by two fixed points called the foci (singular: **focus**). These points (foci) lie on either side of the center of the **major axis**, which is a line through the widest part of an ellipse as shown in Figure 11-6. The orbits of all planets are ellipses with the sun at one focus. It is important to remember that the sun is not at the center of Earth's orbit.

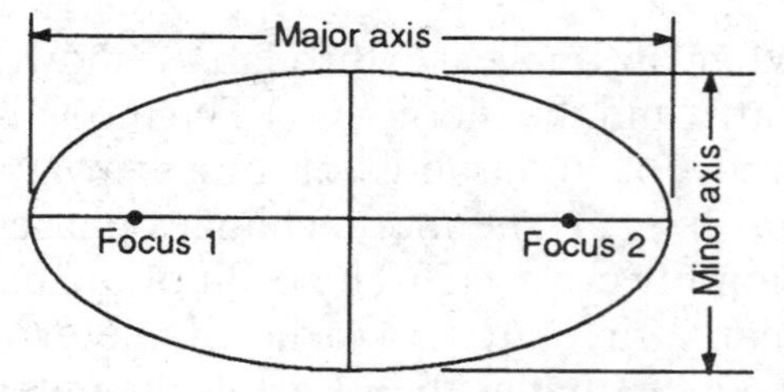

Figure 11-6. The major axis is a straight line through the widest part of the ellipse that passes through the two foci.

Figure 11-7 shows the procedure for drawing an ellipse. You need a piece of string, two pins, and a pencil. The pins are the foci of the ellipse that you draw.

If the two foci are located near the ends of the major axis, an ellipse is long and narrow, like the paths of many comets. At its most extreme elongation, an ellipse becomes a line. As the foci move closer together, the shape of an ellipse becomes circular. A circle is a special kind of an ellipse in which the two foci come together at a single

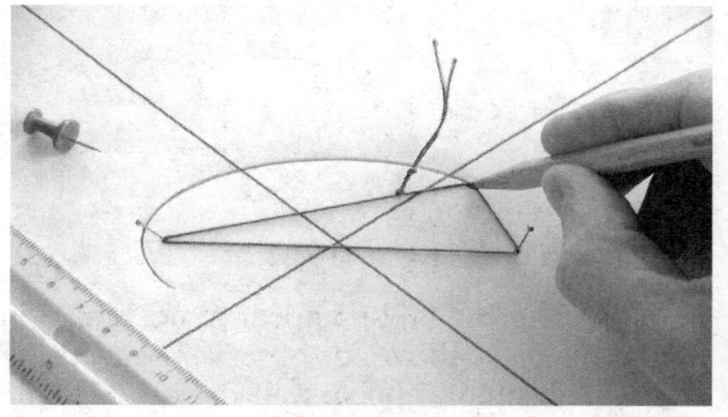

Figure 11-7. You can draw ellipse using a string held by two pins as shown here. Every ellipse has two foci, which in this ellipse, are the positions of the pins.

point. The orbits of the sun's planets are very close to circular, with the two foci close together.

To calculate the **eccentricity** (elongation) of any ellipse, use following formula found on page 1 of the *Earth Science Reference Tables*.

$$\text{eccentricity} = \frac{\text{distance between foci}}{\text{length of major axis}}$$

$$e = \frac{d}{l}$$

Sample Problem

1. Find the eccentricity of the ellipse in Figure 11-8. (The dots are two foci. You will need a centimeter ruler to measure the ellipse.)

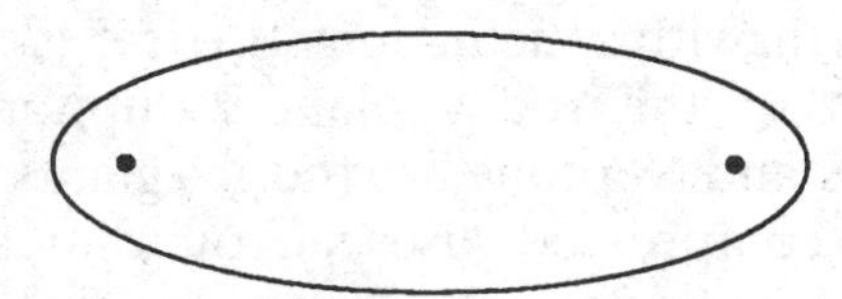

Figure 11-8.

Solution

The distance between the foci is 4 centimeters; the major axis is 5 centimeters.

$$e = \frac{d}{l}$$

$$= \frac{4 \text{ cm}}{5 \text{ cm}} = 0.8$$

(Eccentricity is a ratio and has no units.)

2. Calculate the approximate distance (d) between the foci of Earth's orbit.

Solution

Page 15 of the *Earth Science Reference Tables* lists the eccentricity (e) of Earth's orbit as 0.017 and the radius of Earth's orbit as 149.6 million km. Since the eccentricity is so small (it's nearly circular), we can approximate the length (l) of the major axis as twice the average Earth-sun distance: 149.6×10^6 km $\times 2 = 299.2 \times 10^6$ km.

$e = \frac{d}{l}$ (To isolate d multiply both sides of the equation by l.)

$e \times l = d$

$0.017 \times 299.2 \times 10^6 \text{ km} = 5.09 \times 10^6 \text{ km}$

(This is about 5 million kilometers, which is approximately 2% of the diameter of Earth's orbit.)

THE FORCE OF GRAVITY

Gravity is a force of attraction between objects. This force depends on the mass of each object and the distance between them. Increasing the mass of the objects increases the force. However, increasing the distance between them decreases the force. When you measure your weight, you are measuring the force of attraction between your body and Earth. If you want to lose mass you can eat less and exercise more so that there is less of you. This also would decrease your weight. However, if you could greatly decrease Earth's mass, this would also cause your weight to decrease. Of course, we do not have a way to significantly decrease Earth's mass.

Another way to decrease your weight is to increase your distance from Earth. If you were to become an astronaut and travel into space, this would decrease the gravitational force between you and Earth (your weight). Go far enough away, and you could become virtually weightless, but your body would still have the same mass.

Gravity and the Planets

Gravity is the force that holds the planets and other objects in the solar system in their orbits. Any object that orbits another object in space is a **satellite**. The object it orbits is called the **primary**. For example, Earth is a satellite of the sun; the sun is Earth's primary. The moon is Earth's satellite; Earth is the moon's primary. Because orbits are not perfectly circular, the distance between Earth and the sun changes as Earth revolves around the sun. We generally ignore the gravitational attraction of the planets, because most of the mass of the solar system is found in the sun.

The planets move fastest in their orbit when they are closest to the sun and slowest when they are farthest from the sun. Mercury, the planet closest to the sun, travels about 1.6 times faster than Earth and 10 times the speed of Neptune, which is farthest from the sun. In fact, the orbital velocity of a planet depends only on its distance from the sun. Therefore, if astronomers know a

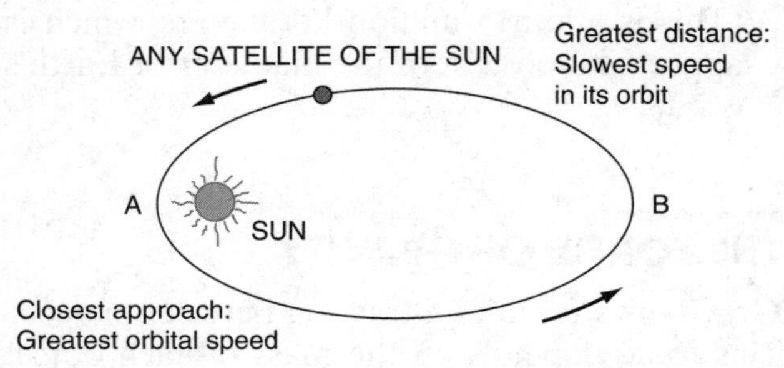

Figure 11-9. The speed of a planet varies with its distance from the sun. This is true for any satellite, including moons. When the satellite is closest, at position *A*, the gravitational attraction is the strongest, so the satellite moves the fastest in its orbit. The satellite slows as it moves toward *B* because its distance from the sun is increasing and the gravitational attraction is decreasing. This is a cyclic change.

planet's average distance from the sun, they can determine its average orbital velocity. The ellipse in Figure 11-9 is greatly exaggerated to show this change in distance. Although Earth's orbit is not a perfect circle, it is so close to one that the changing orbital speed of planet Earth, while measurable, is not noticeable to us.

The elliptical path of any satellite is a result of two factors: inertia and gravity. **Inertia** is the tendency of an object to remain at rest, or if it is moving, to move at constant speed in an unchanging direction. If a satellite has a circular orbit, inertia and the force of gravity are constant. There is no change in speed for this satellite, but there is a constant change in direction, which keeps the satellite in an orbital path. Gravity causes the direction to change continuously as the satellite moves in a circle. If the orbit is elliptical (eccentric), gravity also causes the speed to change; the satellite moves faster when it is near its primary, and slower when it is farther away. For example, the speed of Earth in its orbit around the sun changes slightly on an annual cycle. Earth orbits very slightly faster in the winter when it is a little closer to the sun.

THE SOLAR SYSTEM

Through radiometric analysis, as discussed in Chapter 6, we know that the solar system is about 4.6 billion years old. Although this might seem extremely old, the age of the solar system is only about one-third the age of the universe. The **big bang**, the rapid expansion of concentrated matter that created the universe, did not at first produce heavier elements. These building blocks of the sun's planets (and of our bodies) originated in the explosive deaths of giant stars. As the late astronomer Carl Sagan noted, we are all made of "star-stuff."

Planets of the Solar System

The **solar system** consists of eight planets and other objects that revolve around the sun. The eight planets can be divided into two groups: rocky (terrestrial) planets and gas giants. Mercury, Venus, Earth, and Mars are the rocky planets. Earth, with a mean density of 5.5 g/cm^3, is the densest of the rocky planets. Jupiter, Saturn, Uranus, and Neptune are the gas giants. They are mostly compressed gases surrounding a tiny liquid and/or rocky core. The average density of Saturn is only 0.7 g/cm^3. Figure 11-10 compares the sun and planets by size.

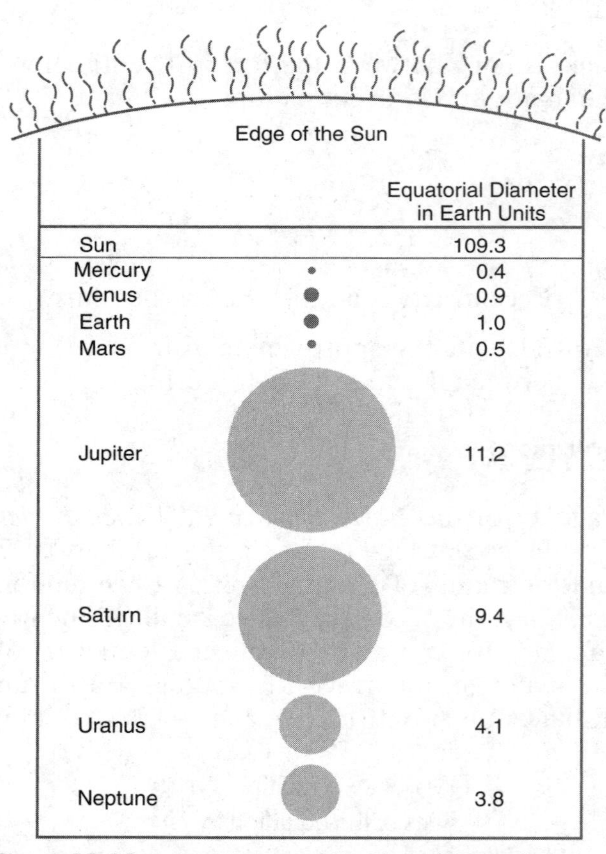

Figure 11-10. A comparison of the size of the major members of the solar system.

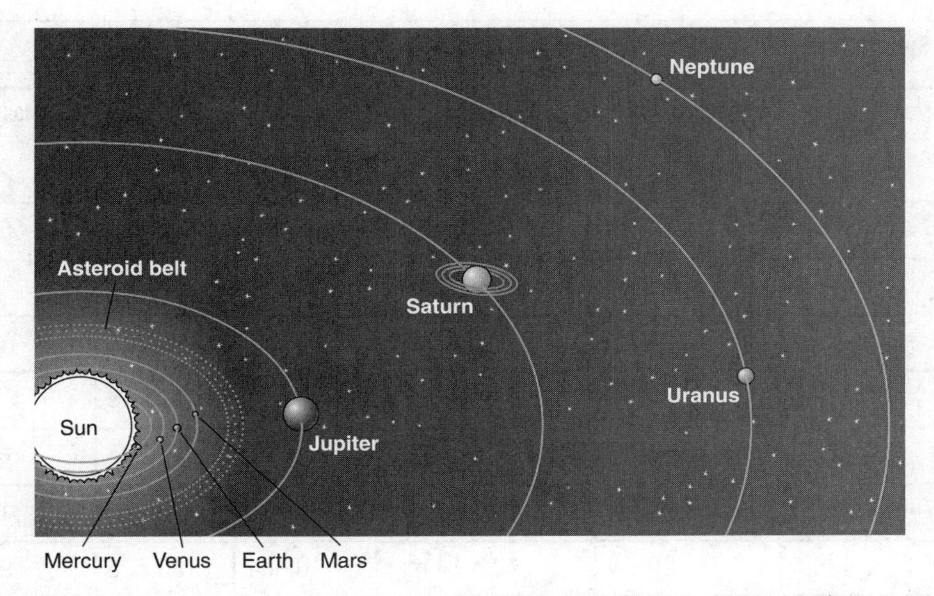

Figure 11-11. The orbits of the terrestrial planets, Mercury, Venus, Earth, and Mars, are closer to the sun than the orbits of the gas giants, Jupiter, Saturn, Uranus, and Neptune. Distances and sizes of the planets are not to scale.

The surface temperatures on the planets are mainly the result of their distance from the sun. For example, the hottest surface temperatures occur on Mercury and Venus. Although Mercury is closer to the sun, Venus's surface temperatures are a little higher than those on the daylight side of Mercury. The high temperatures on Venus are the result of a very dense atmosphere of carbon dioxide: a greenhouse effect run wild. Earth and Mars are cooler because of their greater distance from the sun. (Although the atmosphere of Mars is mostly carbon dioxide, this atmosphere is very thin. Therefore, the greenhouse effect on Mars is small and temperatures are generally cooler than temperatures on Earth.) The dwarf planets, Pluto, Eris, Makemake, and Haumea, are even colder because they are even farther from the sun. Figure 11-11 compares the sizes of the planetary orbits. Note that the inner (mostly rocky) planets have very small orbits compared with those of the outer planets.

Earth Is Unique Our home planet is unique in several ways. Earth is the only planet that has abundant liquid water. Mars shows evidence of erosion by moving water, but liquid water is no longer present on its surface. The moons of Jupiter have abundant water, but in the form of ice. From Chapter 6, you may remember how important liquid water was to the development of life on Earth. The presence of liquid water only on Earth may be the principal reason living organisms have not been detected elsewhere in the solar system. In addition, only our planet has an atmosphere with abundant free oxygen. Oxygen is released when plants, through *photosynthesis*, extract carbon from carbon dioxide in the air.

Page 15 of *The Earth Science Reference Tables* contains a table that lists the orbital characteristics and physical properties of the sun, moon, and the eight planets. (See Table 11-1 on page 282.)

Dwarf Planets, Asteroids, Meteors, and Comets

The eight planets are not the sun's only satellites. A number of smaller objects also orbit the sun, including dwarf planets, asteroids, meteoroids, and comets.

Dwarf Planets For nearly 76 years after its discovery in 1930, Pluto was considered the ninth planet. However, several features of Pluto made it different from other planets. Pluto was the smallest planet and it was especially out of place as the only rocky planet among the outer gas giants. Pluto's orbit is more eccentric than any other planet and its orbit is tilted about 17° from the plain of the other planets. These peculiarities led some astronomers to wonder if it should be called a planet at all.

Table 11-1. Solar System Data

Celestial Object	Mean Distance from Sun (million km)	Period of Revolution (d=days) (y=years)	Period of Rotation at Equator	Eccentricity of Orbit	Equatorial Diameter (km)	Mass (Earth = 1)	Density (g/cm^3)
SUN	—	—	27 d	—	1,392,000	333,000.00	1.4
MERCURY	57.9	88 d	59 d	0.206	4,879	0.06	5.4
VENUS	108.2	224.7 d	243 d	0.007	12,104	0.82	5.2
EARTH	149.6	365.26 d	23 h 56 min 4 s	0.017	12,756	1.00	5.5
MARS	227.9	687 d	24 h 37 min 23 s	0.093	6,794	0.11	3.9
JUPITER	778.4	11.9 y	9 h 50 min 30 s	0.048	142,984	317.83	1.3
SATURN	1,426.7	29.5 y	10 h 14 min	0.054	120,536	95.16	0.7
URANUS	2,871.0	84.0 y	17 h 14 min	0.047	51,118	14.54	1.3
NEPTUNE	4,498.3	164.8 y	16 h	0.009	49,528	17.15	1.8
EARTH'S MOON	149.6 (0.386 from Earth)	27.3 d	27.3 d	0.055	3,476	0.01	3.3

When a slightly larger satellite of the sun, Eris, was discovered beyond Pluto in 2005, many astronomers saw a need for more formal definition of the term "planet." The new definition more closely conforms to the characteristics of the remaining eight planets. Meanwhile, Eris, Pluto and three other objects are now classified as dwarf planets. The other dwarf planets are Ceres, which is in the asteroid belt, Makemake and Haumea, which orbit beyond Neptune. In the future, more dwarf planets are likely to be discovered beyond the orbit of Pluto.

Asteroids Tens of thousands of rocky objects called **asteroids** are located, for the most part, in a belt between the orbits of Mars and Jupiter. Asteroids range in size from roughly 1000 kilometers (600 miles) in diameter down to some the size of pebbles. If all the asteroids were put together, they would form a body about half as big as Earth's moon.

A few asteroids have orbits that can cross Earth's orbit. There is evidence that a large asteroid approximately 10 kilometers (6 miles) in diameter struck Earth in the Yucatán region of Mexico about 65 million years ago. Scientists have found a buried crater in that area and have noted that most sedimentary deposits of that specific time are rich in iridium (a rare element on Earth). In addition, fossils indicate that most living things suddenly became extinct (died out completely) about that time. Scientists think that this catastrophic collision might have thrown tons of dust into the atmosphere, blocking the sunlight, and dramatically cooling Earth's climates for many months. Without sunlight, plants could not grow. Without food, plant-eaters died, which in turn caused meat-eaters to starve. Such a massive food shortage may have caused the extinction of the dinosaurs at the end of the Mesozoic Era.

Meteors Small solid particles from space can be caught by Earth's gravity. As the objects fall, they are heated by friction with Earth's atmosphere and burn. The burning objects produce streaks of light visible at night as **meteors**, or "shooting stars." (Potential meteors in space are called **meteoroids**.) Most meteors are vaporized during their fall through the atmosphere, although some of them are large enough

Figure 11-12. The Willamette Meteorite is a 15-ton iron meteorite that was transported from where it was discovered in Oregon to New York City in 1905. Millions of years of weathering within the soil produced the Swiss-cheese-like cavities in its surface.

Figure 11-13. Meteor Crater in Arizona is Earth's best preserved large impact crater. It was formed about 50,000 years ago by a meteorite estimated to be 4000 times the mass of the meteorite in Figure 11-12.

to hit the ground. (Meteors that survive their fall and hit the ground are called **meteorites.**) Figure 11-12 is a meteorite on display at the American Museum of Natural History in New York City.

The most dramatic structure resulting from a meteor strike on Earth is probably Meteor Crater in northern Arizona. (Figure 11-13.) This crater is more than 1 kilometer (2/3 mile) across and almost 200 meters (600 feet) deep. Although Earth has been struck by meteors many times in its past, it is not covered with impact craters like the moon or Mercury. This is because only the largest meteors survive the fall through Earth's atmosphere. In addition, most meteors strike the oceans, and impact craters on land are eroded away.

Meteor impacts were more common early in the history of the solar system. Over the past 4.5 billion years, most of the interplanetary debris (rocks, dust, and so on) has been swept up by the gravity of the planets.

Comets Comets are icy celestial objects, most of which originate in a region beyond the planets. Some of them travel in highly elliptical orbits, coming close to the sun. Heating by the sun causes them to partially vaporize, producing a tail that we can observe over a period of weeks. Whether a comet is traveling toward or away from the sun, its tail always points away from the sun. Halley's Comet is perhaps the best-known comet. It has a period of approximately 75 years; it will reappear in 2061.

QUESTIONS

Part A

16. How does the planet Mars compare with Earth? (1) Mars is about equal to the diameter of Earth, twice as far from the sun and has a much shorter day. (2) Mars is about twice the diameter of Earth, twice as far from the sun and has twice the day length. (3) Mars is about half the diameter of Earth, twice as far from the sun and has a shorter day length. (4) Mars is about half the diameter of Earth, twice as far from the sun and has a similar day length

17. Which planet has the most circular orbit? (1) Mercury (2) Venus (3) Earth (4) Mars

18. The gravitational attraction between Earth and Mars is greatest when (1) Mars is closest to the sun (2) Mars is farthest from the sun (3) Mars is visible in the night sky (4) Mars is not visible in the night sky

19. Which bar graph below best shows the orbital eccentricities of the planets?

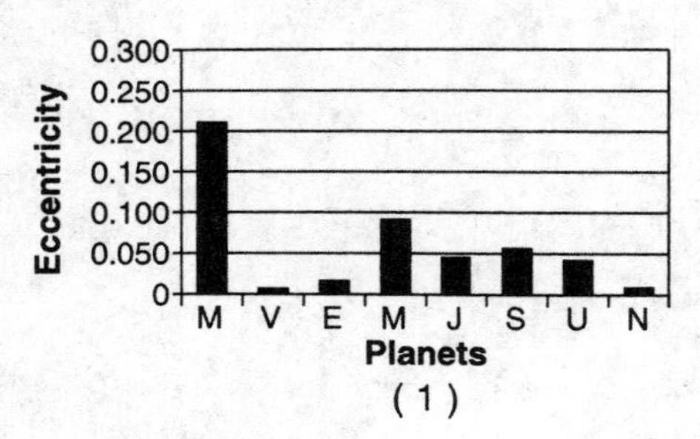

(1)

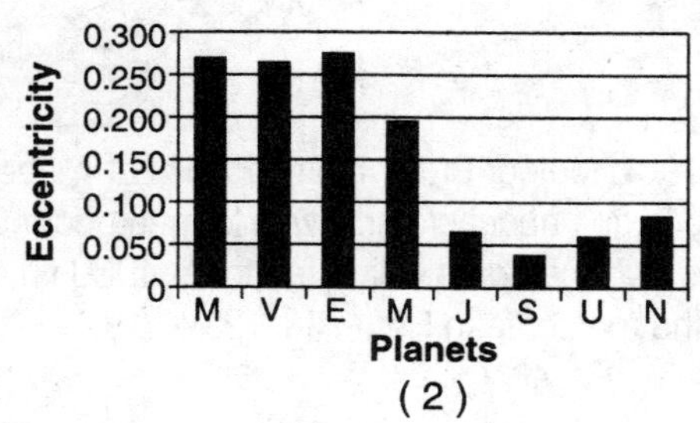

(2)

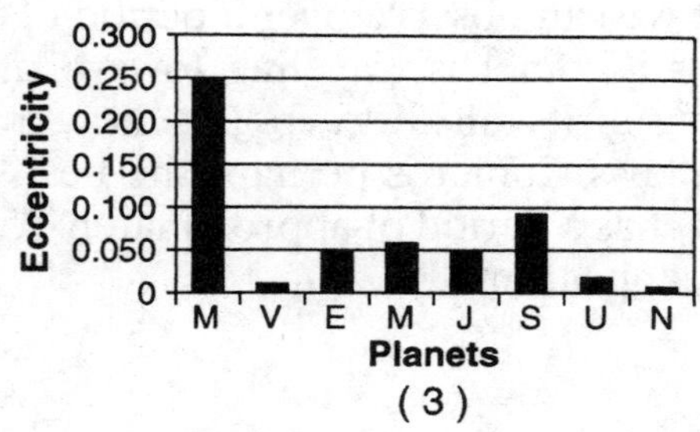

(3)

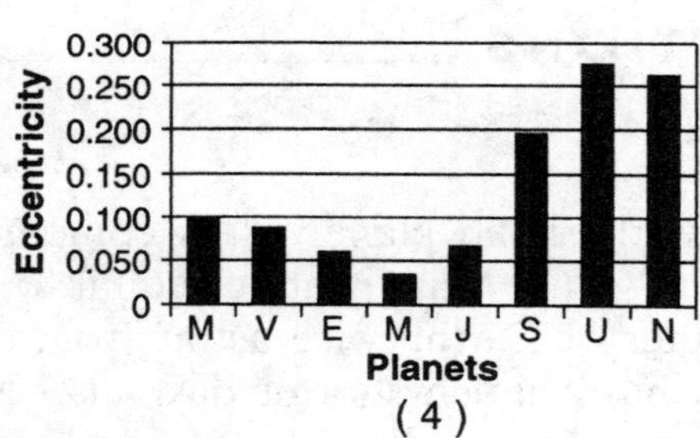

(4)

20. Which object in the solar system has the greatest density? (1) Jupiter (2) the sun (3) the moon (4) Earth

21. Every day the gravitational attraction between Earth and the moon changes by a very small percentage. Why does it change? (1) Earth's axis is tilted $23^1/_2°$. (2) Earth's shape is not quite spherical. (3) The moon's orbit is elliptical. (4) The moon is less dense than Earth.

22. Earth's moon has a volume that is closest the volume of (1) Mercury (2) Mars (3) Jupiter (4) Neptune

23. Which planet has an orbit that is least circular? (1) Mercury (2) Venus (3) Mars (4) Neptune

24. Which lists the gas giant planets in order from largest to smallest? (1) Neptune, Saturn, Uranus, Jupiter (2) Neptune, Jupiter, Saturn, Uranus (3) Jupiter, Saturn, Neptune, Uranus (4) Jupiter, Saturn, Uranus, Neptune

Part B

25. What is the eccentricity of the ellipse below?

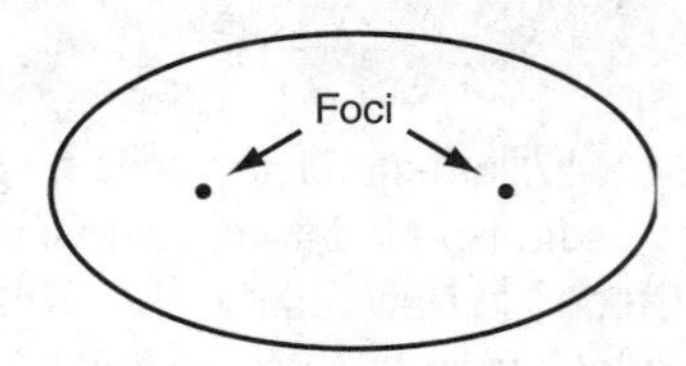

(1) 0.5 (2) 0.8 (3) 1.0 (4) 2.0

Base your answers to questions 26 through 29 on the diagram and table below, which show four motions of Earth and the moon.

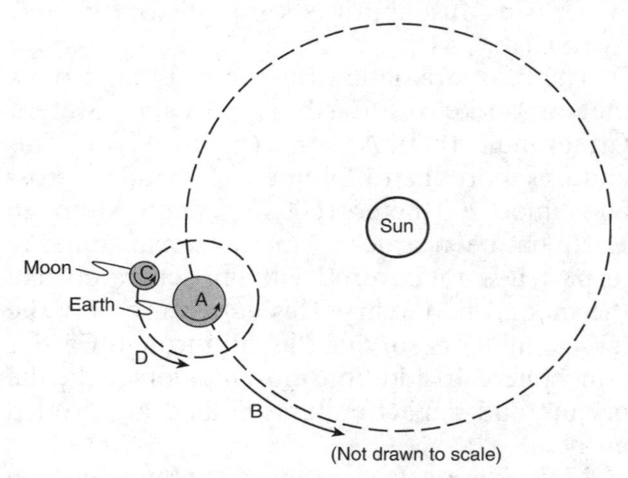

(Not drawn to scale)

Key	
Arrow	**Motion**
A	Earth's rotation on its axis
B	Earth's revolution around the Sun
C	The Moon's rotation on its axis
D	The Moon's revolution around Earth

26. Which two motions indicated in the diagram take about the same time? (1) *A* and *B* (2) *B* and *C* (3) *C* and *D* (4) *A* and *D*

27. Which motion is responsible for the phases of the moon we observe from Earth? (1) *A* (2) *B* (3) *C* (4) *D*

28. How does the orbit of Earth compare with the orbit of the moon? (1) The moon's orbit is more eccentric. (2) Earth's orbit is more eccentric. (3) The moon's orbit is larger. (4) Earth's orbit is smaller.

29. What is the period of the four motions shown in the diagram? (1) A = 1 day, B = 1 week, C = 1 month, D = 1 year (2) A = 1 week, B = 1 year, C = 1 month, D = 1 day (3) A = 1 month, B = 1 day, C = 1 year, D = 1 year (4) A = 1 day, B = 1 year, C = 1 month, D = 1 month

Base your answers to questions 30 through 33 on the text below.

Asteroids

Most of the known asteroids are rocky objects that orbit the sun approximately halfway between the orbits of Mars and Jupiter in the Asteroid Belt. Occasionally, an asteroid is deflected from the Asteroid Belt into a more eccentric orbit that brings it into the region of the terrestrial planets. This may be caused by a collision or by a close approach to Mars or Jupiter.

In 1994, asteroid 1994 XL1 left the Asteroid Belt. It came closer to Earth than the orbit of the moon. When it was first noticed by astronomers, the asteroid was only about 15 hours from Earth and missed a collision by only 52 minutes. Asteroid 1994 XL1 was only 9 meters (30 feet) in diameter and was never bright enough to be visible to the unaided eye.

A 177-kilometer-wide (110-mile-wide) crater was discovered in southern Mexico. The formation of this crater may have led to the extinction of the dinosaurs. Scientists think that climatic changes produced by the dust from the impact were the probable cause of this world-wide extinction. Some scientists think that the crater resulted from the impact of an asteroid that was as large as 10 miles in diameter.

30. What may have caused asteroid 1994 XL1 to leave its orbit within the Asteroid Belt? (1) an eclipse of the sun (2) gravitational force (3) electromagnetic radiation (4) the rotation of the asteroid

31. Why were astronomers not aware of this object until it came so close to Earth? (1) The asteroid was visible only at night. (2) The asteroid was located outside the solar system. (3) The asteroid was relatively dim compared to the stars. (4) The asteroid had a cyclical orbit when it passed close to Earth.

32. Approximately when did the object that may have led to the extinction of the dinosaurs impact Earth? (1) 4.5×10^6 years ago (2) 6.5×10^6 years ago (3) 3.7×10^{10} years ago (4) 6.5×10^{10} years ago

33. Approximately how far from Earth is Venus? (1) 25 million kilometers (2) 50 million kilometers (3) 100 million kilometers (4) 200 million kilometers

The Sun

The sun is by far the nearest star to Earth. A **star** is a large, self-luminous body in space that creates its own radiant energy. Although the sun is smaller than most of the distant stars we see in the night sky, it is larger than most of the nearby stars, which are very dim. We may therefore conclude that the sun is typical of the vast number of stars in the universe. Indeed, the sun, being an average star, appears near the middle of Figure 11-14. The Hertzsprung-Russell diagram (H-R diagram) groups stars by color (an indication of temperature) and **luminosity** (how much light they give off). According to this diagram, most stars can be classified as dwarfs, main sequence stars (such as the sun), giants, or supergiants.

Figure 11-15 illustrates how the Hertzsprung-Russell diagram can be used to describe the life cycle of a star. The path illustrated is for a star about the size of the sun. Different stars have different paths through the H-R diagram depending on their stellar mass. For example, very large stars follow a different path and move beyond the main sequence relatively quickly. They then become unstable and explode as supernovas.

Like all stars, the sun gets its energy from **nuclear fusion**. During fusion under conditions of extreme heat and pressure deep in the sun, the nuclei of four hydrogen atoms join to form a single helium nucleus. This reaction produces a huge amount of energy. Fusion reactions in larger stars such as red giants produce heavier elements such as carbon, oxygen, and iron. The formation of the heaviest natural elements requires the conditions found only in massive stellar explosions.

The nuclear energy that is generated in reactors to produce electricity is the result of nuclear fission. In nuclear fission, large nuclei from elements such as uranium or plutonium are split to form smaller nuclei. Fission does not produce as much energy as fusion and it creates dangerous waste products. However, fission is used because scientists have not been able to sustain controlled nuclear fusion on Earth. Fusion requires extreme temperature and pressure. The only way humans have been able to apply fusion is in the explosion of hydrogen bombs.

The Sun's Rotation If you look at a projected image of the sun, you may see dark spots on its

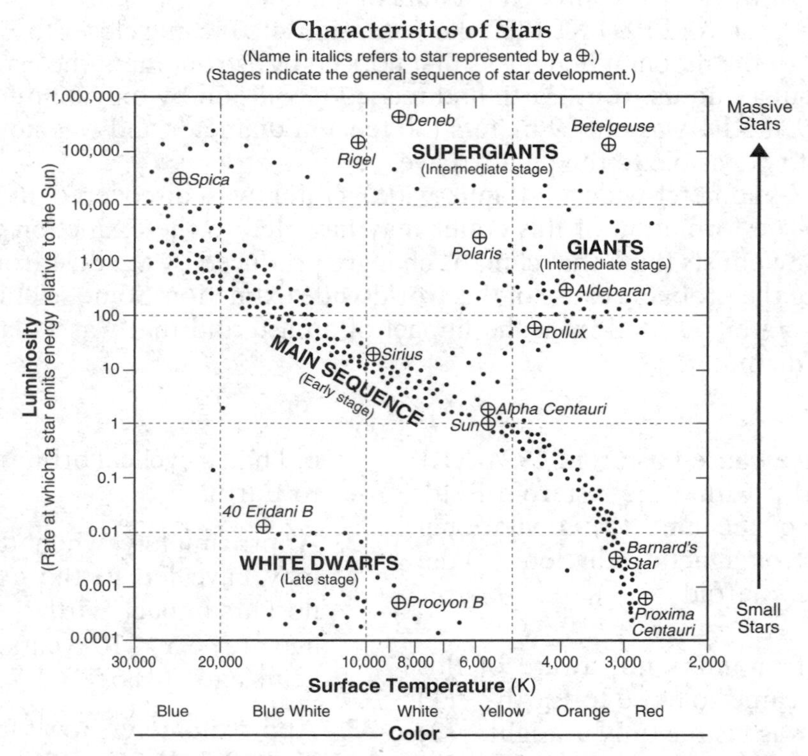

Figure 11-14. The Hertzsprung-Russell (H-R) diagram, named after the scientists who devised it, is used to classify stars by temperature and light output (luminosity). In this scheme, our sun is a typical star. Although the sun is probably brighter than most stars, it is small compared with most of the stars we see at night.

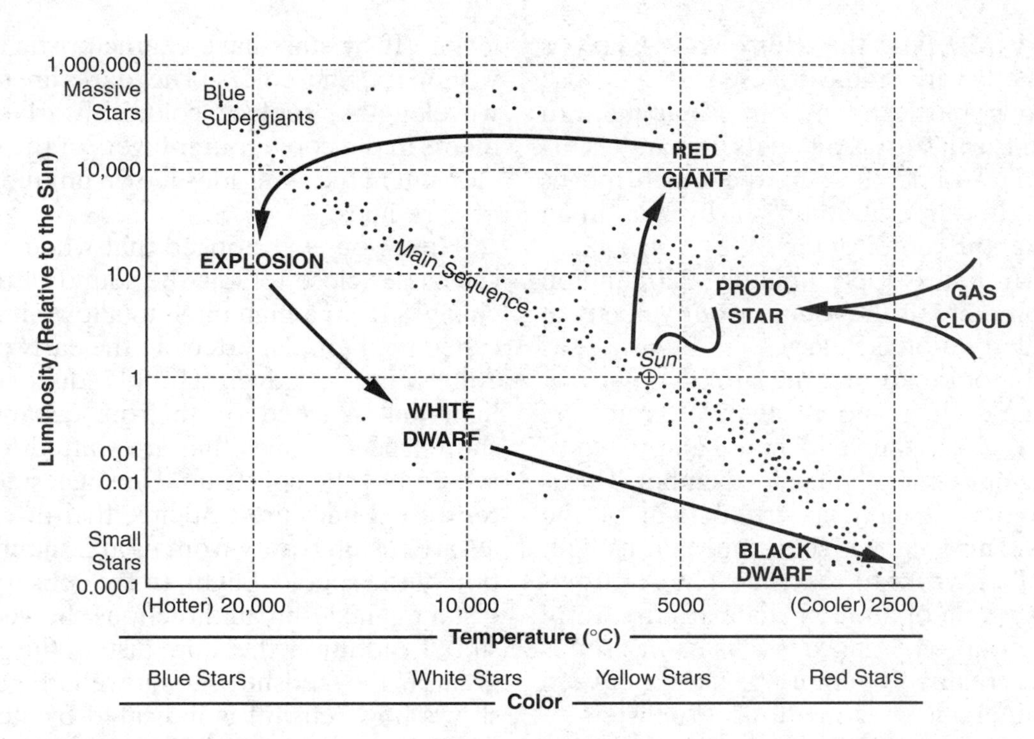

Figure 11-15. The H-R diagram is useful in showing how stars change through time from birth to extinction. Arrows show the life of a star the size of the sun, which is a middle-age star of its type.

surface. *(Caution: Never look directly at the sun, or at the sun through binoculars or a telescope, or without special protective equipment.)* The motion of these spots across the sun over time reveals that the sun rotates on its axis every 27 days. **Sunspots** are temporary storms on the visible surface of the sun. They are a little cooler and dimmer than the normal surface of the sun, and they often occur in pairs, each with opposite magnetic polarity. Sunspots come and go in cycles of about 11 years. The reason for this is not yet known.

GALAXIES OF STARS

When early astronomers looked at the sky on a clear, dark, moonless night, they observed a dim band of light that extends across the sky. They named this band the Milky Way for its faint white color. Telescopic observations have revealed that the Milky Way is actually a collection of distant stars that are too faint to see as individual stars with the unaided eye. When we look at the Milky Way, we are looking along the plane of our own galaxy. A **galaxy** is a huge body of stars and other matter in space. The sun is one of roughly a trillion stars in the Milky Way galaxy.

The Milky Way is classified as a spiral galaxy because of its shape. Our solar system is located in a spiral arm well away from the galactic center. Figure 11-16 shows our position in the Milky

Figure 11-16. Spiral galaxy NGC 3370 is similar in structure to the Milky Way galaxy. The arrow points to the area in NGC 3370 that corresponds to Earth's location in the Milky Way. Earth is located in an arm of the Milky Way far from the more dense, center of the galaxy.

Way galaxy. Mapping the Milky Way galaxy is accomplished with radio telescopes because clouds of dust block the view in visible light. In addition to Earth's rotation on its axis and its orbital motion around the sun, the Earth moves along with the sun and other nearby stars in an orbit around the center of the Milky Way galaxy. Although it takes approximately 220 million years to complete this revolution, our velocity in this galactic rotation is nearly 10 times as fast as our orbital motion around the sun.

On a dark, clear, moonless night, you may also be able to see the very small, faint patch of light that is our nearest galactic neighbor. This is the Andromeda galaxy, a virtual twin of the Milky Way. This galaxy is so distant that its light takes 2 million years to reach us. The Andromeda galaxy is one of about 30 galaxies that form our local group. The galaxies cluster in groups with large, relatively empty spaces between them. A supercluster is made up of clusters of galactic clusters.

EVOLUTION OF THE UNIVERSE

The spectroscope is one of an astronomer's most useful tools. A spectroscope is an instrument that separates light into its component colors, as does the glass prism in Figure 11-17. Spectra of light given off by stars have characteristic dark lines within specific colors. The dark lines (missing wavelengths) replace colors absorbed by elements in the cooler outer layers of the star. Each element in the stars adds it own unique signature of dark lines.

Have you ever noticed that when a speeding car passes close to you, the sound of the engine changes from a high pitch to a low pitch? This is called the Doppler effect. In the early part of the 20th century, Edwin Hubble discovered that light that reached Earth from distant galaxies shows spectral lines that are shifted toward the red end of the spectrum. He suggested that the **redshifted** lines are evidence that distant galaxies are rushing away from us at a significant portion of the speed of light. In fact, observations of distant galaxies in all directions showed the redshift. In addition, the more distant the galaxy, the greater the redshift it showed. Figure 11-18 shows how redshift is indicated by stellar spectra. This indication that distant stars and galaxies are all moving away from us led scientists to the conclusion that the universe is expanding.

If we mentally reverse the expansion, taking us backward in time, we can infer that at one time the universe was a tiny object of primordial matter that had incredible mass and density. This object must have expanded rapidly. This theory of the origin of the universe is now known as the

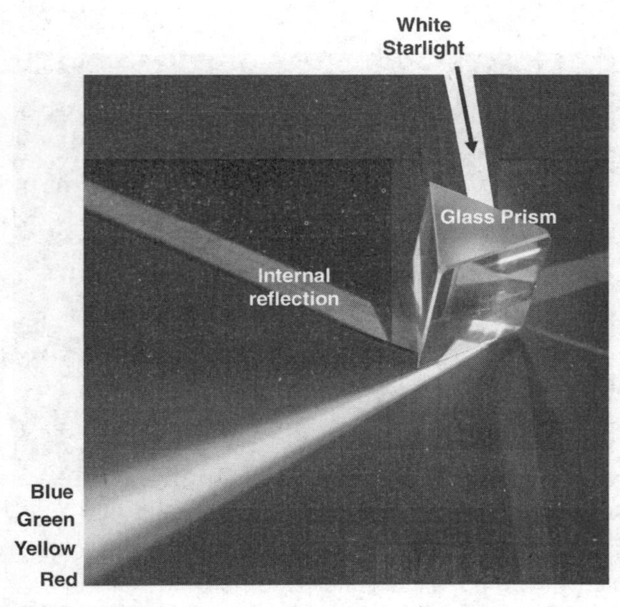

Figure 11-17. A glass prism can be used to separate any white light into its component colors.

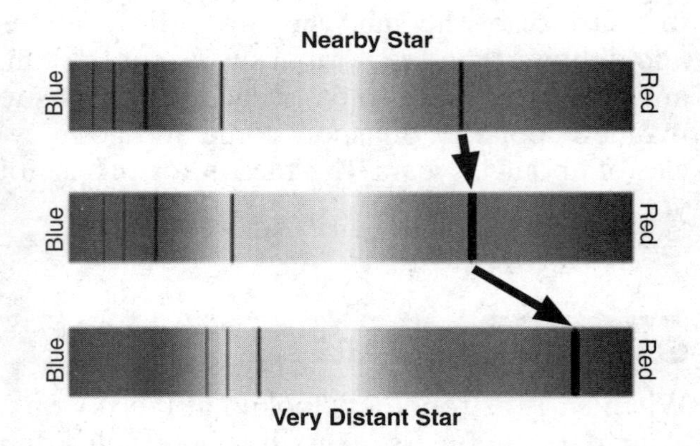

Figure 11-18. Astronomers use light spectra to estimate the distance to distant galaxies by measuring their redshift (Doppler shift). As the universe expands, the more distant a galaxy is the more its spectrum is shifted toward the red (longer wavelength) end of the spectrum. Note the change in the positions of the spectral lines in this diagram as you compare the light from nearby and more distant stars and galaxies.

big bang. In addition, scientists can detect radiation remaining from the big bang.

Astronomers had predicted the existence of radio energy left from the origin of the universe, the big bang. However, it was not detected until two American scientists working on long-distance radio communications observed it in the 1960s. When they pointed their sensitive receiver toward space in any direction, they picked up an annoying background radiation. They eventually realized that they were, in fact, "listening" to the electromagnetic "noise" of the birth of the universe.

How long ago did the big bang occur? In the past several years, discoveries made with the Hubble Space Telescope and other instruments have led to a remarkably precise figure for the age of the universe: 13.7 billion years.

The Size of the Universe

Perhaps the most significant thing for you to understand about the universe is its size. For those who think of a few kilometers, or even a few thousand kilometers, as a great distance, it is difficult to comprehend the size of the universe with a radius on the order of 10^{23} kilometers.

Perhaps this size is easier to understand if you know how fast light travels. About 400 years ago, the Italian scientist Galileo tried to measure the speed of light, but he was unable to do it. Today, we understand why. Light travels so fast that it could circle Earth 7 times in just 1 second. Most scientists believe that the speed of light is the upper limit of velocity. No object or energy can travel faster than light. Light takes about $1^1/_2$ second to get to the moon, and light from the sun takes about 8 minutes to reach Earth. Light from the nearest star visible at night takes about 4 years to reach us. The distance light can travel in 1 year is called a **light-year**. This is a unit of distance. A light-year is about 10 trillion kilometers. The universe is thought to be about 25 billion light-years in diameter.

Astronomers are now investigating superclusters as an even greater level of structure in the universe. Why matter in the universe, which formed from a gigantic "explosion," scientists call the big bang, is distributed so unevenly is one of the most interesting questions facing astronomers today. As astronomers look farther into the universe, they see new levels of structure: the solar system, our galaxy, clusters of galaxies, and even superclusters as shown in Figure 11-19.

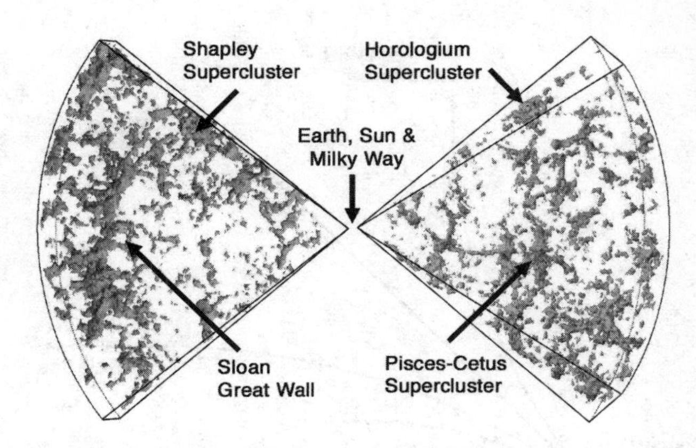

Figure 11-19. Recent mapping of the distant universe has shown gigantic groups of galaxies. Astronomers call these groups superclusters, which seem to be the highest level of structure within the universe.

The Future

The long-term future of the universe is unclear. Some astronomers think that the expansion of the universe will continue forever. Others believe the force of gravity will eventually reverse the expansion, and the universe will fall back together in the **big crunch**. It is even possible that the universe alternates between expansions and contractions.

An analogy may help you to understand this issue. If you throw a baseball upward, you expect gravity to slow its climb until the ball begins to fall back toward the ground. However, if you could throw the ball fast enough, it would have enough speed to overcome Earth's gravitational pull, and the ball would continue into space and never return. This speed is called the escape velocity. Therefore, the ultimate fate of the universe depends on the balance between the rate of expansion and the *escape velocity*. This is determined by the mass and density of the universe. Four models for the fate of the universe are illustrated in Figure 11-20 on page 290.

Recent observations have led astronomers to believe that the expansion rate of the universe is actually accelerating. This is a very surprising finding. Scientists do not understand how this can occur. They have therefore labeled the cause "dark energy." However, what dark energy is and how it works are questions more puzzling

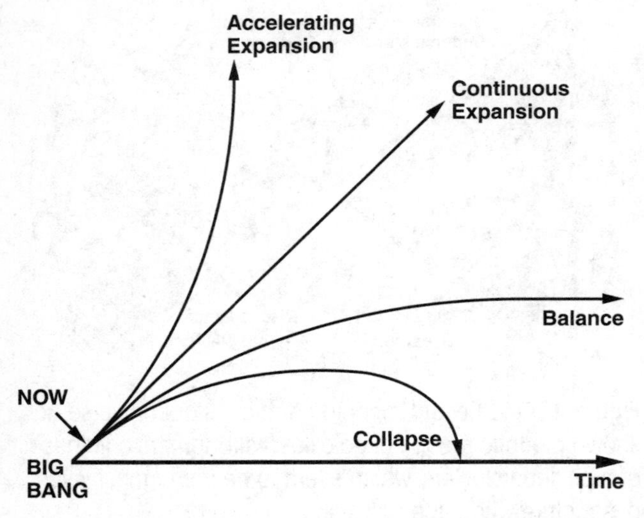

Figure 11-20. Gravity is thought to hold the universe together. Astronomers have therefore suggested three models for the fate of the big-bang universe. They are (1) eventual collapse, (2) balance at a constant size, or (3) continuous expansion. However, a new model has recently been added. Scientists now observe that a mysterious factor they call dark energy seems to be accelerating the expansion.

than any encountered before. Science thrives on unanswered questions, so the future of human scientific inquiry seems secure.

QUESTIONS

Part A

34. Compared with the star Betelgeuse, the sun is (1) smaller, hotter, and less luminous (2) smaller, cooler, and more luminous (3) larger, hotter, and less luminous (4) larger, cooler, and more luminous

35. In which group are the stars the same color and approximately the same temperature? (1) Sirius, 40 Eridani B, Alpha Centauri, and Spica (2) Betelgeuse, Proxima Centauri, and Barnard's Star (3) Rigel, Pollux, and Polaris (4) Procyon B, Spica, Betelgeuse, Polaris, and Aldebaran

36. The following diagram shows the electromagnetic spectrum of the sun and the spectra of four outer stars.

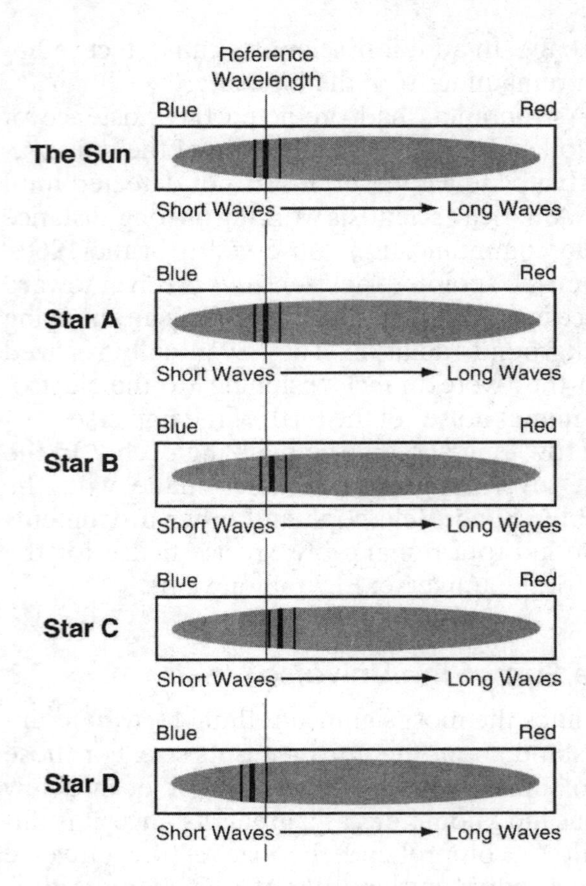

Which star is most likely the greatest distance from Earth? (1) star *A* (2) star *B* (3) star *C* (4) star *D*

Part B

37. Which star has a surface temperature closest to the temperature at the interface between Earth's mantle and core? (1) Sirius (2) Rigel (3) the sun (4) Betelgeuse

Base your answers to questions 38 through 40 on the diagram below, which shows the orbit of Earth and part of the orbit of a comet.

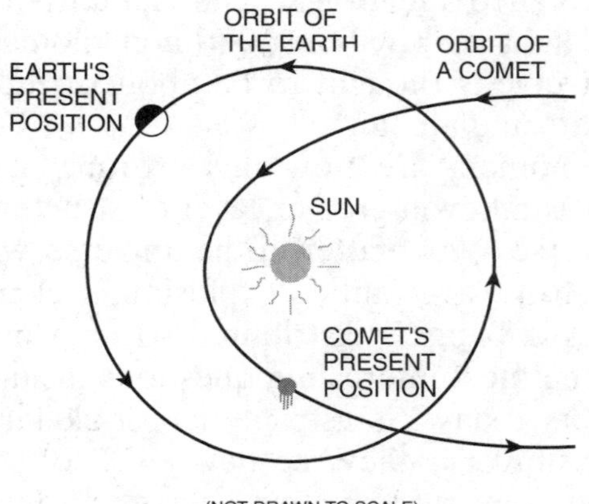

38. Compared to the orbit of Earth, the orbit of the comet is (1) more eccentric and the comet takes longer to complete one orbit (2) more eccentric and the comet takes less time to complete one orbit (3) less eccentric and the comet takes longer to complete one orbit (4) less eccentric and the comet takes less timer to complete one orbit

39. At the time shown in the diagram, an observer on Earth looks into the sky to see this comet. When would the comet be visible? (1) high overhead at midnight (2) near the horizon at midnight (3) high in the sky just before sunrise (4) near the horizon just before sunrise

40. At its present position the comet is (1) speeding up as its gravitation attraction to the sun is increasing (2) speeding up as its gravitation attraction to the sun is decreasing (3) slowing down as its gravitation attraction to the sun is increasing (4) slowing down as its gravitation attraction to the sun is decreasing

Base your answers to questions 41 through 43 on the data table below that shows the history of th universe.

Data Table

Stage	Description of the Universe	Average Temperature of the Universe (°C)	Time From the Beginning of the Universe
1	The size of an atom	?	0 s
2	The size of a grapefruit	?	10^{-43} s
3	"Hot soup" of electrons	10^{27}	10^{-32} s
4	Cooling allows protons and neutrons to form.	10^{13}	10^{-6} s
5	Still too hot to allow the forming of atoms.	10^{8}	3 min
6	Electrons combine with protons and neutrons, forming hydrogen and helium atoms. Light emission begins.	10,000	300,00 years
7	Hydrogen and helium form giant clouds (nebulae) that will become galaxies. First stars form.	−200	1 billion years
8	Galaxy clusters form and first stars die. Heavy elements are thrown into space, forming new stars and planets.	−270	13.7 billion years

41. How soon did protons and neutrons form after the beginning of the universe? (1) 0.0000001 s (2) 0.0000006 s (3) 0.000001 s (4) 0.000006 s

42. What is the most likely temperature for the universe before 10^{-32} s? (1) absolute 0 K (2) 1 million°F (3) 10^{27}°C (4) more than 10^{27}°C

43. Between which two stages did Earth form? (1) 1 and 3 (2) 3 and 5 (3) 6 and 7 (4) 7 and 8

Base your answers to questions 44 and 45 on the graph below, which shows the magnetic activity of the sun and the number of sunspots over a period of approximately 100 years.

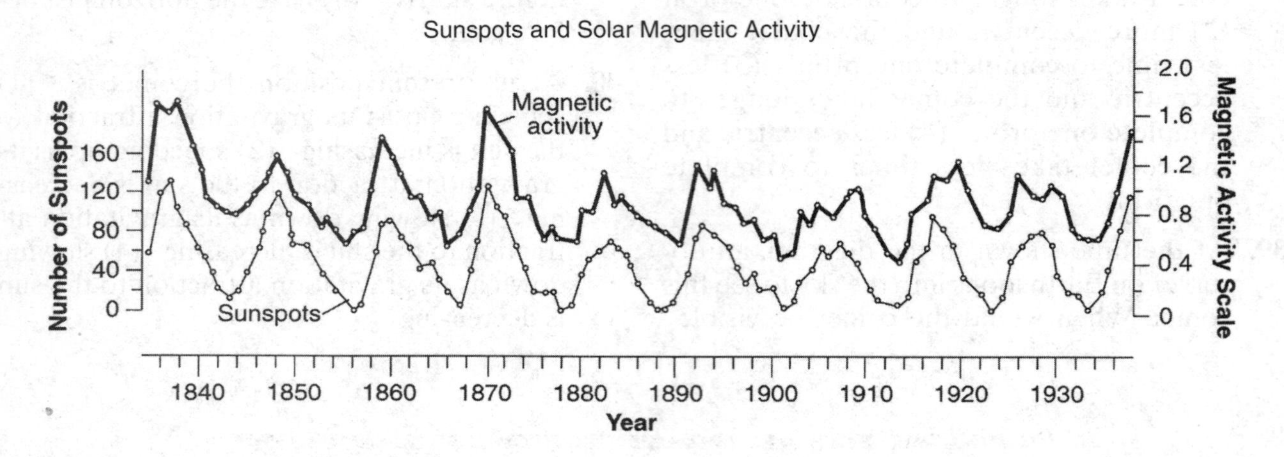

44. The graph indicates that the greatest number of sunspots per year (1) is random and unpredictable (2) occurs at the beginning of each decade (3) is cyclic with a period of approximately 6 years (4) is cyclic with a period of approximately 11 years

45. Which of the following graphs best shows the relationship between magnetic activity and the number of sunspots?

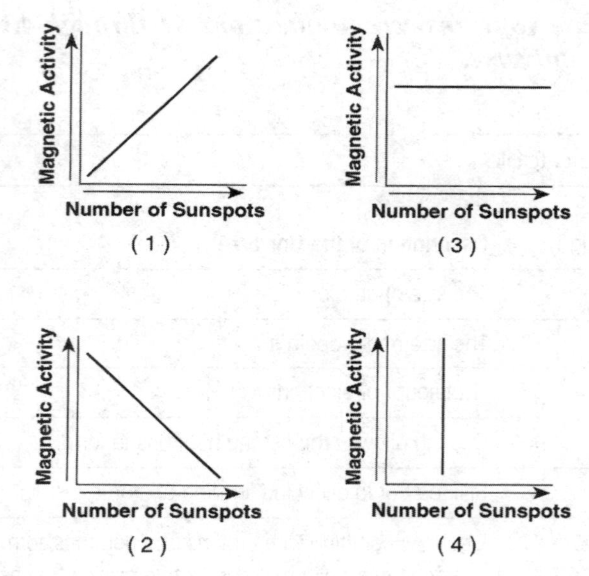

Base your answers to questions 46 through 49 on the text and the diagram below.

The Hubble Classification of Galaxies

In the early 20th century, astronomers did not know that they would find huge groups of stars far beyond the familiar stars in the night skies. At that time, the world's largest telescope was on Mt. Wilson in California. There, Edwin Hubble observed dim, fuzzy objects called "nebulae." Nebula is the Latin word for cloud. He discovered that many nebulae are huge collections of stars and star systems.

Hubble realized that these fuzzy shapes are distant galaxies, similar to our Milky Way. He developed a classification system based on the shapes of the galaxies. He hoped that he could use his classification system to understand how galaxies change through time. Another classification of stars, the H-R diagram, had been devised about a decade earlier. The H-R diagram proved very useful in tracing the evolution of individual stars. (One version of this diagram is on page 15 of the *Earth Science Reference Tables*.)

The "Hubble Tuning Fork," shown in the figure below, led Hubble to hypothesize that galaxies start with an elliptical shape and move through the lenticular stage to become either spiral or barred spiral galaxies. This diagram includes both idealized shapes and photographs of specific examples of these shapes. Since Hubble published his classification scheme in 1926, astronomers have not been able to use it to show a regular evolution of galaxies with age, as Hubble had hoped. However, his system of classification of galaxies based on their shapes is still used today.

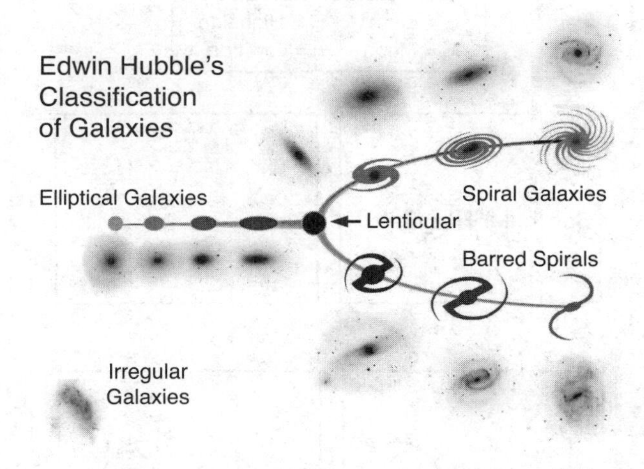

46. Astronomers know that galaxies contain (1) only millions of nearby and distant stars (2) stars, planets, and a variety of other objects (3) only the most distant stars we can observe (4) planets and comets, but no stars

47. The Milky Way is classified as what type of galaxy. (1) elliptical (2) lenticular (3) spiral (4) irregular

48. The nearest galaxy to the Milky Way is the Andromeda galaxy. Which choice below lists these objects from closest to Earth to most distant? (1) sun, Polaris, moon, Andromeda galaxy (2) moon, Polaris, sun, Andromeda galaxy (3) Andromeda galaxy, sun, moon, Polaris (4) moon, sun, Polaris, Andromeda galaxy

49. How does the star Polaris compare with the sun? (1) Polaris appears brighter to us than the sun. (2) Polaris gives off more light than the sun. (3) Polaris is visible in more Earth locations than the sun. (4) Polaris is less massive than the sun.

Problem Solving

Devise a plan to search for intelligent life beyond Earth using only Earth-bound procedures. The whole project must be done here on Earth. Make a list of the equipment you will need and how much these items are likely to cost. Then devise a 5-year plan to maximize your likelihood of finding extraterrestrial civilizations. If you detect such organized life, what should humans do?

Chapter Review Questions

PART A

1. An artificial satellite took the first pictures of the distant side of the moon in 1959. Why had no one observed this side of the moon before these photographs were taken? (1) The moon and Earth rotate around the sun. (2) The far side of the moon is always dark. (3) The rotation and revolution of the moon take equal time. (4) The moon rotates on its axis, which points toward Earth

2. Which planet moves fastest in its orbit? (1) Mercury (2) Earth (3) Jupiter (4) Neptune

3. What is the eccentricity of the orbit shown below? (This diagram is drawn to scale.)

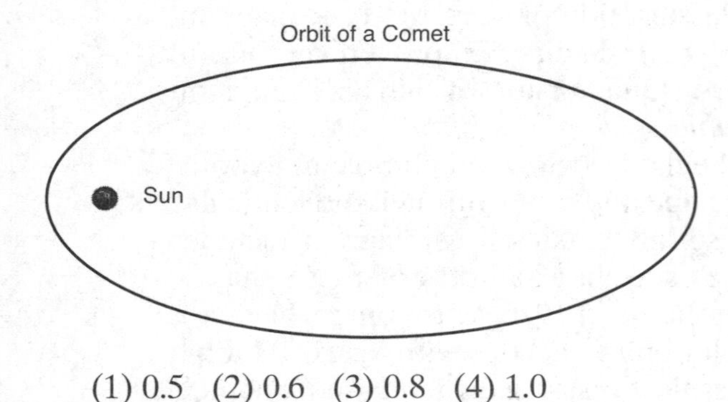

(1) 0.5 (2) 0.6 (3) 0.8 (4) 1.0

4. Compared with nearby galaxies, the most distant galaxies appear (1) more red and are moving closer to Earth (2) more red and are moving farther from Earth (3) more blue and are moving closer to Earth (4) more blue and are moving father from Earth

5. Which two objects are able to orbit closest to each other? (1) Mercury and the sun (2) Mercury and Venus (3) Venus and Earth (4)Earth and Mars

6. The stars in which group have the hottest surface temperature? (1) red dwarfs (2) red giants (3) supergiants (4) blue supergiants

7. The following diagram shows the orbit of a planet revolving around a star.

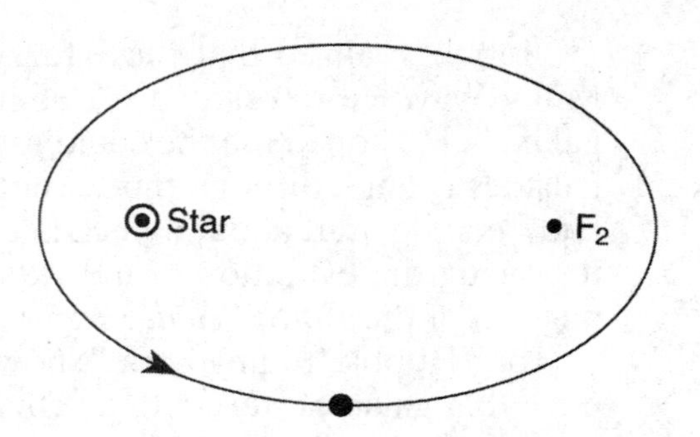

What is the eccentricity of this orbit? (1) 0.22 (2) 0.47 (3) 0.68 (4) 1.47

8. The diagram below shows the sun, moon, and Earth in different positions.

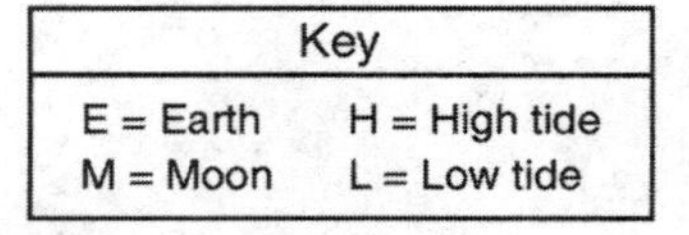

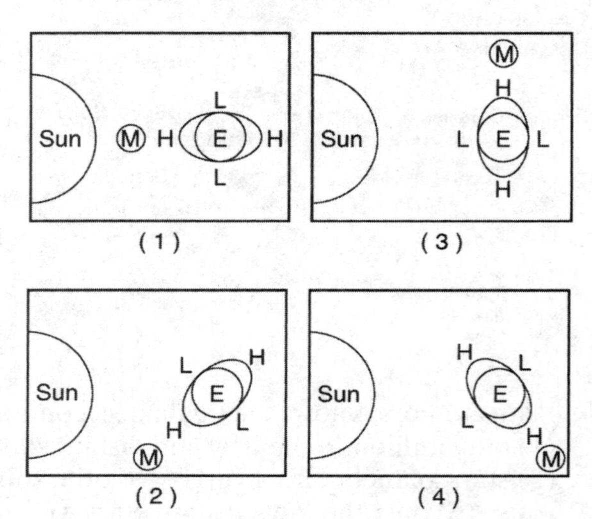

Which diagram best shows these objects when the greatest range of tides occurs?

PART B

Base your answers to questions 9 through 14 on the text and diagram below.

How Old Is the Universe?

In 2008, scientists determined the universe is 13.73 billion years old. They estimate their percent error to be less than 1%. Just a few years ago, the age of the universe was known only within an error of about 40%. This is a remarkable advance for science.

Earlier methods to determine when the big bang occurred depended on observations of the expansion of the universe. The change in the wavelengths of light (redshift) received from the most distant stars is one indication of their age.

Although redshift is not a direct measure of distance, it is clear that the most distant stars and galaxies are also those whose spectra are redshifted the most.

The data used to determine the new age of the universe was gathered by NASA's Wilkinson Microwave Anisotropy Probe (WMAP) satellite. Using a satellite based high above Earth is essential because our atmosphere absorbs microwave radiation. This electromagnetic radiation originated during the big bang as high-energy gamma rays. These rays now have been redshifted into the long-wave part of the spectrum.

In its initial state, the universe gave rise to subatomic particles called neutrinos. For a tiny fraction of a second after the big bang, there was an unimaginably rapid rate of expansion. This expansion is called cosmic inflation. About 400 thousand years after the big bang, the first atoms of hydrogen formed from an ocean of protons and neutrons. The first stars appeared about 400 million years later. Then stars gathered into galaxies and the larger structures we see today. Think about it. All this in less than 14 billion years!

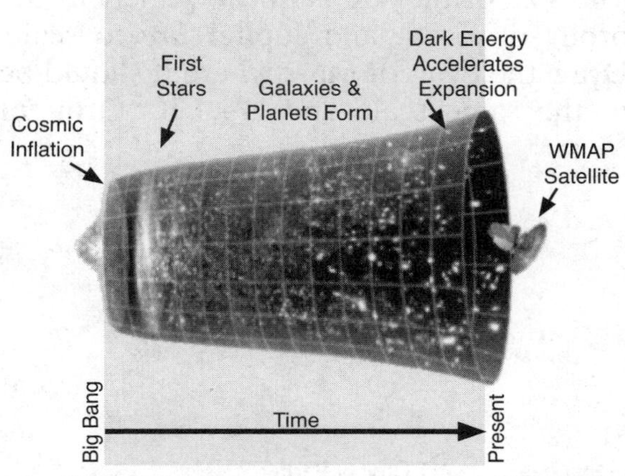

9. How has the radiation given off shortly after the big bang and now detected by the WMAP satellite changed? (1) The wavelength increased. (2) The wavelength decreased. (3) The radiation became stronger. (4) The radiation disappeared.

10. How does Earth's atmosphere affect incoming electromagnetic radiation? (1) All radiation passes through unchanged. (2) The atmosphere blocks all radiation. (3) Most visible light gets through, but not most microwaves. (4) Most visible light is blocked, but microwaves get through.

11. How many years after the big bang did the first stars appear? (1) 400 (2) 400,000 (3) 400,000,000 (4) 4,000,000,000

12. How do the universe, Earth, and the sun compare in age? (1) Earth is approximately a third the age of both the sun and the universe. (2) Earth and the sun are approximately one-third the age of the universe. (3) Earth and the sun are approximately twice the age of the universe. (4) Earth, the sun, and the universe are all the same age.

13. The WMAP satellite orbit takes it 1×10^9 miles from its Florida launch site on Earth's surface. At this distance, where is the satellite? (1) within the troposphere (2) within the mesosphere (3) beyond the atmosphere but closer than the moon (4) beyond the atmosphere and farther than the moon.

14. Which lists the objects in order of increasing size? (1) Earth, sun, universe, galaxy (2) Earth, galaxy, sun, universe (3) Earth, sun, galaxy, universe (4) sun, Earth, galaxy, universe

PART C

15. The photo was taken from an airplane looking down at a remote part of the Colorado Plateau. The straight lines are roads. The curved feature near the top of the picture is a river canyon.

What event created the round hole near the center of the image?

Base your answers to questions 16 through 18 on the diagram below, which shows the moon at eight positions in its orbit around Earth.

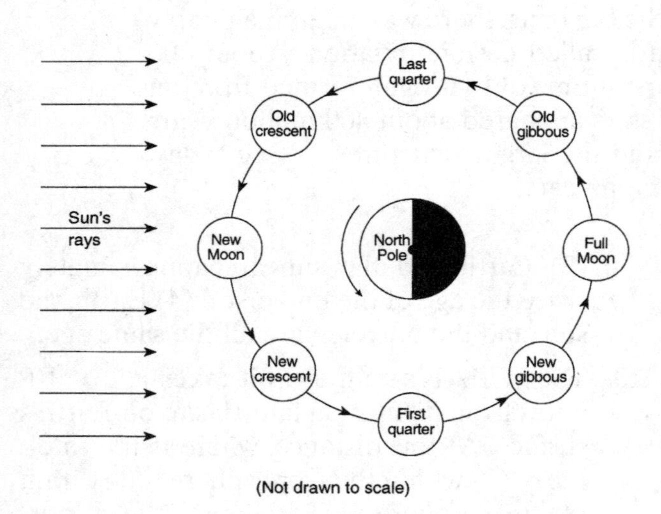

(Not drawn to scale)

16. What causes the moon to change phases as viewed from Earth?

17. In the circle below, shade the portion of the moon that appears dark when the first-quarter phase is viewed from Earth.

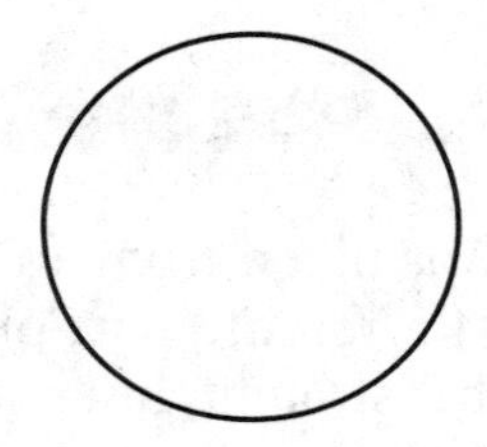

18. Which moon phase do we see approximately 4 days after the full moon?

19. The sun is approximately 400 times the diameter of the moon. Explain why the moon and the sun appear to be almost the same size when we observe them in the sky.

20. Name a star that is hotter than the sun and yet gives off less light than the sun.

21. Why are the phases of the moon called a cycle?

22. The orbits of some planets are more circular than others. What shape describes the orbits of all the planets?

23. When does the moon move fastest in its orbit?

24. Approximately how much more light does the star Aldebaran give off than the sun?

25. When a planet passes in front of the sun as viewed from Earth, the event is called a transit of the planet. Which planets can transit the sun as we observe them from Earth?

26. The diagram below shows the position of the sun and the orbits of Earth and Jupiter. The size of the sun is not to scale, but the orbits of Earth and Jupiter *are* to scale. Draw the orbit of Mars where it should be to the same scale and label it "Orbit of Mars."

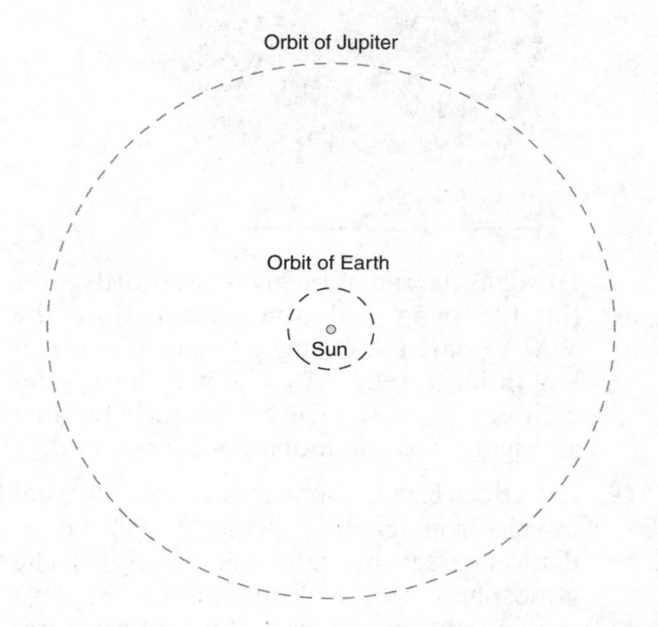

27. Betelgeuse is a prominent red star in the constellation Orion. Although the surface of Betelgeuse is cooler than the sun, Betelgeuse gives off thousands of times more light than the sun. How can a star cooler than the sun give off more light?

Base your answers to questions 28 through 31 on the data table below, which gives the celestial coordinates of the major stars in the constellation Orion.

Location of the Seven Brightest Stars in Orion

Star Number	Celestial Longitude (measured in hours)	Celestial Latitude (measured in degrees)
1	5.9	+7.4
2	5.4	+6.3
3	5.2	-8.2
4	5.8	-9.7
5	5.7	-1.9
6	5.6	-1.2
7	5.5	-0.3

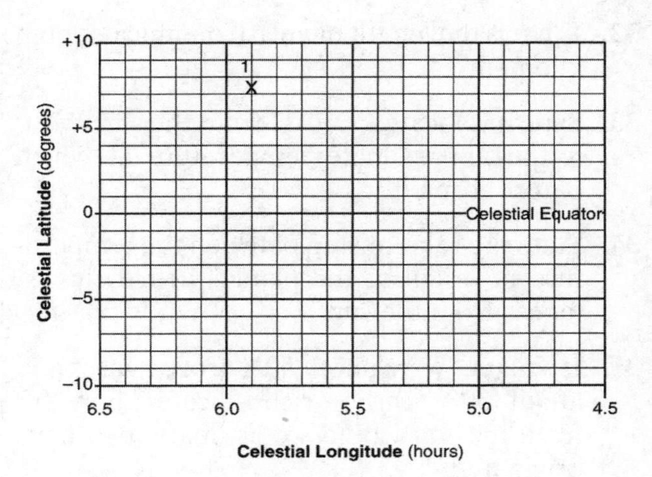

28. On a copy of the grid above draw the constellation Orion by following the steps (*a*) and (*b*).

(*a*) Mark an *X* at the position of each of the seven stars. Write the number of the star beside its *X*. The first star has been plotted for you.
(*b*) Connect the *X*'s in the following order: 5–1–2–7–3–4–5–6–7

29. Star 1 is Betelgeuse and star 3 is Rigel. When you see these stars in the sky, both appear to have about the same brightness. Aside from their different positions, what other difference would you notice?

30. All seven stars are close enough to Earth that they are relatively easy to see in the sky without using a telescope. In what galaxy are these seven stars located?

31. For an observer in New York State, Orion is easy to find in the evening sky in winter. However, it is never visible in New York State in the evening sky in summer. Why is it visible in the evening only in our winter?

Base your answers to questions 32 through 36 on the graph below, which represents the temperature of formation, general composition at the time of formation, and distances from the sun of the planets.

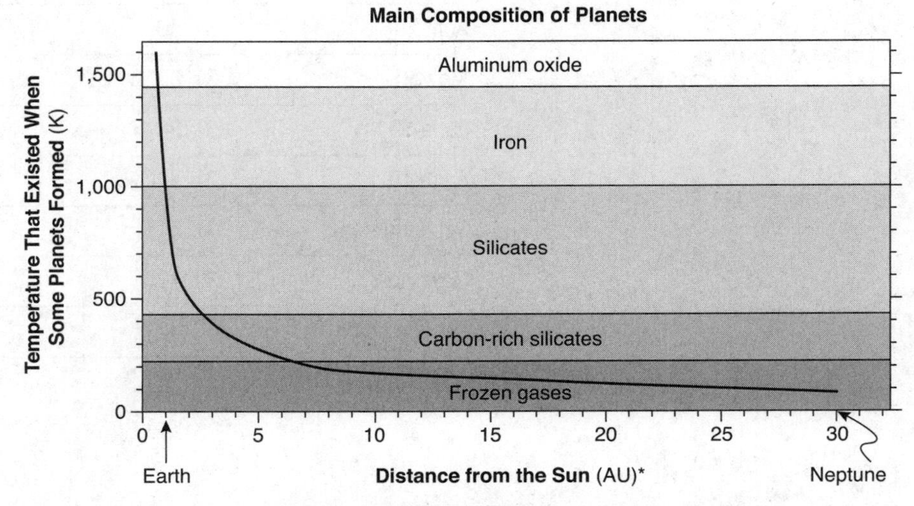

* 1 AU is an astronomical unit: the average distance from Earth to the sun.

32. What is the length of an astronomical unit in kilometers?

33. Saturn is located 9.5 AU from the sun. What is the approximate temperature at which Saturn formed?

34. State the relationship between the temperature of origin of the planets and their distances from the sun.

35. If a planet were about 50% frozen gases and about 50% carbon-rich silicates, how far from the sun would we estimate it to have originated?

36. Name a mineral or mineral family that is common on Earth and also might be common on the planet Mercury.

37. State one finding that scientists use to support the idea that the universe began as a primordial explosion we now call the big bang.

38. In the past century, astronomers have found new and more accurate ways to determine the age of the universe. Using a number of different techniques, the age of the universe recently has been estimated at between 13.5 and 14 billion years. How has the progress in age determination most likely affected the error between scientists' age estimates and the actual age of the universe?

39. Explain why solar and lunar eclipses do not occur with each new moon and full moon.

40. Calculate the eccentricity of the ellipse below to the nearest hundredth. Show your work starting with the appropriate equation.

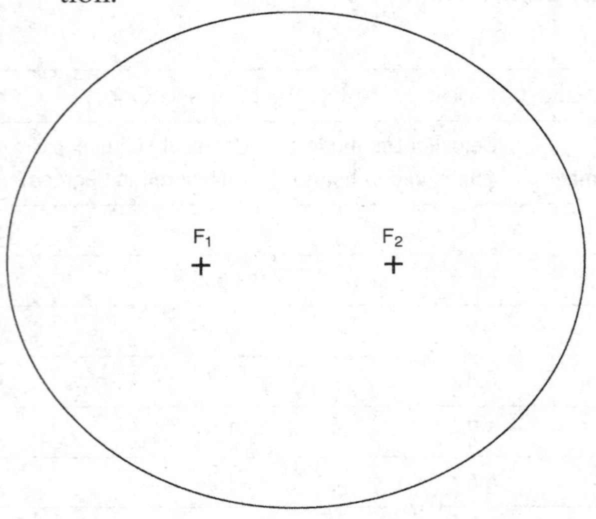

Base your answers to questions 41 through 45 on the table below, which shows changes in the apparent diameter of the sun as observed from New York State over a period of one year.

Apparent Diameter of the Sun During the Year

Date	Apparent Diameter (′ = minutes; ″ = seconds)
January 1	32′ 32″
February 10	32′ 25″
March 20	32′ 07″
April 20	31′ 50″
May 30	31′ 33″
June 30	31′ 28
August 10	31′ 34″
September 20	31′ 51″
November 10	32′ 18″
December 30	32′ 32″

41. On a copy of the grid below, graph the data in the table. Plot each data point as a dot and then connect the dots to make a continuous line.

42. Explain why the apparent diameter of the sun changes for observers on Earth.

43. During this 1-year period, what was the range in the angular diameter of the sun?

44. As Earth orbits the sun, it is closest to the sun each year in January. Explain why residents of New York State generally experience their coldest weather in January.

45. The moon is covered with many visible impact craters. State *two* reasons why there few impact craters on Earth.

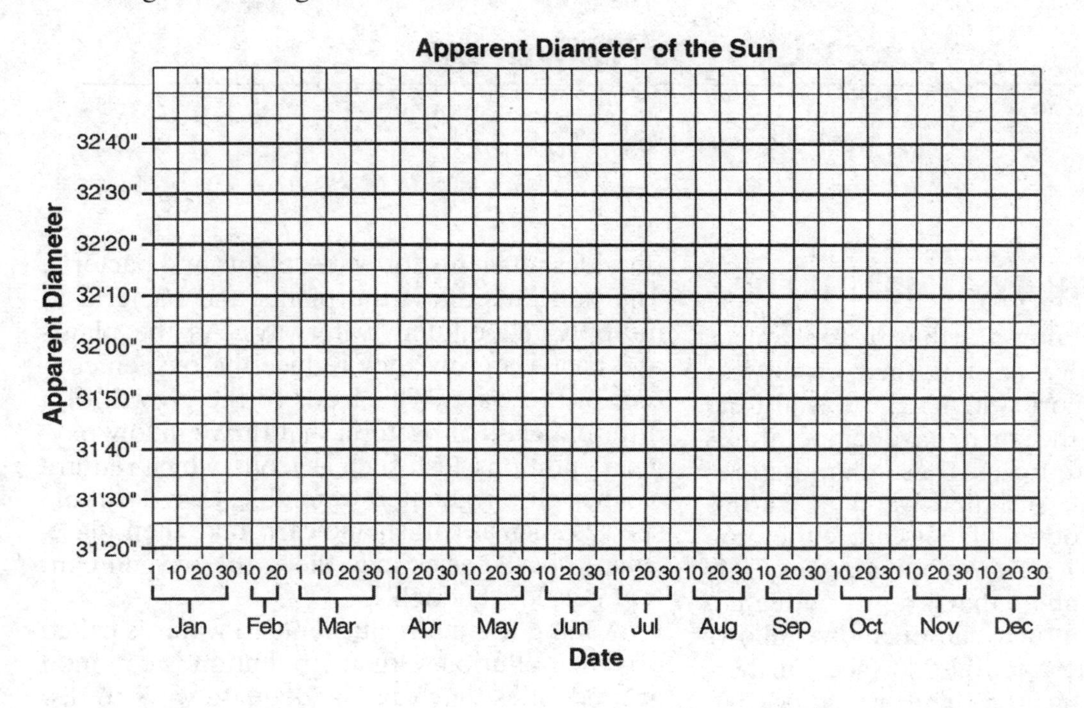

CHAPTER

Environmental Awareness

Vocabulary at a Glance

equilibrium
ozone
ozone layer
photochemical smog
pollution

HOW ARE EARTH SYSTEMS LINKED?

As far as scientists know, Earth is the only place where humans can live without the constant use of artificial life-support systems. Earth is just the right distance from the sun to have temperatures that favor higher life-forms. These temperatures allow large amounts of liquid water at Earth's surface. The sun's other planets are either too hot or too cold to support human life. Only Earth has an atmosphere that is dense enough to protect us from harmful radiation and yet allows solar energy to penetrate to its surface. Furthermore, Earth alone has the right mix of gases to support higher forms of life.

Balance of Nature

A scientific principle states: If part of a system at **equilibrium** (in balance) is disturbed or changed, other parts of the system will change to establish a new balance. As a result, when we change or pollute the environment, other problems are likely to occur as the environment comes to equilibrium at the new conditions. **Pollution** is any substance or form of energy whose concentration is large enough to harm living things or the natural environment. The equilibrium of the environment is upset by pollution.

For example, when you fertilize your lawn, some of the fertilizer will be washed away by rain. It will be carried into nearby streams and lakes. The result may be more significant than just a little fertilizer in the water. The fertilizer provides nutrients for water plants and bacteria. The nutrients allow the plants and bacteria to multiply in polluted waterways. As the plants and bacteria grow, they reduce the oxygen content of the water. This leads to the growth of a different class of bacteria that thrive in low-oxygen conditions. Fish such as trout, which require clean, well-oxygenated water, die. Less desirable species, such as catfish and carp, take their place. Even those species may die if the oxygen content gets low enough.

Adding too many nutrients to water is called organic pollution. Organic pollution is common around cities that have inadequate ways to dispose of human and industrial wastes. It also occurs in farming areas where animal wastes and fertilizers are left on the land. In short, simply fertilizing your lawn can contribute to undesirable changes in other ecosystems.

HOW DOES TECHNOLOGY AFFECT THE ENVIRONMENT?

New technologies often create new sources of pollution. When chemists invented chlorofluorocarbons (CFCs), they thought CFCs were miracle substances. These inexpensive compounds are excellent at carrying away heat, and they are relatively stable. CFCs made air conditioning more affordable in homes, businesses, and motor vehicles. However, in the late 1970s, satellite-based measurements showed that CFCs were drifting into the upper atmosphere where they were destroying Earth's protective **ozone layer**. **Ozone** (O_3) is a form of oxygen that shields Earth from the sun's harmful ultraviolet radiation. CFCs convert ozone into oxygen (O_2), which does not protect us from ultraviolet radiation. In 1992, more than 90 nations, including the United States, agreed to end production of CFCs by 1996. Fortunately, other substances have been found to replace CFCs, and the destruction of Earth's ozone layer seems to be slowing.

Other pollutants are contaminating our waterways. Electrical transformers convert high-voltage electricity, which can be transmitted long distances, into the lower voltage used in our homes. Although transformers are efficient, some of the electrical energy is lost as heat energy. Water could be used to cool the transformers. However, water evaporates easily, leaving the heated parts exposed. The exposed parts could be damaged or cause a fire. Polychlorinated biphenyls (PCBs) disperse waste heat, and do not evaporate easily. In 1979, the United States banned the production of PCBs. Unfortunately, large amounts of PCBs were dumped into the Hudson River at factories where they were manufactured. These PCB wastes settled into river sediments near these factories.

The PCBs have now entered the food chain, and are found in the fish that live in the river. Studies have shown that in high concentrations PCBs may cause birth defects, cancer, liver damage, and nerve disorders in people. There are debates among environmentalists about how best to reduce PCBs in inland waters. Every solution has costs and benefits that must be considered. Is the cleanup worth the expense? What can we do to dispose of the contaminated sediments? Furthermore, many solutions proposed for the disposal of PCBs also have the potential to harm the environment in other locations.

HOW CAN WE CLASSIFY POLLUTION?

We sometimes classify pollution by the part of the environment it affects: air, water, or ground. Pollutants are also classified according to their source. Individuals, households, communities that collect and dispose of pollutants improperly, and industries, some of which use a wide range of hazardous substances and processes, all create pollution. Figure 12-1 shows how pollution may be classified.

Air Pollution

Carbon dioxide (CO_2) is a natural part of Earth's atmosphere. Living organisms that take in oxygen for respiration give off CO_2 as a waste product. Figure 12-2 on page 302 illustrates how CO_2 circulates through the environment. This circulation is the carbon cycle. Scientists know

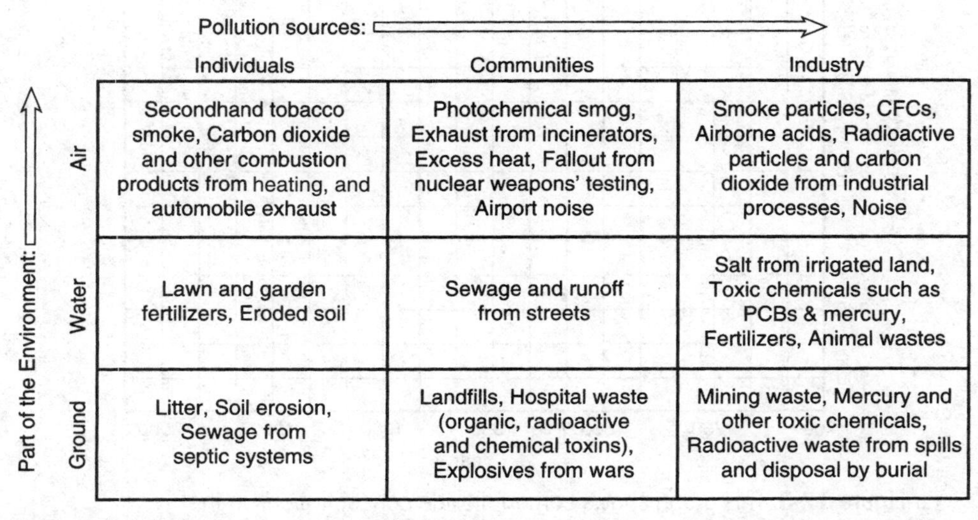

Pollution sources: ⟹

Part of the Environment:	Individuals	Communities	Industry
Air	Secondhand tobacco smoke, Carbon dioxide and other combustion products from heating, and automobile exhaust	Photochemical smog, Exhaust from incinerators, Excess heat, Fallout from nuclear weapons' testing, Airport noise	Smoke particles, CFCs, Airborne acids, Radioactive particles and carbon dioxide from industrial processes, Noise
Water	Lawn and garden fertilizers, Eroded soil	Sewage and runoff from streets	Salt from irrigated land, Toxic chemicals such as PCBs & mercury, Fertilizers, Animal wastes
Ground	Litter, Soil erosion, Sewage from septic systems	Landfills, Hospital waste (organic, radioactive and chemical toxins), Explosives from wars	Mining waste, Mercury and other toxic chemicals, Radioactive waste from spills and disposal by burial

Figure 12-1. Pollution can be classified by the source of the pollutant and the part of the environment it affects. Many forms of pollution belong to more than one class. For example, buried toxins may enter the groundwater and streams (water pollution) or produce an odor that can escape into the air (air pollution).

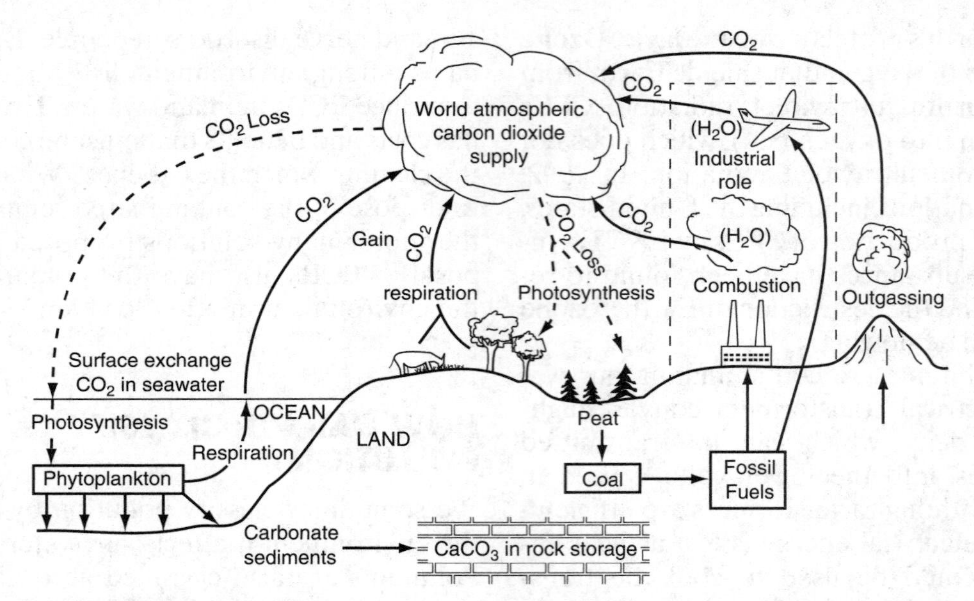

Figure 12-2. Respiration and volcanoes are the primary natural sources of carbon dioxide. Humans also add carbon dioxide (CO_2) to the atmosphere when we burn fossil fuels (oil, coal, and natural gas). CO_2 is absorbed by seawater and taken from the atmosphere by plants through photosynthesis. Large amounts of CO_2 are held in carbonate rocks, primarily limestone.

that burning fossil fuels adds carbon dioxide to Earth's atmosphere. Figure 12-3 shows how the concentration of CO_2 in the atmosphere has increased since the beginning of the Industrial Revolution. In addition, scientists know that carbon dioxide is a "*greenhouse gas*" that causes global warming and an unknown variety of localized climatic changes.

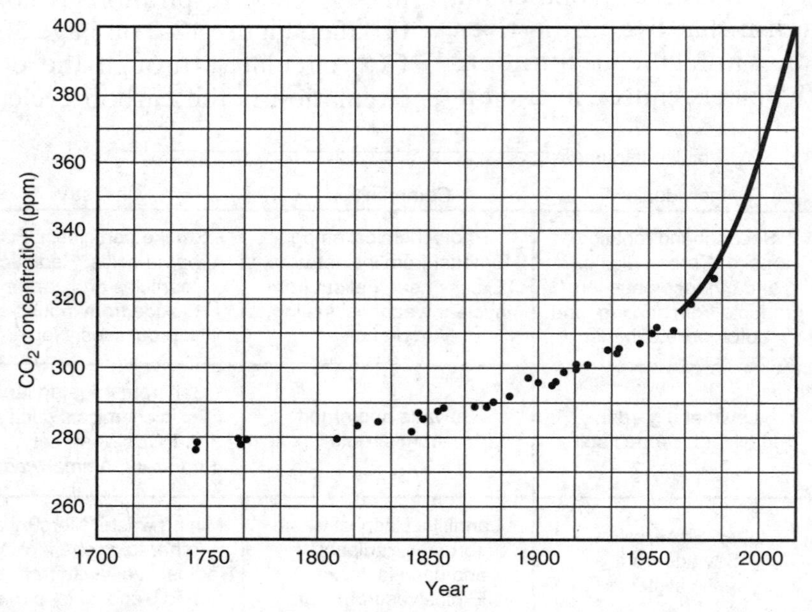

Figure 12-3. This graph shows carbon dioxide concentrations in the atmosphere in parts per million (ppm). The solid line, which begins in 1958, shows continuous measurements of CO_2; the dots represent ice core data. Scientists know the CO_2 content of Earth's atmosphere is increasing. Carbon dioxide is an important greenhouse gas.

In 2009, the U.S. Environmental Protection Agency (EPA) declared that carbon dioxide emissions could be regulated as a source of air pollution. Human-caused emissions of carbon dioxide are considered a major source of greenhouse gas and a cause of global warming. Figure 12-4 shows the most important human-produced greenhouse gases.

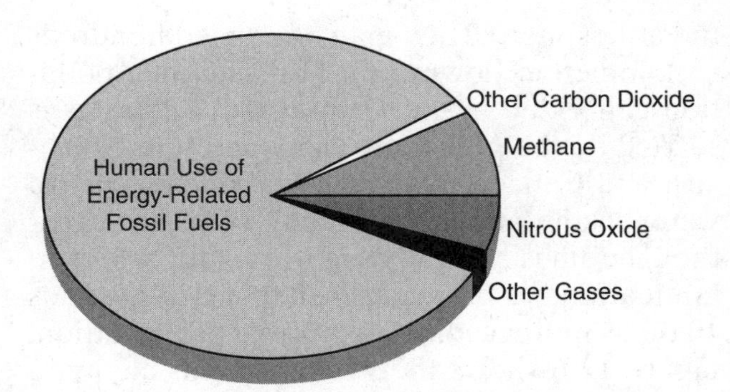

Figure 12-4. Carbon dioxide from the use of fossil fuels (petroleum, coal, natural gas) is the most important contributor to global warming.

It is not clear how global climatic change (global warming) will affect any particular place on Earth. Some places will probably experience more warming than others. In some places, a reduction in rainfall and a change to a desert climate might cause greater problems than the expected increase in temperature.

It is clear that Earth's ice is melting. Most mountain glaciers are melting back, exposing valleys never before seen by people. The polar glaciers are getting thinner. Each summer more sea ice becomes open water. Figure 12-5 shows the Arctic ice cover late in recent summers. Sea ice has recently reached historic lows. A loss of ice is not the most distressing problem. The greater issue is that melting glaciers on Greenland and Antartica could rise sea level hundreds of feet, displacing millions of people now living in coastal areas.

Photochemical Smog Metropolitan Los Angeles is one of the fastest-growing areas in the United States. Los Angeles sits in several valleys located between the San Gabriel Mountains and the Pacific Ocean. The surrounding mountains reduce the free flow of winds from the ocean. Exhausts from motor vehicles, homes, and power plants are trapped and changed by sunlight into the brown haze known as **photochemical smog**. Ozone is a component of photochemical smog. The same ozone we need in the upper atmosphere becomes a pollutant at ground level. Ozone is a concern for people who have respiratory problems.

Oxides of Nitrogen and Sulfur When we burn fossil fuels without proper pollution controls, oxides of nitrogen and sulfur are released into

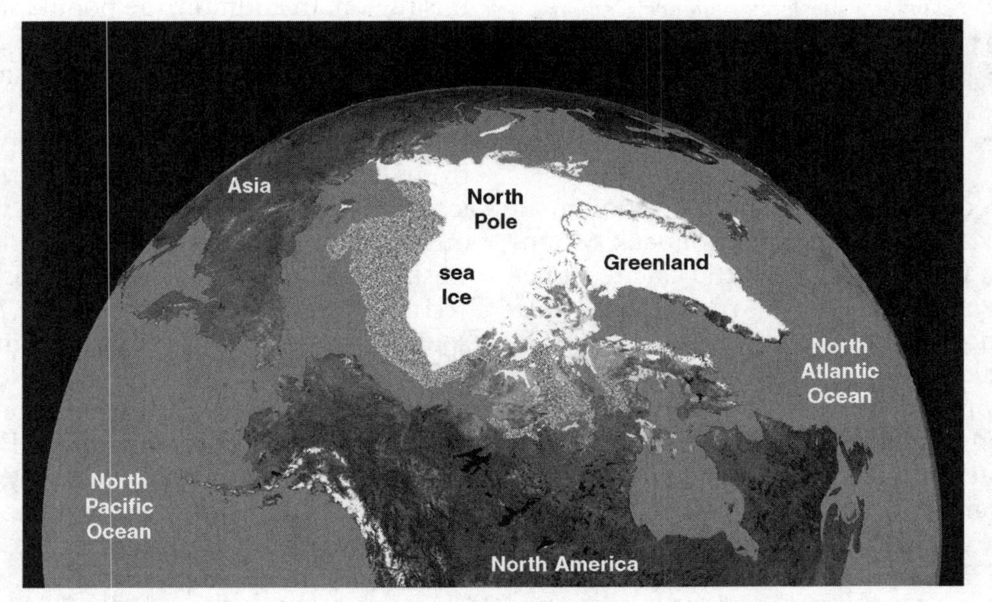

Figure 12-5. In recent years, global warming has caused more Arctic sea ice to melt than ever before. The white areas on this map show sea ice and glaciers in recent summers. The stippled areas show parts of the Arctic Ocean that until recent years did not melt.

the atmosphere. They may be carried hundreds of kilometers downwind. For example, pollution generated in the Central United States is carried by the prevailing winds into the Northeast and Canada. When these oxides mix with water in the atmosphere, they form acids. Although rain is naturally slightly acidic, acid contamination has produced rain that is as much as 40 times more acidic than normal precipitation. Figure 12-6 shows the pH of various common substances. (The pH of a substance is a measure of its acidity. The lower the pH, the more acidic the substance.)

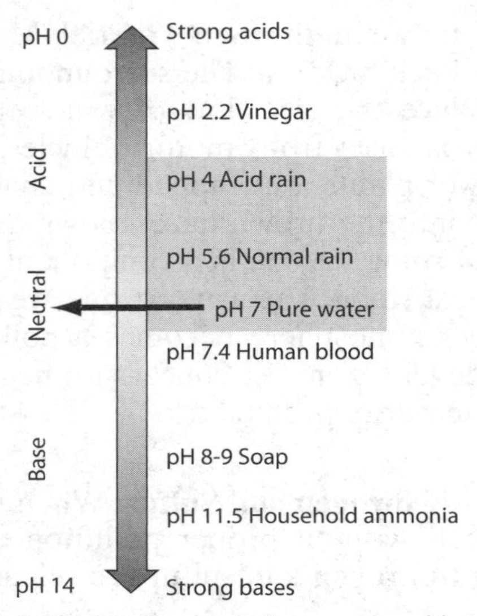

Figure 12-6. Many natural substances are acids, others are bases. The pH scale is used to measure the strength of acids and bases. A pH of 7 indicates that a substance is neither acidic nor basic.

Acid precipitation is a special problem in places such as New York's Adirondack Mountains, where the bedrock has little calcite (limestone or marble). These rocks can help neutralize acid in surface water. The accumulation of acidic winter snow is a special problem because it produces a rush of acid water in the spring, when the snow melts. Unfortunately, fish spawn (mate and lay eggs) in the spring. Acid precipitation harms young fish that are especially sensitive to acidified water. Industrial development and pollution often cause acid deposition to occur downwind. The most important causes of air pollution in the United States are shown in Figure 12-7.

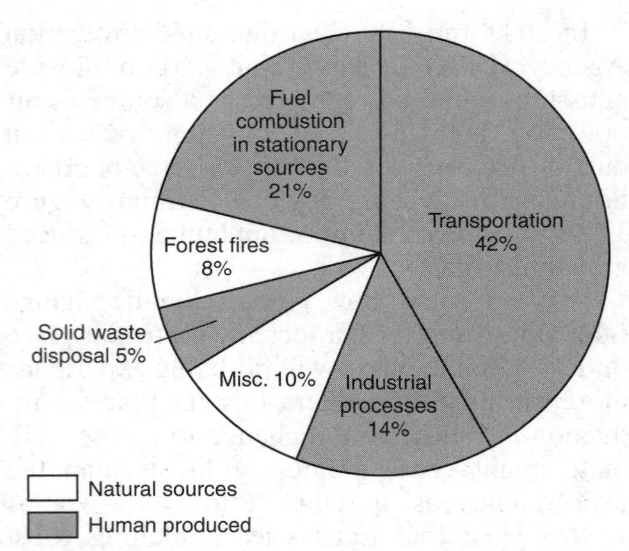

Figure 12-7. Sources of air pollution in the United States (percent by mass).

Water Pollution

The warm, sunny climate and infrequent frosts in Southern California and Arizona make this area an excellent place to grow fruits and vegetables. This is especially true in the winter when fruits and vegetables cannot be grown in other Northern Hemisphere regions because of the cold, winter climate. However, water is in short supply in the Southwest. The Colorado River is used and reused for irrigation as it flows through this region. In addition, the population of this region already uses the entire discharge volume of the Colorado River, and the human population is still growing. During the river's passage through this desert area, the water picks up salts from the soil. By the time the river reaches Mexico, very little water is left, and that water is seriously contaminated with salts and other residue washed from the irrigated fields upstream.

In the late 1950s, many people living in the southern part of Japan along Minamata Bay developed serious health problems. Nearby industries discharged wastes that contained the element mercury into Minamata Bay. Even in very low concentrations, mercury, like other heavy metals, can cause permanent nerve damage. Many people ate fish from the bay and became sick. Some people died. This was a tragic event for local residents. Today scientists know that mercury contamination is a problem in most of the world's oceans. Figure 12-8 presents some

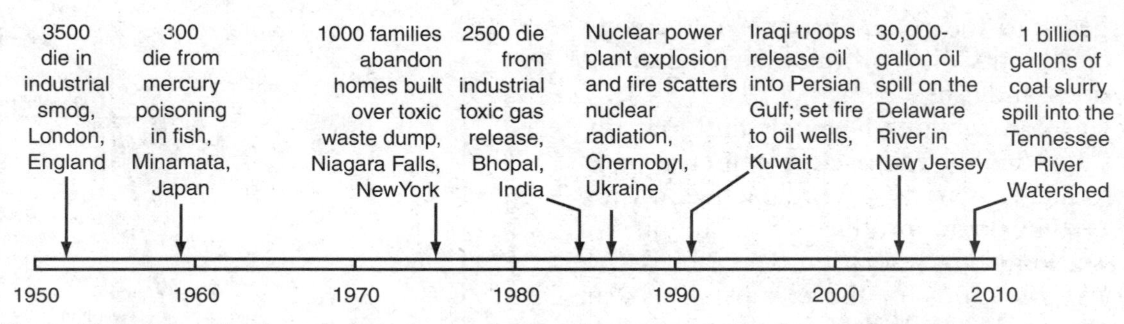

Figure 12-8. Although advances in technology can make our lives better, they can also lead to environmental disasters.

notable environmental disasters in a historical context.

Environmental awareness became an important political issue in the latter part of the 20th century. As a result, new sewage treatment plants were built in regions such as the Hudson River watershed, while dumping of industrial wastes was monitored and reduced. The river looks cleaner today than it has in many years and native fish are returning to a habitat they had once abandoned.

Ground Pollution

We are constantly trying to find better ways to dispose of our household and community wastes. At one time, it was common for garbage to be taken to dumps where it was burned. However, it became clear that the open burning of trash was creating unacceptable levels of air pollution. In addition, these dumps were infested with rats that carried diseases. Many communities developed "sanitary landfills." In a sanitary landfill, the garbage is covered each day with a layer of soil. However, rainwater and groundwater that run through the landfills carry dangerous chemicals into the groundwater and nearby streams.

Today, some communities are building high-temperature incinerators that burn hot enough to decompose most of the dangerous substances that were once released by outdoor burning. Nevertheless, even modern incinerators produce toxic ash. Although burning reduces the volume of the waste, the ash that remains contains a higher concentration of some dangerous substances, such as heavy metals, including mercury. The contaminated ash must be buried in thick plastic sealers to prevent the toxins from escaping into the ground and nearby streams.

In Niagara Falls, New York, local officials allowed the construction of a large housing development and a school on land where toxic chemicals had been buried. This started the Love Canal scandal of the 1970s. After the families moved in, they noticed an above-normal occurrence of diseases and birth defects. In addition, the residents observed toxic liquids seeping out of the ground. About 900 residents were forced to move from their homes. These homes had to be abandoned when the government declared the site to be a toxic waste dump unfit for human habitation. The angry homeowners were not only exposed to dangerous chemicals, but many of them lost the life savings they had invested in their homes. Love Canal because one of four toxic waste sites in New York State designated by the federal government for highest-priority Superfund cleanup.

Energy Pollution

When we think of pollution, we usually think of substances in the environment. However, pollution can also occur in the form of energy. Noise, heat, and nuclear (ionizing) radiation can be harmful to humans.

Noise is easy to observe. People who live near airports or work in noisy places are certainly aware of their exposure to high levels of sound energy. Exposure to loud sounds can cause hearing loss. Most airports try to divert low-flying aircraft from densely populated areas.

Power plants (nuclear and fossil fuel) produce large quantities of excess heat. They use cooling towers that pump water from a nearby river to dispose of the heat. The heated water is discharged into the river, changing its ecology. Fossil fuel plants also produce air pollutants: CO_2, oxides, and ash.

Nuclear power plants, hospitals, and some industries use highly radioactive materials and they produce a variety of radioactive wastes. However, they do not produce air pollutants. We cannot sense nuclear radiation. The radioactive emissions given off by these substances can cause cancer, birth defects, and even death. Some of these waste products will continue to give off deadly quantities of radiation for thousands of years. There is disagreement about the best way to dispose of these substances. Some scientists have suggested burying high-level radioactive wastes deep underground. While others argue that it is safer to store these materials on the sites where they are produced.

Not all exposure to radioactivity is artificial. Harmful levels of radioactive radon gas from the decay of uranium in bedrock can seep into homes built over natural uranium deposits. Some homebuyers now require testing for radioactive radon gas before they purchase a home. Some states require it.

HOW HAS THE HUMAN POPULATION GROWN?

The human population increased gradually from the time of the earliest civilizations until the Industrial Revolution in the 1800s. The industrial revolution allowed farmers to produce more food at lower cost than ever before. At the same time, advances in medicine reduced infant mortality and greatly reduced the occurrence of such diseases as tuberculosis, pneumonia, and smallpox. Now, there are nearly 7 billion people on this planet

Clearly, Earth's human population cannot continue to increase at an accelerating pace. However, the populations of the wealthier nations of Europe and North America have stabilized. In these places, parents no longer need large families to ensure that someone will care for them in their old age. The government can assume that responsibility. The expense of providing an education for their children also helps keep families small in these nations. As wealth spreads from privileged nations to the rest of the world, it is now predicted that the population of

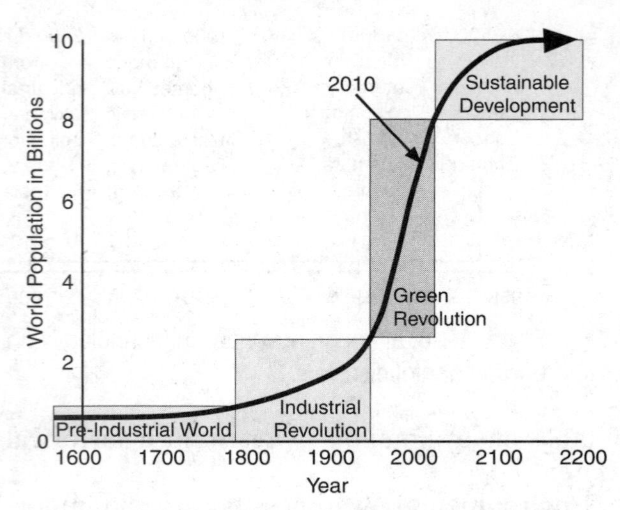

Figure 12-9. The current explosion in world population is primarily due to the benefits of the Industrial Revolution, which dramatically increased production and wealth. In addition, the green revolution has nearly eliminated food shortages. Scientists expect that in the next 200 years the world population will level off, as have the populations of the wealthiest nations.

the planet will stabilize at a sustainable number of roughly 10 billion, as you can see in Figure 12-9. Although this may exceed our present ability to support people, the green revolution (genetic engineering) and technological progress seem likely to make this future quite comfortable for most people.

HOW CAN WE BEST MANAGE OUR RESOURCES?

We depend on Earth for all our physical needs. Many of our resources are renewable. *Renewable resources* circulate through the environment and are replaced, or regenerated, as we use them. For example, as we use water from a stream, it generally is replaced by precipitation and continued stream flow. Trees are cut to make paper and building materials. If we can grow trees at the same rate at which they are cut, our wood reserves remain constant. Wise management will lead us to use our renewable resources no faster than they can be replaced by natural cycles.

Other resources, such as mineral ores, fossil fuels, and soil, are *nonrenewable* because they would take a very long time to regenerate. Most of the world's deposits of iron ore formed billions of years ago before oxygen became abundant in Earth's atmosphere. As we mine iron ore, there is little chance of replacing it through natural cycles. We need to plan wisely so that we can make the best use of these nonrenewable resources. For example, recycling or reusing nonrenewable resources will help extend Earth's limited supply of these materials.

Soil is a critical and nonrenewable resource. As more land is cleared for agriculture, more soil is exposed to erosion. However, farmers have learned to protect soil from erosion by contour plowing, making terraces, and using other conservation practices. They are also able to restore important soil nutrients with fertilizers. Fertilizers and water irrigation are even used to create new farmland.

Many resources are not clearly renewable or nonrenewable. For example, groundwater in a moist climate region is quickly replaced by infiltration from precipitation and streams. However, in desert regions, the rate of water withdrawal may exceed natural replacement. In Arizona and California, "mining" of groundwater thousands of years old is causing the land surface to sink in some places. Is fresh air a renewable resource? Could humans pollute it so badly that fresh air would be no longer available? Clearly this problem already occurs in some urban and industrial areas.

Will We Run Out?

What happens when we run out of resources? History provides us with several examples. In the 1800s, whale oil was used in lamps as a clean-burning source of indoor lighting. As the number of whales decreased, normal effects of "supply and demand" took over; the price of whale oil increased. However, this did not mean that people had to live without interior lighting. At first, whale oil was replaced by kerosene, which was less expensive to produce. The next step was light produced by burning natural gas. Then, in 1878, Thomas Edison invented the electric light bulb, which is our major source of lighting today.

In many places where wood has been used as the main fuel, there are almost no trees left standing. This occurred in much of Europe, Eastern North America, as well as parts of Africa and Latin America. Wood is often replaced by fossil fuels (oil, gas, and coal). As we run low on fossil fuels we will probably need to increase our dependence on nuclear energy and new energy technologies, such as solar, wind power, and geothermal energy. Throughout history, as one source of energy has been used up, another has been developed to take its place. Solar technology, which is nearly pollution free, may provide a large part of our energy in years to come.

The Future

We need to think of our resources as investments in our future. Investments provide the greatest benefit when they are managed properly. If we squander our resources, they will be gone before we can fully appreciate their value. Our goal should be to use resources wisely and with the greatest possible awareness of our future needs.

This is a three-step process. First, scientists need to conduct scientific investigations to find the best ways to use and conserve resources. Second, scientists must explore alternative ways to provide for future needs. Third, we need to educate ourselves and our decision-makers about the need for conservation planning and scientific research.

We can learn to become better citizens of our planet if we understand and appreciate the Native American quotation, "We do not inherit the Earth from our ancestors; we borrow it from our children."

PRESERVING EARTH SYSTEMS

Our planet has many complex natural systems that change through time and affect one another. These systems include living things, the atmosphere, oceans, and continents. No other species has the ability to use and modify Earth in the way humans have. We live in a time when great changes are taking place on Earth. Our planet home is subject to ever-increasing stresses. The products of modern technology demand increasing amounts of limited resources. This places the natural environment under more stress than ever before. However, people can use technology to minimize harmful effects. We can even use technology to help restore the natural environment.

Public pressure plays an important role in changing human behavior. For example, attitudes toward recycling have changed, in part, because of the costs involved with disposing of trash. A market for recycled materials has also helped change attitudes by giving value to materials that were once thrown away. Recent campaigns to recycle materials that were once buried or burned have benefited greatly from public acceptance. This acceptance has, in large measure, been the result of favorable publicity in the media and in schools. In addition, laws have been passed in many cities and towns that require the recycling of some materials. Maintaining a beautiful and productive natural environment is something we must do as individuals, communities, and nations.

Problem Solving

Our national forests are land owned by the federal government for the benefit of all citizens. In some locations, they are grasslands or even desert environments. During the Bill Clinton presidency, large tracts of federal land were set aside purely for preservation. However, when George W. Bush was elected, he decided that certain commercial uses of federal lands, such as cattle grazing, harvest of lumber, and even mining, could be allowed without harmful long-term effects. However, in each case, the land needed to be managed or restored to minimize the environmental impact.

We now have President Obama, who is open to suggestions from citizens. In cooperative groups, make three lists of the ways we can use our federal lands. List *A* will be those uses that have little or no environmental impact. List *B* would be uses for which the increased impact is acceptable or restoration can be readily accomplished. List *C* would be uses for which the environmental impact would be harmful and probably irreversible. Then establish guidelines that your group feels are appropriate for the government to allow commercial use of public lands.

Chapter Review Questions

PART A

1. An environmental scientist needs to prepare a report on the potential effects of a proposed mineral mine. If the scientist wants determine where surface runoff from the mine site will flow, which resource will be most helpful? (1) a topographic map (2) a geologic time scale (3) a map of planetary winds (4) a large-scale road map

2. The photo below shows an airplane flying over central Los Angeles. You can see a thick blanket of photochemical smog over the city. The dark area in the center of the image includes the tall buildings of the downtown area. Which action would *not* help clean the air in major cities such as Los Angeles? (1) Encourage citizens to use

public transportation. (2) Raise taxes on fossil fuels. (3) Build new high-speed lanes on freeways. (4) Pass new laws to control automobile fuel emissions.

3. Which of the following would best reduce greenhouse gases in Earth's atmosphere? (1) Replace oil-burning power plants with coal power. (2) Remove catalytic converter systems from automobile exhaust systems. (3) Raise wages in the developing world to promote a consumer economy. (4) Restore rainforests in tropical and mid-latitude locations.

4. Greenhouse gases such as water vapor and carbon dioxide prevent the escape of infrared radiation from Earth. What wavelength is within the infrared part of the electromagnetic spectrum? (1) 10^{-6} cm (2) 10^{-3} cm (3) 10^{-1} cm (4) 10^{3} cm

5. If human activities continue to add carbon dioxide to Earth's atmosphere at an increasing rate, how would this most likely affect the climate at the North Pole? (1) This would cause warmer summers and warmer winters. (2) This would cause warmer summers and colder winters. (3) This would cause cooler summers and warmer winters. (4) This would cause cooler summers and colder winters.

6. In the late 20th century, trees in the higher parts of the Adirondack Mountains began to die off and aquatic animals in stream and lakes became scarce. Acid precipitation and deposition was the major cause of this change. What is the best way to reduce acid precipitation in the Adirondack Mountains? (1) accelerate global warming (2) stop global warming (3) reduce industrial pollution from the Central United States (4) change world patterns of prevailing winds

7. The graph below shows the percent concentration of copper at various depths in an underground mine in Arizona.

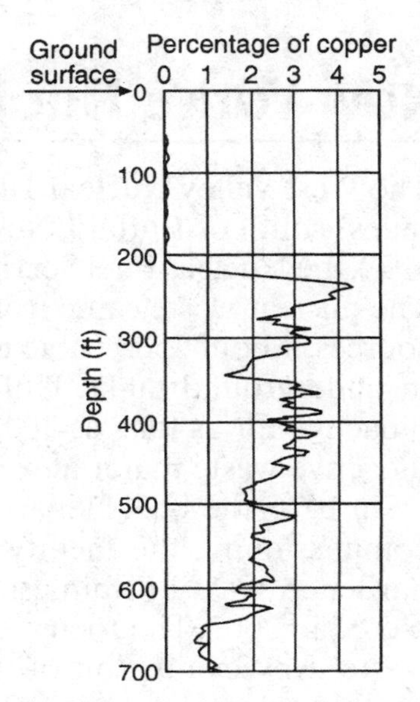

What depth should be mined in order to extract the most productive copper ore? (1) 100–130 ft (2) 230–260 ft (3) 330–360 ft (4) 650–680 ft

8. Which building material is made mostly from the mineral gypsum? (1) plastic pipes (2) window glass (3) drywall panels (4) iron nails

9. Which of the following is generally considered a renewable nonmineral resource? (1) petroleum (2) hematite (3) quartz (4) wind

PART B

Base your answers to questions 10 through 23 on the information and the map below. The map shows two underground storage tanks as gray rectangles 100 meters east of the black dots of the Nuclear Service Center buildings. (The tanks are located just north of the word Western.)

New York's Nuclear Fuel Reprocessing Plant

The West Valley Nuclear Reprocessing Plant is located about 48 kilometers (30 miles) south of Buffalo, New York. In its six years of operation from 1966 until 1972, the Nuclear Fuel Services Center processed 640 metric tons of nuclear fuel. The nuclear waste came from nuclear power plants, research reactors, and other sources. About 2500 cubic meters of highly radioactive liquid waste was stored in underground tanks. Unfortunately, this liquid waste will remain highly radioactive for as long as 200,000 years. In addition, hundreds of tons of solid radioactive waste material were buried on the site.

In 1976, the U. S. Department of Energy began the expensive process of decommissioning the facility. A program to encase the high-level waste within hardened glass in stainless steel canisters was started in 1996. This was completed in 2002. The focus of the cleanup has now shifted to the low-level solid waste. Any evaluation of the West Valley site must include the potential for seismic events, deterioration of containment materials, and long-term soil erosion.

A Portion of the Ashford Hollow USGS Quadrangle, NY
Map region is approximately 42° 30' N, 78° 45' W

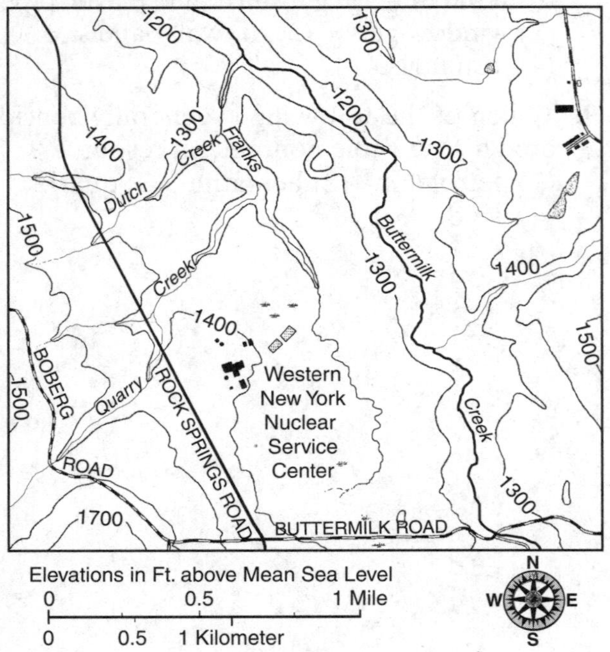

10. In what general direction does Buttermilk Creek flow? (1) northwest (2) northeast (3) southwest (4) southeast

11. If the radioactive liquid waste leaked from the tanks and into the ground, which streams would be most likely to transport radioactive waste? (1) Dutch Creek and Franks Creek (2) Buttermilk Creek and Dutch Creek (3) Franks Creek and Buttermilk Creek (4) Buttermilk Creek, only

12. What is the approximate elevation of the intersection of Rock Springs Road and Buttermilk Road? (1) 1350 feet above sea level (2) 1400 feet above sea level (3) 1450 feet above sea level (4) 1500 feet above sea level

13. If a leak had occurred in the liquid waste tanks, what would be the principal cause of movement of the pollution horizontally through the ground? (1) capillary action (2) stream flow (3) the force of gravity (4) the Coriolis effect

14. In what landscape region is this nuclear waste site located? (1) Tug Hill Plateau (2) Newark Lowlands (3) Taconic Mountains (4) Allegheny Plateau

15. If a major leak occurred at West Valley, which river might eventually become contaminated with radioactive waste through natural drainage processes?
(1) St. Lawrence River (3) Mohawk River
(2) Hudson River (4) Delaware River

16. What is the average gradient of Dutch Creek from where it crosses under Rock Springs Road to where it joins Franks Creek? (1) 150 feet/mile (2) 300 feet/mile (3) 600 feet/mile (4) 1200 feet/mile

17. What is the best way to prevent radioactive seepage from the Nuclear Service Center property? (1) Use high-pressure fire hoses to wash the ground once a week. (2) Build a fence around the radioactive waste site to prevent human access. (3) Build a high dam along Buttermilk Creek where Buttermilk Road crosses the creek. (4) Build a solid wall 50 feet into the ground and cover the site with a waterproof barrier.

18. What form of pollution is all high-level nuclear radioactivity? (1) air pollution (2) ground pollution (3) water pollution (4) energy pollution

19. What is most troubling about the radioactive pollution at the West Valley site?
(1) Radioactive contamination has a very bad odor. (2) Radioactive materials glow in the dark. (3) The radioactivity will last a very long time. (4) The radioactive substances will cause global climate change.

20. Why are seismic events and erosion considered a problem at the West Valley site?
(1) Exposing radioactive substances to the atmosphere makes them decay faster.
(2) When radioactive materials get wet, they stop giving off radiation. (3) Earthquakes and erosion can damage structures made to hold the waste. (4) Earthquakes are likely to prevent erosion and keep the waste at the site.

21. What is the best way for citizens to deal with the issues of radioactive substances?
(1) Learn about the benefits as well as the dangers of radioactive materials. (2) Keep your own supply of radioactive materials for observations and experiments.
(3) Never allow yourself to be exposed to any form of radiation. (4) Pass laws that prevent any use of radioactive substances in the United States.

22. Why did the Department of Energy encase the radioactive materials as a glass in stainless steel containers? (1) Stainless steel is shiny, so it reflects insolation. (2) As a solid, the hazardous waste cannot easily escape. (3) Glass allows electromagnetic energy to pass through. (4) Stainless steel and glass are popular with artists.

23. What is one advantage of nuclear energy over energy production by burning fossil fuels? (1) Nuclear energy does not create waste products. (2) Nuclear fuels are safe to handle and easy to store. (3) Nuclear technology can be shared without risk to people. (4) Nuclear energy does not release large amounts of greenhouse gases.

Base your answers to questions 24 and 25 on the text below.

In 1990, a large container ship ran into a storm, which washed overboard containers holding 80,000 Nike shoes. The spill occurred at 48°N latitude, 161°W longitude. All the shoes had a unique serial number. Over the next several years people found and reported shoes from this accident thousands of miles from the location of the spill.

24. In what body of water did the spill occur?
(1) Atlantic Ocean (3) Indian Ocean
(2) Pacific Ocean (4) Gulf of Mexico

25. Where did most of the shoes wash ashore?
(1) the east coast of North America
(2) the west coast of North America
(3) the east coast of Asia
(4) the east coast of Africa

Base your answers to questions 26 through 29 on the cross-sectional profile below, which represents a location where oil (petroleum), water, and natural gas have moved upward in a permeable sandstone layer to the bottom of a layer of shale.

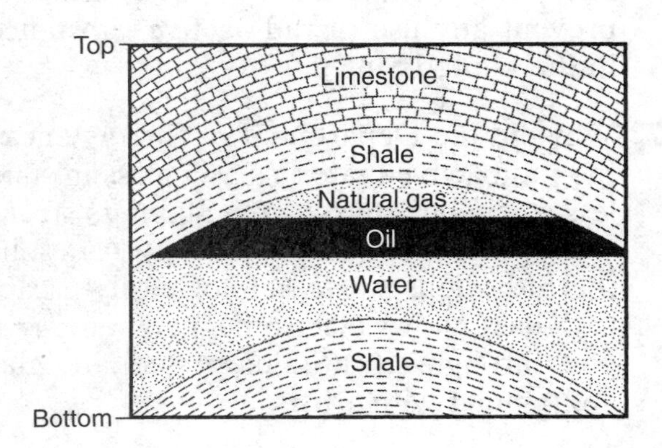

26. The final positions of the natural gas, oil, and water was caused by their differences in
(1) density (2) specific heat (3) relative age
(4) absolute age

27. What stopped the upward movement of these three fluids?
(1) low porosity of the sandstone
(2) low permeability of the sandstone
(3) low porosity of the shale
(4) low permeability of the shale

28. Oil and natural gas are (1) renewable resources that are minerals (2) renewable resources that are not minerals
(3) nonrenewable resources that are minerals (4) nonrenewable resources that are not minerals

29. In the above location, the water, oil, and natural gas are found within
(1) metamorphic rocks
(2) sedimentary rocks
(3) intrusive igneous rocks
(4) extrusive igneous rocks

PART C

Base your answers to questions 30 through 33 on the map below, which shows data points that indicate hydrocarbon contamination levels in the ground in a specific geographic area. A forest, a house, a gas station, a hospital, and a factory are shown in this area. The position of each number shows where that data was collected.

12	13	14	15	15	14	19	21
12	11	12	13	16	17	20	22
11	11	12	14	19	21	24	28
12	12	17	30	36	37	34	32
13	14	30	53	52	45	40	35
12	18	57	45	46	44	39	35
14	15	16	19	27	32	35	33
13	12	12	14	18	23	30	32
12	12	11	14	15	18	23	28
11	12	12	13	13	14	14	15

Key to Features

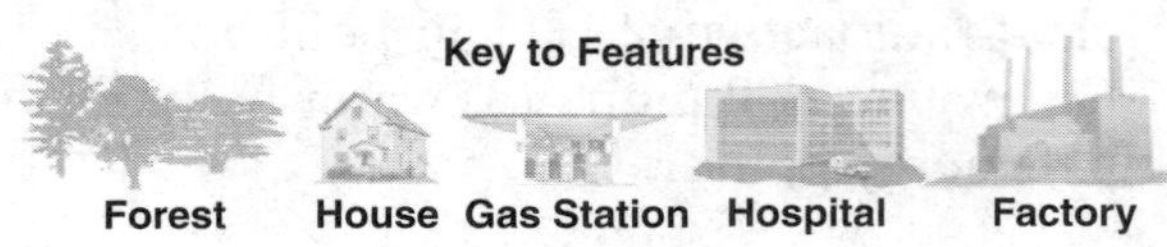

30. Draw the 20- and 40-hydrocarbon contamination unit isolines on the map above.

31. What appears to be the source of this hydrocarbon contamination?

32. North is toward the top of the map. In what direction does the contamination appear to be moving?

33. The people who live in the house get their drinking water from a well in the ground. They are concerned about the potential for contamination of their well. State *two* things they can do to protect themselves from the harmful effects of the contamination of their drinking water.

Base your answers to questions 34 through 37 on the data table below, which shows where carbon dioxide from human activities originates. Some parts of the table have been left blank.

Global Fossil Carbon Emissions*

Emission Source	Metric Tons per year $\times 10^9$	Percent of Total
Petroleum	3300	
Coal	2900	
Natural gas	1400	
Cement production	300	
Global total	7900	100

* Data Source: United States Department of Energy

34. Calculate the four missing values in the "Percent of Total" column.

35. In the blank pie chart below, draw and label the percent contribution of each human-caused carbon dioxide source.

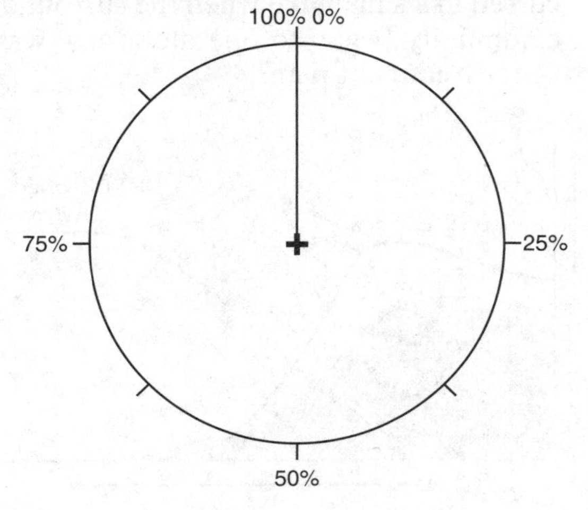

36. Most of us contribute to global warming through our use of fossil fuels. State two things you can do to reduce carbon dioxide emissions into the atmosphere.

37. If humans continue to add more carbon dioxide to Earth's atmosphere, what would probably occur in the Polar Regions that results in the destruction of low-lying coastal cities?

38. The average temperature on Venus is higher than on Mercury. Why is Venus hotter than Mercury?

Base your answers to questions 39 through 42 on the series of maps below, which show the spread of a cloud of volcanic ash from the eruption of El Chichón in Mexico between April 5 and April 25, 1982.

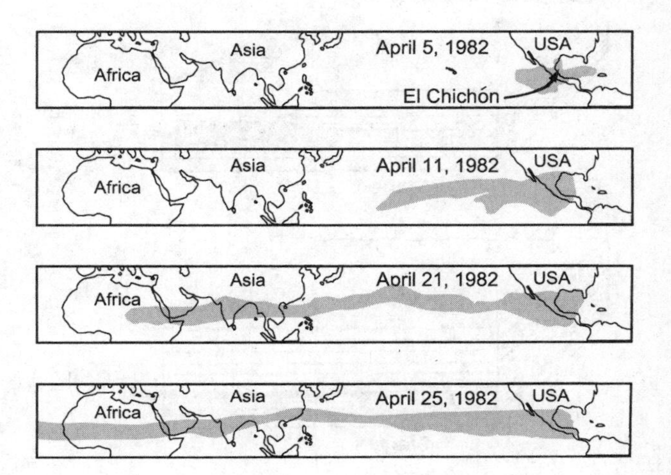

39. Identify the wind direction that is primarily responsible for this movement of the ash cloud.

40. Identify the global wind belt primarily responsible for the movement of this ash cloud.

41. How is this kind of ash cloud likely to affect insolation at Earth's surface?

42. As the ash cloud moved away from the volcano, in what *two* physical characteristics did the ash particles deposited near the volcano differ from those that were transported far out over the oceans.

43. What mineral resource is used in the manufacture of rubber tires?

44. The diagrams below show how coal forms from organic remains. Name *two* processes that convert peat into coal.

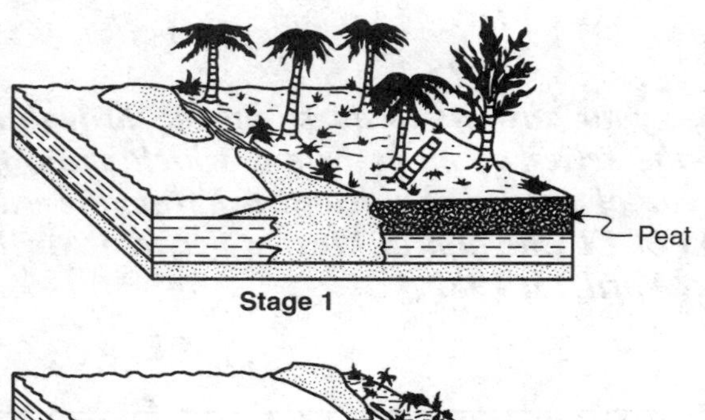

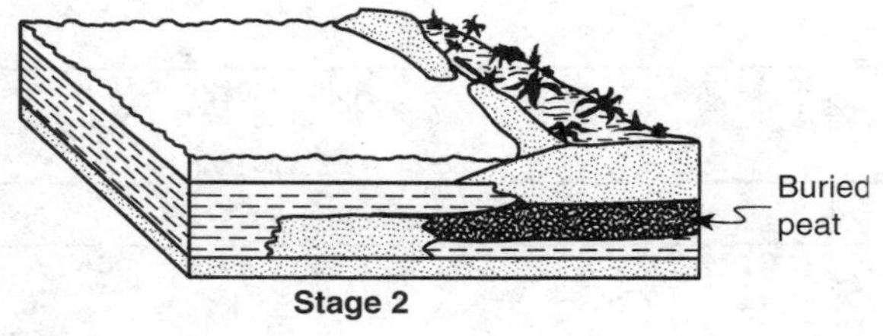

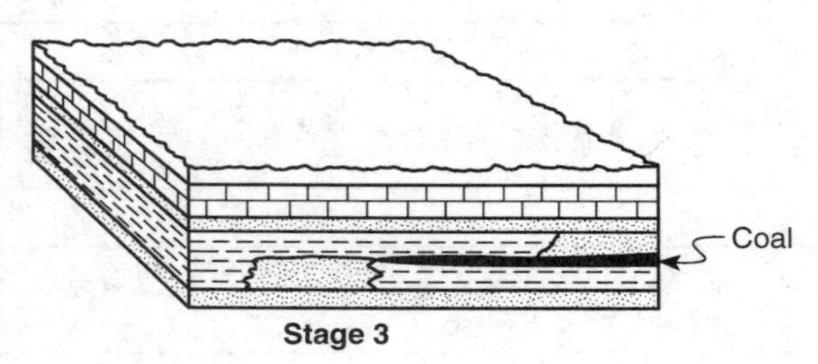

Base your answers to questions 45 through 48 on the table below, which shows estimated water use for a family of four in New York State.

Estimated Water Use for a Family of Four (Liters per Day)	
Irrigation	400
Other outdoor uses	220
Toilets	380
Bathing	320
Laundry	140
Dishes	60
Drinking and cooking	30

45. Using the data in the table, construct a bar graph on the following grid in the following steps:
(*a*) Label the vertical axis with the quantity being graphed, units, and numbers.
(*b*) Draw bars to show the water use for each household purpose.

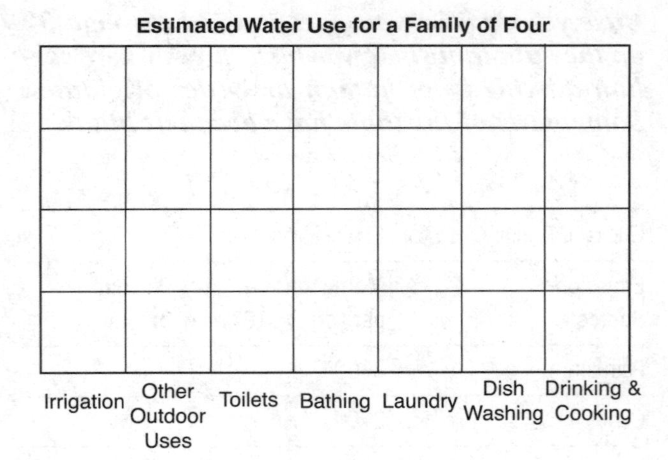

46. State *two* actions that a family could take to reduce water use by one-third without a major impact on safety or lifestyle.

47. What is the general relationship between the number of family members and the use of water?

48. Name three common sources of household water.

49. The graph below shows changes that occurred in a small lake when the surrounding community began to operate a new wastewater treatment plant.

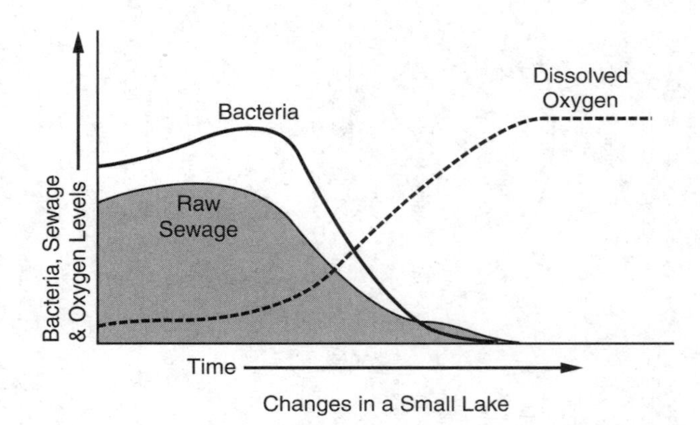

What will be the most likely change in the fish population of the lake over this time period?

50. Global warming is considered a change in Earth's energy equilibrium between incoming solar energy and the energy Earth radiates into space. How can we tell if our planet is in energy equilibrium?

Readings in Science, Engineering, Technology and Society

Geothermal Energy: It's Hot Down There!

For thousands of years, humans have made use of Earth's heat, or geothermal energy. Archeological evidence shows that Native Americans used hot springs for cooking food and bathing more than 10,000 years ago. The ancient Romans charged admission for baths fed by hot springs. That may have been the first commercial use of geothermal energy. During the 1800s, hotels and spas were built near hot springs. In 1892, Boise, Idaho, became the first town in the United States to use water from hot springs to create a municipal heating system.

In 1904, Larderello, Italy, was the first city to use geothermal energy to produce electricity. They used natural steam from vents in the ground to run a turbine. Today, the largest geothermal power installation in the world is in a valley north of San Francisco. It is located in an area called The Geysers. This installation was developed in the 1960s. The power plants at The Geysers generate over 750 megawatts of electricity.

Electricity from geothermal energy has a number of advantages over power generation from oil, coal, or natural gas. Emissions from geothermal power plants consist mainly of water vapor. Geothermal plants produce almost no carbon dioxide (the principal "greenhouse" gas). In addition, they produce no soot particles or other byproducts of combustion. What's more, use of geothermal energy does not require the mining and transport of fossil fuels or radioactive materials. It produces no radioactive wastes. Geothermal energy sources provide power constantly. It is unlike solar power, which cannot be generated at night or wind power which tends to be intermittent. Because Earth is likely to generate internal heat for millions of years to come, geothermal energy is a renewable resource.

However, generating electricity with geothermal energy requires high temperatures—at least 150°C. Such temperatures only occur in places where magma is fairly close to Earth's surface. These places are usually near the boundaries of Earth's tectonic plates. For example, Iceland is an island formed by lava from the Mid-Atlantic Ridge. It generates most of its electricity with geothermal energy.

Countries along the Pacific Ring of Fire (near the subduction zones that surround much of the Pacific Basin) can make good use of geothermal energy. So can places located at "hot spots"—areas where plumes of hot material rise up through Earth's mantle—like the Hawaiian Islands.

What about places located far from the plate boundaries? Is there any way to make use of Earth's inner heat in tectonically quiet areas, such as the northeastern United States? In a word, yes! It turns out that anywhere on the planet, temperatures within Earth increase with increasing depth. This increase in temperature with depth through Earth's crust is the geothermal gradient. The gradient varies from one location to another, but it averages about 25° to 30°C per kilometer.

To take advantage of this, companies drill wells 2 to 3 kilometers into the crust to reach heated groundwater. If the borehole is dry, they can pump water down into the well to absorb heat. However, this approach, known as an enhanced geothermal system, has its dangers. An enhanced geothermal energy project in Switzerland had to be abandoned. Injecting water into deep boreholes triggered earthquakes measuring up to 3.4 on the Richter scale.

Geothermal energy is also available at much shallower depths. Because rock and soil are fairly good insulators, temperatures just a few meters underground remain fairly constant throughout the year. They range from about 10° to 16°C (about 50° to 60°F). These temperatures are cooler than typical summer air temperatures and warmer than winter air temperatures. A geothermal heat pump will heat a home in winter, provide air conditioning in the summer, and produce hot water year-round. Water (or another heat-exchange fluid) pumped into underground pipes absorbs and concentrates the relative warmth of the ground. Pumping the throughout the home provides heat in winter. In summer, the water absorbs heat inside the home to cool the indoor air, and then releases the stored heat into the cooler ground. Geothermal heat pumps are becoming popular for rural and suburban homes.

Geothermal energy is even available in one of the most urban environments you can think of—Manhattan Island in New York City! In early 2008, a school began to drill several wells 450 to 550 meters (1500 to 1800 feet) into the bedrock below Manhattan. Their goal was to reach groundwater that stays at a temperature of about 18°C (about 65°F). They will use the energy in this relatively warm water to heat and cool the school, where about 200 students live and study. When this project is completed, it will be the largest geothermal pumping system in the northeastern United States. There are dozens of other places in Manhattan where wells have been drilled to extract geothermal energy for heating and cooling buildings.

Like any technology, geothermal power has some disadvantages. Drilling deep wells in urban environments requires permission from a variety of state and local city agencies. This can be a complex and frustratingly slow process. In addition, home heating systems using geothermal heat pumps are more expensive to install than conventional oil or gas heating systems. However, since geothermal heat pumps don't rely on the burning of fossil fuels, the systems are much less expensive to operate. Therefore, the higher installation costs are generally recovered after several years. Improvements in deep drilling technology may make widespread development of enhanced geothermal systems easier and more economical in the near future.

In 2006, a panel of experts at the Massachusetts Institute of Technology issued a report suggesting that, with a little help from the federal government, geothermal energy could play a vital role in meeting America's energy needs. It could reduce our dependence on foreign oil and make up for the loss of coal-burning power plants that may need to be retired. As a renewable resource with widespread availability, low emissions, and low operating costs, geothermal energy may well be a major energy source for the future.

QUESTIONS

1. The Philippines generates about 23 percent of its electricity with geothermal energy. What is it about the location of the Philippines that enables it to make use of high-temperature geothermal energy? Support your answer with evidence from the text.

2. The costs for installing a geothermal home heating and cooling system are higher than the costs for installing a conventional system. What are some factors that might influence people to use them? Support your answer with research from the Internet.

3. In terms of environmental impact, what are the advantages of geothermal energy compared with the burning of oil, coal and natural gas? Give examples from current events to clarify your position.

4. Engineers have found that when a dam creates a large reservoir, sometimes there is an increase in low-level earthquake activity in the area. This situation is similar to the enhanced geothermal energy project in Switzerland which also caused earthquakes Based on what you have read here and in the text and in Internet research, what conclusions can you draw?

The Mineral Industry: Where Has It Gone?

Minerals are essential to our lives and our modern society. Almost everything we see or use is related in some way to minerals and other materials extracted from Earth. Concrete used in buildings, sidewalks, and roadways is made of sand and gravel mixed with cement, which is made from limestone. The glass in windows and drinking glasses is made from sand. Ceramic dishes, tiles, and bricks, and porcelain sinks, tubs, and toilets are made from clays. Cars, airplanes, ships, eating and cooking utensils, machinery, tools, and countless other items contain metals from metallic ores mined from the ground. Salt, the most common food seasoning, is a mineral—halite. Even trees used for lumber and plants used for food grow in soil that contains minerals.

Mineral resources are not evenly distributed in Earth's landmasses. Their occurrence is related to mountain building and volcanism, and the landscape features they produce. Nevertheless, large countries such as the United States, Canada, Australia, Russia, and China tend to have large mineral reserves. Small countries such as Japan, must import most of the minerals they use. It is important to remember that a mineral deposit is not considered a resource unless it can be extracted, processed, and sold at a profit.

Materials such as sand and gravel are plentiful and widespread. For a mining operation to be profitable, these materials must be mined close to where they will be used. On the other hand, relatively scarce metals, such as copper, and precious metals, such as gold and silver, are so valuable that they are mined wherever they are found and transported to markets around the world.

Sometimes a mining operations close not because the ore has run out, but for other reasons. For many years, titanium was mined in the Adirondack Mountains in New York State. Titanium is a strong, lightweight metal used in the aerospace industry. Titanium dioxide is a white pigment used in paints, paper, and plastics. Even though there is still plenty of ore left, titanium mining in New York stopped in 1982. Mines elsewhere in the United States and in other countries can produce titanium at a lower cost, so the mine in New York was closed.

In the last few decades, many mines in the U.S. have closed due to "globalization." Simply put, mining companies have found that it is cheaper to explore for and develop mineral deposits in other countries. This is especially true for developing countries. Since 1980, the number of mines in the United States has decreased by more than 60 percent. At the same time, mineral exploration and development in places like Latin America and Africa have increased dramatically.

One reason is that developing countries have a much lower standard of living than we have in the United States. Therefore, workers in those countries will accept lower wages than most American workers are accustomed to.

Another reason is tough environmental restrictions in the United States. Many of the countries to which mining companies are shifting their operations have less strict environmental regulations. To open a mine in the United States, a company must spend millions of dollars on environmental impact studies. In addition, there are federal, state, and local permits that must be paid for before they see any profit. Even then, the company may face fierce opposition from environmental organizations and local civic groups. These groups often have a "not in my backyard" attitude. In contrast, governments and communities in developing nations often welcome the mining companies. The companies bring jobs and pump money into the local economy. Mining companies sometimes build new schools and health clinics to gain the good will of the community.

Environmentalists charge that the mining companies are simply trying to avoid environmental regulations out of greed. To be sure, mining can

have serious environmental consequences. Ores containing valuable metals usually contain other metals that are quite toxic, such as lead, mercury, and arsenic. Extraction and processing of the ores may release these toxic heavy metals into the environment. There they contaminate groundwater and surface water supplies and enter the food chain.

In 2004, residents of a village at Buyat Bay in Indonesia brought a multi-million dollar lawsuit against Newmont Mining Corporation. Newmont is the world's largest producer of gold. The lawsuit charged that the company's mining operations released mercury and arsenic into Buyat Bay. The mercury contaminated the fish that the villagers rely on for food and income. Some villagers complained of strange rashes, large lumps on their necks, and other unusual illnesses. They attribute these illnesses to the heavy metals in the waters and sediment of the bay. The lawsuit was settled in 2006. Newmont agreed to provide $30 million to monitor the environmental and health effects of their mining operations. In addition, they provided funding for enhanced community development programs in the region. Such cases remind us that a mine that is not in your backyard is probably in someone else's backyard. We must also remember that mining activities may have far-reaching social, economic, legal, and even ethical implications.

Another example involves a relatively rare metal called tantalum. This hard, dense metal is a good conductor of electricity. It is used in making capacitors for cell phones. There are large deposits of tantalum in Africa, particularly in the Congo. For years, a civil war has raged between the government of this African nation and several rebel groups. In addition to political motives behind the conflict, there is little doubt that the war is mainly about control of the Congo's tantalum reserves. Companies that import tantalum and use the metal in cell phones often claim not to know if their tantalum came from the Congo.

A similar situation exists regarding the mining of diamonds in parts of Africa. The sale of so-called "blood diamonds" helps to finance armed conflicts in countries such as Uganda and Rwanda. It is sobering to think that by purchasing a cell phone or a diamond engagement ring, you might be helping to fund a civil war in Africa.

QUESTIONS

1. Salt mining is a major industry in New York State. Getting at the salt is challenging because the deposits are located deep below ground. However, salt can be more easily recovered from surface deposits around the shores of the Great Salt Lake in Utah. What is the advantage of using salt mined in New York on the state's roadways, even if mining salt in New York is more expensive than in places farther away?
2. Large areas of the deep seafloor are covered with round nodules and manganese and a number of rare metals. Mineral companies have known about them for many years. Yet, they have not tried to recover them as an industrial resource. Why have the mining companies showed such limited interest?
3. Should the United States make a greater effort to prevent the mining industry from importing mineral resources from abroad, rather than mining them here in the United States? Explain your answer using evidence from the text.
4. To what extent do you think that you, as a consumer, should be concerned with the political and economic conditions in places that make items that you buy and use? Be sure to acknowledge competing views.

Today's Forecast: Light Ash Falls, With a Chance of Lava Flows—Predicting Volcanic Eruptions

Mid-March, 2009—Scientists from the Alaska Volcano Observatory are keeping a close watch on Mt. Redoubt. Although the weather in Alaska is still cold, the scientists know that things could heat up quickly if the mountain—an active volcano—erupts. Mt. Redoubt is part of the Aleutian Range, a chain of volcanic mountains formed where the Pacific Plate is sliding beneath the North American Plate. Since late 2008, seismic activity—earthquakes—has been increasing beneath Mt. Redoubt. This could indicate that magma, molten rock beneath Earth's surface, may be rising within the volcano. This could lead up to an eruption. A small research plane periodically flies over the volcano to measure gas emissions and heat output. This information could provide more clues to the volcano's intentions.

On March 18, the scientists posted the following cautious message on the observatory's website: "The volcano remains restless with abnormally high rates of gas emission and melting of the summit glacier. It is possible for the current period of unrest to result in an eruption. However, the unrest could persist for months or even years and not lead to an eruption. Conditions at the volcano may change rapidly, advancing from low levels of activity to eruption within 24 hours or less." Just four days later, on March 22, Mt. Redoubt erupted, sending a column of steam and ash thousands of feet into the sky.

Fortunately, the March 2009 eruption of Mt. Redoubt caused no injuries or serious property damage. What about volcanoes located in densely populated regions? There are many examples: Mt. Vesuvius in Italy, Mt. Rainier in Washington State, and Mt. Pinatubo in the Philippines, to name a few. In fact, about 500 million people worldwide live near potentially dangerous volcanoes, so the science of predicting volcanic eruptions is not just of academic interest.

Unlike earthquakes, which often occur without warning, volcanoes usually show signs of increased activity leading up to eruptions. Therefore, scientists have had much greater success predicting volcanic eruptions than earthquakes. To evaluate the danger posed by a volcano, scientists begin looking at the volcano's history. They map and date the lava flows and ash deposits around a volcano to determine the time between past eruptions. This can reveal if a volcano is "overdue" for an eruption. Also, volcanoes usually show certain warning signs, or precursors, before an eruption. These include increased seismic activity, swelling of the volcano's slopes, and changes in the gases given off by the volcano. All these are related to the rising of magma within the volcano.

Magma is a mixture of molten rock and dissolved gases, such as sulfur dioxide and carbon dioxide. As magma nears the surface, a volcano may release increased amounts of these gases. Therefore, scientists collect samples of the gases to see how they change over time.

When magma moves upward beneath a volcano, it may force rocks apart, causing sudden, relatively strong earthquakes. On the other hand, magma may flow smoothly through existing channels, producing gentler vibrations. Scientists set up seismographs on and around a volcano to monitor these seismic events. Upwelling magma causes the volcano to swell, much like a balloon being inflated. Therefore, scientists install tiltmeters to measure the angle of the volcano's slopes. (Today's electronic tiltmeters are so sensitive they could detect the tilt produced by placing a dime under one end of a horizontal bar more than half a mile long!)

By carefully monitoring such warning signs, scientists at the Hawaiian Volcano Observatory on Mt. Kilauea have learned to predict its eruptions with great accuracy. Mt. Kilauea is Hawaii's most active volcano. Hawaii's volcanoes are gently sloping shield volcanoes formed by eruptions of very fluid lava. These eruptions tend to be relatively quiet events.

3

In contrast, the steep-sided volcanoes scattered around the Pacific Rim, the so-called "Ring of Fire," tend to erupt much more violently. These volcanoes are located where ocean crust is sliding back into Earth's mantle at subduction zones. The magma in these volcanoes is more viscous (sticky) and contains more trapped gases. This leads to violent, explosive eruptions. Explosive eruptions are harder to predict, and collecting precursor data at such volcanoes can prove deadly. Between 1979 and 2000, 23 scientists were killed while studying subduction zone volcanoes.

Nevertheless, efforts to forecast volcanic eruptions have saved thousands of lives. For example, when Mount St. Helens in Washington State showed signs of an impending eruption early in 1980, the U.S. Forest Service evacuated people from danger zones around the mountain. Although scientists could not pinpoint the exact timing of the powerful eruption that occurred in May of that year, in which 57 people died, the evacuation may have saved about 20,000 lives. On December 18, 2000, scientists predicted a major eruption of Popocatepetl Volcano near Mexico City, leading to the evacuation of 41,000 people; hours later, the volcano had its most violent eruption in over 1,000 years.

Mt. Rainier, Mt. Hood, and the other snow-capped volcanoes of the Cascade Range in the northwestern U.S. give an impression of serene, unchanging beauty. However, research shows that they too have had violent eruptions and spawned devastating lahars in the not-too-distant past. Because of this, the U. S. Geological Survey has set up automated lahar detection systems on Mt. Rainier and other Cascade volcanoes. These systems rely on a network of acoustic flow monitors sensitive to the ground vibrations generated by lahars. A test of one such system at Alaska's Mt. Redoubt volcano in 1990 successfully detected and tracked a lahar. This technology will likely save thousands of lives in the future.

In spite of successes in predicting eruptions and detecting lahars, there are still times when scientists are caught by surprise. In some cases this is because scientists have not had an opportunity to monitor and study a particular volcano. However, volcanoes by nature can be difficult to predict. Sometimes a volcano may erupt without exhibiting the usual precursor events. At other times the precursors are observed, but then the volcano calms down without erupting at all. Time and effort will hopefully bring further improvements in technology and forecasting techniques that will help the scientists who study volcanoes to better protect the millions of people who live in the shadows of these restless giants.

QUESTIONS

1. Why did scientists begin closely monitoring Mt. Redoubt in 2009?
2. What features of tectonic plates are associated with many explosive volcanic eruptions? Cite evidence from the reading.
3. Changes in volcanic gas emissions are often help predict volcanic eruptions. What special danger might be associated with a decrease in the amount of gas venting?
4. If scientists were to forecast a major eruption from a volcano in a densely populated area, and then the eruption did not occur, what problems might such a "false alarm" create? Support your answer with current events.
5. Sketch the profile of a volcano that is ready to erupt. Show the shape of the surface and underground features of the volcano.

On Shaky Ground: Landslide Hazards in New York

About noon on April 27, 1993, in the Tully Valley of central New York State, the Earth moved, or at least a piece of it did. A huge slab of ground—more than a million cubic meters of earth—broke loose from the foot of Bare Mountain. The mountain is located in the town of La Fayette, in Onondaga County. The slab slid eastward across the valley floor. The event affected more than 50 acres of land, destroying three homes and damaging four others. Fortunately, no one was killed or seriously injured. (Three people had to be rescued by helicopter from what was once their backyard.) However, the incident was a dramatic lesson to residents and local officials about the hazards posed by landslides.

As is common in most of New York State, Ice Age glaciers sculpted the landscape around Bare Mountain and Tully Valley. Continental ice sheets that moved south from Lake Ontario scoured their way through north-south-aligned stream valleys. The ice sheets straightened and deepened these valleys. When the ice sheets melted, they left behind piles of sediment—sand, silt, pebbles, boulders, and other debris. Some of these deposits, called glacial moraines, stretched across the valleys. They became natural dams, causing meltwater to back up behind them, forming lakes. New York's famous Finger Lakes were formed in this way. Many such lakes that were formed during the end of the last Ice Age eventually overran or broke through their moraine dams and drained away. They left behind lake deposits consisting mostly of clay. Some of these glacial-lake clay deposits were left on slopes, such as those on Bare Mountain. These clay deposits are unstable, giving those areas a higher than average landslide risk.

If you have ever held wet clay in your hands, you know that it is slippery. Clay minerals consist of tiny platelike particles. These particles slide past one another easily when wet. This characteristic gives clay what engineers and scientists call low shear strength, a low resistance to sideways stresses. In the case of the Tully landslide, the sloping land along the foot of Bare Mountain was resting on a layer of clay that was saturated with water. This layer acted like a banana peel on a freshly waxed wooden floor, allowing the upper soil layers to slide across it. The ground moved more or less as a single chunk, or block, a type of landslide called a slump.

Why was the clay layer so full of water? The water came from the melting of a heavier than usual winter snowpack, plus recent heavy rains. A large amount of water ran down the mountainside, moving over and through the ground. It forced its way into crevices and even into the tiny spaces between the clay particles. This weakened the cohesive forces within the clay layer, setting the stage for the disaster that followed.

Actually, people noticed signs of trouble in the months and years preceding the landslide. As early as 1990, local residents and workers with the New York State Department of Environmental Conservation noticed cracks and bulges in the ground in the area that later slumped. The foundation of one house had been slowly pushed from its proper place. In 1992, it had to be repaired. However, even if these warning signs had been reported to local officials, it is unlikely that they could have predicted the timing and location of the landslide. In addition, there was little they could have done to prevent it or minimize its damage.

After the landslide, geologists working for New York State studied the landslide hazards in the Tully Valley and the surrounding area. They found evidence of earlier landslides. This indicated that the 1993 landslide was not a unique event. Based on what they learned, the geologists produced a color-coded map showing the likelihood of landslides in the area. The map shows zones of low, moderate, and high landslide hazard. The purpose of the map was to help local officials reduce dan-

gers by limiting land uses in landslide-prone terrain. Officials also required special engineering or construction methods for structures built in such areas.

Another area of New York State that has significant landslide hazards is the High Peaks region of the Adirondack Mountains. This area is in the northeastern portion of the state. Most of the high mountains in the Adirondacks are made of the igneous rock anorthosite. Geologists say that anorthosite has a "massive" structure. This means that it lacks the layering typical of sedimentary rocks and the closely spaced fractures found in volcanic rocks such as basalt. There are so few cracks and fractures in the Adirondack's anorthosite bedrock that it is difficult for plant roots to get a good grip on the rock. In addition, the mountainsides were intensely scoured and carved by continental and valley glaciers during the last Ice Age. This steepened slopes and polished the bedrock.

When unusually heavy downpours saturate the thin post-glacial soils on a steep slope, the soils become heavy and lubricated. This causes a strip of forest peels loose from the mountainside. The ground, covered with trees and other vegetation, goes crashing into the valley below. The landslide leaves behind a gleaming stripe of bare bedrock. These slide scars, seen on Whiteface Mountain, Mount Marcy, and elsewhere, are one of the most distinctive features of the Adirondacks. The Tully Valley landslide moved as a slump block. However, the Adirondack landslides are debris avalanches, in which the ground flows in a chaotic jumble of soil, boulders, trees, and plants.

Because of the isolation and lack of development of the Adirondack High Peaks region, the landslides there rarely threaten homes or businesses. Their main effect is on hikers and other recreational users of the mountains. In the autumn of 1999, the remnants of Hurricane Floyd tore through the Adirondacks. The storm released torrential rains that caused new landslides on many of the peaks. In some places, landslide debris wiped out parts of hiking trails. However, the strips of open rock left behind became challenging climbing routes for the adventurous. These routes provide spectacular views of distant mountains.

QUESTIONS

1. Clay is used to make pottery. What do you suppose is the purpose of heating clay pots, after they have been formed into the desired shapes, in a kiln (a special type of high-temperature oven)? Write down your thoughts and share them with your classmates.
2. Suppose you were interested in buying property in western New York; if a landslide hazard map were available to you, how might such a map influence your decision? Be sure to examine competing views.
3. The slope and underlying bedrock in the Tully Valley and the central Adirondacks are very different. What common factor makes these two places especially prone to landslides? Cite examples from the reading.
4. If you were asked to find evidence of prehistoric landslides in Tully Valley, what would you look for? Illustrate your answer with pictures from the Internet.

Hurricanes: The Long Island Express

If you live in the northeastern United States, you probably do not feel threatened by hurricanes. You think that hurricanes mainly affect other areas. Those areas most likely include the islands of the Caribbean Sea, the state of Florida, and other states along the Gulf of Mexico. If and when the remains of a hurricane reach the northeast, they are usually little more than soaking rainstorms with some gusty winds. Still, waves along the coast or flooding caused by heavy rains can cause significant damage. However, it doesn't rise to the catastrophic level of a storm like Hurricane Katrina. That storm devastated New Orleans in 2005. But in September of 1938, New York's Long Island and parts of New England experienced the full wrath of a major hurricane.

A junior forecaster at the U.S. Weather Bureau, Charlie Pierce, predicted that the 1938 hurricane would strike the northeast coast. However, the bureau's chief forecaster and other experts overruled him. They felt certain that the storm would curve out into the Atlantic. At that time, weather forecasters did not have satellites, Doppler radar, computers, and other sophisticated weather tracking technology. They had to base their forecasts on much more limited information.

In addition wind speed within a hurricane, the speed at which the hurricane travels also affects the damage it causes. The 1938 hurricane was traveling at an unusually high speed, about 70 mph. To this day, it is the fastest moving Atlantic hurricane on record. This high speed kept the storm from losing energy as it moved over the cooler waters of the north Atlantic. Atlantic hurricanes spin counterclockwise and generally are traveling to the northwest or north when they strike the United States. Therefore, the speed of the winds just to the right of the storm's eye (called the right eye wall) combines with the traveling speed of the storm.

When the 1938 hurricane made landfall at Suffolk County on Long Island at 3:30 P.M. on September 21st, it had sustained winds of about 120 mph. This combined with the traveling speed of the storm produced some winds up to 186 mph. By unlucky coincidence, the storm struck just a few hours before high tide. To make matters worse, the moon was at new moon phase, when the sun, moon and Earth are lined up, causing higher than normal high tides. All these factors combined to produce an especially powerful storm surge. The hurricane slammed into the coast with such force it registered as an earthquake on a seismograph at Fordham University in New York City. Waves over 30 feet high pounded the shore. Storm tides of 14 to 18 feet rushed inland. The water destroyed houses, wiped out roads, and temporarily turned Montauk Point into an island. New York City was on the weaker, western edge of the hurricane. Even so, the storm knocked out power to the entire city. Winds of up to 75 mph caused the East River to overflow into Manhattan, flooding streets for three blocks inland.

The fast-moving storm quickly tore across Long Island, earning itself the nickname "The Long Island Express." The storm then crossed Long Island Sound and struck the southern coast of New England with equally devastating effects. Storm tides of 18 to 25 feet swept over coastal communities along the south shores of Connecticut, Rhode Island, and Cape Cod. New London, Connecticut was pummeled by waves and wind. A fire in the town's waterfront business area burned out of control for 10 hours. Downtown Providence in Rhode Island was submerged under nearly 20 feet of water. Entire beachfront communities were swept away. The eye of the storm tracked northward through the Connecticut River Valley. It entered central Massachusetts, where the Connecticut River overflowed its banks, causing further destruction. Eventually, the storm continued up through New Hampshire and into Canada, losing strength but still knocking down trees and causing mayhem.

5

When it was all over, more than 700 people had been killed. The storm caused an estimated six billion dollars in damages (in today's dollars). About 4,500 homes and farms had been destroyed, and another 15,000 damaged; 26,000 automobiles were destroyed and 20,000 miles of electric lines and poles were downed. Billions of trees were toppled in New York and New England. The storm also left long-term scars on the coastal landscape. It created and widened several inlets in the barrier islands that line the south shore of Long Island. Out of all the destruction, there may have been some good. The need to rebuild and repair damaged homes, roadways, rail lines, and other infrastructure helped ease unemployment in the region near the end of the Great Depression.

Could it happen again? Powerful hurricanes had struck the northeastern U.S. before the 1938 hurricane. They will likely do so again in the future. However, a peculiar set of circumstances combined to give the "Long Island Express" its unique character. There is little doubt that sooner or later another major hurricane will hit the region. If a storm like the 1938 hurricane were to strike today, it would cause an estimated 39 billion dollars in damages. Since the region is much more densely populated than it was in 1938, there is a potential for great loss of life. On the other hand, advances in weather forecasting technology might give area residents much more advance warning than they had in 1938. This alone could possibly save many lives. Of course, that would require that people take forecasters' warnings seriously and act on them.

QUESTIONS

1. After reading about the rotation of the 1938 hurricane and its northward traveling speed, explain why New York City, located west of the storm's center, experienced much lower wind speeds than areas in Long Island that were east of the storm's center? Cite evidence from the reading.

2. Standard or "normal" air pressure—called one atmosphere—is often defined as equal to 29.9 inches of mercury (in Hg). This means that the downward pressure of the atmosphere can hold up a column of liquid mercury (in a glass tube) that is 29.9 inches tall. Would the air pressure measured at the center of a hurricane be lower or higher than that figure? Support your answer with evidence from the reading. Mercury is about 13.6 times as dense as water. If atmospheric pressure were measured by a column of water, do you think the column of water would be taller or shorter than the column of mercury?

3. Scientists who study climate change are concerned that global warming may increase the frequency or intensity of tropical cyclones. What predicted effect of global warming would increase the destructive power of hurricanes?

4. Atlantic hurricanes tend to occur at a specific time of year, called "hurricane season," with most large storms forming in late summer and early autumn. Explain why major hurricanes happen mainly in August and September. In which months would strong tropical cyclones be most likely to form in the southern hemisphere? (It may help to look at a calendar.) Explain your answer.

Storm Chasers: Taking You on a Whirlwind Tour?

Here is an idea for your next Spring Break family vacation: Fly to Nebraska, and get a hotel room. When the sky turns dark with thunderclouds, tour guides in a van full of sophisticated weather tracking equipment will pick you up. Your guides will drive you straight into a raging storm. There, if you're lucky, you will witness close-up the awesome power of a tornado.

Sound crazy? Perhaps, but this is not a joke. There are, in fact, a number of private "storm chaser" tour companies. For a fee, they will take thrill-seeking tourists in search of tornadoes, thunderstorms, and other extreme weather events.

What are storm chasers, and when did this practice begin? Simply put, storm chasers are people who track (usually by car) tornadoes, thunderstorms, hurricanes, and other types of severe weather. Some are scientists or students trying to gain a better understanding of these storms. Some are photographers or cinematographers who want to capture dramatic weather on film. Some are amateurs with a strong interest in weather, or thrill-seekers looking for an adrenaline rush. No doubt many storm chasers fall into more than one of these categories. And, there are also those who try to earn a living by taking people on storm chases and giving them a chance to experience wild weather at close range.

Many people are fascinated by severe weather. However, no one knows who was first to follow a violent storm, rather run toward safety. Perhaps it was Benjamin Franklin. In the 1750s, he flew a kite during a thunderstorm. He was trying to understand the electrical nature of lightning.

The modern practice of storm chasing, as a hobby and as scientific research, began in the 1950s. Not surprisingly, many of the early chasers came from states in the High Plains of the central U.S. such as the Dakotas, Nebraska, Kansas, and Oklahoma. This was before lap-top computers, global positioning systems, and cell phones provided instant access to up-to-the-minute weather information. Many of the early storm chasers seemed to have a natural talent to pick the "right" storm to follow, meaning a storm likely to generate tornadoes or other impressive effects.

Neil Ward was one of the first to successfully link storm chasing to scientific research. In 1961, Ward hitched a ride with the Oklahoma Highway Patrol. They chased a storm that spawned a large tornado. He used the squad car's radio to relay information to a U.S. Weather Bureau office in Oklahoma City. This allowed the scientists there to correlate what they were seeing on their weather radar screens with his observations. Ward went on to become one of the nation's foremost tornado researchers. In addition to furthering research, the early storm chasers also helped spur the government to develop early warning systems for dangerous weather.

Storm chasers follow many kinds of storms in hopes of catching unusual or spectacular weather phenomena. However, the "Holy Grail" of storm chasing is the tornado. Some of these funnel-shaped whirlwinds are able to lift houses off their foundations. They are among the most violent, localized and short-lived of storms. Tornadoes are usually spawned by huge, rotating thunderclouds called supercells. With wind speeds that can exceed 300 mph, tornadoes may touch down, carve a narrow path of destruction, and disappear within a matter of minutes. This makes close-up observation a rare experience. In addition, tornadoes can be deadly. On average, 60 people are killed by tornadoes every year in the U.S. However, the number can vary greatly from one year to the next; in 2008, the tornado death toll was 125.

The greatest number of tornadoes hit the states of the central U.S. This relatively flat region is nicknamed "tornado alley." Each spring warm, moist air masses from the Gulf of Mexico collide with cold, dry Canadian air masses. This creates the per-

fect conditions for tornado formation. Tornadoes can also be generated by hurricanes. So, they sometimes strike along the Gulf and Southeastern Coasts. For example, 127 tornadoes were associated with Hurricane Ivan. This storm battered the southeastern states in September 2004. The northeastern states also get their share of tornadoes. These are usually spawned by severe thunderstorms or by hurricane remnants passing through the region in the fall. Tornadoes strike all over the world. Bangladesh has the most tornadoes per year after the U.S.

It is very difficult to predict tornadoes. However, advances in technology have brought about improvements in this area. Some advances were made with the help of storm chasers. In 1995, Joshua Wurman, a tornado expert and storm chaser, invented a mobile radar unit called "Doppler on Wheels." In 1999, one of these truck-mounted units tracked a tornado and clocked the fastest wind ever measured—301 mph. Scientists hope that this and other technologies will make it possible to give communities more advance warning for approaching tornadoes. This would help to save lives.

Before the mid-1990s, storm chasing was practiced by a relatively small "lunatic fringe" of hobbyists and scientists. The 1997 release of the motion picture *Twister*, with its memorable airborne cow, created an explosion of interest in storm chasing. Television programs on the Discovery Channel and the Weather Channel also drew attention to storm chasing. As a result, more people began to invade the Plains each spring during tornado season. Many of these people had little or no understanding of severe weather and the risks involved in storm chasing. Some did not always respect the unwritten rules of chasing. These include respecting private property and not interfering with emergency vehicles. This upset the more experienced chasers. They worried that the reckless actions of a few newcomers would tarnish their image in the eyes of the public and law enforcement. Nevertheless, the increased interest also gave rise to storm-chasing tour companies. This allowed some veteran chasers to turn a hobby into an income-generating profession.

If you are thinking about storm chasing, it is worth remembering that it can be an extremely risky activity. Storms are highly unpredictable. Lightning strikes can occur far from storm clouds. Tornadoes can change direction suddenly; the results can be lethal. First timers would be well advised to seek out a reputable private tour group or join tours conducted by university meteorology departments, rather than trying to go it alone.

QUESTIONS

1. Based on the information in this article and your knowledge of Earth science, what are the "key ingredients" involved in producing a tornado?

2. During which season would a tornado be least likely to occur?

3. During storms that can cause flash flooding by releasing a large amount of rain in a short period of time, people in low-lying areas are usually advised to seek higher ground. A hilltop is unlikely to be affected by a flash flood, but what other weather hazards might threaten a person seeking safety there during a storm? Support your answer with evidence from the text.

4. While watching television, you may have noticed weather advisories at the bottom of the T.V. screen called storm watches or storm warnings. Use the Internet to research the difference between a storm watch and a storm warning.

Dinosaurs: Our Fine-Feathered Friends?

Everyone knows that the dinosaurs became extinct 65 millions years ago. Everyone, that is, except a growing number of paleontologists. They believe that the direct descendants of dinosaurs live among us today—as birds! (Paleontologists study ancient life, mainly by collecting and observing fossils.)

The idea that birds are related to dinosaurs is not new. Thomas Henry Huxley was an English naturalist. One evening in the early 1860s, he noticed that the bones of a chicken he had eaten for dinner were very similar to recently discovered dinosaur fossils. Huxley observed that dinosaurs seemed to bridge the gap between reptiles and birds.

In 1861, a spectacular fossil was found in a limestone quarry in Germany. The smooth, fine-grained limestone held a small reptilelike creature. The fossil showed exquisite detail, including the unmistakable impressions of feathers. They named the animal *Archaeopteryx*. It displayed a strange mixture of birdlike and reptilian characteristics. It had a mouthful of teeth, a long, bony tail, and claws on its feathered arms. Today, most scientists consider *Archaeopteryx* to be the first bird.

For many years, most scientists thought that *Archaeopteryx* and the dinosaurs evolved from a common reptilian ancestor along separate paths. However, in 1973 John Ostrom, a scientist from Yale University, pointed out that the skeleton of *Archaeopteryx* bore a strong resemblance to the skeletons the theropod dinosaurs. Theropods were meat-eating dinosaurs that walked upright on their two hind legs. Ostrom listed 22 anatomical features shared by *Archaeopteryx* and theropod dinosaurs. He later argued that some theropods had a primitive wishbone, a bone common to all living birds. These dinosaurs include the *Velociraptors*, made famous in the movie *Jurassic Park*. Since then, paleontologists have discovered about 200 similarities between birds and theropod dinosaurs. In light of this, they consider *Archaeopteryx* and all later birds to be "feathered dinosaurs."

Recent fossil finds from Liaoning Province in northern China provide powerful support for this idea. In 1996, scientists found the fossilized remains of a small theropod dinosaur in lake sediments. These sediments were from the early Cretaceous Period, about 120 to 130 million years ago. (For a clearer picture of where the Cretaceous Period fits into Earth's history, refer to the geologic time scale in the *Earth Science Reference Tables*.) Impressions in the rocks showed that the dinosaur's body was covered in tufts of primitive, downy feathers. Over the next few years, scientists discovered other dinosaur fossils in Liaoning Province. These fossils showed the impressions of barbed feathers more similar to those of modern birds. In 2004, they found the fossil of a small tyrannosaur, a relative of the mighty *T. rex*. This fossil also had the imprints of primitive feathers. In fact, it is not too much of a stretch to imagine that if the *Jurassic Park* films were remade today, the cunning Velociraptors would be covered in feathers rather than scaly, lizardlike skin.

It is now clear that many theropod dinosaurs had feathers, which until recently had been considered a unique feature of birds. However, this does not mean that these dinosaurs could fly. Scientists believe that feathers originally evolved for some other purpose. For example, the strands of primitive downy feathers may have helped dinosaurs retain body heat. The more modern-looking feathers may have served as displays to attract mates, much as they do in today's birds.

In addition to physical features such as feathers and skeletal similarities to birds, dinosaur fossils have been found that appear to preserve bird-like behavior. For instance, some dinosaurs seem to have died while sitting on a nest, protecting their eggs. Some of these dinosaurs were males. This suggests that the parents took turns guarding the nest,

7

just as many birds do today. In 2004, scientists discovered a small theropod dinosaur fossil that apparently was buried while sleeping with its head tucked behind its left arm. Some modern birds also sleep with their heads tucked beneath a wing to conserve heat.

In 2008, scientists analyzed the fossil of a theropod dinosaur found in Argentina. They discovered that it had a birdlike breathing system. It used air sacs to pump air into its lungs. That same year, scientists extracted proteins from soft tissues preserved in the bones of a *T. rex* fossil found in Montana. Tests performed on the proteins showed that the *T. rex* was genetically more similar to ostriches and chickens than to any other creatures, including alligators. This finding provides some of the strongest evidence yet that birds evolved from dinosaurs.

Nevertheless, some scientists are still not convinced that birds descended from dinosaurs. They argue that what appear to be feathers may be the remains of unknown structures in the skin of dinosaurs. And even if some dinosaurs did have feathers, the skeptics point out that these fossils are all about 20 to 30 million years younger than *Archaeopteryx*, the first bird. *Archaeopteryx* has been dated to the late Jurassic Period about 150 million years ago. Defenders of the birds-are-dinosaurs theory reply that evolution is like a branching tree rather than a straight line. Creatures on one branch of that tree may retain primitive features long after other organisms have moved ahead along another branch. For example, primitive mammals like the duck-billed platypus continue to exist today.

The number of doubters is shrinking. Only the discovery of a birdlike dinosaur that is indisputably older than *Archaeopteryx* will convince the last of the skeptics. It may be that the crucial fossil is still waiting to be unearthed somewhere in China, Mongolia, or South America. In the meantime, feel free to look with newfound respect at the sparrows, pigeons, robins, and blue jays that flutter and strut all around us. They are the "feathered dinosaurs" in our midst. Or perhaps you would prefer to dig into a meal of fried chicken and chuckle to yourself about who's having the last laugh!

QUESTIONS

1. Fossils of Cretaceous age dinosaurs recently found in China have provided new evidence that some nonflying dinosaurs are closely related to birds. What is the possible range of absolute (numerical) ages of these fossils? Cite the source of your answer.

2. Rocks are generally classified by their origin. What type of rock is most likely to contain fossils? Why do the other two major types of rocks seldom contain fossils? Explain your answers.

3. Dinosaur fossils are not common in New York State because New York has relatively few surface rocks of Mesozoic age. Where is the most likely place to find dinosaur age fossils in surface bedrock within New York State? Cite the source of your answer.

4. In the 1970s, a paleontologist Robert Bakker proposed that some dinosaurs were fast-moving, warm blooded animals, rather than plodding, cold-blooded reptiles. As evidence, Bakker showed that the internal structure of dinosaur bones is more similar to that of warm-blooded birds and mammals than to the bones of cold-blooded reptiles. What other feature of recently discovered dinosaur fossils suggests that these dinosaurs may have been warm-blooded? Support your answer with evidence from the reading.

Pluto: A Planet No More!

For generations, students were taught that our solar system has nine planets: Mercury, Venus, Earth, Mars, Jupiter, Saturn, Uranus, Neptune, and Pluto. However, in the year 2000, when the Rose Center for Earth and Space opened at the Hayden Planetarium in New York City, Pluto was missing from the family of planets!

This decision did not come out of nowhere. It was the result of years of discoveries about Pluto and the distant region of the solar system it inhabits. These discoveries have changed the way most astronomers think about Pluto.

Clyde Tombaugh, a young astronomer, discovered Pluto in 1930. At the time, astronomers thought Neptune, the eighth planet from the sun, was not following its calculated orbit. They thought that Neptune's motion was affected by the gravity of an undiscovered planet. They called it "Planet X." Scientists later realized that the problems with Neptune's orbit were the result of a mathematical error. Nevertheless, when Tombaugh found Pluto in the area of the sky predicted by the faulty calculations, it was assumed to be Planet X.

Clyde Tombaugh did not discover Pluto by looking through a telescope. Instead, he took photographs several nights apart of the same area of sky. He examined the pair of photos using a technique called "blinking," switching rapidly back and forth from one photo to the other. He was looking for a point of light that from one photo to the next appeared to "jump" or move against the motionless background of stars. This jump indicates the object is much closer than the stars and is likely orbiting the sun. Tombaugh spent many hours painstakingly blinking hundreds of photographic pairs. Today, powerful computers perform this boring procedure much more quickly.

After Pluto was discovered, it soon became clear that it is a rather peculiar object. Scientists realized that Pluto is very small. With an estimated diameter of about 2300 kilometers, Pluto is less than half the size of Mercury, the next smallest planet. It is also smaller than seven moons in the solar system, including Earth's moon. Another odd thing about Pluto is its orbit. The other eight planets revolve around the sun in essentially circular, widely spaced orbits. This suggests that while growing into full-fledged planets, they swept up bits of dust, rock, and ice in their regions of space. Pluto, in contrast, has a more elliptical (oval-shaped) orbit that actually crosses inside part of Neptune's orbit. For about 10 percent of Pluto's orbit, Neptune is farther from the sun than Pluto! The orbit of Pluto is also tilted about 17° with respect to the orbital plane of the other planets.

In 1978, astronomers discovered that Pluto, tiny as it is, has its own moon. Named Charon, the moon is about half the size of Pluto. Some astronomers consider Pluto and Charon to be a "double planet," rather than a planet with an orbiting satellite. In addition, Pluto is thought to be a mixture of rock and ice, similar to comets. Therefore, it is unlike the four inner planets, which are made almost entirely of rock. It is also unlike the four outer giant planets, which consist largely of gases. Nevertheless, for many years, none of these unusual features seriously called into question Pluto's status as a planet.

In 1992, astronomers David Jewitt and Jane Luu were using the 10-meter Keck Telescope in Hawaii. They discovered a small, asteroidlike object orbiting far beyond the orbit of Neptune. It was in a region of space called the Kuiper Belt. Dutch astronomer Gerard Kuiper (rhymes with hyper) had theorized that the region should contain numerous small, icy bodies left over from the formation of the solar system. This region might be a source for comets. Since 1992, more than 1000 other objects have been found in the Kuiper Belt, where Pluto itself resides. Although most of these objects are much smaller than Pluto, several were found to have orbits similar to Pluto's. These ob-

jects were nicknamed Plutinos. To many astronomers, it began to seem that Pluto has more in common with the Plutinos and other "Kuiper Belt Objects" (KBOs) than it does with the other planets.

Defenders of Pluto as a planet pointed out that it is much larger than the first KBOs that were discovered. There is evidence that Pluto even has an atmosphere and seasons. However, in 2002 astronomers discovered a Kuiper Belt Object they named Quaoar. It has a diameter of about 1200 kilometers (roughly the size Charon). In 2004, astronomers found an extremely distant object they named Sedna. They estimated it to be 1600 kilometers wide or about two-thirds the size of Pluto. They determined that it orbits far beyond the outer edge of the Kuiper Belt. Many astronomers feel that as their ability to detect such objects improved, it would only be a matter of time before they found a distant body as large as or larger than Pluto.

They were right. In July 2005, astronomers found a distant object about 2500 kilometers in diameter and 27 percent more massive than Pluto. It has been given the name Eris. Like Pluto, Eris has its own moon. The question then arose: if Pluto is the ninth planet, shouldn't Eris be considered the tenth planet? That was the last straw. In 2006, the International Astronomical Union met and decided that for an object to be considered a planet, the object must not only orbit the sun and be large enough that gravity has made it into a sphere; it must also dominate the zone in which it orbits. Since Pluto does not meet this last condition, they voted to remove Pluto from the list of planets in our solar system. They reclassified Pluto as a "dwarf planet," a status it now shares with Eris, Makemake, Haumea, and Ceres. Ceres is the largest object in the asteroid belt located between the orbits of Mars and Jupiter.

Some scientists favor retaining Pluto's planetary status for the sake of tradition. The astronomers who voted to change Pluto's status did not wish to demote Pluto. They wanted to reclassify it based on their improved understanding of Pluto and the Kuiper Belt. Planet or not, astronomers agree that Pluto is a fascinating object worthy of further study. In January 2006, the New Horizons mission lifted off from Cape Canaveral in Florida. It is on its way to being the first space probe to visit Pluto. The probe carries some of discoverer Clyde Tombaugh's ashes. New Horizons will make its closest approach to Pluto in July 2015. You can bet astronomers will be eagerly awaiting many new discoveries about this distant world—the former ninth planet of our solar system!

QUESTIONS

1. Astronomers have discovered several Kuiper Belt Objects that have satellites of their own. Some people would call them moons. If these Kuiper objects are not true planets, should the objects orbiting them be called moons? Examine competing views. How would you define "moon"?

2. Why do only the nearby objects in space, such as the planets, seem to change their positions among the stars?

3. The definition of a planet is restricted to objects in our solar system. Why do astronomers hesitate to apply their definition of a planet to the satellites of other stars? Use the Internet to research your answer.

4. Some astronomers have suggested that Pluto is the "lost moon" of one of the planets. How could you determine if this is true? What could cause a planet to lose its moon?

Extra-Solar Planets: The Search for Other Earths

Our sun is a star, similar to many of the stars we see in the night sky. Since the sun has family of planets, it seems reasonable to suppose that other stars may also have planets. Some of these planets may be like Earth. Reasonable, perhaps, but until fairly recently was impossible to confirm that conclusion. Distances to other stars are vastly greater than the distances to the other planets of our solar system. In addition, stars are massive, self-luminous objects, far larger and brighter than any planets that may orbit them. Even with the most powerful telescopes, the task of spotting a planet in orbit around a distant star has been compared to trying to see a firefly next to a giant searchlight 3000 miles away!

Faced with such a daunting task, astronomers came up with an indirect way to detect a planet orbiting a distant star. The gravitational force of a planet circling a star tugs the star back and forth. This causes the star to "wobble" in space.

The motion is relatively small. Therefore, measuring it at a great distance in space requires special techniques. Astronomers carefully measure changes in the spectrum of light emitted by the star. They observe alternating redshifts and blueshifts in the starlight: red when the star moves away and blue when the star moves toward us. These changes are small, but astronomers can measure them with great precision.

In 1995, using the "wobble" method two astronomers in France detected a planet orbiting a star in the constellation Pegasus. The discovery of this first extra-solar planet (a planet outside of our solar system) caused a sensation. It confirmed the idea that there are planets orbiting other stars just as the planets of our solar system orbit the sun. However, the newfound planet turned out to be quite an oddball. Calculations indicated that it is about the size of Jupiter, the largest planet in our solar system. However, it orbits its parent star once every 4.2 days. Therefore, it must be much closer to its star than Mercury is to the sun—not a very promising place to find life!

Since 1995, more than 1200 extra-solar planets have been found. Astronomers announce new discoveries every few months. They have found planets around stars as close as 15 light-years and as far as 230 light-years from Earth. Some extra-solar planets have been detected when they passed in front of their parent star. This causes a slight dimming of the star's light. It allows astronomers to determine more accurately the planet's size, and even provides clues to the makeup of the planet's atmosphere. In 2008, astronomers succeeded in directly photographing extra-solar planets through a telescope. This advance opens the door to making spectroscopic observations of planets to determine their composition, temperature, and other properties.

The majority of the extra-solar planets discovered so far are gigantic worlds. Some of them are many times more massive than Jupiter. (Jupiter is about 300 times more massive than Earth and 10 times Earth's diameter.) Astronomers assume that these planets are gas giants like Jupiter and Saturn, rather than rocky planets like Earth and Mars. In addition, most of these newly discovered planets orbit very close to their parent stars. Such planets are the easiest to detect because their great mass and fast orbital period cause their parent star to wobble more noticeably. The wobble occurs at intervals measured in days, weeks, or months, rather than years. (However, that does not mean that they are the most common type of planet out there.)

Some scientists argue that these giant objects are not planets at all, but are actually failed stars or "brown dwarfs." Such objects form in a manner similar to stars but are not quite large enough to sustain the nuclear fusion reactions that cause stars to shine. Other scientists point out that even if they are planets, they could not possibly support life due to their strong surface gravity and extreme surface temperature.

Nevertheless, as detection techniques have improved, astronomers have discovered several planetary systems in which three or more planets orbit a star. In one such system, a Jupiter-size planet orbits its star at a distance similar to that of Mars from our sun. With no solid surface and a thick, choking atmosphere of toxic gases, this planet is unlikely to support life. However, its moons might support life.

Even more encouraging, in 2004 astronomers discovered three extra-solar planets that are closer in size to Earth than to Jupiter. These planets are roughly the mass of the planet Neptune. Astronomers speculate that these could be rocky "super Earths" rather than gas giants. However, these planets orbit too close to their star to allow for the existence of life. Astronomers are confident that rocky, Earth-like planets will soon be found orbiting their parent star at distances within the "comfort zone." This is the area around a star in which planetary temperatures are neither too hot nor too cold. At these temperatures, water can exist as a liquid. Scientists think that this is necessary for life.

Sure enough, in 2007 European astronomers found a "super Earth" orbiting its star in the comfort zone. However, further study indicated that the planet probably suffers from a runaway greenhouse effect. Thus, like Venus in our solar system, it is inhospitable to life.

The true goal of the extra-solar planet hunters is to find an Earth-like planet orbiting its star at the comfort zone that shows signs of biological activity. It is unlikely that Earth-bound telescopes will accomplish this task. In 2006, the French Space Agency launched a satellite called *COROT* to search for extra-solar planets. In February 2009, *COROT* discovered a planet with a diameter just 1.7 times that of Earth, making it the smallest "super Earth" yet found.

In March 2009, NASA launched the *Kepler Mission*. It is a space-based telescope that will survey thousands of stars for the slight, periodic dimming that indicates the presence of an Earth-size planet. There are also plans for other space missions to continue the search. Although currently on hold, the *Terrestrial Planet Finder* mission may be able to detect light reflected by Earth-size planets. It will also search for signs of life, such as the presence of oxygen in the planet's atmosphere. These missions and others may bring us ever closer to finding out if there is life elsewhere in the universe.

QUESTIONS

1. Twirling a ball tied to a string in a circle by making small circles with your hand is a good model for showing why a planet circling a star causes the star to wobble in space. However, there is obviously no string connecting a planet to its star. What is the force that links the planet and star and causes the planet's motion to affect the star?

2. In September 2004, astronomers discovered an extra-solar planet near a dimly shining red dwarf star, orbiting the star at a distance greater than the distance between Pluto and the sun. Although the planet is probably a gas giant, it could have moons with solid, rocky surfaces. Could these moons support life? Explain your answer.

3. The article you have read mentions measurements in light-years. What is a light-year?

4. Most planets detected to date have been giant planets orbiting very close to their parent stars. Why are astronomers especially interested in finding planets similar in size and orbital period to Earth? Use the Internet to research your answer.

Oil Reserves: Is the World Running Dry?

Since the drilling of the first oil well in Titusville, Pennsylvania, in 1859, oil has rapidly become vital to our way of life. What would happen if we suddenly ran out of oil? The effect on our lives would be disastrous.

Almost every major form of transportation—automobiles, planes, trains, and ships—depends on oil, since gasoline, diesel fuel, and jet fuel are made from oil. We produce much of our electricity by burning oil. We heat our homes with oil. Synthetic fabrics such as nylon and polyester are made from petroleum. Petroleum is used in the production of many industrial chemicals, medicines, fertilizers, and even food products. In addition, there are plastics.

During the next 24 hours, try to notice everything around you that is made of plastic. The buttons on your clothes, the handle and bristles of your toothbrush, the water bottle in your backpack, the dashboard of the family car, your cell phone, your music CDs, the housing of your stereo, television, computer, computer monitor and keyboard—all are likely to be made of plastic. Plastics are made from petroleum. What will these things be made of when the oil runs out? In light of these uses, politicians and economists ask the related questions: How much oil is there on Earth, and will the world run out of oil?

Answering these questions is not easy. To begin, it is necessary to understand something about the nature of oil and how it forms. Petroleum (literally, "rock oil") is considered a fossil fuel because it is thought to originate from the remains of tiny organisms that lived in the oceans and died long ago. Their remains sank and settled on the seafloor, where they were covered by sediment. Over time, the layers of sediment and organic remains reached a great thickness. Deep beneath Earth's surface, heat and pressure changed the sediments into rocks, such as shale, sandstone, and limestone, transforming the organic matter into oil. The oil, being less dense than the surrounding rock, migrated upward through the crust.

In some cases, the upward migration of the oil was stopped by underground structures called traps. A trap occurs when a layer of impermeable rock, such as shale, rests above a layer of porous rock, such as sandstone. Liquids cannot pass through impermeable rock. However, porous rock has many empty spaces that can store and transmit fluids. Such rocks, called reservoir rocks, include sandstone and some types of limestone. If enough oil accumulates in the porous rock beneath an impermeable trap, the result can be a usable resource—an oil field that humans can exploit.

The processes that generate oil are occurring today in many parts of the world. However, these processes take millions of years. There is no doubt that humans are using oil at a much faster rate than it can be replaced by natural processes. In other words, oil is a nonrenewable resource. The question, then, is not so much will the world run out of oil, but when will it happen?

To predict when Earth's oil reserves will be depleted, geologists and engineers estimate the amount of oil in oil fields that have already been found. They then try to estimate the number and size of oil fields that will be discovered. The engineers combine these numbers in an effort to predict when the worldwide production of oil will reach its peak. After reaching the peak, production will begin to decline. After that, the price of oil will quickly climb so high that it will become too expensive for many of our present uses. Complicating such predictions is the fact that there are large deposits of oil in the form of tar sands and oil shales. These resources are more difficult and expensive to recover than conventional oil deposits. However, they could become economical if the demand for—and price of—oil rise high enough. The question is whether to include these unconventional deposits when trying to estimate the timing of peak oil production.

The pessimists point out that production from the world's super-giant oil fields is already in decline. The super-giant fields are those of the Arabian Peninsula and Persian Gulf (an area that includes Saudi Arabia, Iran, and Iraq). They also point out that geologists have not discovered any new super-giant fields in the last twenty years. The optimists respond that many large oil deposits, with reserves in the billions of barrels, have been found since then and continue to be found. Huge oil fields might still be found in places that have not been fully explored yet for their oil potential. These places include Polar regions.

Technological developments have made it possible to drill for oil in deeper parts of the ocean and more remote regions than ever before. Advanced exploration technologies such as seismic reflection imaging can produce three-dimensional images of underground geologic structures. This improves the ability of geologists to "see" beneath the surface to locate new oil deposits. New extraction methods have increased the productivity of existing oil fields. For example, water pumped under high pressure into an oil-bearing formation fractures the rocks to increase the flow of oil toward existing wells. Pumping a gas such as nitrogen into an oil reservoir displaces oil trapped in tiny spaces in the rocks so that the oil can be recovered.

From early 2007 through the first half of 2008, prices for oil and gas soared. Oil climbed from $45 up to $147 a barrel. The cost of gasoline reached more than $4.00 a gallon in the U.S. This seemed to confirm fears that oil production had peaked and was beginning to decline. Many factors contributed to these steep price increases. Among them were the rising demand in developing countries, hurricane damage to off-shore drilling platforms and refineries in the Gulf of Mexico, and a drop in the value of the U.S. dollar as compared with other world currencies. By late 2008, the price of oil had dropped considerably, with gasoline selling for less than $2.00 a gallon. Many experts caution that this should not lull us into thinking that oil supplies are inexhaustible.

When oil reserves do run low, we will probably need to increase our reliance on other energy resources such as, wind, solar power, and nuclear energy. That will allow us to extend other uses of "black gold" including plastics and medicines. Perhaps solutions to meeting our energy and raw material needs will be found that will avoid the major social upheaval that some have envisioned for a world that runs short of oil. Let's hope so!

QUESTIONS

1. Major oil deposits are found almost exclusively in sedimentary rocks. Why do most geologists believe that it would be unlikely for significant amounts of oil to accumulate in crystalline igneous or metamorphic rocks?
2. A few scientists believe that oil is generated deep within Earth's mantle by processes that do not involve the remains of living things. The oil migrates up into Earth's crust until it becomes trapped in porous sedimentary rocks. If this "abiotic" theory (without life processes) of oil formation is true, how would this affect estimates of world oil reserves? Use the Internet to research your answer.
3. What are some things that you and your family can do most easily to decrease your use of petroleum resources?
4. How might using less petroleum have a positive affect on our climate? Use the Internet to research your answer. Support your answer with evidence from your research.

Photo Credits

All photographs, except those listed below were provided courtesy of the author, Thomas McGuire.

Figure I-1	Astronaut Karen Nyberg, courtesy of NASA
Figure I-2	Men on buoy, courtesy of U.S. Department of Defense
Figure I-3	Temperature at La Guardia Airport, courtesy of NOAA
Figure I-Q21	Precipitation at Albany, NY, courtesy of NOAA
Figure I-Q24	*Titanic*, courtesy of United States National Archives
Figure I-Q30	Earth's layers, courtesy of USGS
Figure I-CRQ 15	U.S. Weather Deaths, courtesy of NOAA
Figure 1-Q9	Chemistry of Continental Crust, Courtesy of Steven I. Dutch, Professor, Natural and Applied Sciences, University of Wisconsin-Green Bay
Figure 1-Q31	Borrego Palm Canyon, CA, map, courtesy of USGS
Figure 1-CRQ 18	Map, adapted from USGS Popolopen Lake quadrangle, NY
Figure 2-Q65	Map: Solar Energy Potential in the USA, courtesy of the Department of Energy.
Figure 3-13	US seismic hazards map, courtesy of USGS.
Figure 3-CRQ5	Seismic waves and Earth's interior, courtesy of USGS.
Figure 3-CRQ38	Four plate interfaces, courtesy of USGS.
Figure 4-11(center)	Valley eroded by a glacier, courtesy of David G. Vaughan.
Figure 4-CRQ44	Soil horizons, courtesy of United States. Department of Agriculture.
Figure 5-CRQ50	Stromboli volcano, courtesy of Steven Dengler.
Figure 7-Q11	Climber on Mt. Everest, courtesy of Peak Freak Expeditions.
Figure 8-15	Tornado, NOAA, in the public domain.
Figure 8-17	Hurricane paths, adapted from NOAA.
Figure 8-Q39	Hurricane damage, NOAA, in the public domain.
Figure 9-18	Graph: Monthly Carbon Dioxide, courtesy of NOAA.
Figure 9-CRQ38	Major streams and Rivers of New York State, adapted from NOAA map.
Figure 10-2	Astronaut John H. Glenn, Jr. in training, courtesy of NASA.
Figure 10-14	One year at the North Pole, courtesy of NOAA.
Figure 11-16	Galaxy NGC 3370, courtesy of NASA.
Figure 11-17	White starlight passing through a prism © Adam Hart-Davis.
Figure 11-19	Structure of the universe, © Willem Schaap in the public domain.
Figure 12-4	Emission of greenhouse gases in the United States based on graph from the U.S. Department of Energy.
Figure 12-5	Extent of Polar ice based on image from NASA.

Reference Tables for Physical Setting/EARTH SCIENCE

Radioactive Decay Data

RADIOACTIVE ISOTOPE	DISINTEGRATION	HALF-LIFE (years)
Carbon-14	$^{14}C \rightarrow {}^{14}N$	5.7×10^3
Potassium-40	$^{40}K \rightarrow {}^{40}Ar$, $^{40}K \rightarrow {}^{40}Ca$	1.3×10^9
Uranium-238	$^{238}U \rightarrow {}^{206}Pb$	4.5×10^9
Rubidium-87	$^{87}Rb \rightarrow {}^{87}Sr$	4.9×10^{10}

Specific Heats of Common Materials

MATERIAL	SPECIFIC HEAT (Joules/gram • °C)
Liquid water	4.18
Solid water (ice)	2.11
Water vapor	2.00
Dry air	1.01
Basalt	0.84
Granite	0.79
Iron	0.45
Copper	0.38
Lead	0.13

Equations

$$\text{Eccentricity} = \frac{\text{distance between foci}}{\text{length of major axis}}$$

$$\text{Gradient} = \frac{\text{change in field value}}{\text{distance}}$$

$$\text{Rate of change} = \frac{\text{change in value}}{\text{time}}$$

$$\text{Density} = \frac{\text{mass}}{\text{volume}}$$

Properties of Water

Heat energy gained during melting .... 334 J/g
Heat energy released during freezing .... 334 J/g
Heat energy gained during vaporization ..... 2260 J/g
Heat energy released during condensation ... 2260 J/g

Density at 3.98°C .... 1.0 g/mL

Average Chemical Composition of Earth's Crust, Hydrosphere, and Troposphere

ELEMENT (symbol)	CRUST		HYDROSPHERE	TROPOSPHERE
	Percent by mass	Percent by volume	Percent by volume	Percent by volume
Oxygen (O)	46.10	94.04	33.0	21.0
Silicon (Si)	28.20	0.88		
Aluminum (Al)	8.23	0.48		
Iron (Fe)	5.63	0.49		
Calcium (Ca)	4.15	1.18		
Sodium (Na)	2.36	1.11		
Magnesium (Mg)	2.33	0.33		
Potassium (K)	2.09	1.42		
Nitrogen (N)				78.0
Hydrogen (H)			66.0	
Other	0.91	0.07	1.0	1.0

2011 EDITION

This edition of the Earth Science Reference Tables should be used in the classroom beginning in the 2011–12 school year. The first examination for which these tables will be used is the January 2012 Regents Examination in Physical Setting/Earth Science.

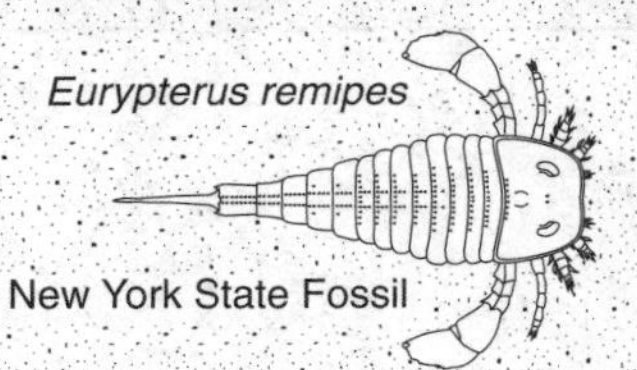

Eurypterus remipes

New York State Fossil

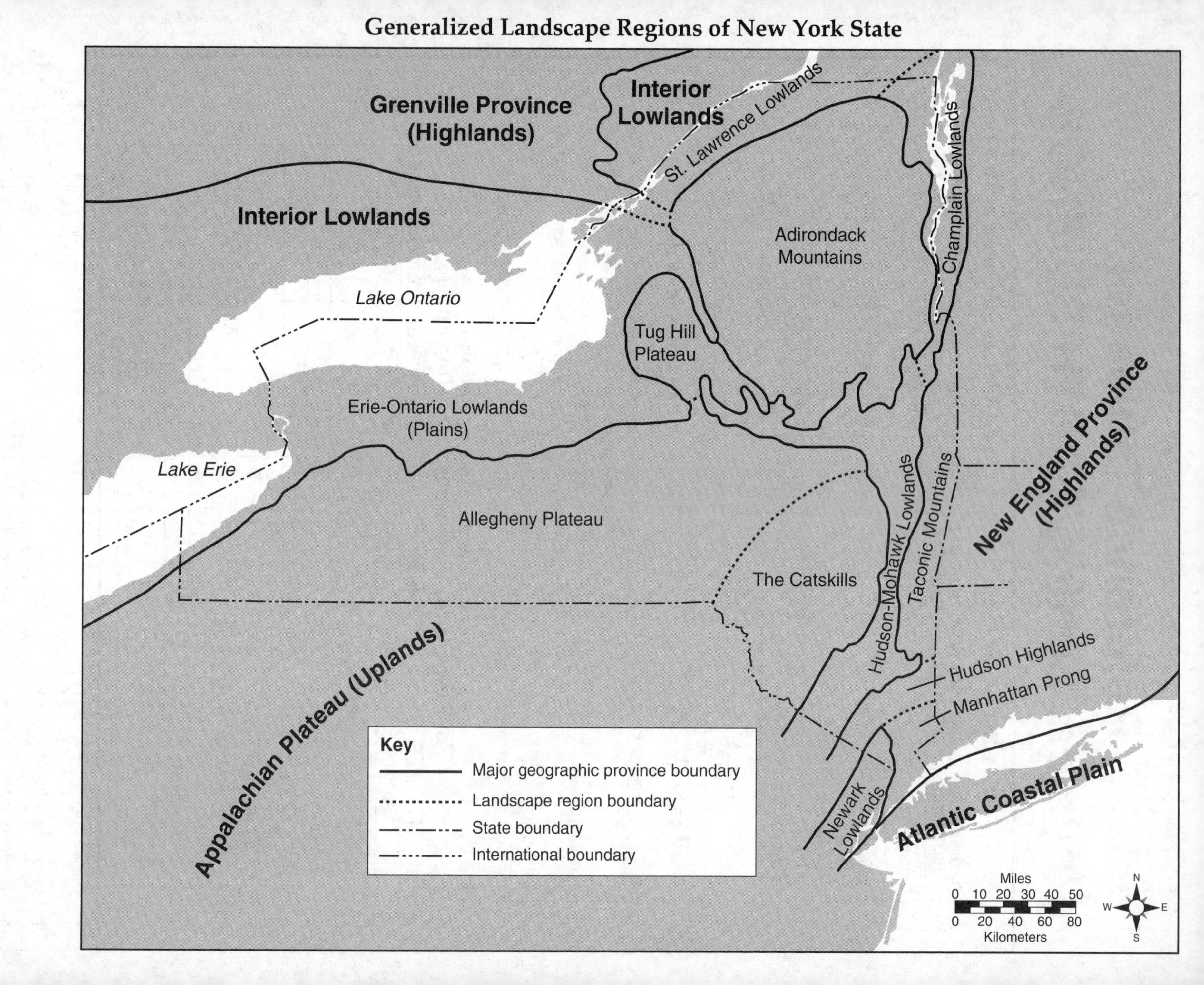
Generalized Landscape Regions of New York State
Grenville Province (Highlands)
Interior Lowlands
St. Lawrence Lowlands
Interior Lowlands
Adirondack Mountains
Champlain Lowlands
Lake Ontario
Tug Hill Plateau
Erie-Ontario Lowlands (Plains)
Lake Erie
Allegheny Plateau
The Catskills
Hudson-Mohawk Lowlands
Taconic Mountains
New England Province (Highlands)
Hudson Highlands
Manhattan Prong
Appalachian Plateau (Uplands)
Newark Lowlands
Atlantic Coastal Plain
Key
Major geographic province boundary
Landscape region boundary
State boundary
International boundary
Miles
0 10 20 30 40 50
0 20 40 60 80
Kilometers
N
W
E
S

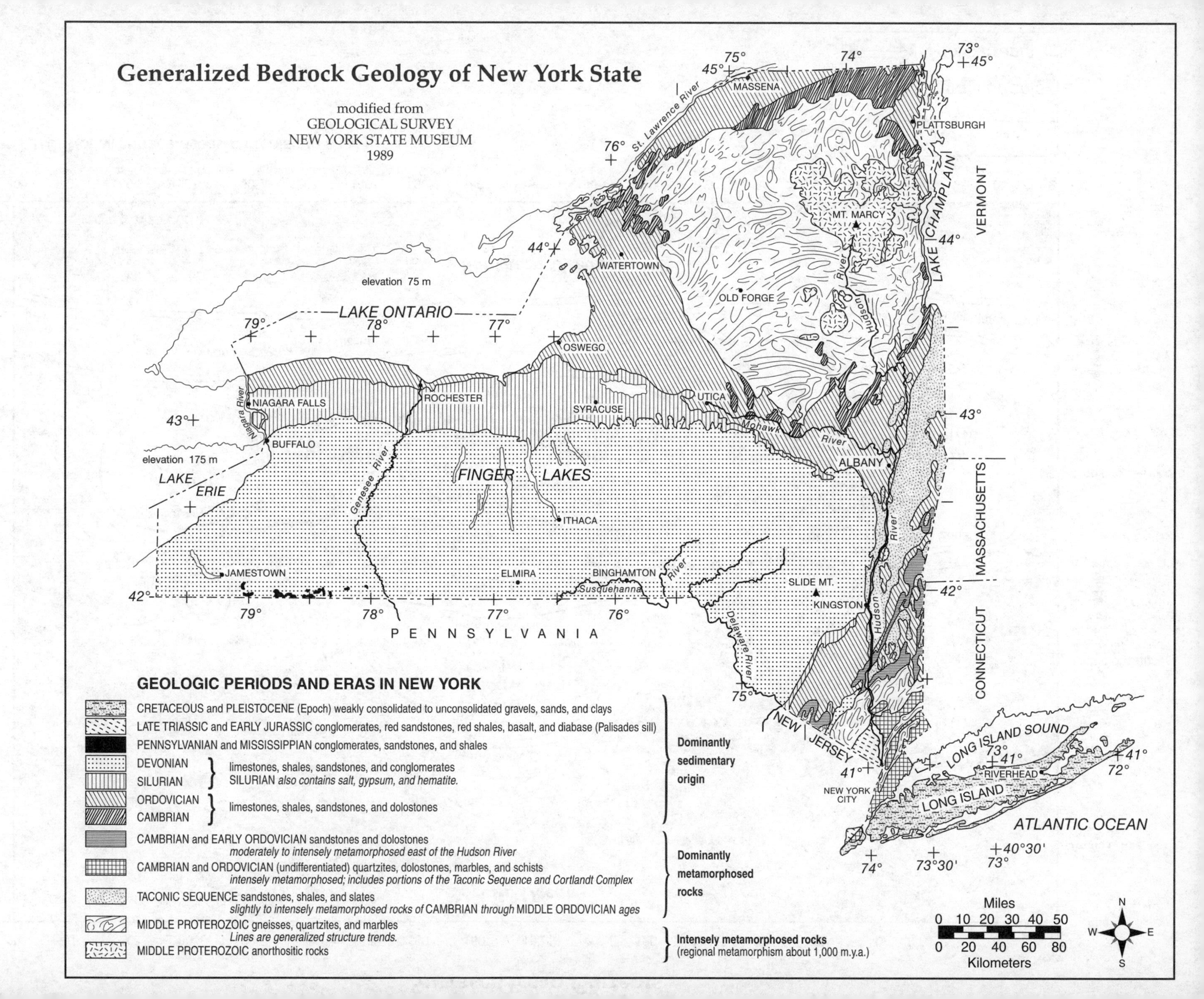
Generalized Bedrock Geology of New York State
modified from
GEOLOGICAL SURVEY
NEW YORK STATE MUSEUM
1989
LAKE ONTARIO
elevation 75 m
LAKE ERIE
elevation 175 m
FINGER LAKES
PENNSYLVANIA
VERMONT
MASSACHUSETTS
CONNECTICUT
NEW JERSEY
LAKE CHAMPLAIN
LONG ISLAND SOUND
LONG ISLAND
ATLANTIC OCEAN
MASSENA
PLATTSBURGH
MT. MARCY
WATERTOWN
OLD FORGE
OSWEGO
NIAGARA FALLS
ROCHESTER
SYRACUSE
UTICA
BUFFALO
ALBANY
ITHACA
JAMESTOWN
ELMIRA
BINGHAMTON
SLIDE MT.
KINGSTON
NEW YORK CITY
RIVERHEAD
St. Lawrence River
Niagara River
Genesee River
Mohawk River
Hudson River
Susquehanna River
Delaware River
GEOLOGIC PERIODS AND ERAS IN NEW YORK
CRETACEOUS and PLEISTOCENE (Epoch) weakly consolidated to unconsolidated gravels, sands, and clays
LATE TRIASSIC and EARLY JURASSIC conglomerates, red sandstones, red shales, basalt, and diabase (Palisades sill)
PENNSYLVANIAN and MISSISSIPPIAN conglomerates, sandstones, and shales
DEVONIAN
SILURIAN
limestones, shales, sandstones, and conglomerates
SILURIAN also contains salt, gypsum, and hematite.
ORDOVICIAN
CAMBRIAN
limestones, shales, sandstones, and dolostones
Dominantly sedimentary origin
CAMBRIAN and EARLY ORDOVICIAN sandstones and dolostones
moderately to intensely metamorphosed east of the Hudson River
CAMBRIAN and ORDOVICIAN (undifferentiated) quartzites, dolostones, marbles, and schists
intensely metamorphosed; includes portions of the Taconic Sequence and Cortlandt Complex
TACONIC SEQUENCE sandstones, shales, and slates
slightly to intensely metamorphosed rocks of CAMBRIAN through MIDDLE ORDOVICIAN ages
Dominantly metamorphosed rocks
MIDDLE PROTEROZOIC gneisses, quartzites, and marbles
Lines are generalized structure trends.
MIDDLE PROTEROZOIC anorthositic rocks
Intensely metamorphosed rocks
(regional metamorphism about 1,000 m.y.a.)
Miles
0 10 20 30 40 50
0 20 40 60 80
Kilometers
N
E
S
W

Surface Ocean Currents

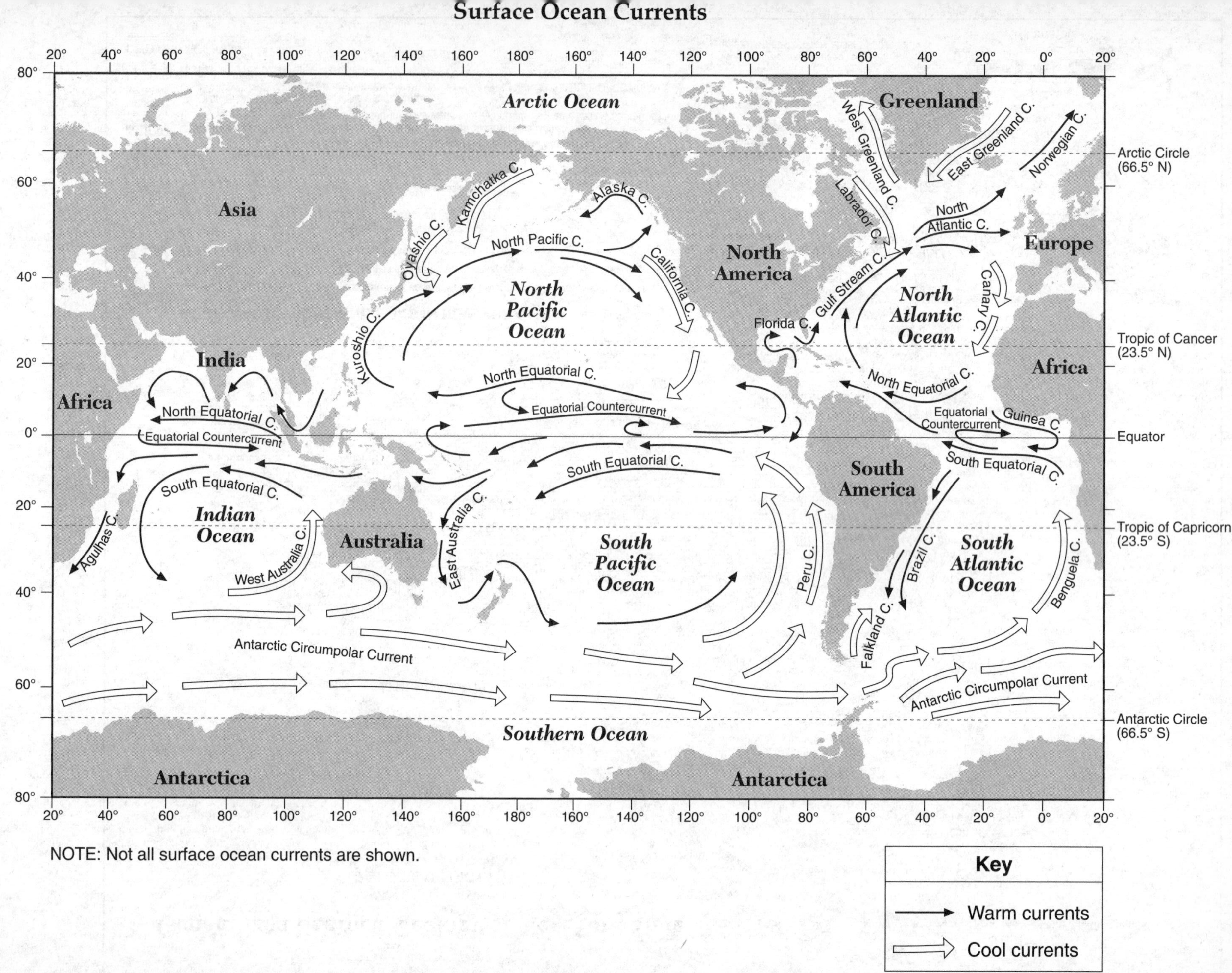

NOTE: Not all surface ocean currents are shown.

Tectonic Plates

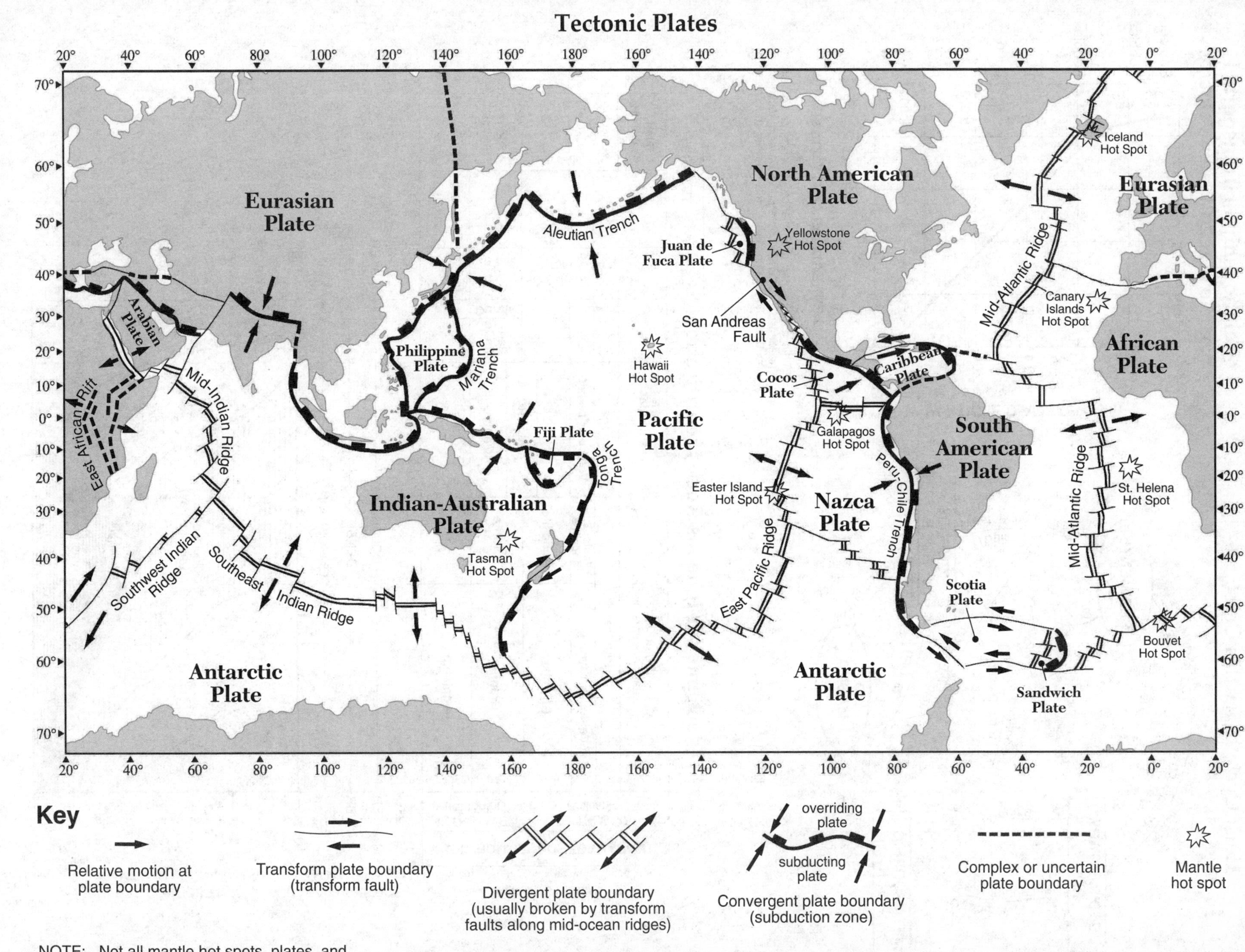

NOTE: Not all mantle hot spots, plates, and boundaries are shown.

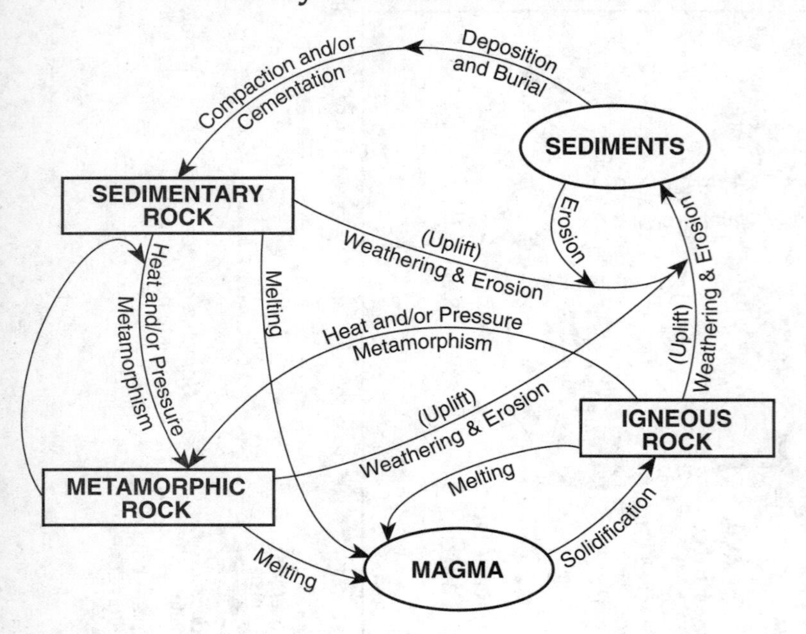

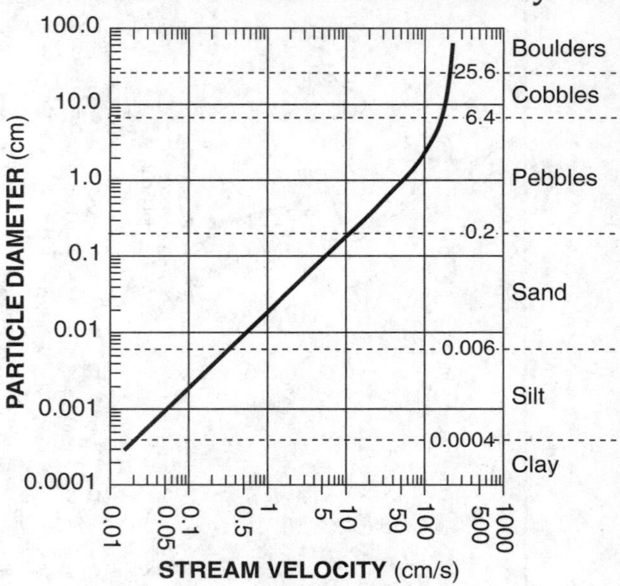

This generalized graph shows the water velocity needed to maintain, but not start, movement. Variations occur due to differences in particle density and shape.

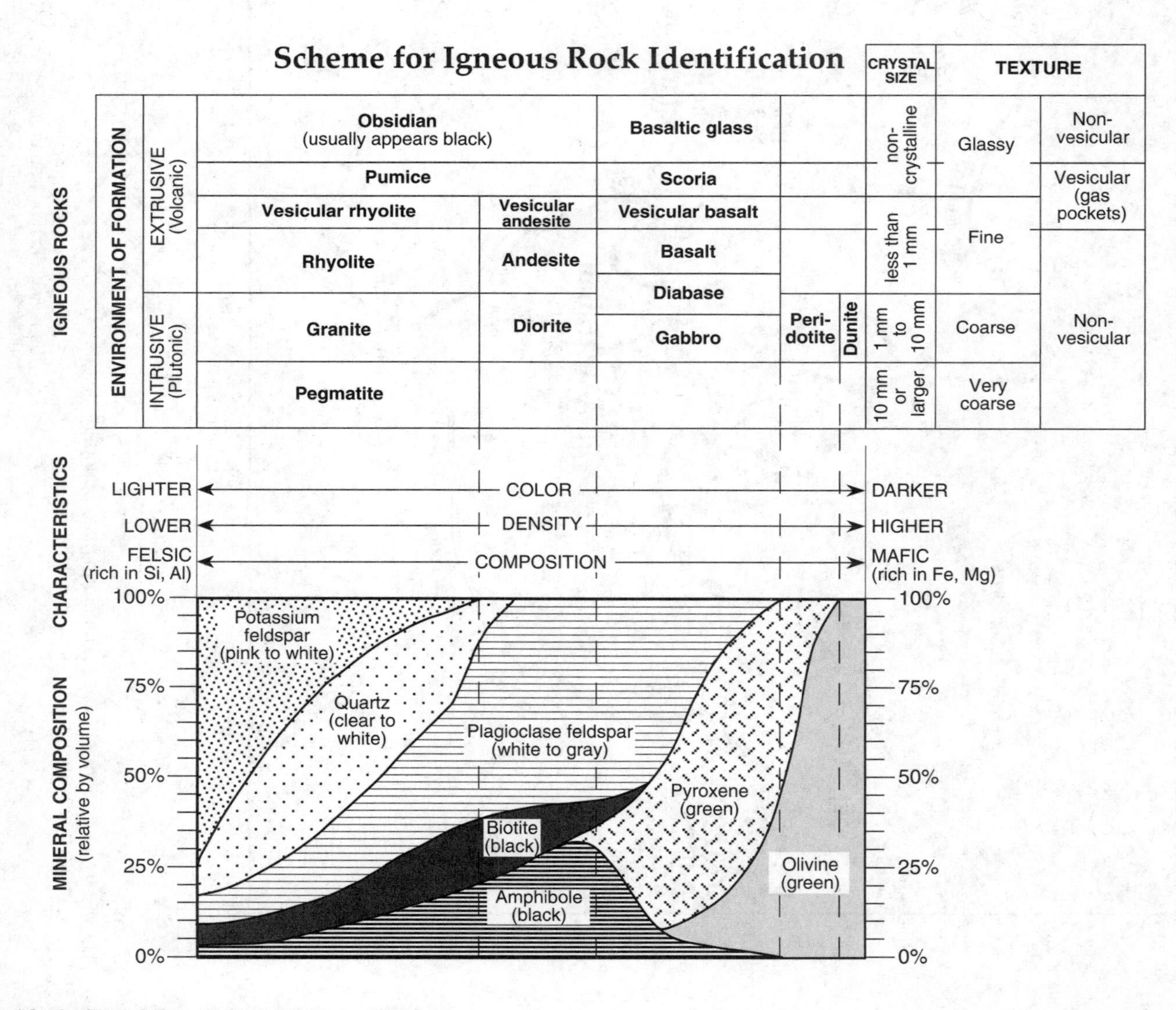

Scheme for Sedimentary Rock Identification

INORGANIC LAND-DERIVED SEDIMENTARY ROCKS					
TEXTURE	GRAIN SIZE	COMPOSITION	COMMENTS	ROCK NAME	MAP SYMBOL
Clastic (fragmental)	Pebbles, cobbles, and/or boulders embedded in sand, silt, and/or clay	Mostly quartz, feldspar, and clay minerals; may contain fragments of other rocks and minerals	Rounded fragments	**Conglomerate**	
Clastic (fragmental)	Pebbles, cobbles, and/or boulders embedded in sand, silt, and/or clay	Mostly quartz, feldspar, and clay minerals; may contain fragments of other rocks and minerals	Angular fragments	**Breccia**	
Clastic (fragmental)	Sand (0.006 to 0.2 cm)	Mostly quartz, feldspar, and clay minerals; may contain fragments of other rocks and minerals	Fine to coarse	**Sandstone**	
Clastic (fragmental)	Silt (0.0004 to 0.006 cm)	Mostly quartz, feldspar, and clay minerals; may contain fragments of other rocks and minerals	Very fine grain	**Siltstone**	
Clastic (fragmental)	Clay (less than 0.0004 cm)	Mostly quartz, feldspar, and clay minerals; may contain fragments of other rocks and minerals	Compact; may split easily	**Shale**	

CHEMICALLY AND/OR ORGANICALLY FORMED SEDIMENTARY ROCKS					
TEXTURE	GRAIN SIZE	COMPOSITION	COMMENTS	ROCK NAME	MAP SYMBOL
Crystalline	Fine to coarse crystals	Halite	Crystals from chemical precipitates and evaporites	**Rock salt**	
Crystalline	Fine to coarse crystals	Gypsum	Crystals from chemical precipitates and evaporites	**Rock gypsum**	
Crystalline	Fine to coarse crystals	Dolomite	Crystals from chemical precipitates and evaporites	**Dolostone**	
Crystalline or bioclastic	Microscopic to very coarse	Calcite	Precipitates of biologic origin or cemented shell fragments	**Limestone**	
Bioclastic	Microscopic to very coarse	Carbon	Compacted plant remains	**Bituminous coal**	

Scheme for Metamorphic Rock Identification

TEXTURE		GRAIN SIZE	COMPOSITION	TYPE OF METAMORPHISM	COMMENTS	ROCK NAME	MAP SYMBOL
FOLIATED	MINERAL ALIGNMENT	Fine	MICA	Regional (Heat and pressure increases)	Low-grade metamorphism of shale	**Slate**	
FOLIATED	MINERAL ALIGNMENT	Fine to medium	MICA, QUARTZ, FELDSPAR, AMPHIBOLE, GARNET	Regional (Heat and pressure increases)	Foliation surfaces shiny from microscopic mica crystals	**Phyllite**	
FOLIATED	MINERAL ALIGNMENT	Fine to medium	MICA, QUARTZ, FELDSPAR, AMPHIBOLE, GARNET, PYROXENE	Regional (Heat and pressure increases)	Platy mica crystals visible from metamorphism of clay or feldspars	**Schist**	
FOLIATED	BANDING	Medium to coarse	MICA, QUARTZ, FELDSPAR, AMPHIBOLE, GARNET, PYROXENE	Regional (Heat and pressure increases)	High-grade metamorphism; mineral types segregated into bands	**Gneiss**	
NONFOLIATED		Fine	Carbon	Regional	Metamorphism of bituminous coal	**Anthracite coal**	
NONFOLIATED		Fine	Various minerals	Contact (heat)	Various rocks changed by heat from nearby magma/lava	**Hornfels**	
NONFOLIATED		Fine to coarse	Quartz	Regional or contact	Metamorphism of quartz sandstone	**Quartzite**	
NONFOLIATED		Fine to coarse	Calcite and/or dolomite	Regional or contact	Metamorphism of limestone or dolostone	**Marble**	
NONFOLIATED		Coarse	Various minerals	Regional or contact	Pebbles may be distorted or stretched	**Metaconglomerate**	

GEOLOGIC HISTORY

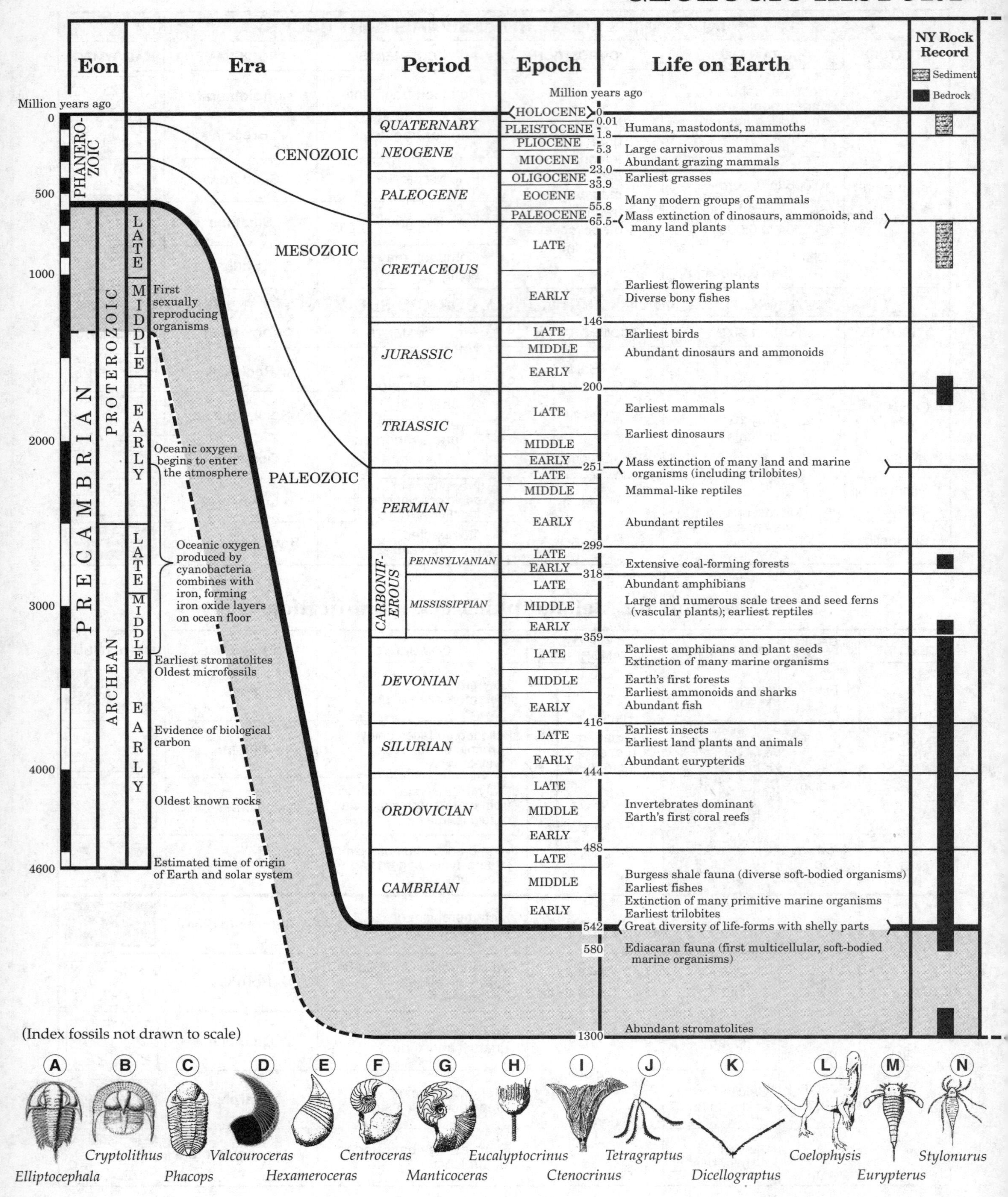

OF NEW YORK STATE

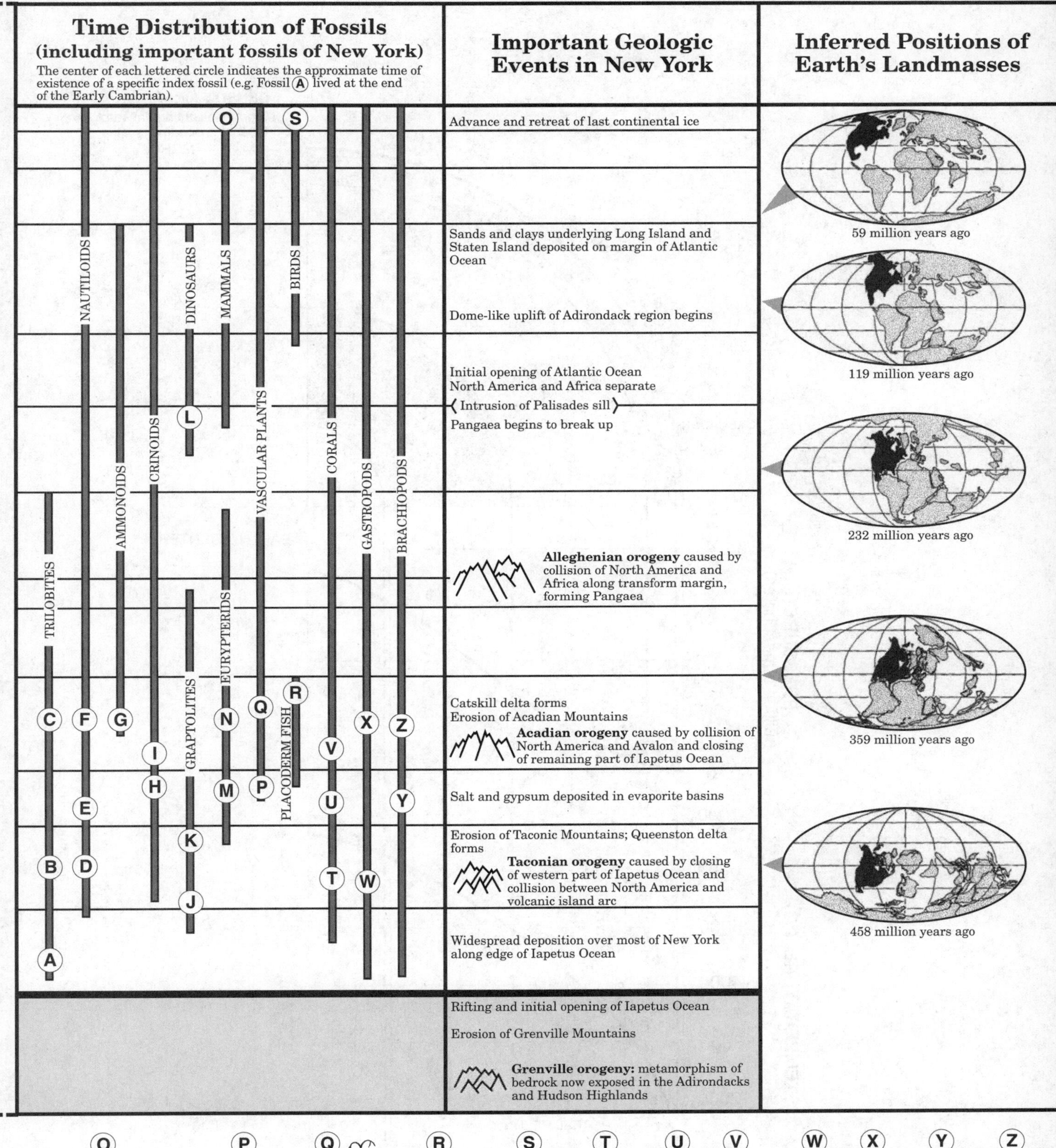

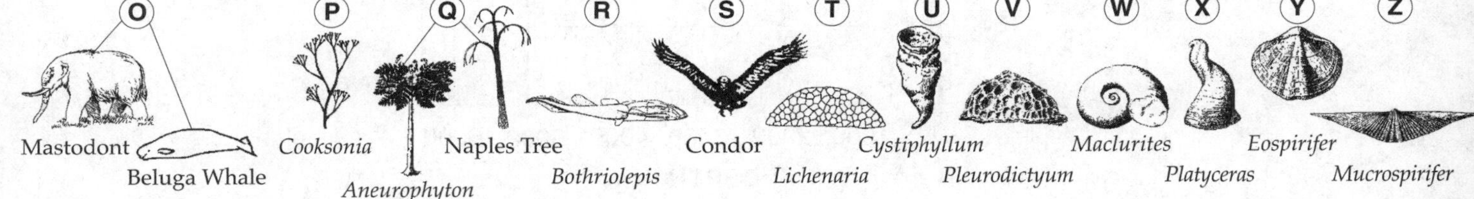

ADU (2011)

Inferred Properties of Earth's Interior

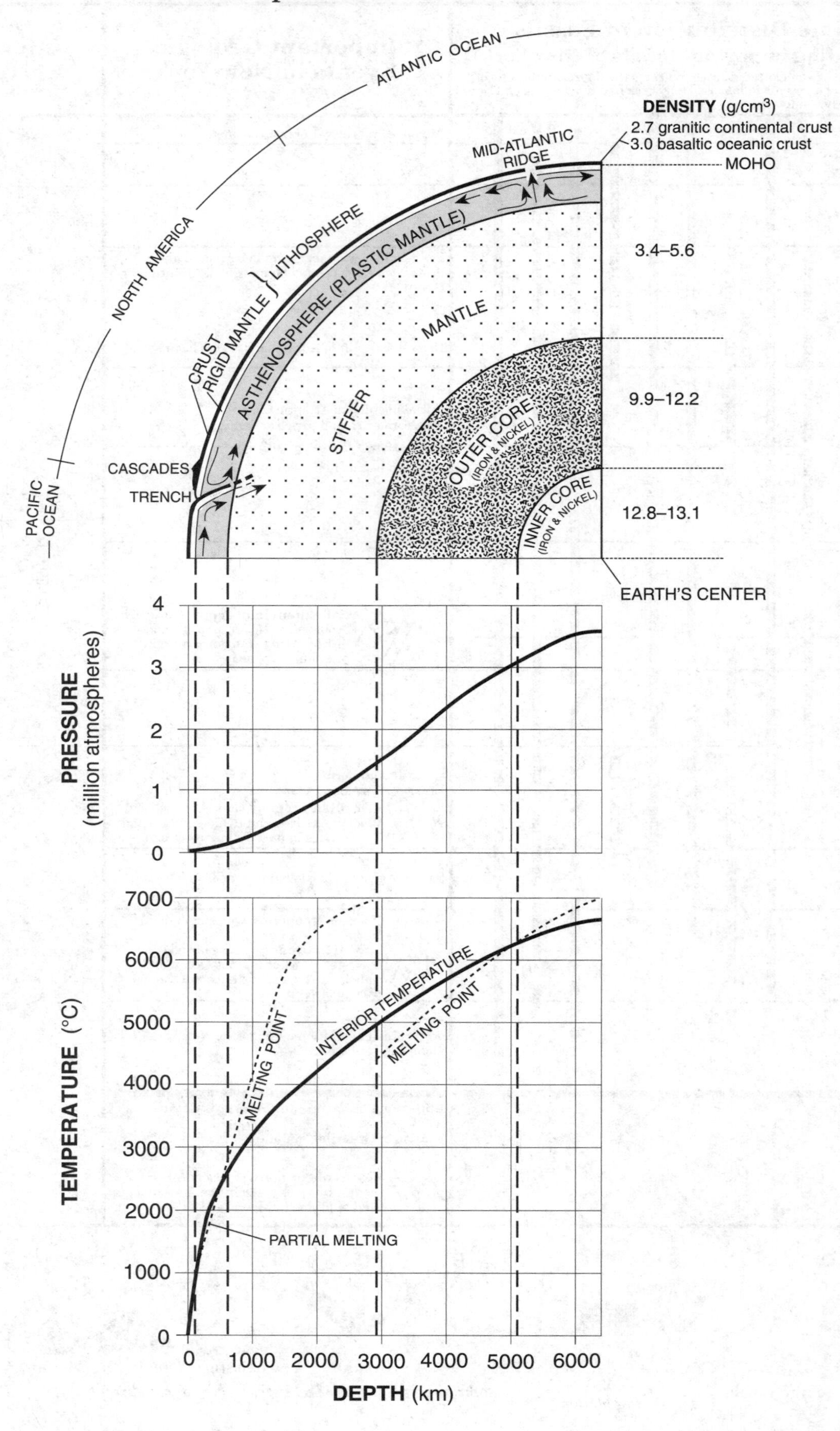

Earthquake P-Wave and S-Wave Travel Time

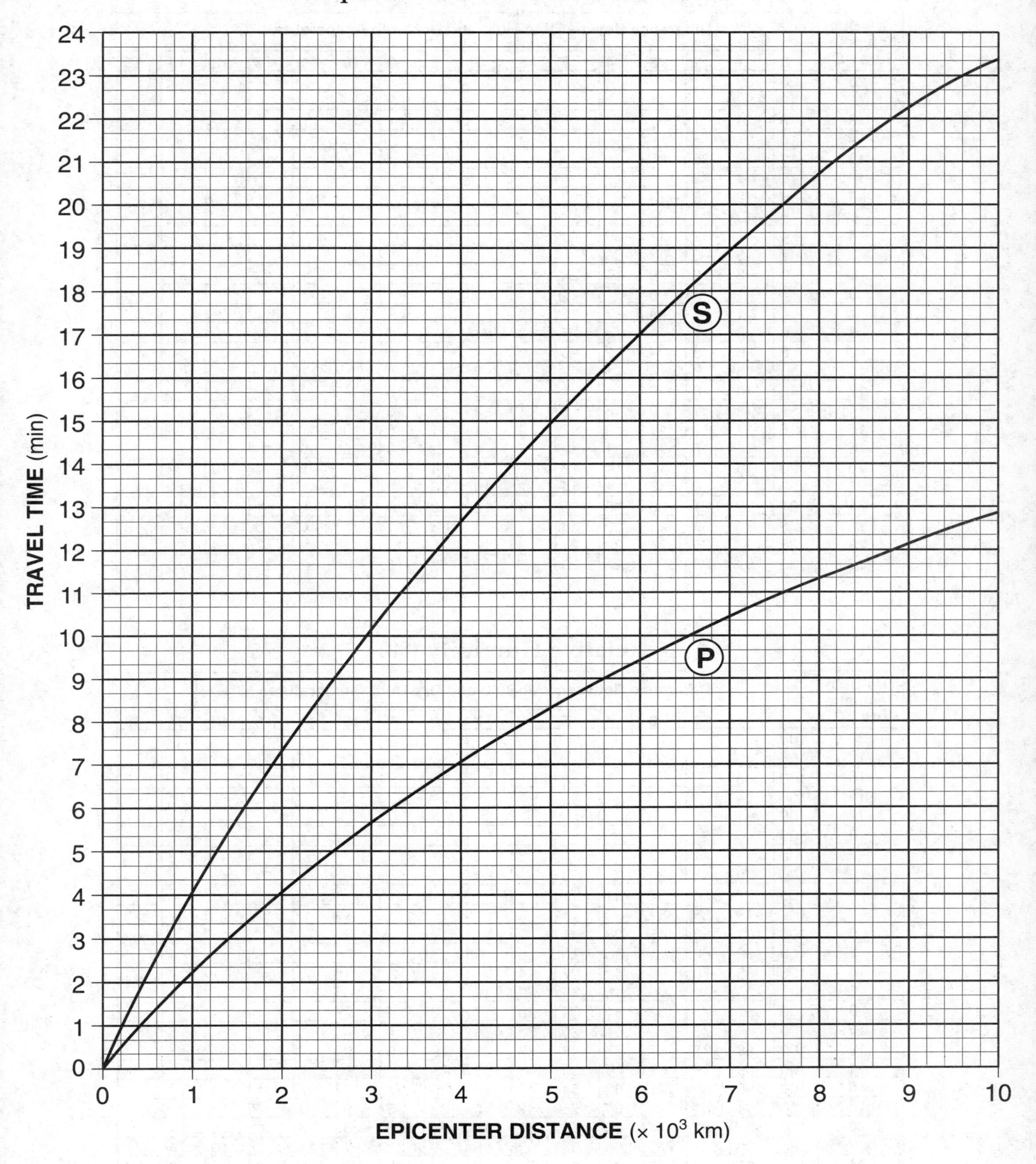

Dewpoint (°C)

Dry-Bulb Temperature (°C)	Difference Between Wet-Bulb and Dry-Bulb Temperatures (C°)															
	0	1	2	3	4	5	6	7	8	9	10	11	12	13	14	15
-20	-20	-33														
-18	-18	-28														
-16	-16	-24														
-14	-14	-21	-36													
-12	-12	-18	-28													
-10	-10	-14	-22													
-8	-8	-12	-18	-29												
-6	-6	-10	-14	-22												
-4	-4	-7	-12	-17	-29											
-2	-2	-5	-8	-13	-20											
0	0	-3	-6	-9	-15	-24										
2	2	-1	-3	-6	-11	-17										
4	4	1	-1	-4	-7	-11	-19									
6	6	4	1	-1	-4	-7	-13	-21								
8	8	6	3	1	-2	-5	-9	-14								
10	10	8	6	4	1	-2	-5	-9	-14	-28						
12	12	10	8	6	4	1	-2	-5	-9	-16						
14	14	12	11	9	6	4	1	-2	-5	-10	-17					
16	16	14	13	11	9	7	4	1	-1	-6	-10	-17				
18	18	16	15	13	11	9	7	4	2	-2	-5	-10	-19			
20	20	19	17	15	14	12	10	7	4	2	-2	-5	-10	-19		
22	22	21	19	17	16	14	12	10	8	5	3	-1	-5	-10	-19	
24	24	23	21	20	18	16	14	12	10	8	6	2	-1	-5	-10	-18
26	26	25	23	22	20	18	17	15	13	11	9	6	3	0	-4	-9
28	28	27	25	24	22	21	19	17	16	14	11	9	7	4	1	-3
30	30	29	27	26	24	23	21	19	18	16	14	12	10	8	5	1

Relative Humidity (%)

Dry-Bulb Temperature (°C)	Difference Between Wet-Bulb and Dry-Bulb Temperatures (C°)															
	0	1	2	3	4	5	6	7	8	9	10	11	12	13	14	15
-20	100	28														
-18	100	40														
-16	100	48														
-14	100	55	11													
-12	100	61	23													
-10	100	66	33													
-8	100	71	41	13												
-6	100	73	48	20												
-4	100	77	54	32	11											
-2	100	79	58	37	20	1										
0	100	81	63	45	28	11										
2	100	83	67	51	36	20	6									
4	100	85	70	56	42	27	14									
6	100	86	72	59	46	35	22	10								
8	100	87	74	62	51	39	28	17	6							
10	100	88	76	65	54	43	33	24	13	4						
12	100	88	78	67	57	48	38	28	19	10	2					
14	100	89	79	69	60	50	41	33	25	16	8	1				
16	100	90	80	71	62	54	45	37	29	21	14	7	1			
18	100	91	81	72	64	56	48	40	33	26	19	12	6			
20	100	91	82	74	66	58	51	44	36	30	23	17	11	5		
22	100	92	83	75	68	60	53	46	40	33	27	21	15	10	4	
24	100	92	84	76	69	62	55	49	42	36	30	25	20	14	9	4
26	100	92	85	77	70	64	57	51	45	39	34	28	23	18	13	9
28	100	93	86	78	71	65	59	53	47	42	36	31	26	21	17	12
30	100	93	86	79	72	66	61	55	49	44	39	34	29	25	20	16

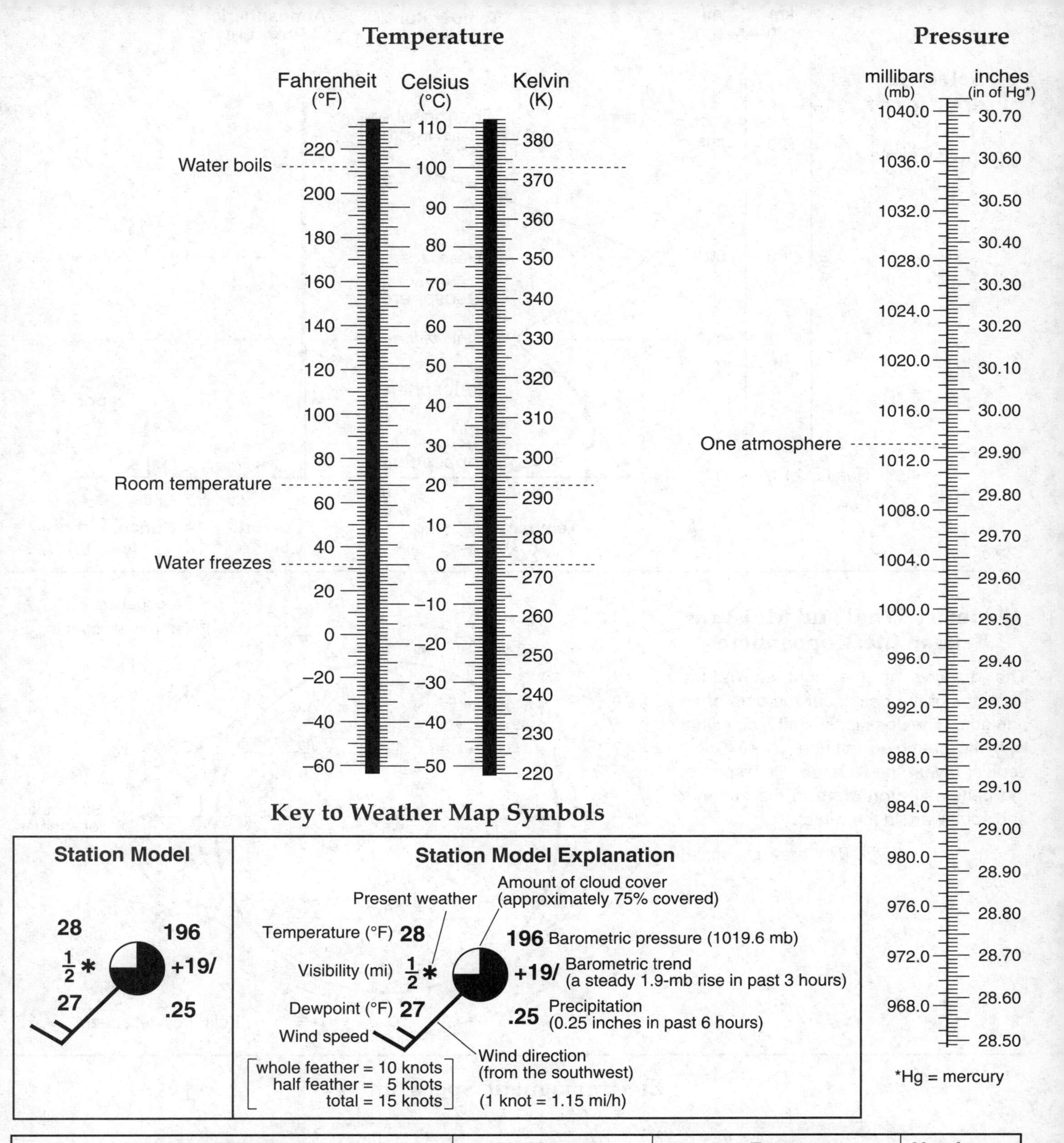

Temperature
Fahrenheit (°F)
Celsius (°C)
Kelvin (K)
220
200
180
160
140
120
100
80
60
40
20
0
–20
–40
–60
110
100
90
80
70
60
50
40
30
20
10
0
–10
–20
–30
–40
–50
380
370
360
350
340
330
320
310
300
290
280
270
260
250
240
230
220
Water boils
Room temperature
Water freezes
Pressure
millibars (mb)
inches (in of Hg*)
1040.0
1036.0
1032.0
1028.0
1024.0
1020.0
1016.0
1012.0
1008.0
1004.0
1000.0
996.0
992.0
988.0
984.0
980.0
976.0
972.0
968.0
30.70
30.60
30.50
30.40
30.30
30.20
30.10
30.00
29.90
29.80
29.70
29.60
29.50
29.40
29.30
29.20
29.10
29.00
28.90
28.80
28.70
28.60
28.50
One atmosphere
*Hg = mercury
Key to Weather Map Symbols
Station Model
28
196
1/2 *
+19/
27
.25
Station Model Explanation
Present weather
Amount of cloud cover (approximately 75% covered)
Temperature (°F) 28
196 Barometric pressure (1019.6 mb)
Visibility (mi) 1/2 *
+19/ Barometric trend (a steady 1.9-mb rise in past 3 hours)
Dewpoint (°F) 27
.25 Precipitation (0.25 inches in past 6 hours)
Wind speed
Wind direction (from the southwest)
whole feather = 10 knots
half feather = 5 knots
total = 15 knots
(1 knot = 1.15 mi/h)

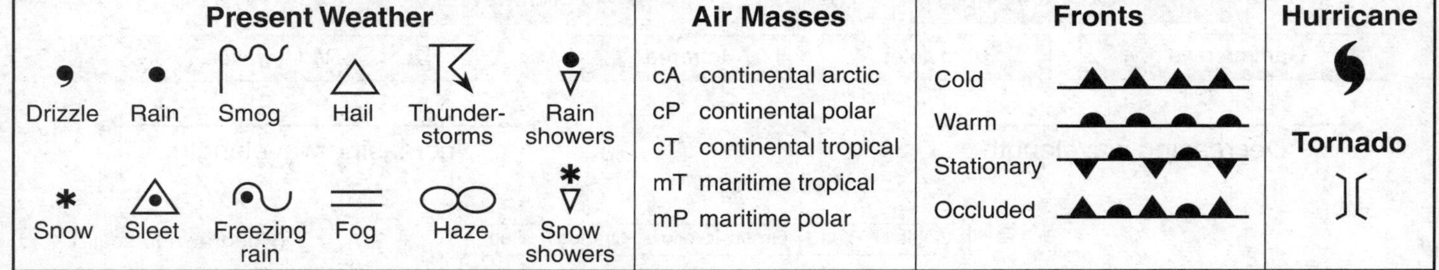

Present Weather
Drizzle
Rain
Smog
Hail
Thunder-storms
Rain showers
Snow
Sleet
Freezing rain
Fog
Haze
Snow showers
Air Masses
cA continental arctic
cP continental polar
cT continental tropical
mT maritime tropical
mP maritime polar
Fronts
Cold
Warm
Stationary
Occluded
Hurricane
Tornado

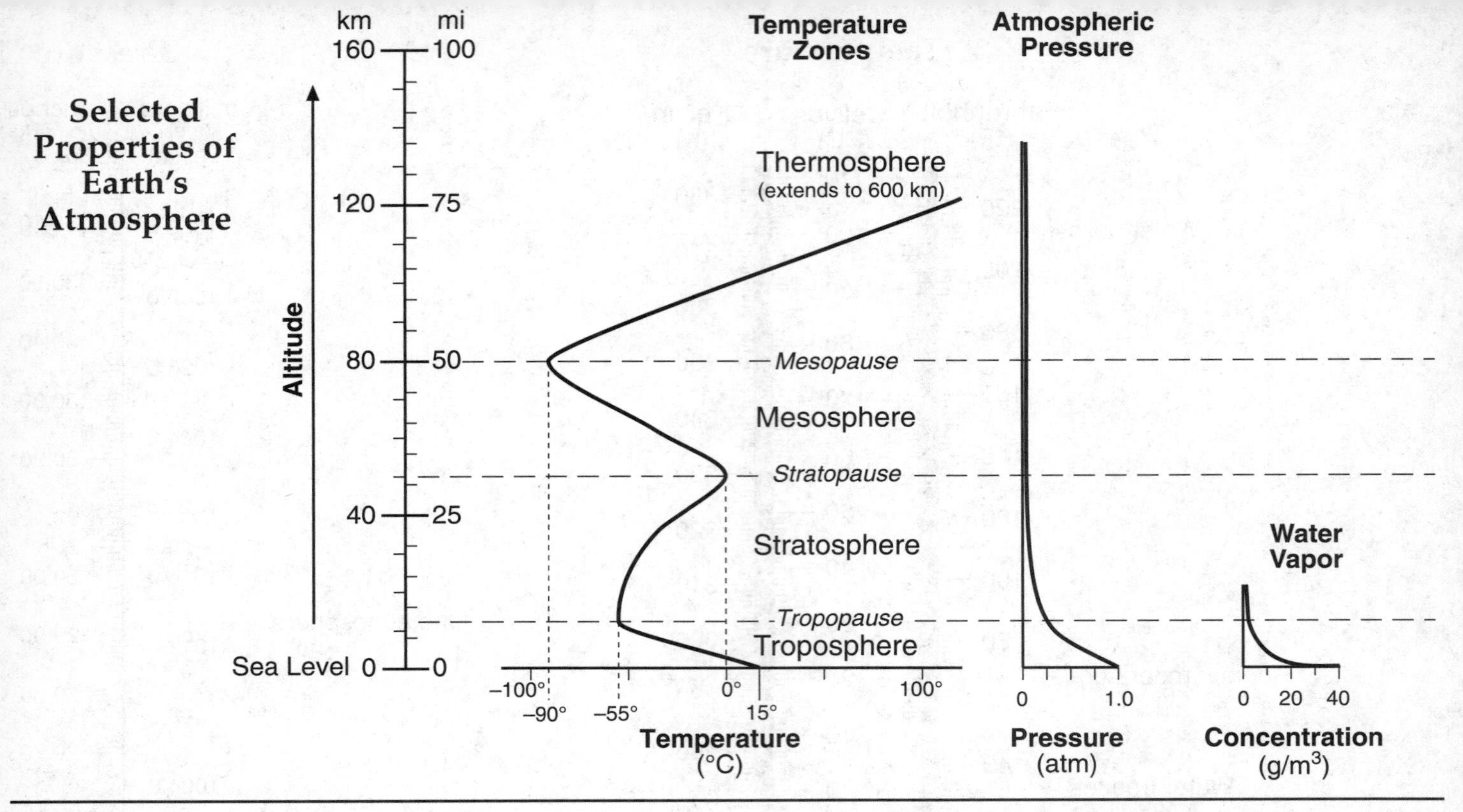

Planetary Wind and Moisture Belts in the Troposphere

The drawing on the right shows the locations of the belts near the time of an equinox. The locations shift somewhat with the changing latitude of the Sun's vertical ray. In the Northern Hemisphere, the belts shift northward in the summer and southward in the winter.

(Not drawn to scale)

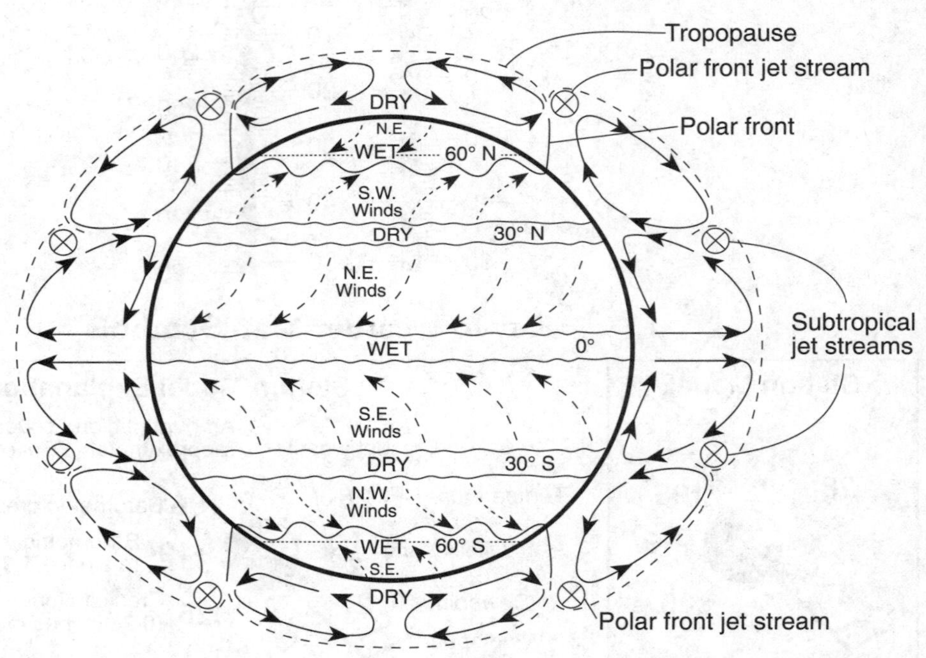

Electromagnetic Spectrum

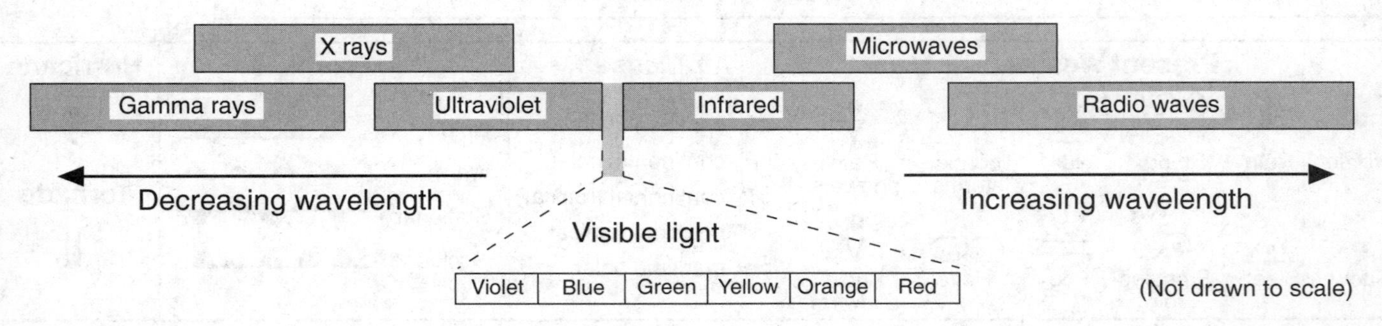

(Not drawn to scale)

Characteristics of Stars

(Name in italics refers to star represented by a ⊕.)
(Stages indicate the general sequence of star development.)

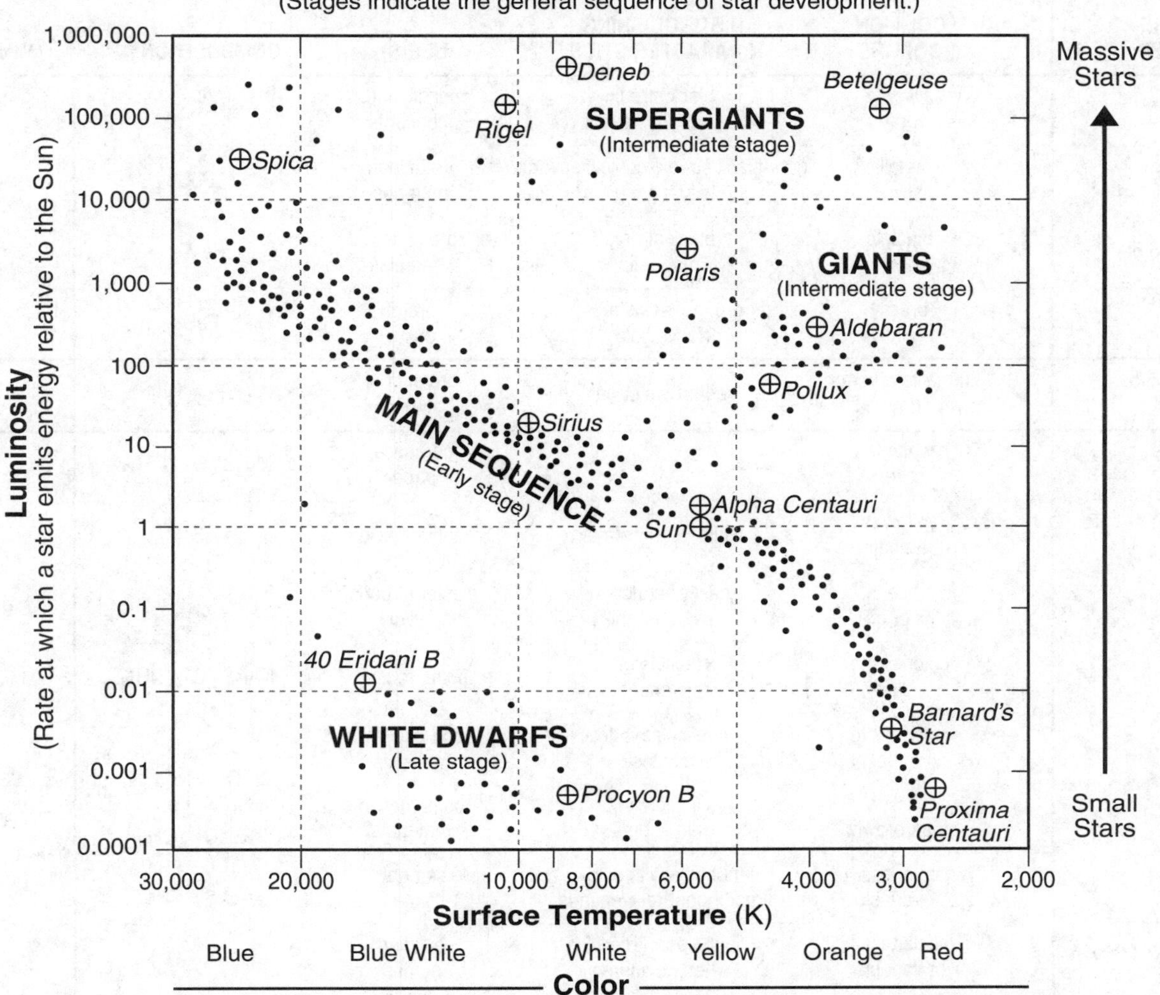

Solar System Data

Celestial Object	Mean Distance from Sun (million km)	Period of Revolution (d=days) (y=years)	Period of Rotation at Equator	Eccentricity of Orbit	Equatorial Diameter (km)	Mass (Earth = 1)	Density (g/cm^3)
SUN	—	—	27 d	—	1,392,000	333,000.00	1.4
MERCURY	57.9	88 d	59 d	0.206	4,879	0.06	5.4
VENUS	108.2	224.7 d	243 d	0.007	12,104	0.82	5.2
EARTH	149.6	365.26 d	23 h 56 min 4 s	0.017	12,756	1.00	5.5
MARS	227.9	687 d	24 h 37 min 23 s	0.093	6,794	0.11	3.9
JUPITER	778.4	11.9 y	9 h 50 min 30 s	0.048	142,984	317.83	1.3
SATURN	1,426.7	29.5 y	10 h 14 min	0.054	120,536	95.16	0.7
URANUS	2,871.0	84.0 y	17 h 14 min	0.047	51,118	14.54	1.3
NEPTUNE	4,498.3	164.8 y	16 h	0.009	49,528	17.15	1.8
EARTH'S MOON	149.6 (0.386 from Earth)	27.3 d	27.3 d	0.055	3,476	0.01	3.3

Properties of Common Minerals

LUSTER	HARDNESS	CLEAVAGE	FRACTURE	COMMON COLORS	DISTINGUISHING CHARACTERISTICS	USE(S)	COMPOSITION*	MINERAL NAME
Metallic luster	1–2	✔		silver to gray	black streak, greasy feel	pencil lead, lubricants	C	**Graphite**
	2.5	✔		metallic silver	gray-black streak, cubic cleavage, density = 7.6 g/cm^3	ore of lead, batteries	PbS	**Galena**
	5.5–6.5		✔	black to silver	black streak, magnetic	ore of iron, steel	Fe_3O_4	**Magnetite**
	6.5		✔	brassy yellow	green-black streak, (fool's gold)	ore of sulfur	FeS_2	**Pyrite**
Either	5.5 – 6.5 or 1		✔	metallic silver or earthy red	red-brown streak	ore of iron, jewelry	Fe_2O_3	**Hematite**
Nonmetallic luster	1	✔		white to green	greasy feel	ceramics, paper	$Mg_3Si_4O_{10}(OH)_2$	**Talc**
	2		✔	yellow to amber	white-yellow streak	sulfuric acid	S	**Sulfur**
	2	✔		white to pink or gray	easily scratched by fingernail	plaster of paris, drywall	$CaSO_4 \cdot 2H_2O$	**Selenite gypsum**
	2–2.5	✔		colorless to yellow	flexible in thin sheets	paint, roofing	$KAl_3Si_3O_{10}(OH)_2$	**Muscovite mica**
	2.5	✔		colorless to white	cubic cleavage, salty taste	food additive, melts ice	$NaCl$	**Halite**
	2.5–3	✔		black to dark brown	flexible in thin sheets	construction materials	$K(Mg,Fe)_3$ $AlSi_3O_{10}(OH)_2$	**Biotite mica**
	3	✔		colorless or variable	bubbles with acid, rhombohedral cleavage	cement, lime	$CaCO_3$	**Calcite**
	3.5	✔		colorless or variable	bubbles with acid when powdered	building stones	$CaMg(CO_3)_2$	**Dolomite**
	4	✔		colorless or variable	cleaves in 4 directions	hydrofluoric acid	CaF_2	**Fluorite**
	5–6	✔		black to dark green	cleaves in 2 directions at 90°	mineral collections, jewelry	$(Ca,Na)(Mg,Fe,Al)$ $(Si,Al)_2O_6$	**Pyroxene** (commonly augite)
	5.5	✔		black to dark green	cleaves at 56° and 124°	mineral collections, jewelry	$CaNa(Mg,Fe)_4(Al,Fe,Ti)_3$ $Si_6O_{22}(O,OH)_2$	**Amphibole** (commonly hornblende)
	6	✔		white to pink	cleaves in 2 directions at 90°	ceramics, glass	$KAlSi_3O_8$	**Potassium feldspar** (commonly orthoclase)
	6	✔		white to gray	cleaves in 2 directions, striations visible	ceramics, glass	$(Na,Ca)AlSi_3O_8$	**Plagioclase feldspar**
	6.5		✔	green to gray or brown	commonly light green and granular	furnace bricks, jewelry	$(Fe,Mg)_2SiO_4$	**Olivine**
	7		✔	colorless or variable	glassy luster, may form hexagonal crystals	glass, jewelry, electronics	SiO_2	**Quartz**
	6.5–7.5		✔	dark red to green	often seen as red glassy grains in NYS metamorphic rocks	jewelry (NYS gem), abrasives	$Fe_3Al_2Si_3O_{12}$	**Garnet**

*Chemical symbols:

Al = aluminum	Cl = chlorine	H = hydrogen	Na = sodium	S = sulfur
C = carbon	F = fluorine	K = potassium	O = oxygen	Si = silicon
Ca = calcium	Fe = iron	Mg = magnesium	Pb = lead	Ti = titanium

✔ = dominant form of breakage

Appendix A: A Guide to the Earth Science Reference Tables

The *Earth Science Reference Tables* (*ESRT*) are a critical resource for class work and examinations in Regents Earth science. This booklet contains 16 pages of tables, graphs, and other information. These tables are sent to teachers at the end of the school year along with the Regents exams. However, most teachers give their students a copy of the tables at the beginning of the course. This document can also be downloaded at the New York State Education Department web site (http://www.emsc.nysed.gov/osa/reftable.html). The *ESRT* are used throughout the year on tests, labs, and homework. No matter what you are doing in Regents Earth science, if you need specific data or graphs, look first in the Reference Tables.

You do not need to memorize the information in the Reference Tables. However, you should have a general idea of the kind of information provided in the *ESRT*, and you should be able to use all of it by the end of the year. In this review book, each table is introduced with the most relevant content area. For example, the use of the "Earthquake P-wave and S-wave Travel Time" graph is explained in Chapter 3. This appendix will serve as a refresher in the use of the whole document.

The complete *ESRT* precedes the appendices. The following explanations follow the arrangement of the 2011 edition of the *ESRT*. A list of vocabulary words is provided for each page. Definitions can also be found in the glossary of this book. The questions are intended to help you understand what information is available in the *ESRT* and how to use it. If possible, for each question, write your answer as well as a brief justification of that answer. If you do not understand any of these examples by the end of the year, ask your teacher to explain the answer to the specific question giving you difficulty.

ESRT Page 1—Physical Constants and Equations

Vocabulary

density: mass per unit of volume, for example g/cm^3

eccentricity: Degree of flattening, or deviation from a perfect circle

gradient: Slope or change in value with distance

half-life: A consistent characteristic of a radioactive isotope that tells the period of time required for half of the radioactive nuclei to break down into the decay product

isotope: A form of a specific element with a set number of neutrons in each atom

radioactive: Describes naturally unstable atoms that radiate energy and/or particles, eventually changing into stable decay products

specific heat: The ability of a particular substance to absorb or give off heat energy as it changes in temperature

NOTE: Water reaches its greatest density at 3.98°C. Water found in nature at any other temperature is less dense. That is why ice floats and the warmer water in a lake is usually near the surface.

Sample Questions

1. What is the half-life of the radioactive substance listed in the *ESRT* that most quickly disintegrates to its decay product?

Answer: 5.7×10^3 *years = 5700 years*

2. If a 10-gram sample were 100% U-238, after 1.35×10^{10} years what decay product is produced and how much?

Answer: 1.35×10^{10} *is 3 half-life periods. After three half-life periods 7/8 of the sample would have changed to lead-206. 7/8* $\times$ *10 g = 8.75 grams of lead-206)*

3. Approximately how much more insolation energy is needed to raise the temperature of a unit mass of ocean water by 1°C than to raise the temperature of the same mass of granite 1°C?

Answer: Note the specific heat value of granite (0.79) and water (4.18). Since the temperature change is the same, 4.18 divided by 0.79 = approximately 5 times the energy.

4. Which process yields more energy to the atmosphere, condensation of water vapor to form a cloud, or the freezing of those cloud droplets to make ice crystals?

Answer: For each gram of water vapor that changes to liquid water, 2260 joules is released. The same mass of liquid water changing to ice crystals only releases 334 joules. Therefore, condensation yields more energy.

ESRT Page 1— Equations

No notes or vocabulary.

Sample Questions

5. The diagram below shows the orbit of a planet around a star to scale, as viewed from a point perpendicular to the plain of the orbit. What is the eccentricity of this orbit of this planet?

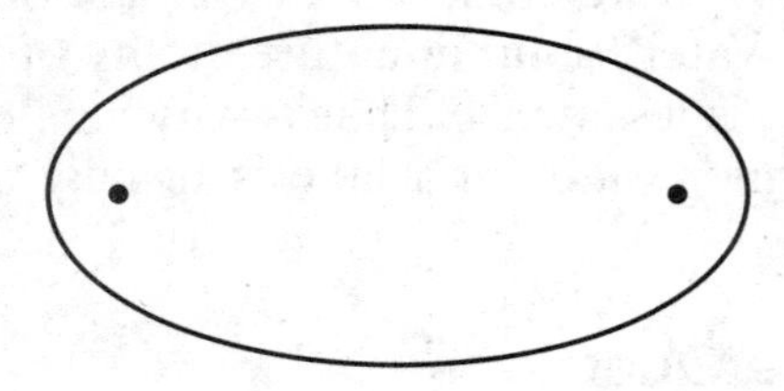

Answer:

$$\text{Eccentricity} = \frac{\text{distance between foci}}{\text{length of major axis}} = \frac{4\text{ cm}}{5\text{ cm}} = 0.8$$

(There are no units; it is a ratio.)

6. What is the average rate of change in temperature between Buffalo, New York, and Oswego, New York, if the temperature in Buffalo is −15°C and the temperature in Oswego is −5°C. (You will need the map on *ESRT* page 3 to find the distance between these two cities. Please show your work and use metric units.)

Answer: The distance between Buffalo and Oswego according to the map on page 3 of the ESRT is 200 km.

Rate of change = change in field value/distance
= 10C°/200 km
= 0.5C°/km

ESRT Page 1—Average Chemical Composition of Earth's Crust, Hydrosphere, and Troposphere

Vocabulary

hydrosphere: The part of Earth that consists of water, mostly the oceans.

troposphere: The lowest, thinnest, most dense, and most massive zone of Earth's atmosphere. (See also ESRT page 14.)

NOTES: For each element that lies within the crust, the chart lists the percent by mass and by volume. The mass percentage of each element within the oceans and atmosphere is not given.

Sample Question

7. What element is relatively common in Earth's crust, hydrosphere, and troposphere?

Answer: According to the "Average Chemical Composition of Earth's Crust, Hydrosphere, and Troposphere" chart, only oxygen is common to all three.

ESRT Page 2—Generalized Landscape Regions of New York State

NOTES: Most landscapes can be classified as one of the following; plains (largely flat lowlands); plateaus (rolling uplands); valley and ridge; or mountains (greatest elevation and topographic relief, or change in elevation).

Sample Questions

8. Describe the landscape of the region near Elmira, New York, in terms of its topographic relief.

Answer: Elmira can be found on the map on page 3. Elmira is located in the Allegheny Plateau, which is a region of moderate elevation change from the hilltops to the valley bottoms.

9. Why was the Erie Canal, which links the Great Lakes to the Atlantic Ocean, constructed from Albany to Utica and Syracuse?

Answer: The canal builders needed to follow a lowland path, the Mohawk River corridor provided an excellent route.

ESRT Page 3-Generalized Bedrock Geology of New York State

NOTES: More information about these geologic periods is given on *ESRT* page 8. Also notice the location of various cities in New York and the latitude/longitude grid along the edges of the map. Rock classifications and specific names are clarified on pages 6-7.

Sample Questions

10. Why is it generally easy to drill or dig water wells on most of Long Island?

Answer: The surface of Long Island is made mostly of sediments (gravel, sand, and clay) rather than solid rock.

11. If you were to travel from Massena, New York, to the central Adirondacks, why would fossils become more difficult to find along your journey?

Answer: There are at least two reasons. Rocks of Precambrian age seldom contain good fossils. In addition, the rocks in the Adirondacks are metamorphosed, which usually destroys fossils.

12. Near which New York State city would you be most likely to find the oldest Placoderm fish?

Answer: To answer this question, you need to find the black vertical bar labeled ® on ESRT page 9. This bar represents the time when Placoderm fish inhabited New York. According to page 9 of the ESRT, this group of animals originated in the late-Silurian Period. Back on page 3, you will notice that Silurian rocks occur at Syracuse and Niagara Falls, although they also might be found close to Buffalo and Rochester.

ESRT Page 4—Surface Ocean Currents

(No notes or vocabulary)

Sample Questions

13. Coastal parts of Los Angeles are well known for their beaches and warm weather. However, people there do not spend as much time in the water as people do along the Gulf of Mexico and South Atlantic resort areas. Why?

Answer: The California Current keeps most coastal waters of California relatively chilly.

14. How are ocean currents in the Southern Hemisphere affected by Earth's rotation?

Answer: Most major ocean currents in the Southern Hemisphere circulate counterclockwise in response to Earth's daily rotation (the Coriolis effect).

ESRT Page 5—Tectonic Plates

Vocabulary

tectonic plate: A relatively rigid geographic section of Earth's surface rocks

hot spot: A place where rising magma or heated rock causes volcanic activity, hot springs, or related surface features

NOTES: The relative motion of the plates affects conditions at the plate boundaries where plates converge (move together), diverge (move apart), or slide past each other (transform boundary).

Sample Questions

15. The subduction of which tectonic plate is most likely responsible for the Yellowstone hot spot?

Answer: The symbol at the edge of the Juan de Fuca Plate indicates that it meets the North American Plate at a convergent boundary. The Juan de Fuca Plate is being subducted (diving) below the North American Plate. After melting within Earth's hot interior, portions of this plate may come back to the surface, causing the Yellowstone Hot Spot.

16. Which major ocean is expanding?

Answer: The Atlantic Ocean is spreading at the Mid-Atlantic Ridge and driving North and South America away from Africa and Europe. The other major oceans are bounded in part by subduction zones where the ocean floor is pulled beneath a continent.

17. What are the approximate terrestrial coordinates (latitude and longitude) of the Hawaii Hot Spot?

Answer: The scale at the edges of the map shows that the Hawaii Hot Spot is located approximately 159°W, 20°N ± 5°.

18. In which direction does the Pacific Plate appear to move along the San Andreas Fault?

Answer: The Pacific Plate appears to move northwest along the San Andreas Fault.

ESRT Page 6—Rock Cycle in Earth's Crust

Vocabulary

magma: Very hot, melted, liquid rock

metamorphism: A process that occurs under extreme heat and/or pressure that can change the mineral composition and observable characteristics of a rock

Sample Questions

19. What name is applied to melted rock?

Answer: On the diagram, all three arrows labeled "Melting" lead to the oval labeled "MAGMA," which is the general name for melted rock.

20. Igneous rock can change directly into three other Earth materials. What are they?

Answer: Arrows on the diagram show that igneous rock can change to sediments by weathering and erosion, to metamorphic rock under the influence of heat and/or pressure, or to magma by remelting.

ESRT Page 6—Relationship of Transported Particle Size to Water Velocity

Vocabulary

velocity: On this page, velocity is used as a synonym for speed

Sample Questions

21. What sediment name is applied to particles finer than sand but coarser than clay?

Answer: According to the graph, the size of silt particles is between sand and clay.

22. In general, how fast must a stream current be to erode pebbles?

Answer: Holding a straight edge vertically on the graph, you can determine that the water velocity needed to move pebbles is at between 10 and 200 centimeters per second.

ESRT Page 6—Scheme for Igneous Rock Identification

Vocabulary

texture: The observable characteristics of a rock's surface such as smoothness, roughness, or layering

intrusive: Describes igneous rocks formed within Earth (plutonic)

extrusive: Describes igneous rocks formed at Earth's surface (volcanic)

grain size: The size of the mineral particles (crystals or fragments) of which a rock is made

vesicular: Containing holes made by the expansion of gases in the magma as it rises toward the surface

felsic: Describes rocks that are dominated by light-colored minerals such as feldspar

mafic: Describes rocks that are dominated by dark-colored minerals

mineral: The substances of which rocks are made, such as quartz and the feldspars

NOTES: Specific rocks, such as granite and basalt, are defined by their position on the igneous rock chart. They will not be defined here.

Sample Questions

23. What relatively common igneous rock contains the largest mineral crystals?

Answer: You will find pegmatite at the bottom of the scheme of rock names; it is the only rock in the "very coarse" section. (No names are shown on the lower right part of the chart because the more mafic magmas seldom form rocks with very large crystals.)

24. In what three ways do granite and rhyolite differ from basalt, diabase, and gabbro?

Answer: The central part of the scheme shows that granite and rhyolite are lighter in color, lower in density, and more felsic (rich in aluminum) than basalt, diabase, and gabbro.

25. What three minerals are almost always found in basalt, diabase, and gabbro?

Answer: According to the scheme, only the shaded zones for plagioclase feldspar, pyroxene, and amphibole extend across the range of compositions for basalt and gabbro.

26. What is the approximate proportion of biotite in granite?

Answer: The thickness of the black (biotite) area in the Mineral Composition section of the scheme indicates that granite is about 5–10% biotite.

ESRT Page 7—Scheme for Sedimentary Rock Identification

Vocabulary

clastic: Composed of fragments of other rocks

crystalline: Composed of intergrown crystals

bioclastic: Composed of the remains of plants and/or animals

fine: Made of small particles, as opposed to medium or coarse (large) particles

NOTES: The map symbols shown on the right side of this page generally correspond to the appearance of the rocks they stand for. These symbols are not universal, and you are likely to see diagrams or maps that use different symbols.

Sample Questions

27. How does a conglomerate differ from a breccia?

Answer: According to the Comments column, conglomerate is composed of large rounded fragments, and breccia is composed of angular fragments.

28. How are most clastic rocks classified?

Answer: Most clastic rocks are classified according to the size of the grains of which they are made.

29. What is shale made from?

Answer: Shale is composed of compact clay-size particles.

ESRT Page 7—Scheme for Metamorphic Rock Identification

Vocabulary

foliated: Containing mineral crystals with a particular alignment

banded: Containing crystalline minerals separated into layers, often appearing as alternating light- and dark-colored layers

metamorphic grade: The degree of change in a rock caused by heat and/or pressure

contact metamorphism: Change due to heating caused by nearness to molten magma

regional metamorphism: Change caused by deep burial within Earth

NOTES: The map symbols shown on the right side of this page generally correspond to the appearance of the rocks. These symbols are not universal, and you are likely to see diagrams or maps that use different symbols.

Sample Questions

30. What mineral family is found in all foliated metamorphic rocks?

Answer: In the Composition column, the shaded boxes show that the mica minerals are common in all foliated (fine to coarse-grained) metamorphic rocks.

31. Name a fine-grained metamorphic rock altered by a nearby igneous intrusion.

Answer: Hornfels is a fine-grained metamorphic rock of variable composition that was changed by contact with magma/lava.

ESRT Pages 8 and 9—Geologic History of New York State

Vocabulary

orogeny: A period of mountain building, such as the Appalachian Orogeny

Iapetus Ocean: An ocean that closed before the opening of the Atlantic Ocean

stromatolites: Small mounds of bacteria especially common in the distant geologic past

fauna: Animals

flora: Plants

NOTES: Divisions of geologic time are listed on the left side of the chart along with their age in millions of years. Notice that the most recent 600 million years has been expanded to fill most of these two pages. The reason for this is rocks older than 600 million years are not very common and most have been greatly changed by metamorphism. The names of the periods are generally related to locations where rocks with the characteristic fossil groups of those ages were originally studied, mostly in Europe.

The black bars in the column at the center of the chart labeled "Rock Record in NYS" indicate the age of rocks found in New York State. The black bars in the column labeled "Time Distribution of Fossils" indicate the period during which certain now fossilized creatures were alive. The letters along the vertical lines match the fossil drawings at the bottom of the page. The column labeled "Important Geologic Events in New York" indicates when mountain building occurred and oceans opened or closed. World maps on the right show the inferred positions of the continents through geologic time.

Sample Questions

32. Which division of geologic time accounts for about 88 percent of the total age of planet Earth?

Answer: Notice the Precambrian on the left side of the chart extends from 4600 million years (4.6 billion years) upward to nearly 500 million years (half a billion years) ago.

33. What fossil group is an excellent index fossil for Paleozoic rocks because it includes fossil remains that extend from early-Cambrian to the end of the Permian period?

Answer: In the "Time Distribution of Fossils," you will find a dark bar that indicates that various species of trilobites existed during the time range in question.

34. The separation of modern North America from Africa started the opening of the North Atlantic Ocean. In absolute, or numerical, time, how long ago did this occur?

Answer: The maps at the right side of page 9 in the column labeled "Inferred Positions of Earth's Landmasses" show that the Atlantic Ocean opened between the Triassic and the Cretaceous. The next column to the left, "Important Geologic Events in New York," tells you the initial opening of the Atlantic Ocean and the separation of North America and Africa began about 200 million years ago.

35. New York State is known for its abundance of Paleozoic fossils. Which Paleozoic period is completely missing from New York bedrock?

Answer: In the "Rock Record in NYS" column, there is a gap in the vertical bar, indicating that rocks of the Permian period are missing.)

ESRT Page 10—Inferred Properties of Earth's Interior

Vocabulary

inferred: Presumed, guessed, or predicted without direct observation

trench: A deep part of the ocean, usually at a location where a tectonic plate is diving into Earth's interior

melting point: The temperature at which a material changes from a solid to a liquid at the current conditions or location.

NOTES: The three parts of this page (the cross section of Earth's layers, pressure graph, and temperature graph) are related by the single scale of depth at the bottom of the page.

Sample Questions

36. What geographic feature between the Americas and Africa and Europe marks a zone of plate divergence?

Answer: In the top section of the diagram, the arrows at the Mid-Atlantic Ridge indicate a spreading plate boundary.

37. What is the general relationship between the density of terrestrial layers and their distance from Earth's center?

Answer: According to the upper right side of the chart, which provides information about density, there is an indirect relationship; that is, the layers become more dense the closer they are to Earth's center.

38. What is the pressure at a depth of 4500 kilometers below the solid surface of Earth?

Answer: From the graph at the center of the diagram, the pressure is 2,700,000 atmospheres, or 2.7×10^6 atm.

39. In what parts of Earth's interior is the rock material inferred to be largely in the liquid state?

Answer: According to the graph at the bottom of the diagram, the actual temperature in the asthenosphere and the outer core is thought to be above the melting point.

ESRT Page 11—Earthquake P-wave and S-wave Time Travel

Vocabulary

P-wave: The fastest energy wave that radiates from an earthquake, primary wave

S-wave: Secondary earthquake wave that travels more slowly than the P-wave.

epicenter: The surface location directly above the focus (interior breakage) of an earthquake

NOTES: The smallest time division on the travel time scale is 1/3 of a minute, or 20 seconds.

Sample Questions

40. How long does it take a P-wave to travel 9000 kilometers?

Answer: Find the line labeled 9 on the axis labeled Epicenter Distance ($\times 10^3$ km) [which means that all numbers on this line are multiplied by 1000]. Read up along this line until it meets the line labeled P, reading across to the left to the Travel Time axis, you find that the answer is 12 minutes, 10 seconds.

41. If the S-wave of a seismic event arrives 6 minutes, 20 seconds after the P-wave, how far away is the epicenter of the event?

Answer: The distance at which the S-wave line is 06:20 higher than the P-wave line is 4600 km, with an acceptable range of answers of $\pm$ 200 km. (See Figure 3-6.)

ESRT Page 12—Dew-Point Temperature (°C)

Vocabulary

dew point: The temperature to which air must be cooled for it to become saturated (100% relative humidity)

wet-bulb temperature: The temperature recorded by a thermometer with the bulb covered by a wet, cloth wick

dry-bulb temperature: The air temperature recorded by the dry bulb of a thermometer

NOTES: The most common error students make using this chart is to forget to subtract the wet-bulb temperature from the dry-bulb temperature to find the temperature difference, which is read along the top row of the chart. The wet-bulb temperature in not used by itself.

Sample Questions

42. What is the dew-point temperature when the temperatures on a sling psychrometer are 21°C and 16°C?

Answer: Unless the temperatures are equal, the lower of the two temperatures is always the wet-bulb temperature. The difference between the wet-bulb and dry-bulb temperature is 5°C (21°C −16°C = 5°C). The value 21°C does not appear along the left side of the table. Therefore, to answer this question, you must interpolate (read between posted values). Reading down the column for a difference of 5C°, you find the values of 12°C and 14°C for 20°C and 22°C on the dry-bulb scale. The interpolated answer is 13°C.

43. What is the dew-point temperature when both thermometers on a sling psychrometer register −8°C?

Answer: The difference between the wet-bulb and dry-bulb temperature is 0. Read down the column for a difference of 0 until you reach −8 in the dry bulb column, reading across the answer is −8. You really do not need the chart when the difference is 0 because the air temperature is the dew point.

ESRT Page 12—Relative Humidity (%)

NOTES: The most common error students make using this chart is to forget to subtract to find the temperature difference, which is read along the top row of the chart. There is no use of the wet-bulb temperature alone.

Sample Question

44. What is the relative humidity when the dry-bulb temperature is 10°C and the wet-bulb temperature is 7°C?

Answer: The difference between the wet-bulb and dry-bulb temperatures is 3°C. Follow the vertical column below 3°C until it meets the horizontal row coming across from 10°C. The relative humidity is 65%.

ESRT Page 13—Temperature

No notes or vocabulary

Sample Question

45. What is the Kelvin equivalent of 212°F?

Answer: This is the boiling temperature of water at or near sea level: 373 K ± 1 K (The degree symbol (°) is not used when the Kelvin scale is used.)

ESRT Page 13—Pressure

NOTE: The inches scale of pressure is based on a mercury-filled barometer. The greater the air pressure, the higher it pushes a column of mercury into a tube. Traditionally, that height is measured in inches.

Sample Question

46. What is normal atmospheric pressure measured in inches of mercury?

Answer: Normal atmospheric pressure is labeled "one atmosphere." It is 29.92 inches of mercury, ± 0.01 inches.

ESRT Page 13—Weather Map Symbols

Vocabulary

knot: One nautical mile per hour (One knot is little more than a standard mile per hour.)
continental: A dry air mass
maritime: A moist air mass
tropical: Warm air mass
polar: Cool air mass
arctic: Very cold air mass
stationary: A weather front that is not moving
occluded: A weather front in which warmer, and often moist, air is isolated well above the ground

NOTES: Weather fronts are named for the kind of weather they bring.

Sample Question

47. To record an air pressure of 998.3 mb on a station model weather map, what number would you write and where?

Answer: The first "9" as well as the decimal point must be dropped. Therefore, "983" is written above and at the right of the circle that indicates the location of the specific weather station.

ESRT Page 14—Selected Properties of Earth's Atmosphere

NOTE: The layers of the atmosphere end in "-sphere," while the top boundaries end in "-pause."

Sample Question

48. What is the typical height of the mesopause in metric units?

Answer: The height of the mesopause is about 80 kilometers.

ESRT Page 14—Planetary Wind and Moisture Belts in the Troposphere

Vocabulary

planetary wind belts: Regions where winds often blow from one general compass direction

jet stream: Fast winds in the upper atmosphere that often help create and steer weather systems

NOTES: Winds are labeled by the direction from which they blow, which is opposite the direction they travel toward.

Sample Questions

49. How do the typical wind directions north of the equator compare with the wind directions south of the equator?

Answer: Southern Hemisphere winds are generally the mirror image of Northern Hemisphere winds.

50. Why do most locations along the equator have a humid climate?

Answer: Rising air near the equator causes cloud formation and precipitation.

ESRT Page 14—Electromagnetic Spectrum

Vocabulary

electromagnetic energy: Radiant energy that includes both invisible radiation such as X rays and infrared (heat rays), as well as the various colors of visible light

Sample Question

51. What name is given to electromagnetic radiation whose wavelength is slightly too long to be visible to humans?

Answer: Infrared (heat) waves

ESRT Page15—Luminosity and Temperature of Stars

Vocabulary

luminosity: A measure of light radiated by a star, *not* how bright it appears.

NOTES: The grouping of stars on the luminosity graph has led astronomers to infer the life cycles of stars.

Sample Question

52. In the night sky, how would the star Betelgeuse appear different from our sun and most other visible stars?

Answer: Although Betelgeuse puts out more light than the sun, due to its distance it appears to be much dimmer. In addition, it is more red in color than the sun and most other night stars.

ESRT Page 15—Solar System Data

Vocabulary

period: Amount of time needed to complete one cycle

rotation: spinning on an axis

revolution: orbiting another object

Sample Questions

53. What planet orbits closest to Earth?

Answer: Earth's orbit is between Venus and Mars. However, subtraction of the orbital values listed in the Mean Distance column shows that Venus is closer to Earth than is Mars.

54. Which planet has an orbit that is closest to a perfect circle?

Answer: The eccentricity of the orbit of Venus is closest to 0, therefore, it has the most circular orbit.

ESRT Page 16—Properties of Common Minerals

Vocabulary

luster, **hardness**, **cleavage**, **fracture**, and **streak** are mineral properties best learned by observing fresh mineral samples. (They are also explained in Chapter 3.)

metallic: Shiny with no light penetration

NOTES: The elements in parentheses in the Composition column have a variable composition. For example, the percentage of calcium and sodium ions in plagioclase is variable. Notice the parentheses in the Mineral Name column. Some minerals listed here are mineral families, such as amphibole and pyroxene. Others, such as gypsum, may be differentiated by different names for the same chemical composition, depending on the observable properties of the sample. (Selenite is a fairly common transparent variety of gypsum.) The chemical symbols of elements commonly found in minerals are listed at the bottom of this table.

Sample Questions

55. Which two minerals in this chart are chemical elements?

Answer: Graphite, a form of carbon (C), and sulfur(S) are elements.

56. Which mineral in this chart is consistently soft?

Answer: Talc, with a Mohs hardness of 1, is a standard of hardness. Some other minerals can be about as soft, but they have a more variable value on the Mohs scale of hardness.

57. How is muscovite most unlike the closely related mineral biotite?

Answer: Muscovite is usually light-colored, while biotite is usually dark, or even black.

58. The following observations were made by a student trying to identify a mineral sample.

COLOR: white

HARDNESS: does not scratch glass

CLEAVAGE/FRACTURE: breaks into geometric pieces

SPECIAL PROPERTY: Fizzes when a drop of dilute hydrochloric acid is placed on a smooth crystal surface

What is the most likely identity of this mineral?

Answer: Only calcite has all these properties, including the reaction with acid. Dolomite is close, but it usually needs to be broken into a powder before it reacts with most acids.

59. Which mineral would you be most likely to use on any Regents exam?

Answer: Graphite is a primary ingredient of pencil lead. (It is often mixed with clay.) Magnetite and hematite are ores of iron, which may be used in making pens and desks.

Reviewing Earth Science: The Physical Setting, Guide to Part D

Dear Teachers and Students,

The New York State Regents Examination in The Physical Setting/Earth Science is composed of four parts. The fourth part, Part D, is a laboratory performance test. As you may know, there was a new version of Part D given for the first time in June 2008 and revised in 2012.

If you are going to take the Earth Science Regents in 2012 or later, the information on the back of this page is important to you. To help students and teachers prepare for the revised Part D, the back of this page contains ***only*** information that the New York State Education Department has made public.

We hope you find this helpful.

Physical Setting/Earth Science
Performance Test – Part D Revised 2012
Materials List

The New York State Regents Examination in Physical Setting/Earth Science consists of two components: a laboratory performance test and a written test. A new form of the laboratory performance test was administered for the first time in June 2008 and revised in 2012. The performance test consists of hands-on tasks set up at three stations. These tasks are designed to measure student achievement of the New York State *Learning Standards for Mathematics, Science, and Technology* included in the Physical Setting/Earth Science Core Curriculum.

The three stations of the performance component of the Regents Examination in Physical Setting/Earth Science 2008 Edition (revised 2012) are shown below along with a materials list for each station. The New York Education Department will provide test booklets, rating guides and other printed administration materials. Schools are responsible for obtaining the performance task materials and assembling them for task performance.

Students should be familiar with the content, concepts, and process skills assessed on the performance tasks and should have performed similar tasks during the normal course of instruction. *However, practice of any of the individual stations before this performance component is administered is not permitted.*

Station 1 . . *Mineral and Rock Identification*

MATERIALS (per setup)

- One hand-sized mineral sample (approximate size: 5 cm × 7 cm × 10 cm) – Any mineral can be used, both familiar and unfamiliar, as long as the properties to be tested are clear and unmistakable. Do not use the same type of mineral at more than one station.
- Three hand-sized rock samples to include one igneous rock, one sedimentary rock, and one metamorphic rock – The rock samples can only be rocks listed on the rock identification charts from the 2011 edition of *Earth Science Reference Tables* and must have unambiguous and unmistakable diagnostic properties. Use different rock combinations or rocks at each station.
- Mineral identification kit containing a penny, a glass scratch plate, a streak plate, and a hand lens.

Station 2 . . *Locating an Epicenter*

MATERIALS (per setup)

- Safe drawing compass

Station 3 . . *Constructing and Analyzing an Elliptical Orbit*

MATERIALS (per setup)

- Cotton string (approximately 30 cm)
- Triple-walled cardboard, foam board or other suitable material (approximately 25 cm × 30 cm)
- Two push pins
- A small container to hold push pins
- One 30-cm metric ruler
- One four-function calculator

ADDITIONAL PREPARATION MATERIALS

- White enamel to label rock and mineral samples
- Page protectors for station directions (approximately 15 per setup)
- Tape
- Scissors

The maximum raw score for the performance test is 16

The maximum total raw score for the written test is 85

Glossary of Earth Science Terms

abrasion: Form of physical weathering caused by friction between rock particles

absolute age: Age of a rock unit, a fossil, or an event expressed in units of time, such as years

absolute time scale: Geologic time that uses numerical time units (years)

adiabatic (temperature change)***:*** Change in the temperature of a gas caused by expansion (cooling) or compression (warming)

agent of erosion: Medium, such as water, wind, or glacial ice, that transports weathered sediments

air mass: Large body of air that has relatively uniform conditions of temperature and pressure

air pressure: Effect of the weight of the atmosphere pressing on a given surface area

alpine glacier: Glacier that forms in high mountains and flows through valleys, giving them a U-shape.

altitude: Height, measured in degrees, of an object above an observer's the horizon

anemometer: A device that measures wind speed

angle of insolation: Angle of the sun above the horizon

angular diameter: Angle formed between opposite sides of an object and an observer's eye

annular drainage pattern: Stream pattern of concentric circles formed on a mountain whose rocks have differing ability to resist weathering

anticyclone: High-pressure system that often brings cool, clear weather as its winds rotate clockwise and away from the center; a zone of divergence

apparent diameter: How large an object looks, which depends on its size and distance from an observer

apparent motion: The way celestial objects appear to move across the sky

arctic air mass: Extremely cold air mass that originated in the far north

arid: Dry; a climate in which there is little precipitation

asteroid: One of the tens of thousands of rocky objects located, for the most part, in a belt between the orbits of Mars and Jupiter

asthenosphere: Part of Earth's interior below the lithosphere that becomes plastic in response to stress

astronomy: Study of the motions and properties of objects in space

atmosphere: Shell of gases that surrounds a planet, for example, the air around the Earth

atmospheric pressure: A measure of the force exerted by the atmosphere

axis: Imaginary line around which an object rotates

banding: Type of layering (foliation) found in some metamorphic rocks that is caused by the movement or growth of minerals into homogeneous layers

barometer: Instrument used to measure air pressure

basaltic: Igneous rock composed mostly of dark-colored, dense minerals containing compounds of iron and magnesium

bedrock: Solid layer of rock that extends into Earth, bedrock is found beneath the soil

big bang: Theory that the universe began as the rapid expansion of an infinitesimal object of incredible mass and density

big crunch: Theory that gravity will reverse the expansion of the universe so that it contracts to a black hole

bioclastic: Sedimentary rock, such as coal and some types of limestone, formed by the accumulation of plant and animal remains

bond: Attachment between atoms and molecules

calorie: A common measuring unit of heat energy

capillarity: Ability of a soil to draw water upward into tiny pores

capillary water: Water held within the aerated zone of the soil above the water table

carbon-14: Radioactive form of the element carbon that has been used to determine the absolute age of recent fossils and geologic events

celestial object: Object in the sky outside Earth's atmosphere; the sun, moon, stars, and planets.

celestial sphere: Imaginary sphere encircling Earth on which all objects in the night sky appear

cement (natural): Mineral or another fine matrix that fills the pores between the grains of sediments in sedimentary rock, holdng it together

chemical weathering: Change in the chemical composition of a rock caused by adjustment to conditions at Earth's surface

circumference: Perimeter of a circle, or the straight-line distance around a sphere

clastic (fragmental): Most common group of sedimentary rocks, made up of different sized particles

classification: Organization by similarities of objects, ideas, or information into groups

cleavage: The way that a mineral splits between layers of atoms that are joined by weak bonds

cleavage planes: Flat surfaces along which some minerals break naturally

climate: Average weather conditions over many years

cloud: Large mass of water droplets or ice crystals suspended in the air

cold front: Boundary between a mass of cold air and the warmer air it is replacing

color: Typically used in Earth science to describe the characteristic color of a mineral

comet: Icy object that most likely originated in a region beyond the planets; some comets travel in highly elliptical orbits

competent rock: Hard rock that resists weathering and erosion

compression waves: (See P-waves.)

condensation: Process by which a gas changes into a liquid; a way in which clouds form

condensation nuclei: Particles suspended in air on which condensation occurs, forming a cloud

conduction: The way heat energy is transferred through matter by the direct contact of molecules

conglomerate: Sedimentary rock composed of cemented gravel, pebbles, or cobbles

constellation: Observed pattern people use to mark the position of stars in the sky; one of the 88 regions of the night sky each associated with a particular group of stars

contact metamorphism: Chemical and physical changes to a rock caused mostly by the heat of a nearby intrusion or extrusion of molten liquid rock

continental air mass: Body of dry air that is low in humidity because it formed over a land area

continental climate: Climate in which there are large seasonal changes in temperature due to the absence of nearby bodies of water that could moderate the climate

continental crust: Rocks within the continents, usually a thin layer of sedimentary rocks over granitic rocks, that are less dense than oceanic crust

continental drift: Idea that the continents move over Earth's surface like rafts on water

continental glacier: A glacier that spreads over a wide geographic area

contour interval: Difference in height between two adjacent contour lines

contour line: Isoline on a map that connects places with the same elevation and shows the shape of the land

contour map: Map that shows the shape of the land using contour lines; a map showing an elevation field

convection: Circulation of a heated fluid (a liquid or a gas) caused by density currents; a form of heat flow in which the heated material moves

convection cell: Circular path of convection flow

convergence: Coming together of winds as they blow into a cyclone (low-pressure system); the coming together of tectonic plates

convergent boundary: Boundary at which crustal plates collide

coordinate system: Grid on which each location has a unique designation defined by the intersecting of two lines; for example, latitude and longitude

core: Innermost layer of Earth, thought to be composed mostly of iron and nickel

Coriolis effect: Apparent curvature of the winds, ocean currents, or objects moving long distances along Earth's surface; caused by Earth's rotation on its axis

correlation: Matching rock layers in different locations by age or by rock types

crust: Thin, outermost layer of the solid Earth

crystal: Solid form of a mineral with a regular shape caused by the internal arrangement of atoms

crystalline: Made of crystals

crystalline sedimentary rocks: Rocks deposited from seawater by chemical precipitation (settling) of minerals

crystallize: To form intergrown crystals as a liquid cools to form a solid

cyclic change: A change that repeats itself, such as the annual cycle of the seasons or the apparent daily motion of the sun through the sky

cyclone: Low-pressure system in which the winds rotate counterclockwise in the Northern Hemisphere and converge to the center

decay product: Element produced by the decay of a radioactive isotope

decay product ratio: Ratio between the mass of a radioactive element and its decay product

day: Average interval of time during which the sun passes from its highest point on one day to its highest point on the next; approximately 24 hours

dendritic: Drainage pattern resembling a tree an its branches

density: Mass per unit volume of a substance

deposition: Settling or release of sediments by an agent of erosion

dew point: Temperature at which the air would be saturated with moisture

diameter: The distance through the center of an object

direct rays: Sunlight that strikes Earth from straight overhead; vertical rays

discharge: Quantity of water flowing past a certain point in a stream per unit of time

divergence: Outward movement of winds from a high-pressure zone (anticyclone)

divergent boundary: Plate boundary at which the plates move apart; an upwelling of material that forms new crust that moves away from the boundary

Doppler shift: (See redshift.)

double refraction: Property of transparent crystals to form a double image of objects viewed through the crystals

drainage pattern: Arrangement of adjoining streams as seen from above

drumlin: Oval-shaped mound of unsorted glacial till

duration of insolation: Length of time from sunrise to sunset that the sun is in the sky

dynamic equilibrium: State of balanced change

earthquake: Natural vibrations, sometimes destructive, that radiate from a sudden movement along a fault zone within Earth or from sudden movements of magma (molten rock) under a volcano

Earth science: Study of Earth's systems and setting in the universe

eccentricity: Degree of elongation of an ellipse

eclipse: Occurs when one celestial object casts its shadow on another celestial object; a lunar eclipse occurs when the Earth lies between the moon and the sun; a solar eclipse occurs when the moon passes between the Earth and the sun

electromagnetic energy: Energy, such as heat waves, visible light, and x-rays, that can radiate through empty space

electromagnetic spectrum: Range of electromagnetic energy from long-wave radio waves to short-wave gamma rays

ellipse: Closed curve around two fixed points known as the foci

El Niño-Southern Oscellation: Condition in the Pacific Ocean when seasonal upwelling of deep equatorial water does not follow its usual pattern. (This can lead to unusual weather conditions over a large area.)

engineering: Application of science and mathematics to solving problems

epicenter: Location along the Earth's surface that is directly above the focus of an earthquake

equator: Imaginary line that circles Earth halfway between the North and South Poles

equilibrium: State of balance between opposing forces in a system

equinox: Time at which the sun is directly above the equator, and all places on the Earth have 12 hours of daylight and 12 hours of darkness (Equinoxes occur near the end of March and September; they mark the beginning of spring and autumn.)

era: Large division of geologic time such as the Mesozoic Era

erosion: Transportation of weathered material (sediments) by water, wind, or ice away from their place of origin

erratic: Large boulder dropped by a glacier

escarpment: Steep slope or cliff where resistant layers of rock overlie weaker layers

esker: Winding ridge of sand and gravel deposited by a stream confined to a tunnel under a glacier

evaporation: Change in phase from a liquid to a gas (vapor); also known as vaporization

evaporite: Sedimentary rock deposited when minerals in a saturated solution settle out of an evaporating body of water

evolution (organic): Principle that through Earth's history, living things have changed in form from a few simple organisms to a great diversity of organisms

exponential notation: (See scientific notation.)

extrusion: Molten, liquid rock (lava) flowing out onto Earth's surface, forming a fine-grained igneous rock

extrusive: Rocks formed from lava that solidifies quickly at Earth's surface

fault: Break in the rock of Earth's lithosphere along which there has been displacement (movement)

felsic: Light-colored rocks composed mostly of feldspar and silica

field: Region of space in which a similar quantity can be measured at every point or location

Finger Lakes: Lakes created in western New York State as the advancing ice deeply scoured former north-south river valleys
flotation: Transportation of sediments along the surface of a stream
fluid: Substance that can flow; liquids and gases
focus: One of two fixed points that determine the shape and position of an ellipse; an earthquake's point of origin within Earth
folds: Layers of rock that have been bent by forces within Earth
foliation: Alignment or segregation of minerals in a metamorphic rock, producing a layered appearance
fossil: Any preserved remains or traces of life
fossil fuels: Fuels, such as coal, oil, and natural gas, that were formed millions of years ago from the remains of ancient organisms
Foucault pendulum: Weight that is free to rotate as it swings back and forth; its slow change in direction of swing is proof of Earth's rotation
fracture: Uneven splitting of a mineral sample
fragmental (clastic): Most common group of sedimentary rocks, made up of different sized particles held by cement
freezing temperature: Temperature at which a liquid will start to change to a solid
front: Boundary, or interface, between different air masses
frost action: Form of physical weathering in climates with seasonal temperature changes alternately above and below 0°C
frost wedging: (See frost action.)

galaxy: Huge body of stars and other matter in space
geocentric model: Early model of the solar system and universe in which Earth is stationary, located at the center of the universe, and around which all celestial objects revolve
geologic time scale: Divisions of Earth's history originally based on observations of fossil evidence; now, through the use of radioactive isotope measurements, it has changed from a relative scale to an absolute scale
geology: study of what Earth is made of and how it changes
glacial erratic: Large rock that has been transported by a glacier
glacial polish: Bedrock surface smoothed by the passage of a glacier
glacier: Large mass of flowing ice.
graded bedding: Layers of sediment that change from coarse particles at the bottom of each layer to progressively finer particles toward the top
gradient: Rate of change in field values between two points in a field; the average slope
granitic: Rocks composed mostly of light-colored, low-density minerals, such as quartz and feldspar
gravity (gravitational force): Attractive force between objects that is directly proportional to their masses and inversely proportional to the distance between them
greenhouse effect: Process by which infrared heat waves are trapped in the Earth's atmosphere by gases such as carbon dioxide and methane
greenhouse gases: Gases, such as carbon dioxide, methane, and water vapor, that trap heat in Earth's atmosphere and affect Earth's climate
Greenwich Mean Time: Time based on observations of the sun along the prime meridian
groundwater: Water that infiltrates the ground

half-life: Time for half the atoms in a radioactive sample to change to the decay product
hardness: In geology, the ability of a mineral to resist scratches
heat energy: Total potential and kinetic energy that an object can release as heat
heat of fusion: Latent potential energy absorbed when a solid melts and released when a liquid freezes
heat of vaporization: Latent potential energy absorbed when a liquid vaporizes and released when a gas condenses
heliocentric model: Modern model of the solar system and universe in which the planets revolve around the sun and Earth undergoes daily rotation
high-pressure system: Anticyclone; a dense mass of air in which the atmospheric pressure is the highest at the center; a zone of divergence
horizontal sorting: Gradual change in the size, density, and shape of particles deposited when a stream slows on reaching calm water (The largest, most dense, and roundest particles settle first while the smaller, least dense, and flatter particles are carried farther out into the calm water.)
hot spot: Location at which a stationary hot plume of magma breaks through the crust
hour: 1/24 of a day
humid: Moist; climate in which there is abundant precipitation
humidity: Measure of the moisture, or water vapor content, of the air
humus: Organic remains that are part of soil
hydrolic cycle: (See water cycle.)
hydrosphere: Earth's liquid water, including oceans, lakes, streams, and groundwater

ice age: Long period of Earth's history when ice sheets covered large areas of the continents

igneous rock: Rock formed by the cooling and crystallization of molten rock (magma or lava)

index fossil: Fossil found over a large geographic area but which existed for a brief period of geologic time

inertia: Tendency of an object to remain at rest, or, if it is moving, to move at constant speed in an unchanging direction

inference: Conclusion based on observations and experiences; something that is thought out, but not directly observed; an interpretation

infiltrate: Water that seeps into the ground

infrared waves: Long-wave heat radiation

inner core: Central portion of Earth's core thought to be composed mostly of solid iron and nickel

inorganic: Substance not formed by or from living things

insolation: Electromagnetic energy that Earth receives from the sun; contraction of the words, "incoming solar radiation"

insolation-temperature lag: Time delay between maximum or minimum insolation and maximum or minimum air temperature

instrument: Device that makes observations or measurements easier to perform or more precise

intensity: Strength; Mercalli scale measure

interface: Boundary between different materials or systems

intrusion: Molten, liquid rock (magma) pushed into cracks within Earth's crust; a body of coarse-grained igneous rock formed by slow cooling within Earth

intrusive (plutonic): Rocks that crystallize slowly inside Earth's crust

island arc: Islands formed by rising magma plumes at a convergent boundary

isobar: Line on a weather map connecting places with the same atmospheric pressure

isoline: Line connecting points having the same value within a field

isosurface: Surface in which all points have the same measured value

isotherm: Line within a temperature field or on a weather map that connects places that have the same temperature

isotope: Form of an element with more or fewer neutrons than other forms of the same element

jet stream: Currents, or very fast winds, that flow in the upper atmosphere

joule: Unit used to measure energy

kame: Delta deposited by a stream at the end of a glacier

kettle: A depression that may be the result of the melting of a buried block of ice

kettle lake: Lake formed when a block of glacial ice melts

kinetic energy: Energy of motion

land breeze: Breeze that blows at night from the land to the sea

landform: Unit of rock of uniform age or composition

landscape: General shape of a region on Earth's surface

latent heat: Energy absorbed or released in a change in state. (Latent heat is so named because it does not show up as a temperature change.)

latitude: Angular distance in degrees north or south of the equator. (It varies from 0° at the equator to 90° at the poles.)

lava: Molten rock at Earth's surface

law of superposition: (See superposition, law of.)

leaching: Process by which groundwater carries dissolved minerals deeper into the soil as the water infiltrates the ground

leeward: Downwind side of a mountain range (Usually the side with less precipitation.)

light-year: Distance light travels in one year

liquid: Phase of matter in which the molecules or atoms are close but free to move independently

lithosphere: Solid portion of Earth below the atmosphere and hydrosphere; a solid layer that includes the crust and the upper portion of Earth's mantle

longitude: Angular distance in degrees east or west of the prime meridian. (Longitude varies from 0° at the prime meridian to 180° near the middle of the Pacific Ocean.)

low-pressure system: Weather system in which the atmospheric pressure is lower than in surrounding areas; cyclone

luminosity: Amount of light given off by a star

luster: Way the surface of a mineral reflects light

mafic: Composed of dark minerals rich in iron and magnesium

magma: Molten rock within Earth

magnitude: Total energy released by an earthquake, measured by the Richter scale

major axis: Line through the widest part of an ellipse

mantle: Portion of Earth below the crust and above the core

maritime air mass: Body of air that is relatively moist because it formed over an ocean

maritime (marine) climate: Climate in which seasonal temperature changes are moderated by a large body of water

mass: Quantity of matter in an object, measured in kilograms or other units of mass

mass extinction: Extinction of a large number of species

mass movement: Downhill movement of rock or sediment without being carried by water, wind, or ice

mass wasting: (See mass movement.)

meander: Natural looping bend, or S-shaped curve, in a stream

measurement: Observation of a quantity made using an instrument

meridian: Imaginary semicircle, drawn around Earth from the North Pole to the South Pole, that represents a constant longitude

metamorphic rock: Rock that has been changed in mineral composition and/or texture by heat and/or pressure without melting

meteor: Solid particle from space caught by Earth's gravity that as they fall are heated by friction with Earth's atmosphere and burn, producing streaks of light visible at night; "shooting stars" (Most meteors vaporize during their fall through the atmosphere, although some of them reach the ground.)

meteorite: Natural object that has fallen to Earth from space

meteoroid: Potential meteors in space

meteorologist: Scientist who studies the weather

meteorology: Study of the changing conditions of the atmosphere, or weather

mid-ocean ridge: System of submerged mountain ranges that encircles Earth and often connects with mountain ranges on the continents, new crust forms here

millibar: Metric unit of atmospheric pressure

mineral: Natural, crystalline inorganic substance of which rock is made

model: Representation of an object or natural event

Moho: Interface between Earth's crust and mantle

Mohs scale: Series of ten minerals used as a scale of hardness

moisture: Presence of water or water vapor, particularly in the atmosphere

moraine: Unsorted mound of sediment deposited directly by moving ice

mountain: Landscape region characterized by nonhorizontal rock structure and great topographic relief; landscape feature usually characterized by high elevation and steep slopes

natural resource: Useful substance that comes from Earth

navigation: Science of locating position on Earth

nonrenewable resource: Resource that cannot be replenished as quickly as they are used

North Star: Polaris; the star located almost directly above the North Pole

nuclear fusion: Under conditions of extreme heat and pressure in the sun, the nuclei of two hydrogen atoms join to form one helium nucleus

oblate spheroid: Nearly spherical shape of Earth, slightly flattened at the poles and slightly bulging at the equator

observation: Information obtained directly from the senses

occluded front: Type of weather front produced when a cold air mass overtakes a warm air mass, isolating the warm air above the ground

ocean trench: Place where tectonic plates slide down into Earth's interior and are destroyed

oceanic crust: Relatively thin, dense layer of basaltic rock that lies under the ocean sediments and on top of the mantle layer

oceanography: Study of the characteristics and dynamics of the blanket of water that covers most of our planet

orbit: Path (usually an ellipse) of any satellite around its primary

orbital speed: Measure of a satellite's orbital motion

organic evolution: (See evolution.)

organic rock: Sedimentary rock formed by the accumulation of plant or animal remains

origin: Method by which rock was formed

original horizontality: Principle that all sedimentary layers were originally flat

origin time: Time an earthquake occurs at its epicenter

orogeny: Process of mountain building caused by forces within Earth

outcrop: Bedrock exposed at the surface because it is not covered by soil

outer core: Outside portion of Earth's core thought to consist mostly of liquid iron and nickel

outwash plain: Layered deposits left by water from a glacier

ozone: Form of oxygen (O_3) that in the upper atmosphere that shields Earth from harmful radiation from space, but at ground level is a pollutant

ozone layer: Protective layer, in the upper atmosphere, that has a high ozone content

Pangaea: Ancient supercontinent that broke apart millions of years ago to form the present continents

parallel: Imaginary line drawn around Earth parallel to the equator that represents a constant latitude

percent deviation: Mathematical method of comparing a measurement with the commonly accepted value for that measurement; percent error from accepted value

period: Subdivision of an era

permeability: Ability of a soil to transmit water

phase: Physical state of matter, solid, liquid, or gas; apparent change in shape of the lighted portion of a celestial object (phases of the moon)

photochemical smog: Air pollution that forms when sunlight changes exhausts from motor vehicles, homes, and power plants into a brown haze

physical weathering: Mechanical breakdown of rocks without a change in chemical composition

plains: Landscape region characterized by horizontal rock structure and low topographic relief

planetary wind belt: Latitude zone of prevailing wind conditions caused by uneven heating of Earth and Earth's rotation

plate boundary: Line along which crustal plates meet and interact

plate tectonics: Unified theory of crustal motion that incorporates continental drift and sea-floor spreading; theory that Earth's surface is composed of about a dozen large, rigid plates that carry the continents

plateau: Landscape region characterized by horizontal rock structure and high topographic relief (A plateau is usually a relatively flat or rolling uplands area deeply cut by stream valleys.)

plutonic: (See intrusive.)

polar air mass: Mass of cool air

Polaris: (See North Star.)

pollution: Substance or form of energy whose concentration is large enough to harm living things or the natural environment

porosity: Portion or percent of empty space within fractured rock or soil; the volume of pore space in a material compared with its volume

potential energy: Energy in storage, energy of position, or energy involved in a change of state

precipitate: Salt crystals that form and settle out of solution

precipitation: Water, in the form of rain, snow, or sleet, falling from the sky; sedimentary process that involves substances settling out of a saturated solution

pressure gradient: Rate of change in air pressure between two points on a map

prevailing wind: Wind that blows more often from one direction than from any other

primary: Object (lying along the major axis of an ellipse) around which a satellite moves. (The sun is the primary of the orbiting Earth, and the Earth is the primary of the moon.)

primary waves: (See P-waves.)

prime meridian: Imaginary line (semicircle) that runs through Greenwich, England, from the North Pole to the South Pole

P-wave: Compression (longitudinal) wave that is the fastest of the seismic waves that radiate from an earthquake (P-waves can travel through solids and liquids; also known as primary waves.)

radial drainage: Stream pattern in which streams radiate like spokes of a wheel down the sides of a central mountain such as a volcano

radiation: Emission and transfer of energy by means of electromagnetic waves, and the only way that energy can travel through empty space; rays or particles given off by an unstable radioactive substance

radiative balance: Equilibrium between absorbed radiant energy and radiant energy given off

radioactive: Emission of energy rays or nuclear particles due to the breakdown of an unstable isotope

redshift: Displacement of spectral lines of distant stars toward the red end of the spectrum; Doppler shift

reflection: Process in which energy waves bounce off a surface or interface

refraction: Process in which energy waves change direction as they move from one medium to another

regional metamorphism: Process by which large masses of rock are changed by deep burial within Earth

relative age: Comparative age; age expressed as before or after other events without specifying the age in units

relative humidity: Ratio between the actual amount of water vapor in the air and the maximum amount of water vapor the air can contain at a given temperature

relative time scale: Time scale that indicates whether the object in question is older or younger than something else

relief (topographic): Change in elevation from one place to another

renewable resources: Natural materials that can be replaced as quickly as they are used

residual soil: Soil that remains on top of the bedrock from which it formed

reversed polarity: Magnetic polarity of an igneous rock that was formed at a time in the geologic past when Earth's north and south magnetic poles had the opposite polarity that they have at the present

revolution: Orbital motion of a satellite around its primary

rift zone: Linear feature of Earth where new crust is being created

Ring of Fire: Zone of volcanoes, earthquakes, and mountain building that surrounds the Pacific Ocean

rock: Natural piece of the solid Earth, usually composed of one or more minerals

rock cycle: Model explaining the natural changes in rocks and sediment

rotation: Spinning of a body around an internal axis (Earth's rotation causes day and night.)

runoff: Precipitation that is unable to infiltrate soil, so it moves overland into streams

satellite: Artificial or natural celestial object that revolves around another object, the primary

saturated: Weather condition in which the air holds as much water vapor as it can at a given temperature; a saturated solution is one that can hold no more of a particular substance dissolved in it

scale: Ratio of the dimensions of the real object to its model

scientific notation: Mathematical shorthand in which numbers are written in the form $M \times 10^n$, where M is a number 1 or greater, but less than 10, and 10^n is a power of 10

sea breeze: Cool breeze that blows off the water during the day

sea-floor spreading: Theory that the oceanic crust has been constructed by material from deep within Earth that rises and spreads apart at the mid-ocean ridges

seasons: Annual cycle of weather conditions as Earth orbits the sun

secondary waves: (See S-waves.)

sediment: Accumulation of particles of weathered rock, organic remains, or both; rock fragments

sedimentary rock: Rock formed by the compression and cementation of particles of sediment

seismic wave: Vibrational energy that radiates through Earth from an earthquake

seismograph: Instrument designed to measure and record the magnitude of an earthquake

seismologist: Scientist who studies earthquakes

seismometer: Instrument that detects vibrations of the ground

sense: Any of the five means by which we directly observe our environment (sight, hearing, smell, taste, and touch)

shear waves: (See S-waves.)

silicates: Large family of minerals that has the silicon-oxygen tetrahedron as its basic structure

slope: (See gradient.)

soil: Weathered rock mixed with organic remains at the top of the lithosphere

soil horizon: Layer within the soil showing a particular stage of soil development

solar system: Planets and other objects that revolve around the sun

solstice: Most hours of daylight in the Northern hemisphere when the sun is directly above the Tropic of Cancer or least hours of daylight when the sun is directly over the Tropic of Capricorn (The solstices occur about June 21 and about December 21, and they mark the beginning of summer and winter in the Northern Hemisphere.)

solution: Mixture in which the particles are so small they cannot be filtered out of the water

sorted: Separated by particle size or other characteristics

source region: Place where an air mass originates

specific gravity: Ratio of the density of a substance to the density of water

specific heat: Amount of heat energy required to raise the temperature of 1 gram of a substance by 1 Celsius degree (Water, with a specific heat of 1, is the common standard.)

spectrum: (See electromagnetic spectrum.)

star: Large, self-luminous body in space that creates its own radiant energy

state of matter: (See phase.)

station model: Standardized form in which weather information is presented at a preset position next to the circle that indicates the geographic location of the reporting station

stationary front: Interface, or boundary, between two air masses that are not moving

streak: Color of the powder of a mineral revealed by rubbing the mineral along a white, unglazed porcelain plate

stream discharge: (See discharge.)

stream velocity: Speed at which a stream flows

striations (glacial): Parallel scratches on the surface of a rock caused by the movement of a glacier

subduction zone: Region in which Earth's crust is destroyed as it is pulled down into the mantle

subsidence: Gradual sinking of a portion of Earth's crust

sunspot: Temporary storm on the visible surface of the sun

superposition (law of): Principle that the lowest layer in a sequence of rock strata must have been deposited before the layers above, unless the rock strata have been turned upside down

suspension: Fluid containing large particles that can be filtered but the particles are too small to settle on their own

S-waves: Transverse earthquake waves that arrive after the P-waves and that cannot travel through a liquid such as Earth's outer core; secondary or shear waves

synoptic weather map: Map showing a variety of field quantities, such as temperature, pressure, and sky conditions, at a particular time and over a large geographic area

tectonics: Study of large-scale deformations of Earth's crust

temperature: Measure of the average vibrational kinetic energy in a substance

terrestrial: Objects that are a part of Earth, such as rocks, oceans, and clouds

texture: Shape or feel of a surface; particularly the shape, arrangement, and size of mineral crystals on a rock surface

thermal energy: (See heat energy.)

tides: Twice daily rise and fall of the oceans caused by the gravitational attraction of the moon and the sun

till: Unsorted sediments deposited directly by glacial ice

tilted sedimentary layers: Beds of rock (usually sedimentary) thought to have been deposited horizontally that have been pushed into a different inclination (angle), usually by motions of Earth's crust

topographic map: (See contour map)

topographic profile: a cross-sectional view that shows the elevation of the land along a particular baseline

topography: Shape of the land surface

transform boundary: Plate boundary at which crustal plates slide past each other

transpiration: Process by which living plants release water vapor to the atmosphere

transported soil: Soil eroded and deposited away from its parent bedrock

trellis drainage: Drainage pattern in which most of the streams occupy parallel valleys; pattern usually develops on folded strata of rocks with differing resistance to erosion

trench (ocean): Ocean floor depression that marks the zone where crust is subducted

Tropic of Cancer: Farthest north the vertical ray of sunlight reaches (at the June solstice); 23 $^1/_2$° north of the equator

Tropic of Capricorn: Farthest south the vertical ray of sunlight reaches (at the December solstice); 23 $^1/_2$° south of the equator

tropical air mass: Warm air mass

troposphere: Lowest layer of air; reaches from Earth's surface up to about 12 km that contains most of the mass of the atmosphere

tsunami: Ocean wave or a series of waves usually associated with an undersea earthquake or landslide

unconformity: Gap in the geologic record caused by the erosion of sediments or rock followed by new deposition

uniformitarianism: Principal that most geologic events of the past are similar to processes that occur in the present

unsorted deposit: Sediment left by melting glacial ice usually containing a mixture of clay, sand, cobbles, and boulders

valley and ridge: Landscape that forms on intensely folded rocks or rocks with parallel bands with high- or low-resistance to erosion

vaporization: (See evaporation.)

vertical ray: Sunlight that strikes Earth from directly overhead (the zenith)

vertical sorting: Sorting of particles from bottom to top in a layer (The roundest, largest, and densest particles are on the bottom and the flattest, smallest, and least dense particles at the top of each layer.)

vesicular: Rock containing many holes created by expanding gas as magma rose to Earth's surface (Scoria and pumice are vesicular.)

volume: Amount of space that matter occupies

warm front: Boundary between a mass of warm air and the colder air it is replacing

water (hydrologic) cycle: Model of the circulation of water between the oceans, atmosphere, and land

watershed: Drainage basin; the geographic area in which water drains into a particular stream or other body of water

water table: Boundary at the top of the saturated zone within fractured bedrock or soil

water vapor: Water in the form of a gas

weather: Short-term condition of the atmosphere, including the changes that occur within hours or days

weathering: Breakdown of rock due to physical or chemical change

wind: Natural movement of air along, or parallel to, Earth's surface; convection within the atmosphere

windward: Side of a mountain range where air is forced to rise, leading to cloud formation and precipitation

year: Time required for a planet (Earth) to complete one orbit around its primary (the sun); $365^1/_4$ days on Earth

zenith: Point on a celestial sphere directly overhead with respect to an observer (90°)

zone of aeration: Soil above the water table in which most of the interconnected pores are filled with air

zone of crustal activity: Area around an ocean ridge or continental mountain range where volcanoes and earthquake epicenters are concentrated

zone of saturation: Fractured rock or soil below the water table in which the pores are filled with groundwater

INDEX

M

N

O

P

Q

R

S

T

U

V

W

Z

Physical Setting/EARTH SCIENCE
June 2016

Part A

Answer all questions in this part.

Directions (1–35): For *each* statement or question, choose the word or expression that, of those given, best completes the statement or answers the question. Some questions may require the use of the *2011 Edition Reference Tables for Physical Setting/Earth Science.* Record your answers on your separate answer sheet.

1 Earth's approximate rate of revolution is

(1) 1° per day
(2) 15° per day
(3) 180° per day
(4) 360° per day

2 Planetary winds in the Northern Hemisphere are deflected to the right due to the

(1) Doppler effect
(2) Coriolis effect
(3) tilt of Earth's axis
(4) polar front jet stream

3 Which star is hotter, but less luminous, than *Polaris*?

(1) *Deneb*
(2) *Aldebaran*
(3) *Sirius*
(4) *Pollux*

4 Which statement best explains why Earth and the other planets of our solar system became layered as they were being formed?

(1) Gravity caused less-dense material to move toward the center of each planet.
(2) Gravity caused more-dense material to move toward the center of each planet.
(3) Materials that cooled quickly stayed at the surface of each planet.
(4) Materials that cooled slowly stayed at the surface of each planet.

5 Which conditions on Earth's surface will allow for the greatest amount of water to seep into the ground?

(1) gentle slope and permeable
(2) gentle slope and impermeable
(3) steep slope and permeable
(4) steep slope and impermeable

6 The photograph below shows a Foucault pendulum at a museum. The pendulum knocks over pins in a regular pattern as it swings back and forth.

This pendulum movement, and the pattern of knocked-over pins, is evidence of Earth's

(1) nearly spherical shape
(2) gravitational attraction to the Sun
(3) rotation on its axis
(4) nearly circular orbit around the Sun

7 Earth's early atmosphere contained carbon dioxide, sulfur dioxide, hydrogen, nitrogen, water vapor, methane, and ammonia. These gases were present in the atmosphere primarily because

(1) radioactive decay products produced in Earth's core were released from Earth's surface
(2) evolving Earth life-forms produced these gases through their activity
(3) Earth's growing gravitational field attracted these gases from space
(4) volcanic eruptions on Earth's surface released these gases from the interior

8 The diagram below represents the apparent positions of the Big Dipper, with respect to *Polaris*, as seen by an observer in New York State at midnight on the first day of summer and on the first day of winter.

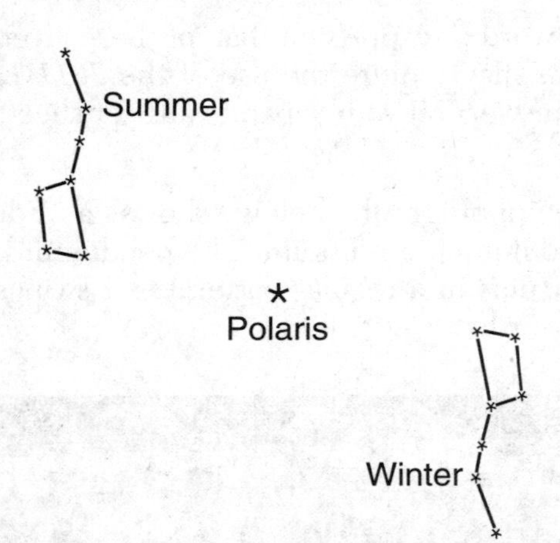

The change in the apparent position of the Big Dipper between the first day of summer and the first day of winter is best explained by Earth

(1) rotating for 12 hours
(2) rotating for 1 day
(3) revolving for 6 months
(4) revolving for 1 year

9 The weather station model shown below indicates that winds are coming from the

(1) southeast at 10 knots
(2) northwest at 10 knots
(3) southeast at 20 knots
(4) northwest at 20 knots

10 Which type of air mass most likely has high humidity and high temperature?

(1) cP (3) mT
(2) cT (4) mP

11 What is the relative humidity if the dry-bulb temperature is 16°C and the wet-bulb temperature is 10°C?

(1) 45% (3) 14%
(2) 33% (4) 4%

12 The table below shows the air temperature and dewpoint at each of four locations, *A*, *B*, *C*, and *D*.

Location	A	B	C	D
Air temperature (°F)	80	60	45	35
Dewpoint (°F)	60	43	35	33

Based on these measurements, which location has the greatest chance of precipitation?

(1) *A* (3) *C*
(2) *B* (4) *D*

13 Which type of electromagnetic radiation has the shortest wavelength?

(1) ultraviolet (3) radio waves
(2) gamma rays (4) visible light

14 Which gas is considered a major greenhouse gas?

(1) methane (3) oxygen
(2) hydrogen (4) nitrogen

15 The diagram below represents Earth and the Sun's incoming rays. Letters *A*, *B*, *C*, and *D* represent locations on Earth's surface.

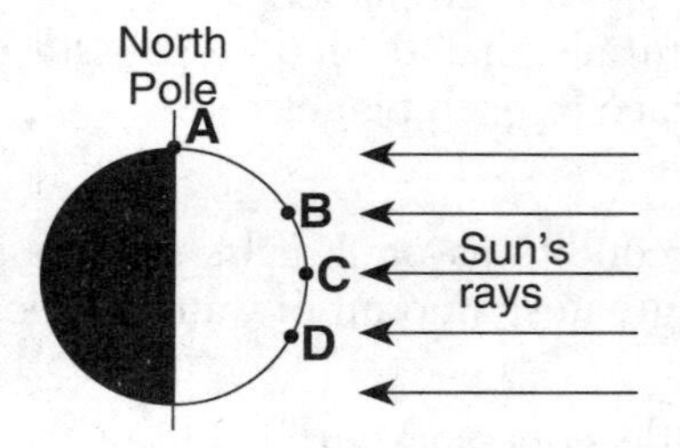

Which two locations are receiving the same intensity of insolation?

(1) *A* and *B* (3) *C* and *D*
(2) *B* and *C* (4) *D* and *B*

16 Most of the sand that makes up the sandstone found in New York State was originally deposited in which type of layers?

(1) tilted
(2) horizontal
(3) faulted
(4) folded

17 The map below shows the current location of New York State in North America.

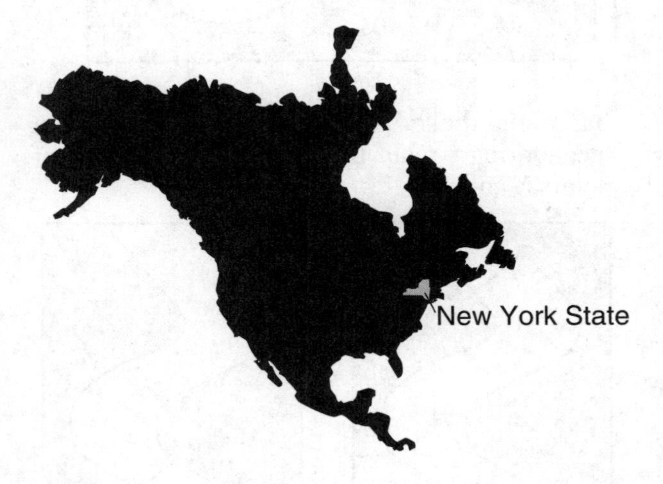

Approximately how many million years ago (mya) was this New York State region located at the equator?

(1) 59 mya
(2) 119 mya
(3) 359 mya
(4) 458 mya

18 Many scientists infer that one cause of the mass extinction of dinosaurs and ammonoids that occurred approximately 65.5 million years ago was

(1) tectonic plate subduction of most of the continents
(2) an asteroid impact that resulted in climate change
(3) a disease spreading among many groups of organisms
(4) severe damage produced by worldwide earthquakes

19 During which geologic epoch do scientists infer that the earliest grasses first appeared on Earth?

(1) Holocene
(2) Pleistocene
(3) Oligocene
(4) Eocene

20 What are the inferred pressure and temperature at the boundary of Earth's stiffer mantle and outer core?

(1) 1.5 million atmospheres pressure and an interior temperature of 4950°C
(2) 1.5 million atmospheres pressure and an interior temperature of 6200°C
(3) 3.1 million atmospheres pressure and an interior temperature of 4950°C
(4) 3.1 million atmospheres pressure and an interior temperature of 6200°C

21 A seismic *P*-wave is recorded at 2:25 p.m. at a seismic station located 7600 kilometers from the epicenter of an earthquake. At what time did the earthquake occur?

(1) 2:05 p.m.
(2) 2:11 p.m.
(3) 2:14 p.m.
(4) 2:36 p.m.

22 A seismic station recorded the *P*-waves, but no *S*-waves, from an earthquake because *S*-waves were

(1) absorbed by Earth's outer core
(2) transmitted only through liquids
(3) weak and detected only at nearby locations
(4) not produced by this earthquake

23 The Catskills of New York State are best described as a plateau, while the Adirondacks are best described as mountains. Which factor is most responsible for the difference in landscape classification of these two regions?

(1) climate variations
(2) bedrock structure
(3) vegetation type
(4) bedrock age

24 An elongated hill that is composed of unsorted sediments deposited by a glacier is called

(1) a delta
(2) a drumlin
(3) a sand dune
(4) an outwash plain

25 Which rock was subjected to intense heat and pressure but did *not* solidify from magma?

(1) sandstone
(2) schist
(3) gabbro
(4) rhyolite

26 The map below shows a stream drainage pattern where the streams radiate outward from the center.

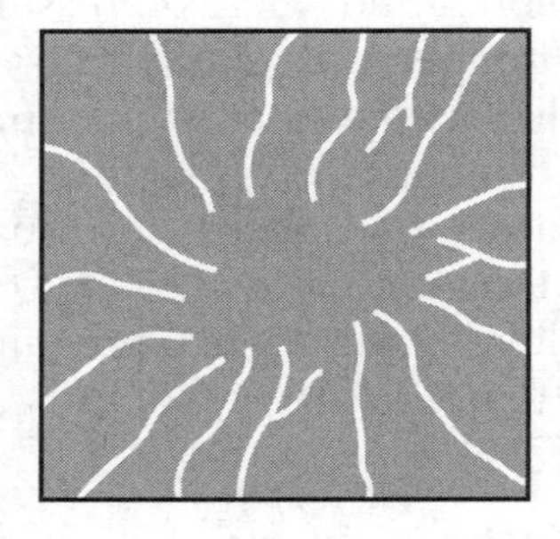

Which landscape feature would produce this stream drainage pattern?

(1) steep cliff
(2) glacial kettle lake
(3) volcanic mountain
(4) flat plain

27 The map below shows the area that, at one time, was covered by ancient Lake Bonneville. Evidence of ancient shorelines indicates that, near the end of the last ice age, Lake Bonneville existed in western Utah and eastern Nevada. The Great Salt Lake in Utah is a remnant of the former Lake Bonneville.

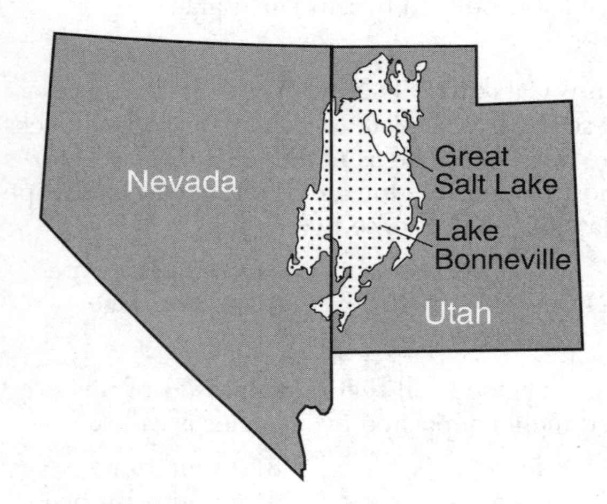

Which material that was formerly on the bottom of Lake Bonneville is most likely exposed on the land surface today?

(1) folded metamorphic bedrock
(2) flat-lying evaporite deposits
(3) coarse-grained coal beds
(4) fine-grained layers of volcanic lava

28 The cross section below represents a portion of a meandering stream. Points *X* and *Y* represent two positions on opposite sides of the stream.

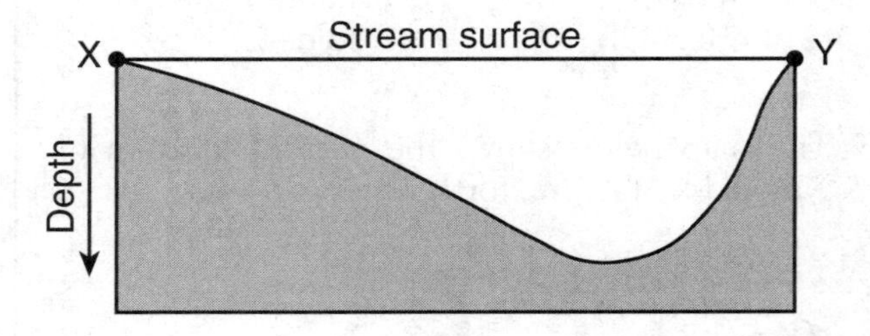

Based on the cross section, which map of a meandering stream best shows the positions of points *X* and *Y*?

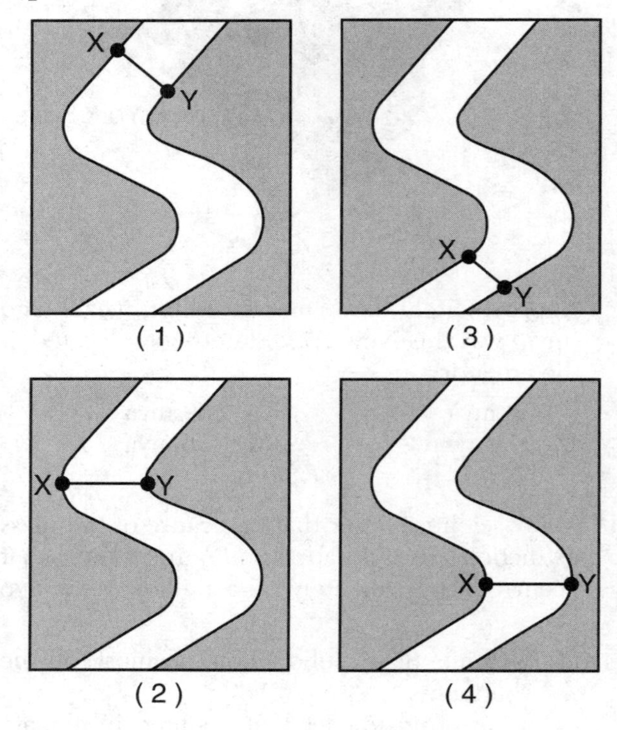

29 When wind and running water gradually decrease in velocity, the transported sediments are deposited

(1) all at once, and are unsorted
(2) all at once, and are sorted by size and density
(3) over a period of time, and are unsorted
(4) over a period of time, and are sorted by size and density

30 The graph below shows ocean water levels for a shoreline location on Long Island, New York. The graph also indicates the dates and times of high and low tides.

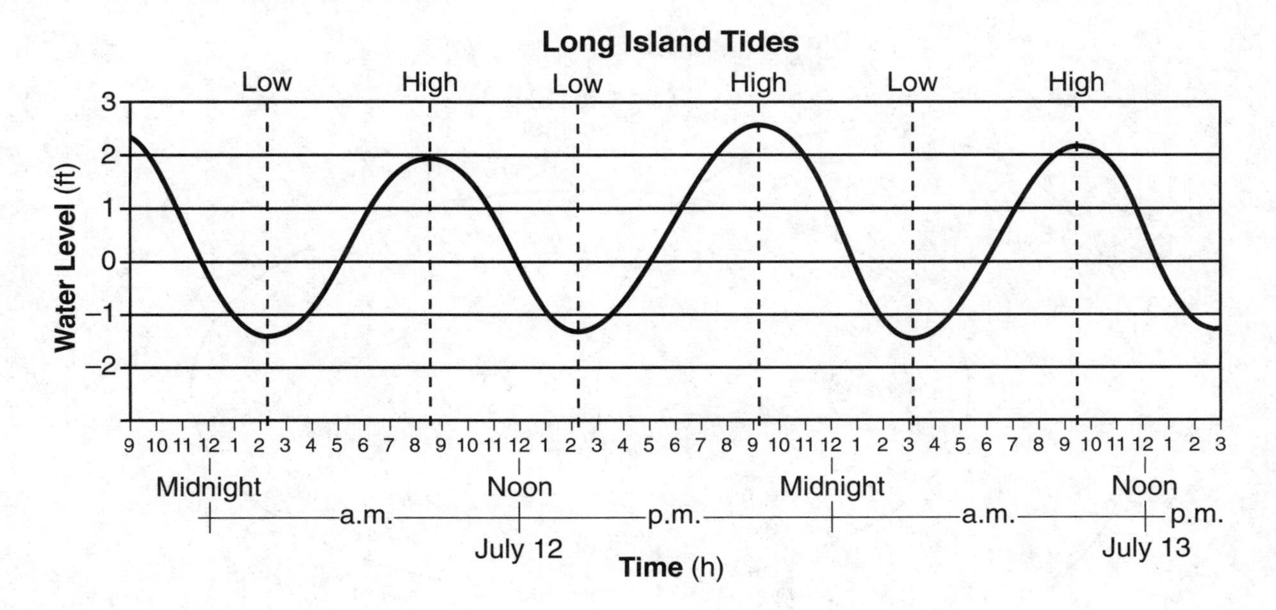

Based on the data, the next high tide occurred at approximately

(1) 4 p.m. on July 13
(2) 10 p.m. on July 13
(3) 4 p.m. on July 14
(4) 10 p.m. on July 14

31 Which diagram best represents heat transfer mainly by the process of conduction?

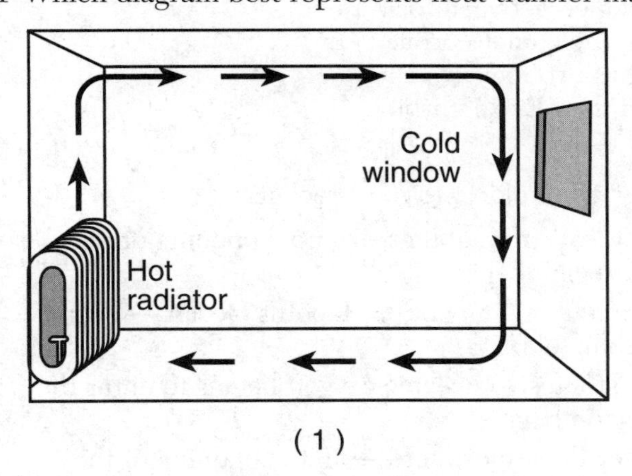

(1)

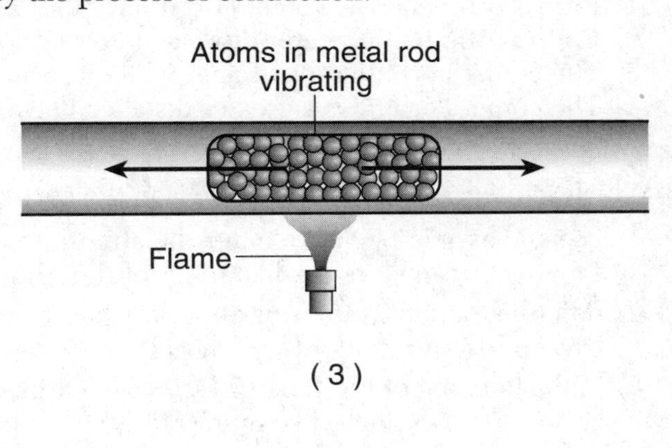

(3)

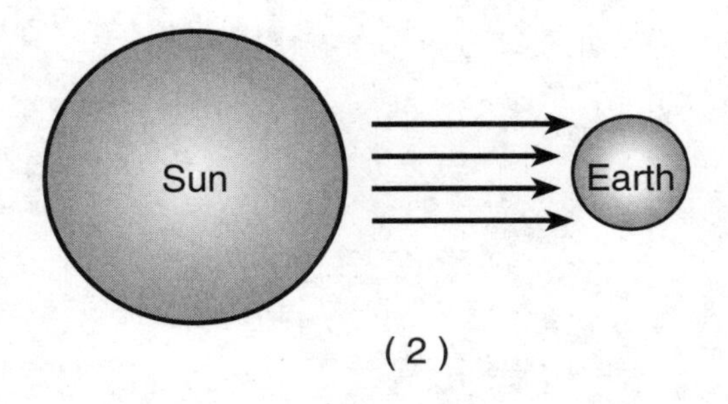

(2)

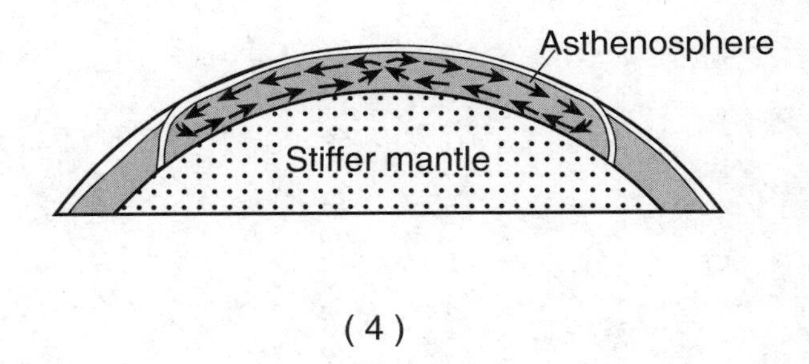

(4)

32 The diagram below represents the position of Earth in its orbit and the position of a comet in its orbit around the Sun.

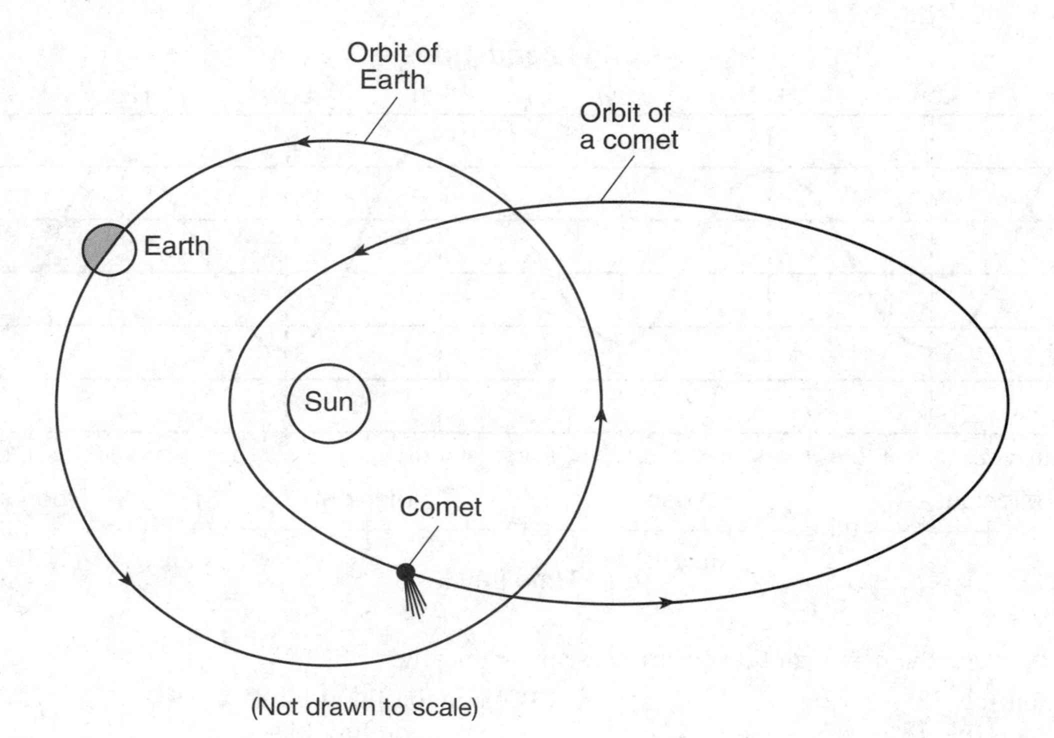

Which inference can be made about the comet's orbit, when it is compared to Earth's orbit?

(1) Earth's orbit and the comet's orbit have the same distance between foci.
(2) Earth's orbit has a greater distance between foci than the comet's orbit.
(3) The comet's orbit has one focus, while Earth's orbit has two foci.
(4) The comet's orbit has a greater distance between foci than Earth's orbit.

33 Which sequence of geologic events is in the correct order, from oldest to most recent?

(1) oceanic oxygen begins to enter the atmosphere → earliest stromatolites → initial opening of the Iapetus Ocean → dome-like uplift of the Adirondack region begins
(2) dome-like uplift of the Adirondack region begins → initial opening of the Iapetus Ocean → oceanic oxygen begins to enter the atmosphere → earliest stromatolites
(3) initial opening of the Iapetus Ocean → earliest stromatolites → oceanic oxygen begins to enter the atmosphere → dome-like uplift of the Adirondack region begins
(4) earliest stromatolites → oceanic oxygen begins to enter the atmosphere → initial opening of the Iapetus Ocean → dome-like uplift of the Adirondack region begins

34 The cross section of the atmosphere below represents the air motion near two frontal boundaries along reference line *XY* on Earth's surface.

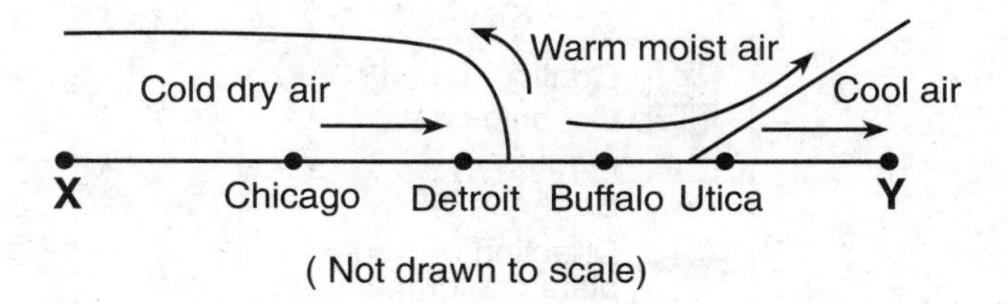

Which weather map correctly identifies these fronts and indicates the direction that these fronts are moving?

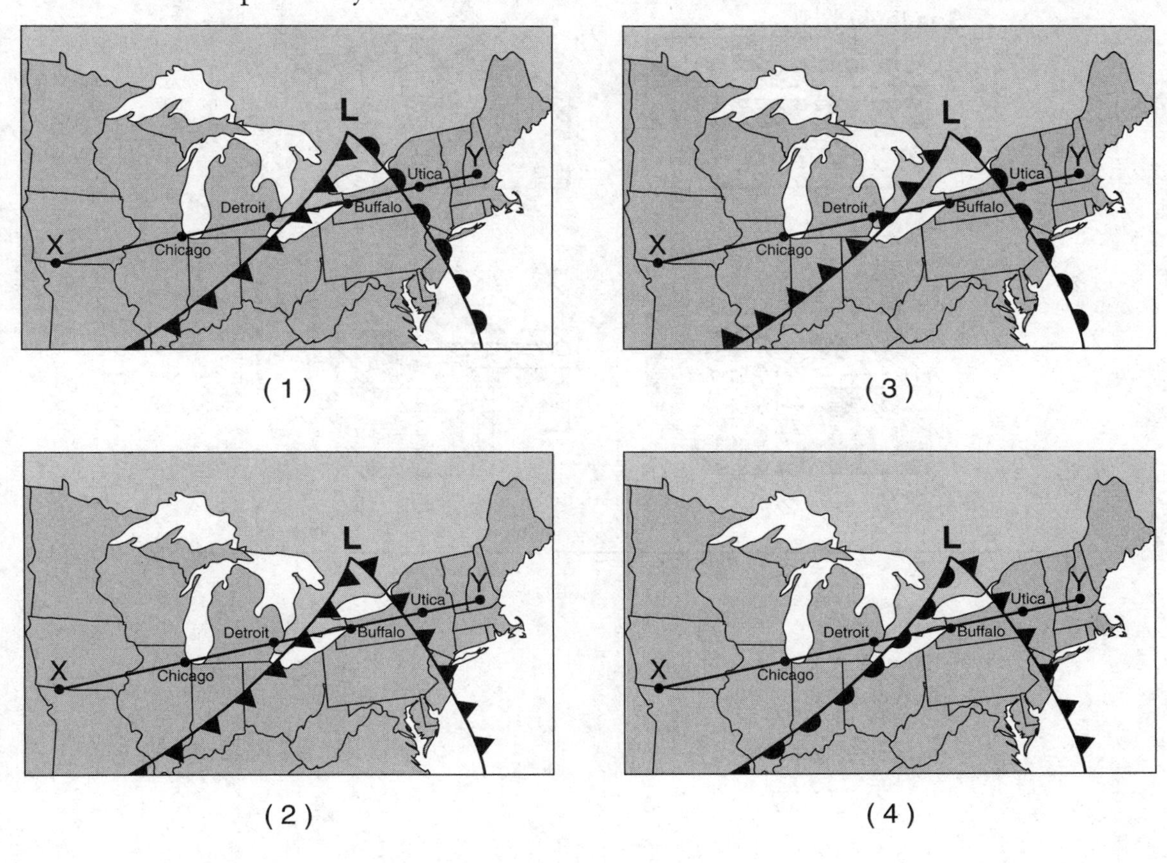

35 Which block diagram represents the plate motion that causes the earthquakes that occur along the San Andreas Fault in California?

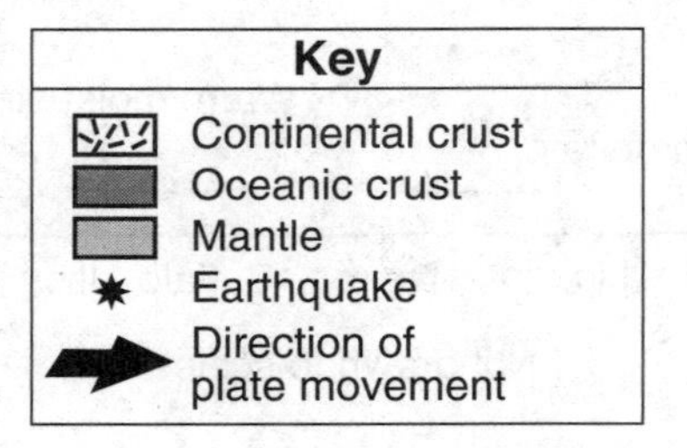

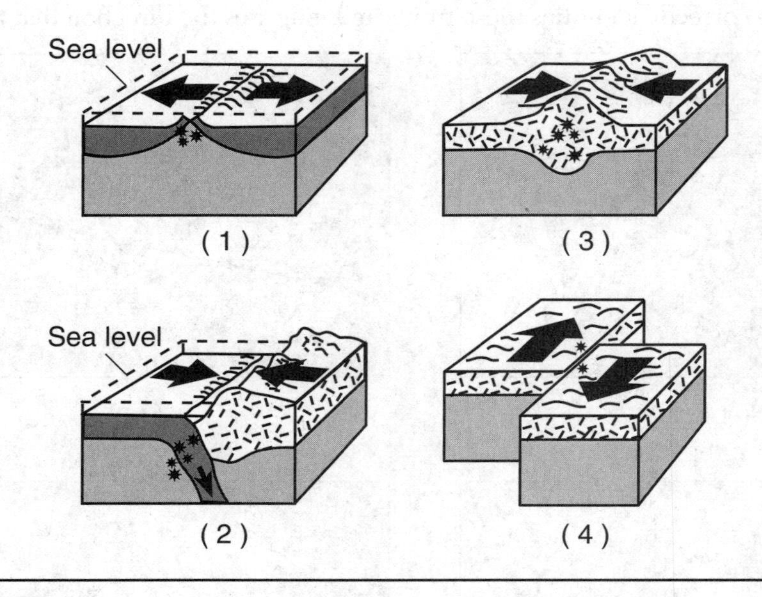

Part B–1

Answer all questions in this part.

Directions (36–50): For *each* statement or question, choose the word or expression that, of those given, best completes the statement or answers the question. Some questions may require the use of the *2011 Edition Reference Tables for Physical Setting/Earth Science.* Record your answers on your separate answer sheet.

Base your answers to questions 36 through 39 on the map and the passage below and on your knowledge of Earth science. The map shows four different locations in India, labeled *A*, *B*, *C*, and *D*, where vertical sticks were placed in the ground on the same clear day. The locations of two cities in India are also shown.

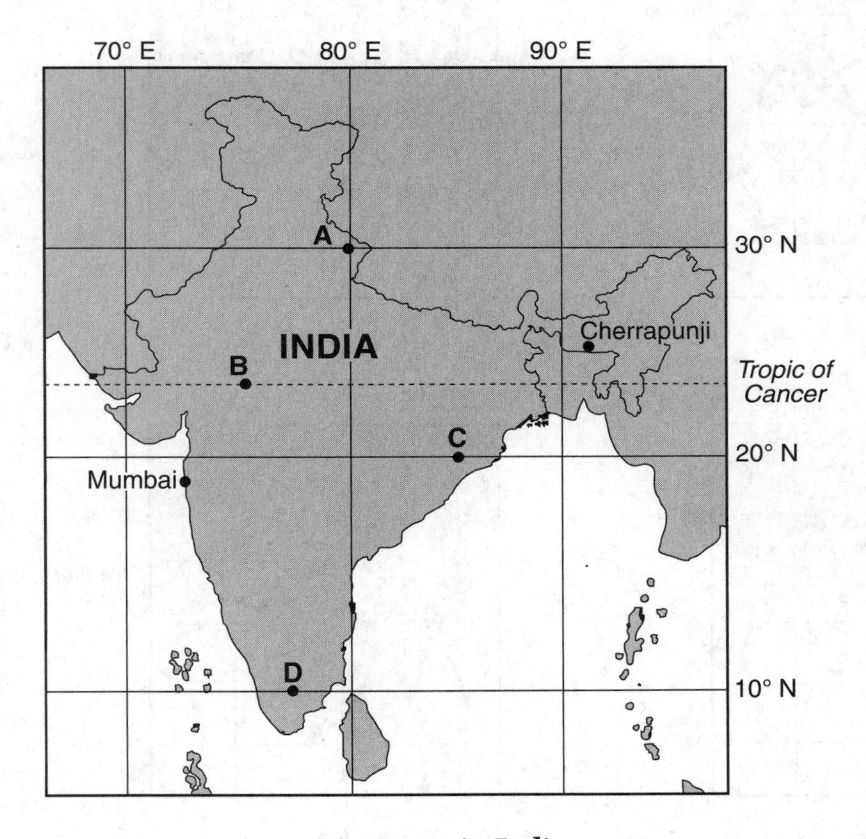

Monsoons in India

A monsoon season is caused by a seasonal shift in the wind direction, which produces excessive rainfall in many parts of the world, most notably India. Cherrapunji, in northeast India, received a record 30.5 feet of rain during July 1861. During the monsoon season from early June into September, Mumbai, India averages 6.8 feet of rain. Mumbai's total average rainfall for the other eight months of the year is only 3.9 inches.

Monsoons are caused by unequal heating rates of land and water. As the land heats throughout the summer, a large low-pressure system forms over India. The heat from the Sun also warms the surrounding ocean waters, but the water warms much more slowly. The cooler air above the ocean is more dense, creating a higher air pressure relative to the lower air pressure over India.

36 At which map location would no shadow be cast by the vertical stick at solar noon on the first day of summer?

(1) *A*
(2) *B*
(3) *C*
(4) *D*

37 Which map shows both the dominant air pressure system that forms over India in the summer and the direction of surface winds around this air pressure system? [High pressure = **H**, Low pressure = **L**]

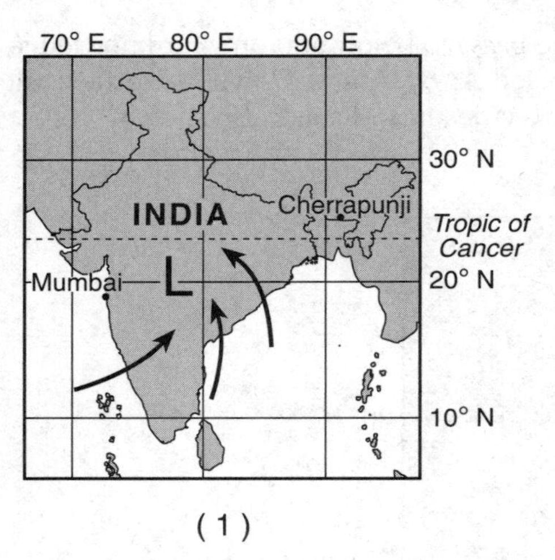

(1)

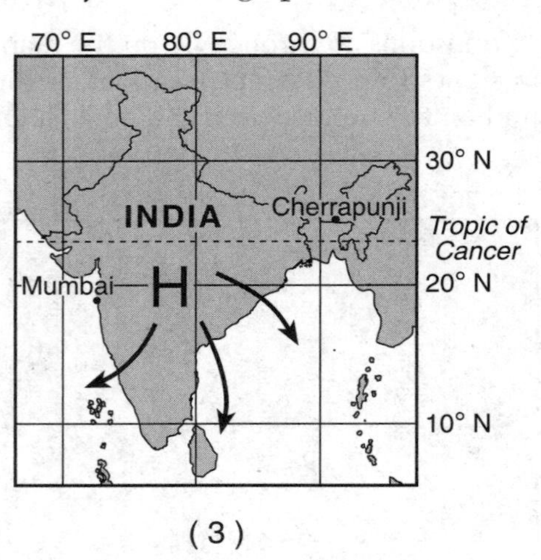

(3)

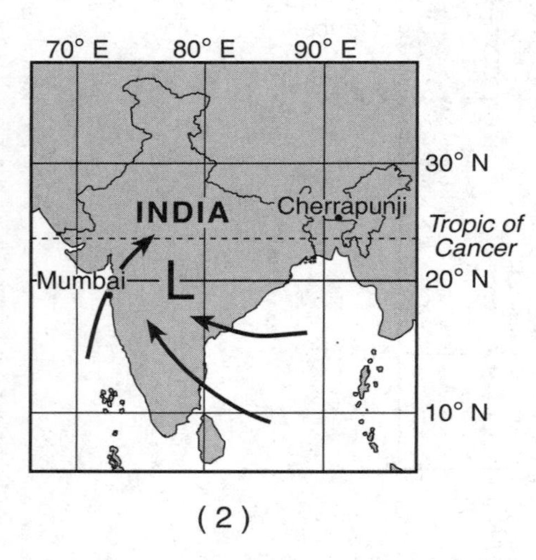

(2)

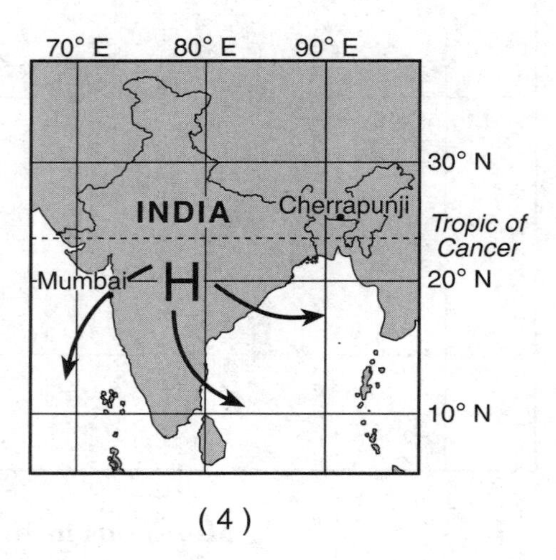

(4)

38 The unequal heating rates of India's land and water are caused by

(1) land having a higher density than water
(2) water having a higher density than land
(3) land having a higher specific heat than water
(4) water having a higher specific heat than land

39 Which processes lead to cloud formation when humid air rises over India?

(1) compression, warming to the dewpoint, and condensation
(2) compression, warming to the dewpoint, and evaporation
(3) expansion, cooling to the dewpoint, and condensation
(4) expansion, cooling to the dewpoint, and evaporation

Base your answers to questions 40 through 42 on the diagram below and on your knowledge of Earth science. The diagram represents the apparent path of the Sun across the sky at a New York State location on June 21. Point *A* represents the position of the noon Sun. Points *A* and *B* on the path are 45 degrees apart.

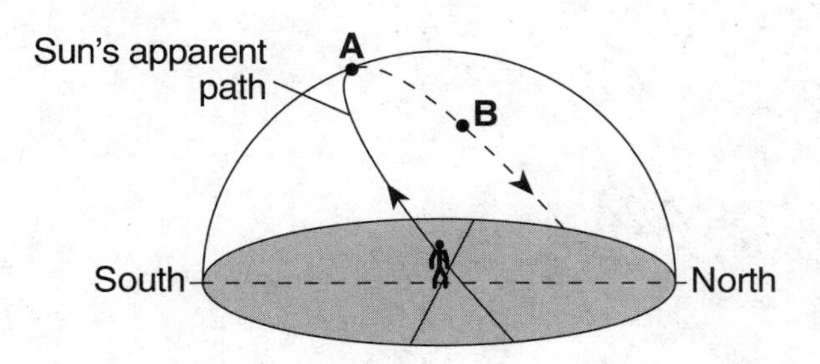

40 How many hours (h) will it take for the apparent position of the Sun to change from point *A* to point *B*?

(1) 1 h
(2) 2 h
(3) 3 h
(4) 4 h

41 Compared to the Sun's apparent path on June 21, the Sun's apparent path on December 21 at this location will

(1) be shorter, and the noon Sun will be lower in the sky
(2) be longer, and the noon Sun will be higher in the sky
(3) remain the same length, and the noon Sun will be lower in the sky
(4) remain the same length, and the noon Sun will be higher in the sky

42 Which diagram represents the correct position of *Polaris* as viewed from this New York State location on a clear night?

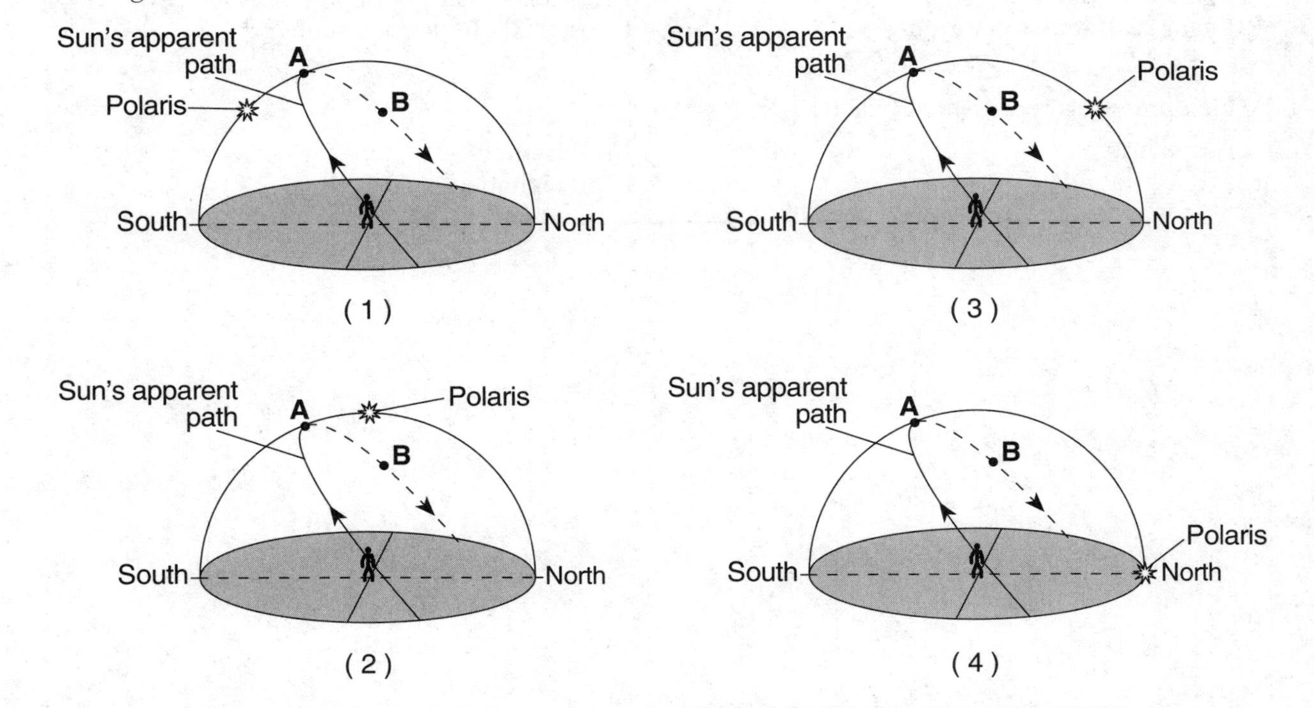

Base your answers to questions 43 and 44 on the diagram below and on your knowledge of Earth science. The diagram represents the water cycle. Letters *A* through *C* represent different processes in the water cycle.

(Not drawn to scale)

43 In order for process *A* to occur, liquid water must

(1) gain 334 Joules per gram
(2) gain 2260 Joules per gram
(3) lose 334 Joules per gram
(4) lose 2260 Joules per gram

44 Which process is represented by letter *B*?

(1) capillarity
(2) transpiration
(3) infiltration
(4) precipitation

Base your answers to questions 45 through 47 on the photograph below and on your knowledge of Earth science. The photograph shows a small waterfall located on the Tug Hill Plateau.

45 During which geologic time period was the surface bedrock at this location formed?

(1) Cretaceous
(2) Triassic
(3) Devonian
(4) Ordovician

46 Compared to the bedrock layers above and below the rock ledge shown at the waterfall, the characteristic that is primarily responsible for the existence of the rock ledge is its greater

(1) resistance to weathering
(2) abundance of fossils
(3) thickness
(4) age

47 Rock fragments that are tumbled and carried over long distances by this stream are most likely becoming

(1) less dense, harder, and smaller
(2) less rounded, jagged, and larger
(3) more dense, angular, and smaller
(4) more rounded, smoother, and smaller

Base your answers to questions 48 through 50 on the rock columns below and on your knowledge of Earth science. The rock columns represent four widely separated locations, *W*, *X*, *Y*, and *Z*. Numbers 1, 2, 3, and 4 represent fossils. The rock layers have *not* been overturned.

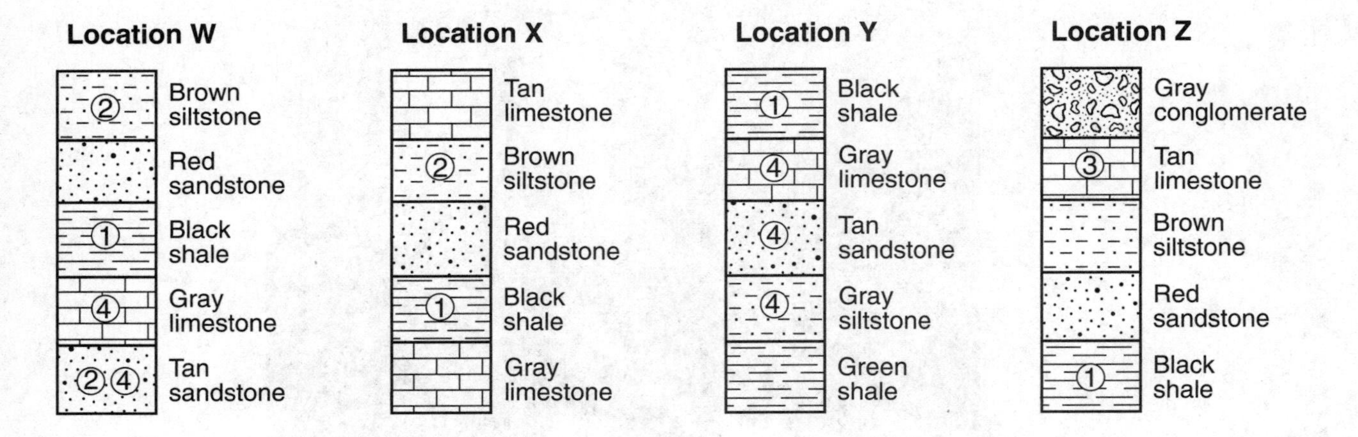

48 Which numbered fossil best represents an index fossil?

(1) 1
(2) 2
(3) 3
(4) 4

49 Which rock layer is the oldest?

(1) tan sandstone
(2) gray limestone
(3) green shale
(4) black shale

50 Which rock layer formed from the deposition of land-derived sediments that had a uniform particle size of about 0.01 cm in diameter?

(1) brown siltstone
(2) black shale
(3) gray conglomerate
(4) red sandstone

Part B–2

Answer all questions in this part.

Directions (51–65): Record your answers in the spaces provided in your answer booklet. Some questions may require the use of the *2011 Edition Reference Tables for Physical Setting/Earth Science.*

Base your answers to questions 51 through 53 on the data table below and on your knowledge of Earth science. The data table lists four constellations in which star clusters are seen from Earth. A star cluster is a group of stars near each other in space. Stars in the same cluster move at the same velocity. The length of the arrows in the table represents the amount of redshift of two wavelengths of visible light emitted by these star clusters.

Data Table

Constellation in which star cluster is seen from Earth	Redshift of two wavelengths of light absorbed by calcium	Distance from Earth (billion light years)	Velocity of star cluster moving away from Earth (km/s)
Ursa Major	Violet → Red	1.0	15,000
Corona Borealis	Violet → Red	1.4	22,000
Boötes	Violet → Red	2.5	39,000
Hydra	Violet → Red	4.0	61,000

Note: One light year is the distance light travels in one year.

51 Describe the evidence shown by the light from these star clusters that indicates that these clusters are moving away from Earth. [1]

52 Write the chemical symbol for the element, shown in the table, that absorbs the two wavelengths of light. [1]

53 Identify the name of the nuclear process that is primarily responsible for producing energy in stars. [1]

Base your answers to questions 54 through 57 on the diagram below and on your knowledge of Earth science. The diagram represents the Moon in eight positions in its orbit around Earth. One position is labeled *A*.

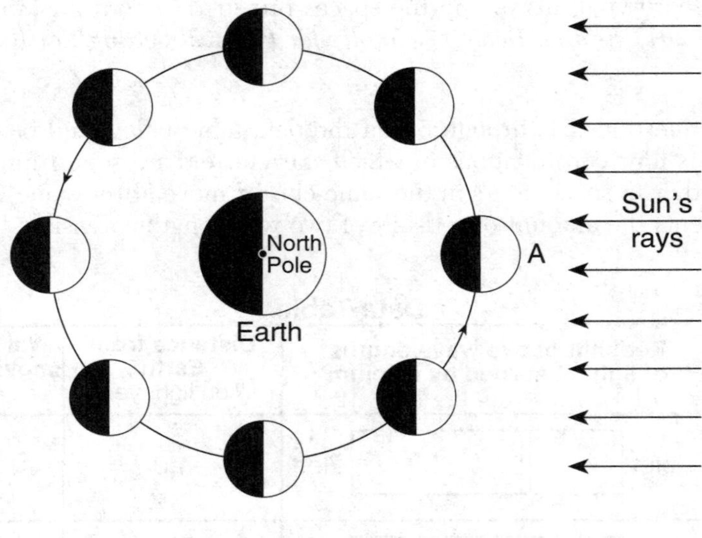

(Not drawn to scale)

54 *In your answer booklet*, circle the type of eclipse that may occur when the Moon is at position *A*. Explain why this type of eclipse may occur when the Moon is at this position. [1]

55 The diagram below represents one phase of the Moon as observed from New York State.

On the diagram *in your answer booklet*, place an **X** on the Moon's orbit to represent the Moon's position when this phase was observed. [1]

56 State the number of days needed for the Moon to show a complete cycle of phases from one full Moon to the next full Moon when viewed from New York State. [1]

57 Explain why the Moon's revolution and rotation cause the same side of the Moon to always face Earth. [1]

Base your answers to questions 58 through 61 on the weather map in your answer booklet and on your knowledge of Earth science. The weather map shows atmospheric pressures, recorded in millibars (mb), at locations around a low-pressure center (**L**) in the eastern United States. Isobars indicate air pressures in the western portion of the mapped area. Point *A* represents a location on Earth's surface.

58 On the weather map *in your answer booklet*, draw the 1012 millibar and the 1008 millibar isobars. Extend the isobars to the east coast of the United States. [1]

59 Identify the weather instrument that was used to measure the air pressures recorded on the map. [1]

60 Identify the compass direction toward which the center of the low-pressure system will move if it follows a typical storm track. [1]

61 Convert the air pressure at location *A* from millibars to inches of mercury. [1]

Base your answers to questions 62 through 65 on the graph below and on your knowledge of Earth science. The graph shows the rate of decay of the radioactive isotope carbon-14 (^{14}C).

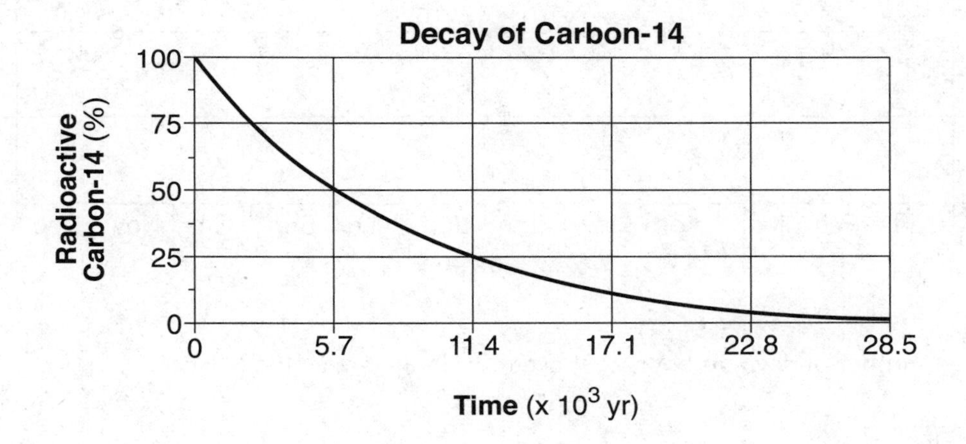

62 Complete the flow chart *in your answer booklet* by filling in the boxes to indicate the percentage of carbon-14 remaining and the time that has passed at the end of each half-life. [1]

63 Identify the decay product formed by the disintegration of carbon-14. [1]

64 Explain why carbon-14 *cannot* be used to accurately determine the age of organic remains that are 1,000,000 years old. [1]

65 State the name of the radioactive isotope that has a half-life that is approximately the same as the estimated time of the origin of Earth. [1]

Part C

Answer all questions in this part.

Directions (66–85): Record your answers in the spaces provided in your answer booklet. Some questions may require the use of the *2011 Edition Reference Tables for Physical Setting/Earth Science.*

Base your answers to questions 66 through 69 on the graph below and on your knowledge of Earth science. The graph shows changes in hours of daylight during the year at the latitudes of 0°, 30° N, 50° N and 60° N.

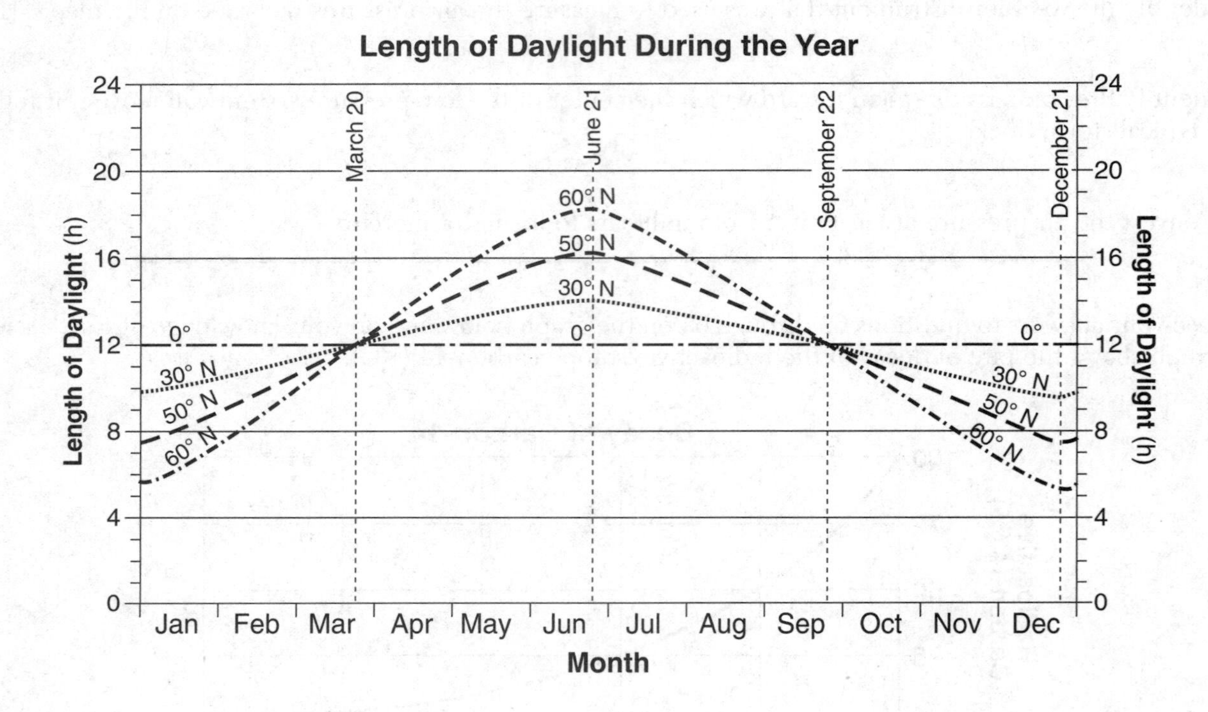

66 Estimate the number of daylight hours that occur on January 1 at 40° N latitude. [1]

67 Identify the latitude shown on the graph that has the earliest sunrise on June 21. Include the units and compass direction in your answer. [1]

68 Explain why all four latitudes have the same number of hours of daylight on March 20 and September 22. [1]

69 The graph *in your answer booklet* shows a curve for the changing length of daylight over the course of one year that occurs for an observer at 50° N latitude. On this same graph in your answer booklet, draw a line to show the changing length of daylight over the course of one year that occurs for an observer at 50° S latitude. [1]

Base your answers to questions 70 through 74 on the passage and data tables below, on the map in your answer booklet, and on your knowledge of Earth science. The data tables show trends (patterns) of two lines of Hawaiian island volcanoes, the Loa trend and the Kea trend. For these trends, ages and distances of the Hawaiian island volcanoes are shown. The map shows the locations of volcanoes, labeled with *X*s, that make up each trend line.

Hawaiian Volcano Trends

The Hawaiian volcanic island chain, located on the Pacific Plate, stretches over 600 kilometers. This chain of large volcanoes has grown from the seafloor to heights of over 4000 meters. Geologists have noted that there appear to be two lines, or "trends," of volcanoes—one that includes Mauna Loa and one that includes Mauna Kea. Loihi and Kilauea are the most recent active volcanoes on the two trends shown on the map.

Loa Trend

Loa Trend Volcanoes	**Volcano Age** (million years)	**Distance from Loihi** (km)
Kauai	4.6	575
Waianae	3.7	465
Koolau	2.2	375
West Molokai	1.7	350
Lanai	1.2	300
Kahoolawe	1.1	250
Hualalai	0.3	130
Mauna Loa	0.2	70
Loihi	0	0

Kea Trend

Kea Trend Volcanoes	**Volcano Age** (million years)	**Distance from Kilauea** (km)
East Molokai	1.7	256
West Maui	1.5	221
Haleakala	0.9	182
Kohala	0.5	100
Mauna Kea	0.4	54
Kilauea	0.1	0

70 The average distance between the volcanoes along the Kea trend is 51.2 kilometers. Place an **X** on the map *in your answer booklet* to identify the location on the seafloor where the next volcano will most likely form as a part of the Kea trend. [1]

71 Identify the *two* volcanoes, one from each trend, that have the same age. [1]

72 State the general relationship between the age of the volcanoes and the distance from Loihi. [1]

73 Identify the tectonic feature beneath the moving Pacific Plate that caused volcanoes to form in *both* the Loa and Kea trends. [1]

74 Identify the compass direction in which the Pacific Plate has moved during the last 4.6 million years. [1]

Base your answers to questions 75 through 79 on the topographic map in your answer booklet and on your knowledge of Earth science. Lines *AB* and *CD* are reference lines on the map. Letter *E* indicates a location in a stream.

75 On the map *in your answer booklet*, draw an **X** on the location with the highest elevation. [1]

76 Using the grid *in your answer booklet*, construct a topographic profile along line *AB* by plotting the elevation of each contour line that crosses line *AB*. Points *A* and *B* have already been plotted on the grid. Connect all plots with a line from *A* to *B* to complete the profile. [1]

77 Calculate the gradient along line *CD*. [1]

78 Describe how the contour lines indicate the direction in which Buck River flows. [1]

79 Determine the velocity of the stream at location *E* where the largest particle being carried at location *E* has a diameter of 10.0 centimeters. [1]

Base your answers to questions 80 through 83 on the passage below and on your knowledge of Earth science.

Dimension Stone: Granite

Dimension stone is any rock mined and cut for specific purposes, such as kitchen countertops, monuments, and the curbing along city streets. Examples of rock mined for use as dimension stone include limestone, marble, sandstone, and slate. The most important dimension stone is granite; however, not all dimension stone sold as granite is actually granite. Two examples of such rock sold as "granite" are syenite and anorthosite. Syenite is a crystalline, light-colored rock composed primarily of potassium feldspar, plagioclase feldspar, biotite, and amphibole, while anorthosite is composed almost entirely of plagioclase feldspar. Like actual granite, both syenite and anorthosite have large, interlocking crystals.

80 Explain why syenite is classified as a plutonic igneous rock. [1]

81 State *one* reason why anorthosite is likely to be white to gray in color. [1]

82 The igneous rock gabbro is sometimes sold as "black granite." Compared to the density and composition of granite, describe how the density and composition of gabbro are different. [1]

83 Identify *one* dimension stone mentioned in the passage that is composed primarily of calcite. [1]

Base your answers to questions 84 and 85 on the map of Australia below and on your knowledge of Earth science. Points *A* through *D* on the map represent locations on the continent.

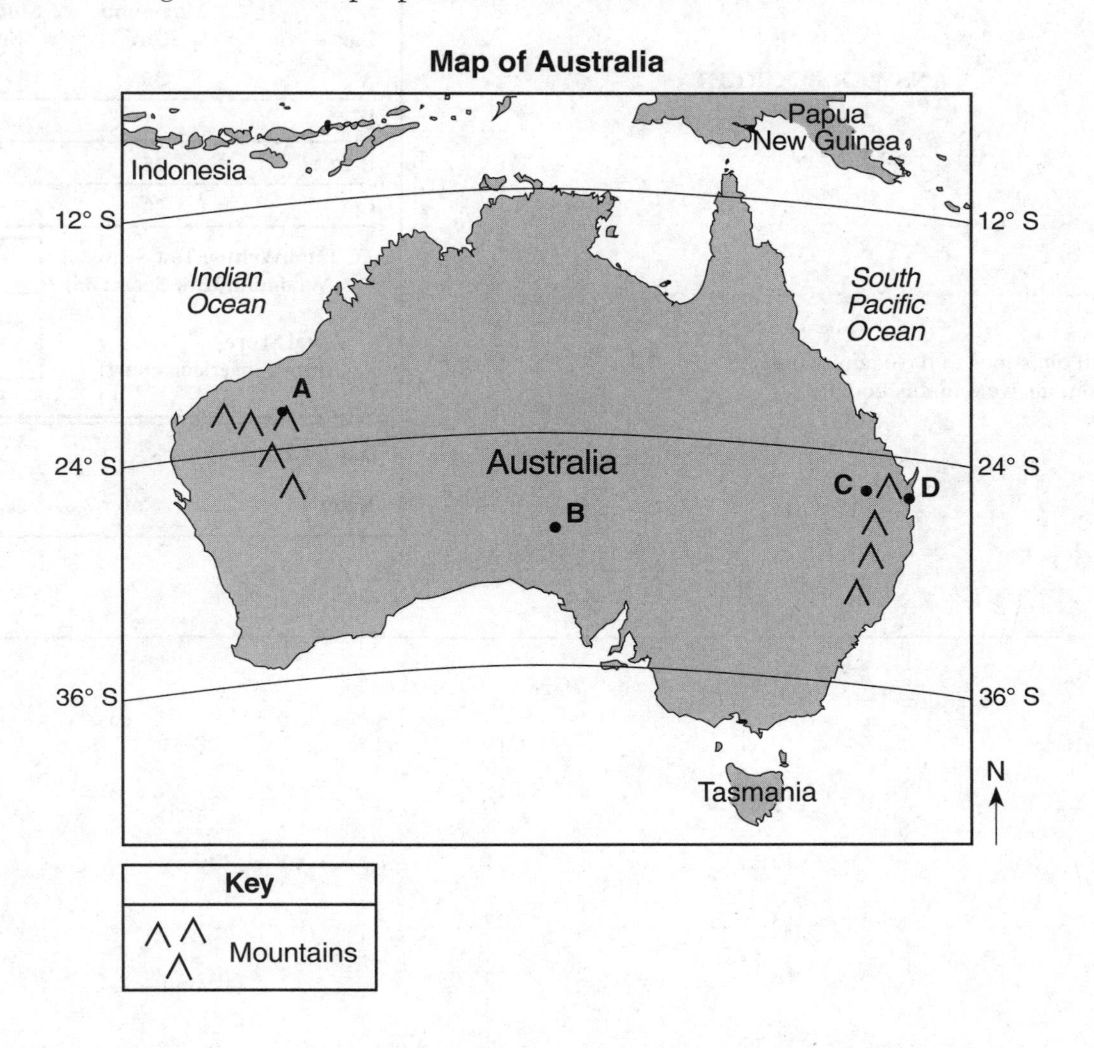

84 Explain why location *A* has a cooler average yearly air temperature than location *B*. [1]

85 The cross section below represents a mountain between locations *C* and *D* and the direction of prevailing winds.

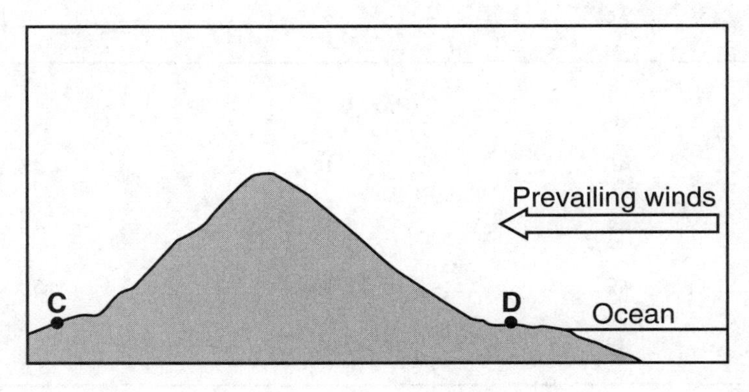

Explain why location *D* has a wetter climate than location *C*. [1]

____

Physical Setting/EARTH SCIENCE
June 2016

	Performance Test Score (Maximum Score: 16)

Part	Maximum Score	Student's Score
A	35	
B-1	15	
B-2	15	
C	20	

Total Written Test Score (Maximum Raw Score: 85)

Final Score (from conversion chart)

Rater's Initials:

Rater 1 .... Rater 2 ....

ANSWER BOOKLET

Student...

Teacher...

School... Grade ....

Answer all questions in the examination.
Record your answers in this booklet.

Part A

1	10.............	19	28.............
2	11.............	20	29.............
3	12.............	21	30.............
4	13.............	22	31.............
5	14.............	23	32.............
6	15.............	24	33.............
7	16.............	25	34.............
8	17.............	26	35.............
9	18.............	27	

Part B

36	40.............	44	48.............
37	41.............	45	49.............
38	42.............	46	50.............
39	43.............	47	

Physical Setting/EARTH SCIENCE
June 2016

ANSWER BOOKLET

☐ Male

Student . . . . Sex: ☐ Female

Teacher . . . .

School . . . . Grade . . . .

Record your answers for Part B–2 and Part C in this booklet.

Part B–2

51 ____

____

52 ____________

53 ______________________________

54 Circle one: **lunar eclipse** **solar eclipse**

Explanation: ____

____

55

56 ____________ **days**

57 ____

____

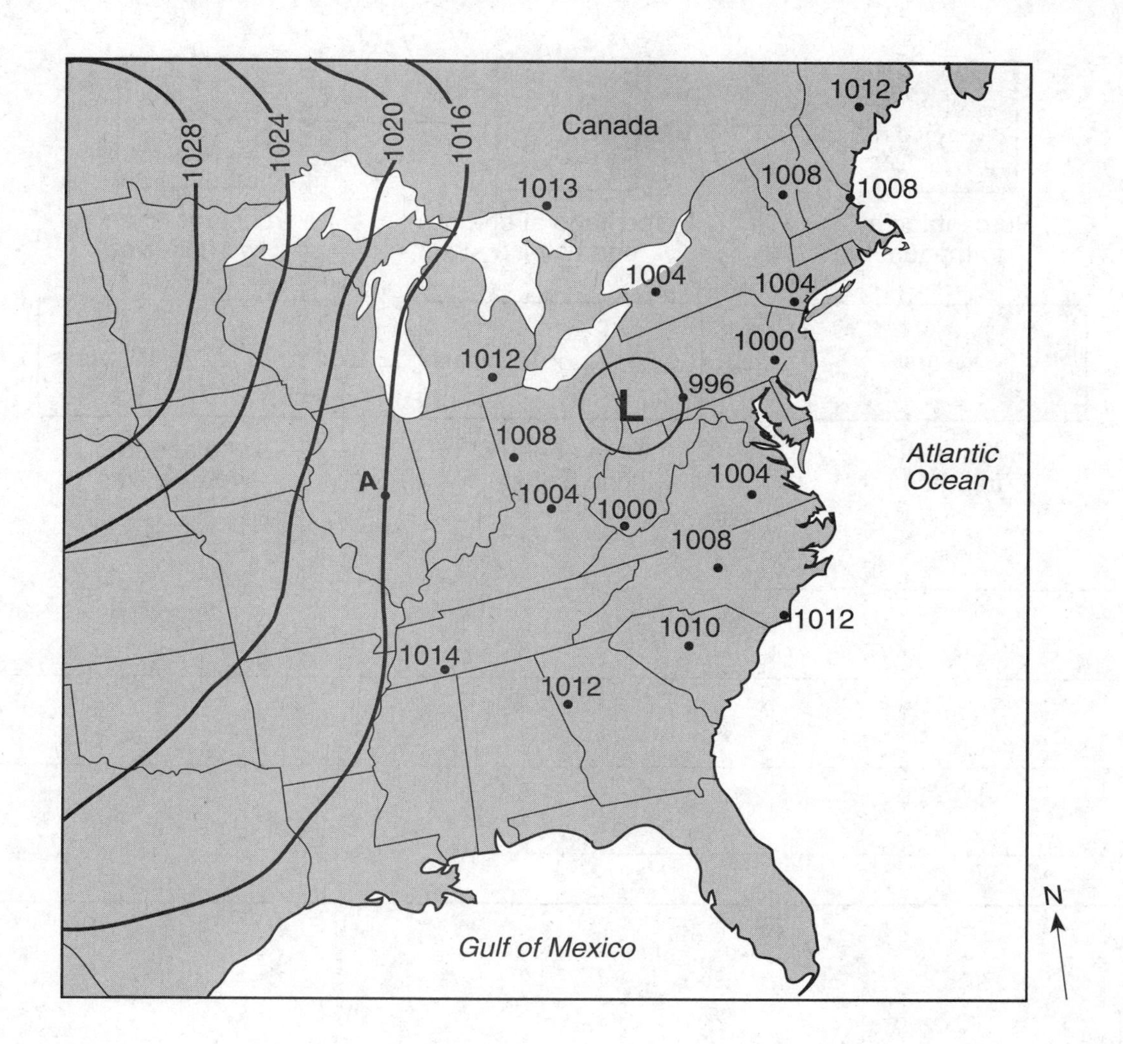

59 ______________________________

60 ______________________________

61 __________ in of Hg

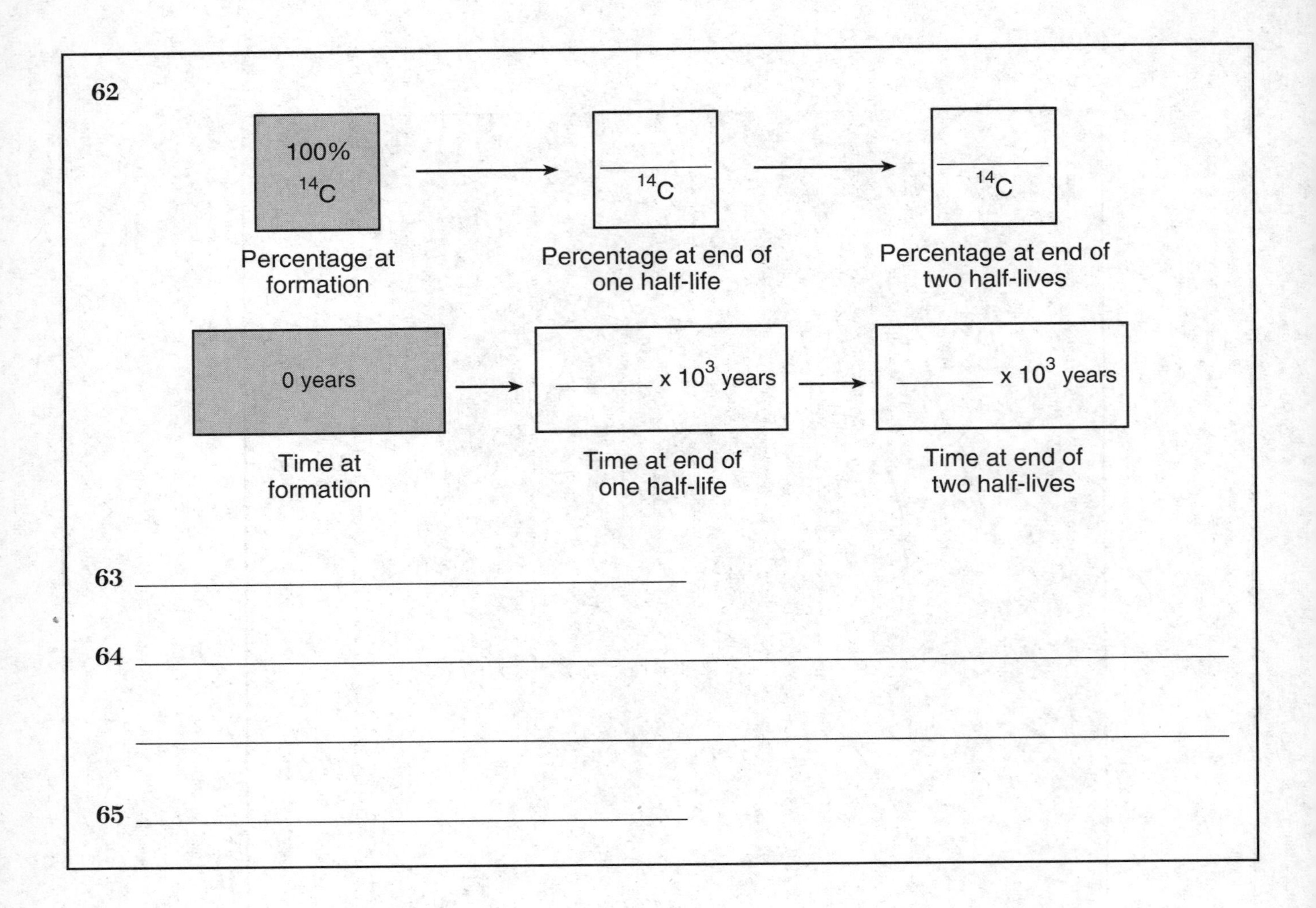
62
100%
^{14}C
^{14}C
^{14}C
Percentage at formation
Percentage at end of one half-life
Percentage at end of two half-lives
0 years
x 10^3 years
x 10^3 years
Time at formation
Time at end of one half-life
Time at end of two half-lives
63
64
65

Part C

66 ____________ h

67 ____________

68 ____

____

69

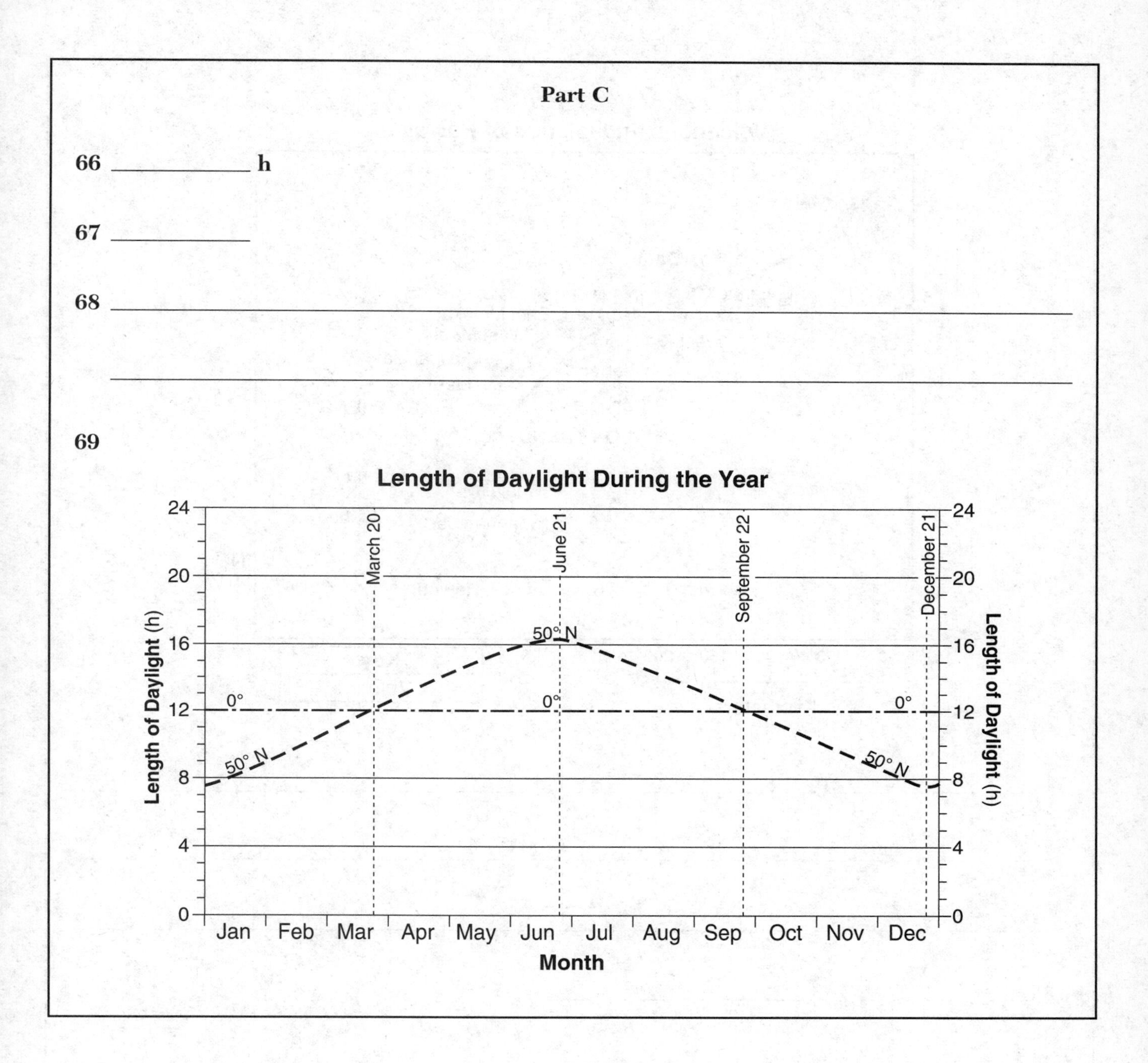

70

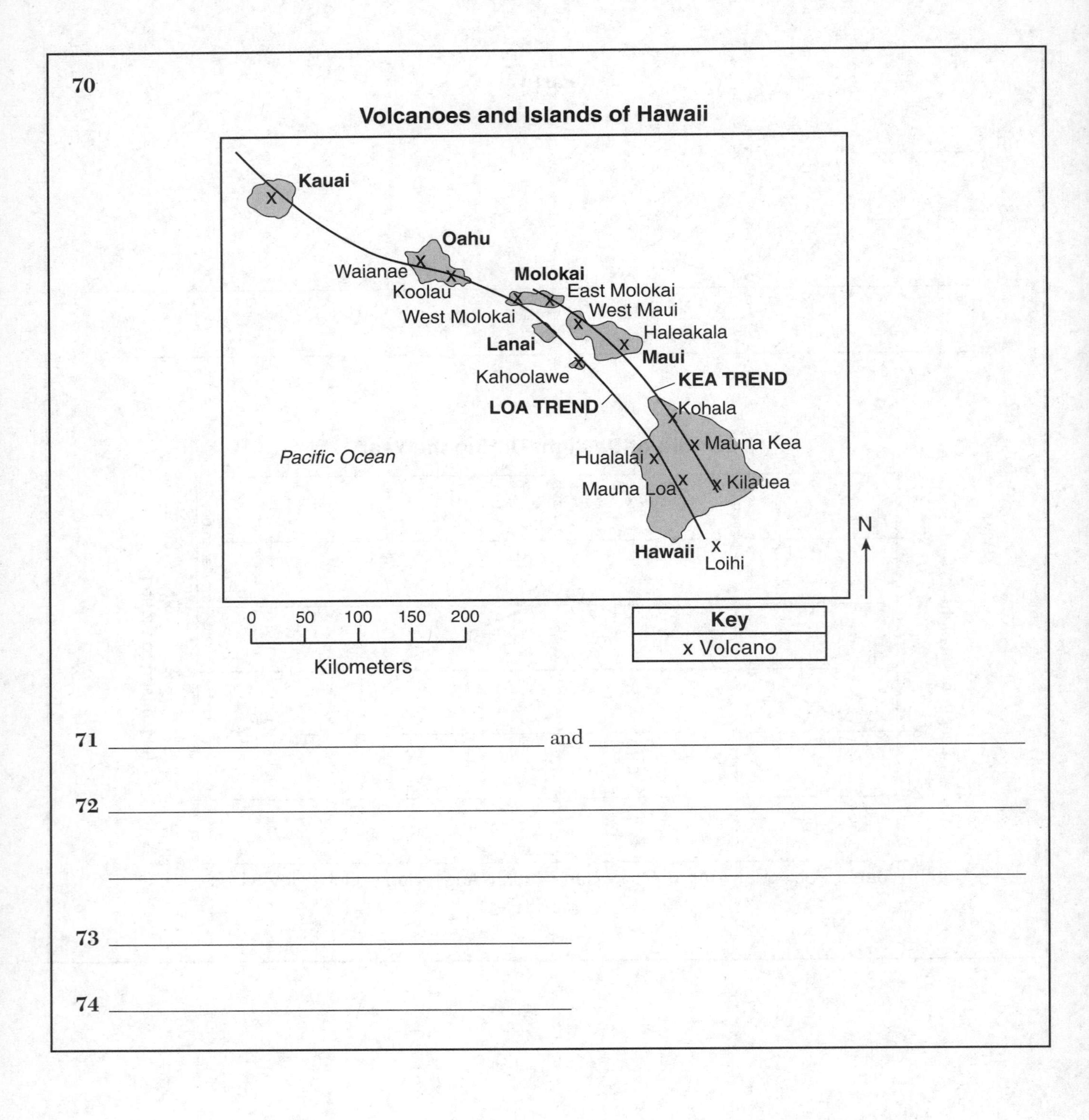

71 ______________________ and ______________________

72 ____

____

73 ______________________

74 ______________________

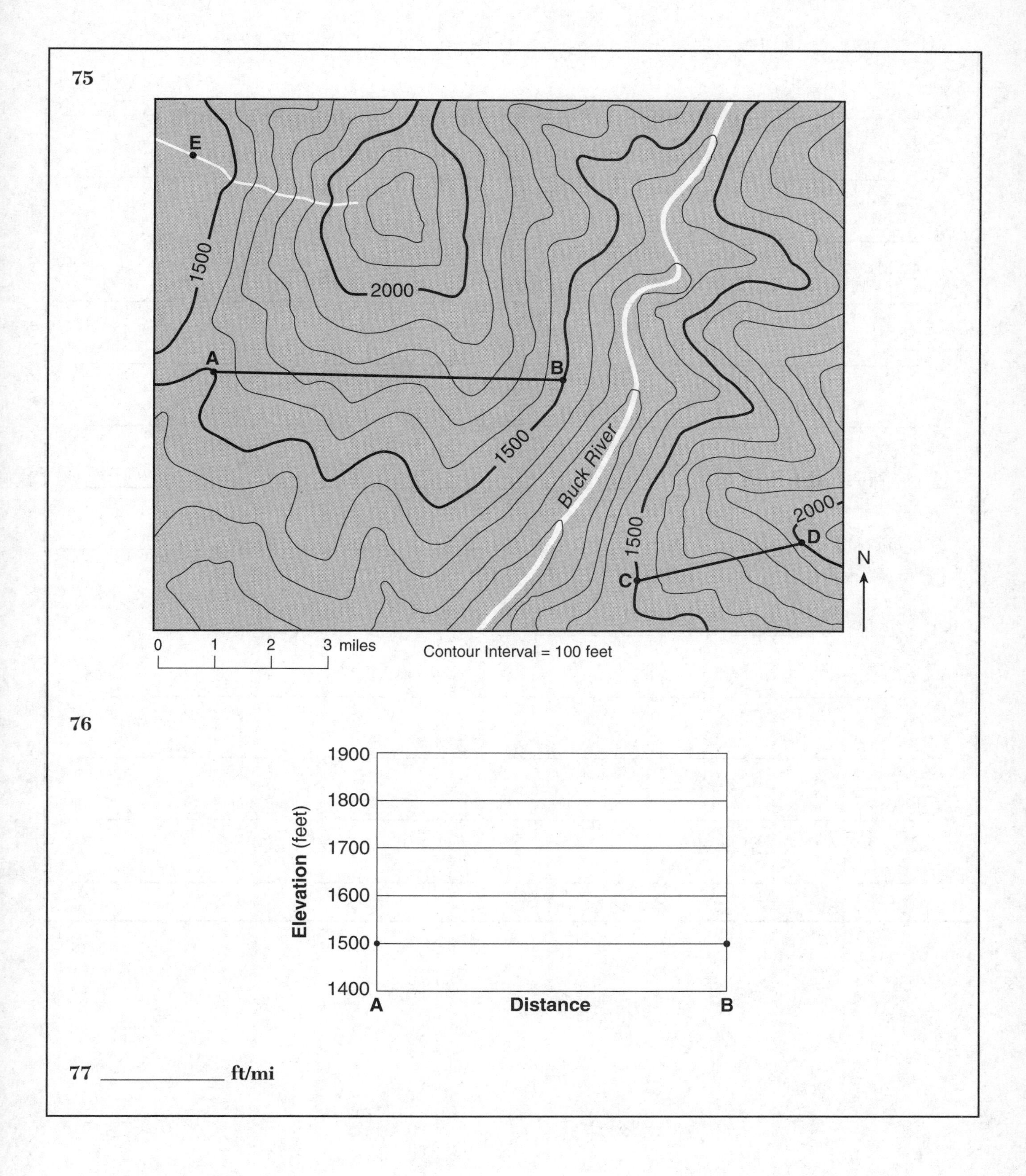

75
E
1500
2000
A
B
1500
Buck River
2000
1500
D
C
N
0
1
2
3 miles
Contour Interval = 100 feet
76
1900
1800
1700
1600
1500
1400
Elevation (feet)
A
Distance
B
77 ____________ ft/mi

78 ____

____

79 __________ **cm/s**

80 ____

____

81 ____

82 Density of gabbro: ____

Composition of gabbro: ____

83 ______________________________

84 ____

____

85 ____

____

Physical Setting/EARTH SCIENCE
January 2017

Part A

Answer all questions in this part.

Directions (1–35): For *each* statement or question, choose the word or expression that, of those given, best completes the statement or answers the question. Some questions may require the use of the *2011 Edition Reference Tables for Physical Setting/Earth Science*. Record your answers on your separate answer sheet.

1 Which statement best explains why stars viewed from the Northern Hemisphere appear to revolve around *Polaris*?

(1) *Polaris* rotates on its axis.
(2) Earth rotates on its axis.
(3) *Polaris* revolves around Earth.
(4) Earth revolves around *Polaris*.

2 The hydrosphere covers approximately what percentage of Earth's lithosphere?

(1) 100% (3) 50%
(2) 70% (4) 25%

3 The deflection of prevailing winds and ocean currents in the Northern Hemisphere is called

(1) eccentricity
(2) refraction
(3) the Coriolis effect
(4) the Doppler effect

4 Earth's rate of revolution is approximately

(1) 1° per day (3) 23.5° per day
(2) 15° per day (4) 360° per day

5 The asteroid Ceres lies at an average distance of 414 million kilometers from the Sun. The period of revolution of Ceres around the Sun is approximately

(1) 438 days (3) 4.6 years
(2) 687 days (4) 12.6 years

6 Which planet has a density that is *less* than the density of liquid water?

(1) Mercury (3) Mars
(2) Earth (4) Saturn

7 The diagram below represents two planets of equal mass, *A* and *B*, revolving around a star. The planets are represented at specific positions in their orbits.

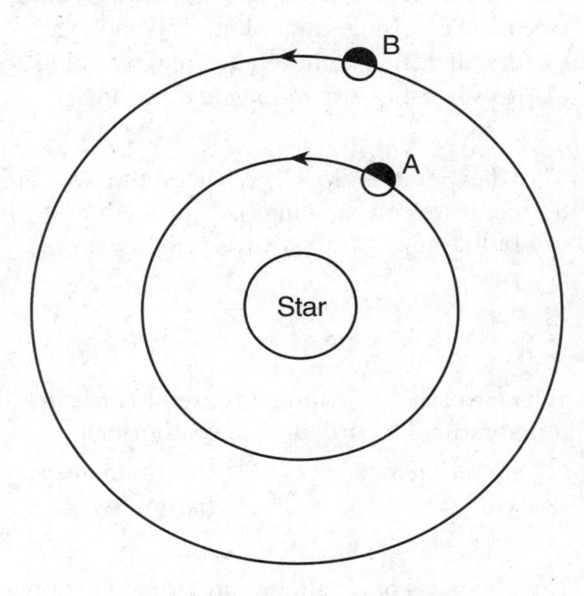

(Not drawn to scale)

When both planets are at the positions represented, planet *B*

(1) can be seen at night from planet *A*, and planet *B* is moving faster in its orbit
(2) can be seen at night from planet *A*, and planet *B* is moving slower in its orbit
(3) cannot be seen at night from planet *A*, and planet *B* is moving faster in its orbit
(4) cannot be seen at night from planet *A*, and planet *B* is moving slower in its orbit

8 Compared to terrestrial planets, Jovian planets have

(1) smaller equatorial diameters and shorter periods of revolution
(2) smaller equatorial diameters and longer periods of revolution
(3) larger equatorial diameters and shorter periods of revolution
(4) larger equatorial diameters and longer periods of revolution

9 Clouds most likely form as a result of

(1) moist air rising, compressing, and warming
(2) moist air rising, expanding, and cooling
(3) dry air rising, compressing, and warming
(4) dry air rising, expanding, and cooling

10 The dewpoint is 15°C. What is the wet-bulb temperature on a sling psychrometer if the dry-bulb temperature is 18°C?

(1) 16°C (3) 3°C
(2) 2°C (4) 20°C

11 Which weather instrument is used to measure air temperatures recorded on a weather map?

(1) anemometer (3) thermometer
(2) wind vane (4) barometer

12 Equal masses of basalt, granite, iron, and copper received the same amount of solar energy during the day. At night, which of these materials cooled down at the fastest rate?

(1) basalt (3) iron
(2) granite (4) copper

13 Equal areas of which type of surface will reflect the most insolation?

(1) light gray rooftop (3) snow-covered field
(2) dark tropical forest (4) black paved road

14 Riverhead, New York, has a smaller average daily temperature range than Elmira, New York, because Riverhead is located

(1) near a large body of water
(2) at a lower latitude
(3) at a higher elevation
(4) near a large city

15 Which diagram best represents the relative wavelengths of visible light, ultraviolet energy, and infrared energy?

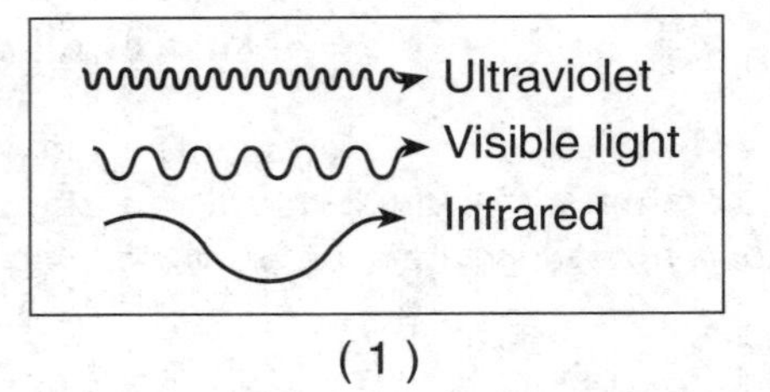

(1)

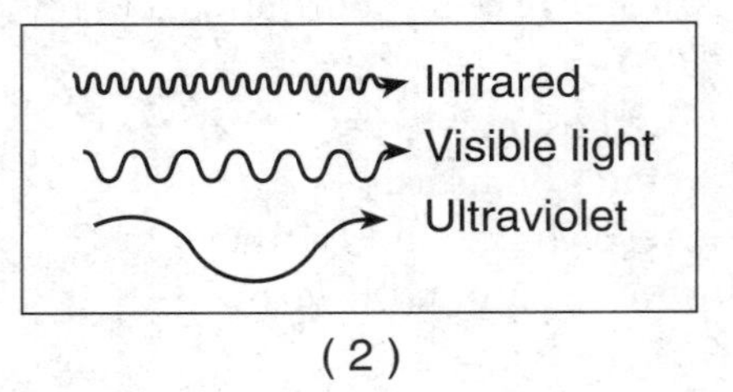

(2)

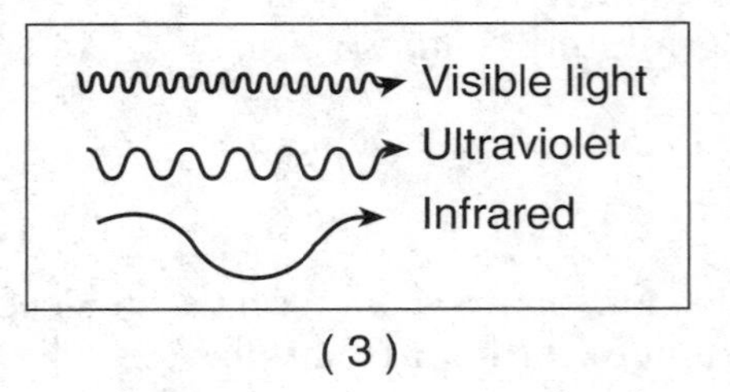

(3)

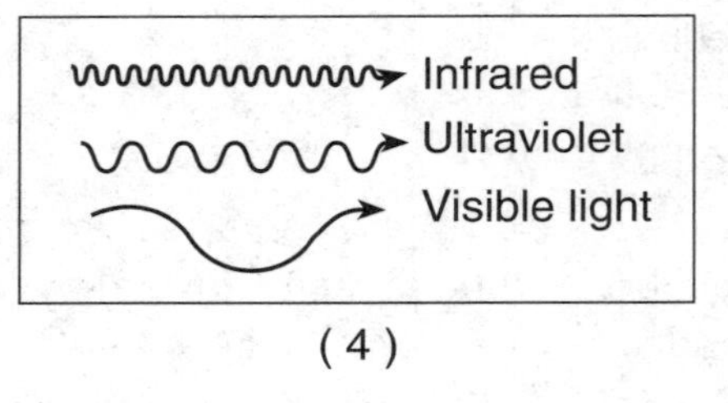

(4)

16 Volcanic ash is a good geologic time marker because the ash

(1) is deposited rapidly over a large area
(2) spreads evenly in all compass directions
(3) is easily weathered and eroded
(4) remains in the atmosphere for millions of years

17 The change in life-forms in the fossil record from less complex organisms to more complex organisms over time is best explained by

(1) extinction
(2) evolution
(3) dynamic equilibrium
(4) original horizontality

18 The graph below shows the yearly air temperature and precipitation of a location on Earth.

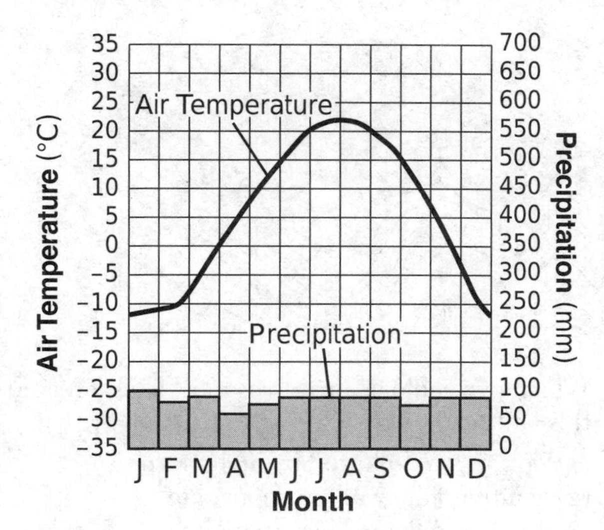

This location would be most likely at a latitude of

(1) 0° (3) 50° N
(2) 35° S (4) 90° N

19 Arrows in the diagram below represent the daytime flow of air over a coastal region.

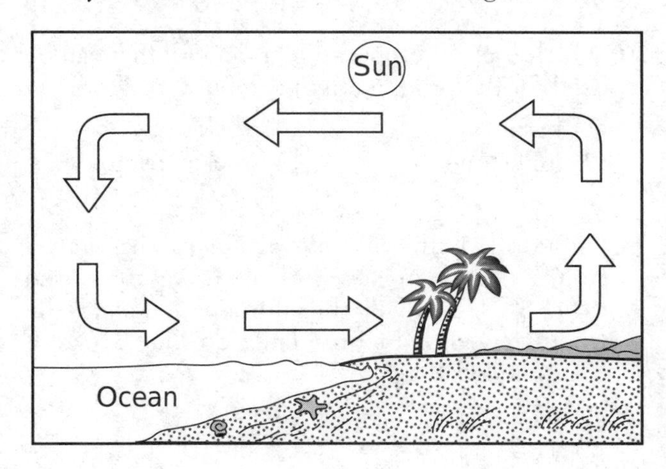

Which process primarily transfers heat by moving air?

(1) conduction (3) radiation
(2) convection (4) transpiration

20 The graph below shows the radioactive decay of rubidium-87.

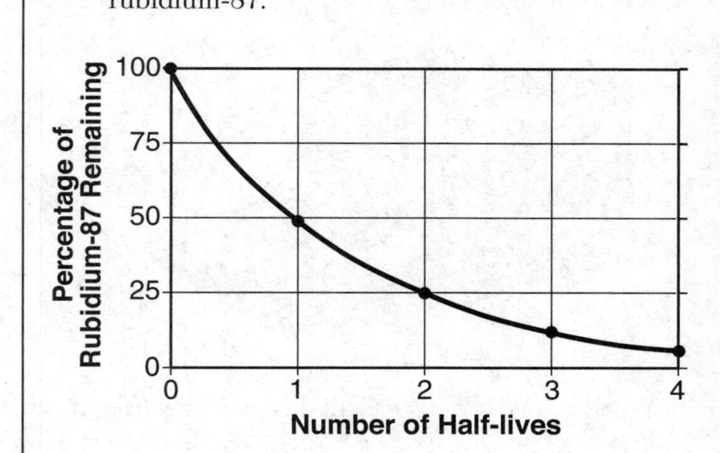

What percentage of rubidium-87 atoms will be left after four half-lives?

(1) 25.0% (3) 6.25%
(2) 12.5% (4) 3.125%

21 The pressure at the interface between Earth's outer core and inner core is inferred to be

(1) 0.2 million atmosphere
(2) 1.5 million atmospheres
(3) 3.1 million atmospheres
(4) 3.6 million atmospheres

22 Which type of tectonic plate boundary is found between the South American Plate and the Scotia Plate?

(1) transform (3) divergent
(2) convergent (4) complex or uncertain

23 The epicenter of an earthquake was located 1800 kilometers from a seismic recording station. If the *S*-wave arrived at the seismic station at 10:06:40 a.m., at what time did the *P*-wave arrive at the same seismic station?

(1) 10:03:00 a.m. (3) 10:09:40 a.m.
(2) 10:03:40 a.m. (4) 10:10:20 a.m.

24 A strong earthquake that occurs on the ocean floor could result in the formation of

(1) a tsunami (3) an El Niño event
(2) a delta (4) an ocean current

25 The block diagram below represents a rapid downslope flow of saturated soil and rock layers.

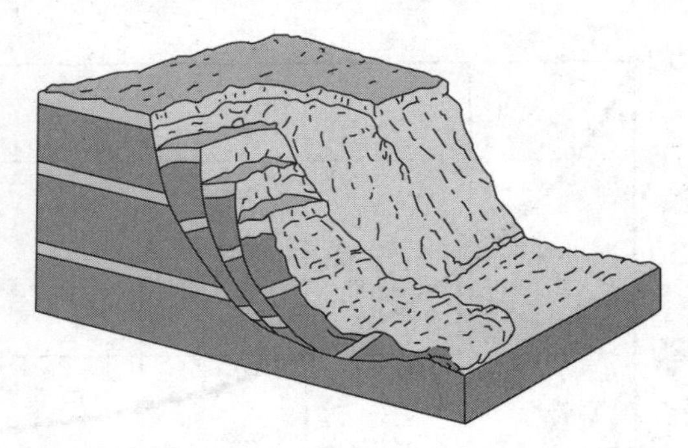

What are two likely causes of this rapid downslope flow?

(1) groundwater and abrasion
(2) groundwater and gravity
(3) prevailing wind and abrasion
(4) prevailing wind and gravity

26 The map below shows a stream. Letters *A*, *B*, *C*, and *D* represent locations on the stream surface. Arrows represent the direction of stream flow.

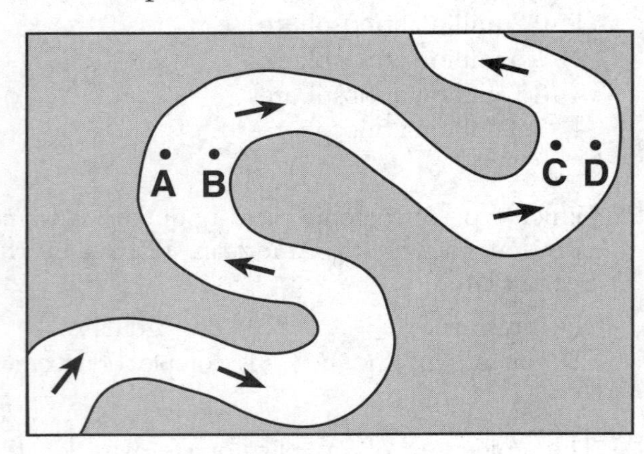

Which two locations have the greatest stream velocities?

(1) *A* and *B*
(2) *B* and *C*
(3) *C* and *D*
(4) *D* and *A*

27 Which climate conditions most likely produce a landscape with rounded hills, large river valleys with many tributaries, and tropical vegetation?

(1) cool and arid
(2) cool and humid
(3) warm and arid
(4) warm and humid

28 The block diagram below represents two parallel mountain ranges.

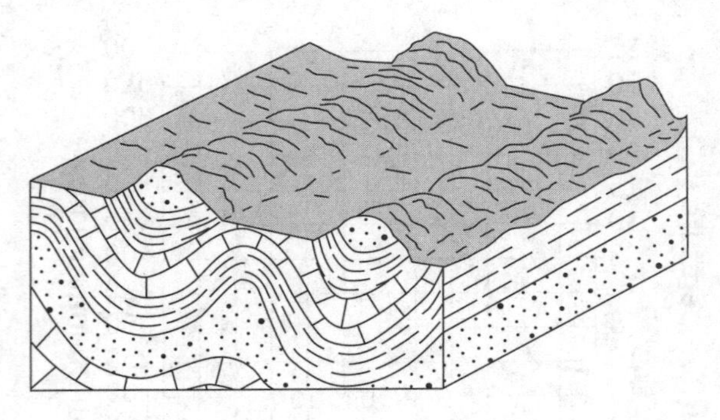

Which two geologic processes most likely created this landscape region?

(1) volcanism, followed by metamorphism
(2) faulting, followed by deposition
(3) folding, followed by erosion
(4) glaciation, followed by rifting

29 Which agent of erosion most likely moves sediments in a sand dune?

(1) wind
(2) glaciers
(3) wave action
(4) running water

30 Which rock is composed of a mineral that can be used for the production of cement?

(1) basalt
(2) limestone
(3) rock salt
(4) rock gypsum

31 On April 21, the altitude of *Polaris*, as viewed from a location in New York State, was measured as 41.3°. What will the altitude of *Polaris* be when viewed one month later, on May 21, from the same location?

(1) 23.5°
(2) 41.3°
(3) 66.7°
(4) 90°

32 The diagrams below represent constellations seen by an observer in New York State facing south at midnight on July 7 and January 3.

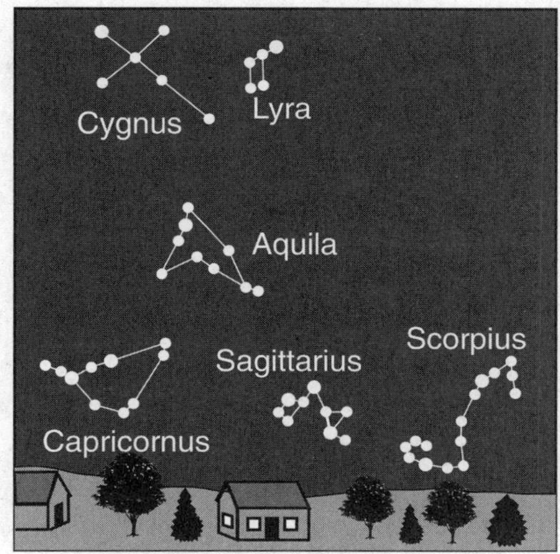

Southern horizon – July 7

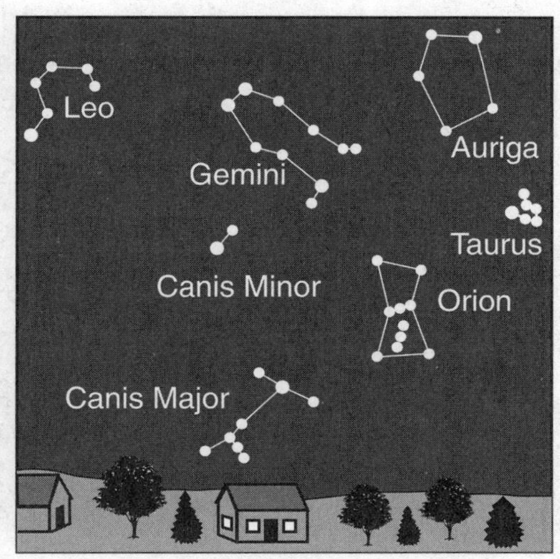

Southern horizon – January 3

Which motion causes the observer to see different constellations at midnight on July 7 compared to midnight on January 3?

(1) revolution of the constellations in their orbits
(2) revolution of Earth in its orbit
(3) rotation of the stars in the constellations
(4) rotation of Earth on its axis

33 The diagram below represents a model of the size of the Sun and indicates the color of the Sun.

Yellow star

Which diagram best represents the relative size and indicates the color of *Polaris* compared to the Sun?

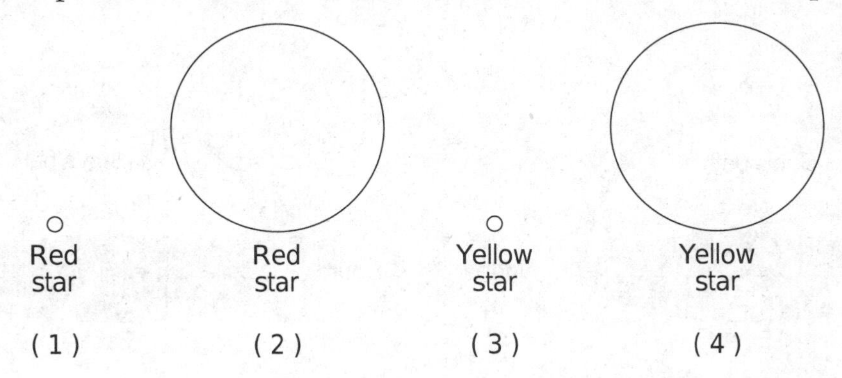

34 The diagram below represents the apparent path of the Sun as seen by an observer on June 21 at a location in New York State.

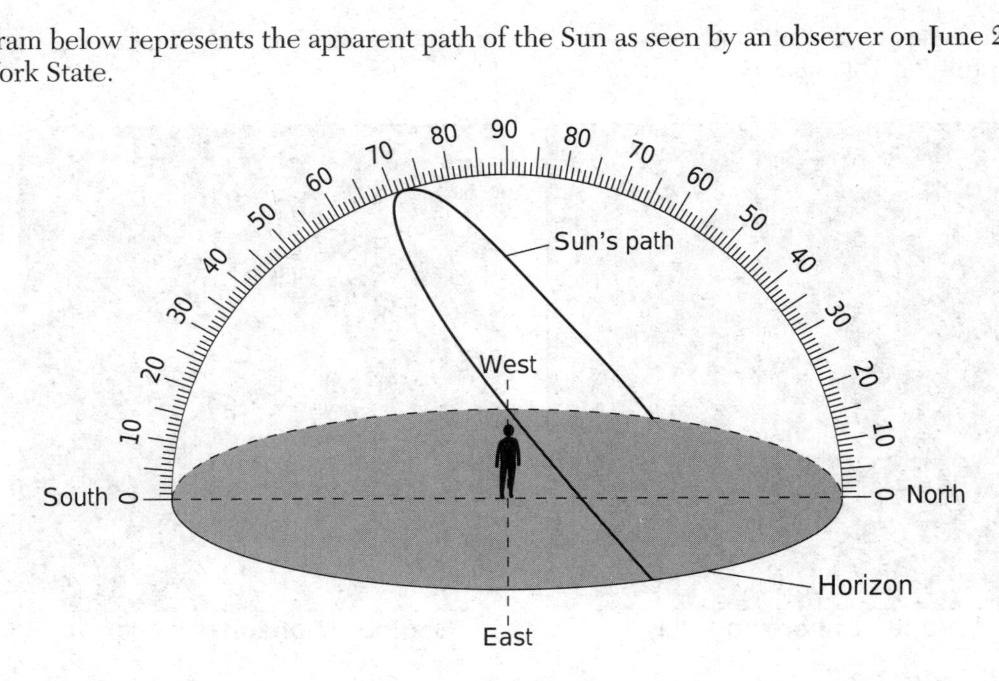

Which diagram best represents the apparent path of the Sun at this same location on December 21?

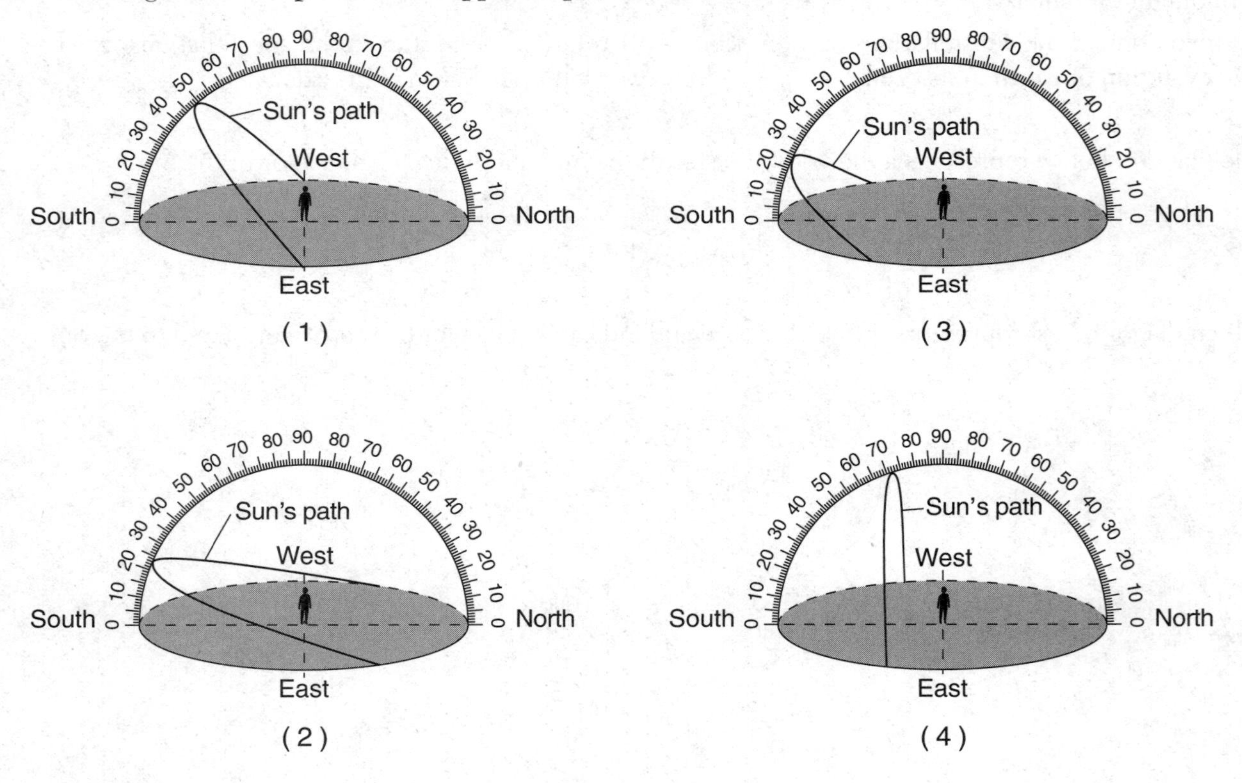

35 The topographic map below shows a portion of the Cayuta Creek that is located in New York State. Points *A*, *B*, *C*, and *D* represent locations on Earth's surface.

Topographic Map

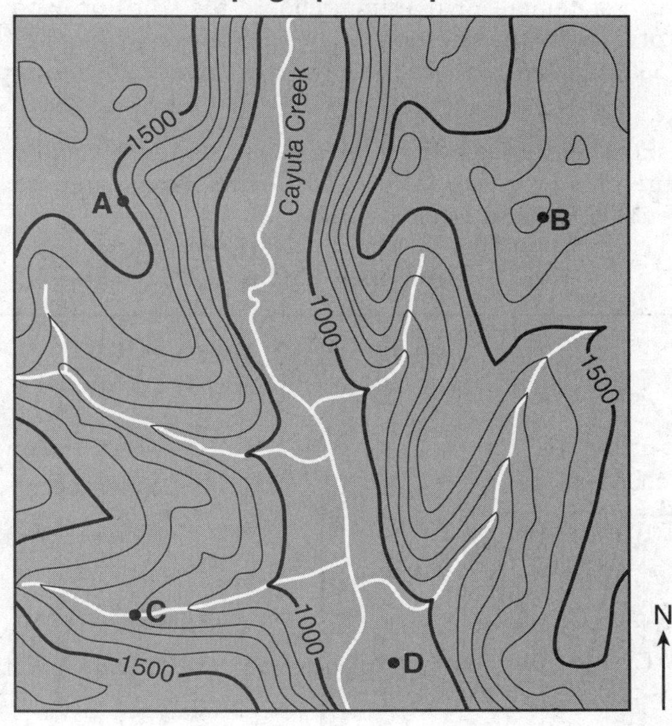

Which point on the map most likely represents a location within the flood plain associated with Cayuta Creek?

(1) *A*
(2) *B*
(3) *C*
(4) *D*

Part B–1

Answer all questions in this part.

Directions (36–50): For *each* statement or question, choose the word or expression that, of those given, best completes the statement or answers the question. Some questions may require the use of the *2011 Edition Reference Tables for Physical Setting/Earth Science*. Record your answers on your separate answer sheet.

Base your answers to questions 36 through 38 on the cross section below and on your knowledge of Earth science. The cross section represents processes in the water cycle. Arrows represent the movement of water. Letters *A*, *B*, *C*, and *D* represent locations on Earth's surface.

The Water Cycle

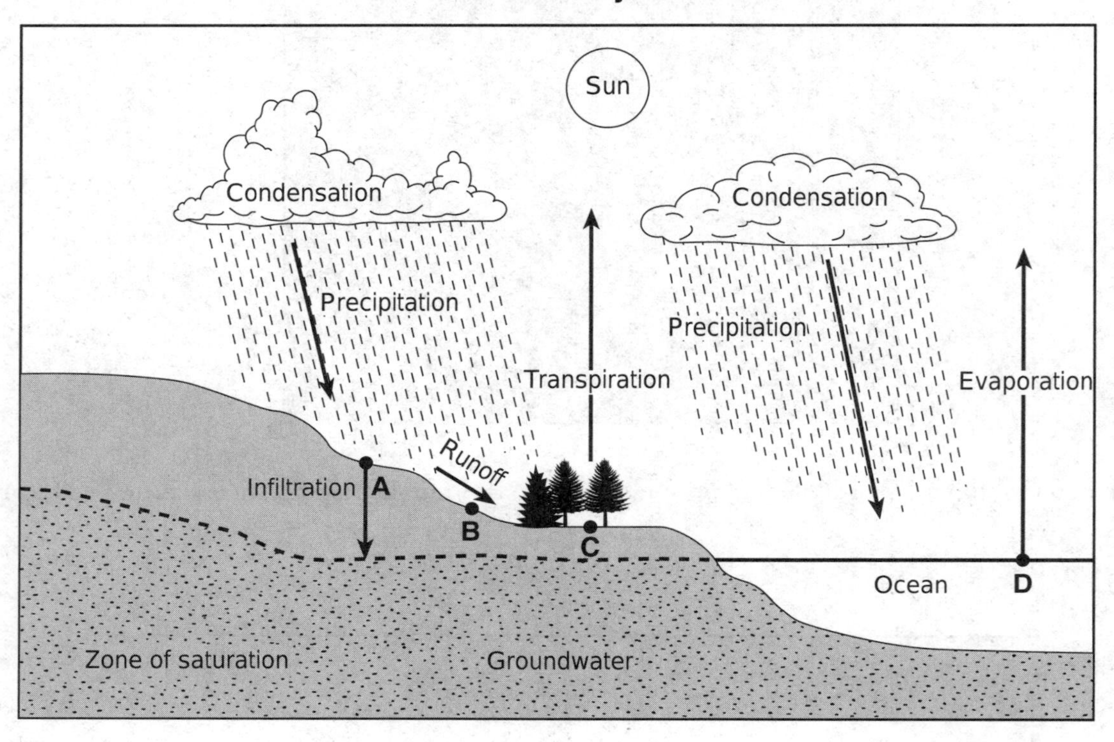

36 The downward movement of water from location *A* will usually be greatest when the soil is

(1) nonporous and the particles are uniformly small in size
(2) nonporous and the particles are uniformly large in size
(3) porous and the particles are uniformly small in size
(4) porous and the particles are uniformly large in size

37 What would most likely reduce the amount of runoff at location *B*?

(1) infiltration occurring faster than precipitation
(2) greater condensation than evaporation
(3) saturated soil below the land surface
(4) a frozen land surface

38 The greatest amount of transpiration and evaporation will occur most likely when the air temperature is

(1) low and the humidity is low
(2) low and the humidity is high
(3) high and the humidity is low
(4) high and the humidity is high

Base your answers to questions 39 and 40 on the graphs below and on your knowledge of Earth science. The graphs show air temperatures and dewpoints in °F, and wind speeds in knots (kt) from 2:00 a.m. to 11:00 p.m. at a certain New York State location.

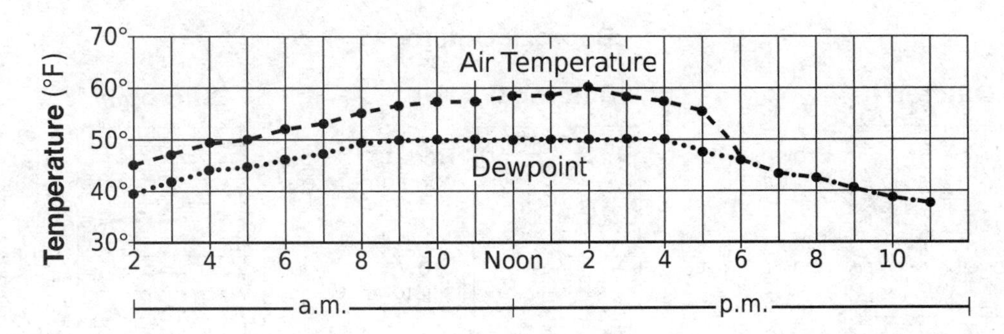

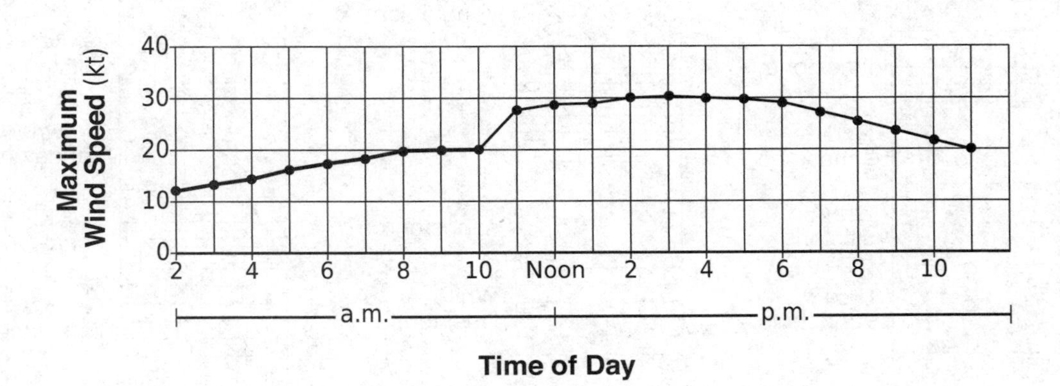

39 Which station model represents the weather data for this location at 4:00 p.m.?

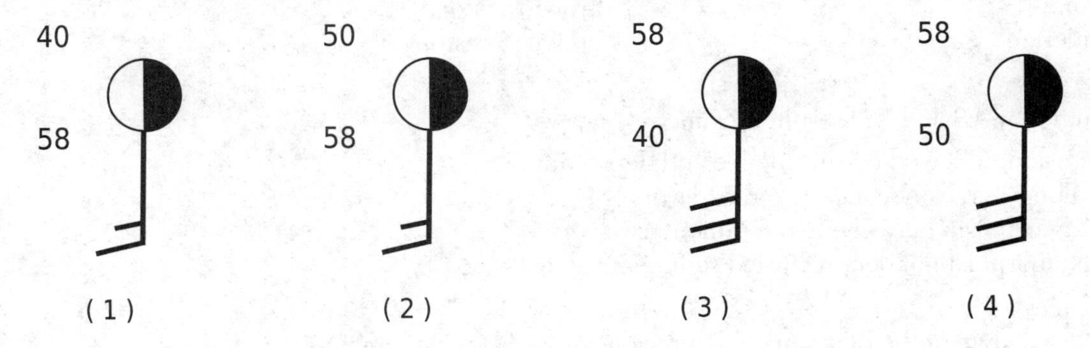

40 What was the relative humidity at 8:00 p.m.?

(1) 30%
(2) 45%
(3) 75%
(4) 100%

Base your answers to questions 41 through 44 on the three bedrock outcrops below and on your knowledge of Earth science. The outcrops, labeled I, II, and III, are located within 15 kilometers of each other. Lines *AB* and *CD* represent unconformities. Line *XY* represents a fault. No overturning of the layers has occurred.

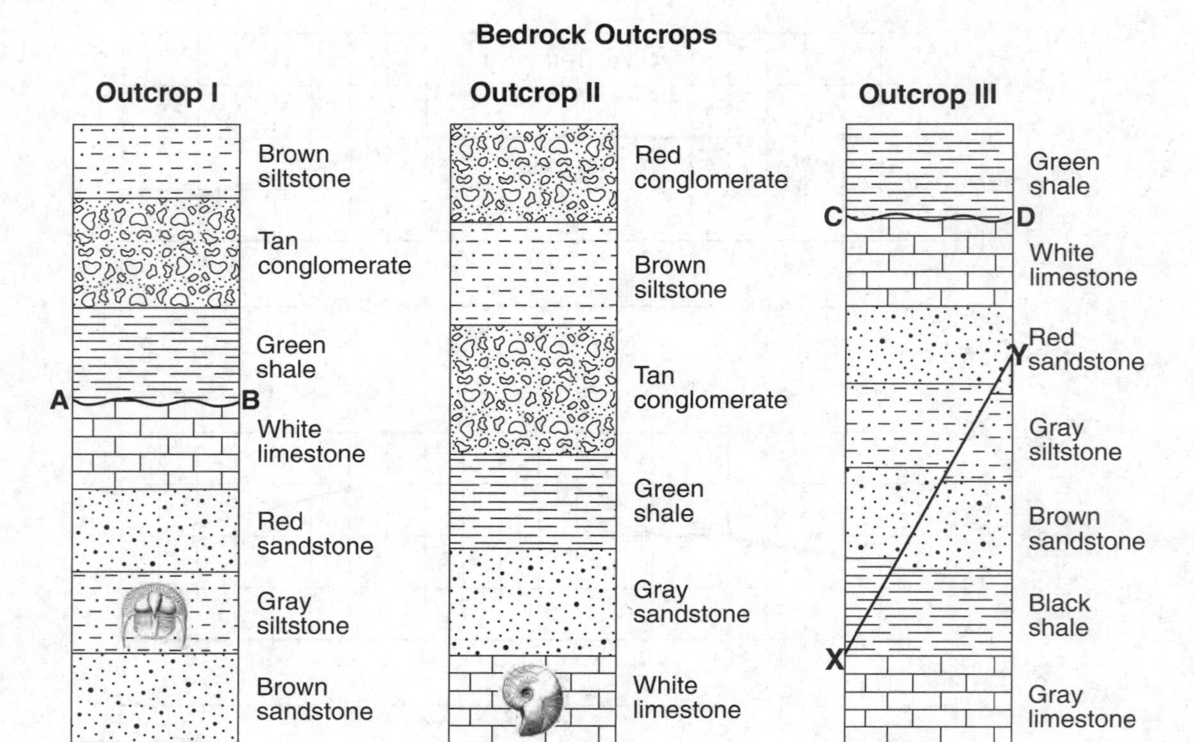

41 Which layer is the youngest?

(1) gray limestone
(2) red conglomerate
(3) brown siltstone
(4) brown sandstone

42 The unconformities at *AB* and *CD* resulted from

(1) uplift and erosion, followed by subsidence and deposition
(2) movement along a crack between two rock layers
(3) contact metamorphism between two sedimentary layers
(4) regional metamorphism of deeply buried sedimentary rocks

43 Based on evidence shown in the diagram, which rock layer is older than fault *XY*?

(1) tan conglomerate
(2) black shale
(3) brown siltstone
(4) white limestone

44 Which processes produced the brown siltstone layer in outcrops I and II?

(1) cooling and solidification of mafic lava at Earth's surface
(2) cooling and solidification of felsic magma deep within Earth
(3) compaction and cementation of rock fragments ranging in size from 0.006 to 0.2 centimeter in diameter
(4) compaction and cementation of rock fragments ranging in size from 0.0004 to 0.006 centimeter in diameter

Base your answers to questions 45 through 47 on the passage below and on your knowledge of Earth science.

Island Arcs

Island arcs are long, curved chains of oceanic islands associated with seismic activity and mountain-building processes at certain plate boundaries. They occur where oceanic tectonic plates collide. Along one side of these island arcs, there is usually a long, narrow deep-sea trench.

At island arcs, the denser plate is subducted and is forced into the partially molten mantle under the less dense plate. The islands are composed of the extrusive igneous rocks basalt and andesite. The basalt originates most likely from the plastic mantle. The andesite originates most likely from the melting of parts of the descending plate and sediments that had accumulated on its surface.

45 An island arc is found along the

(1) East Pacific Ridge
(2) Iceland Hot Spot
(3) Aleutian Trench
(4) Peru-Chile Trench

46 Most of the basalt that forms island arcs comes from the

(1) crust
(2) rigid mantle
(3) asthenosphere
(4) stiffer mantle

47 Which list identifies minerals present in andesite from the greatest percentage by volume to the least percentage by volume?

(1) biotite, plagioclase feldspar, amphibole
(2) biotite, amphibole, plagioclase feldspar
(3) plagioclase feldspar, biotite, amphibole
(4) plagioclase feldspar, amphibole, biotite

Base your answers to questions 48 through 50 on the cross section and data table below and on your knowledge of Earth science. The cross section shows the profile of a stream that is flowing down a valley from its source. Points *A* through *E* represent locations in the stream. The data table shows the average stream velocity at each location. The volume of water in the stream remains the same at all locations.

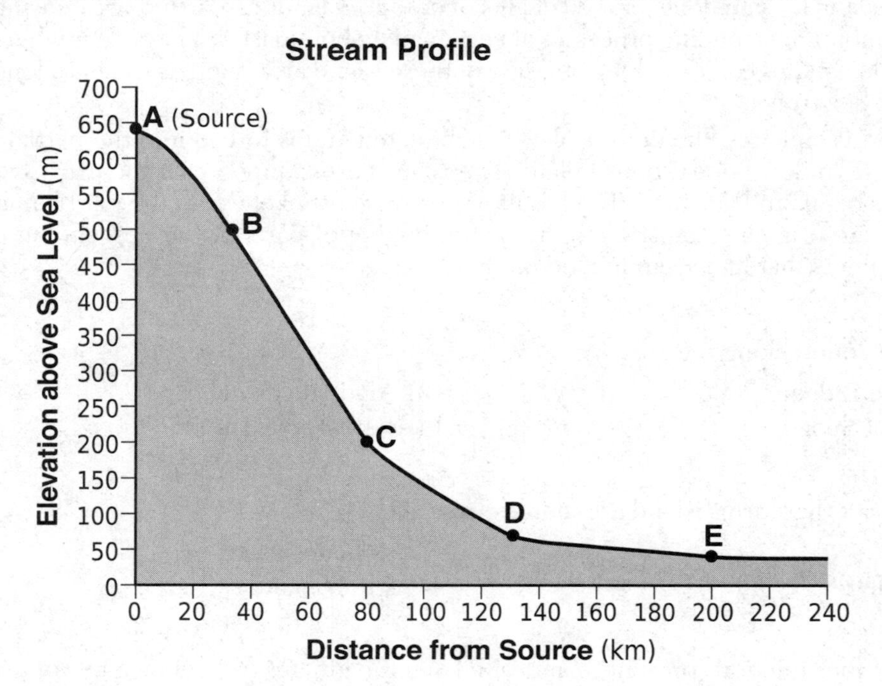

Location in Stream	Average Stream Velocity (cm/s)
A	10
B	110
C	130
D	20
E	15

48 The average stream velocity at each location is controlled primarily by the

(1) elevation above sea level
(2) slope of the land
(3) sediment carried by the stream
(4) distance from the stream's source

49 What is the largest type of sediment that could be transported at location *B*?

(1) silt
(2) sand
(3) pebbles
(4) cobbles

50 Which features could be formed by the stream between locations *D* and *E*?

(1) meanders
(2) kettle lakes
(3) barrier islands
(4) drumlins

Part B–2

Answer all questions in this part.

Directions (51–65): Record your answers in the spaces provided in your answer booklet. Some questions may require the use of the *2011 Edition Reference Tables for Physical Setting/Earth Science*.

Base your answers to questions 51 through 54 on the diagram below and on your knowledge of Earth science. The diagram represents a model of the expanding universe.

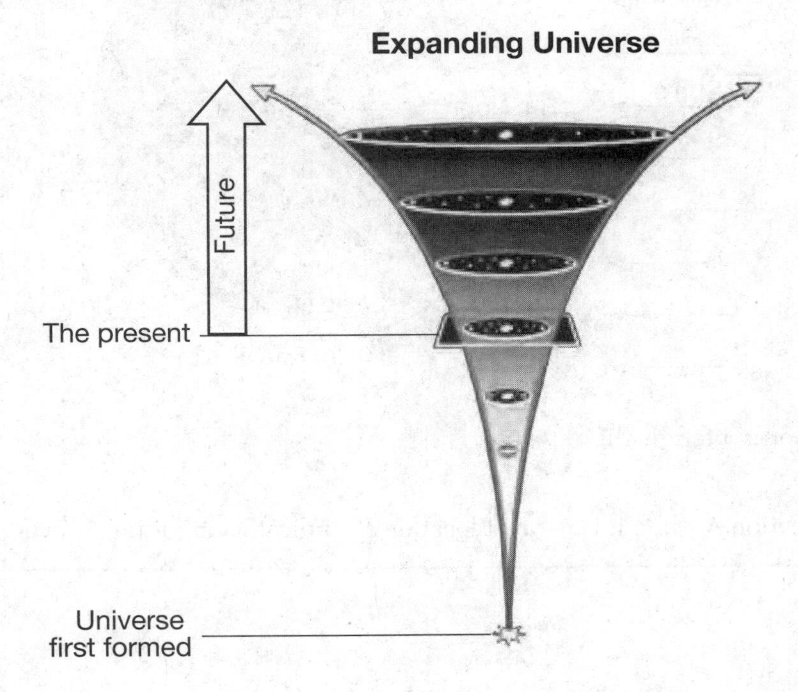

51 Identify the name of the event that is inferred by scientists to have occurred when the universe first formed. [1]

52 Identify *one* piece of evidence that led astronomers to infer that the universe is expanding. [1]

53 Identify the force that caused stars and planets in the universe to become layered according to density differences in their composition. [1]

54 Identify the nuclear process that combines lighter elements into heavier elements to produce the energy radiated by stars. [1]

Base your answers to questions 55 and 56 on the diagram below and on your knowledge of Earth science. The diagram represents a view of Earth from above the North Pole, showing longitude lines at 15 degree intervals. Letters *A* and *B* represent surface locations on the equator.

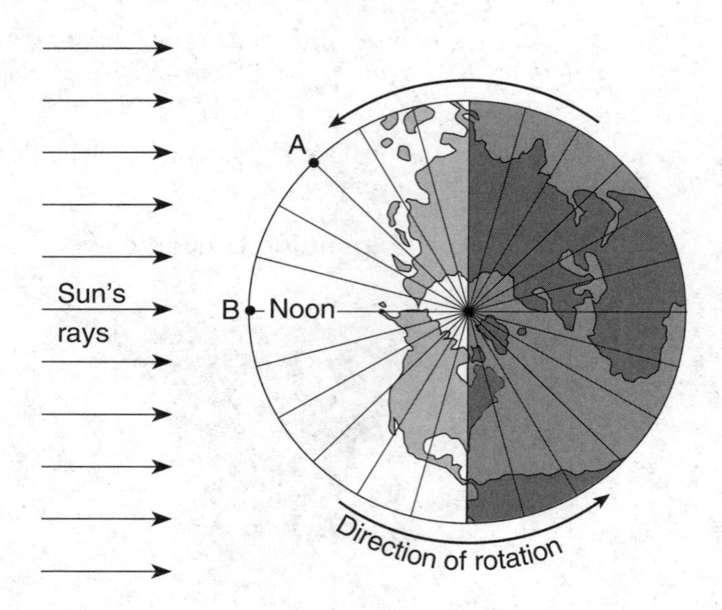

55 Identify *one* date represented by this diagram. [1]

56 State the time at location *A* when it is noon at location *B*. Indicate a.m. or p.m. in your answer. [1]

Base your answers to questions 57 through 60 on the passage and map below and on your knowledge of Earth science. The map shows the positions of the eye (center) of Hurricane Sandy in its path from October 24 to October 31, 2012. A high-pressure center (**H**) is shown on the map.

Hurricane Sandy

In October 2012, Hurricane Sandy produced extreme damage to New York City and the coast of New Jersey due to high winds and a high storm surge. A storm surge is the rise in the level of ocean water along a coast that is caused by strong winds blowing toward land from a severe storm. High ocean tides, occurring at the same time, added to the height of the storm surge. A high-pressure center, located just south of Newfoundland, Canada, affected Hurricane Sandy by altering the path of the jet stream. This change in the jet stream, combined with surface wind circulation around the high-pressure center, caused Hurricane Sandy to curve westward, making landfall along the coast of New Jersey.

Path of Hurricane Sandy from October 24, 2012 to October 31, 2012

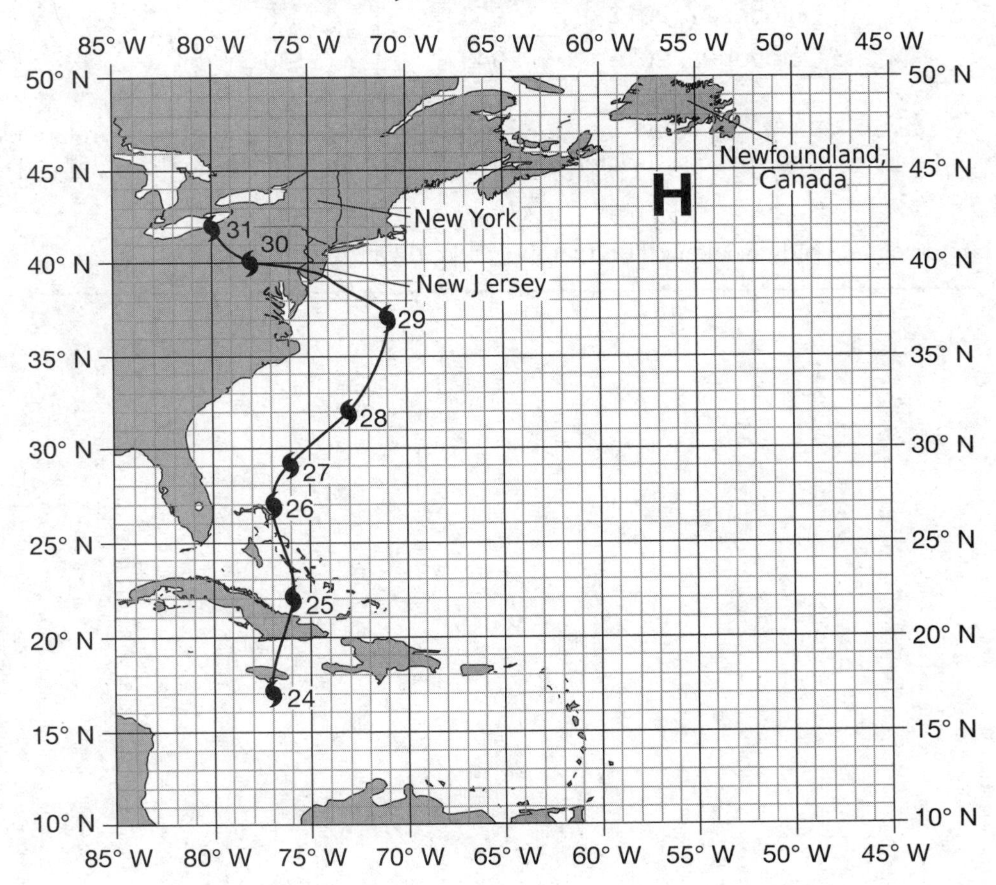

57 Using information from the map, complete the data table *in your answer booklet* by identifying the latitude and longitude positions of the eye of Hurricane Sandy from October 27, 2012 to October 29, 2012. Express your latitude and longitude positions to the nearest whole degree. [1]

58 Describe the surface wind circulation around the high-pressure center (**H**) that is located south of Newfoundland. [1]

59 The data table below shows the air pressure, measured in millibars (mb), and surface wind speed, measured in miles per hour (mi/h), recorded near the center of Hurricane Sandy on three separate days.

Date	**Air Pressure** (mb)	**Surface Wind Speed** (mi/h)
October 24, 2012	973	70
October 27, 2012	958	75
October 29, 2012	943	90

On the set of axes *in your answer booklet*, draw a line to represent the general relationship between air pressure and surface wind speed associated with Hurricane Sandy for these three days. [1]

60 Explain why Hurricane Sandy weakened on October 30 and October 31. [1]

Base your answers to questions 61 through 65 on the geologic timeline below and on your knowledge of Earth science. The geologic timeline, drawn to scale, represents Earth's geologic history. The letters *A* through *H* on the timeline represent the times of occurrence for specific, labeled geologic events. The time of occurrence for letter *A* has been omitted.

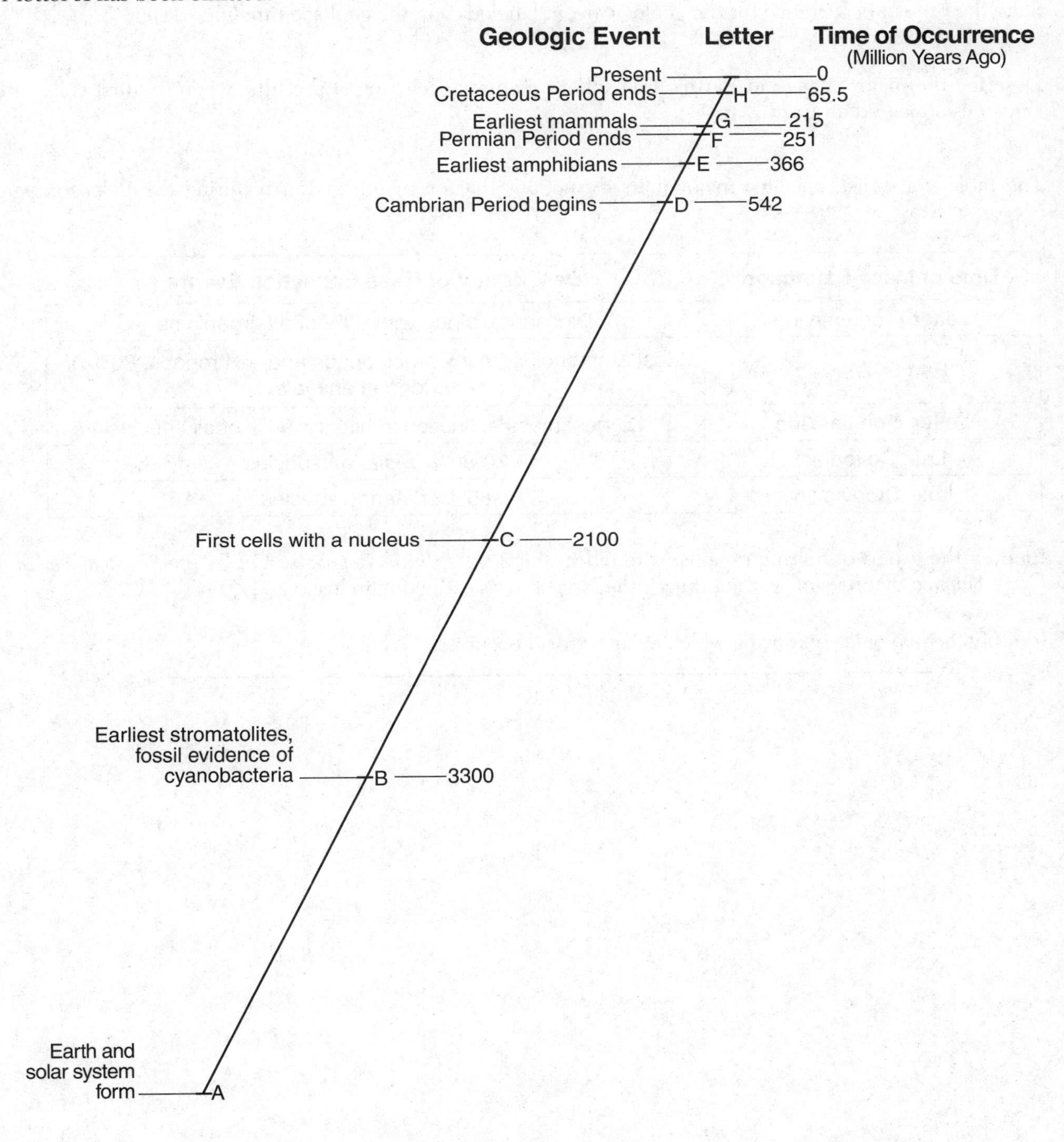

61 Identify the *two* consecutive letters on the timeline that represent the time span within which the earliest insects appeared on Earth. [1]

62 State the time of occurrence for the geologic event labeled *A* on the geologic timeline. [1]

63 Describe the major change in Earth's atmosphere that was occurring at the time when the first cells with a nucleus appeared on Earth. [1]

64 The table below lists the five major mass extinctions that occurred on Earth during the Paleozoic and Mesozoic Eras.

Time of Mass Extinction	Description of Mass Extinction Events
Letter *H* on timeline	Dinosaurs, along with 80% of all organisms
End of Triassic	Most ammonoids, many brachiopods and gastropods, 80% of four-legged animals
Letter *F* on timeline	Largest mass extinction in history, 90% of all species
Late Devonian	70-80% of marine species
Late Ordovician	85% of marine species

Identify the group of marine organisms found in the *2011 Edition Reference Tables for Physical Setting/Earth Science* that became extinct during the largest mass extinction in history. [1]

65 Identify the geologic eon during which event letter *B* occurred. [1]

Part C

Answer all questions in this part.

Directions (66–85): Record your answers in the spaces provided in your answer booklet. Some questions may require the use of the *2011 Edition Reference Tables for Physical Setting/Earth Science*.

Base your answers to questions 66 through 69 on the data table below and on your knowledge of Earth science. The data table shows the average level of atmospheric carbon dioxide (CO_2), measured in parts per million (ppm), for the month of February at the Mauna Loa observatory in Hawaii from 2008 to 2014.

Year	Average February Atmospheric CO_2 Levels (ppm)
2008	386
2009	387
2010	390
2011	392
2012	394
2013	396
2014	398

66 On the grid *in your answer booklet*, construct a line graph by plotting the data for the average February atmospheric carbon dioxide (CO_2) levels for the years 2008 to 2014. Connect the plots with a line. [1]

67 These measurements of atmospheric carbon dioxide were collected at an altitude of 3.4 kilometers. Identify the temperature zone of the atmosphere where these data were collected. [1]

68 Identify *one* major greenhouse gas, other than carbon dioxide. [1]

69 Describe *two* human activities that would *decrease* the amount of carbon dioxide that humans add to Earth's atmosphere. [1]

Base your answers to questions 70 through 73 on the weather map in your answer booklet and on your knowledge of Earth science. The weather map shows the center of a high-pressure system (**H**) and the center of a low-pressure system (**L**) affecting North America. Isobars are drawn for the eastern portion of the map, and one isobar is drawn around the high-pressure center. Air pressures are shown at various points in the western portion of the map. All air pressures were recorded in millibars (mb). Points *A* through *F* represent surface locations.

70 On the map *in your answer booklet*, draw the 1012 mb, 1016 mb, and 1020 mb isobars. Extend the isobars to the edges of the map. [1]

71 Convert the air pressure at location *A* from millibars (mb) to inches of mercury (in of Hg). [1]

72 Calculate the air pressure gradient between locations *A* and *B* in millibars per kilometer. [1]

73 Identify *one* possible air pressure at the center of the low-pressure system. [1]

Base your answers to questions 74 through 77 on the diagram in your answer booklet and on your knowledge of Earth science. The diagram represents the Moon's orbit around Earth as viewed from space above Earth's North Pole (NP). Letter *A* represents one position of the Moon in its orbit.

74 On the diagram *in your answer booklet*, place an **X** on the Moon's orbit to indicate the position of the Moon when a solar eclipse would be observed from Earth. [1]

75 State the number of days that it takes the Moon to orbit Earth once. [1]

76 On the diagram *in your answer booklet*, shade the portion of the Moon that is in darkness as viewed from New York State when the Moon is at position *A*. [1]

77 Describe the actual shape of the Moon's orbit. [1]

Base your answers to questions 78 through 81 on the passage and geologic cross section below and on your knowledge of Earth science. The geologic cross section represents rock layers of a portion of the Niagara Escarpment, and landscape features that are found in the Niagara region. The rock layers have *not* been overturned.

The Niagara Escarpment

A prominent feature found along the shore of Lake Ontario in western New York State is the Niagara Escarpment. This escarpment is the remains of an ancient seabed that was formed when the area was covered by a warm, shallow sea from approximately 450 to 430 million years ago. Erosion of the Taconic Mountains to the east provided the sediments deposited in this basin area. From these sediments, rock layers such as shale, sandstone, and limestone formed. Later, magnesium replaced some of the calcium in the top layer of limestone, turning it into a dolostone layer. When the high ocean levels of the Ordovician Period dropped, the draining of this inland sea caused unequal erosion of the exposed layers. The South Moraine was deposited on the top of the Niagara Escarpment in this region.

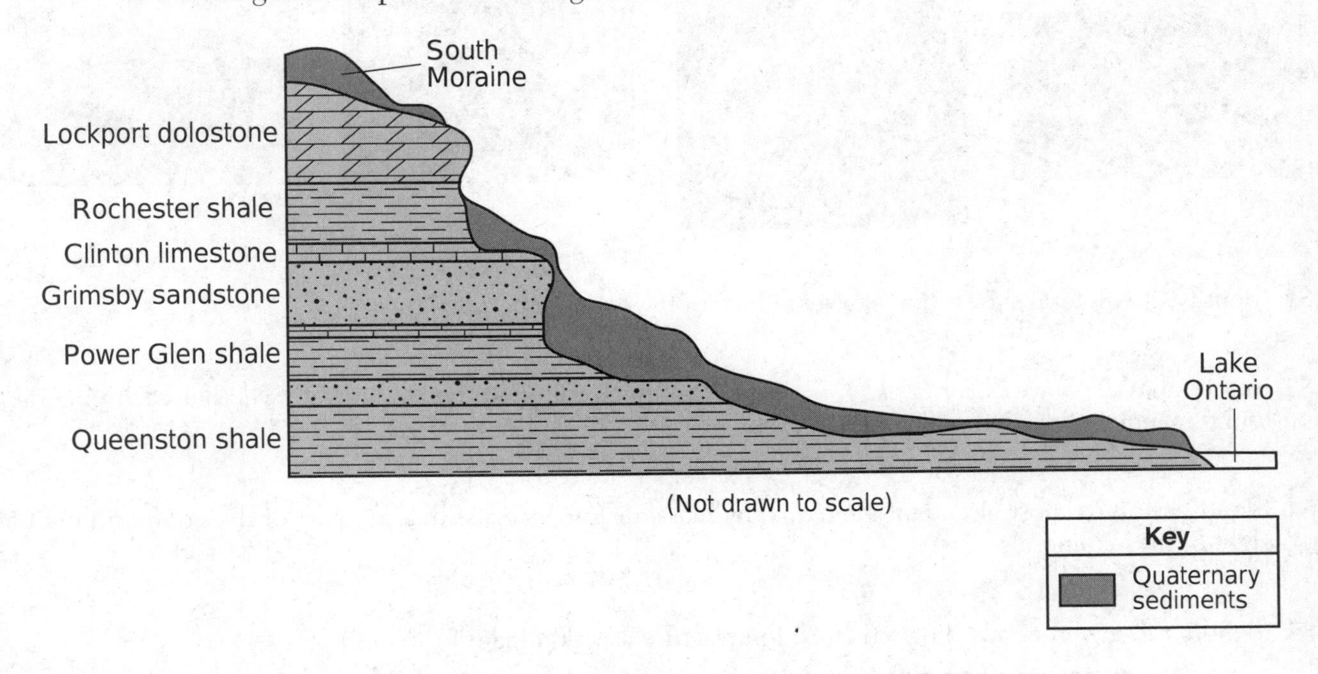

78 Identify the New York State landscape region in which the Niagara Escarpment is located. [1]

79 Identify the mineral composition of the Lockport dolostone. [1]

80 Describe the inferred position of North America when this area was covered by the warm, shallow sea. [1]

81 Describe the tectonic event that caused the Taconian orogeny. [1]

Base your answers to questions 82 through 85 on the photographs below and on your knowledge of Earth science. The photographs show eight common rock-forming minerals.

82 Identify the mineral shown that can scratch all of the other minerals shown. [1]

83 In the table *in your answer booklet*, place an **X** in the appropriate box to indicate whether each mineral is found mainly in felsic or mafic igneous rock. [1]

84 Identify the *two* most abundant elements, by mass, in Earth's crust that are part of the composition of all eight of these minerals. [1]

85 Identify the *two* minerals shown that exhibit fracture as a dominant form of breakage. [1]

Physical Setting/EARTH SCIENCE
January 2017

ANSWER BOOKLET

Student ....

Teacher ....

School .... Grade ....

Answer all questions in the examination.
Record your answers in this booklet.

	Performance Test Score (Maximum Score: 16)

Part	**Maximum Score**	**Student's Score**
A	**35**	
B-1	**15**	
B-2	**15**	
C	**20**	

Total Written Test Score (Maximum Raw Score: 85)

Final Score (from conversion chart)

Rater's Initials:

Rater 1 .... **Rater 2** ....

Part A

1	10..............	19	28..............
2	11..............	20	29..............
3	12..............	21	30..............
4	13..............	22	31..............
5	14..............	23	32..............
6	15..............	24	33..............
7	16..............	25	34..............
8	17..............	26	35..............
9	18..............	27	

Part B

36	40..............	44	48..............
37	41..............	45	49..............
38	42..............	46	50..............
39	43..............	47	

Physical Setting/EARTH SCIENCE
January 2017

ANSWER BOOKLET

☐ Male

Student . . . . Sex: ☐ Female

Teacher . . . .

School . . . . Grade . . . .

Record your answers for Part B–2 and Part C in this booklet.

Part B–2

51 ______________________________

52 ____

53 ______________________________

54 ______________________________

55 ______________________________

56 ______________________________

57

Position of Hurricane Sandy from October 24, 2012 to October 31, 2012

Date	Latitude° (N)	Longitude° (W)
October 24	17	77
October 25	22	76
October 26	27	77
October 27		
October 28		
October 29		
October 30	40	78
October 31	42	80

58 ____

____

59

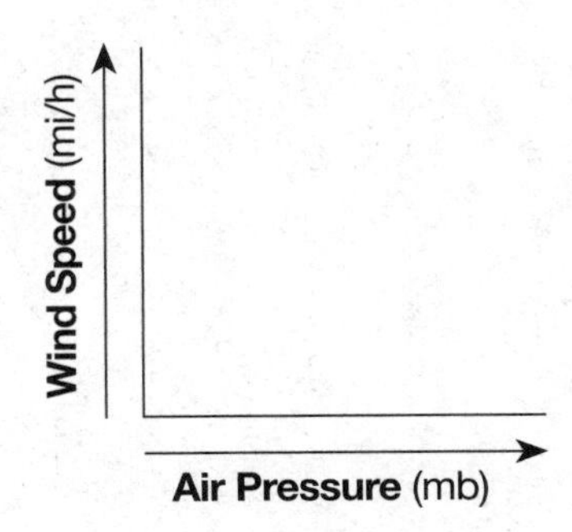

60 ____

____

____

61 ___________ and ___________

62 _____________ **million years ago**

63 ____

____

64 ____

65 ____

Part C

66

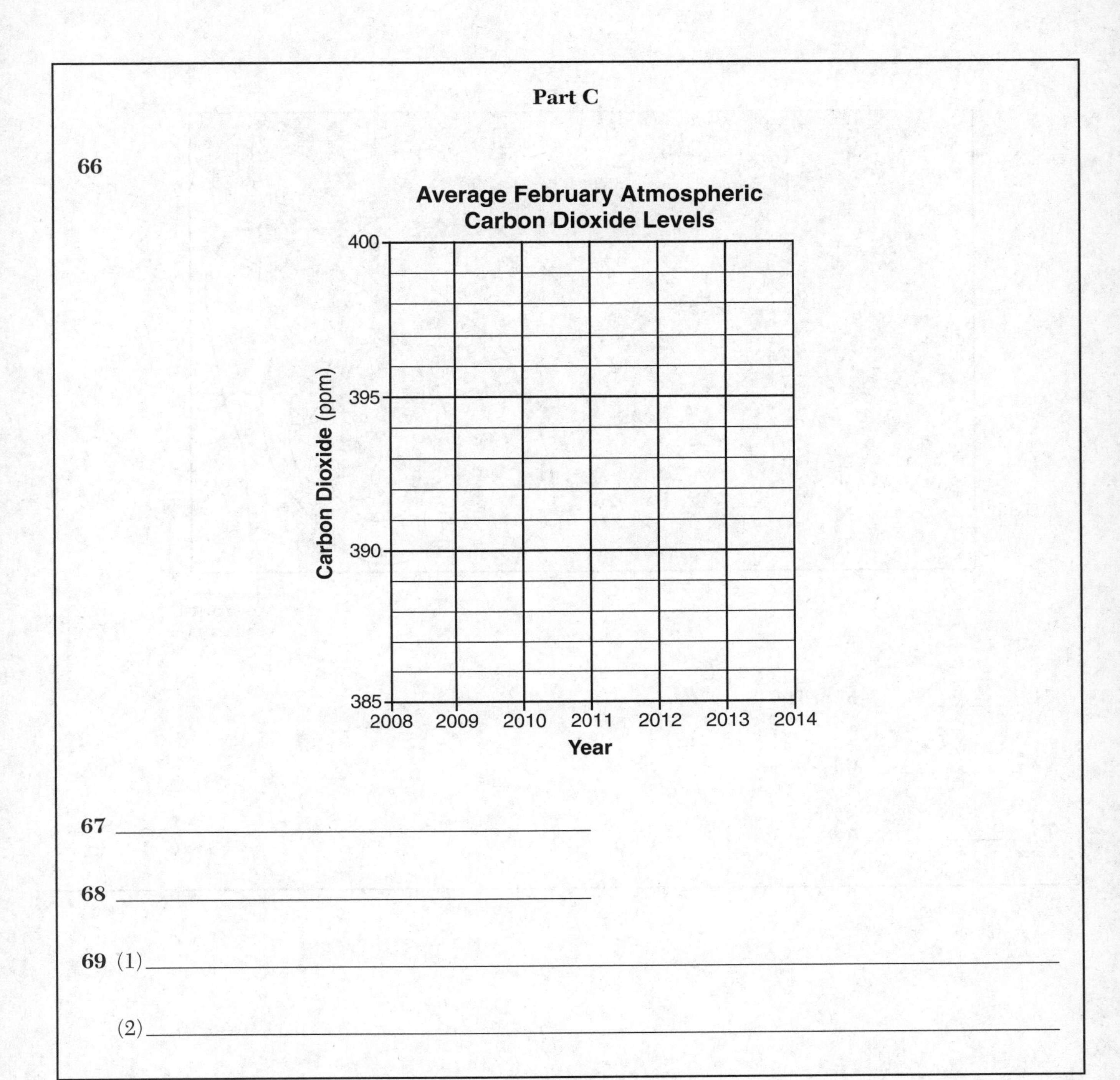

67 ______________________________

68 ______________________________

69 (1) ______________________________

(2) ______________________________

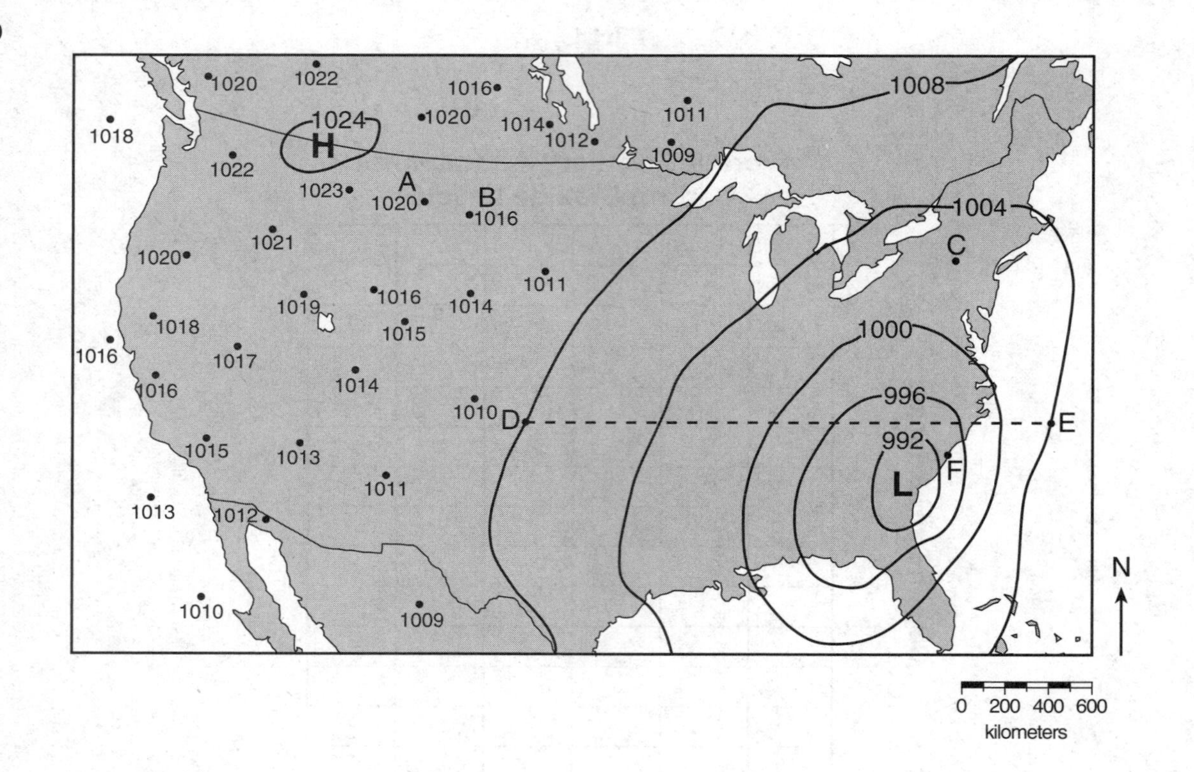

71 ______________ **in of Hg**

72 ______________ **mb/km**

73 ______________ **mb**

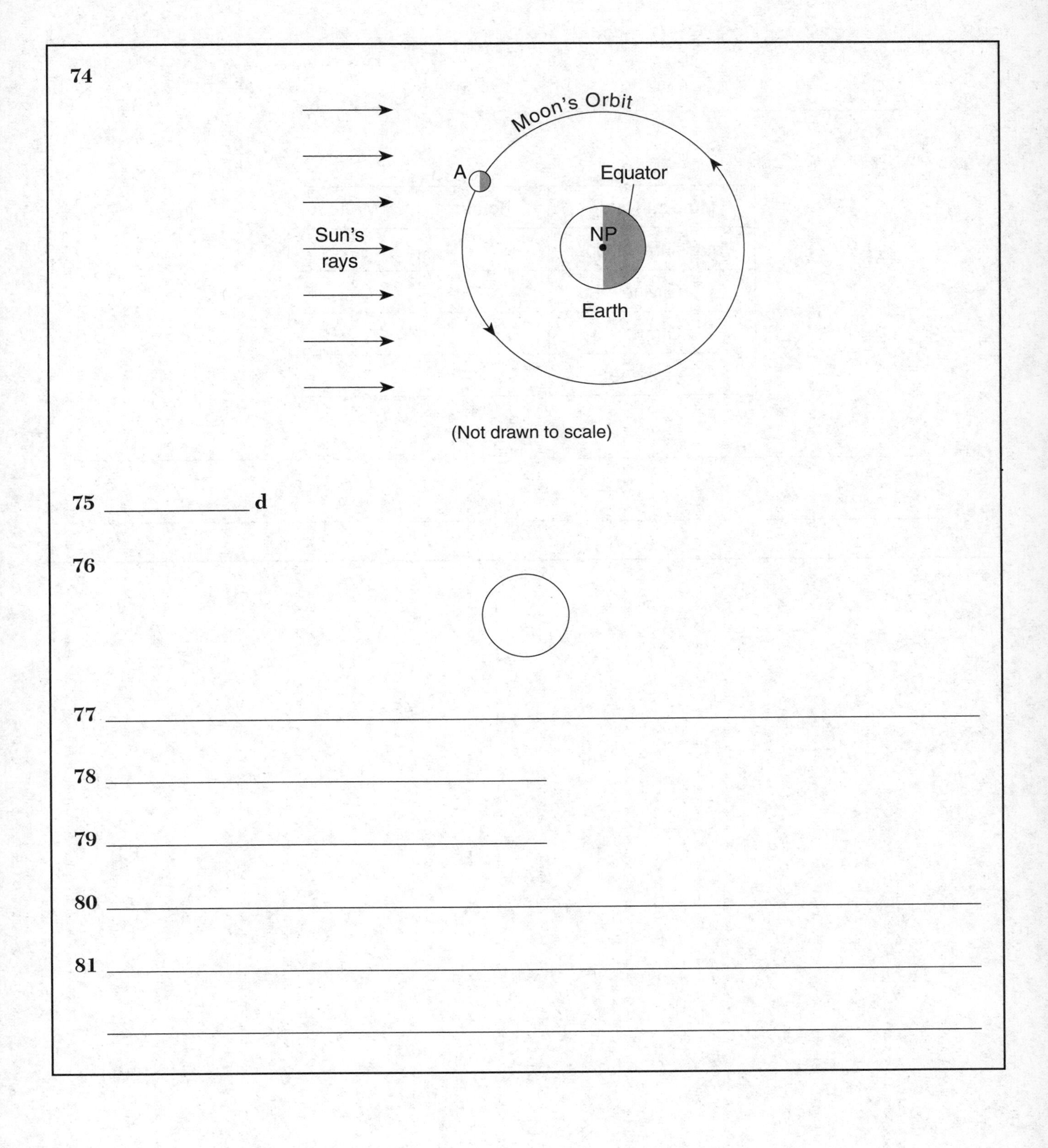

74
Moon's Orbit
A
Equator
Sun's rays
NP
Earth
(Not drawn to scale)
75
d
76
77
78
79
80
81

82 ______________________________

83

Mineral Name	Felsic	Mafic
Potassium feldspar		
Olivine		
Quartz		
Pyroxene		

84 ______________________________ and ______________________________

85 ______________________________ and ______________________________